T0297069

Ecological Climatology

The third edition of Gordon Bonan's comprehensive textbook introduces an interdisciplinary framework to understand the interaction between terrestrial ecosystems and climate change. Ideal for advanced undergraduate and graduate students studying ecology, environmental science, atmospheric science, and geography, it reviews basic meteorological, hydrological, and ecological concepts to examine the physical, chemical, and biological processes by which terrestrial ecosystems affect and are affected by climate. This new edition has been thoroughly updated with new science and references. The scope has been expanded beyond its initial focus on energy, water, and carbon to include reactive gases and aerosols in the atmosphere. This new edition emphasizes Earth as a system, recognizing interconnections among the planet's physical, chemical, biological, and socioeconomic components, and emphasizing global environmental sustainability. Each chapter contains chapter summaries and review questions, and with over four hundred illustrations, including many in color, this textbook will once again be an essential student guide.

Gordon Bonan is senior scientist and head of the Terrestrial Sciences Section at the National Center for Atmospheric Research, Boulder, Colorado. His research focuses on the interactions of terrestrial ecosystems with climate using models of Earth's biosphere, atmosphere, hydrosphere, and geosphere. He has published more than 120 peer-reviewed articles on land-atmosphere coupling and how changes in vegetation alter climate. He is a member of the American Geophysical Union, American Meteorological Society, and Ecological Society of America. He is a Fellow of the American Geophysical Union and has served on advisory boards for numerous national and international organizations and as an editor for several journals.

Ecological Climatology

Concepts and Applications

Third edition

Gordon Bonan
National Center for Atmospheric Research*
Boulder, Colorado

* The National Center for Atmospheric Research is sponsored by the National Science Foundation

CAMBRIDGE
UNIVERSITY PRESS

University Printing House, Cambridge CB2 8BS, United Kingdom

One Liberty Plaza, 20th Floor, New York, NY 10006, USA

477 Williamstown Road, Port Melbourne, VIC 3207, Australia

4843/24, 2nd Floor, Ansari Road, Daryaganj, Delhi - 110002, India

79 Anson Road, #06-04/06, Singapore 079906

Cambridge University Press is part of the University of Cambridge.

It furthers the University's mission by disseminating knowledge in the pursuit of education, learning and research at the highest international levels of excellence.

www.cambridge.org
Information on this title: www.cambridge.org/9781107619050

© Gordon Bonan 2016

First published 2016

A catalogue record for this publication is available from the British Library

Library of Congress Cataloging in Publication data
Bonan, Gordon B.
Ecological climatology : concepts and applications / Gordon Bonan, National Center for Atmospheric Research, Boulder, Colorado. – Third edition.
    pages   cm
Includes bibliographical references and index.
ISBN 978-1-107-04377-0 (hardback) – ISBN 978-1-107-61905-0 (pbk.)
1. Vegetation and climate.   2. Plant ecophysiology.   3. Climatic changes.
I. Title.
QK754.5.B66   2015
581.7′22–dc23   2015012684

ISBN 978-1-107-04377-0 Hardback
ISBN 978-1-107-61905-0 Paperback

Additional resources for this publication at www.cambridge.org/bonan3.

To Amie, who made this possible

To David, Thomas, and Alice, for family

To Milo, Dancer, and Chloe, for hugs and head pats

Contents

Part III Hydrometeorology

Part VI Terrestrial Forcings and Feedbacks

Preface

I began conceiving this book in 1996. At that time, the influence of forests on climate was well-established at the microscale through the study of forest meteorology and biometeorology; that the terrestrial biosphere is essential for climate science and global models of climate was less universally accepted. The first edition was published in 2002. It was an effort to broaden the scope of ecology – to show ecologists the manner in which ecosystems influence climate – and to similarly broaden climate science to recognize the importance of terrestrial ecosystems. The second edition published in 2008 was a marked change from the first edition. It was a complete revision that reflected the expanded scope of the science, improved the organization of the material, and made it more accessible to students.

The third edition is yet another revision and update of the book. The intent has not changed, but the science has so vastly grown. Studies of biosphere–atmosphere interactions at the regional and global scale are now commonplace; all of the major international climate modeling centers include models of the terrestrial biosphere; the carbon cycle and anthropogenic land use and land-cover change are recognized as important facets of climate change; and climate science itself has evolved into a broader perspective of Earth system science. This is seen in the expanded breadth of the book. In the second edition, carbon cycle–climate coupling was still fairly novel. This third edition shows how important that has become to climate science, and additionally includes chapters on the nitrogen cycle, aerosols, and climate change mitigation. A concluding chapter ties together the various topics presented throughout the book. The influences of terrestrial ecosystems on climate must be seen in a larger context of human influences on the global environment and in light of planetary sustainability.

One prominent change over the years has been the extensive growth of the scientific literature. This third edition is not meant to be a survey of all relevant literature; that would be too tedious. Rather, I have highlighted key papers that, with online scholarly databases, provide a springboard to the science. To keep the book manageable, some material had to be deleted from the earlier editions. Many references to scientific studies have been removed or omitted. Nonetheless, this third edition provides a comprehensive survey of the state of the science. The challenge of organizing, synthesizing, and presenting the voluminous material in a comprehensible manner is tempered by the pleasure in seeing the extent to which the science has expanded over the years.

As in the previous editions, this book contains many mathematical equations, but only to illustrate concepts and not with the intent of being a modeling textbook. The book heavily references models, their scientific scope, and their application to understand biosphere–atmosphere interactions. The book also maintains land management, urban planning, and landscape design as a theme. The principles of ecological climatology are applicable to these studies. Unlike global change, land use occurs locally in our communities. It gives substance to environmental issues at spatial and temporal scales to which people can see and respond; we see these changes happen in our communities, often over a period of a few years.

As always, I am indebted to colleagues at the National Center for Atmospheric Research for supporting my efforts to write this third edition, in particular Sam Levis and Keith Oleson, whose long-standing commitment to the development and maintenance of community models, both since 1999, have allowed me to write this book. David Lawrence, too, assumed a leading role in community model development, allowing me to focus my efforts on writing. And new colleagues – Rosie Fisher, Peter Lawrence, Will Wieder, Danica Lombardozzi, Quinn Thomas, Melannie Hartman, and Liz Burakowski – have similarly grown the science and supported community models. Finally, I am indebted to Matt Lloyd at Cambridge University Press, who has supported this endeavor over the many years.

1

Ecosystems and Climate

1.1 Chapter Summary

When viewed from space, Earth is seen as a blue marble. The dominant features of the planet are the blue of the oceans and the white of the clouds traversing the atmosphere. It is an image of fluids – water and air – in motion. Indeed, the study of Earth's climate is dominated by the geophysical principles of fluid dynamics. With closer inspection, however, one can discern land masses – the continents – and the plants that grow on the land. The blue of the oceans gives way to the emerald green of vegetation. Weather, climate, and atmospheric composition have long been known to determine the floristic composition of these plants, their arrangement into communities, and their functioning as ecosystems. Earth system scientists now recognize that the patterns and processes of plant communities and ecosystems not only respond to weather, climate, and atmospheric composition, but also feedback through a variety of physical, chemical, and biological processes to influence the atmosphere. The geoscientific understanding of planet Earth has given way to a new paradigm of biogeosciences. Ecological climatology is an interdisciplinary framework to study the functioning of terrestrial ecosystems in the Earth system through their cycling of energy, water, chemical elements, and trace gases. Changes in terrestrial ecosystems through natural vegetation dynamics and through human

uses of land are a key determinate of Earth's climate.

1.2 Common Science

Ecology is the study of interactions of organisms among themselves and with their environment. It seeks to understand patterns in nature (e.g., the spatial and temporal distribution of organisms) and the processes governing those patterns. Climatology is the study of the physical state of the atmosphere – its instantaneous state, or weather; its seasonal-to-interannual variability; its long-term average condition, or climate; and how climate changes over time. These two fields of scientific study are distinctly different. Ecology is a discipline within the biological sciences and has as its core the principle of natural selection. Climatology is a discipline within the geophysical sciences based on applied physics and fluid dynamics. Both, however, share a common history.

The origin of these sciences is attributed to the Greek scholars Aristotle (ca. 350 BCE) and Theophrastus (ca. 300 BCE) and their books *Meteorologica* and *Enquiry into Plants*, respectively, but their modern beginnings trace back to natural history and plant geography. Naturalists and geographers of the seventeenth, eighteenth, and nineteenth centuries saw changes in vegetation as they explored new regions and laid the foundation for the development of ecology

Table 1.1	Relationship between de Candolle's plant types and Köppen's climate types	
de Candolle plant type	Köppen climate type	Dominant vegetation
Megatherms	Humid tropical	Tropical rainforest
		Tropical savanna
Xerophiles	Dry	Desert
		Grassland
Mesotherms	Moist subtropical mid-latitude	Warm temperate deciduous forest
		Warm temperate coniferous forest
		Mediterranean
Microtherms	Moist continental	Cool temperate deciduous forest
		Cool temperate coniferous forest
		Boreal forest
Hekistotherms	Polar	Tundra

Source: Adapted from Colinvaux (1986, p. 326) and Oliver (1996).

and climatology as they sought explanations for these geographic patterns. Alexander von Humboldt, in the early 1800s, observed that widely separated regions have structurally and functionally similar vegetation if their climates are similar. Alphonse de Candolle hypothesized that temperature creates latitudinal zones of tropical, temperate, and arctic vegetation and in 1874 proposed formal vegetation zones with associated temperature limits. This provided an objective basis to map climatic regions, and in 1884 Wladimir Köppen used maps of vegetation geography to produce climate maps. His five primary climate zones shared similar temperature delimitations as de Candolle's vegetation (Table 1.1). The close correspondence between climate and vegetation is readily apparent, and many secondary climate zones such as tropical savanna, tropical rainforest, and tundra are named after vegetation. Although vegetation is no longer used to map the present climate, it is a primary means to reconstruct past climate from relationships of temperature and precipitation with tree-ring width, pollen abundance, and leaf form.

Despite shared origins, twentieth-century advancement of ecology and climatology proceeded not as an integrated and unified science, but rather in the typical disciplinary framework of science into specialized fields of study that favored reductionism. Plant ecology splintered into topical studies of physiology, populations, communities, ecosystems, landscapes, and biogeochemistry. The study of the atmosphere became organized around spatial scales of micrometeorology, mesoscale meteorology, and global climate and topical fields such as boundary layer meteorology, hydrometeorology, radiative transfer, atmospheric dynamics, and atmospheric chemistry.

With lack of communication across disciplines, ecologists and climatologists can draw different insights from the same observations. Pieter Bruegel the Elder's painting "Hunters in the Snow" exemplifies this (Figure 1.1). The painting has been used in climatology textbooks to illustrate climate change (Lamb 1977, pp. 275–276; Lamb 1995, pp. 233–235). Bruegel painted this scene in the winter of 1565 and it depicts, from a climatologist's perspective, Bruegel's impression of the severe winters of that era. It was the beginning of prolonged artistic interest in Dutch winter landscapes that coincided with an extended period of colder than usual European winters. Ecologists have similarly used this painting to illustrate the ecological concept of a landscape (Forman and Godron 1986, pp. 5–6). Instead of a visual record of an unusually cold climate, the ecological perspective perceives an expression of the core tenets of landscape ecology. From an ecological point of view, the painting depicts heterogeneity of

Fig. 1.1 "Hunters in the Snow" (Pieter Bruegel the Elder). Reproduced with permission of the Kunsthistorisches Museum (Vienna).

landscape elements, spatial scale, and movement across the landscape.

Earth, too, has long been viewed differently by ecologists and climatologists. Ecologists have historically seen weather, climate, and atmospheric composition as external forcings that shape plant communities and ecosystem functions. The manner in which ecosystems influence weather, climate, and atmospheric composition was not examined in classic ecology textbooks. Similarly, climatology textbooks emphasized the physics and fluid dynamics of the atmosphere, not the vegetation at the lowest boundary of the atmosphere.

The advent of global models of Earth's climate in the 1970s and 1980s altered the disciplinary study of ecology and climatology. These models require a mathematical representation of the exchanges of energy, water, and momentum between land and atmosphere. These processes are regulated in part by plants, which with their leaves, stomata, and diversity of life do not conform to the mathematics of fluid dynamics. Atmospheric scientists developing climate models had to expand their geophysical framework to a biogeophysical framework (Deardorff 1978; Dickinson et al. 1986; Sellers et al. 1986). The

ongoing evolution of climate models to models of the Earth system is marked by recognition of the central role of terrestrial ecosystems in regulating Earth's climate through physical, chemical, and biological processes and the critical influence that human appropriation of ecosystem functions has on climate (Pitman 2003; Bonan 2008; Arneth et al. 2010, 2014; Levis 2010; Seneviratne et al. 2010; Mahowald et al. 2011). Models of Earth's land surface, including its terrestrial ecosystems, for climate simulation have expanded beyond their hydrometeorological heritage (with emphasis on surface energy fluxes and the hydrologic cycle) to include carbon, reactive nitrogen, aerosols, anthropogenic land use, and vegetation dynamics. These models are important research tools to study land–atmosphere interactions and climate feedback from ecological processes.

1.3 Deforestation and Climate – Some Early Views

The notion that vegetation affects climate is not new. Over two thousand years ago, Theophrastus

Fig. 1.2 Sketches of land clearing in western New York (clockwise from top left). In the first panel, the initial forest clearing is small, only to fell trees for the small log house and to raise a few livestock. The second panel shows the settler has cleared a few acres of land. Ten years have passed in the third panel. Thirty to forty acres of land have been cleared, and neighbors have cleared their land. The final panel depicts 45 years following the initial clearing. From Turner (1849, pp. 562–566).

wrote that draining marshes removed the moderating effect of water and created a colder climate, while deforestation exposed the ground to the Sun and warmed climate (Glacken 1967, p. 130; Neumann 1985). The concept that forests increase rainfall can be traced back to the Roman natural philosopher Pliny the Elder and his *Natural History*, written in the first century AD (Andréassian 2004). European naturalists in the seventeenth and eighteen centuries, too, believed that the wide-ranging clearing of forests and cultivation of land in Europe had moderated the climate since antiquity (Fleming 1998). Settlers of the New World carried with them a similar sentiment, and a vigorous debate arose about whether the extensive land clearing (Figure 1.2) was indeed changing the climate of America (Thompson 1980; Fleming 1998). Like debates arose in Australia, where much of the native forest and woodland was cleared following British settlement, and similarly with British colonization of India.

One question concerned whether deforestation and cultivation of land created milder winters. A popular view, espoused by the Scottish philosopher David Hume, was that deforestation opened the land to heating by the Sun during winter. In an essay (ca. 1750) he explained that warmer winters occurred because "the land is at present much better cultivated, and that the woods are cleared, which formerly threw a shade upon the earth, and kept the rays of the sun from penetrating to it" (Hume and Miller 1987, p. 451). Because of this, Hume declared that "our northern colonies in America become more temperate, in proportion as the woods are felled." Similar views are seen in the writings of the New England Puritan minister Cotton Mather, who observed that "our cold is much moderated since the opening and clearing of our woods" (Mather 1721, p. 74). Benjamin Franklin, too, believed that climate was warming because "when a country is clear'd of woods, the sun acts more strongly on the face of the earth" (Franklin and Labaree 1966). Franklin's contemporary Hugh Williamson, a physician, scholar, and politician, predicted that with continued clearing of interior lands "we shall seldom be visited by frosts or snows, but may enjoy such a temperature in the midst of winter, as shall hardly destroy the most tender plants" (Williamson 1771). Thomas Jefferson agreed that "a change in our climate … is taking place very sensibly" (Jefferson 1788, p. 88) and urged climate surveys "to show the effect of clearing and culture towards changes of climate" (Jefferson and Bergh 1905). Samuel Williams, a Congregational minister, professor at Harvard, and founder of the University of Vermont, believed that as trees were cut down and settlements increased "the cold decreases, the earth and air become more warm; and the whole temperature of the climate, becomes more equal, uniform and moderate" (Williams 1794, p. 57). This climate change "is so rapid and constant, that it is the subject of common observation and experience."

Not all agreed with this sentiment. In addition to being a lexicographer, Noah Webster of Connecticut was a political writer. He strongly refuted the notion of such changes in climate in an essay published in 1799. Direct observations of climate change were lacking, he noted, and evidence of a warmer climate relied on anecdotes and personal memories. Yet, he, too, admitted to differences between forests and cleared land that altered local climate (Webster 1843, pp. 145). "While a country is covered with trees," he wrote, "the face of the earth is never swept by violent winds; the temperature of the air is more uniform, than in an open country; the earth is never frozen in winter, nor scorched with heat in summer."

The writings of Europeans attested to the prominence of the forest–climate debate, and also to the difference in opinions. Constantin-François Volney published a book in 1803 based on his travels in eastern North America. He observed that "for some years it has been a general remark in the United States, that very perceptible partial changes in the climate took place, which displayed themselves in proportion as the land was cleared" (Volney 1804, p. 266). Alexander von Humboldt responded in 1807 that "the statements so frequently advanced, although unsupported by measurements, that since the first European settlements in New England, Pennsylvania, and Virginia, the destruction of many forests … has rendered the climate more equable, – making the winters milder and the summers cooler, – are now generally discredited" (Humboldt et al. 1850, p. 103). The *Edinburgh Encyclopaedia* asserted that the theory of land-use climate change "is, we fear, to be regarded rather as the birth of a lively fancy, than the offspring of accurate science" (Brewster 1830, pp. 613–614).

Another belief was that forests contributed to the plentiful rainfall in America and that deforestation decreased rainfall. Christopher Columbus developed such a view from his travels to the New World. His son wrote in a biography that he attributed the rainstorms of Jamaica to "the great forests of that land; he knew from experience that formerly this also occurred in the Canary, Madeira, and Azore Islands, but since the removal of forests that once covered those islands, they do not have so much mist and rain as before" (Colón and Keen 1959, pp. 142–143).

Natural scientists developed similar views. John Evelyn declared in 1664 that forests "render those countries and places more subject to rain and mists" (Evelyn 1801, pp. 29–32). John Clayton, an English naturalist and clergyman, attributed the violent thunderstorms of coastal Virginia to the dense forests (Clayton 1693; Berkeley and Berkeley 1965, pp. 48–49). His compatriot John Woodward described how the "great moisture in the air, was a mighty inconvenience and annoyance to those who first settled in America," but that after clearing the forests "the air mended and cleared up apace: changing into a temper much more dry and serene than before" (Woodward 1699). Samuel Williams accounted for the plentiful rainfall because "the immense forests … supply a larger quantity of water for the formation of clouds, than the more cultivated countries of Europe" (Williams 1794, p. 50). Hugh Williamson described a feedback by which forest evaporation enhances rainfall and cools climate (Williamson 1811, pp. 23–25): "The vapours that arise from forests, are soon converted into rain, and that rain becomes the subject of future evaporation, by which the earth is further cooled." In the Caribbean islands, which had undergone widespread forest clearing to grow sugar cane, forest preserves were created to promote rainfall (Anthes 1984).

Settlement of the Great Plains in the 1870s and 1880s shifted the debate from deforestation to afforestation, with the premise that tree planting would increase rainfall (Emmons 1971; Kutzleb 1971; Thompson 1980; Williams 1989). An official in the United States Department of Interior claimed that "the planting of ten or fifteen acres of forest trees on each quarter section [160 acres] will have a most important effect on the climate, equalizing and increasing the moisture" (United States General Land Office 1867, p. 135). That some official believed that "if one-third the surface of the great plains were covered with forest there is every reason to believe the climate would be greatly improved" (United States General Land Office 1868, p. 197). Congress agreed and enacted the Timber Culture Act of 1873 to promote afforestation. Popular science gazettes, too, advocated tree planting to increase rainfall (Oswald 1877; Anonymous 1879). Samuel Aughey, of the University of Nebraska, promoted the notion that plowing the prairie sod was the cause of an increase in rainfall observed at that time. Cultivation allowed the soil to retain more rainfall, which evaporated and rained back onto the land, he theorized (Aughey 1880, pp. 44–45). Charles Wilber popularized this notion with the phrase "rain follows the plow" (Wilber 1881, p. 68). He described how an "army of frontier farmers … could, acting in concert, turn over the prairie sod, and … present a new surface of green, growing crops instead of the dry, hard-baked earth covered with sparse buffalo grass. No one can question or doubt the inevitable effect of this cool condensing surface upon the moisture in the atmosphere."

A sharply divided debate on forest–climate influences continued in the latter half of the nineteenth century and into the twentieth century. Conservationists, botanists, and foresters argued for such influences. George Perkins Marsh devoted a large portion of his treatise *Man and Nature* to forest–climate influences (Marsh 1864). The botanist Richard Upton Piper agreed that "forests trees should be preserved for their beneficial influence upon the climate" (Piper 1855, p. 51). The fledgling forestry division of the United States Department of Agriculture issued reports supportive of the burgeoning field of forest meteorology and forest–rainfall influences (Hough 1878; Fernow 1902; Zon 1927).

Climatologists of the day, however, dismissed the study of forests and climate. A publication on the climate of the United States asserted that "the great differences of surface character which belong to the deserts, woodlands, and other more striking features, are believed to have their *origin* in climate, and not to be agents of causation themselves" (Blodget 1857, p. 482). The geographer Henry Gannett suggested that faulty reasoning was behind the belief that forests increase rainfall. His analysis of precipitation records found no change in rainfall in regions that had undergone increases and decreases in tree cover, and he complained that "a satisfactory explanation of this supposed phenomenon has never … been offered" (Gannett 1888). He further explained that "it may be that in this case an effect has been mistaken for a cause, or rather, since it is universally recognized that rainfall produces forests, the converse has been incorrectly assumed to be also true" (Anonymous 1888). The eminent meteorologist William Ferrel argued in favor of large-scale control of precipitation by atmospheric circulation, not by surface conditions (Ferrel 1889). His colleague Cleveland Abbe wrote that "rational climatology gives no basis for the much-talked-of influence upon the climate of a country produced by the growth or destruction of forests … and the cultivation of crops over a wide extent of prairie" (Abbe 1889). Abbe believed that "the idea that forests either increase or diminish the quantity of rain that falls from the clouds is not worthy to be entertained by rational, intelligent men" (Moore 1910, p. 7).

Foresters and climatologists in the United States Department of Agriculture were sharply divided over forest influences on rainfall. While the foresters issued reports in support of the science (Hough 1878; Fernow 1902; Zon 1927), their colleagues in the department's Weather Bureau (the predecessor of the present-day National Weather Service) resoundingly dismissed these ideas. Mark Harrington, chief of the Weather Bureau, rejected his forestry colleagues' belief that forests affected climate in any manner (Fernow 1902). Willis Moore, Harrington's successor, also rebuffed studies relating forests and climate with the retort that "while much has been written on this subject, but little of it has emanated from meteorologists" (Moore 1910, p. 3). "Precipitation," he explained, "controls forestation, but forestation has little or no effect upon precipitation" (Moore 1910, p. 37). The caustic rhetoric confused a writer in the journal *Nature*, who reported that "the literature on the subject is somewhat bewildering" (Anonymous 1912).

The views about ongoing climate change in colonial America ultimately proved to be false. Climate was not changing; winters were not becoming milder with land clearing; rainfall was not decreasing because of deforestation or increasing because of tree planting and soil cultivation. However, meteorologists of that era, too, were ultimately proved wrong. As they sought physical explanations for geographic variations in climate, they were too quick to dismiss the precept that forests, grasslands, croplands, and other ecosystems do indeed influence climate. Interest in the climatic effects of deforestation, cultivation, and overgrazing reemerged in the 1970s, with recognition that human activities do indeed change climate and that land use is one such mechanism for climate change (Landsberg 1970; Otterman 1974, 1977; Schneider and Dickinson 1974; Sagan et al. 1979). One hundred years after a writer to *Nature* found the debate to be "belwidering," the journal published another paper that found that tropical forests do indeed increase rainfall (Spracklen et al. 2012).

1.4 | Ecological Climatology

Scholars of the eighteenth and nineteenth centuries lacked the scientific tools to properly ascertain forest influences on climate, but scientists in the latter part of the twentieth century had a new tool – global climate models – with which to study how plants and ecosystems affect climate. Scientific interest over the past few decades in the coupling between climate and life has paralleled the trend by atmospheric scientists to recognize the planet as a system of interacting spheres.

Today, scientists identify four main components of the Earth system: atmosphere, air; hydrosphere, water; biosphere, living things; and geosphere, solid portion of Earth. The geosphere can be subdivided into other spheres. The lithosphere is the solid outer layer of Earth including the crust and upper mantle. Its outermost layer is called the pedosphere, or soil. Some scientists separately identify the cryosphere, or frozen portion of Earth. The influence of humans is so prevalent, especially after the Industrial Revolution, that a new sphere, the anthroposphere, has been proposed to describe that part of Earth modified by people for human activities or habitats. Earth's climate must be understood in terms of a system of interacting spheres (atmosphere, hydrosphere, biosphere, geosphere, and anthroposphere); the energy, water, and biogeochemical cycles that link these spheres; and the interactions with human systems that alter these cycles. This parallels a progression in atmospheric sciences from (Figure 1.3): atmospheric general circulation models, which considered atmospheric physics and dynamics; to atmosphere–ocean general circulation models, which included the coupling of the atmosphere with models of ocean and sea-ice physics and dynamics; to global climate models, which additionally accounted for hydrometeorological coupling with land; and now to Earth system models, which also include atmospheric chemistry, terrestrial and marine ecology, and biogeochemistry.

At the intersection of these spheres is an interdisciplinary field of study called biogeoscience that bridges the earth and life sciences (Figure 1.4). Biogeoscience is the study of the interactions between life and Earth's atmosphere, hydrosphere, and geosphere. It has long been synonymous with the study of

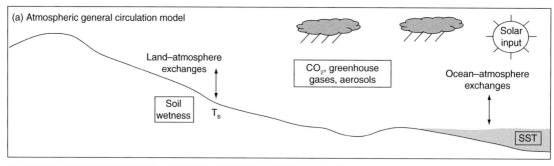

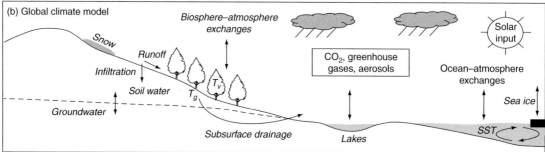

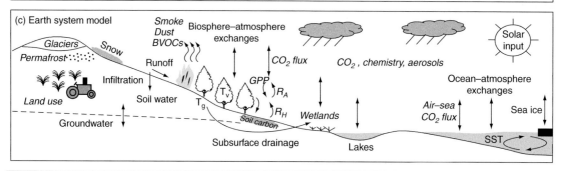

Fig. 1.3 Components of the Earth system, their processes, and interactions as represented in global models. New processes added to each model are highlighted in italics. The distinction among models is not precise, and the transition among models is in fact blurred. (a) Atmospheric general circulation models circa 1970s. These models used prescribed inputs of atmospheric CO₂, other greenhouse gases, and aerosols. They calculated land and ocean physical flux exchanges using prescribed soil wetness and sea surface temperature. (b) Global climate models circa 1990s. These models added the hydrologic cycle on land and plant canopies. They included ocean general circulation models to calculate sea surface temperature, sea ice, and ocean dynamics. (c) Earth system models circa 2010s. These models added the carbon cycle and other biogeochemical processes, anthropogenic land use, wetlands, glaciers and cryospheric processes, atmospheric chemistry, and aerosols.

biogeochemistry, but is broader and includes the interactions between living organisms and the physical environment and the manner in which organisms modify physical systems.

This book examines one element of that science – the commonality between ecological and atmospheric sciences that affect weather, climate, and atmospheric composition. The study of plants and terrestrial ecosystems, and

human appropriation of ecosystem functions, is as essential to the study of Earth's climate as is the study of atmospheric physics and dynamics. This book merges the relevant areas of ecology and climatology, broadly defined to include weather, climate, and atmospheric composition, into an overlapping study of ecological climatology. Ecological climatology is an interdisciplinary framework to understand the functioning of plants and terrestrial ecosystems in the Earth

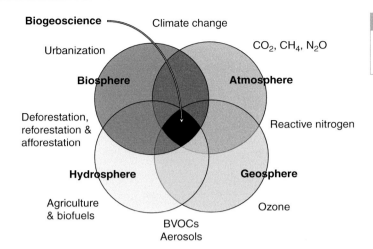

Biogeoscience

Climate change

Urbanization

CO$_2$, CH$_4$, N$_2$O

Biosphere

Atmosphere

Deforestation,
reforestation &
afforestation

Reactive nitrogen

Hydrosphere

Geosphere

Agriculture
& biofuels

Ozone

BVOCs
Aerosols

Fig. 1.4 Biogeoscience as the intersection of the atmosphere, hydrosphere, biosphere, and geosphere. Human influences are shown in the text outside the spheres.

system and the physical, chemical, and biological processes by which the biosphere affects atmospheric processes. A central theme of the book is that plants and terrestrial ecosystems, through their cycling of energy, water, chemical elements, and trace gases, are a critical determinate of climate. Changes in terrestrial ecosystems through natural vegetation dynamics, through human land uses, and through climate change itself significantly affect the trajectory of climate change.

Figure 1.5 illustrates five core areas: the *biogeophysical* and *biogeochemical* processes that regulate the exchanges of energy, water, momentum, and chemical materials with the atmosphere over periods of minutes to hours; *watersheds* and *ecosystems* and the hydrological and ecological processes that regulate these exchanges over periods of days to months; and *landscape dynamics* and the ecological and anthropogenic processes controlling the arrangement of plants into communities, the functioning of ecosystems, and temporal changes in response to disturbance over periods of years to centuries.

Biogeophysics is the study of physical interactions of the biosphere and geosphere with the atmosphere. It considers the transfers of heat, moisture, and momentum between land and atmosphere and the meteorological, hydrological, and ecological processes regulating these exchanges. Momentum is transferred when plants and other rough elements of the land surface interfere with the flow of air. Heat and

moisture are exchanged when net radiation at the surface (R_n) is returned to the atmosphere as sensible heat (H), latent heat (λE), or stored in the ground (G). Biogeophysical feedbacks are understood through the surface energy balance:

$$R_n = \left(S\downarrow - S\uparrow\right) + \left(L\downarrow - L\uparrow\right) = H + \lambda E + G \qquad (1.1)$$

where $S\downarrow$ and $L\downarrow$ are downwelling solar radiation and longwave radiation onto the surface, respectively, and $S\uparrow$ and $L\uparrow$ are the upward radiative fluxes from the surface. Collectively, these four radiative fluxes comprise net radiation. The typical unit of measurement is the flux of energy per unit area (J s^{-1} m^{-2}, or W m^{-2}).

The surface energy balance highlights several important land–atmosphere interactions. One relates to surface albedo (Figure 1.6a). An increase in surface albedo, which can occur with loss of vegetation cover, increases reflected solar radiation, reduces the absorption of solar radiation at the surface, and cools the surface climate. Less energy returns to the atmosphere as sensible and latent heat, which promotes subsidence of air aloft and may reduce precipitation. Such albedo influence on rainfall is particularly important in semiarid climates. In cold, snowy climates, tall trees protrude above the snowpack and reduce surface albedo. Vegetation masking of the high albedo of snow creates a warmer climate than in the absence of trees.

Another important aspect of land–atmosphere coupling is surface roughness (Figure 1.6b).

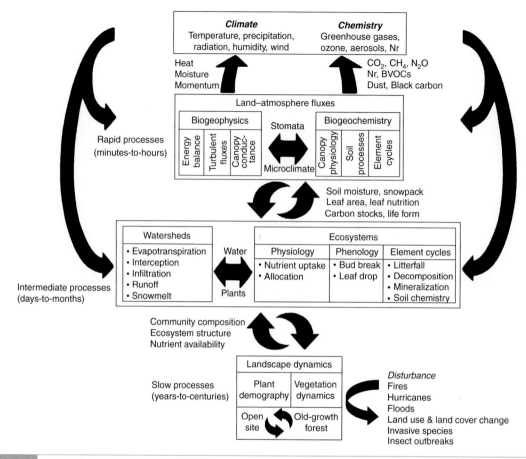

Fig. 1.5 A generalized scope of ecological climatology showing the biogeophysical and biogeochemical processes by which terrestrial ecosystems affect weather, climate, and atmospheric composition, the watershed and ecosystem processes that govern biosphere–atmosphere coupling, and the role of landscape dynamics in initiating change.

Rough surfaces such as forests generate more turbulence and have higher sensible and latent heat fluxes than smoother surfaces such as grasslands, all other factors being equal. A decrease in roughness length, by decreasing turbulence and aerodynamic conductance, can lead to a warmer, drier atmospheric boundary layer.

Biogeochemistry is the study of element cycling among the biosphere, geosphere, and atmosphere. Carbon dioxide (CO_2), methane (CH_4), and nitrous oxide (N_2O) are greenhouse gases regulated in part by terrestrial ecosystems. The net storage of carbon in the biosphere in the absence of fire and other losses, known as net ecosystem production (NEP), is the balance of carbon uptake during gross primary production (GPP), carbon loss during plant respiration (R_A), and carbon loss during decomposition (R_H):

$$NEP = GPP - R_E = (GPP - R_A) - R_H = NPP - R_H$$

$$(1.2)$$

The typical unit of measurement is the mass of carbon exchanged per unit area over some period of time (e.g., g C m^{-2} yr^{-1}). The net carbon uptake by plants ($GPP - R_A$) is known as net primary production (NPP), and the total ecosystem respiration is $R_E = R_A + R_H$. The signature of terrestrial ecosystems is seen in the annual cycle of atmospheric CO_2, which has low concentration during the growing season when plants absorb CO_2 and high concentration during the dormant season. It is also evident in the uptake

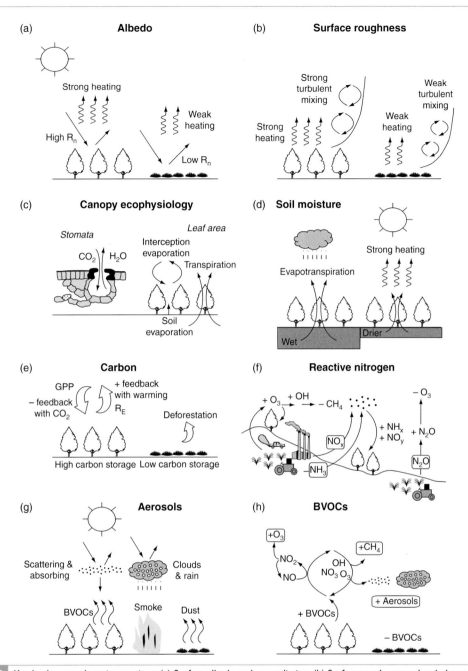

Fig. 1.6 Key land–atmosphere interactions. (a) Surface albedo and net radiation. (b) Surface roughness and turbulent mixing. (c) Canopy ecophysiology, stomata, leaf area, and evapotranspiration. (d) Soil moisture, evapotranspiration, and precipitation. (e) The carbon cycle. (f) Reactive nitrogen, atmospheric chemistry, and aerosols. (g) Aerosols, radiation, and clouds. (h) Biogenic volatile organic compounds, atmospheric chemistry, and secondary organic aerosols.

of anthropogenic CO_2 emissions by terrestrial ecosystems. Only about one-half of current anthropogenic CO_2 emissions remain in the atmosphere. Oceans and terrestrial ecosystems absorb the other half. Two key terrestrial feedbacks that regulate this are the uptake of carbon during photosynthesis with elevated atmospheric CO_2 and the loss of carbon during respiration with a warmer climate. Terrestrial ecosystems differ in these feedbacks, their carbon storage, and their capacity to sequester anthropogenic carbon emissions (Figure 1.6e).

Terrestrial ecosystems similarly regulate the concentrations of CH_4 and N_2O in the atmosphere. Ecosystems are also sources of reactive nitrogen (Nr) that alters atmospheric chemistry and produces aerosols; sources of mineral dust and black carbon (soot) from wildfires, which are important aerosols; and sources of biogenic volatile organic compounds (BVOCs), which produce ozone (O_3), increase CH_4 in the atmosphere, and form aerosols (Figure 1.6f–h).

Biogeophysical and biogeochemical processes do not occur in isolation. For example, stomata open to absorb CO_2 during photosynthesis, but in doing so water diffuses out of the leaf during transpiration (Figure 1.6c). Consequently, water loss during transpiration is tied to carbon uptake during photosynthesis. This is seen in studies that relate leaf photosynthesis, transpiration, and stomatal conductance. The physiology of stomata represents a balance between the conflicting goals of maximizing CO_2 uptake while minimizing water loss.

The exchanges of energy, water, and other materials between biosphere and atmosphere depend on the hydrologic cycle. The fundamental system of study in hydrology is a watershed, or catchment. Over long periods of time, it is commonly assumed that water entering a watershed as precipitation (P) either returns to the atmosphere as evapotranspiration (E) or runs off into streams and rivers (R) so that the annual water balance is:

$$P - E = R \qquad (1.3)$$

The typical unit of measurement is the mass of water flowing per unit area over some period of time (e.g., kg H_2O m^{-2} yr^{-1}).

The hydrologic cycle influences climate in many ways. One prominent means is through latent heat flux, or evapotranspiration. A decrease in leaf area reduces the surface area for transpiration and for the interception of rainfall (Figure 1.6c). Evapotranspiration from the plant canopy decreases, but soil evaporation may increase. In general, a reduction in evapotranspiration, which occurs, for example, with deforestation, produces an increase in runoff. A decrease in vegetation cover that reduces latent heat flux warms surface climate and may reduce precipitation. This is particularly prominent in tropical deforestation. Wet soil can sustain a high latent heat flux and creates a cool, moist atmospheric boundary layer – conditions that may feed back to increase precipitation (Figure 1.6d). In contrast, dry soil decreases latent heat flux and amplifies droughts and heat waves.

Numerous topographic, edaphic, and ecological features control the hydrology of a watershed. Hydrologic processes such as evapotranspiration, interception of precipitation by plants, infiltration of water into soil, runoff into streams and rivers, and snowmelt determine soil moisture, snow pack, and saturated areas within the watershed – conditions that vary with a timescale of days to months and that influence surface fluxes.

Terrestrial ecosystems are an expression of an ecological system. All ecosystems have structure – the arrangement of materials in pools and reservoirs – and function – the flows and exchanges among these pools. For example, the carbon cycle is commonly described by pools such as foliage, stem, and root biomass and decomposing soil organic matter. Functions include carbon uptake during photosynthesis and carbon loss during respiration. A variety of ecological processes operating at timescales of days to months influence ecosystem function. The amount of leaf area is an important determinant of photosynthesis, absorption of solar radiation, heat and momentum fluxes, evapotranspiration, and interception. In many plant communities, the presence of leaves varies seasonally in relation to temperature or moisture stress. Other processes such as litterfall, decomposition, mineralization of organically bound nutrients, nutrient uptake, and the allocation of resources to growth influence carbon storage. Short-term functioning of terrestrial ecosystems is seen in the fluxes of photosynthesis and respiration in relation to the diurnal cycle of solar radiation, temperature, and humidity and day-to-day variability arising from the passage

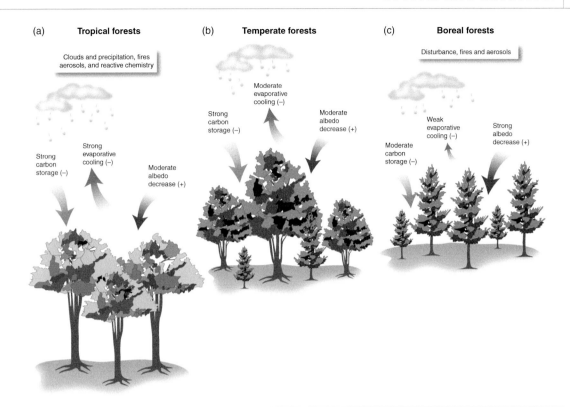

Fig. 1.7 Climate services of (a) tropical, (b) temperate, and (c) boreal forests in terms of albedo, evapotranspiration, and carbon storage. Symbols indicate a positive climate forcing (+, warming) or negative climate forcing (–, cooling). Text boxes indicate other key processes with uncertain climate influences. Adapted from Bonan (2008).

of weather systems. Interannual variability in these functions is seen in the interannual variability of atmospheric CO_2. Long-term functioning manifests in the relationships of climate with net primary production and soil carbon turnover.

Ecosystems are not just static elements of the landscape; they are dynamic. The abundance and biomass of plant species change over periods of years to centuries. Disturbances such as floods, fires, and hurricanes initiate temporal change in ecosystems known as succession. The life history patterns of plants have evolved in part as a result of recurring disturbances. Many plant species are ephemeral members of the landscape, adapted to recently disturbed sites. Others dominate old-growth ecosystems in the late stages of succession. Changes in climate and atmospheric composition affect ecosystems. Long-term changes in temperature, precipitation, atmospheric CO_2, the chemistry

of rainfall, and the deposition of chemical elements onto leaves and soil alter the conditions for vegetation growth. Resulting changes in species composition, ecosystem structure, and nutrient availability each feeds back to affect climate. In particular, forest growth absorbs carbon from the atmosphere while deforestation releases carbon to the atmosphere. Human activities also alter ecosystems through clearing of land for agriculture, farm abandonment, and introduction of invasive species.

The outcome of these physical, chemical, and biological processes can be seen in the influence of the world's forests on climate, shown in Figure 1.7 (Bonan 2008). Tropical rainforests provide a negative climate forcing that cools climate. High rates of carbon storage reduce the accumulation of anthropogenic CO_2 emissions in the atmosphere (reducing greenhouse gas warming). Evaporation of the plentiful rainfall augments this with strong evaporative cooling.

In boreal forests, strong absorption of solar radiation (low surface albedo) may outweigh carbon sequestration so that the boreal forest warms global climate (positive climate forcing) compared with removal of the forest. The net climate forcing of temperate forests is more uncertain. Reforestation and afforestation sequester carbon, but biogeophysical processes augment or diminish the negative biogeochemical forcing of climate. Low surface albedo, especially during winter in snowy climates, contributes to warming, while rates of high evapotranspiration during summer contribute to cooling.

These forest–atmosphere interactions can dampen or amplify anthropogenic climate change. However, an integrated assessment of forest influences entails an evaluation beyond albedo, evapotranspiration, and carbon to include other greenhouse gases, biogenic aerosols, and reactive gases. Forests, in addition to being carbon sinks, also act as sources for aerosol particles. The combined carbon cycle and biogeophysical effect of tropical forests cools climate, but fires, biogenic aerosols, and reactive gases in these forests also affect clouds and precipitation. Biogenic aerosols are also important in boreal forests, where the net forcing from fire must also be considered.

An emerging research frontier is to link the biogeophysical and carbon cycle influences of terrestrial ecosystems with a full depiction of biogeochemical feedbacks mediated through atmospheric chemistry. Terrestrial ecosystems are sources of CH_4 and N_2O. Both are powerful, long-lived greenhouse gases. Additional reactive nitrogen increases N_2O emissions, can produce tropospheric ozone, increase the oxidation capacity of the troposphere (OH radical), decrease CH_4, form aerosols, and increase the deposition of nitrogen onto land (Figure 1.6f). Biomass burning during wildfires injects black carbon (soot) and primary organic aerosols into the atmosphere and also many short-lived gases that affect atmospheric chemistry and air quality (Figure 1.6g). Mineral aerosols (dust) are another important type of aerosol. Plants emit numerous biogenic volatile organic compounds (mostly as isoprene and monoterpenes) that increase the lifetime of CH_4 in the atmosphere and that produce ozone and secondary organic aerosols (Figure 1.6h). Tropospheric ozone is a greenhouse gas that warms climate. The effects of aerosols on climate are complex. Many aerosols scatter solar radiation back to space and cool climate; black carbon absorbs solar radiation and warms climate. Aerosols can also increase cloud brightness (reflecting more solar radiation to space) and suppress rainfall.

1.5 | Timescales of Climate–Ecosystem Interactions

The coupling of ecosystems and climate occurs over a continuum of timescales from minutes to seasons to millennia (Table 1.2). At short timescales, the seasonal emergence and senescence of leaves alters the absorption of radiation, the dissipation of energy into latent and sensible heat, and CO_2 uptake. The effect of these changes can be seen in air temperature, humidity, and the seasonal drawdown of CO_2 in the atmosphere. At seasonal to interannual timescales, photosynthesis, respiration, evapotranspiration, and reactive gas fluxes influence the physical and chemical state of the atmosphere. Interannual variability in temperature and precipitation alter ecosystem metabolism, which is again evident in the concentration of CO_2 in the atmosphere.

Over several decades, people shape the landscape through clearing of land for agriculture, reforestation of abandoned farmland, and through urbanization. These land uses alter surface energy fluxes, biogeochemical cycles, and the hydrological cycle, and they produce a discernible signal in temperature, precipitation, the concentrations of CO_2, CH_4, and N_2O in the atmosphere, and the deposition of atmospheric pollutants such as reactive nitrogen, ozone, and black carbon aerosols onto land. At longer timescales of decades to centuries, slower successional changes in response to disturbances control community composition and ecosystem structure and so alter surface energy fluxes, carbon storage, and trace gas emissions. Coupled climate–ecosystem dynamics are particularly evident over periods of centuries to millennia.

Table 1.2 | Timescales of vegetation change and associated atmospheric impact

Vegetation change	Timescale	Ecological signal	Controlling processes	Atmospheric signal
Leaf emergence	Seasonal to interannual	Leaf area index Canopy conductance	Air temperature Soil moisture Life form	Cooler temperature Higher humidity CO_2 drawdown Lower albedo Greater latent heat Less sensible heat
Ecosystem metabolism	Seasonal to interannual	Leaf area index Carbon storage Water balance Element cycles	Air temperature Soil moisture Humidity Solar radiation Atmospheric CO_2 Nr deposition Ozone	CO_2 drawdown Albedo Latent heat Sensible heat BVOCs Ozone
Land use → Succession → Biogeography →	Decadal Decadal to century Century to millennial	Leaf area index Carbon storage Species composition Nutrient availability Ecosystem structure	People Life history Disturbance Climate Atmospheric CO_2 Nr deposition	Temperature Precipitation Energy balance CO_2, CH_4, Nr BVOCs Aerosols Ozone
Evolution	Millennial	Stomatal density Leaf form	Atmospheric CO_2	Leaf temperature Leaf energy fluxes

Temperature, precipitation, and atmospheric CO_2 are the chief determinants of the geographic distribution of vegetation across the planet. In turn, this biogeography affects climate and atmospheric CO_2 concentration. The outcome of climate–vegetation interactions can also be seen in the evolutionary record. There is a close relationship between leaf shape and climate. Vascular plants introduced numerous biotic feedbacks on climate, primarily related to plant responses to CO_2 that affected stomatal conductance and leaf form.

1.6 | Scientific Tools

The influence of plants and terrestrial ecosystems on the atmosphere can be discerned through environmental monitoring, experimental manipulation, or with the use of numerical models of weather and climate. Intensive field campaigns with ground-based measurements of biosphere–atmosphere flux exchanges and aircraft measurements of atmospheric composition provide datasets to analyze biosphere–atmosphere coupling over short time periods (up to several weeks). Environmental monitoring techniques include eddy covariance flux towers that provide continuous measurements of biosphere–atmosphere exchanges of energy, moisture, and trace gases at fast timescales (subhourly). The longest such sites have been continually operating for over two decades. Ecosystem and watershed studies provide monitoring of carbon and elemental stocks and fluxes and the hydrologic cycle, typically at longer timescales (e.g., annual). Such studies extend over several decades. Satellite observations of leaf area, surface albedo, surface temperature, and other properties provide global coverage at a high spatial resolution for a period extending now for almost three decades. Atmospheric CO_2 observations at numerous locations throughout the world provide information about the seasonal dynamics of land–atmosphere carbon exchange and continental-scale fluxes on timescales of years to decades. The longest such record, at Mauna Loa, Hawaii, dates back to 1958. Ice core measurements reveal the history of CO_2, CH_4, N_2O, and dust in the atmosphere over the past several hundred thousand years and variations with glacial–interglacial cycles.

Whole-ecosystem experimental manipulations provide insight to ecosystem responses to environmental change. Such experiments warm ecosystems or exclude rainfall to study responses to climate change, enrich the air with CO_2 to study responses to elevated atmospheric CO_2 concentrations, and fertilize the soil with nutrients (e.g., nitrogen and phosphorus) to examine responses to perturbed biogeochemical cycles. Watershed manipulation studies that remove vegetation show the biotic control of the hydrologic cycle.

The influence of plants and ecosystems on large-scale climate is difficult to establish directly through observations. Careful examination of climatic data can sometimes reveal an ecological influence, such as the effect of leaf emergence on springtime evapotranspiration and air temperature. Eddy covariance flux towers and field experiments provide local-scale insight to ecosystem–atmosphere interactions, and advances in remote sensing science aid extrapolation to larger spatial scales. More often, however, our understanding of how plants and ecosystems affect climate comes from atmospheric models and their numerical parameterizations of Earth's biosphere. Paired climate simulations, one serving as a control to compare against another simulation with altered vegetation, demonstrate an ecological influence on climate.

1.7 | Overview of the Book

The book describes Earth's climate, the processing shaping climate, how climate changes over time, and the manner in which terrestrial ecosystems influence climate. It is divided into six sections on the Earth system, global physical climatology, hydrometeorology, biometeorology, terrestrial plant ecology, and terrestrial forcings and feedbacks. The first section describes component spheres of the Earth system (Chapter 2) and the energy, water, and biogeochemical cycles that link these spheres (Chapter 3).

The second section reviews climate, climate variability, and climate change. The radiative balance of the atmosphere, its geographic variation, and its annual cycle are an important determinant of climate (Chapter 4). Geographic and seasonal variation in the radiative balance drives the general circulation of the atmosphere (Chapter 5). This gives rise to Earth's macroclimates, and within which mountains, lakes, and vegetation create local climates (Chapter 6). The realized temperature and precipitation in any year can deviate markedly from the long-term climatology because of seasonal-to-interannual atmospheric variability such as the El Niño/Southern Oscillation and North Atlantic Oscillation (Chapter 7). Climate also changes over longer timescales of centuries and millennia in response to changes in insolation, greenhouse gases, and numerous feedbacks within the Earth system (Chapter 8).

The third section on hydrometeorology reviews the hydrologic cycle, surface energy fluxes, and the interactions between the hydrosphere and atmosphere. Soils store vast amounts of energy and water, and this modulates the diurnal and annual cycle of temperature, provides water for evapotranspiration, and regulates the hydrologic cycle on land (Chapter 9). The hydrologic cycle on land is reviewed in terms of point processes (Chapter 10) and watershed processes (Chapter 11). The hydrologic cycle regulates surface energy fluxes. The energy balance at Earth's land surface requires that energy gained from net radiation be balanced by the fluxes of sensible and latent heat to the atmosphere and the storage of heat in soil (Chapter 12). The fluxes of sensible and latent heat occur because turbulent mixing of air transports heat and moisture, typically away from the surface (Chapter 13). Soil moisture exerts a strong control on the partitioning of net radiation into sensible and latent heat fluxes, and through this affects the atmospheric boundary layer (Chapter 14).

The fourth section reviews biometeorology. The exchanges of sensible heat, latent heat, and CO_2 between land and atmosphere are regulated by the physiology and micrometeorology of plant canopies. Individual leaves absorb radiation and exchange sensible heat and latent heat with the surrounding air (Chapter 15). The uptake of CO_2 during photosynthesis is tightly coupled to the loss of water during transpiration (Chapter 16). Both occur through stomatal openings on the leaf surface. The aggregate flux from vegetation is the integral of the individual leaf fluxes over the depth of the canopy (Chapter 17).

The fifth section reviews terrestrial plant ecology. It extends the discussion of plant physiology from the previous section to an overview of whole-plant allocation and plant strategies (Chapter 18), the arrangement of plant species into populations and communities (Chapter 19), and the functioning of ecosystems (Chapter 20). The weathering of rocks, the decomposition of soil organic material, and soil formation are part of the biogeochemical cycling of carbon, nitrogen, and other elements among the atmosphere, biosphere, and geosphere (Chapter 21). Vegetation changes over time in response to recurring disturbance (Chapter 22). Landscapes represent another level of ecological organization, merging the concepts of populations, communities, ecosystems, and plant dynamics. Spatial gradients in the environment combine with natural and anthropogenic disturbances to create a mosaic of plant communities and ecosystems across the landscape (Chapter 23). The structure and composition of vegetation and the functioning of terrestrial ecosystems, which at a local scale are shaped by environmental factors such as temperature and moisture, are also influenced by global climate. This is seen in the biogeography of vegetation and in the global carbon cycle, especially net primary production (Chapter 24).

The final section examines terrestrial forcings and feedbacks in the Earth system, especially how natural and human changes in land cover and land use affect climate. Numerous global climate model experiments have demonstrated the role of land surface hydrology and terrestrial vegetation in determining regional and global climate. Chapter 25 reviews the representation of land surface processes in global models. Soil moisture, snow, and vegetation contribute to climate variability (Chapter 26). Vegetation dynamics in response to climate change alters climate by changing surface albedo, net radiation, and evapotranspiration. The boreal forest–tundra ecotone and the Sahel of North Africa are prominent examples of coupled climate–vegetation dynamics (Chapter 27). Deforestation, reforestation, degradation of drylands, and cultivation of croplands are case studies of how human uses of land alter climate through biogeophysical processes (Chapter 28). In addition, terrestrial ecosystems are coupled to climate through various biogeochemical cycles. The carbon cycle is a prominent feedback with climate change (Chapter 29), as are reactive nitrogen (Chapter 30) and aerosols (Chapter 31). Urbanization also alters climate and the hydrologic cycle (Chapter 32). Because terrestrial ecosystems have such a significant impact on climate through albedo, evapotranspiration, carbon storage, and other processes, they can be managed to mitigate the undesirable effects of anthropogenic climate change (Chapter 33). Greater understanding of Earth and its climate requires that all components of the Earth system – physical, chemical, biological, socioeconomic – be considered. Terrestrial ecosystems – nature's technology – are a critical aspect of planetary habitability in an ever increasing technological world (Chapter 34).

1.8 | Review Questions

1. Scientists have found fossilized remains of tree foliage in Antarctica that date to 100 million years ago. What does this indicate about the climate of that era?

2. A 15 km² region has areas of forest and open land, with an average elevation of 50 m above sea level. Within this region 11 meteorological stations were established in the open land and in small clearings within the forests, all at a height of 1.07 m above the ground. The measurements, obtained in 1886, showed annual precipitation in the open land averaged 383 mm ($n = 4$); the forests averaged 448 mm of rainfall ($n = 7$). Does this support the notion that forests increase rainfall?

3. In the semiarid Sahel region of North Africa, annual rainfall is low and the vegetation is sparse with low annual productivity. Years with high rainfall anomalies have greater vegetation cover and productivity. What does this relationship indicate about rainfall and vegetation?

4. A paired watershed study compares one catchment that is forested and an adjacent catchment that is deforested. Both receive similar annual rainfall, but runoff to streams is greater in the deforested catchment. What does this show about forest–climate influences?

5. Describe changes in ecosystem structure and function with reforestation that affect climate.

1.9 | References

Abbe, C. (1889). Is our climate changing? *The Forum*, 6 (February), 678–688.

Andréassian, V. (2004). Waters and forests: From historical controversy to scientific debate. *Journal of Hydrology*, 291, 1–27.

Anonymous (1879). Rainfall and forests. *Scientific American*, 41(20), 312.

Anonymous (1888). The influence of forests on the quantity and frequency of rainfall. *Science*, 12(303), 242–244.

Anonymous (1912). Forests and rainfall. *Nature*, 89, 662–664.

Anthes, R. A. (1984). Enhancement of convective precipitation by mesoscale variations in vegetative covering in semiarid regions. *Journal of Climate and Applied Meteorology*, 23, 541–554.

Arneth, A., Harrison, S. P., Zaehle, S., et al. (2010). Terrestrial biogeochemical feedbacks in the climate system. *Nature Geoscience*, 3, 525–532.

Arneth, A., Brown, C., and Rounsevell, M. D. A. (2014). Global models of human decision-making for land-based mitigation and adaptation assessment. *Nature Climate Change*, 4, 550–557.

Aughey, S. (1880). *Sketches of the Physical Geography and Geology of Nebraska*. Omaha: Daily Republican Book and Job Office.

Berkeley, E., and Berkeley, D. S. (1965). *The Reverend John Clayton: A Parson with a Scientific Mind: His Scientific Writings and Other Related Papers*. Charlottesville: University Press of Virginia.

Blodget, L. (1857). *Climatology of the United States, and of the Temperate Latitudes of the North American Continent*. Philadelphia: J. B. Lippincott.

Bonan, G. B. (2008). Forests and climate change: Forcings, feedbacks, and the climate benefits of forests. *Science*, 320, 1444–1449.

Brewster, D. (1830). *The Edinburgh Encyclopaedia*, vol. 1, 4th ed. Edinburgh: W. Blackwood.

Clayton, J. (1693). A letter from Mr. John Clayton Rector of Crofton at Wakefield in Yorkshire to the Royal Society, May 12, 1688, giving an account of several observables in Virginia, and in his voyage thither, more particularly concerning the air. *Philosophical Transactions*, 17, 781–795.

Colinvaux, P. (1986). *Ecology*. New York: Wiley.

Colón, F., and Keen, B. (1959). *The Life of the Admiral Christopher Columbus by His Son Ferdinand*. New Brunswick, New Jersey: Rutgers University Press.

Deardorff, J. W. (1978). Efficient prediction of ground surface temperature and moisture, with inclusion of a layer of vegetation. *Journal of Geophysical Research*, 83C, 1889–1903.

Dickinson, R. E., Henderson-Sellers, A., Kennedy, P. J., and Wilson, M.F. (1986). *Biosphere–Atmosphere Transfer Scheme (BATS) for the NCAR Community Climate Model*, Technical Note NCAR/TN-275+STR. Boulder, Colorado: National Center for Atmospheric Research.

Emmons, D. M. (1971). Theories of increased rainfall and the Timber Culture Act of 1873. *Forest History*, 15(3), 6–14.

Evelyn, J. (1801). *Silva: or, A Discourse of Forest-Trees, and the Propagation of Timber in His Majesty's Dominions; as it was delivered in the Royal Society, on the 15th of October 1662*, vol. I. York, United Kingdom: Wilson and Spence.

Fernow, B. E. (1902). *Forest Influences*, United States Department of Agriculture, Forestry Division, Bulletin No. 7. Washington, DC: Government Printing Office.

Ferrel, W. (1889). Note on the influence of forests upon rainfall. *American Meteorological Journal*, 5(10), 433–435.

Fleming, J. R. (1998). *Historical Perspectives on Climate Change*. New York: Oxford University Press.

Forman, R. T. T., and Godron, M. (1986). *Landscape Ecology*. New York: Wiley.

Franklin, B., and Labaree, L. W. (1966). Benjamin Franklin to Ezra Stiles, May 29, 1763. In *The Papers of Benjamin Franklin*, vol. 10, ed. L. W. Labaree. New Haven: Yale University Press, pp. 264–267.

Gannett, H. (1888). Do forests influence rainfall? *Science*, 11(257), 3–5.

Glacken, C. J. (1967). *Traces on the Rhodian Shore: Nature and Culture in Western Thought from Ancient Times to the End of the Eighteenth Century*. Berkeley: University of California Press.

Hough, F. B. (1878) *Report upon Forestry*. Washington, DC: Government Printing Office.

Humboldt, A. von, Otté, E. C., and Bohn, H. G. (1850). *Views of Nature: Or Contemplations on the Sublime Phenomena of Creation; with Scientific Illustrations*. London: H. G. Bohn.

Hume, D., and Miller, E. F. (1987). Of the populousness of ancient nations. In *Essays: Moral, Political, and Literary*, rev. ed., ed. E. F. Miller. Indianapolis: Liberty Fund, pp. 377–464.

Jefferson, T. (1788). *Notes on the State of Virginia*. Philadelphia: Prichard and Hall.

Jefferson, T., and Bergh, A. E. (1905). Thomas Jefferson to Lewis E. Beck, July 16, 1824. In *The Writings of Thomas Jefferson*, vol. 16, ed. A. E. Bergh. Washington, DC: Thomas Jefferson Memorial Association of the United States, pp. 71–72.

Kutzleb, C. R. (1971). Can forests bring rain to the Plains? *Forest History*, 15(3), 14–21.

Lamb, H. H. (1977). *Climate: Present, Past and Future*, vol. 2. *Climatic History and the Future*. London: Methuen.

Lamb, H. H. (1995). *Climate, History and the Modern World*, 2nd ed. London: Routledge.

Landsberg, H. E. (1970). Man-made climatic changes. *Science*, 170, 1265–1274.

Levis, S. (2010). Modeling vegetation and land use in models of the Earth System. *WIREs Climate Change*, 1, 840–856.

Mahowald, N., Ward, D. S., Kloster, S., et al. (2011). Aerosol impacts on climate and biogeochemistry. *Annual Review of Environment and Resources*, 36, 45–74.

Marsh, G. P. (1864). *Man and Nature; or, Physical Geography as Modified by Human Action*. New York: Charles Scribner.

Mather, C. (1721). *The Christian Philosopher: A Collection of the Best Discoveries in Nature, with Religious Improvements*. London: E. Matthews.

Moore, W. L. (1910). *A Report on the Influence of Forests on Climate and on Floods*. Washington, DC: Government Printing Office.

Neumann, J. (1985). Climatic change as a topic in the classical Greek and Roman literature. *Climatic Change*, 7, 441–454.

Oliver, J. E. (1996). Climatic zones. In *Encyclopedia of Climate and Weather*, vol. 1, ed. S. H. Schneider. New York: Oxford University Press, pp. 141–145.

Oswald, F. L. (1877). The climatic influence of vegetation – a plea for our forests. *Popular Science Monthly*, 11(August 1877), 385–390.

Otterman, J. (1974). Baring high-albedo soils by overgrazing: a hypothesized desertification mechanism. *Science*, 186, 531–533.

Otterman, J. (1977). Anthropogenic impact on the albedo of the Earth. *Climatic Change*, 1, 137–155.

Piper, R. U. (1855). *The Trees of America*. Boston: W. White.

Pitman, A. J. (2003). The evolution of, and revolution in, land surface schemes designed for climate models. *International Journal of Climatology*, 23, 479–510.

Sagan, C., Toon, O. B., and Pollack, J. B. (1979). Anthropogenic albedo changes and the Earth's climate. *Science*, 206, 1363–1368.

Schneider, S. H., and Dickinson, R. E. (1974). Climate modeling. *Reviews of Geophysics and Space Physics*, 12, 447–493.

Sellers, P. J., Mintz, Y., Sud, Y. C., and Dalcher, A. (1986). A simple biosphere model (SiB) for use within general circulation models. *Journal of the Atmospheric Sciences*, 43, 505–531.

Seneviratne, S. I., Corti, T., Davin, E. L., et al. (2010). Investigating soil moisture–climate interactions in a changing climate: A review. *Earth-Science Reviews*, 99, 125–161.

Spracklen, D. V., Arnold, S. R., and Taylor, C. M. (2012). Observations of increased tropical rainfall preceded by air passage over forests. *Nature*, 489, 282–285.

Thompson, K. (1980). Forests and climate change in America: Some early views. *Climatic Change*, 3, 47–64.

Turner, O. (1849). *Pioneer History of the Holland Purchase of Western New York*. Buffalo: Jewett Thomas.

United States General Land Office (1867). *Report of the Commissioner of General Land Office, for the Year 1867*. Washington, DC: Government Printing Office.

United States General Land Office (1868). *Report of the Commissioner of General Land Office for the Year 1868*. Washington, DC: Government Printing Office.

Volney, C.-F. (1804). *View of the Climate and Soil of the United States of America: to which are annexed some accounts of Florida, the French colony on the Scioto, certain Canadian colonies, and the savages or natives*. London: J. Johnson.

Webster, N. (1843). On the supposed change in the temperature of winter. In *A Collection of Papers on Political, Literary and Moral Subjects*, ed. N. Webster. New York: Webster and Clark, pp. 119–162.

Wilber, C. D. (1881). *The Great Valleys and Prairies of Nebraska and the Northwest*. Omaha: Daily Republican Print.

Williams, M. (1989). *Americans and Their Forests: A Historical Geography*. Cambridge: Cambridge University Press.

Williams, S. (1794). *The Natural and Civil History of Vermont*. Walpole, New Hampshire: Thomas and Carlisle.

Williamson, H. (1771). An attempt to account for the change of climate, which has been observed in the middle colonies in North-America. *Transactions of the American Philosophical Society*, 1, 272–280.

Williamson, H. (1811). *Observations on the Climate in Different Parts of America, Compared with the Climate in Corresponding Parts of the Other Continent; to Which Are Added, Remarks on the Different Complexions of the Human Race; with Some Account of the Aborigines of America. Being an Introductory Discourse to the History of North-Carolina*. New York: T. & J. Swords.

Woodward, J. (1699). Some thoughts and experiments concerning vegetation by John Woodward, M. D. of the College of Physicians, & R. S. & Professor of Physick in Gresham College. *Philosophical Transactions*, 21, 193–227.

Zon, R. (1927). *Forests and Water in the Light of Scientific Investigation*. Washington, DC: Government Printing Office.

Part I

The Earth System

Components of the Earth System

2.1 | Chapter Summary

Earth's climate is understood in terms of a system of several interacting spheres and the energy, water, and biogeochemical cycles that link these spheres. The main components of the Earth system are: atmosphere, air; hydrosphere, water; cryosphere, frozen portion of Earth; biosphere, living organisms; pedosphere, soil; and anthroposphere, humans. People are important agents of environmental change through land use and land-cover change and co-option of the hydrologic cycle and biogeochemical cycles. Numerous physical, chemical, and biological processes within the Earth system feed back to accentuate or mitigate climate change. Many of these feedbacks relate to terrestrial ecosystems and human activities. Greater understanding of Earth and its climate requires that all components of the Earth system – physical, chemical, biological, socioeconomic – be considered.

2.2 | Atmosphere

The atmosphere is the air that surrounds Earth. It is comprised primarily of nitrogen (N_2) and oxygen (O_2), which together account for 99 percent of the volume of the atmosphere (Table 2.1). Many other gases occur in trace amounts that when combined comprise less than 1 percent of the volume of the atmosphere. Although they occur in minor quantities, some of these gases play an important role in Earth's radiation balance through the greenhouse effect.

Air pressure is a measure of the mass of air above a given point. The total pressure exerted by a parcel of air is the sum of the pressures of all the individual gases in the parcel. Nitrogen, which comprises 78 percent of the air, exerts the most partial pressure, followed by oxygen (21%). Water vapor typically comprises 1–4 percent of air. For example, the atmospheric pressure near sea level is about 1000 hectopascals (hPa, 1 hPa = 100 Pa = 1 millibar). The partial pressure of nitrogen is 780 hPa and oxygen is 210 hPa. If water vapor comprises 1 percent of the parcel, its partial pressure is 10 hPa or 1000 Pa. Because water vapor is only a small constituent of air, vapor pressure is only a small component of total air pressure. Carbon dioxide has a partial pressure of about 40 Pa.

Greenhouses gases are poor absorbers of solar radiation, but are strong absorbers of longwave radiation. As a result, the Sun's radiation passes through the atmosphere and heats the surface, but greenhouse gases in the atmosphere absorb the longwave radiation emitted by the surface. The majority of this longwave radiation is emitted back to the surface, warming the surface. This reemission of terrestrial longwave radiation back to the surface is the greenhouse effect that warms the surface.

The principal greenhouse gases are water vapor (H_2O), carbon dioxide (CO_2), methane (CH_4), and nitrous oxide (N_2O). The amount of water vapor in the atmosphere varies

Table 2.1 | Chemical composition of the atmosphere

Gas	Chemical symbol	Percent (by volume)	
		Current (2011)	Preindustrial (1750)
Nitrogen	N_2	78.08%	
Oxygen	O_2	20.95%	
Argon	Ar	0.93%	
Trace gases		<1%	
Water vapor	H_2O	0–5%	
Carbon dioxide	CO_2	390 ppm	278 ppm
Methane	CH_4	1803 ppb	722 ppb
Nitrous oxide	N_2O	324 ppb	270 ppb

Note: Of the many gases that occur in trace amounts, only the four major greenhouse gases are shown. One part per million (ppm, μmol mol^{-1}) = 0.0001%. One part per billion (ppb, nmol mol^{-1}) = 0.0000001%. CO_2, CH_4, and N_2O are from Hartmann et al. (2013).

geographically and seasonally and can be as high as 5 percent of the atmosphere, with more water vapor in the warm tropics than in colder polar regions and more water vapor in warm seasons than in cold seasons. The concentrations of CO_2, CH_4, and N_2O vary over time. The concentration of CO_2 in the atmosphere averaged for the year 2011 was 390 parts per million (ppm, or μmol mol^{-1}; more precisely, this is the mole fraction, defined as the number of CO_2 molecules in a given number of molecules of dry air) – a 40 percent increase from preindustrial levels, primarily as a result of human activities such as fossil fuel burning and deforestation. The concentrations of CH_4 and N_2O have similarly increased over the past few centuries. Although these gases occur in lower concentration than CO_2, their global warming potential is much greater. Global warming potential measures the effectiveness of gases in absorbing outgoing terrestrial radiation combined with their lifetime in the atmosphere. It is a measure of the total energy added to the climate system relative to that added by CO_2. The 100-year global warming potential of CH_4 is 28 times that of CO_2, and N_2O has a 100-year global warming potential 265 times that of CO_2 (Myhre et al. 2013). Ozone and halocarbons (chlorofluorocarbons (CFCs), hydrofluorocarbons (HFCs), and other carbon compounds containing fluorine, chlorine, bromine, or iodine) are other greenhouse gases.

In addition to these gases, the atmosphere contains microscopic particles ranging in size from a few nanometers to tens of micrometers, known as aerosols (Boucher et al. 2013). Primary aerosols enter the atmosphere directly as dust from land, salts from ocean spray, black carbon (soot) from fires, and volcanic ash. Wind erosion from arid and semiarid environments carries the largest amount of mineral aerosols into the atmosphere. Sea salt produced by breaking waves is a similarly large source of aerosols. Secondary aerosols form when chemical reactions in the atmosphere convert emitted gases to particles. Biological processes on land and in oceans emit sulfate and organic particles. Over land, organic condensates form during chemical reactions involving the emission of nonmethane hydrocarbons from terrestrial vegetation. Human activities produce a wide variety of primary and secondary aerosols, though the total emission is small compared with natural emissions. Sulfate aerosols produced through the combustion of sulfur-containing fossil fuels are particularly important. Airborne particles such as dust and sulfate aerosols directly affect climate by absorbing or scattering solar radiation. In addition, sulfate aerosols alter climate indirectly by influencing the number and size of cloud droplets. In this way, clouds brighten, reducing the sunlight reaching the surface. Black carbon deposited on snow and ice darkens the surface and enhances solar heating of the surface.

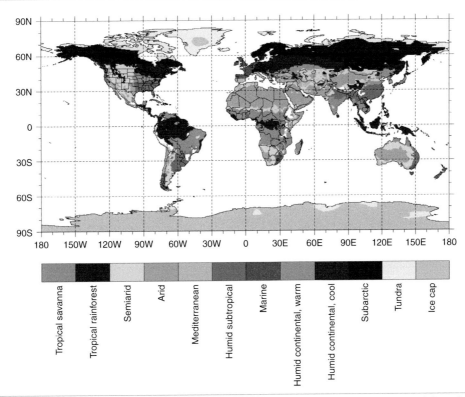

Fig. 2.1 Earth's major climate zones following the Köppen classification as modified by Trewartha (Chapter 6). See color plate section.

The atmosphere is constantly in motion. These motions arise from the heating of the air, land, and oceans by the Sun and the resultant geographic redistribution of heat by atmospheric and oceanic circulations. These motions can be seen over periods of a few minutes in the fluttering of flags in a strong breeze, over the course of a day in the formation of cumulus clouds and afternoon thunderstorms, or in the passage of warm or cold fronts. Atmospheric phenomena at these short timescales are referred to as weather. Whereas weather can be thought of as describing the instantaneous state of the atmosphere, climate describes the average weather, or long-term state of the atmosphere, over periods of many years.

Earth's surface has distinct climate zones defined by temperature and precipitation (Figure 2.1). Tropical climates occur near the equator where monthly mean temperatures are warmer than 18°C. Polar climates form near the poles, where the warmest month of the year is colder than 10°C. Polar climates are categorized as tundra or ice cap based on temperature. In between are the middle latitude climates with both warm and cold seasons. Several such mid-latitude climates form depending on temperature. For example, the subarctic climate zone has less than four months with monthly mean temperature greater than 10°C. Similarly, there are major geographic precipitation patterns. Tropical regions along the equator receive abundant rainfall year-round (tropical rainforest climate). Other tropical regions receive less annual rainfall and have pronounced wet and dry seasons (tropical savanna). Middle latitudes are generally moist, though arid climates, categorized as semiarid or arid based on decreasing moisture, develop along the subtropical high pressures at latitude 30° in both hemispheres and in regions far removed from sources of atmospheric moisture. The Mediterranean climate is a distinct mid-latitude climate with a dry season in summer. Polar regions are

Table 2.2 Water storage on Earth

Pool	Volume (km³)	Percent of total water	Percent of freshwater
Ocean	1,335,040,000	97.0	–
Freshwater	41,984,700	3.0	–
Icecaps, glaciers, and permafrost	26,372,000	1.9	62.8
Groundwater	15,300,000	1.1	36.4
Lakes and rivers	178,000	0.01	0.42
Soil water	122,000	0.01	0.29
Atmosphere	12,700	0.001	0.03

Source: From Trenberth et al. (2007). See also Oki and Kanae (2006).

generally dry because the cold air holds little moisture.

Earth's climate has changed in the past, and is changing still. Global mean planetary temperature increased by 0.85°C over the period 1880–2012 (Hartmann et al. 2013). Human activities that increase greenhouse gases and aerosols in the atmosphere have contributed to this climate change (Bindoff et al. 2013).

2.3 | Hydrosphere

The hydrosphere describes the water on Earth held in rivers, lakes, oceans, the ground, and air. Earth holds about 1377 million km³ of water. This is an amount that if spread over Earth's 510 million km² surface area would be about 2700 m deep. Oceans hold 97 percent of this water (Table 2.2). This is an average depth of about 3700 m spread over the 360 million km² surface area of oceans. Another 1.9 percent is frozen in polar icecaps, glaciers, and permafrost. Only 1.1 percent of Earth's water is liquid on land. Rivers, lakes, and wetlands contain 178,000 km³ of water on the surface. Over 15 million km³ of water is below ground. Deep aquifers that comprise groundwater hold most of this water; the soil near the surface holds very little water (122,000 km³). The atmosphere has the least amount of water, about 13,000 km³ or approximately 25 mm of water spread over Earth's surface area.

Water is an important part of Earth's climate. The oceans store and transport heat, redistributing the uneven geographic heating of Earth by the Sun. Carbon storage in the oceans regulates atmospheric CO_2 concentration. Water vapor is the most important greenhouse gas both in concentration and radiative warming. Water vapor condenses to form clouds, which can precipitate rainfall and which affect the planetary radiation budget by reflecting solar radiation and absorbing and emitting longwave radiation. The latent heat released during condensation provides considerable energy to fuel storms. The hydrologic cycle among the atmosphere, ocean, and land regulates the amount of water vapor in the air. Rates of precipitation and evaporation depend on temperature and other climatic factors so that as climate changes the amount of water vapor in the atmosphere also changes.

2.4 | Cryosphere

The cryosphere is the frozen portion of Earth including glaciers, sea ice, freshwater ice, snow, and permafrost. Glaciers are large, thick masses of ice accumulated from snowfall. They can be geographically small such as mountain glaciers or extensive ice sheets such as those that cover Greenland and Antarctica. Glaciers are the largest store of freshwater (Table 2.2). The melting of glaciers with a warmer climate contributes to sea level rise. Conversely, sea level was lower during the last glacial maximum some 21,000 years

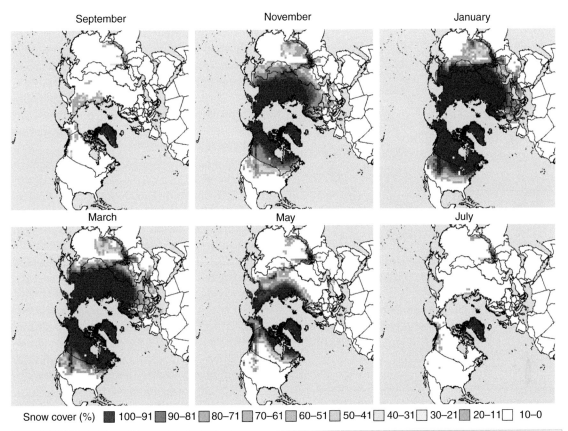

September   November   January

March   May   July

Snow cover (%)   ■ 100–91 ■ 90–81 ▨ 80–71 ▨ 70–61 ▨ 60–51 ☐ 50–41 ☐ 40–31 ☐ 30–21 ▨ 20–11 ☐ 10–0

Fig. 2.2 Monthly snow cover climatology in the Northern Hemisphere for the period 1966–1999. Maps provided courtesy of David Robinson (Global Snow Lab, Rutgers University). See color plate section.

before present when glaciers covered a large portion of North America and Eurasia. Glaciers also affect climate through the planetary energy balance. Glaciers have a much larger albedo than do oceans, soil, or vegetation, which decreases the amount of solar radiation absorbed by the surface and contributes to planetary cooling.

Snow cover and sea ice extent vary seasonally. At its winter maximum, snow covers about 46 million km^2 of land in the Northern Hemisphere (Figure 2.2). Sea ice covers 14–16 million km^2 in the Arctic Ocean and 17–20 million km^2 in the oceans around Antarctica at its seasonal maximum. This sea ice melts during summer to 6–8 million km^2 in the Arctic and 3–4 million km^2 around Antarctica. Similar to glaciers, snow and sea ice affect planetary energetics through high albedo. Fresh snow, for example, can reflect 80–90 percent of incoming solar radiation compared with 10–20 percent for soil or vegetation.

Permafrost is soil or rock where temperatures remain below freezing for two or more years. It is common throughout the Arctic over vast regions of tundra and boreal forest vegetation. Many factors influence the geographic distribution of permafrost, such as air temperature, snow cover, vegetation cover, the presence of an insulating organic layer, and soil water. Soil temperature and permafrost control ecological and biogeochemical processes at high latitudes. Thawing of permafrost with warmer temperatures alters the biogeography, productivity, and carbon balance of arctic ecosystems. Permafrost restricts infiltration and drainage, leading to wet soils and standing surface water, and the degradation of permafrost impacts runoff of freshwater to the Arctic Ocean.

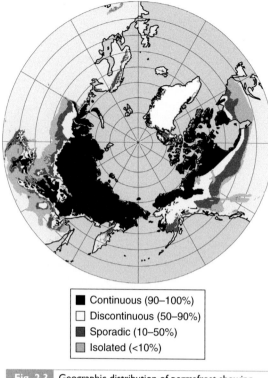

■ Continuous (90–100%)
□ Discontinuous (50–90%)
■ Sporadic (10–50%)
■ Isolated (<10%)

Fig. 2.3 Geographic distribution of permafrost showing the extent of continuous, discontinuous, sporadic, and isolated patches of permafrost in terms of percentage area. Reproduced from Lawrence and Slater (2005).

Permafrost occupies approximately 23 million km², or 24 percent of land in the Northern Hemisphere (Figure 2.3). Continuous permafrost, where permafrost underlies more than 90 percent of the land, extends over some 11 million km² including northern regions of Canada, Alaska, and Russia and is widespread throughout Siberia extending to northern China and Mongolia. In more southern locations, permafrost is discontinuous (50–90% of land), sporadic (10–50%), or occurs in isolated patches (<10%) depending on local climate and site factors.

2.5 | Biosphere

The biosphere describes the plants, animals, fungi, bacteria, algae, and other living organisms that inhabit Earth. Of particular interest in the atmospheric sciences are terrestrial ecosystems. Much of the carbon in biologically active

pools is stored in vegetation and soil (Ciais et al. 2013). It is estimated that plant biomass contains 450–650 Pg (1 Pg = 10^{15} g) of carbon. Soils hold more than three times as much carbon (1500–2400 Pg), with an additional ~1700 Pg C or more locked in permafrost. For comparison, the atmosphere contains about 800 Pg of carbon. The uptake of carbon during photosynthesis and the release of carbon during respiration by plants and soil microorganisms are important parts of the global carbon cycle. In addition, terrestrial ecosystems regulate other biogeochemical cycles that are important to climate (e.g., CH_4, N_2O), the emission of mineral aerosols into the atmosphere, and the emission of chemically reactive gases that affect air quality (e.g., biogenic volatile organic compounds). Plants, by regulating evapotranspiration, infiltration, and runoff, are a key component of the hydrologic cycle.

The natural vegetation of Earth has a distinct geographic pattern that corresponds to climate zones (Figure 2.4). Forests grow in tropical rainforest, humid subtropical, marine, humid continental, and subarctic climates. In these regions, precipitation is abundant year-round. Trees cannot survive in the bitter cold of tundra climates. Instead, small shrubs, herbaceous plants, and mosses grow in the short summers. Extensive grasslands occur in the semiarid and savanna climates of central North America, northern and central South America, central and southern Africa, central Asia, and Australia. Here climate is hot and dry. Short, dense woody bushes form chaparral (also known as Mediterranean) vegetation in the Mediterranean climate where summers are hot and dry and winters are mild and moist. Deserts, with sparse or widely spaced scrubby plants, establish in arid climates. The close correspondence between climate zones and vegetation zones is readily apparent. Climate zones such as tropical savanna, tropical rainforest, Mediterranean, and tundra are named after vegetation.

2.6 | Pedosphere

The lithosphere is the solid outer layer of Earth including the crust and upper mantle. Its

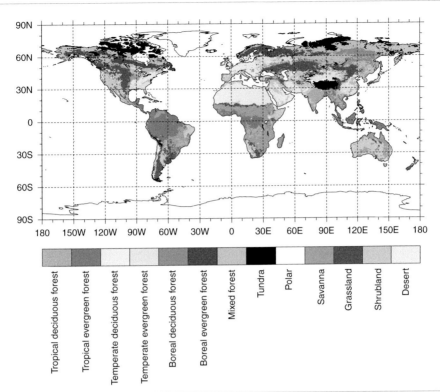

Fig. 2.4 Biogeography of natural vegetation prior to human land use. Data from Ramankutty and Foley (1999). See color plate section.

outermost layer is called the pedosphere, or soil. Soil is the interface in the cycling of energy and materials between the atmosphere and land. It is the matrix through which energy, water, biomass, and nutrients flow. It is the location of large transformations of energy as radiation absorbed by the surface becomes sensible heat, latent heat, or is stored in the ground. Soil is the source of water and nutrients for plant growth. The recycling of nitrogen and other nutrients from vegetation back to the soil is crucial to both soil development and plant growth. Soils store vast amounts of carbon that is slowly released to the atmosphere during decomposition. The transformation of this material to decomposed humus releases nutrients that support plant growth. Numerous microflora and microfauna facilitate the cycling of carbon and nutrients among soil, living biomass, and air.

Soils develop over time in relation to climate, vegetation, and the underlying geologic parent material. There are twelve broad classes of soil, known as soil orders, that differ in color, texture, structure, and chemical and mineralogical properties. Many of these twelve soil orders closely relate to climate and vegetation (Figure 2.5). Entisols and aridisols are soils with little organic matter or soil development and are common in arid climates in association with deserts. Gelisols are cold soils underlain with permafrost and develop throughout the Arctic in association with tundra. Histosols are organic soils with thick peat layers that develop in wet conditions associated with marshes, swamps, and bogs. They are most prevalent in cold climates. Spodosols develop in boreal and cool temperate needleleaf evergreen forests, where the acidic litter enhances leaching. Mollisols are the thick soils with high organic matter content found in association with prairie vegetation. Ultisols are highly weathered clay soils that develop in warm to tropical climates. Oxisols are the most highly weathered soils, with high clay content, and develop where the climate is

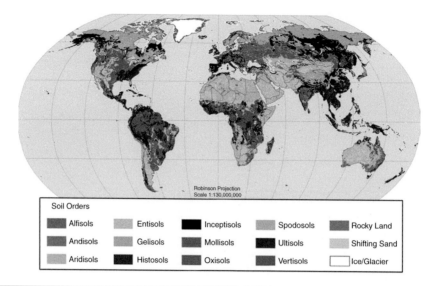

Fig. 2.5 Geographic distribution of the twelve soil orders. Image provided courtesy of the U.S. Department of Agriculture Natural Resources Conservation Service, Soil Survey Division, World Soil Resources, Washington D.C. See color plate section.

hot and wet throughout the year in association with tropical rainforests.

2.7 | Anthroposphere

The world's population in 2005 was estimated to be 6.5 billion people. This is more than a sevenfold increase from that prior to the industrial era, and population is expected to increase further to 8–10 billion by 2050 (Figure 2.6). The ever increasing spread of humanity across the planet has lead to growing recognition of people as agents of environmental change (Vitousek et al. 1997; Foley et al. 2005). Very little of the planet is untouched in some form by human activities. The anthroposphere represents humankind, our socioeconomic systems, and our activities. It is our cities, towns, and villages; the agriculture, energy, and water to sustain the populace; our transportation systems; and our collective influence on the environment. One of the more prominent anthropogenic effects is the emission of CO_2 to the atmosphere from fossil fuel combustion (Figure 2.7). This CO_2 emission has increased to more than 9 Pg C yr^{-1}, and atmospheric CO_2 concentration has increased as a result. The outcome of human activities on planetary functioning is also seen in the elevated levels of atmospheric CH_4, N_2O, and other greenhouse gases over the preindustrial era; by increasing emissions and subsequent deposition of reactive nitrogen; and through land use practices such as agriculture, deforestation, afforestation, and reforestation. By altering biogeochemical cycles, the hydrologic cycle, and energy fluxes, human activities have a significant impact on climate. Inclusion of past and future changes in human activities is an important component of climate change simulations.

Over the past 300 years, human uses of land removed 7–11 million km^2 of forest worldwide (Foley et al. 2005). Managed forests replaced an additional 1.9 million km^2 of natural forest land. Presently, croplands and pastures cover more than one-third of the land surface (Figure 2.8). Croplands cover significant portions of central North America, Europe, and Asia. Pastures and rangelands are extensive throughout the western United States, central Asia, and the tropics. Regionally, large tracts of land have been urbanized, as seen in the nighttime lights of the world (Figure 2.9). It is estimated that about 84,000 km^2 of land in the conterminous United States (an area approximately equal to the state of South Carolina) is covered by buildings, roofs,

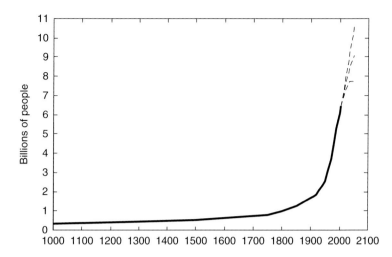

Fig. 2.6 World population from 1000 CE to 2005. Projected population through 2050 is shown for low, medium, and high growth (dashed lines). Data provided courtesy of the Population Division, Department of Economic and Social Affairs, United Nations Secretariat.

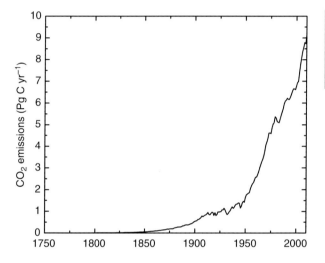

Fig. 2.7 World CO_2 emissions from fossil fuel combustion and cement production. Andres et al. (2012) describe the data. Data provided by the Carbon Dioxide Information Analysis Center (Oak Ridge National Laboratory, Oak Ridge, Tennessee).

roads, parking lots, and other impervious surfaces (Elvidge et al. 2007); globally, impervious surfaces cover an area of about 580,000 km² (larger than France). Other data analysis suggests that the area of cities worldwide may be 660,000 km² (Schneider et al. 2009).

Anthropogenic land-cover change and human uses of land alter ecosystem functions. About 20–30 percent of the net primary production of the world's terrestrial ecosystems is managed by humans (Vitousek et al. 1986; Rojstaczer et al. 2001; Imhoff et al. 2004), and the amount has increased over time (Krausmann et al. 2013). The clearing of forests for agricultural land releases much of the carbon stored in trees and soils to the atmosphere. It is estimated that from 1750 to 2011 an amount of carbon equal to one-half that emitted during the combustion of fossil fuels over the same period was released to the atmosphere as a result of changes in land use (Ciais et al. 2013). Vast sums of available renewable freshwater are withdrawn annually for agricultural, industrial, and municipal uses (Postel et al. 1996; Oki and Kanae 2006).

The production of nitrogen fertilizers, the cultivation of legumes and other nitrogen-fixing crops, and to a lesser extent the combustion of fossil fuels have increased the amount of reactive nitrogen in the Earth system (Galloway et al. 2003, 2004, 2008; Gruber and Galloway 2008; Erisman et al. 2011). The Haber–Bosch process to produce ammonia (NH_3) from N_2 for

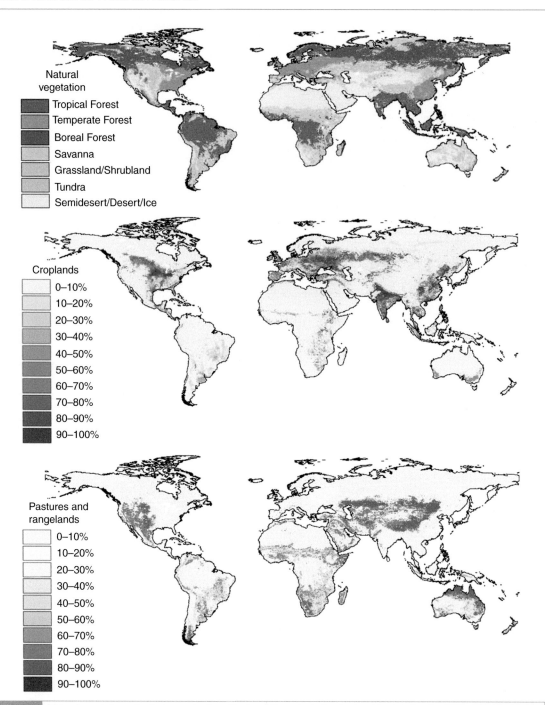

Fig. 2.8 Natural vegetation prior to human land use (top) and the extent of agricultural land during the 1990s. The panels for croplands (middle) and pastures (bottom) show the percentage of the land surface occupied by these land cover types. Reproduced from Foley et al. (2005). See color plate section.

fertilizers and other uses is the single largest source of reactive nitrogen (Figure 2.10). In the late-twentieth century, human production of reactive nitrogen became larger than natural inputs. The Haber–Bosch process produces over 100 Tg N yr^{-1} (1 Tg = 10^{12} g). Cultivation of crops adds additional nitrogen, and fossil fuel combustion adds still more nitrogen. Food production

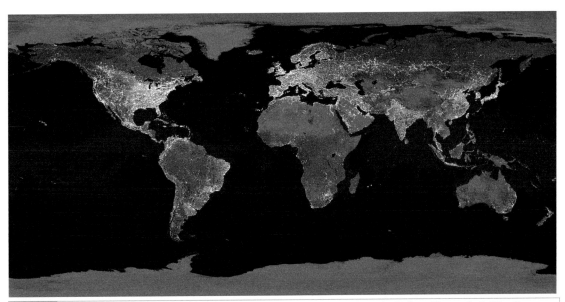

Fig. 2.9 Urban areas as seen by satellite in nighttime lights of the world. Image from the Defense Meteorological Satellite Program and provided by the National Geophysical Data Center (National Oceanic and Atmospheric Administration, Boulder, Colorado) and Goddard Space Flight Center (National Aeronautics and Space Administration, Greenbelt, Maryland) courtesy of Marc Imhoff (GSFC) and Christopher Elvidge (NGDC). The nighttime lights dataset provides a means to monitor urban land cover (Elvidge et al. 1997; Imhoff et al. 1997; Gallo et al. 2004; Small et al. 2005).

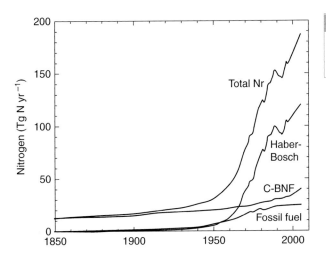

Fig. 2.10 Reactive nitrogen (Nr) creation from fossil fuel combustion, cultivation of biological nitrogen-fixing crops, and the Haber–Bosch process. Adapted from Galloway et al. (2003).

accounts for three-quarter of the reactive nitrogen created by humans. The increased availability of nitrogen contributes to poor air quality and anthropogenic climate change.

Human activities will continue to evolve in the future. Representative concentration pathways (RCPs) are a comprehensive set of four concentration and emission scenarios developed for use as input to models to assess anthropogenic climate change over the twenty-first century (Moss et al. 2010; van Vuuren et al. 2011a). They depict plausible scenarios of how the future may look with respect to population growth, socioeconomic change, technological change, energy consumption, and land use; resulting emissions of greenhouse gases, reactive gases,

Table 2.3 | Main characteristics of each representative concentration pathway (RCP)

Scenario component	RCP2.6	RCP4.5	RCP6.0	RCP8.5
Radiative forcing	2.6 W m^{-2}	4.5 W m^{-2}	6.0 W m^{-2}	8.5 W m^{-2}
Greenhouse gas emissions	Very low	Medium-low	Medium-high	High
Agricultural area	Medium for cropland and pasture	Very low for cropland and pasture	Medium for cropland and very low for pasture	Medium for cropland and pasture
Air pollution	Medium-low	Medium	Medium	Medium-high

Source: From van Vuuren et al. (2011a).

and aerosol precursors; and the concentration of atmosphere constituents (Table 2.3). The four RCPs span a range of radiative forcing values at year 2100 from high to low. (Radiative forcing is the change in the balance between incoming and outgoing radiation at the tropopause due to a change in atmospheric constituents, such as CO_2.) The various radiative forcing pathways are achieved through a range of socioeconomic and technological development scenarios. The RCPs represent a high emission scenario (8.5 W m^{-2}, RCP8.5), two medium stabilization scenarios (6.0 W m^{-2}, RCP6.0; 4.5 W m^{-2}, RCP4.5), and one mitigation scenario leading to low radiative forcing (2.6 W m^{-2}, RCP2.6).

RCP8.5 is a baseline scenario in the absence of climate change policy (Riahi et al. 2011). Increasing greenhouse gas emissions and high greenhouse gas concentrations characterize RCP8.5 (Figure 2.11). Radiative forcing increases to 8.5 W m^{-2} by 2100, with atmospheric CO_2 increasing to 936 ppm, CH_4 to 3751 ppb, and N_2O to 435 ppb. It describes high fossil fuel energy consumption as a result of a large increase in the global population, slow income growth, and modest technological change and energy efficiency improvements in the absence of climate change policies. Global population increases to over 10 billion people in 2050 and to 12 billion people by 2100. Global cropland area increases, primarily in Africa and South America, to meet the growing food demand. Increasing use of fertilizers and intensification of agricultural production give rise to high N_2O

emissions. More livestock and rice production result in high CH_4 emissions. Global forestland decreases, but bioenergy use increases, primarily as a result of wood harvest, giving rise to secondary managed forests. Although RCP8.5 depicts an absence of climate mitigation policies, air quality policies greatly affect the depiction of pollutant emissions. Emissions of sulfur dioxide (SO_2), nitrogen oxides (NO_x), black carbon aerosols, and organic carbon aerosols decline due to clean air legislation and technology improvements.

RCP6.0 is a climate policy intervention scenario (Masui et al. 2011). Total radiative forcing stabilizes at 6.0 W m^{-2} after 2100, with atmospheric CO_2 increasing to 670 ppm, CH_4 decreasing somewhat, and N_2O increasing to 406 ppb (Figure 2.11). Population increases to about 10 billion at 2100, and stabilization is achieved through a variety of technologies and strategies designed to reduce greenhouse gas emissions. However, the degree of greenhouse gas emission mitigation is small compared with RCP4.5 and RCP2.6. Global cropland area increases to feed the growing population while forest area remains relatively constant.

RCP4.5 is another intervention scenario in which total radiative forcing stabilizes at 4.5 W m^{-2} after 2100 (Thomson et al. 2011). Atmospheric CO_2 increases to 538 ppm, CH_4 decreases, and N_2O increases to 372 ppb (Figure 2.11). Global population peaks at 9 billion before declining. The necessary reduction in emissions is achieved through imposition of

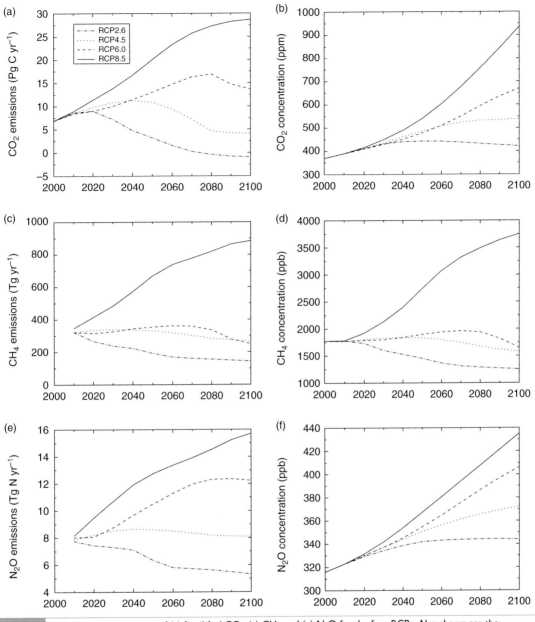

Fig. 2.11 Anthropogenic emissions of (a) fossil fuel CO_2, (c) CH_4, and (e) N_2O for the four RCPs. Also shown are the atmospheric concentrations of (b) CO_2, (d) CH_4, and (f) N_2O. Data from Prather et al. (2013).

climate policies that drive declines in overall energy use and fossil fuel CO_2 emissions with substantial increases in renewable energy and carbon capture systems. Global forest cover increases for carbon storage as part of the overall emissions mitigation strategy. RCP4.5 is the only scenario in which cropland area declines,

because of reforestation policies and assumed increases in crop yield.

RCP2.6 is a scenario that leads to very low greenhouse gas concentrations (van Vuuren et al. 2011b). Radiative forcing peaks at ~3 W m^{-2} before 2100 and declines to 2.6 W m^{-2} by 2100 (CO_2, 421 ppm; CH_4, ~30% lower than

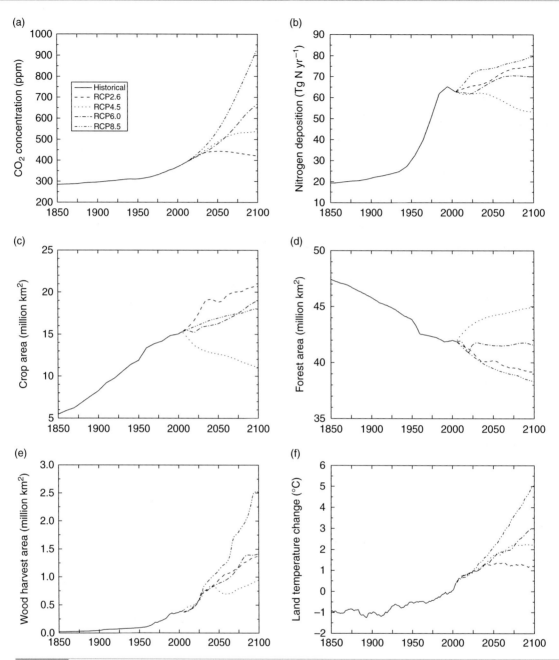

Fig. 2.12 Climate forcings and simulated temperature change in an Earth system model. (a) Atmospheric CO_2 concentration. (b) Annual nitrogen deposition on land. (c) Cropland area. (d) Forest area. (e) Annual wood harvest area. (f) Change in land temperature (relative to 1986–2005). Data from Lawrence et al. (2012).

2000; N_2O, 344 ppb). Greenhouse gas emissions decrease substantially over time in order to reach this low radiative forcing (Figure 2.11), necessitating substantial changes in energy use. The mitigation strategies utilize bioenergy, with

consequences for global land use. RCP2.6 has the largest increase in global cropland area.

Figure 2.12 illustrates some of these historical and future forcings used to simulate climate for the twentieth and twenty-first centuries.

RCP8.5 has the largest warming, followed by RCP6.0 and RCP4.5; RCP2.6 has the smallest warming. A particularly important difference among these is the depiction of future land use (Hurtt et al. 2011).

2.8 | Terrestrial Feedbacks

Numerous physical, chemical, and biological processes within the atmosphere, ocean, and land regulate Earth's climate, and many of these processes feed back to accentuate or mitigate climate change. A positive feedback accentuates a perturbation within the system while a negative feedback mitigates the perturbation. Water vapor is an example of a positive feedback. As temperature increases, the amount of water vapor that can be held in air increases. Because water vapor is a powerful greenhouse gas, any increase in water vapor in the atmosphere as a result of warming will feed back to accentuate the warming. Clouds are another important feedback because they increase planetary albedo (cooling climate) and reduce longwave radiation lost to space (warming climate). The effect depends on the type of cloud.

The amount of snow and ice covering the surface is a key feedback. Sea ice reflects more solar radiation than open water. Glaciers and snow reflect more solar radiation than soil or vegetation. By reflecting more solar radiation, sea ice, glaciers, and snow cool climate because less radiation is available to heat the surface.

Consequently, growth of ice and snow with a colder glacial climate reinforces the cold climate. Conversely, climate warming melts sea ice, glaciers, and snow. This reinforces the warming, because the darker surface absorbs more solar radiation.

Many other climate feedbacks relate to processes that occur on land. Particularly prominent biogeophysical feedbacks relate to albedo, surface roughness, and leaf area (Figure 1.6). Others relate to the hydrologic cycle and biogeochemical cycles. A greater discharge of freshwater into the North Atlantic shuts down the thermohaline circulation. The melting of glaciers provides much of this freshwater but also influences climate through the ice albedo feedback. The geological and biological carbon cycles regulate atmospheric CO_2, and terrestrial ecosystems regulate CH_4 and N_2O concentrations as well. High dust emissions during glacial times may deposit more iron in oceans, which may stimulate phytoplankton growth, reduce atmospheric CO_2, and cool climate. Changes in the geographic distribution of vegetation alter biogeochemical cycles, the hydrologic cycle, and surface energy fluxes. Such feedbacks can be observed in the geological record. In addition, it is widely accepted that glacial–interglacial cycles are a response to orbital forcing, but terrestrial feedbacks within the climate system amplify the response to this forcing. The Arctic is thought to be particularly sensitive to climate change due to numerous feedbacks among climate, the hydrologic cycle, the biosphere, and the cryosphere.

2.9 | Review Questions

1. A change in atmospheric CO_2 concentration of 1 part per million (ppm) is equivalent to about 2.1 Pg C. How much carbon has accumulated in the atmosphere since the preindustrial era?

2. The Amazon basin of South America has a tropical rainforest climate. What type of climate would be expected in the future if annual precipitation decreased and if the seasonality of rainfall increased to distinct wet and dry seasons?

3. By how much would the average ocean depth increase if all of the freshwater in glaciers was put into oceans? Use an ocean area of 360,000,000 km².

The Greenland ice sheet contains about 2,900,000 km³ of frozen water. By how much would sea level rise if all this water melted into oceans?

4. What percentage of the water on Earth is available for human uses?

5. Arctic sea ice is widely expected to decrease in extent with a warmer climate. Why should we care?

6. What type of vegetation would be favored under the climate change scenario of question 2? What type of vegetation would be reduced in extent?

7. By how much would atmospheric CO_2 concentration increase if the biosphere stored no carbon in plants and soil? Assume none of the carbon is taken up by oceans.

8. The Sahel region of Africa lies between the Sahara desert to the north and tropical rainforest to the south. How has human land use altered the vegetation of this region? Use Figure 2.8 for reference.

9. RCP8.5 is considered a business-as-usual scenario in the absence of changes in global energy use. The high energy demand and fossil fuel use in

RCP8.5 implies that achieving climate stabilization will require a substantial reduction of emissions and change in energy use. Discuss ways in which this transformation can be achieved.

10. RCP2.0 is designed to limit global temperature warming to about 2°C. It requires significant reduction in CO_2 emissions. Are these reductions feasible?

2.10 | References

Andres, R. J., Boden, T. A., and Bréon, F.-M., et al. (2012). A synthesis of carbon dioxide emissions from fossil-fuel combustion. *Biogeosciences*, 9, 1845–1871.

Bindoff, N. L., Stott, P. A., AchutaRao, K. M., et al. (2013). Detection and attribution of climate change: from global to regional. In *Climate Change 2013: The Physical Science Basis. Contribution of Working Group I to the Fifth Assessment Report of the Intergovernmental Panel on Climate Change*, ed. T. F. Stocker, D. Qin, G.-K. Plattner, et al. Cambridge: Cambridge University Press, pp. 867–952.

Boucher, O., Randall, D., Artaxo, P., et al. (2013). Clouds and aerosols. In *Climate Change 2013: The Physical Science Basis. Contribution of Working Group I to the Fifth Assessment Report of the Intergovernmental Panel on Climate Change*, ed. T. F. Stocker, D. Qin, G.-K. Plattner, et al. Cambridge: Cambridge University Press, pp. 571–657.

Ciais, P., Sabine, C., Bala, G., et al. (2013). Carbon and other biogeochemical cycles. In *Climate Change 2013: The Physical Science Basis. Contribution of Working Group I to the Fifth Assessment Report of the Intergovernmental Panel on Climate Change*, ed. T. F. Stocker, D. Qin, G.-K. Plattner, et al. Cambridge: Cambridge University Press, pp. 465–570.

Elvidge, C. D., Baugh, K. E., Hobson, V. R., et al. (1997). Satellite inventory of human settlements using nocturnal radiation emissions: A contribution for the global tool chest. *Global Change Biology*, 3, 387–395.

Elvidge, C. D., Tuttle, B. T., Sutton, P. C., et al. (2007). Global distribution and density of constructed impervious surfaces. *Sensors*, 7, 1962–1979.

Erisman, J. W., Galloway, J., Seitzinger, S., Bleeker, A., and Butterbach-Bahl, K. (2011). Reactive nitrogen in the environment and its effect on climate change. *Current Opinion in Environmental Sustainability*, 3, 281–290.

Foley, J. A., DeFries, R., Asner, G. P., et al. (2005). Global consequences of land use. *Science*, 309, 570–574.

Gallo, K. P., Elvidge, C. D., Yang, L., and Reed, B. C. (2004). Trends in night-time city lights and vegetation indices associated with urbanization within the conterminous USA. *International Journal of Remote Sensing*, 25, 2003–2007.

Galloway, J. N., Aber, J. D., Erisman, J. W., et al. (2003). The nitrogen cascade. *BioScience*, 53, 341–356.

Galloway, J. N., Dentener, F. J., Capone, D. G., et al. (2004). Nitrogen cycles: Past, present, and future. *Biogeochemistry*, 70, 153–226.

Galloway, J. N., Townsend, A. R., Erisman, J. W., et al. (2008). Transformation of the nitrogen cycle: Recent trends, questions, and potential solutions. *Science*, 320, 889–892.

Gruber, N., and Galloway, J. N. (2008). An Earth-system perspective of the global nitrogen cycle. *Nature*, 451, 293–296.

Hartmann, D. L., Klein Tank, A. M. G., Rusticucci, M., et al. (2013). Observations: Atmosphere and surface. In *Climate Change 2013: The Physical Science Basis. Contribution of Working Group I to the Fifth Assessment Report of the Intergovernmental Panel on Climate Change*, ed. T. F. Stocker, D. Qin, G.-K. Plattner, et al. Cambridge: Cambridge University Press, pp. 159–254.

Hurtt, G. C., Chini, L. P., Frolking, S., et al. (2011). Harmonization of land–use scenarios for the period 1500–2100: 600 years of global gridded annual land–use transitions, wood harvest, and resulting secondary lands. *Climatic Change*, 109, 117–161.

Imhoff, M. L., Lawrence, W. T., Stutzer, D. C., and Elvidge, C. D. (1997). A technique for using composite DMSP/OLS "city lights" satellite data to map urban area. *Remote Sensing of Environment*, 61, 361–370.

Imhoff, M. L., Bounoua, L., Ricketts, T., et al. (2004). Global patterns in human consumption of net primary production. *Nature*, 429, 870–873.

Krausmann, F., Erb, K.-H., Gingrich, S., et al. (2013). Global human appropriation of net primary production doubled in the 20th century. *Proceedings of the National Academy of Sciences USA*, 110, 10324–10329.

Lawrence, D. M., and Slater, A. G. (2005). A projection of severe near-surface permafrost degradation during the 21st century. *Geophysical Research Letters*, 32, L24401, doi:10.1029/2005GL025080.

Lawrence, P. J., Feddema, J. J., Bonan, G. B., et al. (2012). Simulating the biogeochemical and biogeophysical impacts of transient land cover change and wood harvest in the Community Climate System Model (CCSM4) from 1850 to 2100. *Journal of Climate*, 25, 3071–3095.

Masui, T., Matsumoto, K., Hijioka, Y., et al. (2011). An emission pathway for stabilization at 6 W m^{-2} radiative forcing. *Climatic Change*, 109, 59–76.

Moss, R. H., Edmonds, J. A., Hibbard, K. A., et al. (2010). The next generation of scenarios for climate change research and assessment. *Nature*, 463, 747–756.

Myhre, G., Shindell, D., Bréon, F.-M., et al. (2013). Anthropogenic and natural radiative forcing. In *Climate Change 2013: The Physical Science Basis. Contribution of Working Group I to the Fifth Assessment Report of the Intergovernmental Panel on Climate Change*, ed. T. F. Stocker, D. Qin, G.-K. Plattner, et al. Cambridge: Cambridge University Press, pp. 659–740.

Oki, T., and Kanae, S. (2006). Global hydrologic cycles and world water resources. *Science*, 313, 1068–1072.

Postel, S. L., Daily, G. C., and Ehrlich, P. R. (1996). Human appropriation of renewable fresh water. *Science* 271, 785–788.

Prather, M., Flato, G., Friedlingstein, P., et al. (2013). Annex II: Climate system scenario tables. In *Climate Change 2013: The Physical Science Basis. Contribution of Working Group I to the Fifth Assessment Report of the Intergovernmental Panel on Climate Change*, ed. T. F. Stocker, D. Qin, G.-K. Plattner, et al. Cambridge: Cambridge University Press, pp. 1395–1445.

Ramankutty, N., and Foley, J. A. (1999). Estimating historical changes in global land cover: Croplands from 1700 to 1992. *Global Biogeochemical Cycles*, 13, 997–1027.

Riahi, K., Rao, S., Krey, V., et al. (2011). RCP 8.5: A scenario of comparatively high greenhouse gas emissions. *Climatic Change*, 109, 33–57.

Rojstaczer, S., Sterling, S. M., and Moore, N. J. (2001). Human appropriation of photosynthesis products. *Science*, 294, 2549–2552.

Schneider, A., Friedl, M. A., and Potere, D. (2009). A new map of global urban extent from MODIS satellite data. *Environmental Research Letters*, 4, doi:10.1088/1748-9326/4/4/044003.

Small, C., Pozzi, F., and Elvidge, C. D. (2005). Spatial analysis of global urban extent from DMSP-OLS night lights. *Remote Sensing of Environment*, 96, 277–291.

Thomson, A. M., Calvin, K. V., Smith, S. J., et al. (2011). RCP4.5: A pathway for stabilization of radiative forcing by 2100. *Climatic Change*, 109, 77–94.

Trenberth, K. E., Smith, L., Qian, T., Dai, A., and Fasullo, J. (2007). Estimates of the global water budget and its annual cycling using observational and model data. *Journal of Hydrometeorology*, 8, 758–769.

van Vuuren, D. P., Edmonds, J., Kainuma, M., et al. (2011a). The representative concentration pathways: An overview. *Climatic Change*, 109, 5–31.

van Vuuren, D. P., Stehfest, E., den Elzen, M. G. J., et al. (2011b). RCP2.6: Exploring the possibility to keep global mean temperature increase below 2°C. *Climatic Change*, 109, 95–116.

Vitousek, P. M., Ehrlich, P. R., Ehrlich, A. H., and Matson, P. A. (1986). Human appropriation of the products of photosynthesis. *BioScience*, 36, 368–373.

Vitousek, P. M., Mooney, H. A., Lubchenco, J., and Melillo, J. M. (1997). Human domination of Earth's ecosystems. *Science*, 277, 494–499.

3

Global Cycles

3.1 | Chapter Summary

The functioning of Earth as a system is seen in the global cycling of energy, water, and carbon, and in other biogeochemical cycles. This chapter introduces the fundamental scientific concepts of energy, water, and biogeochemical cycles that regulate climate and link the atmosphere, hydrosphere, cryosphere, biosphere, pedosphere, and anthroposphere. Heat flows between materials due to temperature differences. Heat is transferred in the atmosphere by radiation, conduction, and convection. These flows of heat determine the balance of energy gained, lost, or stored. For the planet as a whole and averaged over the year, the solar radiation absorbed by Earth is equal to the longwave radiation emitted to space. That is, the net radiation absorbed by Earth is zero in the annually averaged planetary mean. The hydrologic cycle describes the cycling of water among land, ocean, and air, principally in terms of precipitation, evaporation, and the runoff of freshwater from land into oceans. The hydrologic cycle regulates the amount of water vapor in the air, which is a key greenhouse gas. The increased capacity of air to hold moisture as temperature increases is an important thermodynamic principle that affects climate. In addition, the change of water among its solid, liquid, and vapor states requires considerable energy. These phase changes provide energy to fuel storms. Atmospheric gases interact with radiant energy flowing through the atmosphere to determine the planetary energy budget. Principal among these are carbon dioxide (CO_2), methane (CH_4), and nitrous oxide (N_2O). These gases cycle among the atmosphere, ocean, and land, regulated in part by biological and geochemical processes. Human activities modify their natural cycles, which can be seen in the rising concentration of these greenhouse gases in the atmosphere.

3.2 | Scientific Units

All units of measurement are derived from four basic units (Table 3.1): length, measured in meters (m); mass, measured in kilograms (kg); time, measured in seconds (s); and temperature, measured in kelvin (K). An additional quantity, mole (mol), is used in chemistry to measure the amount of a substance. Mass and moles are related by the molecular mass of the material (kg mol^{-1}). Force is a quantity that accelerates an object (force = mass × acceleration). The scientific unit for force is the newton (N). One newton is defined as the force needed to accelerate a mass of one kilogram to a speed of one meter per second in one second (1 N = 1 kg m s^{-2}). Work is done when a force acts on an object to move it over a certain distance (work = force × distance). The scientific unit for work is the joule (J). One joule is the work needed to move an object with the force of one newton over a distance of one meter in the direction of the

Table 3.1 Basic and derived scientific units

Quantity name	Unit name	Dimension symbol	Unit symbol	Base units
Length	meter	L	m	–
Mass	kilogram	M	kg	–
Time	second	T	s	–
Temperature	kelvin	K	K	–
Amount	mole	–	mol	–
Area	square meter	L^2	m^2	m^2
Volume	cubic meter	L^3	m^3	m^3
Density	kilogram per cubic meter	$M L^{-3}$	$kg\ m^{-3}$	$kg\ m^{-3}$
Velocity	meter per second	$L T^{-1}$	$m\ s^{-1}$	$m\ s^{-1}$
Acceleration	meter per second per second	$L T^{-2}$	$m\ s^{-2}$	$m\ s^{-2}$
Force	newton	$M L T^{-2}$	N	$kg\ m\ s^{-2}$
Energy	joule	$M L^2 T^{-2}$	J	$kg\ m^2\ s^{-2} = N\ m$
Power	watt	$M L^2 T^{-3}$	W	$kg\ m^2\ s^{-3} = J\ s^{-1}$
Pressure	pascal	$M L^{-1} T^{-2}$	Pa	$kg\ m^{-1}\ s^{-2} = N\ m^{-2}$

Note: $°C = K - 273.15$ so that a change in temperature of $1°C = 1\ K$.

Table 3.2 Metric prefixes

Multiple	Prefix	Symbol	Multiple	Prefix	Symbol
10^{-1}	deci-	d	10^1	deca-	da
10^{-2}	centi-	c	10^2	hecto-	h
10^{-3}	milli-	m	10^3	kilo-	k
10^{-6}	micro-	µ	10^6	mega-	M
10^{-9}	nano-	n	10^9	giga-	G
10^{-12}	pico-	p	10^{12}	tera-	T
10^{-15}	femto-	f	10^{15}	peta-	P

force ($1\ J = 1\ N\ m = 1\ kg\ m^2\ s^{-2}$). Energy is the capacity to do work and has the same units as work. One joule of energy supports $1\ N\ m$ of work. Power is defined as the rate at which work is done. The scientific unit for power is the watt (W), which is equal to the rate of working one joule per second ($1\ W = 1\ J\ s^{-1}$). Pressure is the force per unit area. The scientific unit for pressure is the pascal (Pa), equal to a force of one newton over an area of one square meter ($1\ Pa = 1\ N\ m^{-2} = 1\ kg\ m^{-1}\ s^{-2}$).

These units of measure can have a prefix that indicates multiples or fractions of the units. For example, the prefix kilo- denotes multiplication by one thousand; one kilometer equals one thousand meters ($1\ km = 1000\ m$). The prefix milli- denotes division by one thousand; one millimeter equals one-thousandth of a meter (or conversely, $1000\ mm = 1\ m$). Other common prefixes are: micro- ($µ$, 10^{-6}); hecto- (h, 10^2); mega- (M, 10^6); tera- (T, 10^{12}), and peta- (P, 10^{15}). Table 3.2 lists common scientific prefixes.

3.3 | Energy Fluxes

Energy is the ability to do work. It exists in a variety of forms, but there are two basic categories. Potential energy is stored energy that results from an object's position. It is the work that must be done to move an object from some reference point to another position. For example, a ball at rest on the top of a hill contains gravitational potential energy. A stretched spring has elastic potential energy arising from its deformation. Kinetic energy is the energy of motion. It is the energy an object possesses because of its motion. Kinetic energy is a measure of the amount of work an object in motion can do as a result of its motion. A moving automobile will do work as it hits another vehicle. A ball thrown at a window will do work as it strikes the window.

Temperature is a measure of the energy of motion, or kinetic energy, in the movement of molecules in a substance. All materials are composed of molecules. These molecules are in motion. This motion is most evident for gases or liquids, but the molecules in solids are also in motion through vibrations. As the motion of molecules increases, kinetic energy increases and temperature rises. Temperature is perceived in terms of the relative warmth or coolness of an object.

Heat is a form of kinetic energy that flows from one object to another due to temperature differences between them. In this way, energy transfers from hot to cold materials. In the atmosphere, heat is transferred by radiation, conduction, and convection.

3.3.1 Radiation

All materials with temperature greater than absolute zero (0 K, or −273.15°C) emit energy in the form of electromagnetic radiation. We perceive this energy principally as visible light and in the warmth of the Sun's rays. Terrestrial objects also emit radiation but at wavelengths that are longer than the Sun's and that are not visible to the eye. At lower temperatures, we do not sense this energy because of its low amount. At higher temperatures, such as an electric heater, the radiant energy increases and we feel its warmth. At very high temperatures (1000 K), such as a wood fire, the radiant energy is strong, the emitted radiation shifts to shorter wavelengths in the visible spectrum, and objects glow. Electromagnetic radiation is called radiant energy and has the unit joules (J). Radiant energy emitted or received per unit time is called the radiant flux and has the unit of joules per second, or watts (1 W = 1 J s^{-1}). Radiant flux density is the radiant flux per unit surface area (W m^{-2}). Irradiance is the radiant flux density incident on an object; emittance is that emitted by an object.

Radiant energy travels in waves with peaks and troughs. The distance between successive peaks (or troughs) is the wavelength. Electromagnetic radiation transfers energy in discrete units called quanta or photons. The energy of a photon (e_p, J) is related to wavelength (Λ, m) by:

$$e_p = hc / \Lambda \qquad (3.1)$$

where $h = 6.626 \times 10^{-34}$ J s is Planck's constant and $c = 3 \times 10^8$ m s^{-1} is the speed of light. The longer the wavelength, the lesser the energy of the photon. The energy of a photon of blue light with wavelength 0.450 μm is 4.42×10^{-19} J. A photon of red light (0.680 μm) has 2.92×10^{-19} J.

Planck's law defines the radiant flux density per unit wavelength ($E(\Lambda)$, W m^{-2} m^{-1}) emitted by a blackbody in relation to wavelength and temperature:

$$E(\Lambda) = \frac{2\pi hc^2}{\Lambda^5 \left[\exp(hc / k\Lambda T) - 1 \right]} \qquad (3.2)$$

where $k = 1.38 \times 10^{-23}$ J K^{-1} is the Boltzmann constant and T is temperature (K). Figure 3.1 illustrates this for two bodies with the approximate temperatures of the Sun and Earth. It is readily evident that objects with a higher temperature emit radiation at a greater rate than objects with a lower temperature. It is also evident from Figure 3.1 that the spectral distribution of radiant energy depends on the temperature of the object. The higher the object's temperature, the shorter the wavelength of emitted radiation. The Sun, with a temperature of 6000 K, emits radiation in short wavelengths between 0.2 and

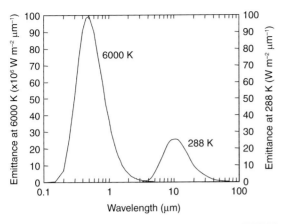

Fig. 3.1 Spectral distribution of radiation emitted by blackbodies at temperatures of 6000 K (Sun, left-hand axis) and 288 K (Earth, right-hand axis). Note that the left-hand axis is larger than the right-hand axis by the factor 10^6.

4 μm. Radiation in these wavelengths is known as solar, or shortwave, radiation. Solar radiation is divided into ultraviolet radiation with a wavelength of less than about 0.4 μm (containing 10% of the Sun's energy), visible radiation between about 0.4 and 0.7 μm (40% of the Sun's energy), and near-infrared radiation at wavelengths of greater than about 0.7 μm (50% of the Sun's energy). Visible radiation is further divided into violet, blue, green, yellow, orange, and red (in order of increasing wavelength). Earth, with an effective temperature of about 288 K, emits less radiation than the Sun and in longer wavelengths from about 3 to 100 μm. Radiation at these high wavelengths is called infrared, or longwave, radiation.

The wavelength at which maximum emission occurs similarly decreases with higher temperature. Maximum emission for the Sun occurs at a wavelength of about 0.5 μm. For Earth, peak emission occurs at a wavelength of 10 μm. Wien's displacement law relates the wavelength of maximum emission to temperature:

$$\Lambda_{max} = 2897 \; \mu m \; K \, / \, T \tag{3.3}$$

This is evident when watching wood burn in a fireplace. As the fire burns out and grows cold, the embers turn from violet and blue when very hot, and more commonly yellow, to orange and then red while continuing to give off heat. The radiation spectrum shifts from shorter wavelengths when the embers are hot to longer wavelengths as they cool.

The Stefan–Boltzmann law relates the radiant flux density emitted by an object to its temperature, obtained by integrating $E(\Lambda)$ over all wavelengths:

$$E = \varepsilon \sigma T^4 \tag{3.4}$$

where E is emittance (W m⁻²), $\sigma = 5.67 \times 10^{-8}$ W m⁻² K⁻⁴ is the Stefan–Boltzmann constant, and ε is the broadband emissivity. This equation is formulated for a blackbody, which is an object that is a perfect absorber of radiation at all wavelengths and that emits the maximum possible energy at all wavelengths for a given temperature. For a blackbody, $\varepsilon = 1$. Most objects are not blackbodies and emit less radiation. Instead, the Stefan–Boltzmann law describes emission of radiation, but blackbody emittance is reduced by the object's emissivity (ε). Emissivity is defined as the ratio of the actual emittance to the blackbody emittance. Most objects have a broadband emissivity of 0.95–0.98 when integrated over all wavelengths.

An object absorbs radiant energy from the Sun. The radiation incident on an object is absorbed, reflected, or transmitted through the material so that:

$$a_\Lambda + r_\Lambda + t_\Lambda = 1 \tag{3.5}$$

Here, a_Λ is the fraction of incident radiation at a specified wavelength that is absorbed (absorptivity, or absorptance); r_Λ is the fraction of incident radiation at a specified wavelength that is reflected (reflectivity, or reflectance); and t_Λ is the fraction of incident radiation at a specified wavelength that is transmitted (transmissivity, or transmittance). These optical properties can vary strongly with wavelength. Green leaves, for example, typically absorb 85 percent of the solar radiation in the visible waveband ($a_{vis} = 0.85$) but absorb less than 50 percent of the radiation in the near-infrared waveband ($a_{nir} < 0.50$). Objects that are sufficiently thick are opaque ($t_\Lambda = 0$), and absorptivity is $a_\Lambda = 1 - r_\Lambda$. The solar radiation absorbed by an object is commonly calculated from this relationship.

Terrestrial bodies also absorb longwave radiation. Equation (3.5) describes absorptance, reflectance, and transmittance of this radiation. Additionally, Kirchhoff's law states that the absorptivity of a material is equal to its emissivity:

$$a_\Lambda = \varepsilon_\Lambda \qquad (3.6)$$

A blackbody is a perfect absorber and emitter of radiation so that $a_\Lambda = \varepsilon_\Lambda = 1$ and $r_\Lambda = t_\Lambda = 0$. An opaque gray body $(t_\Lambda = 0)$ has reflectance $r_\Lambda = 1 - a_\Lambda = 1 - \varepsilon_\Lambda$.

A body with a broadband reflectance r (integrated over all wavelengths) absorbs a fraction of the incoming solar radiation given by $1 - r$. The longwave radiative heat transfer between the body and its surrounding environment is the balance between the energy radiated by the body and that absorbed from the environment. The body absorbs a fraction ε (integrated over all wavelengths) of the energy radiated by its surrounding environment and reflects a fraction $1 - \varepsilon$. The net radiation (R_n, W m^{-2}) absorbed is:

$$R_n = (1 - r)S\downarrow + \varepsilon L\downarrow - \varepsilon\sigma T^4 \qquad (3.7)$$

where $S\downarrow$ is solar radiation (W m^{-2}) and $L\downarrow$ is the radiant energy from the environment (W m^{-2}).

For photosynthesis, the number of photons, not energy, is important. A photon of light with blue wavelength has more energy than a photon of light with red wavelength, but both have the same effect on photosynthesis. Only radiation with wavelengths between 0.4 and 0.7 μm, known also as photosynthetically active radiation, is used during photosynthesis. The energy of a mole of photons is obtained by multiplying the energy per photon by Avogadro's number (6.022×10^{23} mol^{-1}). A mole of photons with wavelength 0.55 μm has the energy:

$$\frac{(6.626 \times 10^{-34} \text{ J s}) (3 \times 10^8 \text{ m s}^{-1})}{0.55 \times 10^{-6} \text{ m}} (6.022 \times 10^{23} \text{ mol}^{-1})$$
$$= 0.218 \times 10^6 \text{ J mol}^{-1}$$

Therefore, photosynthetically active radiation (with average wavelength 0.55 μm) is converted from W m^{-2} (J s^{-1} m^{-2}) to photosynthetic photon flux density with units μmol photon m^{-2} s^{-1} using the factor 4.6 μmol J^{-1}.

3.3.2 Conduction

Conduction is the transfer of heat within a material or between materials arising from molecular vibration without any motion of the material itself. Consider, for example, a metal spoon placed in a pot of hot water. The end of the spoon not in the water becomes hot. As the molecules in the part of the spoon placed in the water absorb heat from the water, they vibrate faster. These molecules cause adjacent molecules to also vibrate faster. This process repeats until all the molecules in the spoon vibrate rapidly. If a person touches the spoon, heat flows from the hot spoon to the skin, causing the molecules in the person's skin to vibrate faster. The person perceives that the spoon is hot. In this way, heat is conducted from the water to the spoon to the skin, flowing from high temperature to low temperature.

The rate of heat transfer in one dimension by conduction (Q, W m^{-2}) is:

$$Q = -\kappa(\Delta T / \Delta z) \qquad (3.8)$$

where κ is the thermal conductivity of the material (W m^{-1} K^{-1}) and $\Delta T / \Delta z$ is the temperature gradient (°C m^{-1}, or K m^{-1}). The rate of heat flow between two points separated by some distance (Δz) is proportional to the temperature difference between the points (ΔT). Thermal conductivity determines the rate of heat transfer for a unit temperature gradient. The negative sign denotes that the flux is positive for a negative temperature gradient.

The type of material affects the rate of heat transfer. Metals are good conductors of heat and have a high thermal conductivity. Wood is a poor conductor and has low thermal conductivity. A metal spoon placed in hot water feels hotter to the touch than does a wooden spoon in the same water because it conducts heat to the hand much more rapidly than the wooden spoon. Materials with low thermal conductivity reduce heat loss by conduction and are effective insulators. Styrofoam is a poor conductor of heat. Air is also a very poor conductor of heat. Double-paned glass windows with an inner layer of air are a very effective insulator.

The thermal conductivity of air, for example, is 0.02 W m^{-1} K^{-1}. A 5°C temperature difference between a body and overlying air over a distance of 10 mm transfers 10 W m^{-2} by conduction. The thermal conductivity of water is 0.57 W m^{-1} K^{-1} and produces a heat flux that is almost thirty times larger (285 W m^{-2}). This is why a body immersed in cold water loses heat rapidly.

3.3.3 Convection

Diffusion is the transport and mixing of heat and mass through the movement of a gas or fluid. Such mixing occurs along a gradient from high to low concentration. The vertical transport and mixing of heat and mass through the movement of air is a form of diffusion termed convection. In the atmospheric sciences, convection refers to vertical motion in the atmosphere. The horizontal transport of heat is termed advection.

Fick's law describes the rate of mass transfer per unit area of a gas (e.g., H_2O, CO_2) in one dimension along a gradient from high to low concentration. The diffusive flux (F_j, kg m^{-2} s^{-1}) between two points relates to the concentration difference ($\Delta\rho_j$, kg m^{-3}) multiplied by a conductance (g'_j, m s^{-1}) or divided by a resistance (r'_j, s m^{-1}):

$$F_j = \Delta\rho_j g'_j = \Delta\rho_j / r'_j \tag{3.9}$$

The conductance accounts for molecular or turbulent motions that mix the fluid.

For diffusive flux calculations, the concentration of a gas (ρ_j) is defined as its mass (m_j, kg) per unit volume of mixture (V, m^3). The ideal gas law describes the volume occupied by a gas. It relates the volume occupied by n moles of a gas at a given pressure (P, Pa = N m^{-2} = J m^{-3}) and temperature (T, K) as:

$$PV = n\Re T \tag{3.10}$$

where $\Re$ is the universal gas constant (8.314 J K^{-1} mol^{-1}). For example, one mole of air occupies a volume $V = 0.0236$ m^3 for a standard atmosphere at sea level. (The standard atmosphere defines sea level as $T = 288.15$ K (15°C) and $P = 1013.25$ hPa, which is denoted STP for standard temperature and pressure.) The inverse (the number of moles per unit volume) is termed molar density (ρ_m, mol m^{-3}), and:

$$\rho_m = \frac{n}{V} = \frac{P}{\Re T} \tag{3.11}$$

Molar density at a given pressure and temperature is constant for all gases and equals 42.3 mol m^{-3} at STP. The density of a gas at a given pressure and temperature is equal to its mass divided by the volume occupied by the gas. An equivalent form of the ideal gas law is:

$$\text{density} = \frac{m_j}{V} = \frac{nM_j}{V} = \frac{P}{\Re T}M_j = \rho_m M_j \tag{3.12}$$

where M_j (kg mol^{-1}) is the molecular mass. For example, dry air (molecular mass, 28.97 g mol^{-1}) has a density of 1.225 kg m^{-3} at STP. If pressure remains constant, any increase in temperature results in a decrease in density. A decrease in temperature results in an increase in density at the same pressure.

Air is comprised of N_2, O_2, Ar, H_2O, CO_2, and other gases in trace amounts (Table 2.1). Each individual gas follows the ideal gas law, and:

$$\rho_j = \frac{m_j}{V} = \frac{n_j M_j}{V} = \frac{P_j}{\Re T}M_j \tag{3.13}$$

with m_j the mass, n_j the number of moles, and M_j the molecular mass of the gas. Here, ρ_j is the mass concentration of the gas and P_j is the partial pressure of the gas. Partial pressure is the pressure that a gas would exert if it alone occupied the same volume as the mixture and at the same temperature.

Volume changes with temperature and pressure so that changes in mass concentration (ρ_j) can occur independent of changes in mass. An alternative measure of concentration, independent of volume, is the mole fraction (c_j, mol mol^{-1}). This is the number of moles of a gas (n_j) in a given volume expressed as a fraction of the total number of moles (n) in the same volume:

$$c_j = \frac{n_j}{n} = \frac{P_j}{P} = \frac{\rho_j}{M_j}\frac{\Re T}{P} = \frac{\rho_j}{M_j\rho_m} \tag{3.14}$$

For example, CO_2 (molecular mass, 44.01 g mol^{-1}) with mole fraction $c_j = 390$ μmol mol^{-1} in the atmosphere at pressure 1013.25 hPa has partial pressure $P_j = 39.5$ Pa and mass concentration $\rho_j = 0.73$ g m^{-3} at 15°C. Inserting Eq. (3.14) into Eq. (3.9), an equivalent form of the diffusive flux equation is:

$$F_j / M_j = \rho_m \Delta c_j g'_j \tag{3.15}$$

Dividing the mass flux (F_j) with units kg m^{-2} s^{-1} by molecular mass (M_j) gives the molar flux with units mol m^{-2} s^{-1}.

The principles of diffusion apply to moisture and heat transfer in the atmosphere. With e the partial pressure of water vapor (Pa), the diffusive flux equation for evaporation (E, kg m^{-2} s^{-1}) is:

$$E \,/\, M_w = \rho_m \frac{\Delta e}{P} g'_w \qquad (3.16)$$

with M_w the molecular mass of water (18.02 g mol^{-1}) and g'_w the conductance for water vapor (m s^{-1}). Heat transfer by convection (H, W m^{-2}) is analogous to diffusion. The term $\rho c'_p \Delta T$ replaces $\Delta \rho_j$ in Eq. (3.9), and:

$$H = \rho c'_p \Delta T g'_h \qquad (3.17)$$

with ρ the density of moist air (kg m^{-3}), c'_p the specific heat of moist air at constant pressure (J kg^{-1} K^{-1}), ΔT (K) the temperature difference, and g'_h the conductance for heat (m s^{-1}). The density of air varies with temperature, pressure, and vapor pressure. A representative value is ρ = 1.22 kg m^{-3} at sea level (1013.25 hPa) and 15°C. The specific heat of air also varies with humidity. A representative value is c'_p = 1010 J kg^{-1} K^{-1}.

The meteorological community expresses diffusive fluxes in terms of a mass flux (F_j) with units kg m^{-2} s^{-1} and conductance (g'_j) with units m s^{-1}, or resistance r'_j (s m^{-1}). Molar units are common in the plant physiological literature, where conductance is preferred because it is directly proportional to the flux and because conductance and flux have the same units. A conductance g'_j with units m s^{-1} is converted to g_j with units mol m^{-2} s^{-1} by multiplying by the molar density (ρ_m) with units mol m^{-3}, given by Eq. (3.11). At STP, 1 mol = 0.0236 m^3 and 1 m s^{-1} = 42.3 mol m^{-2} s^{-1}. The conversion for conductance is:

$$g'_j \left[\frac{m}{s} \right] \times \frac{P}{\Re T} \left[\frac{mol}{m^3} \right] = g_j \left[\frac{mol}{m^2 s} \right] \qquad (3.18)$$

Equivalent forms of the diffusive fluxes (Eq. (3.15), Eq. (3.16), and Eq. (3.17)) are:

$$F_j \,/\, M_j = \Delta c_j g_j \qquad (3.19)$$

$$E \,/\, M_w = \frac{\Delta e}{P} g_w \qquad (3.20)$$

$$H = c_p \Delta T g_h \qquad (3.21)$$

In Eq. (3.19) and Eq. (3.20), the fluxes ($F_j \,/\, M_j$ and $E \,/\, M_w$) and conductances (g_j and g_w) have units mol m^{-2} s^{-1}. Equation (3.21) is the corresponding form of the convective heat flux, with c_p the molar specific heat of moist air at constant pressure (J mol^{-1} K^{-1}) and g_h the molar conductance for heat (mol m^{-2} s^{-1}). A representative value is c_p = 29.2 J mol^{-1} K^{-1}. Molar units and conductances are used in this book, as in Cowan (1977) and Campbell and Norman (1998).

Two types of convection are distinguished in meteorological studies. Free convection occurs due to temperature differences that affect the density, and therefore buoyancy, of air. One means by which convection occurs in the atmosphere is that warm air is less dense than cold air. Warm air, therefore, tends to rise in the atmosphere while cold air sinks. In the process, heat transfers from a warm surface to the colder air above. This transport of heat by vertical motion is called sensible heat because it is heat we can feel. A common example is the warmth felt as warm air rises from a radiator. Convective activity is readily seen in many regions of the world on a hot summer day. Solar radiation heats air near the ground. The warm, moist air rises, where it cools with greater height. Cold air holds less moisture than warm air. As the parcel of air rises and cools, it becomes saturated with moisture. Water vapor condenses, and cumulus clouds form. Forced convection is transport caused by wind. Wind moving across a warm body carries away heat to the cooler air. This is why a breeze is refreshing on a hot summer day. Thick clothes increase the resistance to convective heat loss, diminishing the loss, which is why short-sleeved shirts and shorts are comfortable on hot days. The effect of free and forced convection on mass and heat transport in the atmosphere is represented by the conductances g_j, g_w, and g_h.

3.3.4 Heat Storage

Heat capacity is the amount of energy needed to raise the temperature of a unit volume of material by one degree. A volume with unit area stores the energy:

$$\Delta Q = c_v (\Delta T / \Delta t) \Delta z \qquad (3.22)$$

where ΔQ is the heat absorbed by the volume (W m^{-2}), ΔT is the change in temperature (°C, or K) over the time period Δt (s), Δz is thickness (m), and c_v is heat capacity (J m^{-3} K^{-1}). The heat capacity of water (4.18 MJ m^{-3} K^{-1}) is about twice that of soil. Water has to absorb considerably more energy than soil to warm one degree. The presence of large bodies of waters such as lakes or oceans, therefore, acts to modulate the surrounding climate.

The first law of thermodynamics describes the conservation of energy. It states that in a closed system energy can change from one form to another but it cannot be created or destroyed; the total amount of energy in the system is conserved. Consider, for example, a system in which energy input is balanced by energy output and change in stored energy:

$$\text{energy input} = \text{energy output} + \Delta Q \qquad (3.23)$$

If the system loses the same amount of energy that it gains, there can be no change in storage ($\Delta Q = 0$) and the temperature of the system remains constant ($\Delta T = 0$). If the system gains more energy than it loses, the excess energy is stored in the system as thermal energy, raising the temperature of the system. Conversely, the temperature of the system decreases if it loses more energy than it gains.

3.3.5 Planetary Energy Balance

The principle of energy conservation can be seen in the planetary energy balance (Figure 3.2). Annually, Earth receives approximately 341 W m^{-2} of solar radiation at the top of the atmosphere. Clouds, gases, and aerosols absorb 78 W m^{-2} (23%), and Earth's surface absorbs an additional 161 W m^{-2} (47%). The remainder, 102 W m^{-2} (30%), is reflected back to space by the atmosphere and surface. The absorbed solar radiation warms Earth, which emits longwave radiation. At the surface, Earth emits 396 W m^{-2} of longwave radiation. Clouds, water vapor, CO_2, and other gases in the atmosphere absorb most of this radiation (356 W m^{-2}); only 40 W m^{-2} escapes to space. The gases, particles, and other material suspended in the atmosphere emit longwave radiation. This radiation travels in all directions with some lost to space and some reaching the surface. A total of 199 W

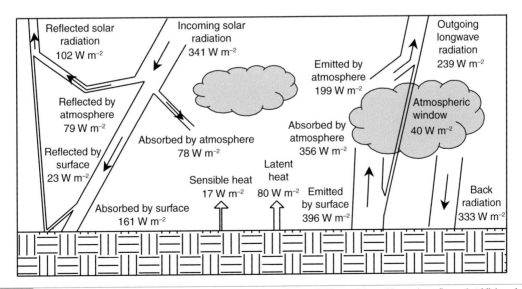

Fig. 3.2 Earth's annual mean global energy budget showing solar radiation (left), sensible and latent heat fluxes (middle), and longwave radiation (right). Data from Trenberth et al. (2009).

m^{-2} escapes to space, which together with the 40 W m^{-2} from the surface balances the 239 W m^{-2} solar radiation absorbed by the atmosphere and surface. That is, the net radiative balance of Earth is zero; the absorbed solar radiation equals the longwave radiation emitted to space.

The fluxes of sensible heat (17 W m^{-2}) and latent heat (80 W m^{-2}), while small compared with radiative fluxes, are important terms in the planetary energy balance (Figure 3.2). Earth's energy budget shows the atmosphere has a deficit of energy while the surface has a surplus. The atmosphere absorbs 78 W m^{-2} of solar radiation and 356 W m^{-2} of longwave radiation from the surface; it emits 199 W m^{-2} of longwave radiation to space and 333 W m^{-2} to the surface. The excess loss of radiation compared with absorption is –98 W m^{-2}. Earth's surface, in contrast, gains 161 W m^{-2} of solar radiation and 333 W m^{-2} of longwave radiation from the atmosphere while emitting 396 W m^{-2} of longwave radiation. This gives the surface a surplus of 98 W m^{-2}. This surplus energy is returned to the atmosphere as sensible heat and latent heat. These fluxes arise as winds carry heat (sensible heat) and moisture (latent heat) away from the surface.

A simple planetary energy balance model provides a descriptor of Earth's temperature and the role of greenhouse gases to warm the surface. Solar radiation heats the planet, and longwave radiation emitted to space cools the planet. This radiative balance determines the mean planetary temperature. The energy balance at the top of the atmosphere (F, W m^{-2}) is:

$$F = \frac{S_c}{4}(1-r) - \sigma T_s^4 = 0 \qquad (3.24)$$

The first term of this equation is the solar radiation absorbed by the atmosphere and surface. In this equation, $S_c = 1364$ W m^{-2} is the amount of radiation emitted by the Sun. The division by four arises because an area of πy^2 intercepts solar radiation (y is the radius of Earth), and $S_c(1-r)\pi y^2$ is the radiant flux (W) received by Earth; but an area of $4\pi y^2$ (i.e., the surface area of a sphere) emits longwave radiation, and $\sigma T_s^4 4\pi y^2$ is the energy flux (W) emitted by Earth. Hence, $S_c / 4 = 341$ W m^{-2} is the incoming solar

radiation at the top of the atmosphere averaged over Earth's surface area. The term r is the planetary albedo, which is the fraction of incoming solar radiation reflected to space ($r = 0.30$). The second term is the outgoing longwave radiation. The emission of longwave radiation to space is given in terms of a global mean surface temperature (T_s, K). The calculated temperature is $T_s = 255$ K (–18°C). Earth's temperature is in fact about 288 K (15°C). The difference in temperature (33°C) is due to the presence of the atmosphere and the absorption within the atmosphere by water vapor, CO_2, and other greenhouse gases of longwave radiation emitted by Earth's surface.

The greenhouse effect can be understood by extending the planetary energy balance model to include an atmosphere (Figure 3.3). In this model, the atmosphere is perfectly transparent to solar radiation (the surface absorbs all the radiation) and also absorbs all the longwave radiation from the surface. The atmosphere radiates up to space and down onto the surface so that an equal amount of energy (239 W m^{-2}) is lost in both directions. To maintain thermal equilibrium, the surface must emit longwave radiation to balance the 239 W m^{-2} from the Sun and an additional 239 W m^{-2} from the atmosphere. Its temperature is 303 K (30°C). The warmer surface temperature with an atmosphere arises because of atmospheric longwave radiation onto the surface; the surface has to be warm enough to emit twice as much radiation as without an atmosphere. The calculated temperature is warmer than the actual temperature of Earth (288 K) because the model neglects the complexity of the atmosphere. However, the model illustrates the basic principle of the greenhouse effect. The surface emits longwave radiation to maintain thermal equilibrium. The atmosphere absorbs this energy and emits some to space and some back onto the surface. The reemission of longwave radiation back to the surface is the greenhouse effect that warms the surface.

The model presented in Figure 3.3b also illustrates how a gain in energy in the system warms the surface. If the system is perturbed such that the surface gains an additional amount of

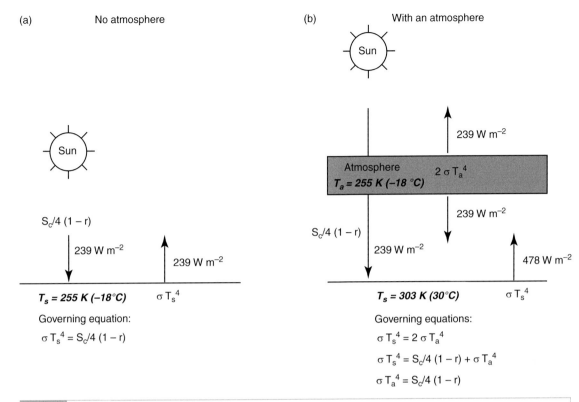

Fig. 3.3 Planetary energy balance and temperature calculation shown (a) without an atmosphere and (b) with an atmosphere.

energy (ΔF), the surface will warm to emit the additional ΔF. One-half of this radiation will be lost to space, and one-half will be emitted back onto the surface. The surface will warm so that it radiates the additional $\Delta F / 2$, but the atmosphere returns one-half of this to the surface. The surface must warm still more to emit another $\Delta F / 4$, and the process continues until thermal equilibrium is achieved. Of the total energy perturbation to the system (ΔF), an amount equal to $\Delta F + \Delta F / 2 + \Delta F / 4 + \Delta F / 8 + \Delta F / 16 + \cdots$ is gained by the surface (equal to $2\Delta F$) and an amount equal to $\Delta F / 2 + \Delta F / 4 + \Delta F / 8 + \Delta F / 16 + \cdots$ (i.e., ΔF) is emitted at the top of the atmosphere.

Equation (3.24) helps to explain the change in planetary temperature to a perturbation of energy in the system. Climate sensitivity is defined as the change in temperature for some change in forcing applied to the system. For Eq. (3.24), and neglecting dependences on other climate processes, the climate sensitivity factor is:

$$\frac{\Delta T_s}{\Delta F} = -\left(\frac{\partial F}{\partial T_s}\right)^{-1} = \left(\frac{S_c}{4}\frac{\partial r}{\partial T_s} + 4\sigma T_s^3\right)^{-1} \qquad (3.25)$$

Ignoring the dependence of albedo on temperature, climate sensitivity is 0.27 K (W m^{-2})$^{-1}$ when Earth is treated as a blackbody. A 1 W m^{-2} gain in energy produces about one-quarter of a degree increase in temperature. This measure of climate sensitivity considers only the blackbody emission of longwave radiation. In fact, the climate sensitivity of Earth is two to three times as large. Feedback mechanisms associated with water vapor, clouds, albedo, and other processes amplify climate sensitivity.

3.4 | Hydrologic Cycle

The hydrologic cycle describes the cycling of water among land, ocean, and air. Evaporation is the physical process by which liquid water in

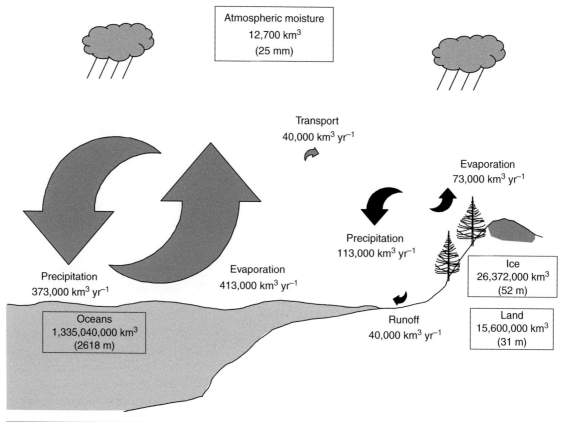

Fig. 3.4 The global hydrologic cycle. Units are km³ of water or, in parentheses, the depth of water spread over Earth's 510 million km² surface area. Data from Trenberth et al. (2007). See also Oki and Kanae (2006).

the oceans or on land changes to vapor in the air. It occurs when unsaturated air comes into contact with a moist surface. Evaporation provides the atmospheric moisture that returns to the surface as rain or snow. Evaporation also consumes an enormous amount of heat, which helps to cool the evaporating surface. Once in the atmosphere, water condenses, forming clouds, and if conditions are right the water falls back to the surface as precipitation. Heat is released as water vapor condenses and changes from vapor to liquid. This heat is a source of energy that drives atmospheric circulation and fuels storms. Oceans are the largest source of water for evaporation. Soils contain less than 1 percent of the unfrozen freshwater on Earth. However, soil water is an important determinant of surface energy fluxes and the climate

near the ground. Additionally, the discharge of freshwater from rivers into oceans prevents oceans from becoming saltier, which in turn influences ocean heat transport.

3.4.1 Global Water Balance

Water flows among the oceans, land, and atmosphere (Figure 3.4). The amount of water and its transfers can be measured by volume (m³) and equivalently by depth (m), mass (kg), or moles (mol) per unit area. These are related by the density of water (1 m³ H$_2$O = 1000 kg) and the molecular mass of water (1 mol H$_2$O = 18.02 g). One kilogram of water spread over an area of one square meter (1 kg m^{-2}) is equivalent to a depth of 1 mm, a volume of 0.001 m³, and 55.5 mol m^{-2}.

Annually, about 486,000 km³ of water falls from the atmosphere as precipitation. The same amount of water returns to the atmosphere annually as evaporation. Although Earth as a whole balances water, oceans and land differ in precipitation and evaporation. Approximately 373,000 km³ of water falls over the oceans each year as precipitation. However, more water evaporates from the oceans (413,000 km³), resulting in an annual surface deficit of 40,000 km³ of water. Runoff from land replenishes this imbalance. Over land, precipitation exceeds evaporation. About 113,000 km³ of water falls on land as precipitation, and 73,000 km³ of water evaporates from land. The surplus water at the land surface (40,000 km³) runs off to streams and rivers where it flows to the oceans to replenish the net loss of water. About 65 percent of the water reaching the land surface as precipitation returns to the atmosphere as evaporation, and 35 percent runs off to the ocean.

The average length of time a parcel of water spends in a reservoir can be calculated from the ratio of water storage to inflow or outflow (inflow and outflow are equal assuming steady state). This ratio is known as residence time or turnover time and measures the time required to replace all the water in the reservoir. For example, the atmosphere holds 12,700 km³ of water but precipitates 486,000 km³ of water per year. The turnover time is 9.5 days. For oceans, the residence time is 3000 years or so. The residence time is days for water in the upper soil to thousands of years for deep aquifers.

3.4.2 Atmospheric Humidity

It is convenient to distinguish water vapor and dry air. The latter is a general term for all gases other than H_2O, and the sum of these two components is the moist air. The total pressure of air (P, Pa) is the sum of dry air and water vapor:

$$P = P_d + e \tag{3.26}$$

where P_d is the partial pressure of dry air (Pa) and e is the partial pressure of water vapor (Pa), commonly called vapor pressure. Dry air with partial pressure $P_d = P - e$ follows the ideal gas law with density (ρ_d, kg m⁻³):

$$\rho_d = \frac{P-e}{\Re T} M_a \tag{3.27}$$

where M_a is the molecular mass of dry air (28.97 g mol⁻¹). The density of water vapor (ρ_v, kg m⁻³) is:

$$\rho_v = \frac{e}{\Re T} M_w = \frac{0.622e}{\Re T} M_a \tag{3.28}$$

with M_w the molecular mass of water (18.02 g mol⁻¹) and $M_w / M_a = 0.622$.

The density of moist air is the sum of that for dry air and water vapor:

$$\rho = \rho_d + \rho_v = \frac{P}{\Re T} M_a \left(1 - 0.378 \frac{e}{P}\right) \tag{3.29}$$

The density of moist air is less than the density of dry air at the same temperature and pressure. For a parcel of air at STP, $\rho = 1.225$ kg m⁻³ with $e = 0$ and $\rho = 1.217$ kg m⁻³ with $e = 1704$ Pa (i.e., saturated). Because moist air is less dense than dry air, water vapor is a source of buoyancy in the atmosphere.

Mass mixing ratio and specific humidity are common measures of the moisture in air. Mass mixing ratio (χ_v, kg kg⁻¹) is defined as the ratio of the mass of water vapor (m_v) in a parcel of air to the mass of dry air (m_d) in the parcel (i.e., excluding the water vapor). It is related to vapor pressure by:

$$\chi_v = \frac{m_v}{m_d} = \frac{\rho_v}{\rho_d} = \frac{0.622e}{P-e} \tag{3.30}$$

Specific humidity (q, kg kg⁻¹) is defined as the ratio of the mass of water vapor in a parcel of air to the total mass of the air. Specific humidity is related to vapor pressure by:

$$q = \frac{m_v}{m_d + m_v} = \frac{\rho_v}{\rho_d + \rho_v} = \frac{0.622e}{P - 0.378e} \tag{3.31}$$

Saturation vapor pressure is the maximum amount of water vapor that a parcel of air can hold. Saturation vapor pressure increases exponentially with warmer temperature, and warm air can hold considerably more water vapor when saturated than can cold air (Figure 3.5). For example, a parcel of air with a temperature of 30°C has a saturation vapor pressure of 4243 Pa; a parcel of air with a temperature of 18.5°C has a saturation vapor pressure of 2129 Pa. One kilogram of air can hold about

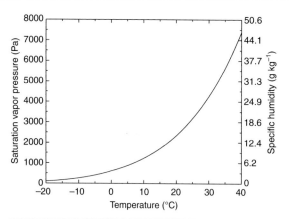

Fig. 3.5 Saturation vapor pressure as a function of temperature. The left-hand axis shows vapor pressure in pascals. The right-hand axis shows specific humidity.

13 g of water when saturated at 18.5°C and 26 g at 30°C.

Relative humidity is a measure of how saturated the air is with water. It is the ratio of e, the actual vapor pressure, to $e_*(T)$, the saturated vapor pressure at temperature T, expressed as a percentage:

$$RH = 100 \, e/e_*(T) \qquad (3.32)$$

For a constant vapor pressure, relative humidity decreases as temperature, and therefore saturated vapor pressure, increases. A parcel of air with a vapor pressure of 2129 Pa is saturated at a temperature of 18.5°C (i.e., $RH = 100\%$), but has a relative humidity of only 50% at 30°C.

The vapor pressure deficit, $e_*(T) - e$, is the difference between the saturation vapor pressure (i.e., the maximum amount of water vapor that can be held in the air) and the actual vapor pressure. It is a measure of the drying potential of air and is an indication of evaporative potential.

3.4.3 Phase Change

The hydrologic cycle is a transfer of water, measured by volume or mass (1 m³ H$_2$O = 1000 kg H$_2$O). It is also an exchange of energy between the surface and atmosphere. Water occurs in three forms: solid (ice), liquid, or gas (vapor). Energy is required to melt ice (solid to liquid) or to evaporate water (liquid to gas). This energy does not change the temperature of the water. Rather, it only changes the molecular state of the water. This energy is stored in the water molecules and released in the reverse process as water vapor condenses to liquid or liquid water freezes to ice. This stored energy is called latent heat because the temperature of water does not change with the gain or loss of heat; only the state of the water molecules changes.

Considerable amounts of energy are required to change water among its solid, liquid, and vapor states (Table 3.3). The energy absorbed in the evaporation of water is called the latent heat of vaporization. At 15°C, 2466 J are required to change one gram of water from liquid to vapor. For the Earth as a whole, 80 W m⁻² are used annually in evaporation – more than three-quarters of the 98 W m⁻² net radiation at the surface (Figure 3.2). For land, more than one-half (39 W m⁻²) of the annual net radiation at the surface (66 W m⁻²) is used to evaporate water (Trenberth et al. 2009). This stored energy is released as latent heat of condensation when water vapor condenses back to liquid. Absorption of energy cools the evaporating surface while condensation releases energy to the surrounding environment. The latent heat of vaporization decreases with warmer temperatures because water molecules contain more internal energy at warmer temperatures and less energy is required for them to evaporate.

The energy required to melt frozen water is called the latent heat of fusion. At 0°C it takes 334 J to melt one gram of water. During melting, this energy is absorbed by the water molecules, changing their phase from solid to liquid rather than warming the environment. The same heat is released when liquid water freezes. This is the reason commercial fruit growers spray orchards with water when a cold freeze is imminent. The change from liquid to ice releases heat, and while the water is freezing its temperature remains at 0°C. The latent heat of fusion decreases with colder temperatures because the water molecules contain less internal energy and less energy is released when they freeze. A third type of phase change, sublimation, occurs when ice changes directly to vapor without passing through the liquid phase.

Table 3.3 Latent heat and saturation vapor pressure in relation to temperature

Temperature (°C)	Latent heat (J g^{-1})			$e_*(T)$ (Pa)	$de_*(T)/dT$ (Pa K^{-1})
	Vaporization	Fusion	Sublimation		
−20	2550	289	2839	103	10
−10	2525	312	2837	260	23
0	2501	334	2835	611	44
5	2490			872	61
10	2478			1227	82
15	2466			1704	110
20	2454			2337	146
25	2442			3167	189
30	2430			4243	243
35	2419			5624	311
40	2407			7378	393

Note: Multiply by the molecular mass of water (18.02 g mol^{-1}) to convert latent heat from J g^{-1} to J mol^{-1}.

3.5 | Biogeochemical Cycles

The chemical composition of the atmosphere is a crucial determinant of climate. Atmospheric gases interact with radiant energy flowing through the atmosphere to affect climate through the greenhouse effect. Chief among these gases are carbon dioxide (CO_2), methane (CH_4), and nitrous oxide (N_2O). The concentration of these gases in the atmosphere is the result of natural and anthropogenic processes.

3.5.1 | Carbon Dioxide

Carbon dioxide is the most widely recognized greenhouse gas. Its average concentration in the atmosphere for the year 2011 was 390 ppm (Table 2.1). This represents the balance of geological processes, biological processes, and human activities. Of the some 10^{23} g of carbon on Earth, all but a small portion is buried in sedimentary rocks. Only about 0.04 percent of the carbon (40,000 Pg) is in biologically active pools near Earth's surface, and oceans hold the vast majority of the biologically active carbon (38,000 Pg C). Soils store most of the biologically active carbon on land. It is estimated that plants contain 450–650 Pg C worldwide while soils hold 1500–2400 Pg C with an additional ~1700 Pg C or more in permafrost. The atmosphere has the least carbon, slightly more than 800 Pg C during the period 2000–2009.

The global carbon cycle represents the interactions of two superimposed cycles: the geological carbon cycle, in which carbon cycles among atmosphere, oceans, and continents in response to the chemical weathering of rocks over a period of millions of years; and the biological carbon cycle, in which carbon cycles among the atmosphere and marine and terrestrial organisms through biological and physical processes over shorter timescales of days, seasons, years, and decades. The annual carbon fluxes in the geological carbon cycle are small compared with the biological fluxes.

In the preindustrial era, the carbon cycle is thought to have been in balance with no net carbon gain by the land or ocean (Figure 3.6a). The biosphere was a small annual sink of carbon (1.7 Pg C yr^{-1}) because photosynthetic uptake exceeded losses from respiration and wildfire. This and additional carbon from rock weathering (0.4 Pg C yr^{-1}) washed into rivers and lakes, where it returned to the atmosphere in freshwater outgassing (1.0 Pg C yr^{-1}), was buried in sediments (0.2 Pg C yr^{-1}), or was

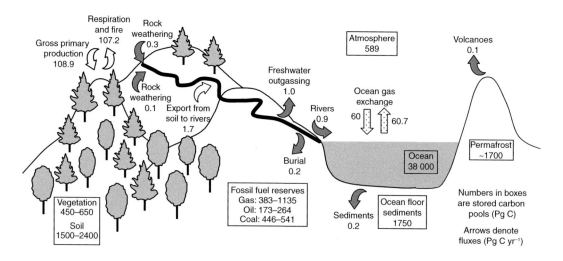

(a) **Preindustrial carbon cycle (~1750)**

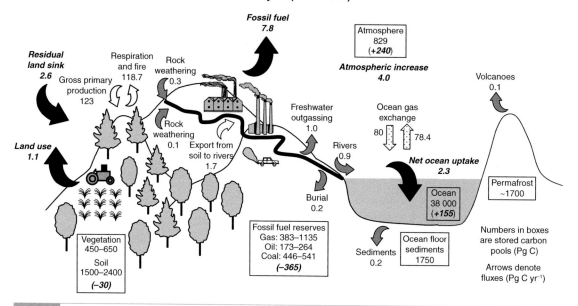

(b) **Modern carbon cycle (2000–2009)**

Fig. 3.6 Global carbon cycle for (a) the preindustrial era (~1750) and (b) the modern carbon cycle (2000–2009). Gray arrows denote fluxes in the geologic cycle. Black arrows and bold italic letters show the major changes in fluxes since 1750. Boxes show carbon pools and the cumulative change in carbon since 1750. Adapted from Ciais et al. (2013).

carried into oceans (0.9 Pg C yr⁻¹). The carbon input balanced a small net loss from air–sea exchange (0.7 Pg C yr⁻¹) and burial on the ocean floor (0.2 Pg C yr⁻¹).

Human activities have significantly impacted the global carbon cycle (Figure 3.6b). The burning of oil and coal to generate heat and electricity,

the combustion of gasoline for transportation, and other industrial processes release CO_2 to the atmosphere. During the period 2000–2009, these activities emitted 7.8 Pg C yr⁻¹. In addition, human uses of land, particularly forest clearing, emitted another 1.1 Pg C yr⁻¹. Slightly less than half of the total anthropogenic emission

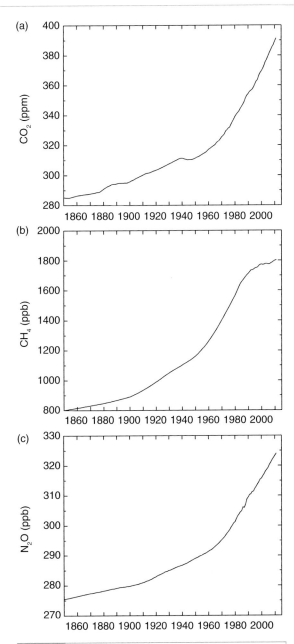

(a)

(b)

(c)

Fig. 3.7 Atmospheric CO_2, CH_4, and N_2O concentration from 1850 to 2011. Data updated from Hansen et al. (1998) and Hansen and Sato (2004) and provided by the NASA Goddard Institute for Space Studies (New York City, New York).

CO_2 in the atmosphere has increased since the preindustrial era (Figure 3.7a).

The land uptake is the balance among several large, but highly uncertain processes. It is estimated that terrestrial plants gain 123 Pg C yr^{-1} during photosynthesis (Beer et al. 2010). It is thought that half of this carbon returns to the atmosphere during autotrophic respiration, by which plants maintain and grow new biomass. The remainder is the annual net primary productivity of terrestrial plants. Leaves, twigs, and other plant debris fall to the ground and decompose. Decomposition of plant debris, wildfire, and other non-respiratory losses return much of this carbon to the atmosphere. From our current understanding of the global carbon cycle, the land must have gained 2.6 Pg C yr^{-1} over the period 2000–2009 after accounting for these carbon losses. Respiration and fire must emit 118.7 Pg C yr^{-1} to balance the photosynthetic carbon gain and export of carbon to rivers. In fact, however, more carbon may be flowing through the aquatic system because of larger carbon export from soils (Raymond et al. 2013; Regnier et al. 2013). The magnitude, geographic location, and causes of the terrestrial carbon sink are the subject of considerable scientific research.

Oceans are thought to have a net uptake of 2.3 Pg C yr^{-1}. Carbon dioxide is exchanged between air and sea through a variety of physical and chemical processes that affect the solubility of CO_2 in water. In addition, growth of phytoplankton absorbs CO_2. The carbon is buried in sediments as the organisms die and settle on the ocean floor. The oceanic biological pump is an important regulator of atmospheric CO_2. Without it, atmospheric CO_2 concentrations would be considerably greater.

3.5.2 Methane
Methane is another important greenhouse gas that cycles among atmosphere, ocean, and terrestrial pools. Its atmospheric concentration, like CO_2, has increased since the beginning of the industrial era (Figure 3.7b). The largest natural source of CH_4 is from wetlands, where anaerobic decomposition in waterlogged soils produces CH_4 (Table 3.4). Termites and other

(4.0 Pg C yr^{-1}) remained in the atmosphere. The rest was taken up by the oceans (2.3 Pg C yr^{-1}) and terrestrial ecosystems (2.6 Pg C yr^{-1}). As a result of these processes, the concentration of

Table 3.4 | Annual production and consumption of methane (CH_4) for 2000–2009

Production	Amount (Tg CH_4 yr^{-1})
Natural sources	
Wetlands	175
Other	43
Total emissions	218
Anthropogenic sources	
Agriculture and waste management	209
Fossil fuels	96
Biomass burning	30
Total emissions	335
Sinks	
Chemical reactions in atmosphere	518
Uptake by soils	32
Total loss	550

Note: These fluxes are highly uncertain, and Ciais et al. (2013) provide other estimates.
Source: From Ciais et al. (2013).

Table 3.5 | Annual sources of nitrous oxide (N_2O) in 2006

Production	Amount (Tg N yr^{-1})
Natural sources	
Soils	6.6
Oceans	3.8
Atmospheric chemistry	0.6
Natural total	11.0
Anthropogenic sources	
Fossil fuel combustion and industrial processes	0.7
Agriculture	4.1
Biomass burning	0.7
Human waste	0.2
Rivers, estuaries, and coastal zones	0.6
Atmospheric deposition	0.6
Anthropogenic total	6.9

Source: From Ciais et al. (2013).

insects also release CH_4 to the atmosphere as a result of anaerobic decomposition of organic matter in their bodies. However, the greater source of CH_4 is from human activities, particularly agriculture, waste management (landfills and sewage), fossil fuel combustion, and biomass burning. Once in the atmosphere, most CH_4 is transformed through a series of chemical reactions with hydroxyl radicals (OH). A much smaller amount is taken up by soils.

3.5.3 Nitrous Oxide

Despite its low concentration in the atmosphere, N_2O is another important greenhouse gas. Its concentration in the atmosphere has also increased since the industrial era (Figure 3.7c). Total sources for 2006 were 17.9 Tg N yr^{-1} (Table 3.5). Human activities contribute about one-third of the annual emission of N_2O to the atmosphere. Agriculture is the single largest anthropogenic source of N_2O. Much of this emission is related to fertilizer application.

3.6 | Review Questions

1. On a winter night, a person sits next to a window with an effective temperature of 6°C. How much longwave radiation does the window emit? With a closed curtain the effective temperature is 18°C. How much longwave radiation is emitted? Assume an emissivity of one. Discuss why the person feels warmer with the curtain closed.

2. The thermal conductivity of glass is about 1 W m^{-1} K^{-1} and is 0.02 W m^{-1} K^{-1} for air. Explain why a double pane window separated by 5 mm of air is more energy efficient than a single pane window.

3. A volume of water and soil with a depth of 5 cm each gains 400 W m^{-2} of energy and loses 350 W m^{-2} over a 10 minute period. What is the change in temperature of each? Heat capacity is 4.18×10^6 J m^{-3} K^{-1} for water and 2×10^6 J m^{-3} K^{-1} for soil.

4. Calculate the mass concentration (μg m^{-3}) of ozone (O_3) with a mole fraction of 80 ppb at STP.

5. Air pressure at sea level is 1013.25 hPa. What is the density of dry air with a temperature of 0°C? How does this compare with density at 25°C?

6. Three parcels of air at sea level (1013.25 hPa) have a temperature of 20°C. One has a mass mixing ratio χ_v = 10 g kg^{-1}, one has a specific humidity q = 10 g kg^{-1}, and one has 75 percent relative humidity ($e_*(T)$ = 2337 Pa). Which has the highest density?

7. The dew-point temperature is another measure of atmospheric moisture. It is the temperature to which a parcel of air must be cooled at constant pressure for it to be saturated. Which parcel of air has higher moisture: one with a high dew-point temperature or one with a low dew-point temperature?

8. A typical vapor pressure gradient between the surface and atmosphere is Δe = 1000 Pa at sea level (P = 1013.25 hPa, T = 15°C) and a representative conductance is g_w = 1 mol H$_2$O m^{-2} s^{-1}. Calculate the latent heat flux. How does this compare with the sensible heat flux between the surface and atmosphere with a temperature difference ΔT = 5°C and a representative conductance g_h = 2 mol m^{-2} s^{-1}? If net radiation is R_n = 650 W m^{-2}, how much energy is available to heat the soil?

9. Calculate the carbon budget of (a) ocean, (b) biosphere, and (c) rivers for the preindustrial carbon cycle. What is the carbon budget of land (biosphere and rivers)? What is the role of rivers in the global carbon cycle?

10. Calculate the change in annual ocean gas exchange and land carbon fluxes since 1750. What terms in the carbon cycle have changed?

11. The export of carbon from soils to rivers may be larger than depicted in Figure 3.6b. What is the consequence of this for our understanding of the terrestrial carbon sink? Most global terrestrial biosphere models exclude the riverine export. What is the implication of excluding this for the terrestrial carbon sink?

12. Why is carbon emission during land use an important term in the global carbon cycle?

13. Contrast the energy balance of the model depicted in Figure 3.3b with the energy balance in Figure 3.2. What processes are neglected that explain some of the discrepancy in the calculated surface temperature (303 K) compared with the actual temperature (288 K)?

14. Calculate the blackbody surface temperature of Earth with a 1 percent increase in the solar constant. Compare this to an increase in planetary albedo of 0.01.

15. Planetary albedo decreases with higher temperature due to changes in sea ice, snow cover, and clouds. In a simple model, planetary albedo (r) depends on global mean surface temperature (T_s) as $r = a - bT_s$. Calculate the blackbody climate sensitivity of Earth when b = 0.01 (so that a 1 K increase in temperature reduces albedo by 0.01).

3.7 | References

Beer, C., Reichstein, M., Tomelleri, E., et al. (2010). Terrestrial gross carbon dioxide uptake: global distribution and covariation with climate. *Science*, 329, 834–838.

Campbell, G. S., and Norman, J. M. (1998). *An Introduction to Environmental Biophysics*, 2nd edn. New York: Springer-Verlag.

Ciais, P., Sabine, C., Bala, G., et al. (2013). Carbon and other biogeochemical cycles. In *Climate Change 2013: The Physical Science Basis. Contribution of Working Group I to the Fifth Assessment Report of the Intergovernmental Panel on Climate Change*, ed. T. F. Stocker, D. Qin, G.-K. Plattner, et al. Cambridge: Cambridge University Press, pp. 465–570.

Cowan, I. R. (1977). Stomatal behaviour and environment. *Advances in Botanical Research*, 4, 117–228.

Hansen, J., and Sato, M. (2004). Greenhouse gas growth rates. *Proceedings of the National Academy of Sciences USA*, 101, 16109–16114.

Hansen, J. E., Sato, M., Lacis, A., et al. (1998). Climate forcings in the Industrial era. *Proceedings of the National Academy of Sciences USA*, 95, 12753–12758.

Oki, T., and Kanae, S. (2006). Global hydrologic cycles and world water resources. *Science*, 313, 1068–1072.

Raymond, P. A., Hartmann, J., Lauerwald, R., et al. (2013). Global carbon dioxide emissions from inland waters. *Nature*, 503, 355–359.

Regnier, P., Friedlingstein, P., Ciais, P., et al. (2013). Anthropogenic perturbation of the carbon fluxes from land to ocean. *Nature Geoscience*, 6, 597–607.

Trenberth, K. E., Smith, L., Qian, T., Dai, A., and Fasullo, J. (2007). Estimates of the global water budget and its annual cycling using observational and model data. *Journal of Hydrometeorology*, 8, 758–769.

Trenberth, K. E., Fasullo, J. T., and Kiehl, J. (2009). Earth's global energy budget. *Bulletin of the American Meteorological Society*, 90, 311–323.

Part II

Global Physical Climatology

4

Atmospheric Radiation

4.1 | Chapter Summary

The balance between absorbed solar radiation and outgoing longwave radiation at the top of the atmosphere is a key determinant of global climate. The Sun's position in the sky, which varies over the course of a day and throughout the year from the geometry of Earth's annual orbit around the Sun and its daily rotation on its axis, determines the intensity of solar radiation. A surface receives the most solar radiation when it is oriented perpendicular to the Sun's rays. At other angles, the Sun's radiation spreads over a larger surface area, with less radiation per unit area. As solar radiation passes through the atmosphere, some is absorbed and some is scattered, both upwards to space and downwards onto the surface. The downward scattered radiation is known as diffuse radiation and emanates from all directions of the sky. Direct beam radiation is not scattered and originates from the Sun's position in the sky. The geographic distribution of net radiation – the difference between solar radiation absorbed and longwave radiation emitted – is unequal. In general, there is an excess of solar radiation gain over longwave radiation loss in the tropics and a deficit at latitudes poleward of 35° to 40°.

4.2 | Solar Geometry

Diurnal and seasonal variation in climate arises from the geometry of Earth's annual orbit around the Sun and its daily rotation on its axis. As Earth rotates over the course of a day, the Sun appears to sweep a broad arc through the sky. Two angles define the Sun's position (Figure 4.1): its altitude above the horizon; and its bearing on the horizon, which is called the azimuth angle. A third angle, the zenith angle, is often used instead of altitude. Whereas altitude is the angular distance above the horizon, zenith angle is the angular distance from a line perpendicular to the surface. Altitude is zero at sunrise and sunset and is greatest at solar noon. At solar noon, the Sun is due south on the horizon in the Northern Hemisphere and is due north in the Southern Hemisphere. In the morning, the Sun is east of south; it is west of south in the afternoon.

Zenith angle and altitude angle vary with latitude, time of year, and time of day:

$$\cos Z = \sin B = \sin \phi \sin \delta + \cos \phi \cos \delta \cos h \qquad (4.1)$$

where Z is the zenith angle, B is the altitude angle, ϕ is latitude, δ is solar declination, and h is the solar hour angle. The hour angle measures time of day. Because Earth rotates on its axis 360° in 24 hours, one hour is equivalent to 15° of longitude, and the hour angle varies by 15° for each hour before or after solar noon. For example, the hour angle is 15° at 1100 hours and –15° at 1300 hours; it is 60° at 0800 hours and –60° at 1600 hours. At solar noon, when $h = 0°$, the zenith angle is at a minimum and the

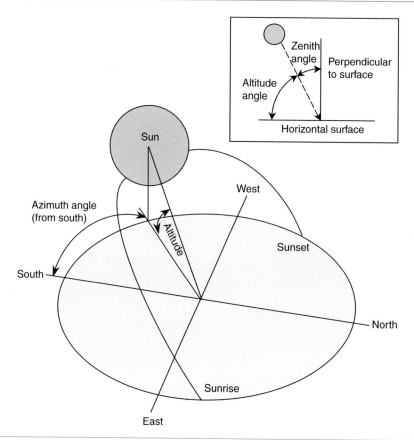

Fig. 4.1 Solar geometry illustrating azimuth, altitude, and zenith angles. Here, azimuth angle is shown from south. The boxed inset shows the difference between the altitude and zenith angles.

altitude angle is at a maximum. At this time, the zenith angle is equal to latitude minus declination, $Z = \phi - \delta$. This is the basis by which ancient mariners navigated the oceans. Measurement of the zenith angle when the Sun is highest in the sky combined with knowledge of the declination angle provides a precise determination of latitude. Solar noon occurs at 1200 hours local time only for the longitude that defines the time zone within which the observer is located. Solar noon occurs earlier in the day east of this longitude and later in the day west of this longitude.

The solar declination angle varies with the day of the year. Earth rotates over the course of a day on an axis tilted at an angle of 66.5° to an imaginary line connecting Earth and the Sun (Figure 4.2). This axis always points to the same location in space. Consequently, as Earth moves in its orbit around the Sun, the direction of tilt relative to the Sun varies. In boreal winter (austral

summer), the Northern Hemisphere tilts away from the Sun while the Southern Hemisphere tilts towards the Sun. In boreal summer (austral winter), the Northern Hemisphere tilts towards the Sun while the Southern Hemisphere tilts away from the Sun. Seasonal changes in orientation towards or away from the Sun are seen in the declination angle. Declination is the angle between a line connecting the centers of the Sun and Earth and the plane of the equator (Figure 4.2). It is the latitude where the Sun is directly overhead at solar noon.

Declination angle varies through the year (Figure 4.3). Four days of the year have special significance. On the winter solstice, which occurs on or about December 21, the declination angle is −23.5°. The Sun is directly overhead (i.e., has a zenith angle of zero) at noon at latitude 23.5° S. This latitude is the farthest point south of the equator where the Sun is directly overhead

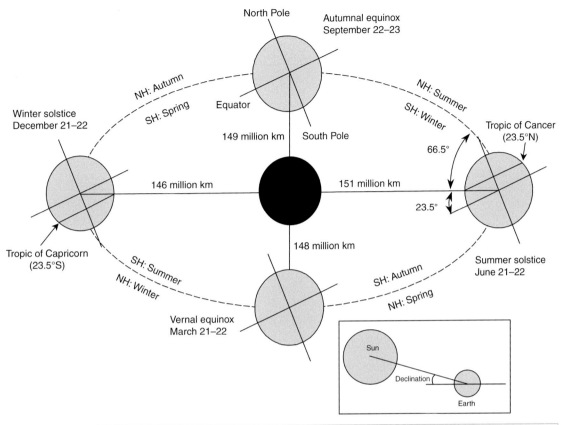

Fig. 4.2 The revolution of Earth around the Sun and the resulting astronomical seasons. The boxed inset illustrates the declination angle. Variation in the time of the solstices and equinoxes occurs because Earth's 365¼ day orbit around the Sun is approximated by 365 days with an extra day every four years.

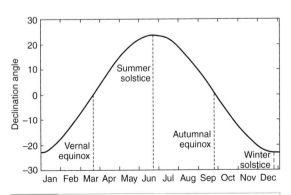

Fig. 4.3 Solar declination angle as a function of day of year.

and is called the Tropic of Capricorn. On this day, which is the shortest day of the year in the Northern Hemisphere, regions north of latitude 66.5° N receive no solar radiation. This latitude defines the Arctic Circle. Daylight is continuous south of latitude 66.5° S. This latitude defines

the Antarctic Circle. On the summer solstice, on or about June 21, the noon Sun is directly overhead at latitude 23.5° N. This latitude defines the Tropic of Cancer – the farthest point north of the equator where the Sun is directly overhead at noon. This day is the longest day of the year in the Northern Hemisphere. Daylight is continuous north of 66.5° N and is absent south of 66.5° S. On the equinoxes (on or about March 21 and September 22), the noon sun is directly overhead at the equator (i.e., the declination is zero). All latitudes receive 12 hours of light. The Sun rises due east at all latitudes on this day.

The azimuth of the Sun is the compass bearing of the Sun on the horizon (Figure 4.1). This direction is in the east in the morning and in the west after noon. The compass bearing at solar noon is either south or north depending on latitude. The azimuth angle (A_{sun}), measured

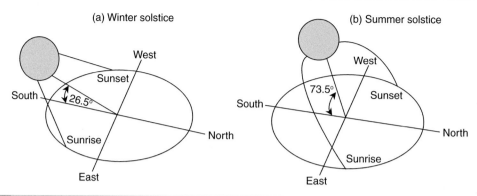

Fig. 4.4 Sun path diagram for latitude 40° N showing azimuth angles at sunrise and sunset and altitude angles at solar noon for (a) the winter solstice and (b) the summer solstice.

as the angular distance from north, varies with time of year, time of day, and latitude:

$$\cos A_{sun} = (\sin \delta \cos \phi - \cos \delta \sin \phi \cos h) / \sin Z \qquad (4.2)$$

This formula gives the angular deviation, up to 180°, from north. It is more convenient to consider azimuth angles in terms of compass bearing. In the morning, when the Sun is east of north, the computed angle is also the compass bearing ranging from 0° (north) to 180° (south). In the afternoon, when the Sun is west of north, the compass bearing is equal to 360° minus the computed angle. Figure 4.4 illustrates the solar geometry at latitude 40° N for the winter and summer solstices. The Sun is higher in the sky and rises and sets further from south in summer than in winter.

The Sun's position in the sky determines the intensity of radiation. A surface receives the most solar radiation when it is oriented perpendicular to the Sun's rays (Figure 4.5). At other angles, the Sun's radiation spreads over a larger surface area leading to less radiation per unit area. For example, when the Sun is directly overhead a unit beam of radiation covers a unit surface area. When the zenith angle is 60° (i.e., the altitude angle is 30°), the same unit beam of radiation spreads over twice as much surface area; the energy per unit area decreases by one-half. The amount of radiation received on a horizontal surface (S_H) relative to that received

on a surface perpendicular to the Sun's rays (S_P) decreases with greater zenith angle:

$$S_H = S_P \cos Z \qquad (4.3)$$

A horizontal surface receives 87 percent of S_P when the zenith angle is 30°, 71 percent at 45°, and 50 percent at 60°.

4.3 | Top of the Atmosphere Solar Radiation

The solar radiation reaching Earth varies over the course of a day and the course of a year according to Earth–Sun geometry. The amount of extraterrestrial solar radiation at the top of the atmosphere on a horizontal surface is:

$$S_H = (S_c / r_v^2) \cos Z \qquad (4.4)$$

where $S_c = 1364$ W m^{-2} is the solar constant and r_v is the radius vector. The cosine of the zenith angle adjusts the radiation that would be received on a surface perpendicular to the solar beam to that received on a horizontal surface. The radius vector accounts for the fact that as Earth moves in an elliptic orbit around the Sun, it is slightly closer to the Sun in January than in July (Figure 4.2). Since the strength of the Sun's radiation decreases with the square of the distance from the Sun, Earth receives slightly more radiation in January than July. This distance,

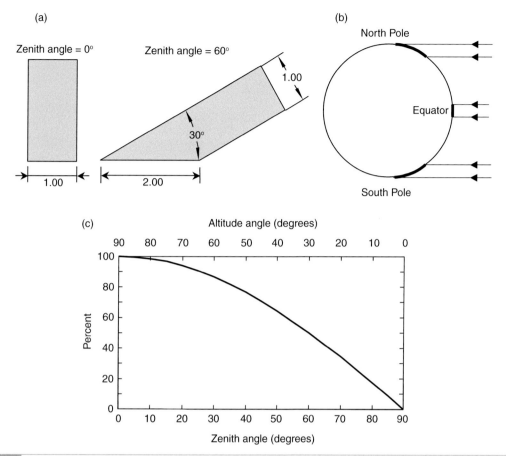

Fig. 4.5 The effect of zenith angle on solar radiation. (a) A beam of solar radiation with unit width spreads over twice as much surface area with a zenith angle of 60° than for a zenith angle of 0°. (b) Radiation spreads over a larger area at the poles than at the equator. (c) Radiation, as a percentage of perpendicular, in relation to zenith angle.

expressed as a fraction of the mean Earth–Sun distance, is called the radius vector and is at a minimum in early January. At this time, Earth receives about 3.5 percent more solar radiation relative to the average Earth–Sun distance. It is at a maximum in early July, when Earth receives about 3.25 percent less solar radiation. However, seasonal changes in radiation due to Earth's tilt are much larger. Indeed, if Earth's axis of rotation were perpendicular to the plane of rotation (i.e., no tilt), there would be no seasons since all points would be illuminated an equal amount of time throughout the year.

Figure 4.6 illustrates the diurnal cycle of solar radiation on June 21 for several latitudes in the Northern Hemisphere. Peak solar radiation at all latitudes occurs at solar noon, when the Sun is at its highest point in the sky. The Sun rises earlier in the day and sets later in the day with higher latitudes. At latitude 75° N, the Sun never sets. The lowest maximum insolation occurs at high latitudes, and insolation increases with southerly latitudes. This is because the Sun's beam is spread over a larger area at high latitudes than at lower latitudes (i.e., the zenith angle is large) so that less radiation is received per unit surface area. However, latitudes 15° N and 30° N receive more solar radiation at solar noon than does the equator. This is because the solar declination angle (23.5°) is far north of the equator, and the Sun has a zenith angle of zero at solar noon at latitude 23.5° N.

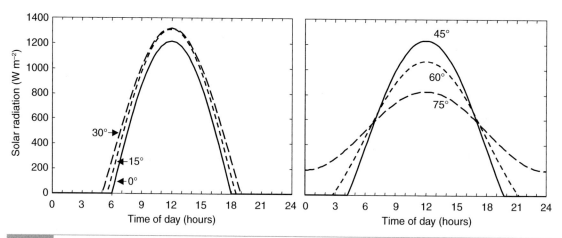

Fig. 4.6 Diurnal cycle of solar radiation at the top of the atmosphere on the summer solstice for latitudes from the equator (0°) to 75° N.

Daylength (in hours), defined as the period during which the Sun is above the horizon, is given by:

$$\frac{24}{\pi}\cos^{-1}\left(-\tan\phi\,\tan\delta\right) \qquad (4.5)$$

Daylength decreases with northern latitudes in winter and increases with northern latitudes in summer (Figure 4.7). Tropical latitudes have relatively little seasonal variation in daylength. Seasonal variation increases with higher latitudes in the Northern and Southern Hemispheres. The most extreme seasonal variation occurs in polar regions, where the Sun is below the horizon in winter and never sets in summer.

Earth's orbit around the Sun over the course of a year drives changes in the apparent motion of the Sun in the sky that affects the amount of solar radiation received. From December 21 to June 21, as declination increases from –23.5° to 23.5°, zenith angle decreases, days become longer, and regions in the Northern Hemisphere receive more solar radiation (Figure 4.8). From June 21 to December 21, declination decreases, days get shorter, and radiation decreases in the Northern Hemisphere. Latitudes near the equator have relatively little seasonal variation in solar radiation. At higher latitudes, the amplitude of the seasonal cycle increases. The most extreme seasonality occurs near the poles, where the Sun is below the horizon until the

equinoxes (March 21, September 22) and where the long summer days result in much radiation. Figure 4.8 also shows a strong poleward decrease in solar radiation during winter, and a much weaker equator-to-pole temperature gradient in summer.

4.4 | Atmospheric Attenuation

As solar radiation passes through the atmosphere, some is absorbed, primarily by water vapor and clouds, and some is scattered, both upwards to space and downwards onto the surface, by clouds, air molecules, and particles suspended in the air. This downward scattered radiation, known as diffuse radiation, emanates from all directions of the sky. In contrast, direct beam radiation is not scattered and originates from the Sun's position in the sky.

Mathematically, the solar radiation on a horizontal surface at the top of the atmosphere (S_H) is attenuated at the surface ($S\downarrow$) as:

$$S\downarrow = S_H\,\tau^m \qquad (4.6)$$

where τ^m is atmospheric transmittance. Typical values for τ range from 0.6 to 0.7 for clear skies. The optical air mass (m) is:

$$m = \frac{1}{\cos Z}\frac{P}{P_s} \qquad (4.7)$$

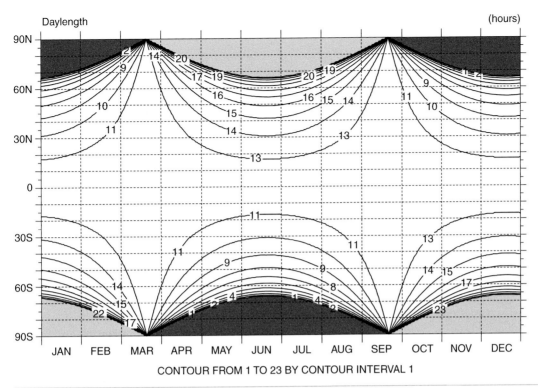

Daylength   (hours)

CONTOUR FROM 1 TO 23 BY CONTOUR INTERVAL 1

Fig. 4.7 Daylength as a function of latitude (vertical axis) and day of year (horizontal axis). Dark shading shows when the Sun never rises. Light shading shows when the Sun is above the horizon for 24 hours.

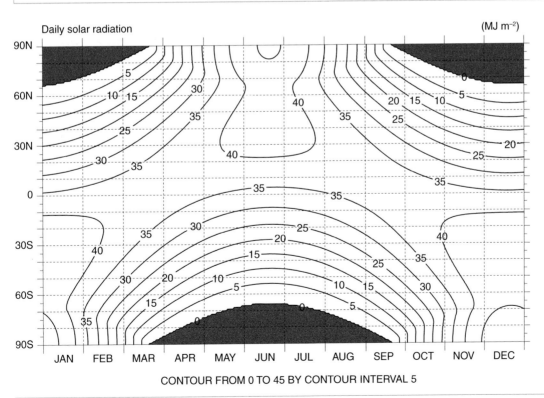

Daily solar radiation   (MJ m⁻²)

CONTOUR FROM 0 TO 45 BY CONTOUR INTERVAL 5

Fig. 4.8 Daily solar radiation at the top of the atmosphere in relation to latitude (vertical axis) and day of year (horizontal axis). Units are megajoules (10⁶ joules) per square meter.

where P is air pressure and P_s is pressure at sea level. Optical air mass increases, so that transmittance decreases, with greater zenith angle. The longer path length through the atmosphere causes less radiation to reach the ground. Altitude also affects solar radiation by reducing the path length that solar radiation travels through the atmosphere. With greater height in the atmosphere, there are fewer air molecules and air pressure decreases. Because there is less air mass at higher elevations, solar radiation is less likely to be scattered or absorbed as it passes through the atmosphere and more radiation reaches the ground. For a given zenith angle, optical air mass decreases with elevation (as pressure decreases). Consequently, atmospheric transmittance increases with higher elevation.

Diffuse radiation is most important when scattering is high. On overcast days, all the radiation is diffuse. The fraction of total radiation that is diffuse is also high when the low solar altitude angle leads to a longer path through the atmosphere, such as in mornings, evenings, winter, and at high latitudes. When the Sun is directly overhead, solar radiation travels through less of the atmosphere than when it is at an angle and hence less is scattered. Consequently, diffuse radiation may account for 25–50 percent of the total radiation when the Sun is low on the horizon, but only 10–20 percent when the Sun is high in the sky. The proportion of the total radiation that is diffuse increases, often by a factor of two, for cloudy, overcast, or polluted skies. Increasing amounts of aerosols in the atmosphere, by altering the proportion of direct and diffuse radiation, can affect plant photosynthesis.

4.5 | Annual Global Mean Energy Budget

Although Earth as a whole balances annual solar and longwave radiation at the top of the atmosphere, the geographic distribution of net radiation is unequal (Figure 4.9). Latitudes near the tropics, between 30° S and 30° N, generally absorb more than 275 W m⁻² solar radiation annually; latitudes closer to the poles absorb less

radiation. The major exception to this is North Africa, where the bright desert soils reflect a large portion of the solar radiation. Tropical latitudes, because they are warmer, generally emit more longwave radiation than high latitudes. Annual outgoing longwave radiation at the top of the atmosphere is generally greater than 250 W m⁻² between latitudes 30° S and 30° N. Regions of high precipitation, which have low longwave fluxes because of the deep, cold clouds associated with precipitation, are an exception. Between latitudes 30° S and 30° N, this heat loss is less than the heat gained from solar radiation. In general, there is an annual excess of solar radiation gain over longwave radiation loss in the tropics and a deficit at latitudes poleward of 35°–40°. This unequal geographic heating is an important determinant of Earth's macroclimate at continental to global scales.

4.6 | Sloped Surfaces

Local and regional climates can deviate from the broad macroscale geography of solar radiation based on topography. The direct beam radiation on a sloped surface depends on the Sun's zenith angle, the angle of slope, the direction of the Sun, and the direction of the sloping surface (Figure 4.10). A surface receives maximum intensity when it is oriented perpendicular to the Sun's rays. On a sloped surface, the angular deviation from perpendicular must be adjusted for the tilt of the surface. The incidence angle is the angle between the Sun's beam and an imaginary line perpendicular to the slope. This angle is:

$$\cos i = \cos s \cos Z \\ + \sin s \sin Z \cos (A_{sun} - A_{slope}) \quad (4.8)$$

where s is the angle of slope and A_{slope} is the azimuth of the slope. The azimuth angle of the slope is the compass direction to which the slope is oriented (e.g., north = 0°, east = 90°, south = 180°, west = 270°). The direct beam radiation onto the surface is:

$$S \downarrow = (S_c / r_v^2) \tau^m \cos i \quad (4.9)$$

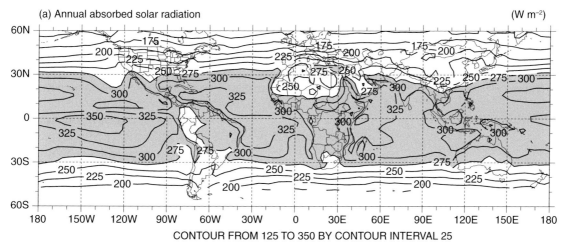

(a) Annual absorbed solar radiation   (W m⁻²)

CONTOUR FROM 125 TO 350 BY CONTOUR INTERVAL 25

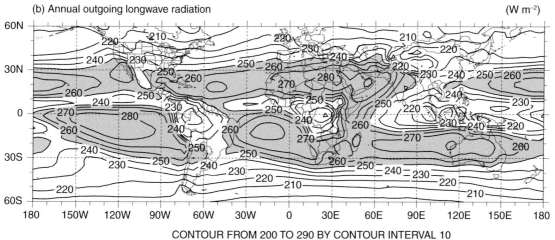

(b) Annual outgoing longwave radiation   (W m⁻²)

CONTOUR FROM 200 TO 290 BY CONTOUR INTERVAL 10

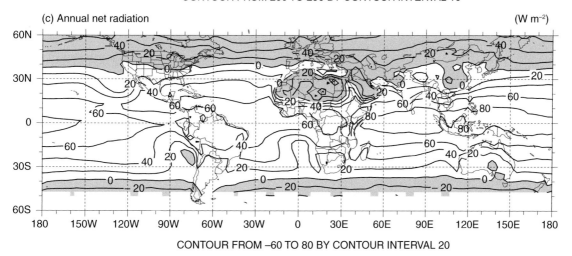

(c) Annual net radiation   (W m⁻²)

CONTOUR FROM −60 TO 80 BY CONTOUR INTERVAL 20

Fig. 4.9 Annually averaged radiative fluxes as observed by satellite. Fluxes poleward of latitude 60° are unreliable and are not shown. (a) Annual absorbed solar radiation. Regions absorbing more than 275 W m⁻² are shaded. (b) Annual outgoing longwave radiation. Regions losing more than 250 W m⁻² are shaded. (c) Annual net radiation (absorbed solar minus outgoing longwave). Regions with a net loss of radiation are shaded. Data provided by the National Center for Atmospheric Research (Boulder, Colorado).

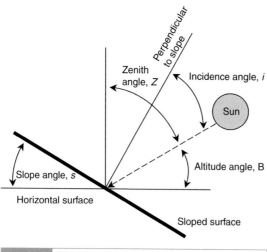

Fig. 4.10 Altitude, zenith, and incidence angles for a sloped surface.

On a horizontal surface, $s = 0°$ and the incidence angle is the zenith angle. On a sloped surface, the incidence angle can be more or less than the zenith angle depending on the angle and orientation of the slope.

The diffuse radiation on a sloped surface also depends on the angle of slope. The sky forms an inverted bowl, or half sphere, above and around a point in the landscape (Figure 4.11). On a horizontal surface, diffuse radiation emanates from all portions of the sky. As the surface tilts at an angle, less of the sky hemisphere is viewed from a point on the surface. The terrain blocks a portion of the sky, from which no sky diffuse radiation is received. With a vertical wall, the sky hemisphere is cut in half and each side of the wall receives diffuse radiation from only one-half of the sky. A sloped surface, therefore, sees less of the sky as the angle of slope

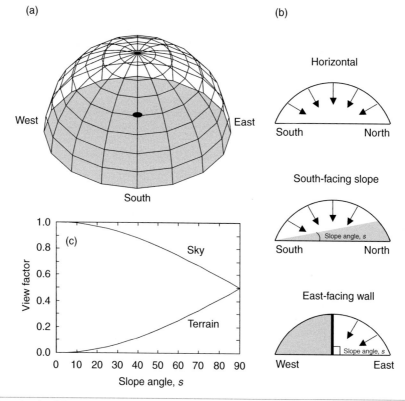

Fig. 4.11 Effect of angle of slope on diffuse radiation. (a) The sky is shown as an inverted bowl, or hemisphere, from which diffuse radiation is received. The point in the center receives diffuse radiation from the entire hemisphere. (b) The three right-hand panels show a cross-section of this hemisphere for a horizontal surface, a south-facing slope, and a wall. The gray area is the portion of the sky that is blocked. (c) Sky and terrain view factors in relation to angle of slope.

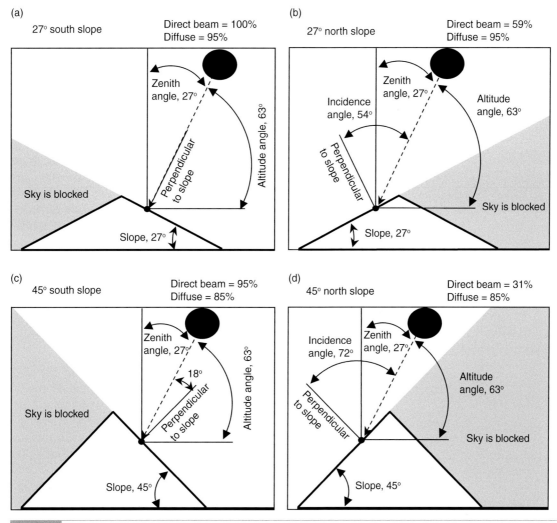

Fig. 4.12 Direct beam and diffuse solar radiation on slopes. In all four panels, the Sun is due south with a zenith angle of 27°. Direct beam radiation varies with angle of slope and aspect in relation to incidence angle. Shown is the percentage of perpendicular radiation. The portion of the sky from which diffuse radiation is received varies with the angle of slope, given as a percentage of sky radiation. (a) 27° south-facing slope. (b) 27° north-facing slope. (c) 45° south-facing slope. (d) 45° north-facing slope.

increases. The fraction of the sky seen, known as the sky view factor, is given by:

$$\psi_{sky} = (1 + \cos\ s)\,/\,2 \qquad (4.10)$$

The portion of the sky that is not blocked, $(1 - \psi_{sky})$, is the fraction of the hemisphere composed of terrain. This terrain is also a source of diffuse radiation, because some of the solar radiation incident on the slope is reflected. As the angle of slope increases, less of the sky contributes diffuse radiation and more of the terrain is viewed.

Figure 4.12 illustrates direct and diffuse radiation on various slopes, ignoring terrain radiation. For all diagrams, the Sun is due south with a zenith angle of 27° (63° elevation above the horizon). On the 27° (51%) south-facing slope, a line perpendicular to the slope is oriented 63° above the horizon. This is the same angle at which the solar beam strikes the surface so that the incidence angle is 0° and the surface

(a)

30°N, 20% southeast slope

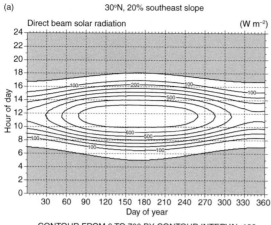

CONTOUR FROM 0 TO 700 BY CONTOUR INTERVAL 100

(b)

45°N, 42% north slope

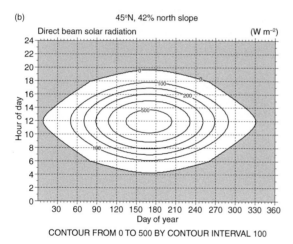

CONTOUR FROM 0 TO 500 BY CONTOUR INTERVAL 100

Fig. 4.13 Clear sky direct beam solar radiation as a function of time of day (vertical axis) and day of year from January 1 to December 31 (horizontal axis) for (a) a 20 percent southeast slope at latitude 30° N and (b) a 42 percent north slope at latitude 45° N. Direct beam radiation is from Eq. (4.9) with $\tau = 0.6$.

95 percent of the sky. Now, however, the solar beam strikes the surface at an angle of 54° from local perpendicular, and the direct beam radiation is only 59 percent. A greater portion of the sky is blocked on the 45° (100%) slopes; the surfaces receive diffuse radiation from only 85 percent of the sky. Both 45° slopes receive less direct beam radiation than the comparable 27° slopes. The incidence angle on the south-facing slope is 18° while that of the north-facing slope is 72°. In these examples, angle of slope has little effect on diffuse radiation. The greatest reduction in diffuse radiation from horizontal is only 15 percent. Angle of slope has relatively minor effect on direct beam radiation for the south-facing slopes. The greatest reduction in radiation comes from the direction of slope. The 27° north-facing slope receives only 59 percent of the direct beam radiation on the south-facing slope. The 45° north-facing slope receives one-third of the direct beam radiation of the south-facing slope.

The angle of slope, its direction of tilt, and latitude interact with time of year and time of day to produce complex patterns of solar radiation on a surface. Figure 4.13 shows the diurnal pattern of direct beam radiation throughout the year on a 20 percent (11°) southeast slope located at latitude 30 °N and a 42 percent (23°) north slope at latitude 45 °N. Noon direct beam radiation is relatively constant throughout the year on the southeast slope, but varies greatly throughout the year on the north slope. The length of day the solar beam illuminates the slope varies by less than three hours over the year on the southeast slope. In contrast, the north slope receives no direct beam radiation from December to mid-January, and daylength increases to 15 hours in June. This topographic variation in solar radiation can create unique microclimates with the general macroclimate.

receives 100 percent of the direct beam radiation. The surface receives diffuse radiation from 95 percent of the sky. On the 27° north-facing slope, diffuse radiation is still received from

4.7 | Review Questions

1. On day July 9, with declination 22°28′, the zenith angle at local solar noon is 13.53°. The time is 1 hour and 37 minutes later than at Greenwich, England. Calculate latitude and longitude.

2. In Mexico City (19° N) on the winter solstice, you need to travel north. You determine the location of the Sun at solar noon. Would you walk towards the Sun or away from the Sun to travel north? What about on the summer solstice?

3. You live in Boulder, Colorado (40° N). By experience, you know that the Sun at solar noon is always due south. You move to Singapore, close to the equator. The azimuth angle of the Sun at solar noon now varies throughout the year. During what period is the Sun in the northern sky? When is it in the southern sky?

4. By experience, you know the following characteristics of solar geometry at Charlottesville, Virginia, USA (38° N) on December 21 (δ = –23.45°) and June 21 (δ = 23.45°). You move to Melbourne, Australia (38° S). Prepare the same table for your new home. How does this compare with your Charlottesville experience?

3000 m, and $P = 0.485\ P_s$ at 5500 m. Calculate the solar radiation received at each of these heights relative to that received at sea level. Use a zenith angle $Z = 20°$ and assume $\tau = 0.7$.

7. Along latitude 40° N, the three-month period May–July has nearly constant daylength, but the three-month periods February–April and August–October have large change in daylength. Explain why.

8. Calculate the direct beam solar radiation for a unit of incoming solar radiation on two hillslopes oriented due south and due north at latitudes 0°, 30° N, and 60° N at local solar noon on the winter solstice (δ = –23.45°) and summer solstice (δ = 23.45°). Each slope is 10 percent. What can be said about the importance of direction of slope in relation to lat-

Charlottesville	Daylength (hours)		Solar azimuth			Noon zenith angle
		Sunrise	Noon	Sunset		
December 21	9.4	120°	180°	240°		61.45°
June 21	14.6	60°	180°	300°		14.55°

5. Compare the amount of extraterrestrial solar radiation received at solar noon on December 21 ($r_v = 0.98372$) and June 21 ($r_v = 1.01630$) at Charlottesville and Melbourne. How should these compare, based solely on zenith angle? Why are they not equal?

6. Air pressure is approximately $P = P_s = 101,325$ Pa at sea level, $P = 0.821\ P_s$ at 1500 m, $P = 0.674\ P_s$ at

itude and season? At 60° N, which slope (north or south) would maximize solar heating of a house during winter?

9. On June 21 would a north-facing wall located at latitude 30° S ever be illuminated by direct beam solar radiation? What about a south-facing wall?

Atmospheric General Circulation and Climate

5.1 | Chapter Summary

Geographic variation in the annual radiative balance at the top of the atmosphere drives the general circulation of the atmosphere and produces the major patterns of climate on Earth. The latitudinal gradient in net radiation results in an equator-to-pole temperature gradient. However, if only radiative processes determined temperatures, the tropics would be tens of degrees warmer than they actually are, and polar regions would be much colder than they actually are. Instead, the uneven geographic distribution of radiation produces winds, set in motion by differences in air pressure, that redistribute heat from the tropics to the poles. Winds are a balance of the pressure gradient force, Coriolis force, and friction acting simultaneously. These forces produce the general circulation of the atmosphere. The continents alter this idealized circulation because landmasses heat and cool faster than oceans. In winter, when the landmasses of the Northern Hemisphere are colder than oceans, high pressure systems form over land while low pressure systems are most pronounced over oceans. The opposite pattern occurs in summer when continents are warmer than oceans. Oceans also influence climate by transporting heat poleward. Prominent ocean circulations are wind-driven surface currents and the density-driven thermohaline circulation. The general circulation of the atmosphere varies over the course of a year in response to seasonal changes in solar radiation. These seasonal changes in surface high and low pressure regions and atmospheric circulation drive seasonal changes in precipitation. Monsoons are one such prominent large-scale seasonal atmospheric circulation.

5.2 | Air Pressure

Air pressure is the force exerted over a given area by the movement of air molecules. Air pressure is also a measure of the mass of air above a given point. Air pressure decreases with greater height above sea level because there are fewer air molecules, and less mass, with greater altitude.

The change in air pressure with height is described by the hydrostatic equation, which relates the upward force on a parcel of air due to the decrease in pressure with height to the downward force from gravity. This balance of forces is:

$$-dP = g\rho dz \tag{5.1}$$

where dP is the change in pressure (pascal, 1 Pa = 1 N m^{-2} = 1 kg m^{-1} s^{-2}), dz is change in height (m), ρ is the density of air (kg m^{-3}), and g = 9.81 m s^{-2} is gravitational acceleration. The term ρdz is the mass of air between heights z and $z + dz$. The downward force due to gravity is $g\rho dz$. This downward force is balanced by an upward force due to the decrease in pressure with height

$(-dP)$. Combining Eq. (5.1) with the ideal gas equation, from Eq. (3.12) and which provides an expression for density, gives:

$$\frac{dP}{P} = \frac{-g}{\Re T / M_a} dz = -\frac{dz}{H} \qquad (5.2)$$

where H is the scale height. Integration of Eq. (5.2) gives the pressure at height z:

$$P = P_s e^{-z/H} \qquad (5.3)$$

where P_s =1013.25 hectopascals (hPa, 1 hPa = 100 Pa = 1 millibar) is surface pressure at sea level ($z = 0$). An approximate value is H = 7600 m so that air pressure is $0.37P_s$ at an altitude of 7.6 km. The force of gravity holds most air molecules near the surface. With greater height in the atmosphere, the number of air molecules and pressure decrease.

Air pressure is related to the mass of air above a given point. This is evident by rearranging the terms in the hydrostatic equation to relate change in mass per unit area (dm, kg m^{-2}) to change in pressure (dP):

$$dm = \rho dz = -dP / g \qquad (5.4)$$

At sea level, where the standard atmospheric pressure is 1013.25 hPa, a column of air covering one square meter of surface area and extending to the top of the atmosphere (i.e., to a pressure of 0 Pa) has mass:

101 325 kg m^{-1}s^{-2} / 9.81 m s^{-2}
= 10 329 kg m^{-2}

Multiplied by Earth's surface area of about 510 million km^2, the total mass of the atmosphere is about 5×10^{18} kg. Approximately fifty percent of the mass of the atmosphere is below a height of 5500 m.

The mass of an atmospheric constituent can be found from its volume mixing ratio. For example, the concentration of CO_2 in the atmosphere is 390 parts per million by volume (Table 2.1). This is converted to total mass of carbon by multiplying this number by the mass of the atmosphere weighted by the relative mass of carbon atoms to the mass of air molecules. The molecular weight of carbon is 12 g mol^{-1}

and the molecular weight of dry air is 28.97 g mol^{-1}. This is equivalent to over 800×10^{15} g of carbon:

$$(390 \times 10^{-6}) \times 12 / 28.97 \times (5 \times 10^{18} \text{ kg})$$
$$= 808 \times 10^{15} \text{g C}$$

5.3 | Wind

Winds are the balance of three forces acting simultaneously: the pressure gradient force, the Coriolis force, and friction. The pressure gradient force is the primary cause of air movement. If the pressure gradient force were the only force acting on air, winds would always flow directly from high to low pressure. However, the Coriolis force, which is the apparent motion caused by Earth's rotation, deflects air as it moves and causes wind to flow parallel to isobars in the absence of friction from the surface. Near the surface, frictional drag slows wind speed and deflects the direction of motion. Surface winds do not flow parallel to the isobars but rather cross them moving from high to low pressure.

5.3.1 | Pressure Gradient Force

Winds are created by horizontal gradients in air pressure. For example, a column of air with a surface pressure of 1020 hPa has a mass of 10,400 kg m^{-2}. Suppose a nearby column of air, separated by a distance of 100 km, has a lower surface pressure of 1012 hPa (10,320 kg m^{-2}). The difference in mass between the two columns is 80 kg m^{-2}, which creates a pressure gradient of 8 hPa per 100 km. The resulting pressure gradient force causes air to flow from high to low pressure. This force is directed from high to low pressure at right angle to lines of equal pressure, or isobars. Closely spaced isobars indicate steep pressure gradients, strong forces, and strong winds; widely spaced isobars indicate weak pressure gradients and weak winds.

Differences in air temperature can create horizontal pressure gradients that initiate air movement. Molecules in warm air move fast

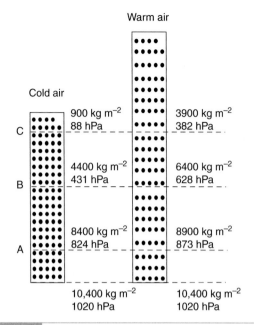

Fig. 5.1 Pressure gradient resulting from temperature differences. In this figure, each dot depicts many air molecules with a combined mass per unit area of 100 kg m^{-2}. The atmosphere is depicted uniformly with height. In fact, mass and pressure decrease rapidly with height. This is omitted from the figure for simplicity and does not invalidate the general conclusions. Shown are mass and pressure at several heights for a column of cold air (left) and a column of warm air (right). Mass at the surface and at heights A, B, and C are the mass of air above each height. Pressure is the corresponding air pressure.

cold air is denser than warm air, more air molecules are closer to the ground in the cold column than in the warm column. For example, air molecules below height **A** in the cold column have a combined mass of 2000 kg m^{-2}. In the warm column, where air molecules spread farther apart, the mass is only 1500 kg m^{-2}. Both columns of air have the same total mass, and therefore the number of molecules (i.e., mass) above height **A** is less in the cold air (8400 kg m^{-2}) than in the warm air (8900 kg m^{-2}); and because there is less mass above height **A** in the cold column, air pressure (i.e., the mass of air above height **A**) is lower than in the warm air column. This is true at other heights.

For any given height above the surface, a column of cold air has fewer air molecules, less mass, and lower air pressure than a column of warm air. Hence, warm air is associated with high air pressure aloft and cold air is associated with low air pressure aloft. Differences in temperature, by creating a horizontal pressure gradient, initiate wind flow aloft from high pressure (warm air) to low pressure (cold air). As air aloft leaves the warm column, the mass of air in the column decreases and surface pressure decreases. Conversely, the addition of air to the cold column increases its mass and surface pressure. As a result, high surface pressure develops under the cold column and low surface pressure develops under the warm column. Surface wind flows from high pressure (cold air) to low pressure (warm air), closing the atmospheric circulation.

This simple model of thermally driven atmospheric circulation begins to explain the geographic redistribution of heat by atmospheric winds. Tropical regions, because they gain radiation, are hot and develop high pressure aloft. Polar regions, because they lose radiation, are cold and develop low pressure aloft. In response to this pressure gradient force, warm tropical air flows towards the poles aloft. A broad band of surface low pressure develops in the tropics, while the poles, where air converges aloft, develop high surface pressure. In response to this surface pressure gradient, cold polar air flows over the surface towards the equator and completes the atmospheric circulation.

and spread apart; the air becomes less dense and the column expands vertically. Conversely, molecules in a column of cold air move slowly and become dense; the column shrinks. That warm air is less dense than cold air is evident from the ideal gas law, from Eq. (3.10). If pressure remains constant, any increase in temperature must result in a corresponding decrease in density. A decrease in temperature results in an increase in density. In other words, a short column of cold, dense air exerts the same surface pressure as a tall column of warm air.

The result is that the mass of air, or air pressure, decreases more rapidly with height in a column of cold air than in a column of warm air. Consider the two columns of air in Figure 5.1. Both have a total mass of 10,400 kg m^{-2} and a surface pressure of 1020 hPa. However, because

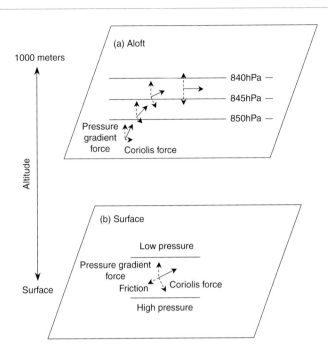

1000 meters

Altitude

Surface

The difference in winds at the surface and aloft in the Northern Hemisphere. (a) Geostrophic wind aloft is balanced by the pressure gradient force and Coriolis force with motion parallel to isobars. (b) Surface wind is balanced by the pressure gradient force, the Coriolis force, and friction. Friction slows wind speed and causes wind to cross isobars typically at a 30° angle.

5.3.2 Geostrophic Wind

This single cell, thermally driven circulation does not fully explain atmospheric circulation. The Coriolis force is the apparent motion caused by Earth's rotation, which pushes winds to the right of their intended path in the Northern Hemisphere and to the left in the Southern Hemisphere. The Coriolis force acts at right angles to the wind's intended direction, influencing its direction but not its speed. The Coriolis force is greater for strong winds than for weak winds and is zero along the equator and greatest at the poles.

The top panel in Figure 5.2 shows how the pressure gradient force and Coriolis force determine wind speed and direction aloft at heights typically greater than 1000 m, where the influence of surface friction is negligible. The pressure gradient force is perpendicular to the isobars in the direction of low pressure. This force accelerates a parcel of air towards the low pressure. As it begins to move, the Coriolis force deflects the air towards the right, curving its path. As the path changes, so does the Coriolis force, which is always directed at right angle to the direction of motion. As the parcel of air increases in speed, the magnitude of the Coriolis force increases. Eventually wind speed and direction are such that the pressure gradient force, acting from high to low pressure, is balanced by the Coriolis force acting in the opposite direction. At this point, the net force acting on the air is zero and the wind flows parallel to the isobars at a constant speed. This flow of air is called geostrophic wind. In the Northern Hemisphere, geostrophic wind flows with low pressure to the left and high pressure to the right with a speed directly related to the pressure gradient. In the Southern Hemisphere, the flow is reversed because the Coriolis force deflects winds to the left of their intended path.

5.3.3 Vertical Air Motions

The frictional drag of objects near the surface slows wind speed and in doing so deflects the direction of motion. Surface winds do not flow parallel to the isobars but rather cross them moving from high to low pressure. Figure 5.2 illustrates the difference in wind between the surface, where friction is important, and aloft, where friction is negligible. The upper air wind is geostrophic, flowing parallel to the isobars with the pressure gradient force balanced by the Coriolis force. At the surface, friction, which acts counter to the direction of motion, reduces wind speed. Because of this, the same pressure gradient produces slower winds at the surface

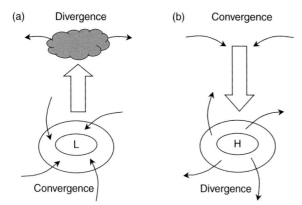

(a) Divergence   (b) Convergence

Convergence   Divergence

Fig. 5.3 Surface winds and vertical air motions in the Northern Hemisphere with (a) low pressure and (b) high pressure.

than aloft. With reduced speed, the Coriolis force decreases, and the weaker Coriolis force no longer balances the pressure gradient force. The winds do not deflect as much towards the right of their intended path and winds flow across isobars towards low pressure.

The deflection of surface winds by friction creates vertical motions in the atmosphere (Figure 5.3). In the Northern Hemisphere, winds flow in a clockwise direction around high pressure cells and counterclockwise around low pressure cells. (In the Southern Hemisphere, where the Coriolis force deflects winds to the left, winds flow clockwise around low pressure cells and counterclockwise around high pressures.) Friction deflects surface winds in towards the center of the low pressure. As the surface air moves inward, the converging air slowly rises, typically to a height of several thousand meters, and diverges aloft. So long as the upper-level outflow of air balances the inflow of surface air, surface pressure remains unchanged. However, if the upper-level divergence exceeds the surface convergence, surface pressure decreases, the pressure gradient increases, and surface winds strengthen. In contrast, surface winds flow outward from the center of a high pressure cell. Air from above converges and descends to replace the diverging surface air.

5.4 | Large-Scale Atmospheric Circulations

Figure 5.4 illustrates the general circulation of the atmosphere as a system of high and low surface pressure regions arising from the unequal heating of the surface. In contrast to the single cell model, each hemisphere has three cells that redistribute heat. In the Northern Hemisphere, warm tropical air with high pressure aloft flows as upper-level wind from the equator towards the North Pole. As it moves north, the Coriolis force deflects the wind to the east. This poleward moving air cools as it moves northwards. The column shrinks, the air becomes denser, and surface pressure increases. The converging masses of air moving from the tropics to middle latitudes further increases surface pressure. This produces a belt of high surface pressure at about latitude 30° N. Air flows along the surface back towards the equator from this high pressure, being deflected to the west by the Coriolis force. These are the northeasterly trade winds. This tropical cell is known as the Hadley circulation. At the North Pole, cold air causes low pressure aloft while warmer air at latitude 60° N creates high pressure aloft. This warm air flows towards the pole as southwesterly wind aloft (northward moving air deflected to the east by the Coriolis force) and back as the polar easterlies along the surface (southward moving air deflected to the west by the Coriolis force). A third cell in the middle latitudes between latitudes 30° N and 60° N, known as the Ferrel cell, circulates in a direction opposite to the Hadley cell and connects the tropical and polar cells. The surface high pressure at latitude 30° N creates subsiding motion and surface divergence. One branch of this air spreads southwards as trade winds.

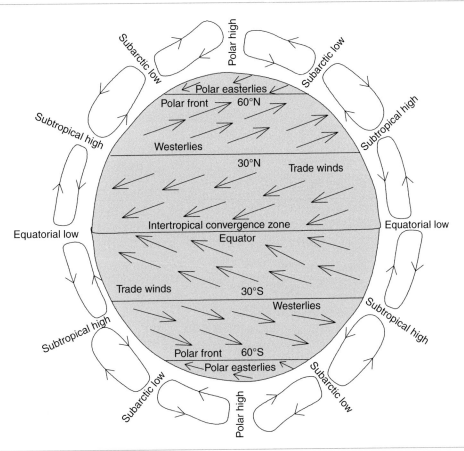

Fig. 5.4 | Idealized atmospheric circulation with resulting surface winds and surface pressure.

The other branch flows northwards, being deflected to the east in middle latitudes. These are the mid-latitude westerlies. Similar circulations develop in the Southern Hemisphere.

Figure 5.4 shows several prominent features characterize the circulation of the atmosphere. Semi-permanent low surface pressure systems develop at the equator and latitude 60°, where air rises. Semi-permanent high surface pressure systems develop at the poles and the subtropics at latitude 30°, where air subsides. Between latitudes 30° N and 30° S, surface trade winds converge along the equator from the northeast in the Northern Hemisphere and the southeast in the Southern Hemisphere in a broad region known as the intertropical convergence zone. Between latitudes 30° and 60° in both hemispheres, surface winds are westerly. At about latitude 60°, these winds clash with polar easterlies along the polar front.

Figure 5.5 shows the actual geographic distribution of surface pressure and surface wind for January and July. The three principal features are: high pressure cells at about 30° latitude in both hemispheres; low pressure along the equator between these subtropical high pressure centers; and low pressure at high latitudes. Wind flows clockwise around high pressure systems in the Northern Hemisphere. Thus, in the Northern Hemisphere the northern edges of the subtropical high pressures drive the mid-latitude westerlies while the southern edges drive the westwardly flowing trade winds. Conversely, winds flow counterclockwise around low pressure centers, driving the polar easterlies to the north and the mid-latitude westerlies to the south. In the Southern Hemisphere, flows are in the opposite direction (counterclockwise around high pressures and clockwise around lows).

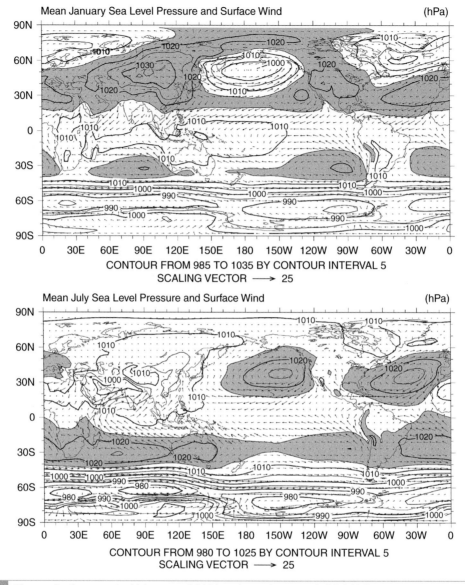

Fig. 5.5 Mean sea level pressure for (a) January and (b) July, with units hectopascals (hPa). High pressure regions greater than 1015 hPa are shaded. Low pressure regions less than 1015 hPa are unshaded. Arrows indicate the direction and magnitude of surface wind in meters per second. Data provided by the National Center for Atmospheric Research (Boulder, Colorado).

The general circulation of the atmosphere accounts for the major climate zones in the world. Temperatures are generally distributed in latitudinal bands with warmest temperatures in the tropics and progressively colder temperatures towards the poles (Figure 5.6). Rainfall is abundant where air rises, such as along the equator, and low where air sinks, such as near latitudes 30° N and 30° S (Figure 5.7). In the tropics, the trade winds converge in the intertropical convergence zone. This convergence and lifting of warm, moist air leads to high annual rainfall. Many tropical regions have wet and dry seasons. This occurs because of seasonal variation in the geographic location of the convergence zone. Regions of subsidence, as occurs in high pressure systems, generally have low rainfall. Many of the world's major deserts – in southwestern United States, North Africa, southern South America, South

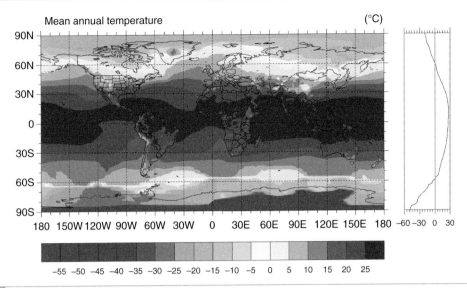

Fig. 5.6 Annual mean air temperature. The chart on the right-hand side shows temperature averaged around the world, from longitude 180° W to 180° E, as a function of latitude. Data provided by the National Center for Atmospheric Research (Boulder, Colorado). See color plate section.

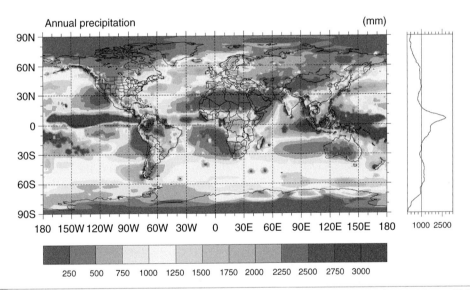

Fig. 5.7 Annual mean precipitation. The chart on the right-hand side shows precipitation averaged around the world, from longitude 180° W to 180° E, as a function of latitude. Data provided by the National Center for Atmospheric Research (Boulder, Colorado). See color plate section.

Africa, and western Australia – occur on the eastern flanks of the subtropical high pressures near latitudes 30° N and 30° S. Rainfall is also high, though not as high as in the tropics, in the middle latitudes between 40° and 60° where warm moist air clashes with cold air along the polar front.

5.5 | Continents

It is evident from Figure 5.5 that the Northern and Southern Hemispheres differ in circulation. The Southern Hemisphere has contiguous bands of high pressure at latitude 30° S and low pressure at latitude 60° S, as expected from Figure 5.4. These bands are intermingled in the Northern Hemisphere. This difference arises because of the different distribution of land in the two hemispheres; 70 percent of all land is in the Northern Hemisphere. Maximum land area is between latitudes 40° N and 75° N, where more than 50 percent of Earth's surface area is land. In the Southern Hemisphere, land is generally less than 25 percent of the surface area. Between latitudes 40° S and 65° S, there is little land. Continents heat and cool faster than oceans. In January, when northern continents are colder than oceans, high pressure systems form over land; low pressure systems are most pronounced over the northern regions of the Pacific and Atlantic Oceans. The opposite pattern occurs in summer when northern continents are warmer than oceans. Strong high pressures develop in the North Pacific and North Atlantic; low pressures develop over Asia and North America.

5.6 | Oceans

Proximity to oceans affects climate. The large heat capacity of oceans creates a thermal inertia that moderates extreme temperature fluctuations compared with interior regions of the continents. Oceans store heat in summer and release heat in winter, damping summertime warming and wintertime cooling. Like the atmosphere, the general circulation of the ocean also transports heat from the tropics to the poles. Oceanic heat transport is the result of two types of circulations: wind-driven surface currents and the density-driven thermohaline circulation.

The trade winds blowing from the southeast in the Southern Hemisphere and northeast in the Northern Hemisphere set in motion westward-flowing equatorial currents to the north and south of the equator. When these currents reach the western edge of the Pacific and Atlantic basins, they deflect and flow as western boundary currents to the north in the Northern Hemisphere and to the south in the Southern Hemisphere. Between latitudes 30°

and 60° in both hemispheres, the prevailing westerly winds push surface water east across the ocean basins. At the eastern edge of the oceans, this water deflects as eastern boundary currents to the south in the Northern Hemisphere and to the north in the Southern Hemisphere. This circulation of surface water creates circular flows in ocean basins known as gyres (Figure 5.8). These gyres are centered in the subtropics at latitudes 30° N in the North Atlantic and the North Pacific and at 30° S in the South Atlantic, South Pacific, and Indian Ocean. These gyres rotate clockwise in the Northern Hemisphere and counterclockwise in the Southern Hemisphere. At high latitudes, polar easterlies drive surface currents in a westward direction to produce subpolar gyres that rotate in a direction opposite to the adjacent subtropical gyres. These subpolar gyres develop in the Atlantic east of Greenland and in the Weddell Sea off Antarctica. In the Southern Hemisphere, the strong surface westerlies near latitude 60° S create an eastward-flowing Antarctic circumpolar gyre.

Western boundary currents are strong poleward-flowing currents that carry warm water from the tropics to middle latitudes. Proximity to these warm currents creates a milder climate than would be expected from latitude. The most prominent are the Gulf Stream current along the western boundary of the Atlantic off the United States coast and the Kuroshio current along the western boundary of the North Pacific. As a result, water off western and northern Europe is much warmer than water off eastern North America at similar latitudes (Figure 5.6). Likewise, water off southeast Alaska and western Canada is warmer than corresponding water north of Japan. In contrast, eastern boundary currents transport cold water towards the tropics. Prominent eastern boundary currents are the California current off the United States west coast, the Peru current off the west coast of South America, the Canary current off northern Africa, and the Benguela current off southern Africa.

In addition to surface currents, deep ocean currents arise from density differences in water. This circulation, called the thermohaline

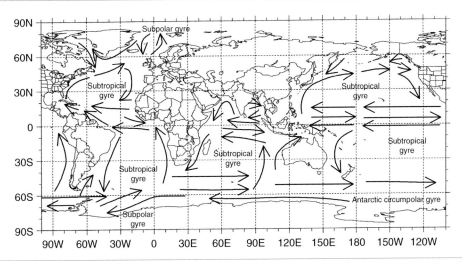

Fig. 5.8 Wind-driven ocean surface currents.

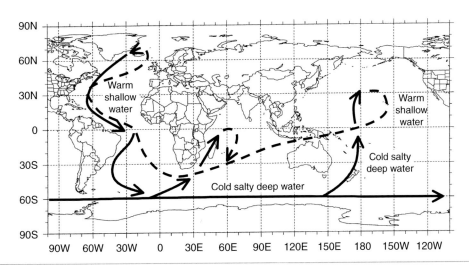

Fig. 5.9 Thermohaline circulation showing cold, salty deep water (solid lines) and warm shallow water (dashed lines).

circulation, arises because variations in the salinity and temperature of water create differences in density. The density of water increases as temperature decreases or salinity increases. Cold, salty water is denser than warm, fresh water. Water involved in the thermohaline circulation initially forms in polar regions at the surface. In the Antarctic Ocean, surface water cools to below freezing because of the cold overlying air. When the water freezes, it forms a layer of sea ice several meters thick. Salt is excluded from the ice as it freezes, and the unfrozen water under the sea ice becomes

saltier. The cold, salty water is dense, sinks to the bottom, encircling Antarctica and flowing northwards along the ocean bottom. Similar masses of cold, dense deep water form in the Arctic Ocean off the coast of Greenland, where it flows southwards along the ocean bottom into the North Atlantic. Figure 5.9 illustrates the full thermohaline circulation. In the North Atlantic, warm surface water flowing northwards becomes more dense as it cools and evaporation increases surface salinity. The dense water sinks and returns southward along the ocean bottom. Near the southern tip of Africa,

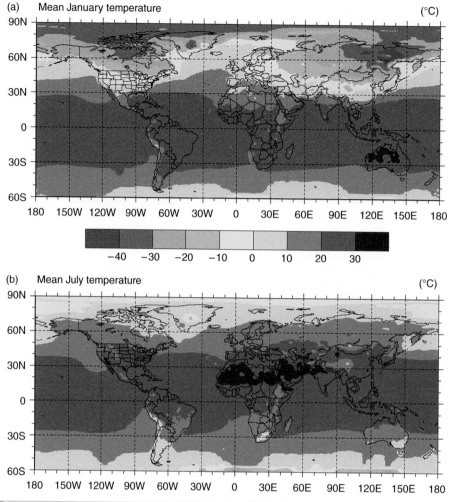

Fig. 5.10 Mean air temperature for (a) January and (b) July. Data provided by the National Center for Atmospheric Research (Boulder, Colorado). See color plate section.

this deep water joins the Antarctic bottom water. This river of bottom water spreads into the Indian and Pacific Oceans. The deep water slowly rises in these oceans and begins its long journey along the surface back to the North Atlantic.

5.7 | Seasons

Changes in solar radiation over the course of a year drive the seasonal heating and cooling of the atmosphere. Air temperatures generally lag by one month compared with solar radiation so that although the least solar radiation is received in December the coldest month is typically January (Figure 5.10). Temperatures generally follow latitudinal bands. In the Northern Hemisphere, which is experiencing winter, the poleward temperature gradient is particularly strong. July is typically the warmest month. In July, the Northern Hemisphere poleward temperature gradient weakens, because the Sun's declination is north of the equator and the latitudinal gradient in solar radiation diminishes. The tropics and subtropics, between latitudes 30° S and 30° N, show the least seasonal variation in temperatures. Temperatures in this region are generally between 20°C and 30°C throughout the year. In contrast, middle and high latitudes have pronounced seasons, with temperatures below freezing in winter and generally from 10°C to 20°C in summer.

The moderating influence of oceans on temperature is evident in Figure 5.10. This is particularly clear in January, when temperatures are colder in the interior regions of North America and Eurasia than at similar latitudes near oceans. Moreover, the continents have a larger seasonal temperature range than corresponding oceans. For example, while the seasonal temperature range in interior North America is about 25°C, the temperature range in the North Atlantic Ocean is only about 10°C. Land has a much lower heat capacity than water so that the continents cool faster than oceans in winter and warm more in summer.

As a result of seasonal changes in solar radiation and temperature, the major surface high and low pressure systems migrate to different locations. In winter, two large subarctic low pressure systems center in the northern regions of the Atlantic and Pacific Oceans – the Icelandic and Aleutian lows, respectively (Figure 5.5). Two high pressure systems – the Bermuda high in the Atlantic and the Pacific high off the California coast – develop in subtropical latitudes near 30° N. High pressure cells form over Asia and North America due to the intense cooling of land. During summer, land warms and low pressure systems replace the continental high pressure cells. The Icelandic and Aleutian lows, so strong in winter, are barely noticeable. Instead, the subtropical Bermuda and Pacific highs dominate.

Seasonal changes in solar radiation and temperature also alter winds aloft. Winds can be partitioned into their south-to-north (meridional) and west-to-east (zonal) components. Figure 5.11 shows the zonal wind as a function of height. The strong westerly winds in the upper atmosphere between 10 and 15 km aloft in the middle latitudes of each hemisphere are the jet streams. They locate along the fronts that separate cold polar air from warm tropical air, where the horizontal temperature and pressure gradients are greatest. The subtropical jet stream is centered at latitude 30°, near the polar extremity of the tropical cell. A second jet, called the polar jet, is located further poleward at the edge of the polar cell. The sharp poleward contrast in temperature produces a rapid horizontal pressure gradient that intensifies winds along the front. Because the poleward temperature gradient is strongest in winter and weakest in summer, the jet streams show seasonal variation. Winds are strongest and dip further from the poles in winter. The jets weaken in summer and migrate to higher latitudes.

These seasonal changes in surface pressure and atmospheric circulation drive seasonal changes in precipitation (Figure 5.12). The intertropical convergence zone, where the convergence of the trade winds produces rising motion and heavy rainfall, migrates from south of the equator in boreal winter (austral summer) to north of the equator in boreal summer (austral winter). For example, tropical regions of South America and Africa between the equator and latitude 10° S receive more rainfall from November to April than from May to October. In contrast, the region north of the equator receives the most rainfall in the boreal summer season.

Monsoons are a prominent large-scale seasonal atmospheric circulation. They arise from the differential heat capacity of land and water and the resulting thermal differences between continents and oceans. Land heats and cools faster than water. In winter, the continents are colder than oceans and have high surface pressure. Oceans are warmer and have low surface pressure. The high pressure over land drives air offshore towards the oceans. The opposite atmospheric circulation – onshore winds – occurs in summer when the continents warm, the surface low pressure intensifies, and moist air flows from oceans (high surface pressure) to warmer land (low surface pressure). Such monsoons can be seen in Asia, Africa, Australia, and southwest North America and bring heavy seasonal rain (Figure 5.5).

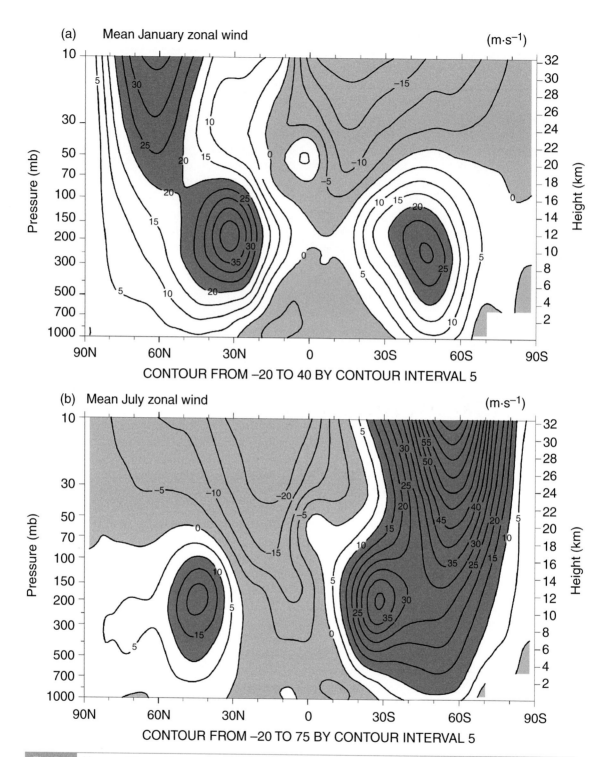

Fig. 5.11 Mean monthly zonal (west-to-east) wind as a function of latitude and height for (a) January and (b) July. Height is in pressure (millibars) on the left axis and altitude (kilometers) on the right axis. One millibar equals 1 hPa. East-to-west winds (negative zonal wind) are lightly shaded. Strong westerly winds are darkly shaded. Winds are averaged around the world, from longitude 180° W to 180° E. As a result, the jet streams are seen as single solid cores. Data provided by the National Center for Atmospheric Research (Boulder, Colorado).

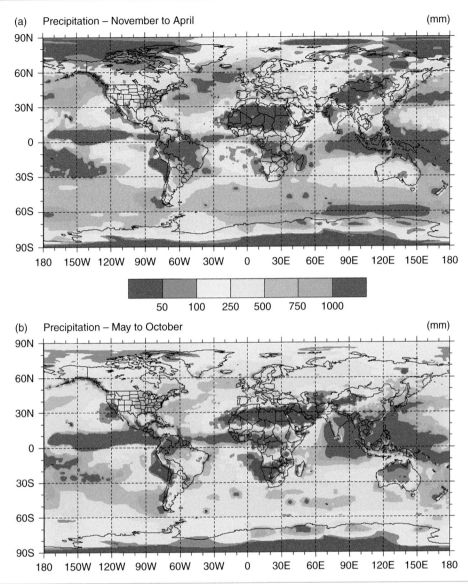

(a) Precipitation – November to April   (mm)

50   100   250   500   750   1000

(b) Precipitation – May to October   (mm)

Fig. 5.12 Precipitation for (a) November–April and (b) May–October. Data provided by the National Center for Atmospheric Research (Boulder, Colorado). See color plate section.

5.8 | Review Questions

1. You plan to ski at Vail, Colorado, USA. Arriving first at Denver, Colorado (elevation 1609 m), what is the decrease in air pressure relative to sea level? The base elevation at Vail is 2475 m and the summit elevation is 3527 m. What is the air pressure relative to sea level at each of these elevations?

2. In climbing Longs Peak, Colorado, USA, the air pressure decreases from 695.02 hPa at the base trailhead (2865 m) to 572.04 hPa at the summit. What is the elevation gain?

3. The concentration of CO_2 in the atmosphere is expected to double at some point in the future from its preindustrial level of 278 ppm. How many grams of carbon must be injected into the atmosphere for this doubling to occur?

4. Explain how the pressure gradient force, Coriolis force, and surface friction affect air flow around low and high pressure centers in the Southern Hemisphere.

5. When English colonists first settled in Massachusetts in the early 1600s, winters were colder than they had expected from experience in England, a more northerly location. Explain why.

6. Contrast temperatures when swimming in the Atlantic Ocean at Virginia Beach, Virginia (36.5 °N) and in the Pacific Ocean off California at similar latitude during summer. Where might you prefer to wear a wetsuit while swimming?

7. Explain why air travel eastward from San Francisco to Washington, DC or from New York City to Munich is faster than the return trip.

8. Southeastern United States typically has hot, humid summers. Temperature and humidity generally decrease to a more pleasant climate in August–September. Explain why.

9. The west coast of the United States has a pronounced seasonality to its precipitation. From Washington to Southern California, much more rain is received from November to April than in the rest of the year. Much of California receives less than 125 mm of rain in the dry season. Describe the changes in atmospheric circulation that produce this.

10. Land with a high surface albedo reflects a high amount of incoming solar radiation. Describe what would happen to the Asian monsoon if the surface albedo of the continent increased during spring and summer.

6

Earth's Climates

6.1 | Chapter Summary

This chapter gives an overview of the various climates found on Earth at the macroscale, mesoscale, and microscale. Macroclimate is the large-scale climate over 2000 km or more resulting from geographic variation in net radiation, the resultant transport of heat by the atmosphere and oceans, and high and low surface pressure belts. Temperature and precipitation distinguish various macroclimate zones. One classification scheme is that of Köppen, which illustrates the major climate zones. Mesoclimates and microclimates are regional and local climates, respectively. Microclimates are climatic features typically smaller than 2 km. A forest has a different microclimate than an adjacent clearing. Mesoscale is between microscale and macroscale, covering atmospheric processes at scales of 2–2000 km. Regional (mesoscale) climates are illustrated in terms of the effect of topography on solar radiation, temperature, and precipitation in mountains. Lakes and oceans also influence regional climate, with generally mild temperatures and reduced temperature variability compared with inland climates. Differential heating between land and ocean results in a local circulation known as a sea breeze.

6.2 | Global Climate Zones

Although no two places experience exactly the same climate, several generalized climate zones can be recognized. Figure 2.1 illustrates one such climate classification – the Köppen classification as modified by Trewartha (Finch et al. 1957; Trewartha 1968). This scheme utilizes five major climate zones based on temperature and precipitation.

1. Humid tropical climate: warm year-round; coldest month 18°C or warmer.
2. Dry climate: deficient precipitation throughout the year; potential evapotranspiration exceeds precipitation.
3. Moist subtropical mid-latitude climate: warm to hot summers with mild winters; coldest month above –3°C but below 18°C; warmest month above 10°C.
4. Moist continental climate: warm summers and cold winters; coldest month below –3°C; warmest month above 10°C.
5. Polar climate: extremely cold winters and cold summers; warmest month below 10°C.

Each of these climate zones has subzones defined by temperature and precipitation (Table 6.1).

Humid tropical climates occur where temperatures are warm throughout the year. A tropical rainforest climate occurs where rainfall is abundant throughout the year; a tropical savanna climate develops where there is a pronounced dry season. In both climate zones, seasonal variation in air temperature is minimal (less than a few degrees). In the tropical rainforest zone, rainfall is abundant throughout the year but may vary seasonally with the position of the intertropical convergence zone (Figure 6.1).

Table 6.1 | Definition of climate zones in Figure 2.1

Climate zone	Major characteristics
Humid tropical climate	warm year-round; coldest month 18°C or warmer
Tropical rainforest	wet all seasons; all months >60 mm rain
Tropical savanna	dry season; rainfall in driest month <60 mm
Dry climate	deficient precipitation throughout year; potential evapotranspiration exceeds precipitation
Semiarid	steppe vegetation
Arid	desert
Moist subtropical mid-latitude climate	warm to hot summers and mild winters; coldest month >−3°C but <18°C; warmest month >10°C
Mediterranean	dry season in summer
Humid subtropical	warmest month >22°C
Marine	all months <22°C
Moist continental climate	warm summers and cold winters; coldest month <−3°C; warmest month >10°C
Humid, warm summer	warmest month >22°C
Humid, cool summer	warmest month <22°C
Subarctic	less than four months >10°C
Polar climate	extremely cold winters and cold summers; warmest month <10°C
Tundra	warmest month <10°C but >0°C
Icecap	warmest month <0°C

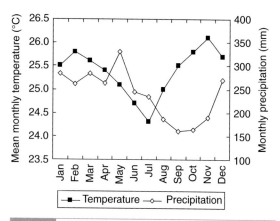

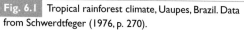

Fig. 6.1 Tropical rainforest climate, Uaupes, Brazil. Data from Schwerdtfeger (1976, p. 270).

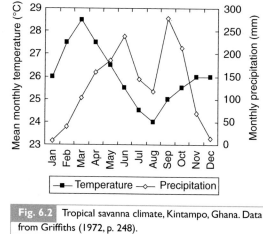

Fig. 6.2 Tropical savanna climate, Kintampo, Ghana. Data from Griffiths (1972, p. 248).

Tropical rainforest climates occur in hot, wet equatorial regions of South America, Africa, southeast Asia, and Indonesia. The tropical savanna climate occurs in tropical regions that are warm year-round but have a pronounced dry season. In the example shown in Figure 6.2, the rainy season is from April to October. Little rainfall occurs in December–February, when a subtropical high pressure influences the region. The double peak in summer precipitation reflects the seasonal migration of the inter-tropical convergence zone across the equator.

Temperatures are warmer during the dry season than in the rainy season when clouds cool the surface. Tropical savanna climates occur in Central America, to the north and south of the Amazon Basin in South America, to the north and south of the Congo Basin in Africa, east Africa, parts of India and Southeast Asia, and northern Australia.

Dry climates, which occur where rainfall is sparse throughout the year, are divided into semiarid and arid climates based on moisture deficiency. The semiarid climate develops in temperate regions, most prominently in the Great Plains of the United States, the steppes of Central Asia, and parts of southern South America, southern Africa, and Australia. Figure 6.3 shows an example climate. In this particular location, most precipitation falls in April and May. Temperatures are hot during June to August when the clear sky and intense solar radiation heat the surface. The arid climate of deserts is not only dry, but also hot as solar radiation readily penetrates the clear, dry skies (Figure 6.4). Desert climates occur on the eastern flanks of the subtropical high pressures near latitudes 30° N and 30° S, chiefly in southwestern United States, North Africa, southern South America, South Africa, and western Australia. They also establish in continental areas of the middle latitudes that are far removed from sources of atmospheric moisture such as central Asia, central Australia, and the Great Basin of western United States.

Moist subtropical middle latitude climates occur in regions with distinct summer and winter seasons, where summers are warm to hot and winters are mild. They are divided into Mediterranean, humid subtropical, and marine zones. Mediterranean climates develop where a summer dry season is pronounced, such as in southern California and along coastal areas of the Mediterranean Sea. These climates have mild, moist winters and hot, dry summers (Figure 6.5). Humid subtropical climates form in southeastern United States, eastern China, Japan, and along the southeastern coasts of South America, Africa, and Australia. These climates occur on the western edge of subtropical high pressure areas, which drives warm,

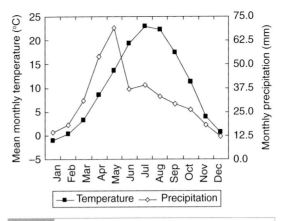

Fig. 6.3 Semiarid climate, Denver, Colorado, USA. Data from Bryson and Hare (1974, p. 277).

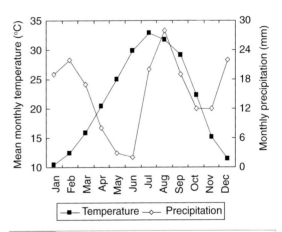

Fig. 6.4 Arid climate, Phoenix, Arizona, USA. Data from Bryson and Hare (1974, p. 268).

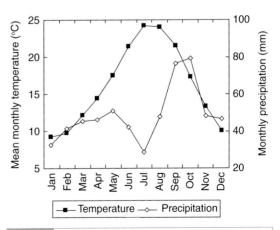

Fig. 6.5 Mediterranean climate, Barcelona, Spain. Data from Wallén (1970, p. 230).

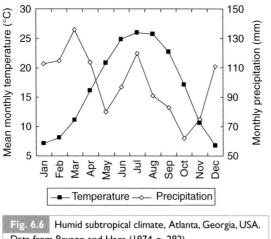

Fig. 6.6 Humid subtropical climate, Atlanta, Georgia, USA. Data from Bryson and Hare (1974, p. 282).

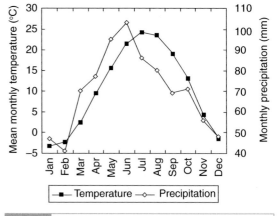

Fig. 6.8 Humid continental, warm summer climate, Chicago, Illinois, USA. Data from Bryson and Hare (1974, p. 285).

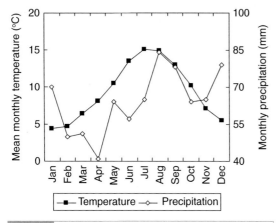

Fig. 6.7 Marine climate, Dublin, Ireland. Data from Wallén (1970, p. 113, p. 114).

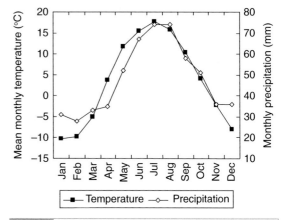

Fig 6.9 Humid continental, cool summer climate, Moscow. Data from Lydolph (1977, p. 394).

moist tropical air towards the middle latitudes, and consequently have hot, humid summers (Figure 6.6). Winters are mild and precipitation is abundant throughout the year. Moderate to pronounced seasonality is a dominant feature of climate. Marine climates occur in the Pacific Northwest region of the United States, western Europe, and western South America in middle latitudes where oceans moderate climate. Marine climates have mild winters, with temperatures rarely below freezing, cool summers, and abundant precipitation year-round (Figure 6.7).

Moist continental climates occur in the northern regions of North America, Europe, and Asia. Large seasonal variation in temperature,

with moderate to cool summers and cold winters, characterizes the climate. The humid continental subzone is divided into warm summer (Figure 6.8) and cool summer (Figure 6.9) regions based on whether the warmest month is above or below 22°C. Farther north, in Alaska, northern Canada, northern Europe, and northern Russia, where the winters are bitterly cold and the summers are cool and short, the climate is subarctic (Figure 6.10). Precipitation is generally light.

Polar climates develop in high latitudes or mountain tops where the warmest month is below 10°C. Tundra climates arise where plants can still survive in the short summers and long, cold winters (Figure 6.11). In the extreme cold

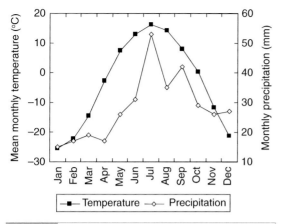

Fig. 6.10 Subarctic climate, Fort Smith, Canada. Data from Bryson and Hare (1974, p. 158).

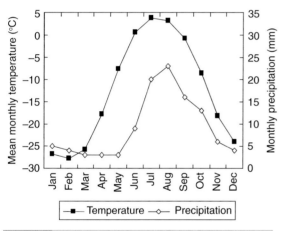

Fig. 6.11 Tundra climate, Barrow, Alaska, USA. Data from Bryson and Hare (1974, p. 144).

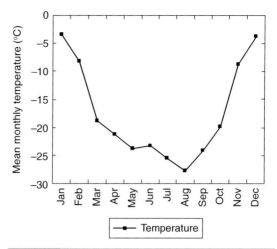

Fig. 6.12 Icecap climate, McMurdo, Antarctica. Data from Orvig (1970, p. 337).

of Greenland and Antarctica, little vegetation grows and permanent glaciers cover the land (Figure 6.12).

At the local and regional scale, climate can deviate markedly from that expected according to macroscale climate classifications. The landscape is a mosaic of patches created by spatial variation in topography, soils, and land cover (Figure 6.13). By altering the cycling of energy and water between land and atmosphere, landscape heterogeneity can create unique microclimates and mesoclimates that differ from the prevailing macroclimate. For example, mountains, with variation in elevation and slope of terrain, have a different mesoclimate from plains. Within mountains, valleys have a different microclimate from ridgetops. Forests have a different microclimate from open rangeland.

6.3 | Hillslopes and Mountains

The direction of slope can be critical to the thermal balance, especially in middle to high latitudes. The effect of radiation loading on air temperature is illustrated in Figure 6.14, which shows air temperature and elevation for a forested site in Germany in summer. At this site, elevation ranges from 340 to 540 m with slopes inclined at 27 percent (15°) to 84 percent (40°). On the particular summer day studied, air temperature varied by as much as 3.5°C depending on aspect. The southwest-facing slope was warmest, with temperature up to 22.5°C. The northeast-facing slope, with temperature as low as 19.0°C, was coldest. On the north side of the hill, temperature increased from 19.0°C in the east to 21.0°C in the west. This is because the Sun was high above the horizon and illuminated the north-facing slopes late in the afternoon. Direction of slope is especially important at high latitudes. North-facing slopes in interior Alaska receive less solar radiation than south-facing slopes and therefore have cold soil underlain with permafrost (Van Cleve et al. 1983, 1986; Viereck et al. 1983).

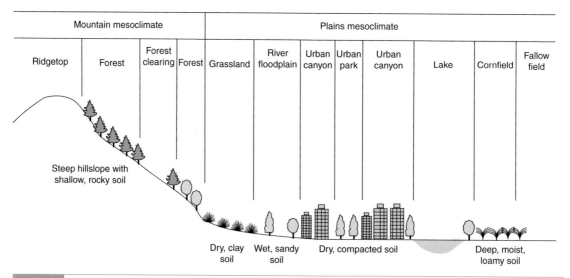

Fig. 6.13 Variation in topography, soils, and land cover across a hypothetical landscape that creates mountain and plains mesoclimates and numerous microclimates within a humid continental macroclimate.

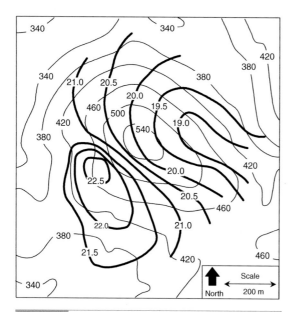

Fig. 6.14 Maximum air temperature in a forest in the Harz mountains of Germany on a warm summer day. Thin lines show elevation (m). Thick lines show temperature (°C). Adapted from Geiger (1965, pp. 426).

Another noticeable feature of mountains is the cool air. So long as air is not saturated with water, it cools at a rate of about 1°C per 100 m. It regains heat at the same rate when it descends. This temperature change is called the dry adiabatic lapse rate. If a low-lying site has

an air temperature of 30°C, a site located 300 m higher will be exposed to 27°C air. At the top of a 1500 m mountain, the temperature is only 15°C (Figure 6.15a).

The amount of water vapor that air can hold without becoming saturated depends on its temperature (Figure 3.5). As moist air rises up a mountain and cools, the amount of water vapor it can hold decreases. So long as the air is not saturated, it cools at the dry adiabatic lapse rate. When the air becomes saturated (i.e., relative humidity is 100 percent), some of the water vapor condenses into droplets, forming fog or clouds. This condensation releases heat (the stored latent heat of vaporization), and the cooling of air as it rises decreases to about 0.5°C per 100 m. This is called the moist adiabatic lapse rate. Figure 6.15b shows changes in air temperature as moist air moves over a mountain. The rising air on the windward slope cools at the dry adiabatic lapse rate until the air becomes saturated, in this example at about 900 m. When saturated, clouds form and the cooling decreases to the moist adiabatic lapse rate. At the summit, the air is 18°C, which is 3°C warmer than if it cooled at the dry adiabatic lapse rate. If, as in this example, precipitation removes water, the air becomes unsaturated and descends on the leeward slope at the dry adiabatic lapse rate.

(a)

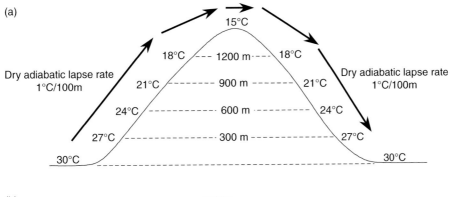

(b)

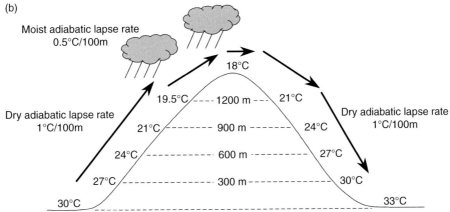

Fig. 6.15 Air temperature in relation to elevation on a 1500-m mountain. (a) Dry adiabatic cooling. (b) Moist adiabatic cooling with precipitation.

Because air is warmed by latent heat of condensation as it moves upslope and by adiabatic heating as it moves downslope, it reaches the bottom warmer than it started on the other side – in this case with a temperature of 33 °C.

Under the right conditions, temperatures increase with elevation rather than decrease. Mountains and hillslopes develop local wind circulations in response to spatial variation in surface heating. Under calm conditions and clear sky, light winds often blow upslope during the day and downslope at night. During the day, mountain slopes absorb solar radiation. These warm surfaces heat air, which becomes less dense and rises. Air flows upslope from low-lying valleys, ravines, or plains to replace the ascending mountain air. These upslope circulations depend on a temperature contrast and develop most strongly on slopes receiving the greatest amount of solar radiation. For example,

an east-facing slope heats up from early morning solar radiation and may develop upslope winds before a west-facing slope. At night, the slopes cool, and cold air near the surface flows downhill and collects in low-lying areas, often forming frost pockets. The effect of cold air drainage on temperature is seen in a study by Hocevar and Martsolf (1971) of early-morning air temperature in relation to elevation in a broad valley in central Pennsylvania. During one particular night, temperature over a 20-km distance varied by as much as 9 °C with low elevations colder than high elevations (air temperature increased 3.4 °C per 100 m). On average for all nights studied, air temperature increased at a rate of 6.2 °C per 100 m.

Warmest nighttime temperatures can be found at mid-slope. Ridgetops are colder due to their high elevation; valleys are cold because of cold air drainage. Figure 6.16 shows the

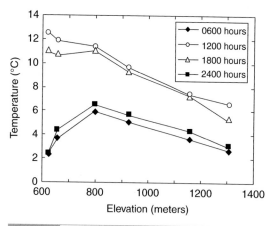

Fig. 6.16 Mid-slope thermal belt in Austrian mountains. Shown is temperature in relation to elevation at 0600, 1200, 1800, and 2400 hours. Data from Geiger (1965, p. 435).

characteristics of this mid-slope thermal belt for Austrian mountains. On the particular mountain studied, daytime temperatures decreased with elevation as expected. Nighttime temperatures increased with elevation up to 800 m; thereafter temperatures decreased with elevation. The warm region at 800 m is the mid-slope thermal belt. Many meteorological factors influence the location of this belt. The thermal belt develops best on clear, calm nights. With high winds or rain, the normal lapse rate occurs. As a result, the elevation with warmest temperature is not constant but varies over time.

As shown in Figure 6.15, the cooling of air with higher elevation often causes condensation and precipitation on the windward side of mountains and dry conditions on the leeward side. This mountain-induced precipitation is called orographic precipitation. The influence of mountains on precipitation is particularly prominent in western United States, along a transect from the Pacific Ocean inland to Colorado. Annual precipitation drops from 1000 mm west of the Pacific Coastal Ranges to about 500 mm on the east side of the mountains. Precipitation on the west side of the Sierra Nevada is 1000 mm, but less than 200 mm on the east side. Similar orographic precipitation occurs in the Rocky Mountains.

Another feature of mountain climates is that mountaintops can often be exceedingly windy. These winds are evident in the twisted, gnarled shapes of exposed trees growing on or near mountaintops. In the United States, the strongest wind recorded (104 m s^{-1}) occurred on the summit of Mount Washington in New Hampshire (Williams 1994, p. 48). As air flows over the land surface, frictional resistance from vegetation, buildings, and the ground slows the wind. This resistance decreases rapidly with height in the atmosphere so that high altitude winds are stronger than surface winds.

6.4 | Lakes and Oceans

Lakes and oceans influence regional climate. In general, milder temperatures with smaller diurnal and seasonal temperature variability characterize seashore and lakeside climates compared with inland climates. Under the right conditions, the climatic effects of lakes and oceans extend far from shoreline. Simulations with a mesoscale climate model show the Great Lakes in North America warmed air temperature by 2°C as far away as Philadelphia, Pennsylvania, during a 2-day cold period in November 1982 (Sousounis and Fritsch 1994; Sousounis 1997, 1998).

The climatic effects of water bodies arise from differences in surface radiation and heat capacity between water and land. Lakes and oceans, when ice-free, generally have a lower albedo than land and absorb more solar radiation. This solar radiation penetrates through the water to considerable depths so that it heats the upper 10 m or so of water rather than merely the surface. In addition, lakes and oceans have a large capacity to store heat. A typical heat capacity of soil is 2 MJ m^{-3} K^{-1} while that of water is 4.2 MJ m^{-3} K^{-1} (Table 9.1). In other words, a cubic meter of water requires twice as much energy to warm 1°C as does the same volume of soil. Conversely, it must lose twice as much energy to cool by 1°C. The large heat capacity of water reduces daytime warming and nighttime cooling compared with land. Likewise, the annual cycle of air temperature is moderated as heat is stored in water during summer and released in winter.

Table 6.2 Cumulative energy fluxes (MJ m^{-2}) during the early and late season calculated assuming the 50,000 km^2 Great Slave Lake study region contains no wetlands and lakes (uplands only) compared with the observed distribution of uplands, wetlands, and lakes

	Early season (April 23–September 2)				Late season (September 2–January 14)			
	R_n	G	λE	H	R_n	G	λE	H
Uplands only	933	57	509	367	161	−57	51	167
Uplands, wetlands, and lakes	973	286	433	254	205	−278	305	178

Note: R_n, net radiation; G, heat storage; λE, latent heat; H, sensible heat. The percentage cover of surface types is: uplands, 55 percent; wetlands, 8 percent; small lakes (< 1 km^2), 7 percent; medium lakes (1–100 km^2), 11 percent; large lakes (> 100 km^2), 19 percent. *Source:* From Rouse et al. (2005).

Studies of the energy balance of upland soil and small, medium, and large lakes in the Great Slave Lake region of Canada covering 50,000 km^2 illustrate the contrast between land and water surfaces (Oswald and Rouse 2004; Rouse et al. 2005). Lakes cover 37 percent of the landscape; wetlands occupy an additional 8 percent of the region. Small, shallow lakes with an area less than 1 km^2 generally thaw rapidly with the onset of air temperature above freezing. Albedo decreases as the snow and ice on the lake melts. Thereafter, solar heating warms the lake and evaporation commences. Autumn freeze-up occurs quickly and limits evaporation. Large, deep lakes with an area greater than 100 km^2 have a much different energy balance. These lakes require longer periods to thaw and warm and remain unfrozen into late autumn or early winter. Ice breakup in Great Slave Lake typically occurs between late May and late June. Ice typically forms between late November and late December.

Energy fluxes vary considerably depending on the size of these lakes. Lakes accumulate more net radiation at the surface over the period April 23 to January 14 compared with uplands, largely due to the lower albedo of water compared with soil. Small, shallow lakes thaw earlier than larger, deeper lakes and therefore accumulate more net radiation earlier in the season. Lakes store considerably more heat than does soil. Large, deep lakes store more heat than do small, shallow lakes. Small lakes reach maximum heat storage earlier (August 1) compared with medium lakes (August 6) and large lakes

(September 8). By late summer, one-quarter to three-quarters of net radiation has been used to heat lakes depending on lake size. Upland soils and small lakes have similar latent heat flux until mid-July, when dry soil limits evaporation. The large heat storage of lakes sustains evaporation into autumn and winter. Small lakes evaporate through September. Evaporation from medium and large lakes begins later than uplands and small lakes. Medium lakes have evaporation rates comparable to small lakes, but extend into November. Evaporation rates are low in large lakes until mid-September, when evaporation increases. Evaporation extends into early January until the water is fully frozen. Evaporation extends over a 19-week season in uplands, 22 weeks in small lakes, 24 weeks in medium lakes, and 30 weeks in large lakes. Overall, the presence of lakes in this region increases net radiation at the surface compared with upland sites (Table 6.2). Maximum heat storage in the early season is increased by a factor of five. Evaporation decreases somewhat during the early season, but increases sixfold in the late season.

The heat stored in lakes is mixed vertically as a result of density differences in water. For freshwater, greatest density occurs at about 4°C. Water warmer than 4°C is less dense. Hence, cold, dense water sinks to the bottom of lakes while warm, light water rises to the surface. This is most evident in summer months, when temperate lakes are stratified into three distinct layers: an upper layer formed by warm, light water (the epilimnion); a transition zone in which temperature changes rapidly with depth (the thermocline); and a deep layer formed by cold, dense water (the hypolimnion). Towards the end of summer and into autumn, surface water cools and sinks to deeper depths. As the surface water mixes with deeper cold water, heat is distributed throughout the vertical profile in a process known as overturning, and the lake develops a uniform temperature profile from surface to bottom. Similar overturning occurs in spring, followed by surface warming and thermal stratification in summer.

In addition to modulating temperature, lakes and oceans can develop local atmospheric circulations. Land and sea breezes develop when the temperature contrast between land and water produces air pressure differences that result in surface breezes onshore during the day and offshore at night (Figure 6.17). During the day, land warms more than water. The air overlying land warms, and the warm surface air rises. The expansion of the air column over land means that pressure aloft is higher over land than over water. The horizontal pressure gradient causes winds aloft to blow offshore from land to water. The offshore flow aloft adds mass to the column of air over water, increasing the pressure at the ocean surface. Surface air flows from high pressure over ocean to the low pressure over land. The result is a local circulation system in which warm air rising over land is replaced with cool ocean air at the surface. These sea breezes generally develop by mid-morning on hot, sunny days with calm weather. At night, land cools more than water. The result is a system of descending air over land and rising air over water driven by offshore surface breezes. Similar breezes can develop for lakes of 200 km² or so (Sun et al. 1997) and for large rivers (Silva Dias et al. 2004).

Another influence of water bodies is the lake effect storms that dump heavy snowfall on the southern and eastern shores of the Great Lakes. Snowfall at lakeside may be twice as deep as inland sites. Lake effect snow develops when cold, dry arctic air moves over the warmer Great Lakes. The cold air is heated and moistened as it travels over the lakes. Heat and moisture make the air unstable. The warm, moist surface air rises, forming clouds that precipitate

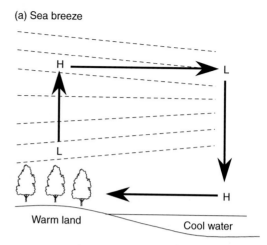

(a) Sea breeze

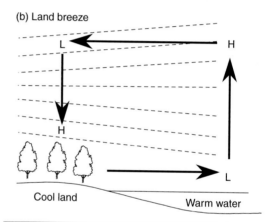

(b) Land breeze

Fig. 6.17 Atmospheric circulation for (a) daytime sea breeze and (b) nighttime land breeze. The dashed lines are isobars of equal pressure. Miller et al. (2003) review sea breeze systems.

snow. In addition, the warmer, more humid air encounters a rough surface as it flows on shore. Winds slow and the convergence of air forces air upwards where it cools, condenses, and precipitates as snow.

6.5 | Forests and Clearings

The effect of vegetation on microclimates is seen by contrasting forested and open locations (Kittredge 1948; Geiger 1965). In general, daytime air temperature during summer is cooler in a forest than above canopy or in clearings. For example, one study found daytime

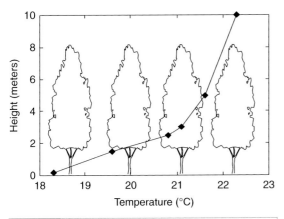

Fig. 6.18 Profile of daily average air temperature above and within a forest canopy. Data from Geiger (1965, p. 321).

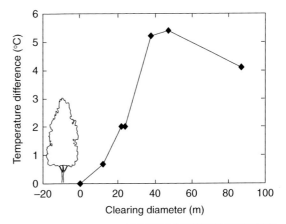

Fig. 6.19 Difference in midday air temperature measured in a clearing and in a forest in relation to clearing size. Data from Geiger (1965, p. 352).

air temperature above the canopy was greater than 22°C while air near the forest floor was 4°C cooler (Figure 6.18). In part, this is because the foliage in the forest canopy absorbs solar radiation so that there is little net radiation to heat the forest floor.

During daytime under periods of strong solar heating, air temperature in a forest is generally a few degrees lower than in adjacent clearings. Many factors such as the orientation of the forest edge, the height of trees, their leaf area, and prevailing winds influence the microclimates of a clearing. The size of a clearing relative to the height of surrounding trees is one critical factor. Figure 6.19 shows that midday air temperature measured at the center of a clearing is much higher than that measured in the surrounding forest. This temperature excess increases with clearing size. This is due to less shading by surrounding trees as the clearing increases in size. With a larger clearing size, however, radiation heating is less important because winds effectively carry heat away from the surface, and the temperature excess decreases.

At night, clearings can be several degrees cooler than within forests because the dense forest foliage blocks the loss of longwave radiation from the surface. Net radiation is an important determinant of nighttime surface air temperature. During night, net radiation is the difference between atmospheric longwave radiation impinging on the surface and longwave radiation emitted by the surface. The surrounding

trees obscure a portion of the sky, blocking some of the atmospheric longwave radiation. For a circular clearing of diameter D surrounded by trees of height H, the portion of the sky seen at the center of the clearing is (Oke 1987, p. 353):

$$\psi_{sky} = \cos^2 \beta \tag{6.1}$$

where β is the angle formed between the center of the clearing and treetop ($\beta = \tan^{-1} 2H / D$). The remainder $(1 - \psi_{sky})$ is the terrain view factor. The total longwave radiation incident at the center of the clearing is the sum of that from the sky (L_{sky}) and that from the surrounding trees (L_{veg}). The incoming longwave radiation is:

$$L \downarrow = L_{sky} \psi_{sky} + L_{veg}(1 - \psi_{sky}) \tag{6.2}$$

As the diameter of a clearing increases in size, a greater fraction of the longwave radiation comes from the sky and less comes from the surrounding trees. Since atmospheric radiation is considerably less than terrestrial radiation at night, there is less energy at the surface as the clearing size increases (Figure 6.20). Hence, one would expect smaller clearings to be warmer than larger clearings at night.

The presence of trees can also alter local winds. Patches of trees arranged closely together generally decrease wind to a distance of 20–30 times their height. This windbreak, or shelterbelt, effect has been widely used in rural settings

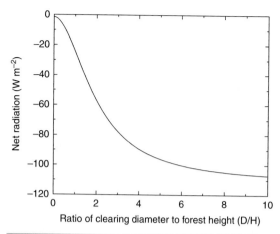

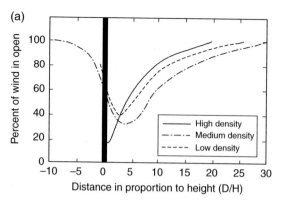

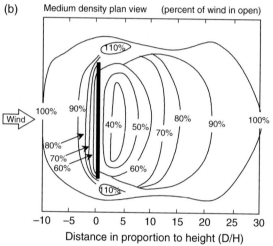

Fig. 6.20 The effect of clearing size (D/H) on nighttime radiation balance illustrated assuming L_{sky} = 250 W m^{-2}, L_{veg} = 360 W m^{-2} (i.e, the radiative temperature is about 10°C), and the longwave radiation emitted by the ground in the clearing is also 360 W m^{-2}. Net radiation is $R_n = 250\,\psi_{sky} + 360(1 - \psi_{sky}) - 360$.

Fig. 6.21 Effects of trees on wind. The thick solid line represents a windbreak. Distance is measured upwind and downwind of the windbreak as a proportion of tree height. (a) Wind speed in relation to distance for low, medium, and high density vegetation. (b) Spatial pattern of wind speed around a medium density windbreak. Adapted from Oke (1987, p. 244, p. 245).

to reduce winds blowing across open farmlands, thereby reducing erosion of soil and heat loss from isolated buildings (Oke 1987). Figure 6.21 illustrates the general aspects of windbreaks. The greatest reduction in wind occurs within a distance that is less than five times tree height, where winds decrease to 20–40 percent of those in open areas. The penetrability of barriers greatly affects wind reduction. A dense barrier has the largest reduction in wind speed, but the abrupt blockage of wind produces strong turbulence downwind so that the shelter effect is limited to only a short distance. The shelter effect is also small in sparse vegetation, where the winds readily penetrate the barrier. In contrast, the shelter effect extends further downwind in medium density windbreaks. Phenology influences the shelter effect. Winds are generally 20–30 percent lower when deciduous trees are in leaf than when leaves are off the trees (Landsberg 1981, p. 130).

6.6 | Review Questions

1. Describe each of the following locations by its Köppen climate zone. (a) Uaupes, Brazil (Figure 6.1) if dry season precipitation decreases by 120 mm per month. (b) Fort Smith, Canada (Figure 6.10) if temperature increases by 5°C in summer. (c) How much summer warming would be needed for Barrow, Alaska (Figure 6.11) to become a subarctic climate? (d) How would climate have to change for the semiarid climate of Denver, Colorado (Figure 6.3) to become a humid continental, warm summer climate comparable to Chicago, Illinois (Figure 6.8)?

2. On a hot summer day in which the high temperature at Denver, Colorado, USA (elevation 1609 m) is 35°C, you travel into Rocky Mountain National Park and ascend Longs Peak (elevation 4345 m). Assuming air cools at the dry adiabatic lapse rate, what is the temperature at the summit compared with that at Denver?

3. Use Eq. (5.3) to determine the surface pressure at Denver and Longs Peak. Then use Eq. (4.6) and Eq. (4.7) to compare the amount of solar radiation received at Denver and Longs Peak. Assume the zenith angle is $Z = 30°$ and $\tau = 0.6$.

4. Two homeowners would like to add solar panels to their house. The roof of one house faces in several directions, including due east and due south. In the other house, the roof is aligned along a north-to-south axis; one-half faces due west and the other half faces due east. Which home has better solar access, assuming no obstructions?

5. How much energy is needed to warm a cubic meter of water and a cubic meter of soil by 2°C?

6. The state of North Carolina in the United States has a humid subtropical macroclimate. The state extends from the Blue Ridge and Appalachian Mountains in the west to the Atlantic Ocean on the east. Describe two mesoclimates found in the state that allow residents in the city of Greensboro, located in the interior of the state, to escape the hot summer climate. What are the main characteristics of these mesoclimates?

7. What is the sky view factor for a forest clearing with a diameter of 25 m surrounded by trees with a height of 25 m? What is the sky view factor if the clearing is increased in diameter to 100 m? Which clearing is expected to be colder at night?

8. In the early growing season, sudden frosts can kill new plant growth. This can be a critical factor in successful forest regeneration. Which is more likely to have successful seedling establishment: a small clearing or a large clearcut?

9. What are two ways in which trees can be used to create a more favorable microclimate around a house?

6.7 | References

Bryson, R. A., and Hare, F. K. (1974). *Climates of North America. World Survey of Climatology*, vol. 11. Amsterdam: Elsevier.

Finch, V. C., Trewartha, G. T., Robinson, A. H., and Hammond, E. H. (1957). *Elements of Geography: Physical and Cultural*, 4th ed. New York: McGraw-Hill.

Geiger, R. (1965). *The Climate Near the Ground*. Cambridge, Massachusetts: Harvard University Press.

Griffiths, J. F. (1972). *Climates of Africa. World Survey of Climatology*, vol. 10. Amsterdam: Elsevier.

Hocevar, A., and Martsolf, J. D. (1971). Temperature distribution under radiation frost conditions in a central Pennsylvania valley. *Agricultural Meteorology*, 8, 371–383.

Kittredge, J. (1948). *Forest Influences: the Effects of Woody Vegetation on Climate, Water, and Soil, with Applications to the Conservation of Water and the Control of Floods and Erosion*. New York: McGraw-Hill.

Landsberg, H. E. (1981). *The Urban Climate*. New York: Academic Press.

Lydolph, P. E. (1977). *Climates of the Soviet Union. World Survey of Climatology*, vol. 7. Amsterdam: Elsevier.

Miller, S. T. K., Keim, B. D., Talbot, R. W., and Mao, H. (2003). Sea breeze: structure, forecasting, and impacts. *Reviews of Geophysics*, 41, 1011, doi:10.1029/2003RG000124.

Oke, T. R. (1987). *Boundary Layer Climates*, 2nd ed. London: Routledge.

Orvig, S. (1970). *Climates of the Polar Regions. World Survey of Climatology*, vol. 14. Amsterdam: Elsevier.

Oswald, C. J., and Rouse, W. R. (2004). Thermal characteristics and energy balance of various-size Canadian Shield lakes in the Mackenzie River basin. *Journal of Hydrometeorology*, 5, 129–144.

Rouse, W. R., Oswald, C. J., Binyamin, J., et al. (2005). The role of northern lakes in a regional energy balance. *Journal of Hydrometeorology*, 6, 291–305.

Schwerdtfeger, W. (1976). *Climates of Central and South America. World Survey of Climatology*, vol. 12. Amsterdam: Elsevier.

Silva Dias, M. A. F., Silva Dias, P. L., Longo, M., Fitzjarrald, D. R., and Denning, A. S. (2004). River breeze circulation in eastern Amazonia: Observations and modelling results. *Theoretical and Applied Climatology*, 78, 111–121.

Sousounis, P. J. (1997). Lake-aggregate mesoscale disturbances. Part III: Description of a mesoscale aggregate vortex. *Monthly Weather Review*, 125, 1111–1134.

Sousounis, P. J. (1998). Lake-aggregate mesoscale disturbances. Part IV: Development of a mesoscale aggregate vortex. *Monthly Weather Review*, 126, 3169–3188.

Sousounis, P. J., and Fritsch, J. M. (1994). Lake-aggregate mesoscale disturbances. Part II: A case study of the effects on regional and synoptic-scale weather systems. *Bulletin of the American Meteorological Society*, 75, 1793–1811.

Sun, J., Lenschow, D. H., Mahrt, L., et al. (1997). Lake-induced atmospheric circulations during BOREAS. *Journal of Geophysical Research*, 102D, 29155–29166.

Trewartha, G. T. (1968). *An Introduction to Climate*, 4th ed. New York: McGraw-Hill.

Van Cleve, K., Dyrness, C. T., Viereck, L. A., et al. (1983). Taiga ecosystems in interior Alaska. *BioScience*, 33, 39–44.

Van Cleve, K., Chapin, F. S., III, Flanagan, P. W., Viereck, L. A., and Dyrness, C. T. (1986). *Forest Ecosystems in the Alaskan Taiga: A Synthesis of Structure and Function.* New York: Springer-Verlag.

Viereck, L. A., Dyrness, C. T., Van Cleve, K., and Foote, M. J. (1983). Vegetation, soils, and forest productivity in selected forest types in interior Alaska. *Canadian Journal of Forest Research*, 13, 703–720.

Wallén, C. C. (1970). *Climates of Northern and Western Europe. World Survey of Climatology*, vol. 5. Amsterdam: Elsevier.

Williams, J. (1994). *The Weather Almanac 1995.* New York: Vintage Books.

7

Climate Variability

7.1 | Chapter Summary

The previous chapters focused on the mean state of the atmosphere and climate zones. However, the realized temperature and precipitation in any year can deviate markedly from the long-term climatology. Some years are warmer or colder than normal; some are wetter or drier than normal. This is a realization of climate variability at seasonal-to-interannual timescales. Major modes of climate variability include the El Niño/Southern Oscillation and the North Atlantic Oscillation. El Niño is a large-scale warming of sea surface temperature in the eastern tropical Pacific. It is associated with a reduced Walker circulation that weakens the trade winds, with below-normal rainfall in the western tropical Pacific and above-normal rainfall along the equator east of the dateline. The changes in tropical sea surface temperature and atmospheric circulation affect temperature and precipitation worldwide. The North Atlantic Oscillation is associated with changes in the strength of the Icelandic low and subtropical high in the North Atlantic during winter. This affects the strength of middle latitude westerlies, with consequences for winter temperature and precipitation from eastern North America to Europe and Asia.

7.2 | Floods, Droughts, and Heat Waves

Climate is often described in terms of climatic means, typically defined by 30-year averages. For example, the mean temperature for the month of July in central England during the period 1971–2000 was 16.5°C (Figure 7.1a). However, the monthly temperature varied considerably from this 30-year mean in any given year. The minimum monthly temperature during the 30-year period was 14.7°C; the maximum was 19.5°C. The average temperature was realized in only two years. The July temperature was below average in 16 years and above average in 12 years. Temperature in five years was one or more standard deviations (1.3°C) below average and more than one standard deviation above average in another five years. An extended temperature record for central England over a 347-year period further illustrates the nature of climate variability (Figure 7.1b). Between 1659 and 2005, July temperature averaged 16.0°C but ranged from a high of 19.5°C to a low of 13.4°C. As seen in the 30-year running mean temperature, there have been distinct periods of warming and cooling in central England.

Precipitation also varies from year to year. The summer of 1993 brought severe flooding in

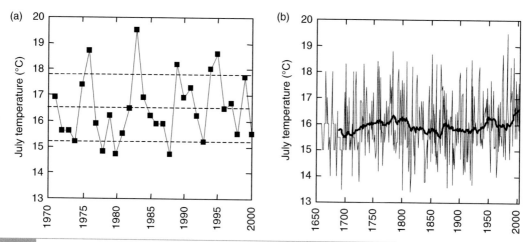

Fig. 7.1 Mean monthly July temperature measured in central England. (a) Temperature for 1971–2000, with dashed lines denoting the mean and one standard deviation. (b) Temperature for 1659–2005. The thick solid line shows the 30-year mean temperature. Data provided by the Hadley Centre (Met Office, Exeter, UK). Manley (1974), Parker et al. (1992), and Parker and Horton (2005) describe this temperature record.

Midwest United States (Changnon 1996). Over 7 million hectares (ha) of land (1 ha =10,000 m²) were flooded, causing 15–20 billion U.S. dollars in damage and forty-eight deaths. At St. Louis, Missouri, the Mississippi River was above flood stage for eighty days and above record flood stage for twenty-three consecutive days. At Davenport, Iowa, the river was above flood stage for forty-three days. Many of the large rivers that flow into the Mississippi also experienced major flooding. During this period, the northern and central Plains states and Midwest states received much more precipitation than normal (Figure 7.2). Precipitation from June to August was more than 100–200 mm above normal in a wide region and was more than 300 mm above normal in a smaller region centered on Iowa.

In addition to floods, central and western regions of the United States are prone to recurring droughts (Figure 7.3). The summers of 1934, during the infamous Dust Bowl, and 1956 were two of the most extreme drought years in the 1900s. In these years, extensive portions of the central Plains and West experienced extremely dry soils with a Palmer Drought Severity Index of –4 or less. The summer of 1988 was also a severe drought that caused 40 billion U.S. dollars in damage and 5000–10,000 heat-related deaths

(Riebsame et al. 1990). These three drought years contrast with the widespread surplus of moisture in the summer of 1993. A 1930s magnitude Dust Bowl drought has occurred once or twice a century over the past three to four hundred years (Karl and Koscielny 1982; Woodhouse and Overpeck 1998), and multiyear droughts are common in western United States over the past millennium (Cook et al. 2004; Herweijer et al. 2007). Droughts persist longer in interior portions of the country than in areas closer to the coasts (Karl and Koscielny 1982; Diaz 1983; Karl 1983).

Variation in temperature and precipitation from year to year are a manifestation of naturally occurring climate variability. The atmosphere is a chaotic system in which small-scale atmospheric events may have large-scale consequences. This chaotic behavior has been aptly characterized by the infamous butterfly effect by which a butterfly flapping its wings in Asia can cause events affecting weather over the United States (Lorenz 1993). Chaos theory puts a limit on the predictability of weather beyond several days because forecasters can never precisely know all the conditions affecting weather. However, over longer timescales of seasons some general patterns are evident. In particular,

(a) Total precipitation (mm)

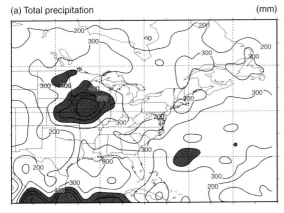

(b) Anomaly (mm)

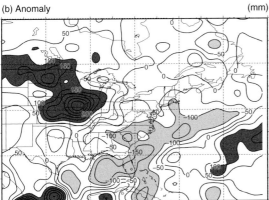

Fig. 7.2 Precipitation for the June–August 1993 summer season. (a) Total precipitation. Regions with greater than 400 mm precipitation are shaded. (b) Departure from normal. Regions more than 100 mm above normal are darkly shaded. Regions more than 100 mm below normal are lightly shaded. Data provided by the National Center for Atmospheric Research (Boulder, Colorado).

tropical atmospheric circulation and precipitation are strongly linked to the temperature of the underlying sea surface in a mode of variability known as the El Niño/Southern Oscillation (ENSO). These changes in tropical circulation and precipitation influence seasonal temperature and precipitation throughout the world. The changes in climate and weather created by El Niño and La Niña can have devastating socioeconomic consequences. Events in the North Atlantic greatly affect winter temperature and precipitation in a broad region from North America to Europe and Asia through the North Atlantic Oscillation (NAO).

7.3 El Niño/Southern Oscillation

In contrast to the meridional circulations depicted in Figure 5.4, the Walker circulation is a prominent zonal circulation across the tropical Pacific Ocean (Figure 7.4). It is characterized by low-level easterly winds (trade winds) and upper-level westerly winds across the tropical Pacific, ascending motion in the western basin and descending motion in the eastern basin. This zonal circulation is associated with a strong temperature contrast across the Pacific basin.

The trade winds drive warm surface water westward across the tropical Pacific with the result that surface water in the western tropical Pacific is normally several degrees warmer than water in the eastern tropical Pacific (Figure 7.5). The western Pacific near Australia and Indonesia has an extensive pool of warm surface water with temperatures in excess of 28°C. In contrast, water in the eastern Pacific off the Peruvian and Ecuadorian coast of South America is several degrees cooler, because deep cold water replaces the warm surface water. Surface air in the western tropical Pacific, heated by the warm seas, rises (low surface pressure). The ascending warm, moist air forms deep clouds that produce heavy rains over northern Australia, Indonesia, and the Philippines. Colder air in the eastern tropical Pacific descends (high surface pressure). Rainfall is light in this region of subsidence.

Every few years this pattern changes, and there is a large-scale warming of sea surface temperature in the tropical Pacific east of the dateline known as El Niño. This period of higher than normal sea surface temperature is often followed by a cold phase (La Niña) in which waters in the eastern Pacific have below normal temperature. El Niño, or warm episodes, and La Niña, or cold episodes, are opposite extremes of the El Niño/Southern Oscillation cycle. El Niño refers to the warming of surface waters while Southern Oscillation refers to changes in the Walker circulation. El Niños typically last 6 to 12 months and recur every few years. Historical records show El Niño has occurred

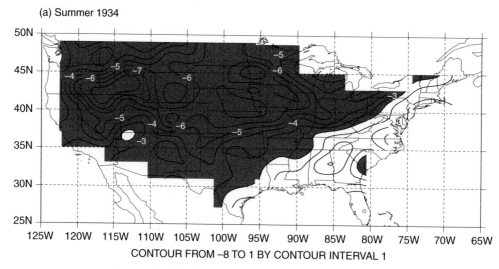

(a) Summer 1934

CONTOUR FROM −8 TO 1 BY CONTOUR INTERVAL 1

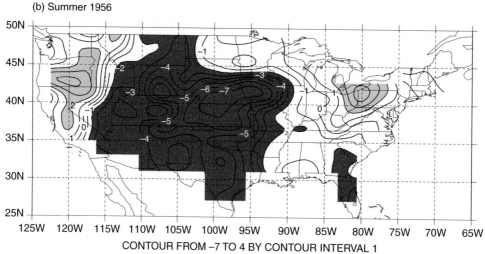

(b) Summer 1956

CONTOUR FROM −7 TO 4 BY CONTOUR INTERVAL 1

Fig. 7.3 Summertime Palmer Drought Severity Index for the years (a) 1934, (b) 1956, (c) 1988, and (d) 1993. Values between −2 and +2 indicate near normal moisture conditions. Values less than −2 indicate increasing severity of drought and are darkly shaded. Values greater than +2 indicate increasing moisture surplus and are lightly shaded. Data from Cook et al. (1999) and provided by the National Geophysical Data Center (National Oceanic and Atmospheric Administration, Boulder, Colorado).

periodically over the past several thousand years (Masson-Delmotte et al. 2013).

Abnormally warm sea surface temperatures across the eastern equatorial Pacific characterize El Niño. During a strong El Niño such as in 1998, the 28°C isotherm extends well east of the dateline and ocean temperatures are more than 2–3°C warmer than normal between the dateline and the west coast of South America (Figure 7.6a,b). During El Niño episodes, the normal contrast between high pressure over the eastern tropical

Pacific and low pressure over the west, which drives the easterly trade winds, diminishes. This is seen in the Southern Oscillation Index, which quantifies the difference in sea level pressure between Tahiti and Darwin, Australia. Prolonged periods of a negative phase, with below normal air pressure at Tahiti and above normal air pressure at Darwin, coincide with abnormally warm water in the eastern tropical Pacific. This reflects a reduced strength of the Walker circulation (Figure 7.7a). Higher than normal air pressure in

(c) Summer 1988

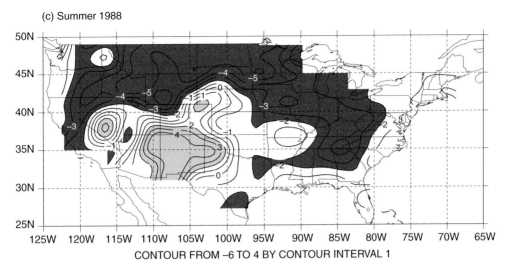

CONTOUR FROM −6 TO 4 BY CONTOUR INTERVAL 1

(d) Summer 1993

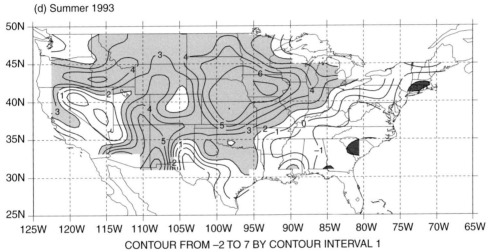

CONTOUR FROM −2 TO 7 BY CONTOUR INTERVAL 1

Fig. 7.3 (*continued*)

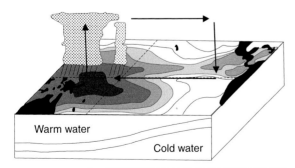

Fig. 7.4 Main features of the Walker circulation, from December to February, during normal conditions. Shading indicates progressively warmer sea surface temperature.

the west over Indonesia and northern Australia and lower than normal air pressure in the eastern tropical Pacific weakens the trade winds; warm water drifts further east, from the dateline to South America. Rainfall patterns follow the warm water, with dry conditions over northern Australia, Indonesia, and the Philippines and heavy rains along the equator from the dateline east to the South American coast where sea surface temperatures reach 28°C or more (Figure 7.8a).

La Niña episodes feature abnormally cold surface water. During a strong La Niña such

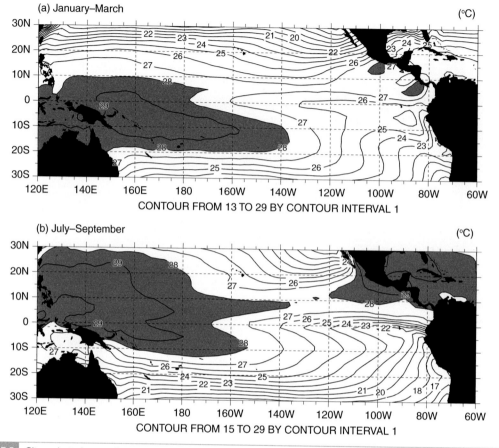

Fig. 7.5 Climatological sea surface temperature (1950–1998) for the equatorial Pacific between latitudes 30° S and 30° N during (a) January–March and (b) July–September. Temperatures greater than 28°C are shaded. Data provided by the National Center for Atmospheric Research (Boulder, Colorado).

as in 1989, the 28°C isotherm is restricted to west of the dateline along the equator and sea surface temperatures are a few degrees below normal between the dateline and western South America (Figure 7.6c,d). During La Niña episodes, air pressure is lower than normal over Indonesia and higher than normal over the eastern tropical Pacific (a positive Southern Oscillation Index). This reflects an enhanced Walker circulation (Figure 7.7b). The strong east-to-west pressure difference drives strong easterly surface winds across the Pacific from the Galapagos Islands to Indonesia. The warm pool of surface water does not drift eastwards across the dateline. As a result, the equatorial Pacific from the dateline to South America is extremely dry while northern Australia,

Indonesia, and the Philippines receive heavy rainfall (Figure 7.8b).

El Niño is a cyclically recurring event in which small changes in the strength of the easterly surface winds along the equator lead to changes in oceanic circulation. This affects sea surface temperatures and rainfall, which feed back to affect the strength of the easterlies until a mature El Niño is established. The El Niño cycle represents a coherent, large-scale fluctuation in ocean temperature, rainfall, air pressure, and atmospheric circulation across the tropical Pacific. The shifts in tropical atmospheric circulation alter global atmospheric circulation and influence temperature and precipitation worldwide. These changes are most apparent between December and February, though there

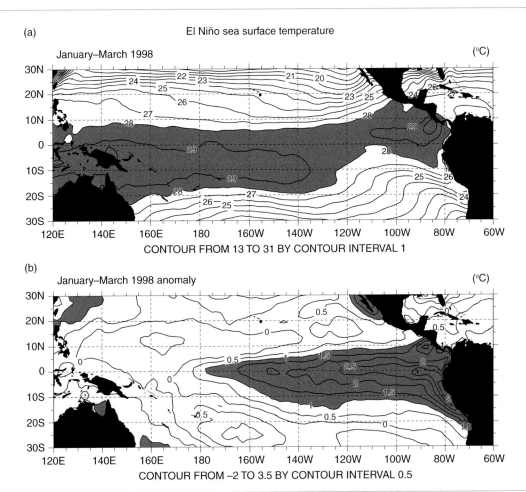

Fig. 7.6 Sea surface temperatures (January–March) in the Pacific for the 1998 El Niño and 1989 La Niña. (a) Observed sea surface temperature during the 1998 El Niño. Temperatures greater than 28°C are shaded. (b) Difference from the climatological mean. Temperature anomalies greater than 1°C are shaded. (c) Observed sea surface temperature during the 1989 La Niña. Temperatures greater than 28°C are shaded. (d) Difference from the climatological mean. Temperature anomalies less than −0.5°C are shaded. Data provided by the National Center for Atmospheric Research (Boulder, Colorado).

are important patterns throughout the year (Figure 7.9a). During this time, wetter than normal conditions occur along the northwest coast of South America, southern Brazil and central Argentina along the southeast coast of South America, and central east Africa. California and the Gulf Coast of the United States also tend to be wetter than normal. Drier than normal conditions occur over northeast South America and southern Africa. Temperatures are warmer than normal across southern Africa, southeast Asia, southeastern Australia, Japan, southern Alaska, much of Canada, and southeastern Brazil. Cooler than normal temperatures occur in southeastern

United States. The opposite patterns generally occur during La Niña (Figure 7.9b).

7.4 North Atlantic Oscillation

The seesaw behavior of El Niño and La Niña is a naturally recurring pattern of climate variability. A similar type of oscillation occurs in the North Atlantic in winter, where north-to-south variation in surface pressure affects climate in a pattern called the North Atlantic Oscillation (Hurrell et al. 2003). During summer, a strong subtropical high pressure system centered

(c)

La Niña sea surface temperature

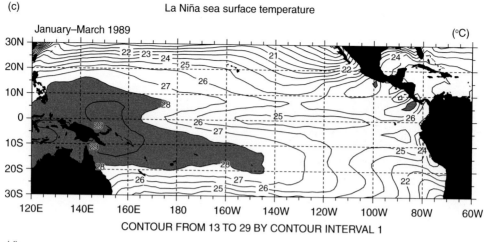

January–March 1989   (°C)

CONTOUR FROM 13 TO 29 BY CONTOUR INTERVAL 1

(d)

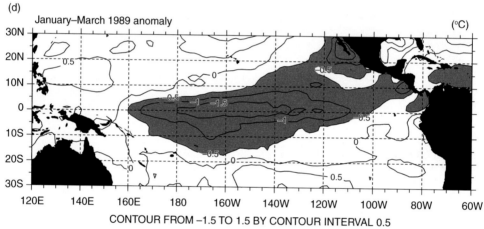

January–March 1989 anomaly   (°C)

CONTOUR FROM −1.5 TO 1.5 BY CONTOUR INTERVAL 0.5

Fig. 7.6 (continued)

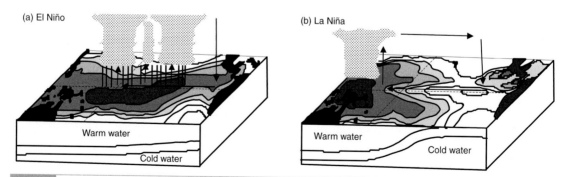

(a) El Niño   (b) La Niña

Warm water

Cold water

Warm water

Cold water

Fig. 7.7 Changes in the Walker circulation during (a) El Niño and (b) La Niña. Shading indicates progressively warmer sea surface temperatures.

El Niño precipitation

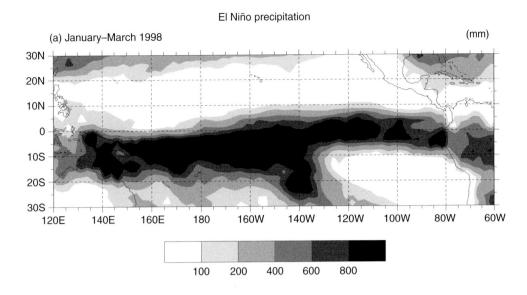

La Niña precipitation

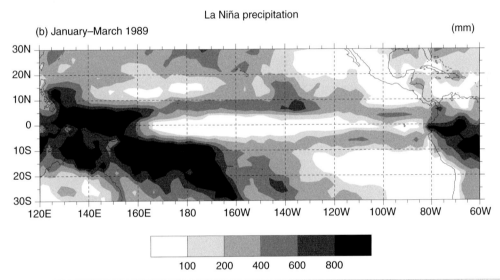

Fig. 7.8 Contrast in precipitation (January–March) across the tropical Pacific between (a) the 1998 El Niño and (b) the 1989 La Niña. Data provided by the National Center for Atmospheric Research (Boulder, Colorado).

near the Azores dominates much of the North Atlantic (Figure 5.5). The Azores high weakens during winter, and the Icelandic low dominates the North Atlantic. A weaker subtropical high migrates towards the equator. The Icelandic low and Azores high drive the westerly flow of air across the North Atlantic and guide storms onto Europe. The meridional gradient in winter surface pressure between Iceland and the Azores changes over time, with great consequences for eastern North America, Europe, and northern Asia. The normalized difference in surface pressure between Iceland and some southerly location (e.g., Lisbon, Portugal) is an index of the North Atlantic Oscillation. In the positive phase, pressure is lower than normal over the region of the Icelandic low and higher than normal across the subtropical Atlantic. Positive values are associated with stronger than average westerlies across the North Atlantic. In the negative phase, both pressure centers are weakened.

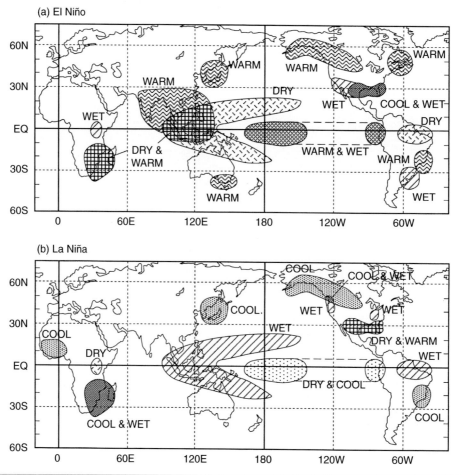

Fig. 7.9 Changes in temperature and precipitation for the December–February season during (a) El Niño and (b) La Niña. Source: Climate Prediction Center (National Centers for Environmental Prediction, National Oceanic and Atmospheric Administration, Washington, DC).

The climatic effects of the North Atlantic Oscillation are most pronounced in winter and are seen in temperature (Figure 7.10) and precipitation (Figure 7.11). During the positive phase, stronger than average middle latitude westerlies cause more frequent and stronger winter storms to cross the North Atlantic on a more northerly track. Counterclockwise flow around the deepened Icelandic low drives cold polar air onto Greenland and Labrador, where temperature and precipitation are below normal. Strong westerlies force maritime air onto northern Europe, where winters are warmer and wetter than normal. Clockwise flow around

the strengthened subtropical high forces warm subtropical air into eastern United States (mild, wet winter with fewer snow days) and cool northerly air into the Mediterranean (below normal-precipitation). The reverse patterns occur in the negative phase. Storms are less frequent and weaker and cross the Atlantic on a more zonal track. The Mediterranean has wetter than normal conditions while winter across northern Europe is colder and drier than normal. Winter is colder than normal across eastern United States with more snow storms while Greenland experiences milder than normal temperature.

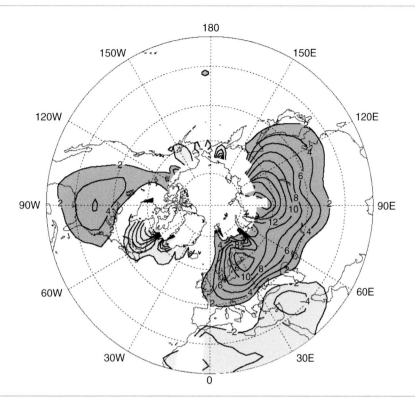

Fig. 7.10 Change in mean winter (December–March) surface temperature (× 0.1°C) due to a unit increase in the North Atlantic Oscillation index. Contours are in 0.2°C increments with contours > 0.2°C darkly shaded and those < –0.2°C lightly shaded. Reproduced from Hurrell et al. (2003).

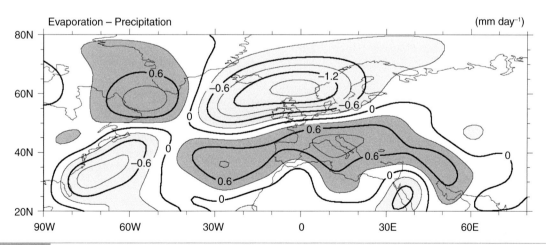

Fig. 7.11 Difference in mean winter (December–March) evaporation minus precipitation (E – P) between years with high and low North Atlantic Oscillation index. Contours are in 0.3 mm per day increments with contours > 0.3 mm day⁻¹ (dry conditions) darkly shaded and those < –0.3 mm day⁻¹ (wet conditions) lightly shaded. Reproduced from Hurrell et al. (2003).

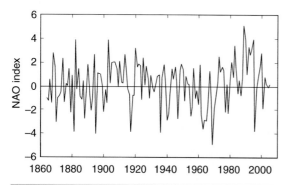

Fig. 7.12 Winter (December–March) index of the North Atlantic Oscillation (NAO) from 1864 to 2005 based on the difference in normalized sea level pressure between Lisbon, Portugal, and Stykkisholmur/Reykjavik, Iceland (Hurrell 1995; Hurrell et al. 2003). Data provided by James Hurrell (National Center for Atmospheric Research, Boulder, Colorado).

The North Atlantic Oscillation is observed in historical records over several centuries (Hurrell and van Loon 1997; Jones et al. 1997, 2003; Cook 2003; Masson-Delmotte et al. 2013). Figure 7.12 shows a time series of the NAO index since 1864. The period 1900–1930 had predominantly positive index values while the 1960s were characterized by negative index values. The North Atlantic switched again to positive index during the 1980s and 1990s. The upward trend in the NAO index and its relationship with the observed warming of Northern Hemisphere surface temperature over the same period is the subject of considerable research (Hurrell 1995; Hurrell et al. 2003).

7.5 | Other Modes of Variability

The ENSO and NAO are two of many modes of interannual climate variability, and they are second only to the change in seasons in terms of

their impact on global climate. Other modes of variability occur, and these climate fluctuations impact marine and terrestrial ecosystems worldwide (Stenseth et al. 2002).

The Arctic Oscillation (AO), also known as the Northern Annular Mode (NAM), is characterized by winds moving counterclockwise around the Arctic. In its positive phase, surface pressure in polar regions is lower than normal. The upper-level westerly winds are strong and restrict cold arctic air to polar regions. Much of the United States and northern Eurasia experience above-normal temperatures. These winds weaken during the negative phase of the AO, when pressure is higher than normal in polar regions. The weaker circumpolar circulation allows cold arctic air masses to move southward. Winters are colder and stormier than normal.

The Atlantic Multi-decadal Oscillation (AMO) is another mode of variability occurring in the North Atlantic Ocean. It is characterized by a coherent pattern of warm or cool sea surface temperature anomalies with a period of several decades. The AMO affects temperature and precipitation over North America and Europe.

The Pacific Decadal Oscillation (PDO) is a long-lived (20–30 years) ENSO-like event in the Pacific Ocean. Its phases are characterized by warm or cool ocean temperature anomalies, and the effects of the PDO on climate are seen primarily in the North Pacific and in North America. The positive phase is characterized by anomalously cool sea surface temperatures in the North Pacific and warm ocean temperatures along the Pacific Coast. A deeper wintertime Aleutian low pushes warm, humid air onto the Pacific Northwest, Alaska, and British Columbia, which experience above-normal temperatures. Southeastern United States has below-normal winter temperatures. These patterns are reversed in the negative phase.

7.6 | Review Questions

1. The average July temperature in central England for the period 1911–1940 is 15.93°C, for 1941–1970 is 16.0°C, and for 1971–2000 is 16.48°C. With the information in the following table, use a t-test to determine if temperature during the period 1911–1940 is statistically different from temperature during the period 1971–2000. Compare 1941–1970 with 1971–2000.

	1911–1940	1941–1970	1971–2000
n	30	30	30
Σx	477.9	480	494.4
Σx^2	7659.27	7705.56	8193.22
s	1.26386	0.93882	1.25269

2. Graph and compare the cumulative frequency distribution of central England July temperature for the periods 1911–1940, 1941–1970, and 1971–2000.

Midpoint temperature (°C)	Frequency		
	1911–1940	1941–1970	1971–2000
13.75	2	0	0
14.25	1	2	0
14.75	2	0	3
15.25	8	7	2
15.75	3	6	8
16.25	6	5	3
16.75	1	5	6
17.25	2	4	2
17.75	2	1	1

Midpoint temperature (°C)	Frequency		
	1911–1940	1941–1970	1971–2000
18.25	2	0	2
18.75	1	0	2
19.25	0	0	0
19.75	0	0	1

3. Sea level pressure is above normal at Tahiti and below normal at Darwin. How do sea surface temperatures and rainfall at the Galapagos Islands compare with normal conditions?

4. Conditions are such that during a particular December–February period: much of Canada has temperatures above normal; the Rocky Mountains receive above normal precipitation; the northeast region of Brazil receives below normal precipitation. These patterns are characteristic of what mode of climate variability? What has happened to sea surface temperatures in the tropical Pacific and to the Walker circulation?

5. Above normal sea level pressure at Lisbon and below normal sea level pressure at Iceland is characteristic of what mode of climate variability? Describe the expected winter conditions over northern Europe. Contrast this with conditions over Greenland and Labrador.

7.7 | References

Changnon, S. A., Jr. (1996). *The Great Flood of 1993: Causes, Impacts, and Responses.* Boulder, Colorado: Westview Press.

Cook, E. R. (2003). Multi-proxy reconstructions of the North Atlantic Oscillation (NAO) index: A critical review and a new well-verified winter NAO index reconstruction back to AD 1400. In *The North Atlantic Oscillation: Climatic Significance and Environmental Impact*, ed. J. W. Hurrell, Y. Kushnir, G. Ottersen, and M. Visbeck. Washington, DC: American Geophysical Union, pp. 63–79.

Cook, E. R., Meko, D. M., Stahle, D. W., and Cleaveland, M. K. (1999). Drought reconstructions for the continental United States. *Journal of Climate*, 12, 1145–1162.

Cook, E. R., Woodhouse, C. A., Eakin, C. M., Meko, D. M., and Stahle, D. W. (2004). Long-term aridity changes in the western United States. *Science*, 306, 1015–1018.

Diaz, H. F. (1983). Some aspects of major dry and wet periods in the contiguous United States, 1895–1981. *Journal of Climate and Applied Meteorology*, 22, 3–16.

Herweijer, C., Seager, R., Cook, E. R., and Emile-Geay, J. (2007). North American droughts of the last millennium from a gridded network of tree-ring data. *Journal of Climate*, 20, 1353–1376.

Hurrell, J. W. (1995). Decadal trends in the North Atlantic Oscillation: Regional temperatures and precipitation. *Science*, 269, 676–679.

Hurrell, J. W., and van Loon, H. (1997). Decadal variations in climate associated with the North Atlantic Oscillation. *Climatic Change*, 36, 301–326.

Hurrell, J. W., Kushnir, Y., Ottersen, G. and Visbeck, M. (2003). An overview of the North Atlantic Oscillation. In *The North Atlantic Oscillation: Climatic Significance and Environmental Impact*, ed. J. W. Hurrell, Y. Kushnir, G. Ottersen, and M. Visbeck. Washington, DC: American Geophysical Union, pp. 1–35.

Jones, P. D., Jonsson, T., and Wheeler, D. (1997). Extension to the North Atlantic Oscillation using early instrumental pressure observations from Gibraltar and south-west Iceland. *International Journal of Climatology*, 17, 1433–1450.

Jones, P. D., Osborn, T. J., and Briffa, K. R. (2003). Pressure-based measures of the North Atlantic Oscillation (NAO): A comparison and an assessment of changes in the strength of the NAO and in its influence on surface climate parameters. In *The North Atlantic Oscillation: Climatic Significance and Environmental Impact*, ed. J. W. Hurrell, Y. Kushnir, G. Ottersen, and M. Visbeck. Washington, DC: American Geophysical Union, pp. 51–62.

Karl, T. R. (1983). Some spatial characteristics of drought duration in the United States. *Journal of Climate and Applied Meteorology*, 22, 1356–1366.

Karl, T. R., and Koscielny, A. J. (1982). Drought in the United States: 1895–1981. *Journal of Climatology*, 2, 313–329.

Lorenz, E. N. (1993). *The Essence of Chaos*. Seattle: University of Washington Press.

Manley, G. (1974). Central England temperatures: Monthly means 1659 to 1973. *Quarterly Journal of the Royal Meteorological Society*, 100, 389–405.

Masson-Delmotte V., Schulz, M., Abe-Ouchi, A., et al. (2013). Information from paleoclimate archives. In *Climate Change 2013: The Physical Science Basis. Contribution of Working Group I to the Fifth Assessment Report of the Intergovernmental Panel on Climate Change*, ed. T. F. Stocker, D. Qin, G.-K. Plattner, et al. Cambridge: Cambridge University Press, pp. 383–464.

Parker, D., and Horton, B. (2005). Uncertainties in central England temperature 1878–2003 and some improvements to the maximum and minimum series. *International Journal of Climatology*, 25, 1173–1188.

Parker, D. E., Legg, T. P., and Folland, C. K. (1992). A new daily central England temperature series, 1772–1991. *International Journal of Climatology*, 12, 317–342.

Riebsame, W. E., Changnon, S. A., and Karl, T. R. (1990). *Drought and Natural Resource Management in the United States: Impacts and Implications of the 1987–1989 Drought*. Boulder, Colorado: Westview Press.

Stenseth, N. C., Mysterud, A., Ottersen, G., et al. (2002). Ecological effects of climate fluctuations. *Science*, 297, 1292–1296.

Woodhouse, C. A., and Overpeck, J. T. (1998). 2000 years of drought variability in the central United States. *Bulletin of the American Meteorological Society*, 79, 2693–2714.

8

Climate Change

Chapter Summary

Climate has changed over the course of Earth's history and will change in the future. Just 18,000 years before present (18 kyr BP), Earth was in the grips of a prolonged cold period in which ice covered vast tracts of the Northern Hemisphere. Over the past few million years there have been numerous such ice ages separated by shorter, warm interglacial periods. The geologic record also reveals numerous rapid climate changes over periods as short as decades or centuries. This climate change is the result of changes in the external forcing of the Earth system by the Sun and internal physical, chemical, and biological feedbacks among the atmospheric, oceanic, and terrestrial components of the Earth system. Plate tectonics and changes in the geometry of Earth's orbit around the Sun influence climate at timescales of millennia or longer. Changes in the concentration of greenhouse gases or in the runoff of freshwater to oceans affect climate at timescales of centuries to millennia. Changes in solar irradiance or volcanic eruptions that emit aerosols into the atmosphere influence climate at timescales of years to decades. Climate change over the twentieth and twenty-first centuries is reviewed in the context of greenhouse gases and anthropogenic influences on climate.

8.2 Glacial Cycles

Over the past several hundred thousand years, glaciers have cyclically advanced and retreated in the Northern Hemisphere with a period of about 100 kyr. Glaciation typically takes 90 kyr to complete while deglaciation occurs over 10 kyr. Ice cores extracted deep below the surface in Greenland and Antarctica record Earth's climate history and reveal these ice ages. Atmospheric gases trapped in ice as glaciers grow provide a record of the temperature and chemical composition of the atmosphere at the time the ice formed. A 2,755-m deep record of the Vostok ice core extracted from Antarctica illustrates climate change over the past 250 kyr BP. The record shows two distinct cold periods beginning about 190 kyr BP and lasting until 140 kyr BP and again beginning about 115 kyr BP until 18 kyr BP (Figure 8.1). During these glacial periods, temperature was several degrees colder than present. The record also shows two warm periods following these ice ages from about 140 kyr to 125 kyr BP and presently beginning about 15 kyr BP. During these interglacial periods, climate warmed by several degrees. The EPICA Dome C ice core in Antarctica covers the past 800 kyr and reveals eight glacial cycles characterized by low atmospheric concentrations of CO_2, CH_4, and N_2O (Figure 8.2). Indeed, ice cores reveal a close relationship between greenhouse

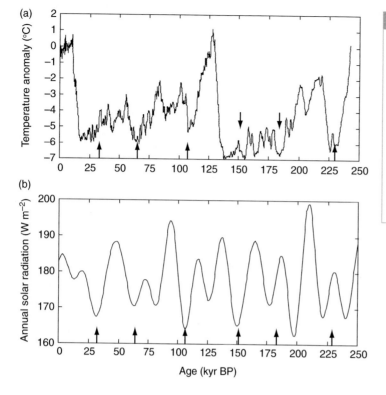

(a)

(b)

Fig. 8.1 Climate history over the past 250 kyr reconstructed from the Vostok ice core. (a) Temperature deviation from present (Jouzel et al. 1996). (b) Annual solar radiation at latitude 60° N (Berger 1978; Berger and Loutre 1991). This latitude is used because solar radiation at high latitudes in the Northern Hemisphere is critical to glacier dynamics. Arrows show periods of low solar radiation. Data provided by the National Geophysical Data Center (National Oceanic and Atmospheric Administration, Boulder, Colorado).

gases and temperature over glacial–interglacial cycles (Masson-Delmotte et al. 2013).

The onset of the last glaciation began about 115 kyr BP, when temperatures rapidly dropped by several degrees. A cold period about 75 kyr BP ushered in the main glacial phase. The buildup of glaciers culminated in a glacial maximum about 18–21 kyr BP (Figure 8.3). At this time, the Cordilleran ice sheet covered northern North America in the west, and the Laurentide ice sheet covered the northern continent in the east. Ice covered Greenland, Iceland, and much of northern Europe and Russia. These ice sheets were at least one kilometer thick and in many regions were two or more kilometers thick. Then climate warmed rapidly and glaciers melted over a period of about 6 kyr. Plants colonized newly exposed soil, and vegetation that had been restricted to southern locations migrated northward.

The warming arose from changes in insolation (Kutzbach and Guetter 1986; COHMAP 1988; Huntley and Webb 1988; Wright et al. 1993). At 18 kyr BP, the amount of solar radiation in the Northern Hemisphere was similar to present. Over the next several thousand years, summer

solar radiation in the Northern Hemisphere increased while winter radiation decreased (Figure 8.4). The increased summer radiation warmed continents and melted glaciers. The greatest increase in summer radiation occurred between 12 and 6 kyr BP, when the seasonality of Northern Hemisphere radiation increased compared with present. Summer solar radiation in the Northern Hemisphere increased by about 8% (30 W m^{-2}) compared with present, and winter radiation decreased by a similar amount. Regions of the Northern Hemisphere were warmer than present. The climate of northern Africa was wetter than present. Since then, summer radiation decreased and winter radiation increased to present conditions.

Ice cores and other data reveal rapid climate changes since the last glacial maximum (Alley et al. 2002, 2003; Masson-Delmotte et al. 2013). Changes in temperature of several degrees occurred throughout the Northern Hemisphere numerous times in periods as short as years to decades. One such climate change was the Younger Dryas cold period, which occurred from about 12.8 to 11.5 kyr

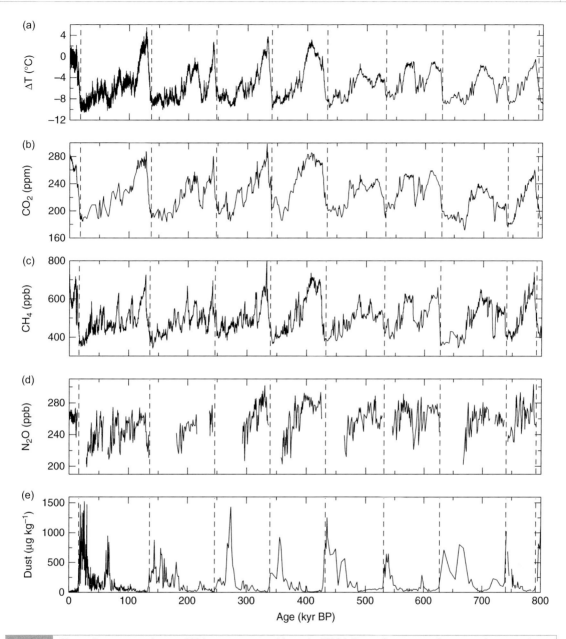

Fig. 8.2 Climate history over the past 800 kyr reconstructed from the EPICA Dome C ice core in Antarctica showing eight glacial cycles. (a) Temperature deviation (Jouzel et al. 2007). (b) CO_2 concentration (Lüthi et al. 2008). (c) CH_4 concentration (Loulergue et al. 2008). (d) N_2O concentration (Schilt et al. 2010). (e) Dust concentration (Lambert et al. 2008). Dashed vertical lines indicate approximate glacial terminations. Data provided by the National Climatic Data Center Paleoclimatology program (National Oceanic and Atmospheric Administration, Boulder, Colorado).

BP. At that time, following prolonged warming that brought an end to the ice age, temperature abruptly decreased in North America and Europe. Newly emerged forests reverted to glacial tundra. Glaciers advanced southwards and down mountains. This cold period lasted for about 1300 years before warming began again. Another cold period occurred 8.2 kyr BP, when temperature cooled by 2–6°C in Greenland, with lower temperatures extending to North

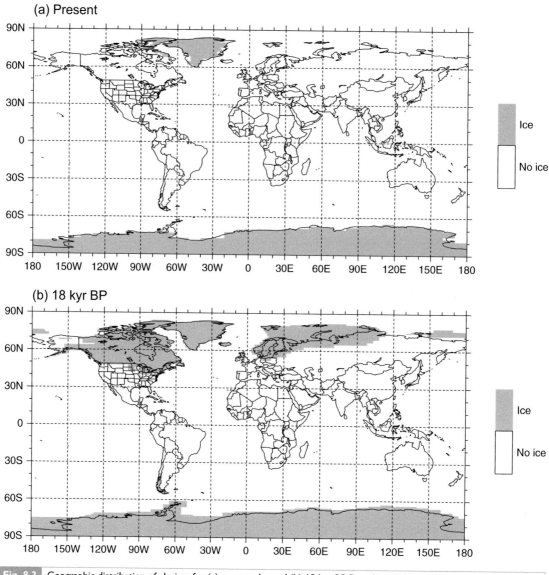

Fig. 8.3 Geographic distribution of glaciers for (a) present-day and (b) 18 kyr BP. Data from Peltier (1994) and provided by the National Geophysical Data Center (National Oceanic and Atmospheric Administration, Boulder, Colorado).

America and Europe (Alley and Ágústsdóttir 2005; Masson-Delmotte et al. 2013).

Climate also changes over shorter timescales (Masson-Delmotte et al. 2013). Figure 8.5 shows temperature for the Northern Hemisphere over the past 1500 years. Climate was relatively mild from about AD 950 to 1250 in an era known as the Medieval Warm Period. Then climate began to fluctuate, with cold winters interspersed among warm winters. Beginning about 1550, climate entered a prolonged period of cold temperature known as the Little Ice Age that lasted until about 1850, with a main phase from 1550 to 1700. Winters were long and cold; summers were short. Alpine glaciers advanced to lower elevations. After about 1700, temperature warmed in an erratic recovery from the Little Ice Age. Temperature has increased substantially since the mid-1800s, though not continuously.

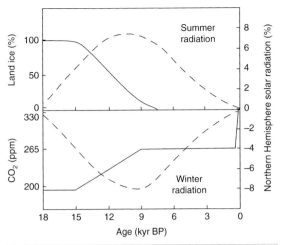

Fig. 8.4 Solar radiation, atmospheric CO_2, and glacier volume over the past 18 kyr. Land ice is expressed as a percentage of the ice volume 18 kyr BP. Northern Hemisphere solar radiation (dashed lines) is shown for summer (June–August) and winter (December–February) as a percentage difference from present values. Adapted from Kutzbach and Guetter (1986), COHMAP (1988), and Kutzbach and Webb (1993).

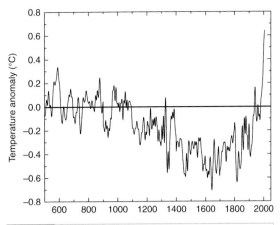

Fig. 8.5 Annual mean temperature of the Northern Hemisphere over the past 1500 years (AD 500–2006) as a deviation from the 1961–1990 mean. Temperature was reconstructed from tree rings, ice cores, corals, sediments, and other records. Data from Mann et al. (2009) and provided by the National Climatic Data Center Paleoclimatology program (National Oceanic and Atmospheric Administration, Boulder, Colorado).

8.3 | Mechanisms of Climate Change

Climate changes because of many processes, both natural and anthropogenic. Plate tectonics alters the continents and oceans and creates mountains. The geometry of Earth's orbit around the Sun changes insolation. The chemical composition of the atmosphere, with increasing concentrations of CO_2, CH_4, N_2O, and other gases, warms climate through the greenhouse effect. Changes in the runoff of freshwater to oceans alter the thermohaline circulation. The amount of radiation emitted by the Sun varies in relation to sunspot activity. Aerosols in the atmosphere alter Earth's radiative balance.

8.3.1 Plate Tectonics

Plate tectonics, or continental drift, refers to the slow movement of continents at rates of a few centimeters per year. About 540–500 million years ago, the continents were widely dispersed along the equator. They drifted together and collided over time so that by 300 million years ago a large supercontinent called Pangaea had formed. Pangaea began to break apart 200 million years ago, and since then the continents have slowly drifted to their current locations. The current continental geography is part of a cycle in which continents assemble into a supercontinent that slowly breaks apart only to reassemble again. This cycle of supercontinent formation and destruction takes about 500 million years to complete.

The period about 80 million years ago during the late Cretaceous illustrates the climate-altering influences of plate tectonics (Crowley and North 1991). Three continental blocks formed large contiguous land areas: North America–Greenland–Eurasia; South America–Antarctica–India–Australia; and Africa (Figure 8.6). The oceans consisted of a single large Pacific basin. North America and Europe were closer together as were South America and Africa, and the Atlantic Ocean did not yet exist. India and Australia were attached to Antarctica, and the Indian Ocean had not yet formed. Sea level was about 100–200 m higher than at present. This flooded portions of western Europe,

Land distribution 80 million years ago

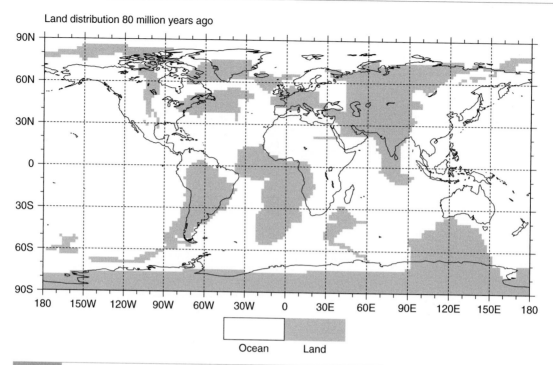

northern Africa, and interior North America with shallow seas. The southerly location of Africa, India, and Australia precluded development of a circumpolar Antarctic current. Instead, open passageways between North and South America and Eurasia and Africa led to the development of a shallow equatorial seaway known as the Tethys Sea that allowed circumglobal ocean circulation. Climate was several degrees warmer than present, especially at high latitudes. Atmospheric CO_2 levels were two to nine times higher than present, which contributed to the warming, as did changes in oceanic heat transport.

Since that time, the continents have slowly moved to their modern geographic locations. The westward drift of North America and South America opened the Atlantic Ocean. Africa moved northwards to converge with Europe, and the Indian subcontinent moved northwards to converge with Asia. This, combined with northward movement of Australia and the emergence of the Central American isthmus,

closed the Tethys equatorial seaway. The formation of the North Atlantic, South Atlantic, and Indian Oceans allowed development of subtropical gyres at these latitudes. The opening of the Drake Passage between South America and Antarctica and the northward movement of India and Australia led to a circumpolar ocean circulation around Antarctica. Resultant changes in ocean circulation altered heat transport and climate. Formation of the Antarctic circumpolar current may have reduced poleward heat transport to Antarctica, thereby facilitating the glaciation of Antarctica 34 million years ago.

Mountain building also changes climate. Locally, temperatures cool due to higher elevation of the land. Regions downwind from mountains typically are in a rain shadow and receive only sparse precipitation, while upwind slopes receive heavy precipitation. In addition, regional and hemispheric climate change occurs because the presence of mountains alters atmospheric circulation. High mountains and plateaus block the west-to-east flow of the jet streams in middle latitudes. The eastward flow of air diverts around the high terrain, enhancing meanders in the jet streams. Changes in summer heating and winter cooling over uplifted terrain also alters seasonal monsoon circulations. In summer, the Sun heats high plateaus. The warm air becomes less dense and rises, creating a surface low pressure. Air flows in from adjacent regions, flowing along the surface in towards the low pressure cell. The opposite circulation occurs in winter. Cold, dense air sinks over high plateaus, creating a high surface pressure and outward surface flow.

Geologic evidence shows that uplifting over the past 10–40 million years altered regional temperature and precipitation (Kutzbach et al. 1989; Ruddiman and Kutzbach 1989; Ruddiman et al. 1989, 1997; Prell and Kutzbach 1992; Zhisheng et al. 2001). The Tibetan Plateau covers more than 2 million km^2 with an average elevation of about 4.5 km. The Himalayan Mountains form a narrow range along the southern edge of the plateau. The high topography of this region is a result of the collision of India with Asia 40–50 million years ago and subsequent uplifting of land. Formation of the Himalayas and the Tibetan Plateau changed the climate of Southeast Asia and the strength of the monsoon. Before India collided with Asia, the small size of the Asian continent and its low elevation prevented a strong land–sea temperature contrast, which dampened the monsoon. Increased elevation following uplifting strengthened the monsoon and brought more precipitation onto the continent. In western United States, uplifting over the past 15 million years formed the Sierra Nevada and Rocky Mountains and the high Great Basin and Colorado Plateau between them, altering regional climate.

8.3.2 Orbital Changes

One reason for the recurring waxing and waning of glaciers is that the amount of solar radiation received on Earth varies in a process known as the Milankovitch cycles (Hays et al. 1976; Berger 1988; Berger et al. 1993; Shackleton 2000). Changes in orbital geometry affect the amount of solar radiation on Earth. The Vostok ice core reveals a close correspondence between temperature and solar radiation in which several well-marked temperature minima correspond to minima in solar radiation (Figure 8.1).

The eccentricity of Earth's orbit around the Sun varies from elliptical to nearly circular and back to elliptical with a period of about 100 kyr (Figure 8.7a). The current orbit results in about ±3% variation in solar radiation between when Earth is closest and farthest from the Sun. Greater eccentricity produces greater variation and changes the lengths of seasons. Earth's orbit is currently only modestly elliptical and is becoming even less elliptical over time.

The angle of tilt (also known as obliquity), which gives us seasons, varies from 22° to 24.5° with a period of about 41 kyr (Figure 8.7b). Smaller tilt results in less seasonal variation between winter and summer in middle and high latitudes; winters are milder and summers are cooler. Larger tilt amplifies the seasons at high latitudes. The current tilt of 23.5° is slowly decreasing.

The third cycle, known as the precession of the equinoxes, changes the distance between Earth and Sun during the seasons (Figure 8.8). The location of the equinoxes and solstices shift celestial location in Earth's orbit around the Sun with a period of about 22 kyr. Currently, Earth is closest to the Sun in January and farthest in July. In about 11 kyr, Earth will be closer in July than in January. The planet will receive more radiation than present in July and less in January. Compared with the present climate, the Southern Hemisphere winter will be warmer and summer cooler while the Northern Hemisphere winter will be colder and summer warmer.

8.3.3 Greenhouse Gases

Carbon dioxide cycles among atmosphere, oceans, and continents in response to the chemical weathering of rocks (Figure 8.9). The dominant form of chemical weathering involves carbonic acid. Carbon dioxide (CO_2) in soil reacts with water (H_2O) to form carbonic acid (H_2CO_3), which weathers and disintegrates rocks through a series of chemical reactions. In carbonate rocks such as calcium carbonate ($CaCO_3$), H_2CO_3 dissolves minerals to yield one calcium (Ca^{2+}) ion

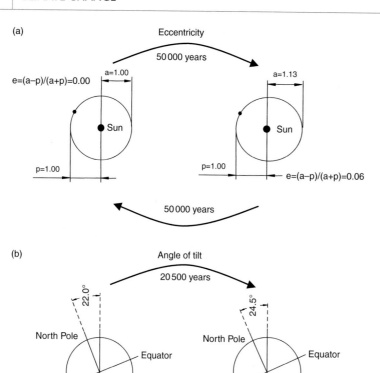

(a)

Eccentricity

50 000 years

$e=(a-p)/(a+p)=0.00$ a=1.00   a=1.13

Sun   Sun

p=1.00   p=1.00 $e=(a-p)/(a+p)=0.06$

50 000 years

(b)

Angle of tilt

20 500 years

22.0°   24.5°

North Pole   North Pole

Equator   Equator

South Pole   South Pole

20 500 years

Fig. 8.7 Eccentricity and obliquity in the Milankovitch cycles. (a) Changes in the eccentricity of Earth's orbit over the course of 100 kyr. (b) Changes in angle of tilt over the course of 41 kyr.

and two bicarbonate ions (HCO_3^-). Silicate minerals such as calcium silicate ($CaSiO_3$) dissolve to yield Ca^{2+}, two HCO_3^-, and additionally silica (SiO_2). Runoff carries the dissolved products into rivers and deposits it in oceans, where marine organisms use Ca^{2+} and HCO_3^- to build skeletons or shells of $CaCO_3$. This releases CO_2 back to the atmosphere. For carbonate rocks, the overall reaction is balanced, and the CO_2 used during weathering returns to the atmosphere. One molecule of CO_2 dissolves $CaCO_3$ to yield two HCO_3^-. Marine life transforms one HCO_3^- into $CaCO_3$; the other becomes CO_2 released to the atmosphere.

The weathering of silicate rocks requires two molecules of CO_2, but only one returns to the atmosphere through marine organisms. Instead, the second CO_2 molecule returns to the atmosphere over geologic timescales through volcanic eruptions. When marine organisms die, $CaCO_3$ contained in skeletons and shells deposits on the sea floor and buries underground. At high temperatures deep in the Earth, $CaCO_3$ reacts with SiO_2 to form $CaSiO_3$ and CO_2. Volcanic eruptions and soda springs vent this CO_2 into the atmosphere. Thus, metamorphism of rocks deep underground and venting to the atmosphere through volcanoes is required before the second molecule of CO_2 returns to the atmosphere. The weathering of silicate rocks results in a net loss of atmospheric CO_2 in the absence of volcanoes. Annual carbon fluxes in the geologic cycle are small (Figure 3.6), but can result in large perturbations of atmospheric CO_2 over tens to hundreds of millions of years (Berner 1991, 1998, 2003).

The biological carbon cycle superimposes large, rapid changes on the small, slow changes of the geological cycle (Figure 3.6). Over a timescale of decades to centuries, changes in ecosystem productivity and ocean biology alter atmospheric CO_2, but over timescales of

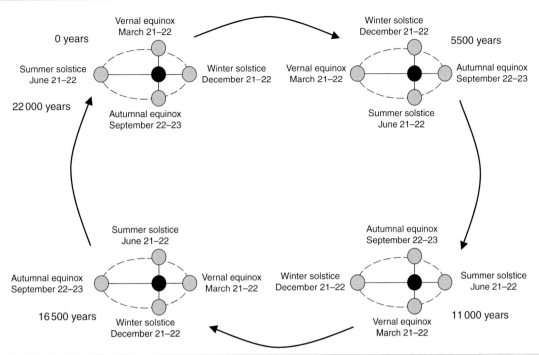

Fig. 8.8 Precession of the equinoxes in the Milankovitch cycles. Diagrams show changes in the celestial location of the solstices and equinoxes over the course of 22 kyr. The eccentricity of Earth's orbit is highly exaggerated.

millions of years, the geological carbon cycle is most important. It is the geological carbon cycle that has maintained the relatively low concentration of CO_2 in the atmosphere throughout Earth's history.

The concentrations of CO_2, CH_4, and N_2O in the atmosphere have varied over the past several hundred thousand years (Figure 8.2). Atmospheric concentrations of these gases were lower during ice ages than during interglacials. Ice-core records extending back 800 kyr indicate CO_2 varies naturally from 180 ppm (glacial maximum) to 300 ppm (interglacial); CH_4 ranges from 350 to 800 ppb; and N_2O varies from 200 to 300 ppb (Masson-Delmotte et al. 2013). The co-occurrence in time of low atmospheric concentrations of CO_2, CH_4, and N_2O and cold temperatures is not coincidental. Low concentrations of these gases cool climate by reducing the greenhouse effect while high concentrations warm climate. In addition, the rates of biological uptake and release of these gases depend on climate.

Following recovery from the last ice age, atmospheric CO_2 concentrations remained at about 260–280 ppm for several thousand years (Masson-Delmotte et al. 2013). Since the mid-1800s, the concentration of CO_2, CH_4, and N_2O in the atmosphere has increased (Figure 3.7). The atmospheric concentrations in 2011 of CO_2 (390 ppm), CH_4 (1803 ppb), and N_2O (324 ppb) (Table 2.1) are higher than observed over the past 800 kyr. The current concentration of atmospheric CO_2 is more than 100 ppm above its pre-industrial value. Although these gases naturally cycle among atmosphere, land, and oceans, they have emissions that are directly attributable to human activities such as fossil fuel combustion, land use, and agriculture.

Human activities emitted little CO_2 prior to 1850. Since then, as society industrialized and population increased, the annual emission rate has increased. The two major anthropogenic sources of CO_2 are fossil fuel burning and land-use practices such as deforestation (Figure 3.6). The concentrations of CH_4 and N_2O in the atmosphere have also increased from human activities. Anthropogenic sources of CH_4 include fossil fuel use, decomposition of wastes

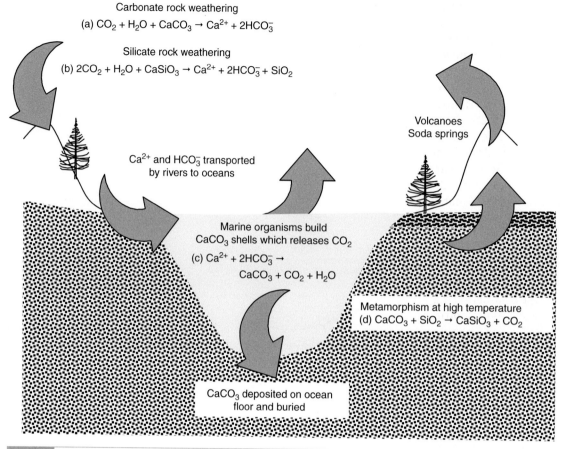

Carbonate rock weathering

(a) $CO_2 + H_2O + CaCO_3 \rightarrow Ca^{2+} + 2HCO_3^-$

Silicate rock weathering

(b) $2CO_2 + H_2O + CaSiO_3 \rightarrow Ca^{2+} + 2HCO_3^- + SiO_2$

Volcanoes
Soda springs

Ca^{2+} and HCO_3^- transported
by rivers to oceans

Marine organisms build
$CaCO_3$ shells which releases CO_2

(c) $Ca^{2+} + 2HCO_3^- \rightarrow$
    $CaCO_3 + CO_2 + H_2O$

Metamorphism at high temperature
(d) $CaCO_3 + SiO_2 \rightarrow CaSiO_3 + CO_2$

$CaCO_3$ deposited on ocean
floor and buried

Fig. 8.9 Geologic carbon cycle arising from the weathering of rocks, sedimentation, and volcanoes. Adapted from Berner and Lasaga (1989).

in landfills, and agriculture practices such as rice paddies, biomass burning, and digestion by grazing animals (Table 3.4). Agriculture, industry, and combustion of fossil fuels emit N_2O to the atmosphere (Table 3.5). Conversion of forest to cultivated lands and the application of fertilizer or manure are important anthropogenic sources of N_2O.

8.3.4 Freshwater Runoff and Thermohaline Circulation

Oceans transport vast quantities of heat poleward from the tropics in the thermohaline circulation (Figure 5.9). A weak thermohaline circulation reduces heat transport and causes colder winters. The formation of deep water in the North Atlantic, where cold, salty water sinks to the ocean bottom and flows southward, maintains the circulation. Weakening of the thermohaline circulation as a result of climate change is an important climate feedback (Broecker et al. 1985; Broecker 1997; Manabe and Stouffer 1999). Increases in freshwater input to the North Atlantic due to changes in continental runoff or glacial melt affect the poleward transport of heat by making the surface water less salty, so that the water is less dense and does not sink to the ocean bottom.

Climate records throughout the North Atlantic region show frequent large and rapid climate changes over the past several thousand years related to the reorganization of the North Atlantic thermohaline circulation (Alley et al. 2002, 2003; Masson-Delmotte et al.

2013). A shutdown of the thermohaline circulation as a result of a diversion of runoff from the Mississippi River drainage basin to the St. Lawrence River may have caused the Younger Dryas cold period about 12.8 kyr BP and lasting for about 1300 years (Broecker et al. 1989; Clark et al. 2001; Broecker 2003, 2006; Meissner and Clark 2006; Carlson and Clark 2012). Prior to that time, runoff from melting glaciers in Canada drained into the Gulf of Mexico through the Mississippi River. As the glaciers retreated northward, melt water was channeled into the Great Lakes and flowed through the St. Lawrence River into the North Atlantic. The northward transport of heat in the North Atlantic Ocean diminished and climate cooled. Massive outflow of freshwater from Hudson Bay into the Labrador Sea may have caused the abrupt cold period 8.2 ky BP (Clark et al. 2001; Alley and Ágústsdóttir 2005; Meissner and Clark 2006; Masson-Delmotte et al. 2013). Prior to this cold episode, remnant ice occupied Hudson Bay, creating a large ice dam for glacial lakes draining northwards. Rapid melting of this ice cleared the Hudson Strait and allowed the glacial lakes to drain several tens of thousands of cubic kilometers of water northward through the Hudson Strait into the Labrador Sea and the North Atlantic.

8.3.5 Solar Variability

The Sun is often described as producing a constant amount of radiation. Indeed, the term solar constant embodies this concept. This number is the amount of radiation received at the mean Earth–Sun distance by one square meter of surface area at the top of the atmosphere oriented perpendicular to the Sun. In fact, however, the amount of radiation has varied by about 2 W m^{-2} over the past several hundred years (Figure 8.10). The appearance of dark spots on the Sun's surface causes this variation in solar radiation. During times of maximum sunspot activity, the Sun emits more radiation. There is greater solar radiation emission during times of high sunspot number and less solar radiation emission during times of few sunspots. Because they are easily observed, a long record of sunspot

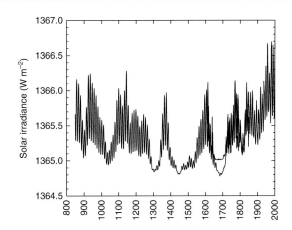

Fig. 8.10 Total solar irradiance for the years AD 850–2000 using the VSK reconstruction (850–1849) and the WLS reconstruction (1610–2000). An estimated 11-year solar cycle was superimposed on the VSK data. Data from Schmidt et al. (2012).

occurrence exists. This record shows an irregular cycle in sunspot abundance with a period of about 11 years. Sunspots were virtually absent during the period 1645–1715, during which time Earth received less solar radiation. This decrease in solar radiation, known as the Maunder Minimum, corresponds to the Little Ice Age and may explain why temperatures were abnormally cold during this time (Eddy 1976). Since the Maunder Minimum, solar radiation has generally increased.

8.3.6 Aerosols

Aerosols enter the atmosphere in many forms, including mineral dust from land, volcanic ash, and sulfate aerosols from volcanic gases and human activities. The occurrence of these aerosols in the atmosphere alters climate by absorbing or scattering atmospheric radiation, by altering clouds, and by influencing biogeochemical cycles. Their net effect in the atmosphere is to cool climate (Myhre et al. 2013). The amount of aerosols in the atmosphere changes considerably over time. For example, ice cores show large changes in dust deposition during the past 800 kyr (Figure 8.2e). Large dust spikes occurred during glacial periods. These high dust fluxes are related to the expansion of deserts.

Volcanoes are a particularly important source of short-term climate change. Volcanoes inject dust, debris, and gases high into the atmosphere at altitudes of 15–25 km where they can remain for many months. Strong winds at these altitudes transport the volcanic material around the world. This material alters the radiation balance and cools surface temperatures worldwide by reflecting more solar radiation to space. In particular, sulfur dioxide (SO_2), the primary gas emitted by volcanoes, forms sulfuric acid (H_2SO_4). This condenses into sulfate aerosols that scatter incoming solar radiation back to space.

The last millennium has had several large volcanic eruptions, such as the AD 1258 eruption that injected 250 Tg of sulfate aerosol into the stratosphere, the 1452 Kuwae eruption (~140 Tg), and the 1815 Tambora eruption (110 Tg) (Figure 8.11). One of the more devastating eruptions occurred in April 1815 when Mount Tambora in Indonesia exploded in the largest known volcanic eruption. Volcanic ash, estimated to be 100 km^3 of debris, extended into the upper atmosphere, where it blocked the Sun's rays from reaching Earth's surface for the next 18 months. Temperatures in the following year were below normal over much of the Northern Hemisphere. The period from 1600 to 1900 saw repeated volcanic eruptions. The 1600s were an especially active time for volcanism. Major volcanic eruptions cool global annual mean temperature by about 0.1–0.3°C for up to three years following eruption (Masson-Delmotte et al. 2013). Recurring volcanic activity in conjunction with decreased solar radiation has been implicated in the cooling of the Little Ice Age (Free and Robock 1999) and contributed to climate variability over the past millennia (Ammann et al. 2007; Landrum et al. 2013; Masson-Delmotte et al. 2013).

The combustion of fossil fuels also emits SO_2 to the atmosphere. Through a variety of chemical reactions, sulfur dioxide becomes sulfate aerosols, with greatest concentration in the Northern Hemisphere in summer. Sulfate aerosols cool climate, both directly by reflecting more solar radiation to space and indirectly by altering the radiative properties of clouds (Myhre et al. 2013).

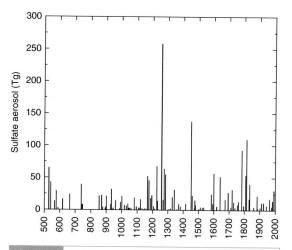

Fig. 8.11 Global stratospheric volcanic sulfate aerosol injection for the years AD 501–2000. Data from Gao et al. (2008).

8.4 Climate of the Twentieth Century

The period since the mid-1800s has seen a prominent warming of Earth's surface (Figure 8.12a). Observations show Earth's annual global mean surface temperature increased by 0.85°C over the period spanning 1880–2012 (Hartmann et al. 2013). The warming over the last 62 years (1951–2012) was 0.72°C. The rate of warming over this period (0.12°C per decade) was twice that over the entire 133-year record (0.06°C per decade). The rate of warming was still higher over the period 1979–2012 (0.16°C per decade). Each of the past three decades (1980s, 1990s, 2000s) was warmer than the previous decades since 1850. All ten of the warmest years since 1850 occurred after 1997. This warming was not continuous from year to year, but rather occurred in two distinct periods. Temperature was relatively stable prior to about 1915 and warmed thereafter until about 1945. Subsequent decades through the 1970s saw a small temperature decrease, with a significant warming trend thereafter. This warming is particularly prominent on land (Figure 8.12b). Land temperature increased at a rate of 0.18°C per decade over the period 1951–2012 and by 0.26°C per decade for

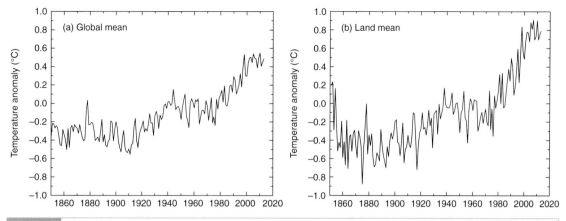

Fig. 8.12 Surface temperature for 1850–2013 based on station, ship, and buoy observations. Temperature is the anomaly from the 1961–1990 mean. (a) Annual global mean over ocean and land from the HadCRUT4 dataset (Morice et al. 2012). (b) Annual global mean air temperature over land from the CRUTEM4 dataset (Jones et al. 2012). Data provided by the Climatic Research Unit (University of East Anglia, Norwich). See Hartmann et al. (2013) for other datasets.

1979–2012 (Hartmann et al. 2013). The occurrence of cold days and nights has decreased while the occurrence of warm days and nights has increased.

Air temperature is not the only indicator of Earth's changing climate (Hartmann et al. 2013; Rhein et al. 2013; Vaughan et al. 2013). Oceans have warmed, and ground temperatures have increased. The consequences of this warming are especially noticeable in the cryosphere. Spring snow cover in the Northern Hemisphere has decreased, Northern Hemisphere lakes and rivers are freezing later in autumn and thawing earlier in spring, glaciers and permafrost are melting, and Arctic sea ice is shrinking. The hydrologic cycle, too, has changed. Annual precipitation has increased in the extratropics, as has the occurrence of extreme rainfall events.

One key issue in the scientific debate is the extent to which the warming of the twentieth century reflects natural climate variability. The instrumental record dates back to the mid to late 1800s. Prior to that, tree rings, coral, ice cores, and other temperature-sensitive proxy data are used to reconstruct hemispheric and global temperature (Figure 8.5). Though such temperature reconstructions are subject to methodological uncertainties, they show that the late 1900s stands out as an exceptionally warm period with an unprecedented rate of warming (Masson-Delmotte et al. 2013).

Climate warming can be understood in terms of various radiative forcings, which constitute the changes in energy available to the climate system (Myhre et al. 2013). A positive radiative forcing means more incoming energy remains in the system, and planetary temperature increases; a negative radiative forcing means more outgoing energy, and planetary temperature decreases. Figure 8.13 shows changes in these forcings between 1750 and 2011. Increases in greenhouse gases (CO_2, CH_4, N_2O, and halocarbons) are a positive radiative forcing that has warmed climate. The combined radiative forcing is 2.83 W m^{-2}, of which CO_2 (+112 ppm) contributes 1.82 W m^{-2}; CH_4 (+1081 ppb), 0.48 W m^{-2}; halocarbons (chlorofluorocarbons, hydrochlorofluorocarbons, and chlorocarbons), 0.36 W m^{-2}; and N_2O (+54 ppb), 0.17 W m^{-2}. Ozone is another important greenhouse gas. Increasing amounts of ozone in the troposphere is a positive radiative forcing (0.40 W m^{-2}) while ozone depletion in the stratosphere is a small negative forcing (–0.05 W m^{-2}). Human activities have increased stratospheric water vapor, because chemical destruction of anthropogenic CH_4 in the stratosphere produces a small amount of water vapor. This provides a positive radiative forcing (0.07 W m^{-2}).

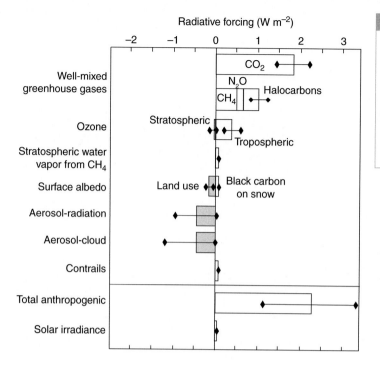

Fig. 8.13 Change in global mean radiative forcing between 1750 and 2011. Anthropogenic forcings are greenhouse gases, ozone, water vapor, surface albedo, aerosols, and contrails. Solar irradiance is a natural forcing that increased during this period. This is compared with the net anthropogenic forcing. Bars indicate uncertainty estimates. Adapted from Myhre et al. (2013).

Aerosols affect climate by absorbing and scattering radiation (aerosol–radiation interactions) and by altering cloud albedo (aerosol–cloud interactions). Both produce a negative radiative forcing. The net aerosol–radiation interactions forcing from sulfate aerosols, fossil fuel organic and black carbon aerosols, biomass burning aerosols, secondary organic aerosols, nitrate aerosols, and mineral dust is a negative radiative forcing (–0.45 W m⁻²), primarily from sulfate aerosols, though black carbon emitted during combustion of fossil fuel absorbs solar radiation and is a significant positive radiative forcing. The aerosol–cloud interactions forcing is comparable (–0.45 W m⁻²). Additionally, aerosols deposited onto snow and ice decrease surface albedo, which provides a small positive radiative forcing (0.04 W m⁻²). Clearing of land for agriculture has increased surface albedo over large regions of the world. This is a negative radiative forcing (–0.15 W m⁻²). Contrails in the atmosphere produce a small positive radiative forcing (0.06 W m⁻²).

The net anthropogenic radiative forcing between 1750 and 2011 is 2.3 W m⁻². In contrast, the natural radiative forcing from volcanoes and solar variability is about 2% of this value. Emission of volcanic aerosols, while important, is episodic, and sulfate aerosols from SO₂ are relatively short lived (less than one year). Solar irradiance has produced a small positive radiative forcing (0.05 W m⁻²). The preponderance of evidence shows that the net anthropogenic radiative forcing is positive and vastly exceeds that from natural processes. It is highly unlikely that natural processes have had a warming influence comparable to that of the net anthropogenic radiative forcing.

The transient response of climate to changes in external forcings is a complex outcome of physical, chemical, and biological feedbacks within the Earth system that amplify or dampen the response to forcings (Figure 8.14). For example, water vapor, clouds, surface albedo, and ocean processes provide strong positive and negative physical feedbacks in response to warming. Other processes such as soil moisture, the greening or dieback of vegetation, the carbon cycle, and the cycling of nitrogen relate to ecosystems and their responses to climate change. Natural climate variability internal to the system modulates the system response to these

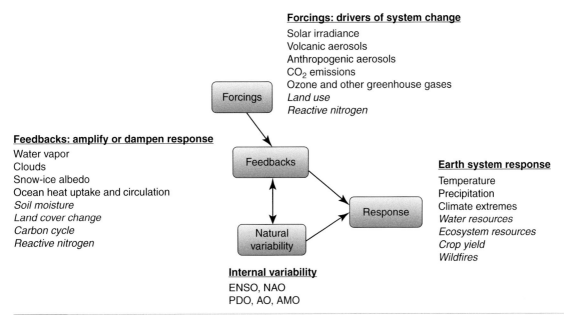

Forcings: drivers of system change
Solar irradiance
Volcanic aerosols
Anthropogenic aerosols
CO_2 emissions
Ozone and other greenhouse gases
Land use
Reactive nitrogen

Feedbacks: amplify or dampen response
Water vapor
Clouds
Snow-ice albedo
Ocean heat uptake and circulation
Soil moisture
Land cover change
Carbon cycle
Reactive nitrogen

Earth system response
Temperature
Precipitation
Climate extremes
Water resources
Ecosystem resources
Crop yield
Wildfires

Internal variability
ENSO, NAO
PDO, AO, AMO

Fig. 8.14 Depiction of climate response to forcings as modulated by physical, chemical, and biological feedbacks within the system and natural variability internal to the system. Ecological processes are highlighted in italics.

feedbacks. The El Ñino/Southern Oscillation, North Atlantic Oscillation, and other modes of variability create internally generated fluctuations in temperature and precipitation superimposed on the long-term response to forcings. The realized outcome is the response to the imposed forcings, feedbacks, and internal variability.

Global climate models are used to attribute temperature trends over the twentieth century to particular forcings, and to study how climate might change in the future with altered forcings. These models of the atmosphere–ocean–land–sea ice system simulate the climate of the preindustrial era, often taken as the mid-1800s (e.g., 1850), given appropriate forcings for that era. A long, multi-century simulation provides an estimate of climate in the absence of temporal trends in forcings. Then, the models simulate the time evolving climate through the twentieth century (e.g., 1850–2005) with prescribed concentrations of greenhouse gases and other forcings. Particular natural and anthropogenic forcings are included or excluded to test which forcings produce the best match between simulated and observed temperature trends over the twentieth century.

Climate models focus on physical processes coupling the atmosphere–ocean–land–sea ice system. In contrast, Earth system models represent the physics, chemistry, and biology of the Earth system and include atmospheric chemistry and terrestrial and marine ecology and biogeochemistry. A prominent use of Earth system models is to simulate the global carbon cycle and its feedback with climate change. While climate model simulations are driven by prescribed CO_2 concentrations, Earth system models are driven by anthropogenic CO_2 emissions and simulate atmosphere CO_2 as the balance among emissions and terrestrial and marine ecosystems processes.

Model simulations that include only natural forcings (solar irradiance, volcanic aerosols) do not reproduce the warming in the latter half of the twentieth century while models that also include anthropogenic forcings (greenhouse gases, aerosols, ozone, land-use change) do simulate the warming (Bindoff et al. 2013). Figure 8.15 shows results from such simulations for one climate model (Meehl et al. 2004). The model replicates the early twentieth century warming with natural forcings (chiefly

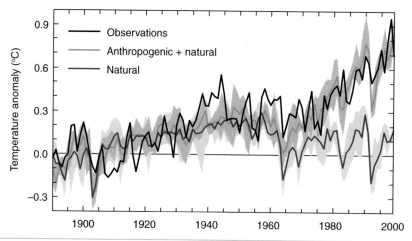

solar irradiance), but reproduces the late twentieth century warming only when anthropogenic forcings (primarily greenhouse gases) are included. That the observed warming of the twentieth century is only simulated when anthropogenic forcings are included suggests an influence of humans on climate. However, not all anthropogenic emissions warm climate. Sulfate aerosol particles produced by industrial activities have lowered global temperature and have decreased the temperature warming below that expected from higher greenhouse gas concentrations alone.

Figure 8.15 illustrates the three key components of climate change simulations: forcings (perturbations to the system); model response (also called the forced response); and natural climate variability (the unforced variability internal to the system). Each of these produces uncertainty in the simulations. The importance of natural variability is assessed through a multi-member ensemble of simulations with a single model. Because of the chaotic and nonlinear nature of climate, small differences in initial conditions produce different climate trajectories, each of which is an equally plausible realization of climate. This is seen in the ensemble

spread for the two forcing simulations. Various models differ in their forced response, due to their spatial resolution, parameterization of processes, and other factors. While Figure 8.15 shows results for a single model, climate change assessment is typically given in terms of multi-model ensembles. Uncertainty in the forcing occurs because direct observations are not available to reconstruct the various forcing agents.

Climate model simulations can be used to partition the temperature increase to particular forcings. Earth's annual global mean surface temperature increased by about 0.65°C over the period 1951–2010, and multi-model simulations show that most of this warming came from greenhouse gases (Bindoff et al. 2013). The models attribute 0.5–1.3°C (midpoint, 0.9°C) of the temperature increase to greenhouse gases. Other anthropogenic forcings (aerosols, ozone, land-use change) contributed between −0.6°C and 0.1°C (midpoint, −0.25°C). The net anthropogenic effect was a warming of 0.6–0.8°C, consistent with the observed warming over this period. Natural forcings had a minor effect on temperature (−0.1 to 0.1°C), as did internal variability (−0.1 to 0.1°C).

8.5 | Climate of the Twenty-First Century

Climate models simulate climate change for the twenty-first century using greenhouse gas concentrations and other forcings derived from representative concentration pathways (RCPs). These depict four different scenarios of the future with respect to population growth, socioeconomic change, technological change, energy consumption, and land use; resulting emissions of greenhouse gases, reactive gases, and aerosol precursors; and the concentration of atmosphere constituents (Table 2.3, Figure 2.11). They span a range of radiative forcing values at year 2100, with a high emission scenario (8.5 W m^{-2}, RCP8.5), two medium stabilization scenarios (4.5 W m^{-2}, RCP4.5; 6.0 W m^{-2}, RCP6.0), and one

mitigation scenario with low radiative forcing (2.6 W m^{-2}, RCP2.6).

Figure 8.16a shows a multi-model synthesis of climate simulations through 2050 for the four RCPs using 25–42 models depending on scenario (Kirtman et al. 2013). The spread across models and RCPs is about 0.5°C and increases over time. Over the near-term (2016–2035), the increase in annual global mean surface air temperature is simulated to be 0.3–0.7°C relative to the 1986–2005 reference period. Over this time period, the spread across the four RCPs is small (0.2°C) relative to the spread among models (0.4°C).

The long-term warming is larger (Collins et al. 2013). Figure 8.16b shows a multi-model synthesis of climate simulations through 2100 and extended for an additional 200 years until 2300, and Table 8.1 summarizes results.

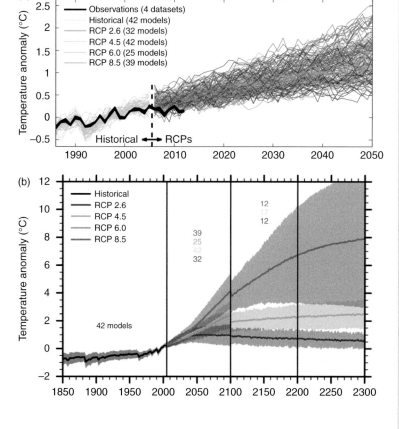

Fig. 8.16 Transient climate model simulations for the historical era and with the four representative concentration pathways (RCPs). Shown are the annual global mean surface air temperature anomalies (relative to 1986–2005) for the historical period (prior to 2005) and for the RCPs. Only one ensemble member was used from each model, and numbers in the figure indicate the number of models contributing to the different time periods. (a) Near-term climate change for 1986–2050. Each individual line is a single model realization for the historical period and the RCPs. The thick black line denotes observed temperature based on four datasets. Adapted from Kirtman et al. (2013). (b) The full simulations for the historical period (1850–2005) and the RCPs (2005–2300). Solid lines show the multi-model mean and the shading denotes the spread among the individual models. The discontinuity at 2100 for RCP8.5 is an artifact of the model sample size. Adapted from Collins et al. (2013). See color plate section.

Table 8.1 Multi-model ensemble mean global temperature change for the middle and late twenty-first century and the late twenty-second and twenty-third centuries in the four representative concentration pathways

Scenario	Global mean warming (°C)			
	2046–2065	2081–2100	2181–2200	2281–2300
RCP2.6	1.0 ± 0.3	1.0 ± 0.4	0.7 ± 0.4	0.6 ± 0.3
RCP4.5	1.4 ± 0.3	1.8 ± 0.5	2.3 ± 0.5	2.5 ± 0.6
RCP6.0	1.3 ± 0.3	2.2 ± 0.5	3.7 ± 0.7	4.2 ± 1.0
RCP8.5	2.0 ± 0.4	3.7 ± 0.7	6.5 ± 2.0	7.8 ± 2.9

Note: Data are annual global mean surface air temperature change relative to 1986–2005. Shown is the multi-model mean ± one standard deviation across the individual models.

Source: From Collins et al. (2013).

Annual global mean surface air temperature increases during the twenty-first century for all scenarios, and the warming directly relates to the magnitude of the radiative forcing. The low radiative forcing scenario (RCP2.6) has a warming of 1.0°C at mid-century and 1.0°C at the end of the century. The stabilization scenario (RCP4.5) has a warming of 1.4°C at mid-century and 1.8°C at the end of the century. RCP6.0 has still larger warming (1.3°C and 2.2°C, respectively). The largest warming is with RCP8.5, where temperature increases by 2.0°C at mid-century and 3.7°C at the end of the century. Warming continues unabated in the business-as-usual scenario with high radiative forcing (RCP8.5).

Even if concentrations are stabilized without further increases, climate will continue to warm for hundreds of years (Collins et al. 2013). Past anthropogenic emissions commit us to long-term warming because of the large thermal reservoir of the ocean and the slow mixing of the radiative forcing energy perturbation into the ocean. This climate change commitment is seen in the simulation with RCP4.5 extended through 2300. In this scenario, atmospheric CO_2 and radiative forcing stabilize before 2100. However, the warming continues through 2300 (Figure 8.16b). Global temperature at the end of the simulation (averaged for the period 2281–2300) is 0.7°C warmer than that of the period 200 years earlier, at the end of the twenty-first century (Table 8.1).

The atmospheric CO_2 concentration at a given point in time depends on the total amount of anthropogenic CO_2 released in the atmosphere (the cumulative carbon emission), the resulting climate change, and feedbacks that alter the accumulation of anthropogenic emissions by the terrestrial biosphere and oceans. Despite these complexities, cumulative total anthropogenic CO_2 emissions and the change in global mean surface temperature are approximately linearly related (Allen et al. 2009; Matthews et al. 2009, 2012; Collins et al. 2013). The exact relationship is model dependent, but each model shows a near linear relationship between temperature change and cumulative emissions, independent of the exact emissions scenario. Figure 8.17 shows this relationship as the multi-model mean of many different climate simulations over the period 1870–2100. If the warming caused by anthropogenic CO_2 emissions is to be limited to less than 2°C (relative to the period 1861–1880), cumulative emissions from all anthropogenic sources must be less than 1000 Pg C since that period (Collins et al. 2013). About one-half of that carbon has already been emitted by 2011. This illustrates the difficulty of achieving climate targets as CO_2 mitigation is delayed (Stocker 2013).

Uncertainty in climate change projections arises from the forcings (scenarios), the model response, and internal (natural) variability (Hawkins and Sutton 2009; Meehl et al. 2009). Forcing uncertainty is assessed through

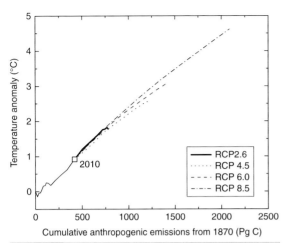

Fig. 8.17 Annual global mean surface air temperature increase in relation to cumulative anthropogenic CO_2 emissions. Temperature is the anomaly relative to 1861–1880. Shown is the multi-model mean for the historical period (until 2010) and each of the four RCPs until 2100. Each line is a temporal trajectory of the temperature anomaly at a given period in relation to the cumulative emissions to that period. Adapted from IPCC (2013).

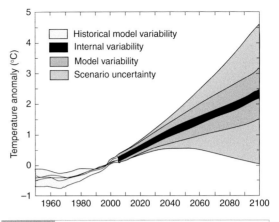

Fig. 8.18 Sources of uncertainty in climate simulations through 2100. Shown is the annual global mean surface air temperature anomaly (relative to 1986–2005). The shading denotes uncertainty from internal variability, model variability, and scenario variability. Also shown are model variability over the historical period (1850–2005) and three datasets of global mean temperature (solid lines). Adapted from Kirtman et al. (2013).

different emission scenarios (i.e., the four RCPs). Model response uncertainty is assessed through multi-model ensembles. Natural variability is assessed through a multi-member ensemble of simulations for each model. For temperature projections at near-term decadal timescales (10–30 years), the particulars of the forcing (i.e., scenario uncertainty) are less important than model response uncertainty and internal variability (Figure 8.18). Scenario uncertainty is important at multi-decadal lead times.

Distinguishing between the internally generated climate variations (i.e., natural variability) and the anthropogenically forced climate change is a key requirement in climate change detection and attribution analyses (Deser et al. 2012; Hawkins and Sutton 2012; Mahlstein et al. 2012). Natural variability manifests as interannual-to-decadal climate variability, seen in observations and an individual model realization, as well as ensemble variability within a model. Climate change detection and attribution requires determining the time when the signal of the forced temperature change becomes large relative to its natural variability. The time at which the forced climate change signal emerges from the noise of natural climate variability is known as the time of emergence. The time of emergence is defined as the year at which the forced climate change signal (S, e.g., change in annual temperature) exceeds the noise (N, e.g., standard deviation of annual temperature) by a particular threshold (e.g., $S/N > 1$ or > 2). The time of emergence for temperature can range from a few decades in the mid-latitudes to several decades in regions with high natural variability.

8.6 | Review Questions

1. Describe trends in orbital forcing over the next several thousand years associated with the Milankovitch cycles. In the absence of other climate forcings, how might climate change?

2. Ice cores reveal a close coupling between temperature and atmospheric CO_2 concentration. What physical and biological processes affect this coupling?

3. Describe what would happen to atmospheric CO_2 in the absence of volcanic activity.

4. The rate of rock weathering generally increases with warmer temperature. Is the weathering of rocks a positive or negative feedback on rising atmospheric CO_2?

5. Discuss terrestrial feedbacks that accentuate or mitigate cooling during ice ages.

6. Describe how long-term warming and increased precipitation in the Arctic can lead to glaciation.

7. Discuss two anthropogenic activities that have reduced the temperature increase that is expected from rising concentrations of greenhouse gases. Is it feasible to use these in geoengineering practices to reduce global warming?

8. Assume policymakers would like to devise socioeconomic growth scenarios that will limit global warming in the middle of the twenty-first century to less than 0.4°C compared with the end of the twentieth century. Is this a feasible target?

9. In developing scenarios of plausible greenhouse gas emissions, should policymakers focus on short- or long-term socioeconomic growth? Explain why.

10. The radiative forcing from land-use change is generally considered to be negative, because of an increase in surface albedo. Discuss other processes by which land use affects climate.

8.7 | References

Allen, M. R., Frame, D. J., Huntingford, C., et al. (2009). Warming caused by cumulative carbon emissions towards the trillionth tone. *Nature*, 458, 1163–1166.

Alley, R. B., and Ágústsdóttir, A. M. (2005). The 8k event: Cause and consequences of a major Holocene abrupt climate change. *Quaternary Science Reviews*, 24, 1123–1149.

Alley, R. B., Marotzke, J., Nordhaus, W., et al. (2002). *Abrupt Climate Change: Inevitable Surprises*. Washington, DC: The National Academies Press.

Alley, R. B., Marotzke, J., Nordhaus, W. D., et al. (2003). Abrupt climate change. *Science*, 299, 2005–2010.

Ammann, C. M., Joos, F., Schimel, D. S., Otto-Bliesner, B. L., and Tomas, R. A. (2007). Solar influence on climate during the past millennium: results from transient simulations with the NCAR Climate System Model. *Proceedings of the National Academy of Sciences USA*, 104, 3713–3718.

Berger, A. L. (1978). Long-term variations of daily insolation and Quaternary climatic changes. *Journal of the Atmospheric Sciences*, 35, 2362–2367.

Berger, A. (1988). Milankovitch theory and climate. *Reviews of Geophysics*, 26, 624–657.

Berger, A., and Loutre, M. F. (1991). Insolation values for the climate of the last 10 million years. *Quaternary Science Reviews*, 10, 297–317.

Berger, A., Loutre, M. F., and Tricot, C. (1993). Insolation and Earth's orbital periods. *Journal of Geophysical Research*, 98D, 10341–10362.

Berner, R. A. (1991). A model for atmospheric CO_2 over Phanerozoic time. *American Journal of Science*, 291, 339–376.

Berner, R. A. (1998). The carbon cycle and CO_2 over Phanerozoic time: the role of land plants. *Philosophical Transactions of the Royal Society B*, 353, 75–82.

Berner, R. A. (2003). The long-term carbon cycle, fossil fuels and atmospheric composition. *Nature*, 426, 323–326.

Berner, R. A., and Lasaga, A. C. (1989). Modeling the geochemical carbon cycle. *Scientific American*, 260(3), 74–81.

Bindoff, N. L., Stott, P. A., Achuta Rao, K. M., et al. (2013). Detection and attribution of climate change: from global to regional. In *Climate Change 2013: The Physical Science Basis. Contribution of Working Group I to the Fifth Assessment Report of the Intergovernmental Panel on Climate Change*, ed. T. F. Stocker, D. Qin, G.-K. Plattner, et al. Cambridge: Cambridge University Press, pp. 867–952.

Broecker, W. S. (1997). Thermohaline circulation, the Achilles heel of our climate system: Will man-made CO_2 upset the current balance? *Science*, 278, 1582–1588.

Broecker, W. S. (2003). Does the trigger for abrupt climate change reside in the ocean or in the atmosphere? *Science*, 300, 1519–1522.

Broecker, W. S. (2006). Abrupt climate change revisited. *Global and Planetary Change*, 54, 211–215.

Broecker, W. S., Peteet, D. M., and Rind, D. (1985). Does the ocean–atmosphere system have more than one stable mode of operation? *Nature*, 315, 21–26.

Broecker, W. S., Kennett, J. P., Flower, B. P., et al. (1989). Routing of meltwater from the Laurentide Ice Sheet during the Younger Dryas cold episode. *Nature*, 341, 318–321.

Carlson, A. E., and Clark, P. U. (2012). Ice sheet sources of sea level rise and freshwater discharge during the last deglaciation. *Reviews of Geophysics*, 50, RG4007, doi:10.1029/2011RG000371.

Clark, P. U., Marshall, S. J., Clarke, G. K. C., et al. (2001). Freshwater forcing of abrupt climate change during the last glaciation. *Science*, 293, 283–287.

COHMAP (1988). Climatic changes of the last 18,000 years: Observations and model simulations. *Science*, 241, 1043–1052.

Collins, M., Knutti, R., Arblaster, J., et al. (2013). Long-term climate change: Projections, commitments and irreversibility. In *Climate Change 2013: The Physical Science Basis. Contribution of Working Group I to the Fifth Assessment Report of the Intergovernmental Panel on Climate Change*, ed. T. F. Stocker, D. Qin, G.-K. Plattner, et al. Cambridge: Cambridge University Press, pp. 1029–1136.

Crowley, T. J., and North, G. R. (1991). *Paleoclimatology*. New York: Oxford University Press.

Deser, C., Knutti, R., Solomon, S., and Phillips, A. S. (2012). Communication of the role of natural variability in future North American climate. *Nature Climate Change*, 2, 775–779.

Eddy, J. A. (1976). The Maunder Minimum. *Science*, 192, 1189–1202.

Free, M., and Robock, A. (1999). Global warming in the context of the Little Ice Age. *Journal of Geophysical Research*, 104D, 19057–19070.

Gao, C., Robock, A., and Ammann, C. (2008). Volcanic forcing of climate over the past 1500 years: An improved ice corebased index for climate models. *Journal of Geophysical Research*, 113, D23111, doi:10.1029/2008JD010239.

Hartmann, D. L., Klein Tank, A. M. G., Rusticucci, M., et al. (2013). Observations: atmosphere and surface. In *Climate Change 2013: The Physical Science Basis. Contribution of Working Group I to the Fifth Assessment Report of the Intergovernmental Panel on Climate Change*, ed. T. F. Stocker, D. Qin, G.-K. Plattner, et al. Cambridge: Cambridge University Press, pp. 159–254.

Hawkins, E., and Sutton, R. (2009). The potential to narrow uncertainty in regional climate predictions. *Bulletin of the American Meteorological Society*, 90, 1095–1107.

Hawkins, E., and Sutton, R. (2012). Time of emergence of climate signals. *Geophysical Research Letters*, 39, L01702, doi:10.1029/2011GL050087.

Hay, W. W., DeConto, R. M., Wold, C. N., et al. (1999). Alternative global Cretaceous paleogeography. In *Evolution of the Cretaceous Ocean–Climate System*, ed. E. Barrera and C. C. Johnson. Boulder, Colorado: Geological Society of America, pp. 1–47.

Hays, J. D., Imbrie, J., and Shackleton, N. J. (1976). Variations in the Earth's orbit: Pacemaker of the Ice Ages. *Science*, 194, 1121–1132.

Huntley, B. J., and Webb, T., III (1988). *Vegetation History*. Dordrecht: Kluwer Academic Publishers.

IPCC (2013). Summary for policymakers. In *Climate Change 2013: The Physical Science Basis. Contribution of Working Group I to the Fifth Assessment Report of the Intergovernmental Panel on Climate Change*, ed. T. F. Stocker, D. Qin, G.-K. Plattner, et al. Cambridge: Cambridge University Press, pp. 3–29.

Jones, P. D., Lister, D. H., Osborn, T. J., et al. (2012). Hemispheric and large-scale land-surface air temperature variations: An extensive revision and an update to 2010. *Journal of Geophysical Research*, 117, D05127, doi:10.1029/2011JD017139.

Jouzel, J., Waelbroeck, C., Malaize, B., et al. (1996). Climatic interpretation of the recently extended Vostok ice records. *Climate Dynamics*, 12, 513–521.

Jouzel, J., Masson-Delmotte, V., Cattani, O., et al. (2007). Orbital and millennial Antarctic climate variability over the past 800,000 years. *Science*, 317, 793–796.

Kirtman, B., Power, S. B., Adedoyin, J. A., et al. (2013). Near-term climate change: projections and predictability. In *Climate Change 2013: The Physical Science Basis. Contribution of Working Group I to the Fifth Assessment Report of the Intergovernmental Panel on Climate Change*, ed. T. F. Stocker, D. Qin, G.-K. Plattner, et al. Cambridge: Cambridge University Press, pp. 953–1028.

Kutzbach, J. E., and Guetter, P. J. (1986). The influence of changing orbital parameters and surface boundary conditions on climate simulations for the past 18 000 years. *Journal of the Atmospheric Sciences*, 43, 1726–1759.

Kutzbach, J. E., and Webb, T., III (1993). Conceptual basis for understanding late-Quaternary climates. In *Global Climates since the Last Glacial Maximum*, ed. H. E. Wright, Jr., J. E. Kutzbach, T. Webb, III, et al. Minneapolis: University of Minnesota Press, pp. 5–11.

Kutzbach, J. E., Guetter, P. J., Ruddiman, W. F., and Prell, W. L. (1989). Sensitivity of climate to late Cenozoic uplift in southern Asia and the American west: numerical experiments. *Journal of Geophysical Research*, 94D, 18393–18407.

Lambert, F., Delmonte, B., Petit, J. R., et al. (2008). Dust–climate couplings over the past 800,000 years from the EPICA Dome C ice core. *Nature*, 452, 616–619.

Landrum, L., Otto-Bliesner, B. L., Wahl, E. R., et al. (2013). Last millennium climate and its variability in CCSM4. *Journal of Climate*, 26, 1085–1111.

Loulergue, L., Schilt, A., Spahni, R., et al. (2008). Orbital and millennial-scale features of atmospheric CH_4 over the past 800,000 years. *Nature*, 453, 383–386.

Lüthi, D., Le Floch, M., Bereiter, B., et al. (2008). High-resolution carbon dioxide concentration record 650,000–800,000 years before present. *Nature*, 453, 379–382.

Mahlstein, I., Hegerl, G., and Solomon, S. (2012). Emerging local warming signals in observational data. *Geophysical Research Letters*, 39, L21711, doi:10.1029/2012GL053952.

Manabe, S., and Stouffer, R. J. (1999). The role of thermohaline circulation in climate. *Tellus B*, 51, 91–109.

Mann, M. E., Zhang, Z., Rutherford, S., et al. (2009). Global signatures and dynamical origins of the Little Ice Age and Medieval Climate Anomaly. *Science*, 326, 1256–1260.

Masson-Delmotte V., Schulz, M., Abe-Ouchi, A., et al. (2013). Information from paleoclimate archives. In *Climate Change 2013: The Physical Science Basis. Contribution of Working Group I to the Fifth Assessment Report of the Intergovernmental Panel on Climate Change*, ed. T. F. Stocker, D. Qin, G.-K. Plattner, et al. Cambridge: Cambridge University Press, pp. 383–464.

Matthews, H. D., Gillett, N. P., Stott, P. A., and Zickfeld, K. (2009). The proportionality of global warming to cumulative carbon emissions. *Nature*, 459, 829–832.

Matthews, H. D., Solomon, S., and Pierrehumbert, R. (2012). Cumulative carbon as a policy framework for achieving climate stabilization. *Philosophical Transactions of the Royal Society A*, 370, 4365–4379.

Meehl, G. A., Washington, W. M., Ammann, C. M., et al. (2004). Combinations of natural and anthropogenic forcings in twentieth-century climate. *Journal of Climate*, 17, 3721–3727.

Meehl, G. A., Goddard, L., Murphy, J., et al. (2009). Decadal prediction: Can it be skillful? *Bulletin of the American Meteorological Society*, 90, 1467–1485.

Meissner, K. J., and Clark, P. U. (2006). Impact of floods versus routing events on the thermohaline circulation. *Geophysical Research Letters*, 33, L15704, doi:10.1029/2006GL026705.

Morice, C. P., Kennedy, J. J., Rayner, N. A., and Jones, P. D. (2012). Quantifying uncertainties in global and regional temperature change using an ensemble of observational estimates: The HadCRUT4 data set. *Journal of Geophysical Research*, 117, D08101, doi:10.1029/2011JD017187.

Myhre, G., Shindell, D., Bréon, F.-M., et al. (2013). Anthropogenic and natural radiative forcing. In *Climate Change 2013: The Physical Science Basis. Contribution of Working Group I to the Fifth Assessment Report of the Intergovernmental Panel on Climate Change*, ed. T. F. Stocker, D. Qin, G.-K. Plattner, et al. Cambridge: Cambridge University Press, pp. 659–740.

Peltier, W. R. (1994). Ice age paleotopography. *Science*, 265, 195–201.

Prell, W. L., and Kutzbach, J. E. (1992). Sensitivity of the Indian monsoon to forcing parameters and implications for its evolution. *Nature*, 360, 647–652.

Rhein, M., Rintoul, S. R., Aoki, S., et al. (2013). Observations: Ocean. In *Climate Change 2013: The Physical Science Basis. Contribution of Working Group I to the Fifth Assessment Report of the Intergovernmental Panel on Climate Change*, ed. T. F. Stocker, D. Qin, G.-K. Plattner, et al. Cambridge: Cambridge University Press, pp. 255–315.

Ruddiman, W. F., and Kutzbach, J. E. (1989). Forcing of late Cenozoic Northern Hemisphere climate by plateau uplift in southern Asia and the American west. *Journal of Geophysical Research*, 94D, 18409–18427.

Ruddiman, W. F., Prell, W. L., and Raymo, M. E. (1989). Late Cenozoic uplift in southern Asia and the American west: Rationale for general circulation modeling experiments. *Journal of Geophysical Research*, 94D, 18379–18391.

Ruddiman, W. F., Kutzbach, J. E., and Prentice, I. C. (1997). Testing the climatic effects of orography and CO_2 with general circulation and biome models. In *Tectonic Uplift and Climate Change*, ed. W. F. Ruddiman. New York: Plenum Press, pp. 203–235.

Schilt, A., Baumgartner, M., Blunier, T., et al. (2010). Glacial–interglacial and millennial-scale variations in the atmospheric nitrous oxide concentration during the last 800,000 years. *Quaternary Science Reviews*, 29, 182–192.

Schmidt, G. A., Jungclaus, J. H., Ammann, C. M., et al. (2012). Climate forcing reconstructions for use in

PMIP simulations of the Last Millennium (v1.1). *Geoscientific Model Development*, 5, 185–191.

Shackleton, N. J. (2000). The 100,000-year ice-age cycle identified and found to lag temperature, carbon dioxide, and orbital eccentricity. *Science*, 289, 1897–1902.

Stocker, T. F. (2013). The closing door of climate targets. *Science*, 339, 280–282.

Vaughan, D. G., Comiso, J. C., Allison, I., et al. (2013). Observations: Cryosphere. In *Climate Change 2013: The Physical Science Basis. Contribution of Working Group I to the Fifth Assessment Report of the Intergovernmental Panel on Climate Change*, ed. T. F. Stocker, D. Qin, G.-K. Plattner, et al. Cambridge: Cambridge University Press, pp. 317–382.

Wright, H. E., Jr., Kutzbach, J. E., Webb, T., III, et al. (1993). *Global Climates since the Last Glacial Maximum*. Minneapolis: University of Minnesota Press.

Zhisheng, A., Kutzbach, J. E., Prell, W. L., and Porter, S. C. (2001). Evolution of Asian monsoons and phased uplift of the Himalaya–Tibetan plateau since Late Miocene times. *Nature*, 411, 62–66.

Part III

Hydrometeorology

9

Soil Physics

9.1 | Chapter Summary

Soils store a considerable amount of heat and water. The diurnal cycle of soil temperature and seasonal variation in soil temperature over the course of a year are important determinants of the land surface climate. The amount of water held in soil regulates evapotranspiration. This chapter reviews the physics of soil heat transfer and soil water relations. Heat flows from high to low temperature through conduction. Important soil properties that determine heat transfer are thermal conductivity and heat capacity. Two forces govern water movement in soil. Gravitational potential represents water movement due to the force of gravity. The second force, called matric potential, occurs because water is bound to soil particles. Water flows from high to low potential as described by Darcy's law. The Richards equation combines Darcy's law with principles of water conservation to describe the change in soil water content over time. Key hydraulic properties are porosity, matric potential, and hydraulic conductivity. These latter two properties vary with soil water. Soils differ in hydraulic properties in relation to the size and arrangement of pores. The pores in sandy soil are large, water loosely adheres to soil particles, water movement is rapid, and the soil drains rapidly. Pores are smaller in clay soil, water is tightly bound to soil particles, movement is slow, and drainage is impeded. Loams are intermediate, draining more slowly than sands and retaining more water.

9.2 | Soil Texture and Structure

Soils are composed of organic material, mineral particles, air, and water. A typical mineral soil is 55 percent solid particles and 45 percent air and water. Most soils have relatively low organic matter content, ranging from 1 percent to 10 percent. These soils are known as mineral soils. In contrast, organic soils are those in which more than 80 percent of the material is organic matter. These soils develop in swamps, bogs, and marshes where waterlogged conditions inhibit decomposition. The type, abundance, and arrangement of mineral and organic particles determine heat and water flow.

Mineral particles consist of three types determined by size. Sand particles are the largest, ranging in size from 0.05 mm (very fine sand) to 2 mm (very coarse sand). They are rounded or irregular, which creates large pore spaces between particles. Sand particles have a low capacity to hold water. Clay particles are the smallest mineral particles and are less than 0.002 mm in size. Clay particles are generally flat, plate-like, and fit closely together. They have the largest surface area, which facilitates the adsorption of water to the clay particles. In between, ranging in size from 0.002 to 0.05 mm, are silt particles.

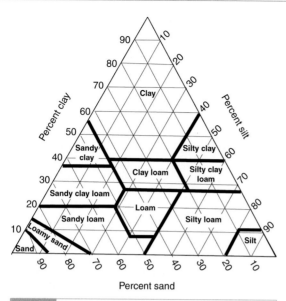

Fig. 9.1 Soil texture triangle showing the relationship between the 12 soil texture classes and percentage of sand, silt, and clay. The percentage of particles that are clay is determined by the lines running parallel to the sand side of the triangle. The percentage silt is determined by the lines running parallel to the clay side. The percentage sand lines run parallel to the silt side. For example, a soil that is 60 percent sand, 30 percent clay, and 10 percent silt is a sandy clay loam.

The relative abundance of sand, silt, and clay particles determines soil texture (Figure 9.1). There are three broad texture classes: sands – soils in which sand particles comprise more than 70 percent of the material by weight; clays – soils in which clay particles comprise at least 35–40 percent of the material; and loams – soils that are a mixture of sand, silt, and clay. A sandy soil has loose, individual grains that can be seen and felt. It readily falls apart when dry. If squeezed when dry, a loamy soil will form a molded shape that stays together with careful handling. When wet, the molded form is more durable and can be handled without breaking. A clay soil forms hard clumps when dry and is sticky when wet. Wet clay will form a long, flexible ribbon when squeezed.

Soil texture considers the size distribution of mineral particles. The arrangement of soil particles into large recognizable units, known as soil structure, is equally important. Loose, granular

soil is much more porous than compacted soil. The development of good soil structure requires some agent that coheres and cements individual mineral particles together. Clay particles, because of their large surface area, are one such binding agent. Lack of clay is one reason why sandy soils crumble so easily. Organic matter is another cementing agent that allows individual particles to adhere to one another. Through this, organic matter has a great influence on the ability of soil to store water.

The sand, silt, clay, and organic particles that form soil lie in close contact, but because of their irregular shapes they do not fit evenly together. Instead, voids exist around the individual particles. The total volume of voids is known as pore space, or porosity. It is the volume of soil that is occupied by air and water. These pores can fill with water, such as happens when water infiltrates into the soil. Or when dry, the pores consist mainly of air. Most field conditions are in between, and the soil is a mix of solid particles, water, and air. In a typical soil, about 55 percent of the soil volume comprises solids and 45 percent of the soil volume is pore space comprising air and water.

9.3 | Soil Temperature

Soils are a large source or repository of heat that moderates the diurnal and seasonal range in surface temperature. During the day, when solar radiation heats the surface, the surface is warmer than the underlying soil and heat flows into the soil. This transfer of heat away from the surface cools the surface. At night, the surface is cooler than the soil and heat flows out of the soil. This gain of energy at the surface warms the surface. As a result, surface air temperature shows less of a diurnal range than if no heat were stored in the soil. The same behavior occurs annually, when soil stores heat in warm months and releases heat during cold months.

Figure 9.2a illustrates a typical summer temperature profile during the day and night. At night, temperatures increase with depth. Because heat flows from high to low temperatures, heat flows upward from deeper depths

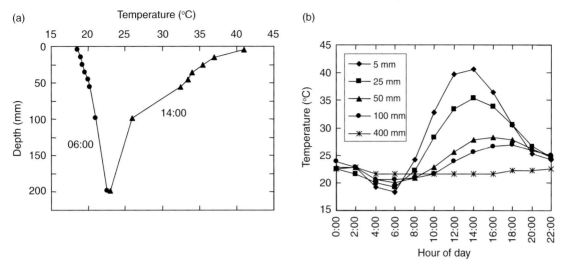

Fig. 9.2 Summer soil temperatures. (a) Typical night and day soil temperature profiles during summer. (b) Diurnal cycle of soil temperature at several depths on a typical summer day. Adapted from Hartmann (1994, p. 86).

to the surface. During the day, the soil profile warms, but the warming decreases with greater depth. The deep soil hardly warms at all from its nighttime temperature. As a result, daytime soil temperatures decrease with depth, and heat flows downward from the surface towards the deep soil. The diurnal cycle of soil temperature at different depths further illustrates this behavior (Figure 9.2b). The soil close to the surface, at a depth of 5 mm, warms rapidly as the Sun's radiation heats the surface, increasing from 18°C at 0600 hours to 41°C at 1400 hours. This upper soil also cools rapidly at night. Deeper soil layers are cooler than upper layers during the day (e.g., from 1200 to 1600 hours) and warmer at night (e.g., from 0400 to 0600 hours). The diurnal range in temperature decreases with depth, and maximum temperatures occur later in the day.

Two soil properties (thermal conductivity, heat capacity) determine the temperature profile for a given heat flux at the surface. First, heat flows from high temperature to low temperature. The rate at which heat flows between two points separated by a distance of Δz meters is equal to soil thermal conductivity times the temperature gradient:

$$F = -\kappa(\partial T / \partial z) \tag{9.1}$$

with F heat flux (W m^{-2}), κ thermal conductivity (W m^{-1} K^{-1}), and $\partial T / \partial z$ (K m^{-1}, or °C m^{-1}) the temperature gradient. This latter term is the change in temperature with depth in soil, approximated numerically by $\Delta T / \Delta z$. The negative sign denotes that the heat flux is positive for a negative temperature gradient; the heat flux out of the soil is positive, and the flux into the soil is negative. Thermal conductivity determines the heat flow in unit time by conduction through a unit thickness of a unit area of material across a unit temperature gradient.

Second, if more heat enters a volume of soil than exits, the soil gains heat and warms. Conversely, net loss of heat cools the soil volume. Heat capacity is a measure of the temperature change arising from this change in heat storage. It is the amount of heat required to change the temperature of a unit volume of material by 1°C. Energy conservation requires that the difference between heat coming into the top of a slab of soil at depth z and heat exiting the bottom of the slab at depth $z + \Delta z$ equal the rate of heat storage (Figure 9.3):

$$\rho c(\Delta T / \Delta t)\Delta z = -(F_z - F_{z+\Delta z}) \tag{9.2}$$

where ρc is heat capacity (J m^{-3} K^{-1}), $\Delta T / \Delta t$ is the change in temperature with time (K s^{-1},

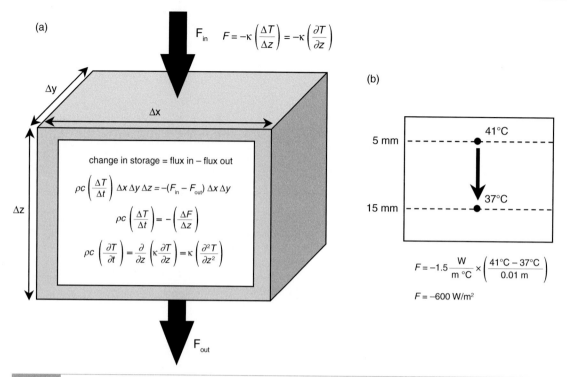

Fig. 9.3 Heat transfer in soil. (a) Heat balance of a volume of soil. Equations are shown in both their numerical finite difference form and in the notation of calculus. (b) Example of vertical heat transfer between two points with a temperature difference of 4°C separated by a distance of 10 mm with a thermal conductivity of 1.5 W m⁻¹ K⁻¹. Hanks (1992) and Hillel (1998) review soil heat transfer.

or °C s⁻¹), and F_z and $F_{z+\Delta z}$ are the heat flux (W m⁻²) into and out of the soil slab. The negative sign ensures that temperature increases when there is a net gain of heat.

Combining equations for heat flux, given by Eq. (9.1), and storage, given by Eq. (9.2), and assuming heat capacity and thermal conductivity do not change with depth gives the change in temperature over time (Figure 9.3), or in the notation of calculus:

$$\frac{\partial T}{\partial t} = \frac{1}{\rho c}\left[\frac{\partial}{\partial z}\left(\kappa\frac{\partial T}{\partial z}\right)\right] = \frac{\kappa}{\rho c}\left(\frac{\partial^2 T}{\partial z^2}\right) \qquad (9.3)$$

The change in soil temperature over time is directly proportional to the thermal conductivity and inversely proportional to the heat capacity. Thermal conductivity determines the rate of heat transfer and heat capacity determines the temperature change as a result of this heat transfer. Soils with a high thermal conductivity gain and lose energy faster than soils with a low thermal conductivity. Soils with a low heat capacity warm and cool faster, for a given heat flux, than soils with a high heat capacity.

Thermal conductivity and heat capacity vary depending on mineral composition, porosity, organic matter content, and the water content of soils (Table 9.1). Soils consist of solid particles, air, and water. The overall thermal conductivity of a soil is a weighted average of the conductivity of its solid, air, and water fractions. Quartz has a very high thermal conductivity, and soils with high quartz content (e.g., sandy soils) have a high thermal conductivity. Clay minerals have a lower thermal conductivity, and clay soils have a lower thermal conductivity than sandy soils. Organic material has an extremely low thermal conductivity, and soils with high organic matter content have a thermal conductivity that is one-quarter to one-third that of mineral soils. Air and water

Table 9.1 Thermal conductivity and heat capacity for soil components and for sand, clay, and peat soils in relation to soil water

	Thermal conductivity (W m^{-1} K^{-1})	Heat capacity (MJ m^{-3} K^{-1})
Soil component		
Quartz	8.80	2.13
Clay minerals	2.92	2.38
Organic matter	0.25	2.50
Water	0.57	4.18
Air	0.02	0.0012
Sandy soil (porosity = 0.4)		
0%	0.30	1.28
50%	1.80	2.12
100%	2.20	2.96
Clay soil (porosity = 0.4)		
0%	0.25	1.42
50%	1.18	2.25
100%	1.58	3.10
Peat soil (porosity = 0.8)		
0%	0.06	0.50
50%	0.29	2.18
100%	0.50	3.87

Note: Soil water content is expressed as a percentage of saturation.
Source: From Monteith and Unsworth (2013, p. 281).

occur in the voids, or pore space, around soil particles. Air and water have a lower thermal conductivity than mineral particles. Consequently, soils with a high pore space have a lower thermal conductivity, all other factors being equal, than soils that are less porous. Sandy soils are less porous than clay soils, which is another reason why they have a higher thermal conductivity. Organic soils are often extremely porous. Thermal conductivity of soil increases greatly with increasing soil water content because the thermal conductivity of water is more than 20 times that of air. Similarly, the heat capacity of water is 3500 times that of air, and the heat capacity of soil increases with water content.

The practical implications of these differences in thermal properties are clearer under idealized conditions. The diurnal and annual cycles of temperature at the soil surface can be represented as a sine wave in which surface temperature varies periodically between some maximum and minimum values. Mathematically, the temperature at the soil surface at some time t is:

$$T_s(t) = \bar{T}_s + A_s \sin\left(2\pi t \,/\, p\right) \qquad (9.4)$$

where p is the period of oscillation (e.g., 86,400 seconds for a diurnal cycle), $\bar{T}_s$ is the average temperature over this period, and A_s is the amplitude (i.e., one-half the difference between maximum and minimum temperatures) over the same time period. This periodic behavior is seen for the near-surface temperature in Figure 9.2b, which has a minimum early in the morning, a maximum in early afternoon, and decreases again during the night.

With a periodic surface temperature and if thermal properties are constant with depth, temperature at a depth of z meters is:

$$T_z(t) = \bar{T}_s + A_s e^{-z/D} \sin\left(2\pi \frac{t}{p} - \frac{z}{D}\right) \qquad (9.5)$$

where $D = \sqrt{\alpha p / \pi}$ and $\alpha = \kappa / \rho c$ is the thermal diffusivity ($m^2\ s^{-1}$). The term $A_s e^{-z/D}$ describes the decrease in surface temperature amplitude with depth. At depth $z = D$ the amplitude is $0.37 A_s$. This depth is called the damping depth. The amplitude at depth $z = 2D$ is $0.14 A_s$ and at depth $z = 3D$ is $0.05 A_s$. In other words, the temperature amplitude decreases with depth, exactly as seen in Figure 9.2b.

A typical soil diffusivity is $\alpha = 7 \times 10^{-7}\ m^2\ s^{-1}$. Over the course of a day ($p = 86,400$ seconds), the damping depth is 14 cm. That is, the diurnal range in temperature at a depth of 14 cm is 37 percent that at the surface. Over the course of a year ($p = 86,400 \times 365$ seconds), the damping depth is 2.65 m. At a depth equal to three times the damping depth, the range in temperature is 5 percent that at the surface. So at a depth of 42 cm, the temperature is approximately equal to the average daily temperature. At a depth of about 8 m, the temperature is approximately equal to the average annual temperature.

The term z / D represents the shift in time with depth when maximum and minimum temperatures occur. For example, maximum temperature at the surface occurs at $t = 0.25 p$, but maximum temperature at depth $z = D$ occurs at $t = 0.41 p$. That is, over the course of a day the maximum temperature at the damping depth occurs almost 4 hours later than the maximum temperature at the surface. With deeper depths, maximum temperature occurs later. At depth $z = \pi D$, temperature is at a maximum when surface temperature is at a minimum. Again, this is exactly what is seen in Figure 9.2b.

Soils are often covered by organic material or snow. Forests, in particular, are typically covered with a layer of decomposing leaf litter several centimeters thick. Organic material has a heat capacity similar to mineral soil, but a much lower thermal conductivity (Table 9.1). As a result, decomposing organic material acts as an insulator, preventing soil from warming in the day and from cooling at night. Snow, with its low thermal conductivity (e.g., 0.34 W m^{-1} K^{-1}), has a similar insulating effect. A deep snow pack early in winter can keep soil warmer than if no snow was present.

In seasonally frozen soils, it is necessary to account for the different thermal properties of water and ice (Farouki 1981; Lunardini 1981). The heat capacity of ice (approximately 2 MJ m^{-3} K^{-1}) is one-half that of water, while its thermal conductivity (2.2 W m^{-1} K^{-1}) is almost four times that of water. Additionally, the change in phase of water consumes and releases heat. At 0°C, 334 joules are needed to melt one gram of water (Table 3.3). This energy (latent heat of fusion) changes the phase of water from ice to liquid rather than warming the soil. The same heat is released when water freezes. While water is freezing, its temperature remains at 0°C. The importance of phase change is seen in simulations of soil temperature with and without phase change. Without phase change, an unfrozen soil undergoing freezing cools too rapidly.

9.4 | Soil Water

The upper meter of soil typically holds 10–45 cm of water. One measure of soil moisture is the volumetric water content, which is defined as the volume of water per unit volume of soil (i.e., the fraction of the soil volume that is water). Figure 9.4 shows typical water contents for a loam soil. The maximum amount of water held in soil occurs when all the pore space is filled with water. The soil is a mixture of water and solid particles. For loam, pores comprise 45.1 percent of the volume of soil, and this is also the water content at saturation. In other words, 1 m^3 of loam holds 0.451 m^3 of water when saturated. Volumetric water content is also the depth of water per unit depth of soil; 1 m of loam holds 45.1 cm of water at saturation. This water is loosely held in the soil, and it quickly drains due to the force of gravity. The amount of water held in the soil after gravitational drainage is its field capacity. The soil is no longer saturated, but rather is a mixture of water, air, and solids. At field capacity, 39.3 percent of the soil volume (87.1% of the pore space) is water. As the soil dries, water movement becomes more difficult, and at some critical water content the water is so strongly bound to soil particles that it can no

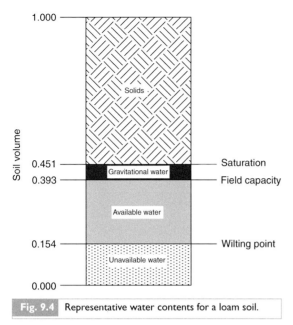

Fig. 9.4 Representative water contents for a loam soil.

point. For wet soil, gravity is the dominant force causing water to move.

The second force, called the matric potential, occurs because water is attracted to the solid surfaces of mineral and organic particles. This attraction creates a negative pressure, or suction, that binds water to the soil matrix. Matric potential depends on water content. For saturated soil, relatively weak matric forces are exerted on water. As the soil dries, its matric potential decreases and strong pressures bind water to the soil matrix (i.e., suction increases). Water movement is more difficult because of this binding force. This can be described mathematically (Clapp and Hornberger 1978):

$$\psi = \psi_{sat} \left(\theta / \theta_{sat} \right)^{-b} \tag{9.6}$$

where ψ is matric potential, ψ_{sat} is matric potential at saturation, θ is volumetric water content, θ_{sat} is water content at saturation (porosity), and b is an empirical parameter. Soils differ in hydraulic properties and matric potential (Table 9.2). The same amount of water is held more strongly in clay than in sand (Figure 9.5). van Genuchten (1980) provides another commonly used soil water retention function.

Field capacity typically occurs at a suction of 1000 mm. Because water is held more tightly by clay than by sand, field capacity occurs at higher water content for clay than for sand (Figure 9.5). The water content at which wilting occurs depends on plant physiology, but it generally occurs at a suction of 150,000 mm (Figure 9.5). Soil water contents at saturation, field capacity, and wilting point vary with texture (Table 9.2). At wilting point, clay holds more water than loam or sand and water fills more of the pore space. Sand and clay have low available water, and loam has high available water.

The sum of gravitational and matric potential governs water movement, and water flows from high to low potential. Except when wet, the adsorptive force binding water to the soil matrix exceeds the gravitational force pulling water downward. For example, consider a column of loam one meter deep. The difference in gravitational potential between water at the top of the column and water at the bottom is

longer be removed. This water content is known as the wilting point because a plant is likely to lose turgor and wilt if the soil is not replenished with water. The soil consists primarily of air and solid particles, and the pores are relatively devoid of water. At wilting point, 15.4 percent of the soil volume (more than one-third of the pore space) is water, but this water is unavailable to plants because it is tightly bound to the soil particles. The difference between the water content at field capacity and the water content at wilting point is the water available to support plant growth. The available water capacity is 23.9 percent, meaning that water can be extracted from only about one-quarter the volume of a loam soil. One meter of loam soil, therefore, only has at most 23.9 cm of available water. When dry, it still contains 15.4 cm of water, but plants cannot extract this water.

Two forces, or potentials, govern water movement in soil. Gravitational potential represents water movement due to the force of gravity. Gravitational potential is the elevation above some arbitrary reference height. A point one meter higher than another point has a gravitational potential that is 1000 mm greater, and because water flows from high potential to low potential water flows downhill to the lower

Table 9.2 Hydraulic properties in relation to soil texture

Soil texture	Porosity, θ_{sat} (fraction)	Percentage of saturation			Hydraulic conductivity at saturation, K_{sat} (mm hr^{-1})	Matric potential at saturation, ψ_{sat} (mm)	Exponent b
		Field capacity	Wilting point	Available water			
Sand	0.395	59	17	42	634	−121	4.05
Sandy loam	0.435	73	26	47	125	−218	4.90
Loam	0.451	87	34	53	25	−478	5.39
Clay loam	0.476	95	52	43	9	−630	8.52
Clay	0.482	92	59	33	5	−405	11.4

Note: See also Cosby et al. (1984) for relationships with sand and clay content.
Source: From Clapp and Hornberger (1978).

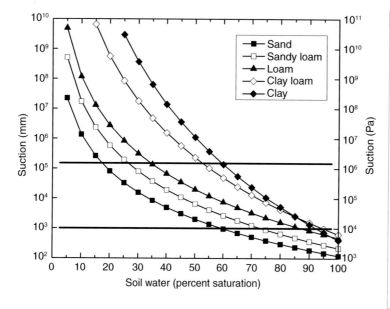

Fig. 9.5 Adsorptive forces binding water to soil particles as a function of water content for sand, sandy loam, loam, clay loam, and clay soils. Suction is a negative pressure (i.e., suction is the negative of matric potential). Suction and matric potential can be expressed in units of pressure (Pa = kg m^{-1} s^{-2}) or height of a column of water under that pressure (mm). These units are related by the density of water (ρ_w, 1000 kg m^{-3}) and the gravitational constant (g, 9.8 m s^{-2}) as 1 mm = (0.001 m) × $\rho_w g$ = 9.8 Pa. Field capacity generally occurs at a suction of 1000 mm. Wilting point typically occurs at a suction of 150,000 mm. Data from Clapp and Hornberger (1978). See also van Genuchten (1980) for another soil water retention function.

1000 mm. When the soil is more than 85 percent saturated, the force binding water to the soil matrix (Figure 9.5) is less than the gravitational force and the soil readily drains. At a water content less than about 85 percent of saturation, however, the matric suction exceeds the gravitation potential.

Water flow is governed by Darcy's law:

$$F = -K\left[\frac{\partial(\psi+z)}{\partial z}\right] \quad (9.7)$$

where F is water flux (mm s^{-1}), K is hydraulic conductivity (mm s^{-1}), ψ is matric potential (mm), and z is gravitational potential (mm), which is defined as the height above some arbitrary height. The term $(\psi+z)$ is the total potential, and the term $\partial(\psi+z)/\partial z$ is the change in total potential with depth, approximated numerically by $\Delta(\psi+z)/\Delta z$. The negative sign ensures that downward water flow is negative and upward water flow is positive. Darcy's law is analogous to soil heat transfer. Vertical water

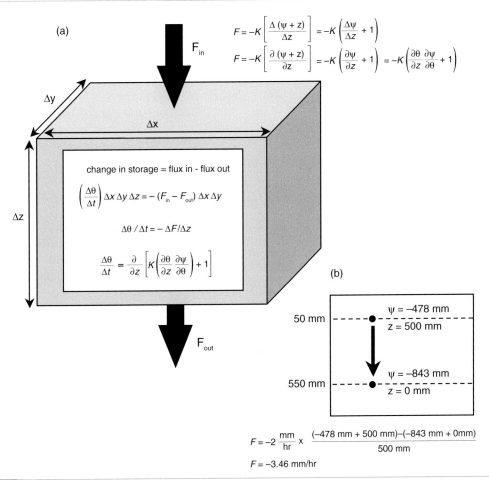

(a)

F_{in}

$$F = -K \left[\frac{\Delta (\psi + z)}{\Delta z} \right] = -K \left(\frac{\Delta \psi}{\Delta z} + 1 \right)$$

$$F = -K \left[\frac{\partial (\psi + z)}{\partial z} \right] = -K \left(\frac{\partial \psi}{\partial z} + 1 \right) = -K \left(\frac{\partial \theta}{\partial z} \frac{\partial \psi}{\partial \theta} + 1 \right)$$

Δy

Δx

Δz

change in storage = flux in - flux out

$$\left(\frac{\Delta \theta}{\Delta t} \right) \Delta x \, \Delta y \, \Delta z = -(F_{in} - F_{out}) \, \Delta x \, \Delta y$$

$$\Delta \theta / \Delta t = -\Delta F / \Delta z$$

$$\frac{\Delta \theta}{\Delta t} = \frac{\partial}{\partial z} \left[K \left(\frac{\partial \theta}{\partial z} \frac{\partial \psi}{\partial \theta} \right) + 1 \right]$$

F_{out}

(b)

$\psi = -478$ mm

50 mm — — — —

$z = 500$ mm

$\psi = -843$ mm

550 mm — — — —

$z = 0$ mm

$$F = -2 \frac{mm}{hr} \times \frac{(-478 \text{ mm} + 500 \text{ mm}) - (-843 \text{ mm} + 0 \text{mm})}{500 \text{ mm}}$$

$$F = -3.46 \text{ mm/hr}$$

Fig. 9.6 Water flow in soil. (a) Water balance of a volume of soil. Equations are shown in both their numerical finite difference form and in the notation of calculus. (b) Example of vertical water flow between two points with a matric potential difference of 365 mm separated by a distance of 500 mm with a hydraulic conductivity of 2 mm hr^{-1}. Hanks (1992) and Hillel (1998) review unsaturated soil water flow.

flow is related to a hydraulic conductivity times a potential gradient, flowing from high to low potential.

The change in soil water over time is given by combining Darcy's law with principles of water conservation (Figure 9.6). The change in water storage in a volume of soil equals the difference between the flux of water into the volume and the flux of water out of the volume. The volume of water in a cube with the dimensions $\Delta x \Delta y \Delta z$ is $\theta \Delta x \Delta y \Delta z$. At the top of the cube, F_{in} mm s^{-1} of water flows into the soil across a cross-sectional area $\Delta x \Delta y$ mm^2. Likewise, $F_{out} \Delta x \Delta y$ mm^3 s^{-1} of water flows out at the bottom. Combining

equations for water flux, given by Eq. (9.7), and change in storage gives the water balance $\Delta \theta / \Delta t = -\Delta F / \Delta z$, or in the notation of calculus:

$$\frac{\partial \theta}{\partial t} = \frac{\partial}{\partial z} \left[K \left(\frac{\partial \theta}{\partial z} \frac{\partial \psi}{\partial \theta} \right) + 1 \right] \qquad (9.8)$$

This equation, known as the Richards equation, relates the change in water content over time to hydraulic conductivity and matric potential. The left-hand side of the equation is the rate of change of water in a volume of soil and the right-hand side is the difference between inflow and outflow rates, each expressed on a per unit volume basis. The term $\partial \psi / \partial z = (\partial \theta / \partial z)(\partial \psi / \partial \theta)$

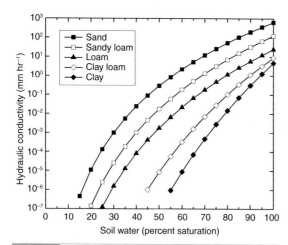

Fig. 9.7 Hydraulic conductivity in relation to soil water for sand, sandy loam, loam, clay loam, and clay soils. Data from Clapp and Hornberger (1978). See also van Genuchten (1980) for another function.

arises because water flux is given in terms of matric potential (ψ) whereas water storage is in terms of volumetric water content (θ). The relationship between ψ and θ provides common terms, but makes the Richards equation difficult to solve numerically because it is non-linear (Figure 9.5).

An additional complication is that hydraulic conductivity is a non-linear function of soil water. Hydraulic conductivity decreases as soil dries, described mathematically (Clapp and Hornberger 1978):

$$K = K_{sat}\left(\theta / \theta_{sat}\right)^{2b+3} \tag{9.9}$$

where K_{sat} is hydraulic conductivity at saturation (Figure 9.7). When saturated, a loam has a hydraulic conductivity of 25 mm hr^{-1}. At a water content of 90 percent of saturation, the hydraulic conductivity is only 6 mm hr^{-1}, and at 80 percent of saturation it is only 1 mm hr^{-1}. As soil dries, not only is water held more tightly to the solid particles (i.e., suction increases and matric potential becomes more negative), but the pore space filled with water becomes smaller and discontinuous. As a result, movement of water is hindered and hydraulic conductivity decreases. In general, sand has the highest hydraulic conductivity for a given water content, loam has

moderate hydraulic conductivity, and clay has extremely low conductivity.

Table 9.2 compares hydraulic properties for several soil textures. In addition to total pore space, physical properties of soil such as hydraulic conductivity and the soil moisture retention curve determine how fast water drains in a soil, how much water it can hold at field capacity, and how much water is available to plants. Total pore space, expressed on a volumetric basis, ranges from 40 percent in sand to 48 percent in clay. Although porosity is low in sand, the individual pores are large because of the large mineral particles. Water loosely adheres to mineral particles (high matric potential when saturated), water movement is rapid (high saturated hydraulic conductivity), and sand drains rapidly. As a result, sands have low water content at field capacity. On the other hand, fine-textured soil such as clay has high porosity. Although the pores are smaller than those of sand, clay has many more pores. Strong adsorptive forces bind water to the soil matrix (low matric potential when saturated), movement is slow (low saturated hydraulic conductivity), and drainage is impeded (high field capacity). Loams are intermediate, draining more slowly than sands and retaining more water.

Infiltration is the vertical flow of water into soil. The maximum amount of water that can infiltrate is the infiltration capacity. This depends on soil wetness at the start of infiltration and the length of time. In general, infiltration rates are high initially and decrease as the soil becomes wet. Many methods are available to estimate infiltration. The Green–Ampt equation uses the notion of a horizontal wetting front that moves downward in the soil (Hillel 1998). From Darcy's law, the infiltration rate (i, mm s^{-1}) is:

$$i = -K_{sat}\left(\frac{0 - \psi_{sat}}{z_{sat}} + 1\right) = -K_{sat}\left(\frac{-\psi_{sat} + z_{sat}}{z_{sat}}\right) \tag{9.10}$$

where z_{sat} is the depth (mm) of the wetting front. The surface is assumed to have a matric potential of zero, and the wetting front is saturated with a matric potential equal to ψ_{sat}. The total cumulative infiltrated water (I, mm) is given by the depth of the wetting front multiplied by the difference between saturated water content (θ_{sat}) and initial water content (θ_i):

$$I = z_{sat}(\theta_{sat} - \theta_i) \tag{9.11}$$

The infiltration rate is the negative of the time derivative of I (the negative occurs because a negative water flux means downward motion):

$$i = -\frac{dI}{dt} = -(\theta_{sat} - \theta_i)\frac{dz_{sat}}{dt} \tag{9.12}$$

Combining Eq. (9.10) and Eq. (9.12) gives an equation for z_{sat} with respect to time:

$$(\theta_{sat} - \theta_i)\frac{dz_{sat}}{dt} = K_{sat}\left(\frac{-\psi_{sat} + z_{sat}}{z_{sat}}\right) \tag{9.13}$$

Solution of this equation, substituting Eq. (9.11) for z_{sat}, provides the time (t, seconds) required for a given cumulative infiltration (I):

$$t = \frac{I}{K_{sat}} + \frac{\psi_{sat}(\theta_{sat} - \theta_i)}{K_{sat}}\ln\left[1 + \frac{I}{-\psi_{sat}(\theta_{sat} - \theta_i)}\right] \tag{9.14}$$

Equation (9.14) relates cumulative infiltration (I) to time (t) given the soil properties K_{sat}, ψ_{sat}, and θ_{sat} (Table 9.2) and initial water content (θ_i). The infiltration rate (i) at time t can be found from Eq. (9.10) and Eq. (9.11):

$$i = -K_{sat}\left[\frac{-\psi_{sat}(\theta_{sat} - \theta_i)}{I} + 1\right] \tag{9.15}$$

For example, if K_{sat} = 0.007 mm s^{-1}, ψ_{sat} = –478 mm, θ_{sat} = 0.451, and θ_i = 0.392 (values

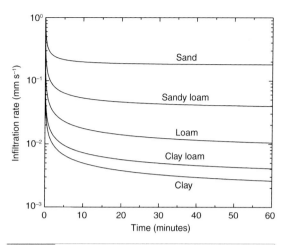

Fig. 9.8 Infiltration over a 60-minute period for sand, sandy loam, clay loam, and clay soils at field capacity using the Green–Ampt equation with hydraulic properties from Table 9.2.

representative of a loam soil at field capacity), the time required for I = 25 mm of water to infiltrate into the soil is about 17 minutes. At this time, the depth of the wetting front is z_{sat} = 424 mm, meaning the top 424 mm of soil is saturated. Calculation of infiltration at different times shows a high initial infiltration rate that decreases towards saturated hydraulic conductivity over time (Figure 9.8). Sand has the greatest infiltration capacity, loam an intermediate value, and clay has the lowest infiltration capacity. The same soils have higher initial infiltration rates and accumulate more water over one hour when drier.

9.5 | Review Questions

1. On a cold winter day, which floor feels colder when walking on it barefoot – a wood floor with a thermal conductivity of 0.15 W m^{-1} K^{-1} or a tile floor with a thermal conductivity of 2 W m^{-1} K^{-1}?

2. The heat flux measured in soil at depth 50 mm is –150 W m^{-2}. The heat flux is –120 W m^{-2} at depth 100 mm. Calculate the time rate of change of temperature ($\Delta T / \Delta t$). The heat capacity is ρc = 2.9 MJ m^{-3} K^{-1}.

3. Calculate the heat flux in Figure 9.3b using the thermal conductivity of sand (2.0 W m^{-1} K^{-1}), clay (1.4 W m^{-1} K^{-1}), peat (0.5 W m^{-1} K^{-1}), and snow (0.34 W m^{-1} K^{-1}). How does the heat flux in organic soil and snow compare with that of mineral soil? Which material is likely to warm more?

4. Geothermal heating takes advantage of soil heat storage to heat and cool a house. A heat pump placed at a specified depth in soil transfers heat from the warm soil in the winter and conversely cools a building in summer. If the annual mean temperature is 11°C and the temperature range is 24°C, find the depth at which soil temperature does not drop below freezing throughout the year. Use $\alpha = 7 \times 10^{-7}$ m² s⁻¹. At what depth does soil temperature not vary by more than ±1°C from the annual mean?

5. Explain why the temperature in a stone cathedral is less than the outside air temperature on a hot summer day.

6. A volume of soil 10 cm wide × 10 cm long × 10 cm deep has a mass of 1.5 kg when wet. The soil mass is 1.25 kg when dry. Calculate the volumetric water content.

7. Find the available water holding capacity of sandy clay loam with $\psi_{sat} = -299$ mm, $\theta_{sat} = 0.42$, and $b = 7.12$. How much water is available for evapotranspiration in a sandy clay loam soil that is 100 cm deep?

8. The flux of water into a volume of soil 50 mm deep is –2.5 mm hr⁻¹. The flux of water out of the soil column at the bottom is –1.0 mm hr⁻¹. Calculate the time rate of change in water content ($\Delta\theta / \Delta t$).

9. In the example problem given in Figure 9.6, calculate the water flux if the flow is horizontal rather than vertical. Both points are at the same height and are separated by a horizontal distance of 500 mm.

10. A soil 50 cm deep has a volumetric water content of 0.17. What is the depth of water that must be applied to the soil to raise the water content to 0.33?

11. A soil has an initial volumetric water content of 0.12 and water content at saturation of 0.42. How deep will 25 mm of rainfall saturate the soil?

12. From the Green–Ampt equation, what is the infiltration rate at infinity?

13. A column of soil is saturated with water. Use the data in Table 9.2 to calculate the initial rate of drainage from the bottom of the column for (a) sandy loam, (b) loam, and (c) clay loam. Which soil has the highest initial drainage rate?

9.6 | References

Clapp, R. B., and Hornberger, G. M. (1978). Empirical equations for some soil hydraulic properties. *Water Resources Research*, 14, 601–604.

Cosby, B. J., Hornberger, G. M., Clapp, R. B., and Ginn, T. R. (1984). A statistical exploration of the relationships of soil moisture characteristics to the physical properties of soils. *Water Resources Research*, 20, 682–690.

Farouki, O. T. (1981). The thermal properties of soils in cold regions. *Cold Regions Science and Technology*, 5, 67–75.

Hanks, R. J. (1992). *Applied Soil Physics*, 2nd ed. New York: Springer-Verlag.

Hartmann, D. L. (1994). *Global Physical Climatology*. San Diego: Academic Press.

Hillel, D. (1998). *Environmental Soil Physics*. San Diego: Academic Press.

Lunardini, V. J. (1981). *Heat Transfer in Cold Climates*. New York: Van Nostrand Reinhold.

Monteith, J. L., and Unsworth, M. H. (2013). *Principles of Environmental Physics*, 4th ed. Amsterdam: Elsevier.

van Genuchten, M. T. (1980). A closed-form equation for predicting the hydraulic conductivity of unsaturated soils. *Soil Science Society of America Journal*, 44, 892–898.

10

Water Balance

Chapter Summary

This chapter introduces the hydrologic cycle on land. The overall hydrologic cycle is reviewed and simplified into an expression where the change in soil moisture is the balance between precipitation input and losses from evapotranspiration and runoff. The various terms in this equation are discussed. Some precipitation is intercepted by the plant canopy and evaporates back to the atmosphere. Evaporation is the physical process by which liquid water changes to vapor. Transpiration is the evaporation of water held internal to plants as water moves from the soil through plants into the atmosphere along a continuum of decreasing water potential. The basic meteorological and biological processes controlling evapotranspiration are introduced, but Chapters 12 and 13 provide a more in-depth discussion of meteorological processes, and Chapters 15–17 review the biological control of transpiration by leaves and plant canopies. Runoff is discussed in separate sections on infiltration and overland flow. The section on overland flow reviews the effects of soil texture, soil water, and land cover on runoff. A simple model of the water balance illustrates geographic patterns of water availability.

10.2 Cycling of Water on Land

The cycling of water over land depicted in Figure 10.1 involves much more detail than that shown in Figure 3.4. Solar energy drives the hydrologic cycle, evaporating water from soil, lakes, and rivers. When conditions are favorable, this water vapor condenses to form clouds and eventually *precipitation*, which replenishes soil water and renews the cycle. Precipitation occurs when air cools and water vapor condenses to form cloud droplets or ice crystals. Cooling is caused by air rising in altitude, such as when air is lifted over mountains. This type of precipitation is known as orographic precipitation. Precipitation also occurs through frontal convergence, when a warm air mass rises over a colder air mass. In summer, thunderstorms develop when strong solar radiation heats the ground, warming the surface air and causing the less dense air to rise. This type of rainfall is known as convective precipitation.

Not all of the precipitation reaches the ground. Leaves, twigs, and branches of plants intercept rain and snow. *Interception* is the process by which precipitation is temporarily stored on plant surfaces. This water quickly evaporates and never replenishes the soil. The water that is not intercepted falls to the ground as throughfall or stemflow. *Throughfall* is water reaching the ground directly through openings in the plant canopy or by dripping down from leaves, twigs, and branches. *Stemflow* is water that reaches the ground by flowing down plant stems and tree trunks.

In many regions, a significant portion of winter precipitation falls as *snow*. If temperatures are cold enough, this water accumulates and is stored for periods of hours, days, or months

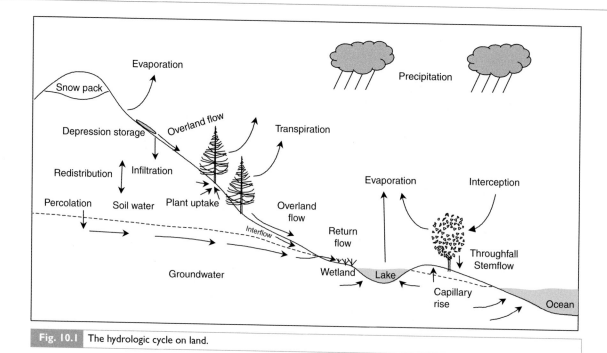

The hydrologic cycle on land.

before melting. Winter storage of precipitation in snow and subsequent spring snowmelt is a large source of water in seasonally-cold climates.

Liquid water reaches the ground as rainfall where vegetation is absent, as throughfall and stemflow under vegetation, or from snowmelt. Some of this water infiltrates into the soil. *Infiltration* is the physical process by which water moves into the soil. When the infiltration capacity of soil is exceeded, water collects as puddles in small depressions (*depression storage*). When these are filled, water runs off over the ground surface as overland flow. *Overland flow* is runoff generated when the infiltration capacity of soil is exceeded by the rainfall intensity, resulting first in ponding of water on the soil surface and then flow across the surface. It moves downhill, first in small rills and gullies and then into creeks and streams that feed large rivers.

The water that infiltrates into the soil wets the soil and is stored as soil water. *Soil water* is water held in the unsaturated zone between the soil surface and the water table. Soil water returns to the atmosphere through evaporation from bare ground and transpiration from plants. *Evaporation* is the physical process by which water changes from a liquid to vapor in the air. *Transpiration* is evaporation of water held inside plants.

Within the soil, water is removed during evaporation and by plant roots (*plant uptake*) when plants replenish water lost during transpiration. Water also moves vertically and horizontally due to internal forces determined by how wet or dry the soil is and by gravity. In most cases, gravity is the greatest force, causing water to flow downwards. This movement is known as *redistribution*, or more commonly percolation. If the water movement is deep, the percolating water will *recharge* the groundwater. On very shallow soils underlain with impermeable material (e.g., bedrock), the infiltrating water may move downhill as subsurface interflow. *Interflow* is the lateral movement of water in upper soil layers. For most landscapes, interflow is not thought to be important relative to overland flow.

Groundwater is the subsurface region that is saturated with water. The top is defined by the water table, which separates the saturated and unsaturated zones, and the bottom is defined by an impermeable layer (e.g., bedrock). Water moves horizontally within these aquifers, typically at a rate of about one-half to one meter per day. This lateral water flow recharges rivers, lakes, wetlands, and oceans and provides the base flow to maintain riverflow in the absence of rainfall. When this flowing water reaches the surface, it is known as return flow. *Return flow* is the process by which groundwater re-emerges from the soil in a saturated area and flows downslope as overland flow. Groundwater is typically recharged by water percolating through the soil column. Water also flows upward, wetting the area of soil immediately above the water table (*capillary rise*). As a result, the upper boundary of groundwater (i.e., the water table) fluctuates seasonally as water enters and leaves the aquifer.

Insights to the hydrologic cycle can be gained by reducing the full cycle shown in Figure 10.1 to a more simple form, $\Delta S = P - E - R$. In this water budget, the change in water storage on land (ΔS) is the difference between precipitation input (P), evapotranspiration loss to the atmosphere (E), and runoff to the oceans (R). The next sections examine the terms in this equation in more detail.

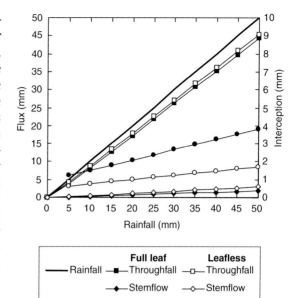

Fig. 10.2 Typical throughfall and stemflow (left axis) and interception (right axis) in relation to rainfall for a deciduous forest in full leaf and without leaves. Data from Zinke (1967). See also Helvey and Patric (1965).

10.3 | Interception and Throughfall

Precipitation falls on foliage, branches, and bark, where it collects and readily evaporates back to the atmosphere. The amount collected depends on rainfall intensity, frequency, and duration. Most plants have a small capacity to store water. Early studies of interception suggested that trees, shrubs, and grasses can store about 1.3 mm of liquid water on their foliage (Zinke 1967). A slightly larger amount of water (3.8 mm) can be stored as snow in trees. This storage capacity varies with the leaf area index (L) of the canopy. Studies find that storage capacity (mm) ranges from $0.15L$ to $0.3L$, and may be as large as $0.5L$ in some forests (Carlyle-Moses and Gash 2011). The interception capacity and amount of water held in the canopy has a significant effect on climate because it readily evaporates (Davies-Barnard et al. 2014).

During brief, moderate storms, this storage capacity may not be exceeded and most of the rainfall is intercepted. In long, intense storms, the storage capacity is quickly exceeded and water drips to the ground. Figure 10.2 shows typical relationships among rainfall amount and throughfall and stemflow. Stemflow is typically minor. Throughfall is the dominant component. For example, a deciduous forest in full leaf may allow 18 mm of rainfall to reach the ground during a 20 mm storm. Of this, 17 mm reaches the ground as throughfall and 1 mm is stemflow. The remaining water (2 mm) is intercepted. Intercepted water readily evaporates because it is held externally on foliage and wood. One study found that in pine forests the rate of evaporation of intercepted water was three times the rate of transpiration under the same radiation conditions (Stewart 1977).

Plant canopies generally intercept about 10–20 percent of annual precipitation (Carlyle-Moses and Gash 2011). Lowland tropical forests intercept, on average, 15 percent of annual precipitation (Krusche et al. 2011). Satellite-based estimates of interception loss range from 13 percent of annual rainfall for broadleaf evergreen forest to 19 percent for broadleaf deciduous forest and 22 percent for needleleaf forests (Miralles et al. 2010). One determinant of interception is vegetative cover; the denser the foliage the greater the amount of water stored. The seasonal emergence and senescence of leaves also matters. Deciduous forests intercept less rainfall when leafless than during the growing season when leaves are present (Figure 10.2). When in full leaf, grasses have the same storage capacity as trees, but because they have no woody material when dormant they intercept considerably less precipitation annually than trees. The type of leaf also determines the amount of interception. Broad-shaped leaves, such as found on deciduous trees, allow water droplets to run together, forming larger drops that readily drip off the leaf. Needle-shaped foliage, such as on coniferous trees, does not facilitate dripping because the separated small needles do not allow water to coalesce into large droplets.

10.4 | Evapotranspiration

Evaporation occurs when a moist surface is exposed to drier air. As air parcels move away from the surface, they carry with them moisture from the surface. Water evaporates from the surface, increasing the amount of water vapor in the surrounding air. When the air is saturated with water vapor, evaporation ceases. Transpiration is evaporation of water from plant leaves as it moves from the soil through plants and out through leaves to the air. Plants consume large amounts of water during growth. A field of corn covering 4000 m² can consume 10,000–15,000 liters of water (2.5–3.75 mm) in a day. A single well-watered tree can transpire 100–150 liters of water per day. Meteorological processes near the surface control evaporation and transpiration. Transpiration is also regulated by the physiology of plants. When plants cover a small portion of the surface, evaporation is the dominant flux. Transpiration becomes more important as plant cover increases. However, it is difficult to distinguish evaporation from transpiration, and the two terms are often combined into evapotranspiration.

Global evapotranspiration from Earth's land surface is about 550 mm of water per year (Jung et al. 2010). Other estimates range from 544 to 631 mm per year (Mueller et al. 2011). Transpiration accounts for 80–90 percent of global evapotranspiration (Jasechko et al. 2013). This water loss varies considerably among various biomes (Figure 10.3). Monthly evapotranspiration in tropical rainforest averages 2.5–3.5 mm per day and has little seasonal variation. Annual water loss is 1100 mm. Other biomes have comparable peak monthly rates, but strong seasonal variation yields much less annual water loss (200–500 mm). Seasonally cold biomes (temperate deciduous forest, boreal forest, and tundra) have a distinct annual cycle with high evapotranspiration during the warm summer months and low rates in the cold winter season. Temperate deciduous forest has peak monthly rates of 2–2.5 mm per day during the growing season. Boreal forest and tundra have lower maximum rates (1.5–2 mm per day) and a shorter growing season. Annual water loss declines from deciduous forest (380 mm) to boreal forest (230 mm) to tundra (170 mm). Ponderosa pine and grassland have distinct annual cycles related to precipitation. Peak rates in ponderosa pine (2.5 mm per day) and grassland (>2 mm per day) are comparable to other sites, but decline markedly in the dry season. Annual water loss is, however, similar to, or greater than, other sites (300 mm, grassland; 480 mm, ponderosa pine).

The rate of evapotranspiration depends on the availability of energy to evaporate water and the ability of water vapor to diffuse into the atmosphere. At 15 °C, 2466 J are needed to change one gram of water to vapor (Table 3.3). Tropical climates, with a surplus of net radiation, have more energy to evaporate water than arctic climates. The capacity of air to remove water from the evaporating surface is also important. This is related to the humidity of the air. Dry air has a greater evaporative demand than humid air. It is also related to wind speed. As water evaporates

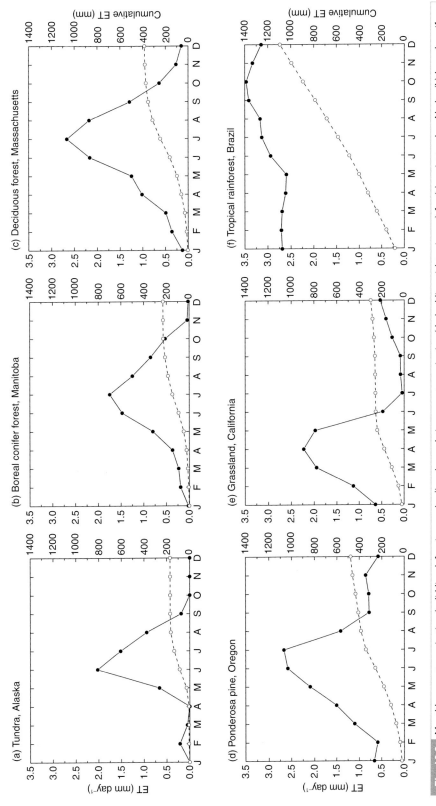

Fig. 10.3 Monthly evapotranspiration (solid line, left axis, mm day⁻¹) and cumulative evapotranspiration (dashed line, right axis, mm) for (a) moist tundra, Alaska, (b) boreal conifer forest, Manitoba, (c) temperate deciduous forest, Massachusetts, (d) ponderosa pine, Oregon, (e) grassland, California, and (f) tropical rainforest, Brazil. See Figure 12.3 for monthly energy fluxes and site details.

from a surface, parcels of air near the surface become more humid. Under calm conditions, evapotranspiration decreases as the air becomes saturated with water vapor. With windy conditions, these parcels of air are carried away and replaced by less humid parcels.

The type of soil and its water content also regulate evapotranspiration. The rate of evaporation is determined by the rate at which water is supplied to the surface. An insufficient rate of soil water flow upward to the evaporating surface decreases the rate at which water can evaporate. The hydraulic properties of soil and the extent of drying determine the rate of replenishment (Chapter 9). A dry soil or a soil with low hydraulic conductivity provides less water for evapotranspiration than does a wet soil or one with high hydraulic conductivity.

The type of vegetation is also important. Leaves have microscopic pores called stomata that open to allow the plant to absorb CO_2 during photosynthesis. The plant cannot grow if stomata are not open, but when stomata are open, water inside the leaf diffuses out to the surrounding drier air during transpiration. If too much water is lost, the plant becomes desiccated and will die if its internal water is not replenished from water in the soil. Plants have evolved compromises between the need to open stomata to take up CO_2 and the need to close them to prevent water loss (Chapter 16).

Water lost from leaves during transpiration must be replenished from the soil. As transpiration increases during the day, water is first drawn from internal plant storage and then from soil near the roots. The movement of water from the soil through plants into the atmosphere occurs along a continuum of decreasing water potential. Water potential is a negative suction. The atmosphere exerts the most suction and has the lowest water potential. When wet, soil particles exert minimal suction on water and have a high water potential. Water flows from soil (on the order of –0.01 MPa when wet) into roots (–0.1 MPa) through the plant and out through foliage (–1 MPa) into the surrounding air (–100 MPa), moving from high to low water potential. At night, plant water uptake replenishes water depleted during the day. By morning,

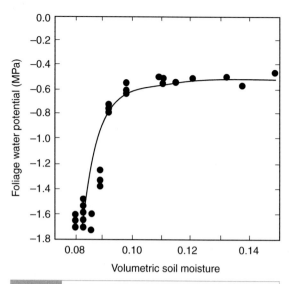

Fig. 10.4 Pre-dawn foliage water potential in red pine trees in relation to volumetric soil water content for loamy sand. Volumetric water content is based on the depth of water in the upper 46 cm of soil. Adapted from Sucoff (1972).

before transpiration begins, water in soil near the roots, water in plant storage, and water in foliage are again nearly equal in potential. The gradient in water potential re-establishes during the day as transpiration increases.

As a plant extracts water from the soil, some critical water content is reached at which further decrease in soil water increases plant stress. In trees, water stress is seen in the pre-dawn water potential of foliage, which is a reliable indicator of soil moisture. Pre-dawn foliage water potential decreases as soil moisture drops below some threshold (Sucoff 1972; Hinckley and Ritchie 1973; Running et al. 1975). In the data shown in Figure 10.4, for example, pre-dawn foliage water potential of red pine trees growing on loamy sand is invariant of soil moisture at high water contents and decreases linearly with volumetric soil water less than about 0.1 m³ m⁻³.

Several classes of models are used to estimate evapotranspiration (Fisher et al. 2011). One method uses air temperature as a surrogate for the energy available to evaporate water. The Thornthwaite equation exemplifies this approach (Thornthwaite 1948). In this formulation, monthly potential evapotranspiration (E_p, mm) is:

$$E_p = 16\left(\frac{L}{12}\right)\left(\frac{N}{30}\right)\left(\frac{10T}{I}\right)^a \qquad (10.1)$$

where L is daylength (hours), N is the number of days in a month, T is mean monthly air temperature (°C), a is defined as:

$$a = 6.75 \times 10^{-7} I^3 - 7.71 \times 10^{-5} I^2 \\ + 1.79 \times 10^{-2} I + 0.49 \qquad (10.2)$$

and I is summed for months with $T > 0°C$ as:

$$I = \sum (T/5)^{1.514} \qquad (10.3)$$

Evapotranspiration from Thornthwaite's method is a potential evapotranspiration because it does not account for the reduction in evapotranspiration as a result of soil drying. However, simple relationships can be used to decrease potential evapotranspiration to the extent that soil water is limiting. One approach is to assume that evapotranspiration proceeds at its potential rate until the soil is depleted of water. However, this ignores the tighter binding of water to soil particles as the soil dries. An alternative is to assume a linear decrease in evapotranspiration as the soil becomes drier, scaled to give the potential rate when the soil is fully wet and zero when the soil is dry.

Another class of models relates evapotranspiration to available energy, given by net radiation. The Priestley–Taylor equation exemplifies this type of model (Priestley and Taylor 1972). This equation relates potential evapotranspiration (E_p, mm day⁻¹) to net radiation (R_n) as:

$$E_p = \alpha \frac{s}{s+\gamma} \frac{R_n}{\lambda} \qquad (10.4)$$

Dividing by the latent heat of vaporization (λ, MJ kg⁻¹) converts R_n from an energy flux (MJ m⁻² day⁻¹) to a mass flux (kg m⁻² day⁻¹), equivalent to a depth of water (mm day⁻¹) because the density of water is 1000 kg m⁻³ (1 kg m⁻²/1000 kg m⁻³ = 0.001 m). In this equation, s (kPa K⁻¹) is the change in saturation vapor pressure with respect to temperature (Table 3.3) and γ is the psychrometric constant (a representative value is 0.0665 kPa K⁻¹, Chapter 12). The coefficient α equals 1.26 for a wet surface, but is lower for

vegetation. For example, α equals 0.82 in tropical and temperate broadleaf forests and equals 0.65 and 0.55 in temperate and boreal conifer forests, respectively (Komatsu 2005; Baldocchi and Ryu 2011).

The Penman equation is a combination equation that includes both available energy and diffusion (Penman 1948). As given by Shuttleworth (1993, 2007), potential evapotranspiration (E_p, mm day⁻¹) over open water is:

$$E_p = \frac{s}{s+\gamma} \frac{R_n}{\lambda} + \frac{\gamma}{s+\gamma} \frac{6.43(1+0.536u)D}{\lambda} \qquad (10.5)$$

This equation is similar to the Priestley–Taylor equation, but additionally includes wind speed (u, m s⁻¹) and vapor pressure deficit (D, kPa). Evapotranspiration is a weighted linear combination of available energy and vapor pressure deficit. Evapotranspiration increases as more energy is available and as the atmospheric demand (i.e., vapor pressure deficit) increases, all other factors being equal. The Penman–Monteith equation (Chapter 12) is an extension of the Penman equation and illustrates the thermodynamic, aerodynamic, and biological processes controlling evapotranspiration. It can be applied to calculate evapotranspiration for a reference crop (Shuttleworth 1993, 2007). The Penman–Monteith equation is also used with satellite remote sensing to estimate continental-scale evapotranspiration (Zhang et al. 2010; Mu et al. 2011; Ryu et al. 2011). Radiation, atmospheric humidity, and wind speed are critical determinants of evapotranspiration, and temperature-based estimates (such as Thornthwaite's method) give different (and poorer) estimates of the water balance than do radiation-based and combination-based methods (Fisher et al. 2009; Sheffield et al. 2012).

10.5 | Runoff

The rate at which water infiltrates depends in part on the rate at which it is supplied to the soil surface. When the rainfall rate is less than the infiltration capacity, all the water infiltrates into the soil. Water delivered in excess of infiltration capacity initially accumulates as puddles in small depressions on the surface. Once this depression storage capacity is exceeded, the excess water flows downhill as overland flow, or surface runoff. The time when this occurs, known as time to ponding, depends on soil texture, antecedent soil water, and delivery rate. High infiltration capacity does not allow ponding on sand or sandy loam except under extremely high precipitation rates. In general, infiltration rate decreases, time to ponding decreases, and runoff increases from sand to loam to clay or with initially wetter soil.

The Green–Ampt equation (Chapter 9), and other such formulations, represents infiltration into idealized soil columns. In addition to micropores arising from the shape, arrangement, and aggregation of mineral and humus particles, soils have macropores formed by plant roots, earthworms, ants, and other burrowing organisms. These macropores can increase infiltration rates. Additionally, soil properties vary spatially, and other methods must be used to account for the effect of soil heterogeneity on infiltration (Chapter 11). Because of its importance to stormflow, several empirical formulas have been devised to determine runoff for application in landscape and urban planning. These equations illustrate the environmental controls of runoff.

One simple means to estimate runoff is the Rational method, which is commonly used in urban planning (Strom and Nathan 1993; Ferguson 1998). Runoff (R, m³ s⁻¹) is:

$$R = 0.278cPA \tag{10.6}$$

where P is rainfall intensity (mm per hour), A is the drainage area (km²), and c is a coefficient that varies with land cover. The factor 0.278 converts units to m³ s⁻¹. The equation states that peak runoff is equal to the fraction of the rainfall that runs off (cP) multiplied by the size of the drainage area (A). The runoff coefficient ranges from zero for a completely pervious surface to one for a completely impervious surface (Table 10.1). Urban landscapes generally generate more runoff than vegetated landscapes. Vegetated landscapes generate less runoff than bare ground.

Table 10.1 Runoff coefficients for use with the Rational method

Land cover	Range
Natural landscapes	
Forest, 0–5% slope	0.10–0.40
Forest, 5–10% slope	0.25–0.50
Forest, 10–30% slope	0.30–0.60
Grass, 0–5% slope	0.10–0.40
Grass, 5–10% slope	0.16–0.55
Grass, 10–30% slope	0.22–0.60
Bare ground, 0–5% slope	0.30–0.60
Bare ground, 5–10% slope	0.40–0.70
Bare ground, 10–30% slope	0.52–0.82
Urban landscapes	
Suburban residential	0.25–0.40
Row houses	0.60–0.75
Industrial	0.50–0.90
Central business district	0.70–0.95

Source: From Strom and Nathan (1993, p. 106).

The United States Soil Conservation Service developed a method for estimating runoff based on soil type, land use, land cover, and antecedent soil moisture (SCS 1985, 1986). This method is also used in urban planning (Strom and Nathan 1993; Ferguson 1998). Runoff (R, mm) is:

$$R = \frac{(P - I_a)^2}{P - I_a + S_{max}} \tag{10.7}$$

where P (mm) is rainfall over a 24-hour interval, I_a (mm) is the initial loss of water to infiltration and in surface depressions before runoff begins (known as the initial abstraction), and S_{max} is the potential maximum retention after runoff begins. This latter term is related to a curve number (CN) that depends on soil type, land use, land cover, and antecedent soil moisture. For fluxes in millimeters:

$$S_{max} = \left(\frac{1000}{CN} - 10\right)25.4 \tag{10.8}$$

The initial abstraction is $I_a = 0.2S_{max}$.

Figure 10.5a illustrates runoff in relation to precipitation for a variety of curve numbers.

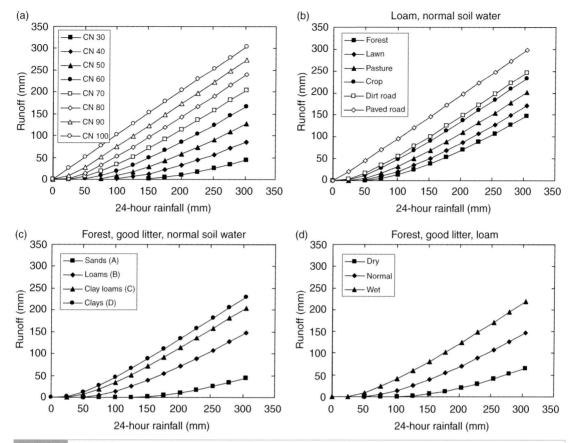

Fig. 10.5 U.S. Soil Conservation Service runoff in relation to precipitation. Curve numbers are given in Table 10.2. (a) Effect of curve number from 30 to 100 on runoff. (b) Effect of land cover for a loam with normal soil water. (c) Effect of soil texture for a forest with good litter and normal soil water. (d) Effect of soil water for a forest with good litter and loam soil.

No surface runoff occurs when the duration of the storm is less than the time required to saturate the soil or if the intensity of rainfall is less than the soil's infiltration capacity. Runoff increases with curve number until $CN = 100$, when there is a one-to-one relationship between precipitation and runoff. The initial detention of precipitation prior to runoff decreases with curve number. For $CN = 30$, runoff does not begin until rainfall exceeds 150 mm. For $CN = 60$, runoff begins with rainfall in excess of 50 mm. Runoff begins almost immediately for $CN \geq 90$.

Table 10.2 shows curve numbers for several land cover types and soils with normal soil water, and Figure 10.5 shows resulting runoff in relation to storm rainfall for a variety of conditions. For a given soil type (e.g., loam), a forest with a good litter cover generates the least runoff. The litter cover retards surface water flow, giving the water additional time to enter the soil. In addition, large, extensive tree roots make the soil more porous, allowing more water to enter the soil. Crops and dirt roads generate high runoff. Paved roads generate the most runoff. Sands have high infiltration rates and low runoff potential. Clays have low infiltration rates and high runoff potential. Loams are intermediate soils, with moderate to low infiltration rates. Curve numbers must be adjusted for antecedent moisture conditions (SCS 1985, pp. 4.10–4.12, p. 10.7). Dry soils have lower curve numbers than wet soils so that dry soils have less runoff than wet soils.

Table 10.2 Soil Conservation Service curve numbers (CN) in relation to soil texture and land cover for normal soil moisture conditions

Land cover	Soil texture			
	Sands (A)	Loams (B)	Clay loams (C)	Clays (D)
Woods, good litter	30	55	70	77
Grass lawn, good condition	39	61	74	80
Pasture, good condition (moderately grazed)	49	69	79	84
Cropland, good condition	67	78	85	89
Dirt road	72	82	87	89
Paved road	98	98	98	98

Source: From SCS (1985, pp. 9.1–9.11) and SCS (1986).

10.6 | Soil Water

In most locations, the water table, which defines the depth at which soil is saturated, is well below the ground surface. Wetlands, where the water is perched above the ground, are an exception. Between the ground surface and the water table is a region where soil is less than saturated. This is known as the unsaturated, or vadose, zone, and water in this region is known as soil water. Plant roots are typically restricted to the unsaturated zone, which supplies plants with necessary moisture.

The unsaturated zone is divided into three zones related to the distribution of water above the water table (Figure 10.6). The first 50–100 cm of soil, where plants typically have most of their roots, is known as the rooting zone. This zone is often saturated during rainfall when water infiltrates into the soil. However, the near-surface soil quickly dries as some of the water drains downward due to the force of gravity and some evaporates to the atmosphere. In addition, plant roots extract water to meet transpiration needs. Consequently, water contents in the root zone range from saturation during infiltration to wilting point in dry periods. Immediately below the root zone is the intermediate zone. This zone is recharged when water in excess of field capacity percolates down the soil column, eventually reaching a zone of saturation bounded by an impermeable layer (e.g., bedrock). Water contents can reach saturation during heavy storms, but mostly the water content is near field capacity. Below the intermediate zone is the groundwater, and immediately above this is a small zone called the capillary fringe, which is kept saturated by water rising from groundwater.

Figure 10.7 illustrates the drainage and wetting of the unsaturated zone over a 16-day period in a Canadian pine forest. On May 30, the overall water content was high, though the near-surface soil was dry. By June 13, the upper soil had dried as a result of evapotranspiration while the deeper soil had dried from drainage. Two days later, heavy rainfall wetted the upper soil. A distinct wetting front at a depth of 50 cm is apparent.

Plant roots extract water from the soil to replenish water lost through transpiration. In some instances, plant roots can also redistribute water within the soil profile. Such activity is known as hydraulic lift or more generally hydraulic redistribution. Through this process, drier upper soil layers are moistened by water from wetter deeper layers. Hydraulic redistribution is widespread among plant species and has been observed in grasses, shrubs, and trees in deserts, temperate forests, and tropical savannas (Richards and Caldwell 1987; Caldwell and Richards 1989; Dawson 1993a, 1996; Burgess et al. 1998; Caldwell et al. 1998; Jackson et al. 2000; Meinzer et al. 2004; Domec et al. 2010; Neumann and Cardon 2012). By keeping

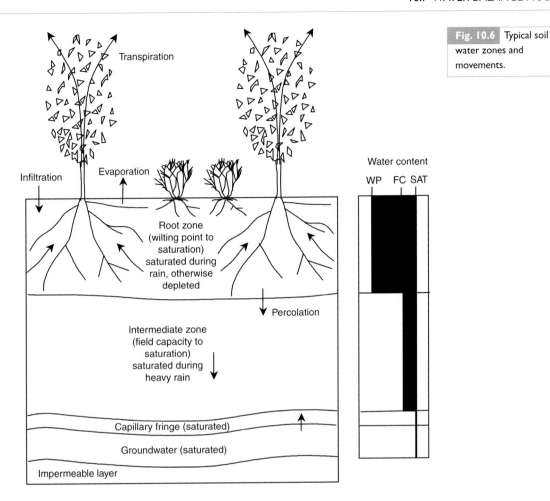

Fig. 10.6 Typical soil water zones and movements.

Transpiration

Infiltration

Evaporation

Root zone
(wilting point to
saturation)
saturated during
rain, otherwise
depleted

Percolation

Intermediate zone
(field capacity to
saturation)
saturated during
heavy rain

Capillary fringe (saturated)

Groundwater (saturated)

Impermeable layer

Water content

WP   FC SAT

upper soils moist, hydraulic redistribution can enhance water availability and sustain transpiration, with important effects on climate (Lee et al. 2005; Wang 2011).

10.7 Water Balance Model

The complexities of the hydrologic cycle on land can be reduced to a simple form in which the change in soil water (ΔS) is the balance between water input from precipitation (P), water loss from evapotranspiration (E), and water lost as runoff (R). Mathematically, $\Delta S = P - E - R$. Thornthwaite and Mather (1955, 1957) used this equation in a simple bucket model of monthly root zone soil water (Figure 10.8a). The soil is treated as a bucket with a maximum

water-holding capacity. Precipitation fills the bucket, and evapotranspiration depletes the bucket. Any water in excess of the maximum water-holding capacity overflows the bucket and is lost as runoff. Potential evapotranspiration is calculated from Eq. (10.1) and reduced to actual evapotranspiration based on the ratio of soil water to maximum water-holding capacity.

Figure 10.9 illustrates this methodology applied over a 12-month period using the bucket model of Mintz and Walker (1993). From January to May and again from September to December, soil water does not limit evapotranspiration so that actual evapotranspiration equals potential evapotranspiration. In these months, evapotranspiration is less than precipitation and the excess precipitation runs off. In June, July, and August, soil water limits evapotranspiration to

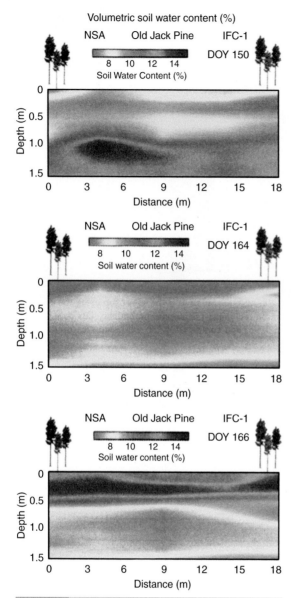

Volumetric soil water content (%)

NSA   Old Jack Pine   IFC-1

DOY 150

8 10 12 14
Soil Water Content (%)

NSA   Old Jack Pine   IFC-1

DOY 164

8 10 12 14
Soil water content (%)

NSA   Old Jack Pine   IFC-1

DOY 166

8 10 12 14
Soil water content (%)

Fig. 10.7 Soil water content (%) with depth across an 18 m transect in a Canadian jack pine forest in late spring 1994. Top: moist conditions on May 30. Middle: dry down on June 13. Bottom: wetting front on June 15. Reproduced from Cuenca et al. (1997). See color plate section.

less than the potential rate. There is no runoff in these months, because water loss from evapotranspiration balances water input from precipitation.

Figure 10.10 shows annual $P - E$ for the United States calculated by this method. The eastern half of the United States, except southern Florida, has a large surplus of water. The Pacific Northwest, where annual precipitation is high, and parts of the mountainous West, where cold temperatures reduce evapotranspiration, also have large water surplus. The least surplus water occurs in the hot, arid Southwest.

Other water balance models follow the same principles, but represent more physical processes. For example, the monthly water balance model of McCabe and Wolock (2011a,b) additionally temporarily stores water as snow (Figure 10.8b). Precipitation is partitioned as rain or snow based on temperature. This snowfall accumulates and melts at a rate determined by temperature. Runoff is generated from infiltration-excess overland flow and additionally from surplus soil water. Monthly temperature and precipitation are climate inputs to the model. Seven site-specific inputs are: daylength (used in potential evapotranspiration); rain and snow threshold temperatures; the maximum snow melt rate; the fraction of rainfall that becomes direct runoff; the fraction of surplus water that becomes runoff; and soil water storage capacity.

10.8 | Isotope Hydrology

Isotopes are variants of a particular element that differ in their number of neutrons. Stable isotopes do not undergo radioactive decay and are commonly used in hydrological and biogeochemical analyses. Carbon, for example, has three naturally occurring isotopes: two stable isotopes (^{12}C and ^{13}C) and one radioactive isotope (^{14}C). Hydrogen has two stable isotopes, ^{1}H and ^{2}H (commonly called deuterium and abbreviated by the symbol D). Two stable isotopes of oxygen used in geochemical studies are ^{16}O and ^{18}O. The lighter isotopes are much more common than the heavier isotopes; ^{12}C, ^{1}H, and ^{16}O have natural abundances of 98.89 percent, 99.98 percent, and 99.76 percent, respectively. The heavier isotopes (^{13}C, ^{2}H, and ^{18}O) are much less abundant.

The isotopic composition of a sample (given by the ratio of the heavy isotope to the light

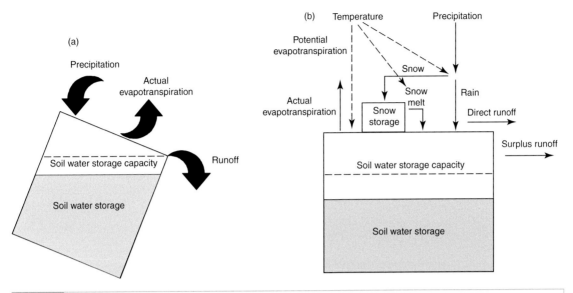

Fig. 10.8 Schematic representation of (a) a bucket model with the water balance $\Delta S = P - E - R$ (Thornthwaite and Mather 1955, 1957) and (b) a more detailed water balance model (McCabe and Wolock 2011a).

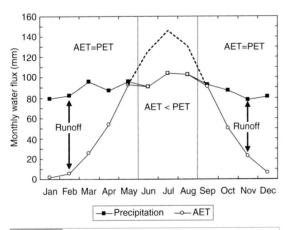

Fig. 10.9 Monthly water balance. Runoff is the difference between precipitation and actual evapotranspiration (AET). The dashed line shows potential evapotranspiration (PET). In other months, AET = PET.

isotope) is measured relative to a standard. For ^{18}O:

$$\delta^{18}O = \left[\frac{\left(\frac{^{18}O}{^{16}O}\right)_{sample}}{\left(\frac{^{18}O}{^{16}O}\right)_{standard}} - 1 \right] 1000 \qquad (10.9)$$

and is reported as per mil (‰). The standard, by definition, has $\delta = 0$‰. A negative value indicates a ratio of heavy-to-light isotope that is less in the sample compared with the standard. The sample is said to be lighter or depleted relative to the standard. A positive value indicates a ratio of heavy-to-light isotope that is greater in the sample compared with the standard. The sample is said to be heavier or enriched. The δ^2H is based on the ratio $^2H/^1H$, and the $\delta^{13}C$ is based on $^{13}C/^{12}C$.

The isotopic composition of a product can differ from the source material. The process by which this occurs is termed isotope fractionation. In general, lighter isotopes are favored in evaporation and photosynthesis, leaving the source material heavier. This isotopic fractionation, or natural variation in the isotopic composition of substances, provides mechanistic understanding of geochemical cycles. For example, three isotopic types of water are $^1H_2^{16}O$ (>99%) and the heavier water $^1H^2H^{16}O$ (commonly abbreviated HDO) and $^1H_2^{18}O$. Isotopic fractionation occurs naturally through evaporation and condensation and imparts a discernible signature to the hydrologic cycle (Figure 10.11). Lighter isotopes evaporate more easily than heavier isotopes and

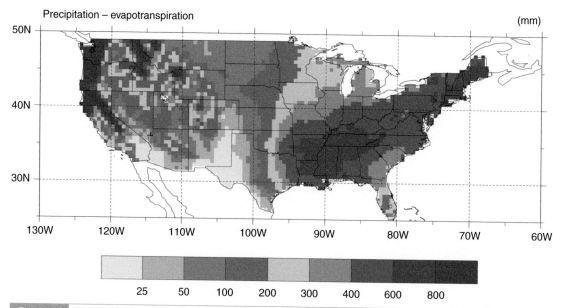

Precipitation – evapotranspiration

(mm)

25 50 100 200 300 400 600 800

Fig. 10.10 Geographic distribution of the difference between annual precipitation and evapotranspiration in the United States. Evapotranspiration is based on the water balance model of Figure 10.9 using monthly temperature and precipitation climatologies (Legates and Willmott 1990a,b) and observed soil water-holding capacity (Rosenbloom and Kittel 1996). See color plate section.

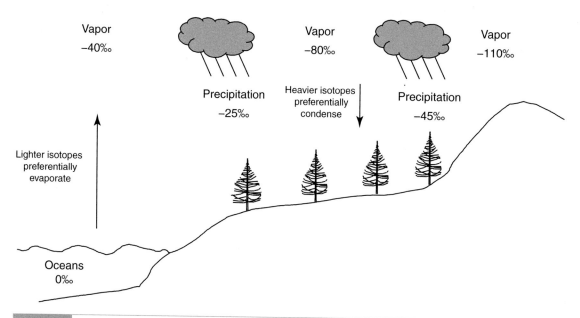

Fig. 10.11 Generalized representation of isotopes in the hydrologic cycle. Shown are representative values for δ^2H with evaporation from tropical oceans and subsequent precipitation over land. Adapted from Dawson (1993b).

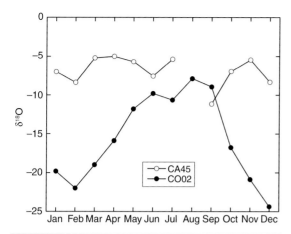

Average monthly $\delta^{18}O$ of precipitation for a low elevation site near the California coast (CA45) and an inland site in the Colorado Rocky Mountains (CO02). Data from Vachon et al. (2010).

^{18}O or ^2H (higher $\delta^{18}O$ and δ^2H) compared with water vapor. The air mass itself becomes further depleted in ^{18}O and ^2H as it moves inland. Temperature affects the isotopic composition of precipitation. With increasing temperature, precipitation is enriched in heavier isotopes (^{18}O and ^2H). Thus, precipitation is depleted of heavy isotopes in polar regions (low $\delta^{18}O$ and δ^2H). The effect of temperature and source region on the isotopic signature of precipitation is seen in comparison of $\delta^{18}O$ for coastal California and inland Colorado sites (Figure 10.12). The coastal site has higher $\delta^{18}O$. The inland site is more depleted in ^{18}O (lower $\delta^{18}O$), especially in winter.

The isotopic composition of water can identify the source of atmospheric water (Henderson-Sellers et al. 2004; Noone et al. 2013). Isotopes can additionally discriminate among processes within a plant canopy, especially partitioning evapotranspiration into soil evaporation and transpiration (Yakir and Sternberg 2000; Yepez et al. 2003; Williams et al. 2004; Dawson and Simonin 2011; Jasechko et al. 2013). This is because evaporation enriches soil water in ^{18}O and ^2H, but transpiration does not produce similar isotopic fractionation.

occur preferentially in the vapor phase. Heavier isotopes preferentially condense and occur preferentially in the liquid phase. Thus, water vapor that evaporates from tropical oceans is depleted in ^{18}O or ^2H (low $\delta^{18}O$ and δ^2H). Heavier isotopes condense and precipitate preferentially to lighter isotopes, so that rainfall is enriched in

10.9 | Review Questions

1. Daily precipitation at a particular site is 40 mm; interception is 3 mm; transpiration is 4 mm; soil evaporation is 1 mm; and runoff is 6 mm. How much water is lost in evapotranspiration? How much water infiltrates into the soil? What is the change in soil water?

2. The data in Figure 10.2 show that a deciduous forest in full leaf intercepts 2 mm of rainfall in a 20 mm storm; 3 mm in a 35 mm storm; and 3.8 mm in a 50 mm storm. Why does the percentage of rainfall intercepted decrease with greater rainfall?

3. Calculate monthly potential evapotranspiration (E_p, mm) using Thornthwaite's method for a site with the following monthly temperature (T, °C), precipitation (P, mm), and daylength (L, hours):

	Jan	Feb	Mar	Apr	May	Jun	Jul	Aug	Sep	Oct	Nov	Dec
T	−10.6	−9.4	−3.9	5.0	12.8	18.3	20.6	19.4	15.0	8.3	0.6	−7.2
P	94	81	94	61	79	89	97	86	89	84	86	94
L	9.6	10.6	11.8	13.2	14.3	15.0	14.8	13.8	12.6	11.3	10.0	9.4
Days	31	28	31	30	31	30	31	31	30	31	30	31

4. Does the site in question 3 gain or lose water over a year? During what months might plants become water stressed?

5. A tropical rainforest has annual evapotranspiration equal to 1100 mm. A temperate deciduous forest has annual evapotranspiration equal to 380 mm. Using the Priestley–Taylor equation, what is the average daily net radiation at these sites? Assume λ = 2466 J g^{-1}, s = 110 Pa K^{-1}, and γ = 66.5 Pa K^{-1}.

6. Use the Penman equation to calculate daily potential evapotranspiration for a site with R_n = 10 MJ m^{-2} day^{-1} with a wind speed 2 m s^{-1} and vapor pressure deficit (a) 1000 Pa and (b) 2000 Pa. What is potential evapotranspiration if wind speed increases to 5 m s^{-1}? Use λ, s, and γ as in question 5.

7. 50 mm of rain falls over a 24-hour period. Use the United States Soil Conservation Service method to calculate runoff for a pasture on clay loam, forest on clay, and crop on sand. Use the data in Table 10.2 and assume normal soil moisture. Which site absorbs the most water prior to runoff generation? Which site has the greatest runoff?

8. Based on the United States Soil Conservation Service method, for what soil texture types will a forest generate more runoff than a pasture? Use the data in Table 10.2 and assume normal soil moisture.

9. A bucket of soil water has an initial water depth W = 120 mm and water holding capacity W_{max} = 150 mm. It receives 105 mm of rainfall and potential evapotranspiration is 145 mm. Calculate actual evapotranspiration, runoff, and the change in soil water. Assume soil water decreases potential evapotranspiration linearly in relation to W / W_{max}. What is the runoff if precipitation is 155 mm?

10. A forest receives 1000 mm precipitation in a year and generates 100 mm of runoff. A grassland receives 800 mm rainfall annually and generates 300 mm of runoff. Which site has the higher annual evapotranspiration?

11. Describe how transpiration affects the relative abundance of ^{18}O and ^{2}H of leaf water.

10.10 | References

Baldocchi, D. D., and Ryu, Y. (2011). A synthesis of forest evaporation fluxes – from days to years – as measured with eddy covariance. In *Forest Hydrology and Biogeochemistry: Synthesis of Past Research and Future Directions*, ed. D. F. Levia, D. Carlyle-Moses, and T. Tanaka. Dordrecht: Springer, pp. 101–116.

Burgess, S. S. O., Adams, M. A., Turner, N. C., and Ong, C. K. (1998). The redistribution of soil water by tree root systems. *Oecologia*, 115, 306–311.

Caldwell, M. M., and Richards, J. H. (1989). Hydraulic lift: water efflux from upper roots improves effectiveness of water uptake by deep roots. *Oecologia*, 79, 1–5.

Caldwell, M. M., Dawson, T. E., and Richards, J. H. (1998). Hydraulic lift: consequences of water efflux from the roots of plants. *Oecologia*, 113, 151–161.

Carlyle-Moses, D. E., and Gash, J. H. C. (2011). Rainfall interception loss by forest canopies. In *Forest Hydrology and Biogeochemistry: Synthesis of Past Research and Future Directions*, ed. D. F. Levia, D. Carlyle-Moses, and T. Tanaka. Dordrecht: Springer, pp. 407–423.

Cuenca, R. H., Stangel, D. E., and Kelly, S. F. (1997). Soil water balance in a boreal forest. *Journal of Geophysical Research*, 102D, 29355–29365.

Davies-Barnard, T., Valdes, P. J., Jones, C. D., and Singarayer, J. S. (2014). Sensitivity of a coupled climate model to canopy interception capacity. *Climate Dynamics*, 42, 1715–1732.

Dawson, T. E. (1993a). Hydraulic lift and water use by plants: Implications for water balance, performance and plant-plant interactions. *Oecologia*, 95, 565–574.

Dawson, T. E. (1993b). Water sources of plants as determined from xylem-water isotopic composition: Perspectives on plant competition, distribution, and water relations. In *Stable Isotopes and Plant Carbon–Water Relations*, ed. J. R. Ehleringer, A. E. Hall, and G. D. Farquhar. San Diego: Academic Press, pp. 465–496.

Dawson, T. E. (1996). Determining water use by trees and forests from isotopic, energy balance and transpiration analyses: the roles of tree size and hydraulic lift. *Tree Physiology*, 16, 263–272.

Dawson, T. E., and Simonin, K. A. (2011). The roles of stable isotopes in forest hydrology and biogeochemistry. In *Forest Hydrology and Biogeochemistry: Synthesis of Past Research and Future Directions*, ed. D. F. Levia, D. Carlyle-Moses, and T. Tanaka. Dordrecht: Springer, pp. 137–161.

Domec, J.-C., King, J. S., Noormets, A., et al. (2010). Hydraulic redistribution of soil water by roots affects whole-stand evapotranspiration and net ecosystem carbon exchange. *New Phytologist*, 187, 171–183.

Ferguson, B. K. (1998). *Introduction to Stormwater: Concept, Purpose, Design*. New York: Wiley.

Fisher, J. B., Malhi, Y., Bonal, D., et al. (2009). The land–atmosphere water flux in the tropics. *Global Change Biology*, 15, 2694–2714.

Fisher, J. B., Whittaker, R. J., and Malhi, Y. (2011). ET come home: potential evapotranspiration in geographical ecology. *Global Ecology and Biogeography*, 20, 1–18.

Helvey, J. D., and Patric, J. H. (1965). Canopy and litter interception of rainfall by hardwoods of eastern United States. *Water Resources Research*, 1, 193–206.

Henderson-Sellers, A., McGuffie, K., Noone, D., and Irannejad, P. (2004). Using stable water isotopes to evaluate basin-scale simulations of surface water budgets. *Journal of Hydrometeorology*, 5, 805–822.

Hinckley, T. M., and Ritchie, G. A. (1973). A theoretical model for calculation of xylem sap pressure from climatological data. *American Midland Naturalist*, 90, 56–69.

Jackson, R. B., Sperry, J. S., and Dawson, T. E. (2000). Root water uptake and transport: Using physiological processes in global predictions. *Trends in Plant Science*, 5, 482–488.

Jasechko, S., Sharp, Z. D., Gibson, J. J., et al. (2013). Terrestrial water fluxes dominated by transpiration. *Nature*, 496, 347–351.

Jung, M., Reichstein, M., Ciais, P., et al. (2010). Recent decline in the global land evapotranspiration trend due to limited moisture supply. *Nature*, 467, 951–954.

Komatsu, H. (2005). Forest categorization according to dry-canopy evaporation rates in the growing season: comparison of the Priestley–Taylor coefficient values from various observations sites. *Hydrological Processes*, 19, 3873–3896.

Krusche, A. V., Ballester, M. V. R., and Leite, N. K. (2011). Hydrology and biogeochemistry of terra firme lowland tropical forests. In *Forest Hydrology and Biogeochemistry: Synthesis of Past Research and Future Directions*, ed. D. F. Levia, D. Carlyle-Moses, and T. Tanaka. Dordrecht: Springer, pp. 187–201.

Lee, J.-E., Oliveira, R. S., Dawson, T. E., and Fung, I. (2005). Root functioning modifies seasonal climate. *Proceedings of the National Academy of Sciences USA*, 102, 17576–17581.

Legates, D. R., and Willmott, C. J. (1990a). Mean seasonal and spatial variability in global surface air temperature. *Theoretical and Applied Climatology*, 41, 11–21.

Legates, D. R., and Willmott, C. J. (1990b). Mean seasonal and spatial variability in gauge-corrected, global precipitation. *International Journal of Climatology*, 10, 111–127.

McCabe, G. J., and Wolock, D. M. (2011a). Century-scale variability in global annual runoff examined using a water balance model. *International Journal of Climatology*, 31, 1739–1748.

McCabe, G. J., and Wolock, D. M. (2011b). Independent effects of temperature and precipitation on modeled runoff in the conterminous United States. *Water Resources Research*, 47, W11522, doi:10.1029/2011WR010630.

Meinzer, F. C., Brooks, J. R., Bucci, S., et al. (2004). Converging patterns of uptake and hydraulic redistribution of soil water in contrasting woody vegetation types. *Tree Physiology*, 24, 919–928.

Mintz, Y., and Walker, G. K. (1993). Global fields of soil moisture and land surface evapotranspiration derived from observed precipitation and surface air temperature. *Journal of Applied Meteorology*, 32, 1305–1334.

Miralles, D. G., Gash, J. H., Holmes, T. R. H., de Jeu, R. A. M., and Dolman, A. J. (2010). Global canopy interception from satellite observations. *Journal of Geophysical Research*, 115, D16122, doi:10.1029/2009JD013530.

Mu, Q., Zhao, M., and Running, S. W. (2011). Improvements to a MODIS global terrestrial evapotranspiration algorithm. *Remote Sensing of Environment*, 115, 1781–1800.

Mueller, B., Seneviratne, S. I., Jimenez, C., et al. (2011). Evaluation of global observations-based evapotranspiration datasets and IPCC AR4 simulations. *Geophysical Research Letters*, 38, L06402, doi:10.1029/2010GL046230.

Neumann, R. B., and Cardon, Z. G. (2012). The magnitude of hydraulic redistribution by plant roots: a review and synthesis of empirical and modeling studies. *New Phytologist*, 194, 337–352.

Noone, D., Risi, C., Bailey, A., et al. (2013). Determining water sources in the boundary layer from tall tower profiles of water vapor and surface water isotope ratios after a snowstorm in Colorado. *Atmospheric Chemistry and Physics*, 13, 1607–1623.

Penman, H. L. (1948). Natural evaporation from open water, bare soil and grass. *Proceedings of the Royal Society of London*, 193A, 120–145.

Priestley, C. H. B, and Taylor, R. J. (1972). On the assessment of surface heat flux and evaporation using large-scale parameters. *Monthly Weather Review*, 100, 81–92.

Richards, J. H., and Caldwell, M. M. (1987). Hydraulic lift: Substantial nocturnal water transport between soil layers by *Artemisia tridentata* roots. *Oecologia*, 73, 486–489.

Rosenbloom, N., and Kittel, T. G. F. (1996). *A User's Guide to the VEMAP Phase I Database*, Technical Note NCAR/TN-431+IA. Boulder, Colorado: National Center for Atmospheric Research.

Running, S. W., Waring, R. H., and Rydell, R. A. (1975). Physiological control of water flux in conifers: A computer simulation model. *Oecologia*, 18, 1–16.

Ryu, Y., Baldocchi, D. D., Kobayashi, H., et al. (2011). Integration of MODIS land and atmosphere products with a coupled-process model to estimate gross primary productivity and evapotranspiration from 1 km to global scales. *Global Biogeochemical Cycles*, 25, GB4017, doi:10.1029/2011GB004053.

SCS (1985). *National Engineering Handbook. Section 4: Hydrology*. Washington, DC: Soil Conservation Service, U.S. Department of Agriculture.

SCS (1986). *Urban Hydrology for Small Watersheds*, Technical Release No. 55. Washington, DC: Soil Conservation Service, U.S. Department of Agriculture.

Sheffield, J., Wood, E. F., and Roderick, M. L. (2012). Little change in global drought over the past 60 years. *Nature*, 491, 435–438.

Stewart, J. B. (1977). Evaporation from the wet canopy of a pine forest. *Water Resources Research*, 13, 915–921.

Shuttleworth, W. J. (1993). Evaporation. In *Handbook of Hydrology*, ed. D. R. Maidment. New York: McGraw-Hill, pp. 4.1–4.53.

Shuttleworth, W. J. (2007). Putting the "vap" into evaporation. *Hydrology and Earth System Sciences*, 11, 210–244.

Strom, S., and Nathan, K. (1993). *Site Engineering for Landscape Architects*, 2nd ed. New York: Van Nostrand Reinhold.

Sucoff, E. (1972). Water potential in red pine: Soil moisture, evapotranspiration, crown position. *Ecology*, 53, 681–686.

Thornthwaite, C. W. (1948). An approach toward a rational classification of climate. *Geographical Review*, 38, 55–94.

Thornthwaite, C. W., and Mather, J. R. (1955). *The Water Balance*, Publications in Climatology Volume 8, Number 1. Centerton, New Jersey: Drexel Institute of Technology.

Thornthwaite, C. W., and Mather, J. R. (1957). *Instructions and Tables for Computing Potential Evapotranspiration and the Water Balance*, Publications in Climatology Volume 10, Number 3. Centerton, New Jersey: Drexel Institute of Technology.

Vachon, R. W., Welker, J. M., White, J. W. C., and Vaughn, B. H. (2010). Monthly precipitation isoscapes ($\delta^{18}O$) of the United States: Connections with surface temperatures, moisture source conditions, and air mass trajectories. *Journal of Geophysical Research*, 115, D21126, doi:10.1029/2010JD014105.

Wang, G. (2011). Assessing the potential hydrological impacts of hydraulic redistribution in Amazonia using a numerical modeling approach. *Water Resources Research*, 47, W02528, doi:10.1029/2010WR009601.

Williams, D. G., Cable, W., Hultine, K., et al. (2004). Evapotranspiration components determined by stable isotope, sap flow and eddy covariance techniques. *Agricultural and Forest Meteorology*, 125, 241–258.

Yakir, D., and Sternberg, L. da S. L. (2000). The use of stable isotopes to study ecosystem gas exchange. *Oecologia*, 123, 297–311.

Yepez, E. A., Williams, D. G., Scott, R. L., and Lin, G. (2003). Partitioning overstory and understory evapotranspiration in a semiarid savanna woodland from the isotopic composition of water vapor. *Agricultural and Forest Meteorology*, 119, 53–68.

Zhang, K., Kimball, J. S., Nemani, R. R., and Running, S. W. (2010). A continuous satellite-derived global record of land surface evapotranspiration from 1983 to 2006. *Water Resources Research*, 46, W09522, doi:10.1029/2009WR008800.

Zinke, P. J. (1967). Forest interception studies in the United States. In *Forest Hydrology*, ed. W. E. Sopper and H. W. Lull. Oxford: Pergamon Press, pp. 137–161.

Watershed Hydrology

11.1 Chapter Summary

The flow of water in streams and rivers is a key measure of the hydrologic cycle integrated over large areas. A watershed is the geographic area that contributes to water flow in a stream or river. Building upon concepts introduced in the previous chapter, this chapter introduces the study of watersheds. The overall hydrologic balance of a watershed is discussed, and three cases studies (Hubbard Brook, Coweeta, and Walker Branch) illustrate the hydrologic balance of watersheds. Surface runoff, or overland flow, is generated within a watershed when water reaching the ground exceeds the soil's capacity to gain water during infiltration (infiltration-excess runoff) or when rain falls on saturated areas of the watershed (saturation-excess runoff). The processes that generate runoff are reviewed and illustrated by numerical models of watershed hydrology. The spatial distribution of precipitation, spatial variability in infiltration capacity, antecedent soil moisture, and topography are important determinants of runoff at the watershed scale. Riverflow is an integrator of runoff, and the processes regulating riverflow, especially flooding, are discussed and illustrated. The chapter concludes with a discussion of global drainage basins and observed riverflow for major river systems. Comparison of simulated versus observed riverflow is one means to test the hydrologic cycle of climate models.

11.2 Watersheds

The cycling of water depicted in Figure 10.1 can be applied to particular geographic regions to calculate the water balance. One such area is a watershed or drainage basin. A watershed is the geographic area that contributes to flow in a stream or river. It can be hundreds of thousands of square kilometers for a large river such as the Mississippi or Amazon or a few square kilometers for a small creek. A watershed is topographically defined; it is bounded along its edges by divides formed from high elevation points. A drop of water on the streamward side of the divide flows downslope to the stream; a drop of water on the other side of the divide flows into another stream.

Figure 11.1 depicts a typical watershed, bounded on its sides by elevation, on the bottom by bedrock, and above by the atmosphere. The water balance extending from the divides to groundwater is:

$$\Delta S = P + G_{in} - \left(E + q_{over} + q_{base}\right) \qquad (11.1)$$

where ΔS is the change in water storage, P is precipitation, G_{in} is groundwater flowing into the watershed, E is water loss during evapotranspiration, q_{over} is surface runoff, and q_{base} is subsurface groundwater flow. The total lateral outflow of water is also the total runoff ($R = q_{over} + q_{base}$), and because watersheds are topographically defined by high elevations, groundwater inflow is typically negligible. Consequently, if there is

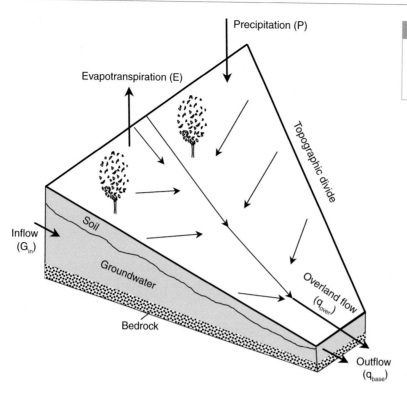

Precipitation (P)

Evapotranspiration (E)

Topographic divide

Inflow
(G_{in})

Soil

Groundwater

Overland flow
(q_{over})

Bedrock

Outflow
(q_{base})

Fig. 11.1 A topographically defined watershed. The thick black line shows the topographic divide. Arrows indicate water fluxes.

no change in water storage, water input from precipitation is balanced by evapotranspiration and runoff ($P - E = R$).

11.3 | Watershed Studies

The term $P - E = R$ averaged over long time periods (e.g., annually) is the amount of water flowing in streams and rivers. Streamflow, therefore, provides a basis for monitoring and diagnosing the hydrologic balance of a watershed. In watershed studies, precipitation is typically measured by a network of rain gauges and spatially averaged across the watershed. Streamflow is monitored by gauging stations. Annual evapotranspiration is estimated as the difference between precipitation and streamflow.

One watershed where the water budget has been studied in detail is the Hubbard Brook Experimental Forest in the White Mountains of New Hampshire. The Hubbard Brook watershed extends over 3076 ha (1 ha = 10,000 m²) and is covered by northern hardwood forest (Likens et al. 1977; Bormann and Likens 1979; Likens

and Bormann 1995; Likens 2004). Between 1956 and 1974, annual precipitation averaged 1300 mm with 500 mm (38%) being returned to the atmosphere in evapotranspiration and 800 mm (62%) leaving the watershed as runoff in streams (Figure 11.2a). Over the period studied, there was considerable interannual variability in annual precipitation, which ranged from 950 to 1860 mm. Evapotranspiration remained relatively constant from year to year, ranging from 418 to 542 mm. Evapotranspiration did not increase in wetter years or decrease in drier years. Annual streamflow increased in wetter years, and there was a linear increase in annual streamflow in response to precipitation. This suggests that precipitation is first used to replenish water lost during evapotranspiration. Any excess water then contributes to streamflow.

Similar relationships occur at the Walker Branch watershed near Oak Ridge, Tennessee. This is a 98 ha watershed with mixed deciduous forest (Johnson and Van Hook 1989). For the 35-year period (1969–2003), the average annual precipitation was 1331 mm (Figure 11.2b).

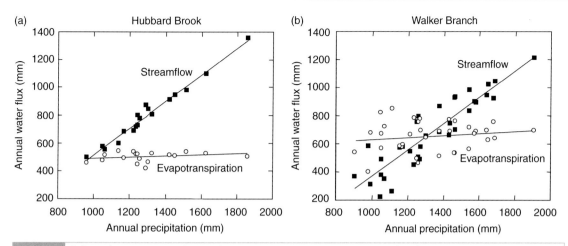

Fig. 11.2 Relationships among annual streamflow, evapotranspiration, and precipitation for (a) the Hubbard Brook Experimental Forest during 1956–1974 and (b) the Walker Branch watershed during 1969–2003. Data from Likens et al. (1977, p. 22) and the Oak Ridge National Laboratory (Oak Ridge, Tennessee).

Table 11.1 Physiographic and hydrologic characteristics of six watersheds at the Coweeta Hydrologic Laboratory

	Watershed					
	2	14	18	34	27	36
Area (ha)	12.3	61.0	12.5	32.7	39.0	48.6
Minimum elevation (m)	709	707	726	852	1061	1021
Maximum elevation (m)	1004	992	993	1184	1455	1542
Annual precipitation (mm)	1772	1876	1939	2009	2451	2222
Annual runoff (mm)	854	988	1034	1175	1737	1675
Annual evapotranspiration (mm)	918	887	905	835	713	547
Runoff ratio	0.48	0.53	0.53	0.58	0.71	0.75

Source: From Swift et al. (1988).

Streamflow for the watershed, monitored for two subcatchments, averaged 679 mm (51% of annual precipitation). Evapotranspiration, taken as the residual precipitation, averaged 652 mm (49%). This estimate of evapotranspiration derived from the catchment water balance is similar in magnitude to eddy covariance estimates (Wilson et al. 2001). The Walker Branch watershed has higher annual evapotranspiration than Hubbard Brook despite similar annual precipitation. The Walker Branch data show considerable more variability in their relationship with precipitation than the Hubbard Brook data. This is due to the large water-holding capacity of Walker Branch soils compared with Hubbard Brook soils (Luxmoore and Huff 1989).

The 2185 ha Coweeta Hydrologic Laboratory in the mountains of southwestern North Carolina illustrates the wide range in the hydrologic cycle than can be found even within a small region (Swank and Crossley 1988). In six subcatchments of mixed deciduous forest that vary in elevation, annual runoff ranges from about 50 percent of annual precipitation at low elevations to 70–75 percent at high elevations (Table 11.1).

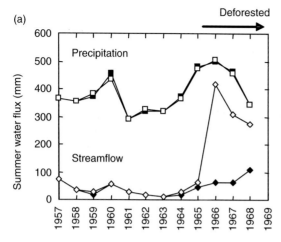

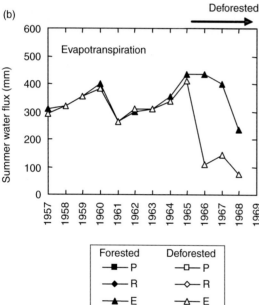

Fig. 11.3 Effect of deforestation on (a) summer precipitation (P) and streamflow (R) and (b) evapotranspiration (E) for two watersheds in the Hubbard Brook Experimental Forest. One watershed was deforested in 1965–66 and vegetation regrowth was suppressed for three years. The other watershed was not deforested. Adapted from Bormann and Likens (1979, p. 85).

In general, annual precipitation increases with higher elevation. Conversely, soil depth decreases with higher elevation so that water-holding capacity is greater at lowland than at upland sites. High elevations sites also have less evapotranspiration demand than lowland sites. Consequently, watersheds 2, 14, and 18 at low elevations, where more of precipitation is stored in deep soils, have a lower proportion of annual precipitation as runoff than do watersheds 27 and 36 at high elevations, where shallow soils limit water storage and promote runoff.

The importance of vegetation in regulating the water balance can be demonstrated by experimentally clearing a watershed. Such studies routinely show decreased evapotranspiration and increased streamflow in deforested watersheds compared with forested watersheds (Bosch and Hewlett 1982; Hornbeck et al. 1993; Zhang et al. 2001; Andréassian 2004; Brown et al. 2005). Comparison of forested and deforested watersheds at Hubbard Brook illustrates this response to clearing (Hornbeck et al. 1970, 1997; Bormann and Likens 1979). Prior to deforestation, both watersheds had similar evapotranspiration and streamflow (Figure 11.3). After clearing in 1965–66, evapotranspiration decreased and streamflow increased in the deforested watershed compared with the forested watershed. Greatest increase in streamflow occurred during the growing season when evapotranspiration decreased due to forest clearing. The Coweeta study shows a similar response to clearing. During a 7-year period in which regrowth was cut annually, streamflow increased compared with mature forest (Figure 11.4a). Clearcutting had minor effect in late winter and early spring when the soil was recharged and greater effect in the growing season. Conversely, afforestation of grassland and shrubland reduces streamflow (Farley et al. 2005; Jackson et al. 2005).

The type of vegetation also influences runoff. Conversion of Coweeta watersheds from mature deciduous forest to young white pine trees reduced annual streamflow by 20 cm (20%) below that of deciduous forest (Swank and Miner 1968; Swank and Douglass 1974). This was a result of greater interception and subsequent evapotranspiration of rainfall by evergreen pines than deciduous trees during the dormant season. The reduction in streamflow occurred every month and was greatest before and during leaf emergence (March, April, May) (Figure 11.4b).

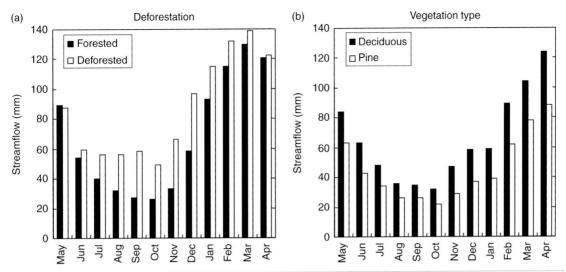

Fig. 11.4 Hydrologic response of (a) Coweeta watershed 17 to clearcutting and (b) Coweeta watershed 1 to conversion from deciduous trees to evergreen white pine. Data for watershed 17 are shown for the period before cutting (forested) and for a 7-year period after cutting (deforested). Data for watershed 1 are shown for mature deciduous forest and a 4-year period after planting pine. Adapted from Swank et al. (1988).

11.4 | Runoff Processes

Surface runoff, or overland flow, is generated within a watershed when the amount of water reaching the ground exceeds the soil's capacity to gain water during infiltration. This type of overland flow is known as infiltration-excess runoff or Horton runoff. The local runoff rate (R) at a particular point in a watershed is the rainfall (P) in excess of infiltration capacity (i):

$$R = \begin{cases} \text{infiltration-excess} \\ P - i & \text{for} \quad P > i \\ 0 & \text{for} \quad P \leq i \\ \text{saturation-excess} \\ P \end{cases} \quad (11.2)$$

The total runoff from a catchment is an integration of local runoff across all points within the watershed.

Infiltration-excess runoff typically occurs on upslope areas of a watershed or on soils with low infiltration capacity. In many temperate forests, soils have a high infiltration capacity and precipitation rarely generates infiltration-excess overland flow. In contrast, this can be the dominant form of runoff in watersheds where the vegetation has been cleared, on construction sites where the soil is compacted, and in urban watersheds where much of the surface is impervious.

Overland flow is also generated when rain falls on saturated areas, and the water immediately runs off. This type of overland flow is called saturation-excess runoff. It is common in wetlands and low-lying areas along stream channels with shallow water tables. The amount of water from saturation-excess runoff equals the precipitation rate, and total runoff from the watershed is the precipitation rate times the area of the watershed that is saturated. Saturation-excess runoff is an important contributor to storm flow in watersheds where infiltration capacity is high and little infiltration-excess runoff is generated.

The saturated areas in a watershed expand in size during a rainstorm and contract during extended dry periods. The area of the watershed that contributes to streamflow through saturation-excess overland flow therefore changes over the course of a storm. Figure 11.5 illustrates this concept of variable contributing area, showing the growth of saturated areas during a rainstorm. In this example, saturated zones are restricted to the immediate stream

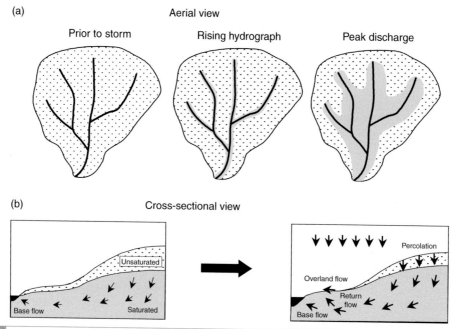

(a)   Aerial view

Prior to storm   Rising hydrograph   Peak discharge

(b)   Cross-sectional view

Fig. 11.5 Growth of saturated areas (shaded) along streams (black) during a rainstorm, shown as (a) aerial view and (b) cross-sectional view. Stippling denotes unsaturated areas.

banks prior to the storm. No runoff occurs, and streamflow is maintained by groundwater discharge. As rainfall occurs, areas along the banks begin to get saturated. All rainfall on the saturated area runs off and the streams begin to rapidly rise. At time of peak discharge, the saturated zones are extensive.

11.5 | Catchment Runoff

The total runoff from a catchment is an integration of a number of runoff generation processes within the watershed. One such factor is the rate of precipitation, which can vary spatially depending on storm characteristics. Figure 11.6 shows the spatial distribution of rainfall over the 154 km² Walnut Gulch watershed in Arizona during two storms. Both storms delivered a similar amount of rainfall, but they differed greatly in their spatial characteristics. The storm on September 8, 1970 was locally concentrated. The average precipitation across the watershed was 6.4 mm, but this rain fell over only 68.5 percent of the watershed with a local precipitation

maximum of 39 mm. Ten percent of the watershed received more than 25 mm of rainfall. In contrast, the storm on August 25, 1972 was more evenly distributed. The spatial average precipitation was 7.3 mm. This rain fell over the entire basin. Ninety percent of the watershed received between 3 and 10 mm of rain. None of the watershed received more than 14 mm of rainfall.

In addition, soil water varies spatially depending on soil texture. Figure 11.7 shows soil water content for the Little Washita watershed near Chickasha, Oklahoma. This watershed drains an area of 610 km² and is covered by pasture rangeland and crops. Heavy rain fell across the watershed on June 5 followed by moderate rainfall until June 9. Thereafter there was no rain for several days. On June 10 volumetric soil water content in the first 5 cm of soil ranged from 0.15 to 0.35 m³ m⁻³. The near-surface soil was close to saturation in the west and east, where the soil is primarily silt loam and loam. Central areas, where the soil is sandy loam and sand, were drier. This general spatial pattern was maintained as the soil drained over the next few days.

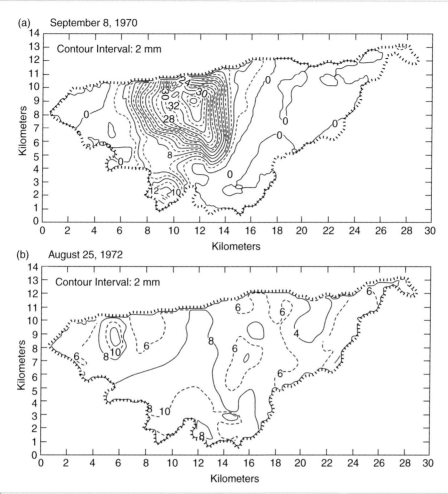

Fig. 11.6 Rainfall distribution across the 154 km² Walnut Gulch watershed near Tucson, Arizona, for (a) September 8, 1970 and (b) August 25, 1972. Redrawn from Fennessey et al. (1986).

11.5.1 Exponentially Distributed Precipitation

The example shown in Table 11.2 illustrates how spatial variation in precipitation affects runoff. In this particular example, the average precipitation is 1.85 mm, but the actual amount received varies with four sub-catchments. For example, 25 percent of the watershed receives only 0.2 mm of rainfall; 10 percent receives 7 mm. The total runoff from the watershed is found by applying Eq. (11.2) to each sub-area, weighted by the appropriate fractional area. In this example, 0.65 mm of runoff is generated by the watershed assuming a spatially invariant infiltration capacity of 2 mm. In contrast, no

runoff is generated by the watershed if Eq. (11.2) is evaluated with the average precipitation. This is because local precipitation exceeds infiltration capacity for 40 percent of the watershed. The remaining 60 percent of the watershed generates no runoff.

One approach commonly used to account for spatial variability in precipitation is to represent the precipitation rate at a point as an exponential probability density function (Shuttleworth 1988; Pitman et al. 1990; Dolman and Gregory 1992; Eltahir and Bras 1993). Rain is assumed to fall over a fraction of the surface (μ) and the remainder $(1 - \mu)$ receives no rainfall. Within the raining area, the local precipitation

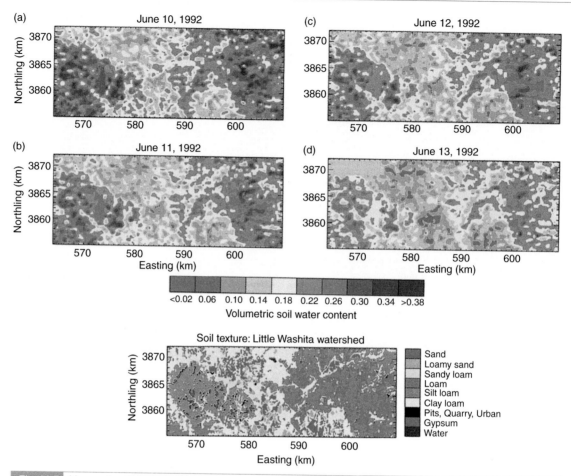

Fig. 11.7 Near-surface (0–5 cm depth) volumetric soil water content for June 10–13, 1992 in the Little Washita watershed, Oklahoma. Data have a spatial resolution of 200 m. Reproduced from Mattikalli et al. (1998). See color plate section.

Table 11.2 Example runoff calculation for a catchment divided into four sub-areas each receiving point precipitation (P) and generating runoff (R) as the precipitation in excess of infiltration capacity (i)

Fractional area	Point precipitation (mm)	Infiltration capacity (mm)	Runoff (mm)
0.25	0.2	2.0	0.0
0.35	1.0	2.0	0.0
0.30	2.5	2.0	0.5
0.10	7.0	2.0	5.0
1.00	1.85	2.0	0.65

rate at a given point in space (P) is exponentially distributed with the probability density function:

$$f(P) = \frac{\mu}{\bar{P}} \exp\left(\frac{-\mu P}{\bar{P}}\right) \qquad (11.3)$$

where $\bar{P}$ is the precipitation rate averaged over the entire surface. The average precipitation rate over the raining area is $\bar{P}/\mu$, and the rain covered portion of the surface receives higher precipitation rates as rainfall is concentrated into

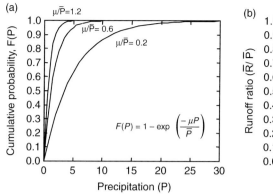
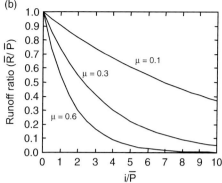

Fig. 11.8 Influence of precipitation variability on runoff. (a) Cumulative probability distribution for $\mu = 0.1, 0.3,$ and 0.6 with $\bar{P} = 0.5$ mm min^{-1}. (b) Runoff ratio ($\bar{R}/\bar{P}$) in relation to $i/\bar{P}$ for $\mu = 0.1, 0.3,$ and 0.6.

a smaller area (i.e., μ decreases). This is evident from the cumulative probability distribution, which gives the probability that local precipitation is less than a particular value. As shown in Figure 11.8a, the occurrence of extreme high local rainfall rates increases as μ decreases, resulting in an increase in the mean and median rates over the area μ that receives rainfall.

Equation (11.3) can be used to scale runoff at a particular point in the watershed to the entire catchment. If infiltration capacity is spatially invariant, the average runoff is obtained by integrating Eq. (11.2) with respect to P (which provides the runoff rate from the rain covered fraction μ) and recognizing that the fractional area $1 - \mu$ receives no rainfall:

$$\bar{R} = \mu \int_i^\infty (P - i) f(P) dP = \bar{P} \exp\left(\frac{-\mu i}{\bar{P}}\right) \quad (11.4)$$

Figure 11.8b illustrates the behavior of Eq. (11.4). The amount of precipitation that becomes runoff ($\bar{R}/\bar{P}$) decreases as infiltration capacity increases relative to precipitation (i.e., as $i/\bar{P}$ increases). For a given $i/\bar{P}$, the runoff ratio increases as the fractional area of precipitation (μ) decreases. That is, locally concentrated rainfall increases runoff.

11.5.2 Variable Infiltration Capacity

Spatial variability in soil water can be represented through heterogeneity in soil water-holding capacity. This approach treats the soil as a bucket

with a specified water-holding capacity. The soil fills with water until this capacity is reached and excess water becomes runoff. Because the water-holding capacity is the maximum amount of water that the soil can store, it can also be thought of as the infiltration capacity of the soil. The water balance of the bucket is:

$$R = \begin{cases} P - (i - W_0) & \text{for } P > i - W_0 \\ 0 & \text{for } P \leq i - W_0 \end{cases} \quad (11.5)$$

where R is runoff, P is precipitation, i is water-holding capacity (i.e., infiltration capacity), and W_0 is soil water, all with units mm of water. A watershed can be represented by many individual buckets that vary in infiltration capacity. In this approach, runoff only occurs from buckets that are filled to capacity and are saturated. That is, rainfall infiltrates where there is capacity to absorb water and runs off where the infiltration capacity is exceeded.

The Xinanjiang runoff formulation uses this concept to represent spatial variability in soil water storage capacity and its effect on runoff (Zhao et al. 1980; Wood et al. 1992; Zhao 1992; Zhao and Liu 1995). The spatial variation of water-holding (or infiltration) capacity (i, mm) is described by:

$$i = i_m \left[1 - (1 - A)^{1/B} \right] \quad (11.6)$$

or equivalently:

$$A = 1 - (1 - i/i_m)^B \quad (11.7)$$

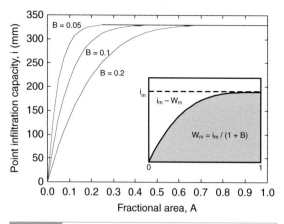

Fig. 11.9 Xinanjiang infiltration capacity for i_m = 330 mm and B = 0.1. For comparison, curves for B = 0.05 and B = 0.2 are also illustrated. The inset panel figure shows the area above and below the infiltration capacity curve

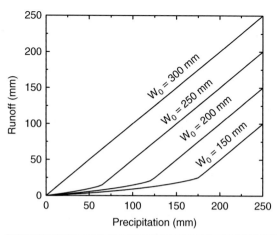

Fig. 11.10 Runoff for the Xinanjiang formulation in relation to precipitation for W_m = 300 mm, B = 0.1, and four different values of initial water storage (W_0).

where i_m (mm) is the maximum infiltration capacity, A is the fractional area for which infiltration capacity is less than or equal to i, and B is a shape parameter with typical values of ~0.1. As $B \to 0$, Eq. (11.6) becomes a constant value i_m and there is no spatial variability in infiltration capacity. Spatial variation in infiltration capacity increases as B increases. Figure 11.9 illustrates the infiltration capacity curve for i_m= 330 mm and B= 0.1. About 40 percent of the basin has infiltration capacity less than 330 mm; 10 percent, less than 215 mm; and 5 percent, less than 132 mm.

The maximum amount of water that can be held in the basin is the area under the infiltration capacity curve, given by:

$$W_m = i_m/(1+B) \qquad (11.8)$$

The area above the curve $(i_m - W_m)$ is the excess water that cannot be stored in the soil and is lost as runoff. If infiltration capacity is spatially invariant across the watershed (i.e., B= 0), the maximum amount of water held in the basin is i_m. As B becomes larger the area-integrated water storage capacity decreases from i_m and more water is lost to runoff.

The saturated fraction of the basin (A_s) with infiltration capacity filled is defined for a particular soil water (W_0) by:

$$A_s = 1 - (1 - W_0/W_m)^{B/(1+B)} \qquad (11.9)$$

This W_0 and A_s have an associated infiltration capacity (i_0) found by evaluating Eq. (11.6) with A_s:

$$i_0 = i_m \left[1 - (1 - A_s)^{1/B} \right] \qquad (11.10)$$

W_0 is the area under the infiltration capacity curve from A= 0 to A= 1 and below the horizontal line defined by i_0. The amount of water $i_0 - W_0$ cannot be held in the soil and is lost as runoff.

These equations are used to calculate runoff over the basin. Some portion of the rainfall (P) infiltrates into the soil (ΔW) and the remainder $(P - \Delta W)$ runs off from the saturated area. The change in soil water storage (ΔW) is given by the area under the infiltration capacity curve from A_s to 1 and bounded by i_0 to $i_0 + P$, and runoff is:

$$R = \begin{cases} P - W_m + W_0 + W_m \left[1 - \dfrac{(i_0 + P)}{i_m} \right]^{1+B} & \text{for } i_0 + P < i_m \\ P - W_m + W_0 & \text{for } i_0 + P \geq i_m \end{cases}$$

$$(11.11)$$

Runoff is the amount of rain that falls on the saturated fraction where the infiltration capacities are less than $i_0 + P$ and have been filled. Figure 11.10 illustrates calculated rainfall–runoff relationships. The shape parameter B influences runoff generation through the saturation-excess process. Small values of B lead

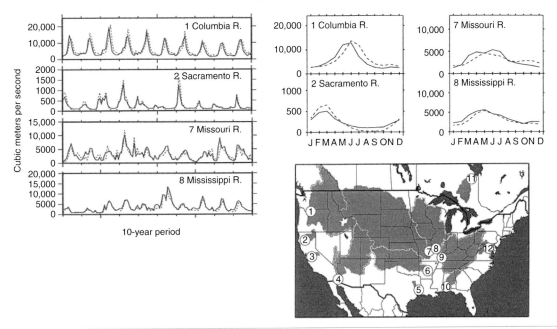

Fig. 11.11 Comparison of observed (solid line) and simulated (dashed line) runoff (m³ s⁻¹) over a 10-year period using the VIC model for the Columbia, Sacramento, Missouri, and Mississippi River basins. Also shown is average monthly flow. Adapted from Maurer et al. (2002).

to a small fraction of the basin with low infiltration capacity, producing a small saturated area and low runoff. A higher value of B leads to a larger fraction of the basin with low infiltration capacity, a larger fraction of the basin that is saturated, and greater runoff for a given amount of rainfall. The two model input parameters are W_m and B.

The Xinanjiang runoff formulation is used in the variable infiltration capacity (VIC) model (Liang et al. 1994, 1996). This is a macroscale hydrologic model of the surface energy and water balance. Distinguishing hydrologic features of the model are its representation of the effect of spatial variability in soil water storage on surface runoff and its representation of base flow, which occurs from the deepest soil layer as a non-linear function of soil water. In this way the model separates runoff to rivers into a fast response from saturation-excess surface runoff and a slower response via subsurface flow. The model has been implemented and evaluated for particular river basins as well as continental and global simulations (Nijssen et al. 2001; Mauer et al. 2002). The model reproduces observed

peak streamflow, the low flows dominated by base flow, and interannual variability, as shown in Figure 11.11 for several major drainage basins in the United States. This suggests that the model adequately captures the partitioning of precipitation into runoff and evapotranspiration. The Xinanjiang runoff formulation has also been used to represent runoff in the land component of climate models (Dümenil and Todini 1992; Stamm et al. 1994; Ducharne et al. 1998; Hagemann and Gates 2003; Wang et al. 2008; Balsamo et al. 2009; Li et al. 2011). A challenge with the model, however, is that the runoff parameters are not directly observable and instead are estimated from calibration with observed riverflow.

11.5.3 TOPMODEL

Topography influences the amount of runoff generated by a watershed by controlling the development of saturated areas. Saturated areas occur when the water table rises to the surface. This typically happens in regions of gentle slopes and convergent water flow. Watersheds having a large percentage of their total area

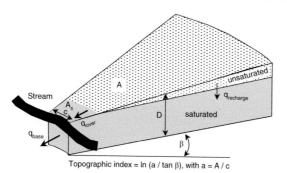

Topographic index = ln (a / tan β), with a = A / c

Fig. 11.12 Topographically defined variable contributing area in a watershed. Shaded areas are saturated. Stippled areas are unsaturated.

Saturation-excess runoff occurs over the saturated fraction of the watershed:

$$R = (A_s/A)P \qquad (11.14)$$

Similar to the VIC model, TOPMODEL separates total runoff to rivers into saturation-excess overland flow and subsurface base flow.

High values of the topographic index indicate large contributing areas and shallow slopes, typically at the base of hillslopes and near streams. Low values indicate relatively little upslope contributing area and steep slopes, typically along ridges and hilltops. For example, a segment of a watershed with an upslope area A= 500 m², contour length at the base c = 25 m, and a local slope $\tan \beta$ = 0.08 has a topographic index of 5.5. A smaller upslope area (A = 250 m²) gives a smaller topographic index (4.8), indicating greater depth to water table and less chance of saturation. A shallower slope ($\tan \beta$ = 0.02) produces a larger topographic index (6.9), indicating a greater likelihood for saturation. A smaller contour length (c = 5 m) indicates more convergent flow, also giving a higher topographic index (7.1).

The spatial distribution of the topographic index approximates the spatial distribution of depth to the water table, and a map of the topographic index throughout a watershed shows areas where saturation-excess overland flow is likely to occur. Figure 11.13 shows the application of this topographic index to a small 36 ha watershed in the Vosges Mountains near Strasbourg, France. Elevation ranges from 1000 m along the top of ridges to 748 m along the valley bottom. The slopes are steep, with an average of 20° and a maximum slope of 35°. Soils are mostly sand and gravel. The vegetation is a mixture of grassland on granite bedrock and forest on sandstone along upper slopes. Two streams near the center of the basin and one smaller stream near the basin outlet drain the watershed. These streams and nearby saturated areas are clearly indicated by high values of the topographic index. Upslope areas have low topographic index and hilltops have the lowest index.

The TOPMODEL framework is a distributed modeling approach in that hydrologic

with gentle slopes and convergent flow produce more runoff in the form of saturation-excess overland flow than watersheds without those characteristics.

One approach that links the variable contributing area concept to catchment topography is the watershed hydrology model TOPMODEL (Beven and Kirkby 1979; Beven et al. 1995; Hornberger et al. 1998). This model divides a catchment into discrete blocks defined by the topographic index $\ln(a/\tan \beta)$, where a is the upslope area per unit contour width and $\tan \beta$ is the slope of the surface (Figure 11.12). The upslope contributing area per unit contour width is $a = A / c$, where here A is the area of the hillslope segment and c is the hillslope width. The local water table depth (z) at any location is:

$$z = \bar{z} + \frac{1}{f}\left[\lambda - \ln(a/\tan \beta)\right] \qquad (11.12)$$

where $\bar{z}$ is the watershed average depth to water table and λ is the watershed average value of the topographic index. The parameter f describes the change of saturated hydraulic conductivity with depth. Saturated hydraulic conductivity is assumed to decrease exponentially with depth (z):

$$K_{sat}(z) = K_{sat} \exp(-fz) \qquad (11.13)$$

where K_{sat} is saturated hydraulic conductivity at the surface. Saturated areas in the watershed are those with $z \le 0$. Equation (11.12) is applied to every location in the watershed to calculate the saturated fraction of the watershed (A_s / A).

(a)

Ringelbach catchment 3D elevation

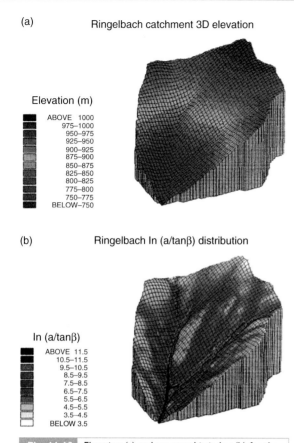

Elevation (m)

ABOVE 1000
975–1000
950–975
925–950
900–925
875–900
850–875
825–850
800–825
775–800
750–775
BELOW–750

(b)

Ringelbach ln (a/tanβ) distribution

ln (a/tanβ)

ABOVE 11.5
10.5–11.5
9.5–10.5
8.5–9.5
7.5–8.5
6.5–7.5
5.5–6.5
4.5–5.5
3.5–4.5
BELOW 3.5

Fig. 11.13 Elevation (a) and topographic index (b) for the 36 ha Ringelbach watershed. Data have a spatial resolution of 5 m. Reproduced from Ambroise et al. (1996). See color plate section.

calculations are performed for increments of the topographic index over the entire catchment. By representing topography within the watershed as parcels of discrete topographic index values or statistical distributions of similar index, runoff can be modeled and routed over the surface into streams during a storm. The TOPMODEL approach has been applied to many watersheds (Beven and Wood 1983; Beven et al. 1984; Hornberger et al. 1985, 1994; Wolock et al. 1989, 1990; Band et al. 1993; Famiglietti and Wood 1994; Wolock and Price 1994; Wolock 1995). It is also being used to represent hydrologic processes in the land component of climate models (Stieglitz et al. 1997; Koster et al. 2000; Chen and Kumar 2001; Gedney and Cox 2003; Niu et al. 2005; Decharme and Douville 2006; Clark

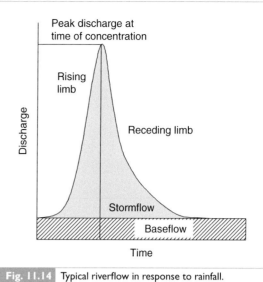

Fig. 11.14 Typical riverflow in response to rainfall.

and Gedney 2008). By altering soil moisture, the VIC and TOPMODEL runoff parameterizations have a significant effect of the global terrestrial carbon cycle (Lei et al. 2014).

11.6 Riverflow

Total riverflow is derived from two components. Base flow is water that enters a river from persistent sources, usually groundwater discharge. It maintains riverflow between storms and is why many rivers still flow during prolonged droughts, although at lower levels. Storm flow is surface runoff that enters rivers in response to hydrologic events such as rainstorms or snowmelt. Overland flow from saturated areas provides most of the stormflow.

Figure 11.14 shows the typical response of a river to rainfall. As rain falls onto the watershed, river discharge slowly increases above base flow. The response is initially small because the discharge is augmented only by rain falling onto the channel itself and nearby areas. Over time, the discharge rises rapidly as the watershed becomes saturated and more distant regions of the basin contribute runoff. At peak discharge, all of the watershed that can contribute runoff is producing runoff. The time for peak discharge to occur is known as the time of concentration.

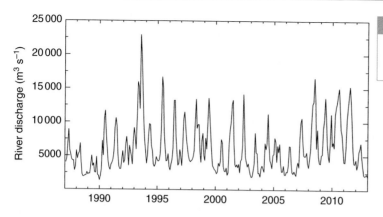

Fig. 11.15 Monthly discharge for the Mississippi River at St. Louis, Missouri, for 1987–2012. Data provided by the U.S. Geological Survey.

So long as rainfall continues, the discharge remains constant. Then, as rainfall ceases, river-flow slowly returns to pre-storm levels. The geomorphology and geology of a watershed control the path, magnitude, and timing of water flow at the outlet. Small basins, basins with impermeable soils, and basins with large saturated areas attain peak discharge quicker than large basins, basins with permeable soils, or basins with little saturation.

Figure 11.15 shows observed monthly riverflow for the Mississippi River at St. Louis, Missouri, during the period 1987–2012. The Mississippi, which at St. Louis drains an area of 1.8 million km^2, has an average flow of 6100 m^3 s^{-1}, but has distinct periods of floods and droughts. For example, monthly flow rates were less than 2000 m^3 s^{-1} in the summer of 1988 during a prolonged dry period. A particularly large flood occurred in the summer of 1993, with flow in excess of 20,000 m^3 s^{-1}. Other periods of floods and droughts are evident in the riverflow.

11.7 | Global Drainage Basins

Figure 11.16 shows the amount of water flowing in major rivers worldwide. With an annual flow of 155,430 m^3 s^{-1}, the Amazon River, which enters the Atlantic Ocean in Brazil along the equator, discharges the most water. The Congo River in western equatorial Africa, the Orinoco River in northern South America, and the Changjiang River in China are the next largest, each carrying from 25,000 to 33,000 m^3 s^{-1}.

Three of the 12 largest rivers (Ob, Yenisei, and Lena) drain Siberia and flow northward into the Arctic Ocean. The Mississippi, which drains much of the United States east of the Rocky Mountains, carries 14,703 m^3 s^{-1} to the Gulf of Mexico and ranks ninth worldwide in terms of annual flow. Comparison with observed riverflow is an important means to test the hydrologic cycle of climate models (Coe 2000; Lawrence et al. 2011).

Figure 11.17 divides the continents into source regions of runoff for each major ocean basin. For example, 89 percent of South America contributes runoff to the Atlantic Ocean. Only a small region west of the Andes Mountains (7% of the continent) generates runoff for the Pacific Ocean. North American rivers feed the Pacific Ocean west of the Rocky Mountains (23% of continental area), the Atlantic Ocean east of the Rocky Mountains (56%), and the Arctic Ocean in the north (19%). European land area is also split among three oceans, with 35 percent of the land area contributing to the Atlantic Ocean, 30 percent to the Mediterranean Sea, and 15 percent to the Arctic Ocean. Africa and Asia similarly drain into three oceans, with Asian land area roughly evenly apportioned to the Pacific (31%), Arctic (25%), and Indian (22%) oceans. In Africa, water from nearly one-half (45%) of all land area flows into the Atlantic while 24 percent and 21 percent of the land area contributes to the Mediterranean and Indian Ocean, respectively. More than one-half (58%) of the land area of Australia drains into the Indian Ocean. Only 14 percent drains into the Pacific while

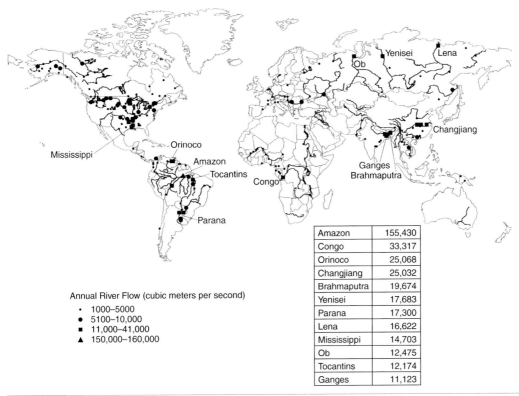

Amazon	155,430
Congo	33,317
Orinoco	25,068
Changjiang	25,032
Brahmaputra	19,674
Yenisei	17,683
Parana	17,300
Lena	16,622
Mississippi	14,703
Ob	12,475
Tocantins	12,174
Ganges	11,123

Annual River Flow (cubic meters per second)
- · 1000–5000
- ● 5100–10,000
- ■ 11,000–41,000
- ▲ 150,000–160,000

Fig. 11.16 Annual riverflow for 241 gauging stations worldwide. The 12 largest rivers and their annual flow at the station closest to their mouth are identified. Data from Coe (2000).

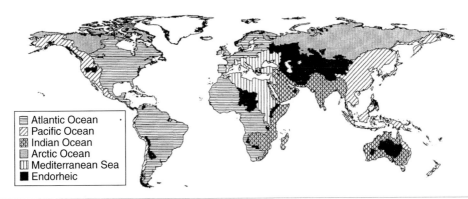

☰ Atlantic Ocean
▨ Pacific Ocean
▦ Indian Ocean
▥ Arctic Ocean
▥ Mediterranean Sea
■ Endorheic

Fig. 11.17 Continental runoff source regions for the Atlantic Ocean, Pacific Ocean, Indian Ocean, Arctic Ocean, and Mediterranean Sea. Endorheic regions are closed basins with internal drainage. Reproduced from Vörösmarty et al. (2000).

28 percent of the land area has no connection to the sea.

Rivers in northern Alaska, the Canadian Arctic, and Russia replenish water in the Arctic Ocean. Asia provides two-thirds (66%) of the total contributing basin area. North America comprises one-quarter (25%) of the contributing area while Europe is only 9 percent. The Atlantic Ocean is fed primarily by land area from South America (35% of total contributing basin area), Africa (29%), and North America (28%). Europe comprises only 8 percent of the

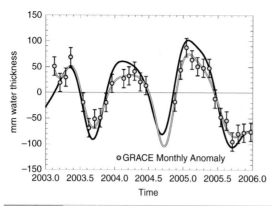

Fig. 11.18 Observed total water storage anomalies (soil water and groundwater) for a network of sites in Illinois (black line) and total water storage anomalies derived from GRACE (circles and gray line) for the period 2003–2005. Reproduced from Swenson et al. (2006).

America and Australia are only minor components of the Pacific Ocean drainage basin (each with 6% of total area).

Continental-scale variations in the water budget can also be monitored from space. Terrestrial water storage affects Earth's gravitational field. Satellite-based measurements of month-to-month variation in Earth's gravitational field from the Gravity Recovery and Climate Experiment (GRACE) give an indication of changes in surface and subsurface water storage on land (Ramillien et al. 2008). The GRACE data can be used to infer the water balance (Swenson and Wahr 2006; Rodell et al. 2009; Syed et al. 2009; Famiglietti et al. 2011). For example, Swenson et al. (2006) showed that GRACE data compare well with observed total water storage (soil water plus groundwater) in Illinois (Figure 11.18). Maximum storage occurs in spring, with minimum water storage in autumn. GRACE estimates of seasonal variation in water storage provide a means to assess the terrestrial hydrologic cycle simulated by global climate models (Niu and Yang 2006; Swenson and Milly 2006).

total drainage basin, but has a greater relative role in the North Atlantic. One-half (47%) of the drainage area of the Indian Ocean is in Asia, one-third (31%) in Africa, and one-quarter (22%) in Australia. Two-thirds (64%) of the drainage area of the Pacific Ocean originates in Asia and one-quarter (24%) in North America. South

11.8 | Review Questions

1. Fill in the missing data in the following table. Which watershed has the highest annual evapotranspiration? Which has the lowest annual evapotranspiration? Which watershed is likely to have deep soils? Which is likely to have shallow soils?

Watershed	Precipitation, P (mm)	Runoff, R (mm)	R / P	E (mm)
1	2000		0.40	
2	1700	1020		
3	1800			900
4	1600		0.55	
5	1500	675		

2. The watershed that supplies water to a city is covered with a mature pine forest. The forest is cleared during logging. How will water supply be affected? The forest then regrows in deciduous trees. As the deciduous forest matures, how will water supply compare with the prior pine forest?

3. Use Eq. (11.2) to numerically calculate catchment runoff from the data in the following table. How does this compare with runoff calculated using the average precipitation and infiltration capacity?

4. Two storms have the same area mean precipitation, but differ in their spatial characteristics. One

Fractional area	Point precipitation (mm)	Infiltration capacity (mm)
Unsaturated		
0.10	0.3	0.1
0.20	1.0	1.2
0.25	2.0	0.7
0.20	4.0	1.4
Saturated		
0.25	1.5	1.8

has precipitation evenly distributed across the watershed. The other has precipitation locally concentrated in only 20 percent of the watershed. Which storm is likely to generate more catchment runoff? Why?

5. Two watersheds experience the same storm. One has an average volumetric soil water content, relative to saturation, $s = \theta / \theta_{sat} = 0.3$, and the other has $s = 0.6$. Both watersheds have the same soil

texture. Which watershed is likely to have more runoff? Why?

6. Use the Xinanjiang formulation to calculate runoff for a watershed with $P = 50$ mm, $W_m = 125$ mm, $B = 0.1$, and $W_0 = 50$ mm.

7. Compare the answer to question 6 to the runoff produced in each of the following three scenarios: (a) wetter soil, $W_0 = 75$ mm; (b) higher maximum water, $W_m = 150$ mm; and (c) greater spatial variation in infiltration capacity, $B = 0.2$.

8. Which area within a watershed is more likely to produce runoff during a storm: where the topographic index is 6.5 or where the topographic index is 10.0? Why?

9. For what values of $\ln(a / \tan \beta)$ is a watershed saturated?

10. Use Eq. (11.12) to calculate the depth to the water table $(z - \bar{z})$ for an area within a watershed with $\ln(a / \tan \beta) = 15$. Assume $\lambda = 8$ and $f = 3$. In another area, the topographic index is 5. What is the depth to the water table there? Which area is more likely to generate runoff during a storm?

11.9 | References

Ambroise, B., Beven, K., and Freer, J. (1996). Toward a generalization of the TOPMODEL concepts: Topographic indices of hydrological similarity. *Water Resources Research*, 32, 2135–2145.

Andréassian, V. (2004). Waters and forests: From historical controversy to scientific debate. *Journal of Hydrology*, 291, 1–27.

Balsamo, G., Viterbo, P., Beljaars, A., et al. (2009). A revised hydrology for the ECMWF model: Verification from field site to terrestrial water storage and impact in the integrated forecast system. *Journal of Hydrometeorology*, 10, 623–643.

Band, L. E., Patterson, P., Nemani, R., and Running, S. W. (1993). Forest ecosystem processes at the watershed scale: incorporating hillslope hydrology. *Agricultural and Forest Meteorology*, 63, 93–126.

Beven, K. J., and Kirkby, M. J. (1979). A physically based variable contributing area model of basin hydrology. *Hydrological Sciences Bulletin*, 24, 43–69.

Beven, K., and Wood, E. F. (1983). Catchment geomorphology and the dynamics of runoff contributing areas. *Journal of Hydrology*, 65, 139–158.

Beven, K. J., Kirkby, M. J., Schofield, N., and Tagg, A. F. (1984). Testing a physically-based flood forecasting model (TOPMODEL) for three U.K. catchments. *Journal of Hydrology*, 69, 119–143.

Beven, K. J., Lamb, R., Quinn, P. F., Romanowicz, R., and Freer, J. (1995). TOPMODEL. In *Computer Models of Watershed Hydrology*, ed. V. P. Singh. Highlands Ranch, Colorado: Water Resources Publications, pp. 627–668.

Bormann, F. H., and Likens, G. E. (1979). *Pattern and Process in a Forested Ecosystem*. New York: Springer-Verlag.

Bosch, J. M., and Hewlett, J. D. (1982). A review of catchment experiments to determine the effect of vegetation changes on water yield and evapotranspiration. *Journal of Hydrology*, 55, 3–23.

Brown, A. E., Zhang, L., McMahon, T. A., Western, A. W., and Vertessy, R. A. (2005). A review of paired catchment studies for determining changes in water yield resulting from alterations in vegetation. *Journal of Hydrology*, 310, 28–61.

Chen, J., and Kumar, P. (2001). Topographic influence on the seasonal and interannual variation of water and energy balance of basins in North America. *Journal of Climate*, 14, 1989–2014.

Clark, D. B., and Gedney, N. (2008). Representing the effects of subgrid variability of soil moisture on runoff generation in a land surface model. *Journal of Geophysical Research*, 113, D10111, doi:10.1029/2007JD008940.

Coe, M. T. (2000). Modeling terrestrial hydrological systems at the continental scale: testing the accuracy of an atmospheric GCM. *Journal of Climate*, 13, 686–704.

Decharme, B., and Douville, H. (2006). Introduction of a sub-grid hydrology in the ISBA land surface model. *Climate Dynamics*, 26, 65–78.

Dolman, A. J., and Gregory, D. (1992). The parametrization of rainfall interception in GCMs. *Quarterly Journal of the Royal Meteorological Society*, 118, 455–467.

Ducharne, A., Laval, K., and Polcher, J. (1998). Sensitivity of the hydrological cycle to the parameterization of soil hydrology in a GCM. *Climate Dynamics*, 14, 307–327.

Dümenil, L., and Todini, E. (1992). A rainfall-runoff scheme for use in the Hamburg climate model. In *Advances in Theoretical Hydrology: A Tribute to James Dooge*, ed. J. P. O'Kane. Amsterdam: Elsevier, pp. 129–157.

Eltahir, E. A. B., and Bras, R. L. (1993). A description of rainfall interception over large areas. *Journal of Climate*, 6, 1002–1008.

Famiglietti, J. S., and Wood, E. F. (1994). Multiscale modeling of spatially variable water and energy balance processes. *Water Resources Research*, 30, 3061–3078.

Famiglietti, J. S., Lo, M., Ho, S. L., et al. (2011). Satellites measure recent rates of groundwater depletion in California's Central Valley. *Geophysical Research Letters*, 38, L03403, doi:10.1029/2010GL046442.

Farley, K. A., Jobbágy, E. G., and Jackson, R. B. (2005). Effects of afforestation on water yield: A global synthesis with implications for policy. *Global Change Biology*, 11, 1565–1576.

Fennessey, N. M., Eagleson, P. S., Qinliang, W., and Rodriguez-Iturbe, I. (1986). *Spatial Characteristics of Observed Precipitation Fields: A Catalog of Summer Storms in Arizona, Volume I*, Ralph M. Parsons Laboratory Report No. 307. Cambridge: Massachusetts Institute of Technology.

Gedney, N., and Cox, P. M. (2003). The sensitivity of global climate model simulations to the representation of soil moisture heterogeneity. *Journal of Hydrometeorology*, 4, 1265–1275.

Hagemann, S., and Gates, L. D. (2003). Improving a sub-grid runoff parameterization scheme for climate models by the use of high resolution data derived from satellite observations. *Climate Dynamics*, 21, 349–359.

Hornbeck, J. W., Pierce, R. S., and Federer, C. A. (1970). Streamflow changes after forest clearing in New England. *Water Resources Research*, 6, 1124–1132.

Hornbeck, J. W., Adams, M. B., Corbett, E. S., Verry, E. S., and Lynch, J. A. (1993). Long-term impacts of forest treatments on water yield: A summary for northeastern USA. *Journal of Hydrology*, 150, 323–344.

Hornbeck, J. W., Martin, C. W., and Eagar, C. (1997). Summary of water yield experiments at Hubbard Brook Experimental Forest, New Hampshire. *Canadian Journal of Forest Research*, 27, 2043–2052.

Hornberger, G. M., Beven, K. J., Cosby, B. J., and Sappington, D. E. (1985). Shenandoah watershed study: Calibration of a topography-based, variable contributing area hydrological model to a small forested catchment. *Water Resources Research*, 21, 1841–1850.

Hornberger, G. M., Bencala, K. E., and McKnight, D. M. (1994). Hydrological controls on dissolved organic carbon during snowmelt in the Snake River near Montezuma, Colorado. *Biogeochemistry*, 25, 147–165.

Hornberger, G. M., Raffensperger, J. P., Wiberg, P. L., and Eshleman, K. N. (1998). *Elements of Physical Hydrology*. Baltimore: Johns Hopkins University Press.

Jackson, R. B., Jobbágy, E. G., Avissar, R., et al. (2005). Trading water for carbon with biological carbon sequestration. *Science*, 310, 1944–1947.

Johnson, D. W., and Van Hook, R. I. (1989). *Analysis of Biogeochemical Cycling Processes in Walker Branch Watershed*. New York: Springer-Verlag.

Koster, R. D., Suarez, M. J., Ducharne, A., Stieglitz, M., and Kumar, P. (2000). A catchment-based approach to modeling land surface processes in a general circulation model, 1: Model structure. *Journal of Geophysical Research*, 105D, 24809–24822.

Lawrence, D. M., Oleson, K. W., Flanner, M. G., et al. (2011). Parameterization improvements and functional and structural advances in version 4 of the Community Land Model. *Journal of Advances in Modeling Earth Systems*, 3, doi:10.1029/2011MS000045.

Lei, H., Huang, M., Leung, L. R., et al. (2014). Sensitivity of global terrestrial gross primary production to hydrologic states simulated by the Community Land Model using two runoff parameterizations. *Journal of Advances in Modeling Earth Systems*, 6, 658–679, doi:10.1002/2013MS000252.

Li, H., Huang, M., Wigmosta, M. S., et al. (2011). Evaluating runoff simulations from the Community Land Model 4.0 using observations from flux towers and a mountainous watershed. *Journal of Geophysical Research*, 116, D24120, doi:10.1029/2011JD016276.

Liang, X., Lettenmaier, D. P., Wood, E. F., and Burges, S. J. (1994). A simple hydrologically based model of land surface water and energy fluxes for general circulation models. *Journal of Geophysical Research*, 99D, 14415–14428.

Liang, X., Lettenmaier, D. P., and Wood, E. F. (1996). One-dimensional statistical dynamic representation of subgrid spatial variability of precipitation in the two-layer variable infiltration capacity model. *Journal of Geophysical Research*, 101D, 21403–21422.

Likens, G. E. (2004). Some perspectives on long-term biogeochemical research from the Hubbard Brook ecosystem study. *Ecology*, 85, 2355–2362.

Likens, G. E., and Bormann, F. H. (1995). *Biogeochemistry of a Forested Ecosystem*, 2nd edn. New York: Springer-Verlag.

Likens, G. E., Bormann, F. H., Pierce, R. S., Eaton, J. S., and Johnson, N. M. (1977). *Biogeochemistry of a Forested Ecosystem*. New York: Springer-Verlag.

Luxmoore, R. J., and Huff, D. D. (1989). Water. In *Analysis of Biogeochemical Cycling Processes in Walker Branch Watershed*, ed. D. W. Johnson and R. I. Van Hook. New York: Springer-Verlag, pp. 164–196.

Mattikalli, N. M., Engman, E. T., Jackson, T. J., and Ahuja, L. R. (1998). Microwave remote sensing of temporal variations of brightness temperature and near-surface soil water content during a watershed-scale field experiment, and its application to the estimation of soil physical properties. *Water Resources Research*, 34, 2289–2299.

Maurer, E. P., Wood, A. W., Adam, J. C., Lettenmaier, D. P., and Nijssen, B. (2002). A long-term hydrologically based dataset of land surface fluxes and states for the conterminous United States. *Journal of Climate*, 15, 3237–3251.

Nijssen, B., O'Donnell, G. M., Lettenmaier, D. P., Lohmann, D., and Wood, E. F. (2001). Predicting the discharge of global rivers. *Journal of Climate*, 14, 3307–3323.

Niu, G.-Y., and Yang, Z.-L. (2006). Assessing a land surface model's improvements with GRACE estimates. *Geophysical Research Letters*, 33, L07401, doi:10.1029/2005GL025555.

Niu, G.-Y., Yang, Z.-L., Dickinson, R. E., and Gulden, L. E. (2005). A simple TOPMODEL-based runoff parameterization (SIMTOP) for use in global climate models. *Journal of Geophysical Research*, 110, D21106, doi:10.1029/2005JD006111.

Pitman, A. J., Henderson-Sellers, A., and Yang, Z.-L. (1990). Sensitivity of regional climates to localized precipitation in global models. *Nature*, 346, 734–737.

Ramillien, G., Famiglietti, J. S., and Wahr, J. (2008). Detection of continental hydrology and glaciology signals from GRACE: A review. *Surveys in Geophysics*, 29, 361–374.

Rodell, M., Velicogna, I., and Famiglietti, J. S. (2009). Satellite-based estimates of groundwater depletion in India. *Nature*, 460, 999–1002.

Shuttleworth, W. J. (1988). Macrohydrology – the new challenge for process hydrology. *Journal of Hydrology*, 100, 31–56.

Stamm, J. F., Wood, E. F., and Lettenmaier, D. P. (1994). Sensitivity of a GCM simulation of global climate to the representation of land-surface hydrology. *Journal of Climate*, 7, 1218–1239.

Stieglitz, M., Rind, D., Famiglietti, J., and Rosenzweig, C. (1997). An efficient approach to modeling the topographic control of surface hydrology for regional and global climate modeling. *Journal of Climate*, 10, 118–137.

Swank, W. T., and Crossley, D. A., Jr. (1988). *Forest Hydrology and Ecology at Coweeta*. New York: Springer-Verlag.

Swank, W. T., and Douglass, J. E. (1974). Streamflow greatly reduced by converting deciduous hardwood stands to pine. *Science*, 185, 857–859.

Swank, W. T., and Miner, N. H. (1968). Conversion of hardwood-covered watersheds to white pine reduces water yield. *Water Resources Research*, 4, 947–954.

Swank, W. T., Swift, L. W., Jr., and Douglass, J. E. (1988). Streamflow changes associated with forest cutting, species conversions, and natural disturbances. In *Forest Hydrology and Ecology at Coweeta*, ed. W. T. Swank and D. A. Crossley Jr. Springer-Verlag, New York, pp. 297–312.

Swenson, S. C., and Milly, P. C. D. (2006). Climate model biases in seasonality of continental water storage revealed by satellite gravimetry. *Water Resources Research*, 42, W03201, doi:10.1029/2005WR004628.

Swenson, S., and Wahr, J. (2006). Estimating large-scale precipitation minus evapotranspiration from GRACE satellite gravity measurements. *Journal of Hydrometeorology*, 7, 252–270.

Swenson, S., Yeh, P. J.-F., Wahr, J., and Famiglietti, J. (2006). A comparison of terrestrial water storage variations from GRACE with in situ measurements from Illinois. *Geophysical Research Letters*, 33, L16401, doi:10.1029/2006GL026962.

Swift, L. W., Jr., Cunningham, G. B., and Douglass, J. E. (1988). Climatology and hydrology. In *Forest Hydrology and Ecology at Coweeta*, ed. W. T. Swank and D. A. Crossley Jr. Springer-Verlag, New York, pp. 35–55.

Syed, T. H., Famiglietti, J. S., and Chambers, D. P. (2009). GRACE-based estimates of terrestrial freshwater discharge from basin to continental scales. *Journal of Hydrometeorology*, 10, 22–40.

Vörösmarty, C. J., Fekete, B. M., Meybeck, M., and Lammers, R. B. (2000). Global system of rivers: its role in organizing continental land mass and defining land-to-ocean linkages. *Global Biogeochemical Cycles*, 14, 599–621.

Wang, A., Li, K. Y., and Lettenmaier, D. P. (2008). Integration of the variable infiltration capacity model soil hydrology scheme into the community land model. *Journal of Geophysical Research*, 113, D09111, doi:10.1029/2007JD009246.

Wilson, K. B., Hanson, P. J., Mulholland, P. J., Baldocchi, D. D., and Wullschleger, S. D. (2001). A comparison of methods for determining forest evapotranspiration and its components: Sap-flow, soil water budget, eddy covariance and catchment water balance. *Agricultural and Forest Meteorology*, 106, 153–168.

Wolock, D. M. (1995). Effects of subbasin size on topographic characteristics and simulated flow paths in Sleepers River watershed, Vermont. *Water Resources Research*, 31, 1989–1997.

Wolock, D. M., and Price, C. V. (1994). Effects of digital elevation model map scale and data resolution on a topography-based watershed model. *Water Resources Research*, 30, 3041–3052.

Wolock, D. M., Hornberger, G. M., Bevin, K. J., and Campbell, W. G. (1989). The relationship of catchment topography and soil hydraulic characteristics to lake alkalinity in the northeastern United States. *Water Resources Research*, 25, 829–837.

Wolock, D. M., Hornberger, G. M., and Musgrove, T. J. (1990). Topographic effects on flow path and surface water chemistry of the Llyn Brianne catchments in Wales. *Journal of Hydrology*, 115, 243–259.

Wood, E. F., Lettenmaier, D. P., and Zartarian, V. G. (1992). A land-surface hydrology parameterization with subgrid variability for general circulation models. *Journal of Geophysical Research*, 97D, 2717–2728.

Zhang, L., Dawes, W. R., and Walker, G. R. (2001). Response of mean annual evapotranspiration to vegetation changes at catchment scale. *Water Resources Research*, 37, 701–708.

Zhao, R.-J. (1992). The Xinanjiang model applied in China. *Journal of Hydrology*, 135, 371–381.

Zhao, R.-J., and Liu, X.-R. (1995). The Xinanjiang model. In *Computer Models of Watershed Hydrology*, ed. V. P. Singh. Highlands Ranch, Colorado: Water Resources Publications, pp. 215–232.

Zhao, R.-J., Zuang, Y.-L., Fang, L. R., Liu, X.-R., and Zhang, Q.-S. (1980). The Xinanjiang model. In *Hydrological Forecasting: Proceedings of the Oxford Symposium, 15–18 April 1980*, Publication No. 129. International Association of Hydrological Sciences, Wallingford, United Kingdom, pp. 351–356.

12

Surface Energy Fluxes

12.1 | Chapter Summary

The energy balance at Earth's land surface requires that the energy gained from net radiation be balanced by the fluxes of sensible and latent heat to the atmosphere and the storage of heat in soil. These energy fluxes are a primary determinant of surface climate. The annual energy balance at the land surface varies geographically in relation to incoming solar radiation and soil water availability. Over land, annual evapotranspiration is highest in the tropics and generally decreases towards the poles. Geographic patterns of evapotranspiration are explained by Budyko's analysis of the control of evapotranspiration by net radiation and precipitation. Energy fluxes vary over the course of a day and throughout the year, also in relation to soil water availability and the diurnal and annual cycles of solar radiation. The various terms in the energy budget (net radiation, sensible heat flux, latent heat flux, and soil heat flux) are illustrated for different climate zones and for various vegetation types. The Penman–Monteith equation illustrates relationships among net radiation, latent heat flux, sensible heat flux, and surface temperature. Soil experiments that alter surface albedo, surface conductance to evapotranspiration, and thermal conductivity illustrate the importance of these properties in regulating surface temperature and energy fluxes.

12.2 | Surface Energy Budget

The solar and longwave radiation that impinges on Earth's surface heats the surface. The surface reflects some of the incoming solar radiation and also emits outgoing longwave radiation. The remaining radiation is the net radiation at the surface. Net radiation is dissipated in three ways.

Movement of air transports heat in a process known as convection. A common example is the cooling effect of a breeze on a hot summer day. This heat exchange is called sensible heat. Greenhouse microclimates are an example of the warm temperatures that can arise in the absence of convective heat exchange (Avissar and Mahrer 1982; Mahrer et al. 1987; Oke 1987). It is generally thought that greenhouses provide a warm environment to grow plants because glass or other translucent coverings allow solar radiation to penetrate and warm the interior of the greenhouse while longwave radiation emitted by the interior surfaces is trapped within the greenhouse. Although this can happen, the daytime warmth in greenhouses is largely a result of negligible convective heat exchange with the outside environment. The sensible heat from the warm interior surfaces is trapped within the greenhouse, warming the interior air.

Evapotranspiration is another way in which heat is dissipated at the surface. The change in phase of water from liquid at the surface to vapor in air requires considerable energy. Evapotranspiration, therefore, involves a transfer of mass and energy to the atmosphere. Transfer of mass is seen as wet clothes dry on a clothesline. Heat loss is why a person may feel cold on a hot summer day when wet but hot after being dried with a towel. When water changes from liquid to gas (vapor), energy is absorbed from the evaporating surface without a rise in temperature. This latent heat of vaporization varies with temperature, but is 2.466 MJ kg^{-1} at 15°C (Table 3.3). A typical summertime rate of evapotranspiration is a depth of 5 mm of water per day, which with a density of 1000 kg m^{-3} is equivalent to a mass of 5 kg of water per square meter (0.005 m × 1000 kg m^{-3} = 5 kg m^{-2}). A water loss of 5 kg m^{-2} day^{-1} is equivalent to a heat loss:

$$5\frac{kg}{m^2 day} \times \frac{1\ day}{86\ 400\ s} \times 2\ 466\ 000\frac{J}{kg} = 143\frac{W}{m^2}$$

This heat is transferred from the evaporating surface to the air, where it is stored in water vapor as latent heat. It is released when water vapor condenses back to liquid.

Some heat is exchanged with the underlying soil through conduction. Conduction is the transfer of heat along a temperature gradient from high temperature to low temperature but in contrast to convection due to direct contact rather than movement of air. The heat felt when touching a hot object is an example of conduction.

Remembering that the change in energy storage in a system is equal to the difference between energy input and energy output, given by Eq. (3.22), the overall energy balance of a volume of soil with surface area $\Delta x \Delta y$ (m²) and depth Δz (m) is:

$$c_v\left(\Delta T / \Delta t\right)\Delta z = \left(S\downarrow - S\uparrow + L\downarrow - L\uparrow\right) - H - \lambda E = G$$

$$(12.1)$$

The left-hand side of this equation represents the change in storage (W m^{-2}), where c_v is the volumetric heat capacity (J m^{-3} K^{-1}) and $\Delta T / \Delta t$ is the change in temperature with time (K s^{-1}). The right-hand side of the equation represents

the energy (W m^{-2}) gained from incoming solar radiation ($S\downarrow$) and longwave radiation ($L\downarrow$) minus energy losses from reflected solar radiation ($S\uparrow$), outgoing longwave radiation ($L\uparrow$), sensible heat (H) and latent heat (λE). Latent heat flux is the product of the evaporative water flux (E, kg m^{-2} s^{-1}) times the latent heat of vaporization (λ, J kg^{-1}). The difference among net radiation, sensible heat flux, and latent heat flux is the heat flux (G) stored in the soil by conduction. The term Δz appears on the left-hand side of Eq. (12.1) because energy is stored in the soil volume with dimensions $\Delta x \Delta y \Delta z$ (m³) while surface fluxes are exchanged over a surface area $\Delta x \Delta y$ (m²); the common term ($\Delta x \Delta y$) drops out.

More commonly, the surface energy balance is written:

$$R_n = \left(S\downarrow - S\uparrow\right) + \left(L\downarrow - L\uparrow\right) = H + \lambda E + G \quad (12.2)$$

Net radiation is the total solar and longwave radiation absorbed by the surface after accounting for reflection of solar radiation and emission of longwave radiation. The net radiation absorbed by the surface is balanced by energy lost or gained by sensible heat, latent heat, and change in heat storage.

Consider, for example, a volume of soil 50 cm thick, deep enough so that no heat is conducted out of the bottom of the soil column over the course of a day. Typical energy fluxes at midday for a moist soil are R_n = 650 W m^{-2}, H = 200 W m^{-2}, λE = 350 W m^{-2}, and G = 100 W m^{-2}. With a heat capacity c_v = 2.5 MJ m^{-3} K^{-1}, typical of moist soil, the soil warms at a rate:

$$\Delta T/\Delta t = G/(c_v \Delta z) = 0.29\ °C\ \text{per hour}$$

The Bowen ratio (defined as the ratio of sensible heat flux to latent heat flux, $\beta = H / \lambda E$) is a key metric of the surface energy balance. With this definition:

$$\lambda E = \frac{R_n - G}{1 + \beta} \quad (12.3)$$

For a given amount of available energy ($R_n - G$), latent heat flux decreases as the Bowen ratio increases.

12.3 | Energy Balance of Earth's Surface

Although small compared with other fluxes, evapotranspiration is an important term in Earth's planetary energy budget and hydrologic cycle (see Wang and Dickinson (2012) for a review). Annual evapotranspiration over land is about 60 percent of net radiation, when the global energy budget analysis of Figure 3.2 is extended to land only (Trenberth et al. 2009). Evapotranspiration is about 65 percent of the annual precipitation over land (Figure 3.4).

The magnitude of sensible and latent heat fluxes varies geographically, but some general patterns are evident. Annual evapotranspiration over land is highest in the tropics and generally decreases towards the poles. Annual evapotranspiration is a balance between radiative heating, which provides energy to evaporate water, and precipitation, which provides water to evaporate. The tropics have high net radiation to evaporate water and ample precipitation year-round to sustain evapotranspiration. In other regions, there may still be sufficient energy to evaporate water, but low annual precipitation reduces the availability of water to be evaporated. Typical values of the Bowen ratio are: 0.1–0.3 for tropical rainforests, where high annual rainfall keeps soil wet year-round; 0.4–0.8 for temperate forests and grasslands, where less rainfall causes drier soils; 2.0–6.0 for semiarid regions with extremely dry soils; and greater than 10.0 for deserts (Oke 1987, p. 70).

Budyko (1974, pp. 322–327, 1986, pp. 76–79) proposed a relationship to describe geographic variation in annual evapotranspiration (E, mm yr^{-1}) in relation to annual precipitation (P, mm yr^{-1}) and annual net radiation (R_n, MJ m^{-2} yr^{-1}). A key index is the ratio of net radiation to the amount of energy required to evaporate the annual precipitation. This latter term is given by λP, where λ (MJ kg^{-1}) converts P from mm yr^{-1} (equivalent to kg m^{-2} yr^{-1}) to an energy flux (MJ m^{-2} yr^{-1}), and the ratio $R_n / \lambda P$ is the radiative dryness index. Where the soil is dry, Budyko assumed that runoff is zero and all the

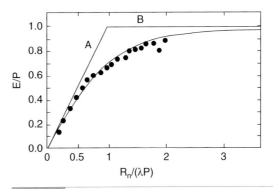

Fig. 12.1 Relationship between the ratio of annual evapotranspiration to precipitation (E / P) and the ratio of net radiation to the amount of energy required to evaporate the annual precipitation ($R_n / \lambda P$). Redrawn from Budyko (1974, p. 325).

precipitation evaporates provided there is sufficient net radiation to evaporate the water:

$$E \to P \text{ for } R_n >> \lambda P, \text{ or, } \frac{E}{P} \to 1 \text{ as } \frac{R_n}{\lambda P} \to \infty \quad (12.4)$$

This is represented by line B in Figure 12.1. Where soil water is plentiful, evapotranspiration is limited not by precipitation but rather by net radiation:

$$E \to \frac{R_n}{\lambda} \text{ for } R_n << \lambda P, \text{ or, } \lambda E \to R_n \text{ as } \frac{R_n}{\lambda P} \to 0$$
$$(12.5)$$

This is represented by line A in Figure 12.1. These two lines set the bounds on evapotranspiration. Observations show, in fact, a smooth transition between these two limiting cases. The fitted relationship is:

$$E = \left[\frac{R_n P}{\lambda} \left(\tanh \frac{\lambda P}{R_n} \right) \left(1 - \cosh \frac{R_n}{\lambda P} + \sinh \frac{R_n}{\lambda P} \right) \right]^{1/2}$$
$$(12.6)$$

With $R = P - E$, Eq. (12.6) can be used to derive runoff. Figure 12.2 illustrates these relationships over a range of annual precipitation and net radiation.

Analyses of climate model data show the generality of this relationship (Koster et al. 1999). The Budyko relationship describes evapotranspiration and runoff across large spatial scales. The

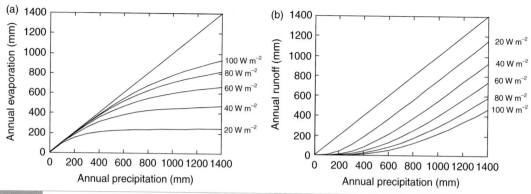

Fig. 12.2 Relationships of (a) annual evapotranspiration and (b) annual runoff with annual precipitation for annual net radiation ranging from 20–100 W m⁻². See also Budyko (1974, p. 327).

constraint imposed by net radiation and precipitation (or instead soil moisture) on latent heat flux and runoff provide an important check on those simulated by climate models (Koster and Mahanama 2012). Deviations from the Budyko relationship provide an indication of the effects of vegetation on evapotranspiration (Zhang et al. 2001; Donohue et al. 2007, 2010; Oudin et al. 2008; Gentine et al. 2012; Williams et al. 2012).

12.4 | Annual Cycle

The energy balance at the land surface varies seasonally due to changes in incoming solar radiation and precipitation. Figure 12.3 illustrates observed monthly surface fluxes for six types of vegetation located in different climate zones. In the Amazon region of Brazil, the climate is tropical rainforest. Solar radiation is relatively constant throughout the year. High amounts of rainfall keep the soil wet, and much of the available energy is used to evaporate water.

Solar radiation has a pronounced annual cycle in middle and high latitudes. Arid climates have high dry season solar radiation because the clear skies allow much of the solar radiation at the top of the atmosphere to reach the ground. The California grassland has a Mediterranean climate with wet, mild winters and dry, hot summers. Sensible heat flux is the dominant flux during summer, when the grasses wither under heat and moisture stress and latent heat

flux is negligible. The opposite occurs during winter, when latent heat flux exceeds sensible heat flux. The Oregon pine forest also experiences wet, cool winters and dry, warm summers. In this locale, sensible heat flux exceeds latent heat flux in most months.

In climates with pronounced cold seasons, latent heat flux is low during winter when there is little energy available to evaporate water and increases markedly in summer. In the temperate continental climate of Massachusetts, the deciduous forest has an annual cycle to latent heat flux that tracks solar radiation. Sensible heat flux exceeds latent heat flux in spring, before budbreak and leaf emergence. Thereafter, latent heat flux exceeds sensible heat flux during the summer months as the foliage sustains high rates of transpiration. A similar seasonal trend occurs in the boreal forest at Manitoba. Here, the cold winter with snow and frozen soil restricts latent heat flux until warmer months. The arctic tundra site similarly has low latent heat flux restricted to a few summer months.

The concepts embodied in Budyko's analysis of net radiation and precipitation as limiting controls of evapotranspiration can be extended to the monthly timescale. Monthly evapotranspiration can be assumed to be limited by the lesser of net radiation (energy to evaporate water) or precipitation (water supply). This, indeed, is the case where there is little inter-seasonal storage of water in soil. In the example shown in Figure 12.4, evapotranspiration closely follows

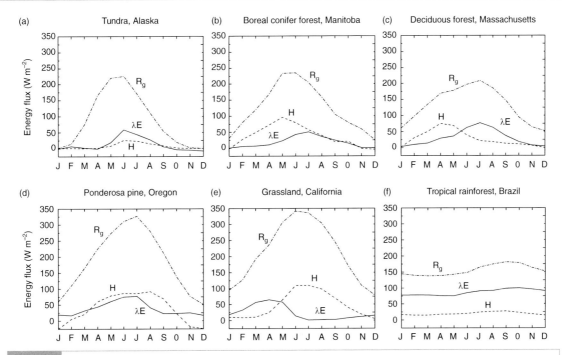

Fig. 12.3 Monthly surface energy fluxes of solar radiation (R_g), latent heat (λE), and sensible heat (H) for (a) moist tundra, Alaska (Schwalm et al. 2010), (b) boreal conifer forest, Manitoba (Goulden et al. 2006), (c) temperate deciduous forest, Massachusetts (Urbanski et al. 2007), (d) ponderosa pine forest, Oregon (Thomas et al. 2009), (e) grassland, California (Ma et al. 2007), and (f) tropical rainforest, Brazil (Miller et al. 2011).

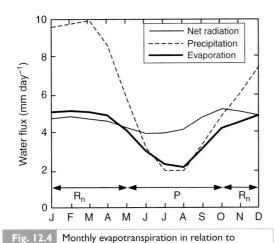

Fig. 12.4 Monthly evapotranspiration in relation to precipitation and net radiation.

precipitation exceeds evapotranspiration sustains evapotranspiration in months when there is energy available to evaporate water but precipitation is low (Milly 1994).

12.5 | Diurnal Cycle

Surface energy fluxes vary over the course of a day in response to the diurnal cycle of solar radiation. In early morning, the land surface typically has a negative radiative balance because no solar radiation is absorbed but longwave radiation is lost. Sensible and latent heat fluxes are small. During daylight hours, the absorption of solar radiation increases and there is a net gain of radiation at the surface. The land warms and some of this energy is returned to the atmosphere as sensible and latent heat. The remainder warms the soil. Fluxes typically are strongest in early to middle afternoon and decrease late in the afternoon when solar radiation diminishes.

the rate imposed by net radiation during the period November–May when precipitation is plentiful, but precipitation limits evapotranspiration during the dry season of June–October. With storage, however, water gained when

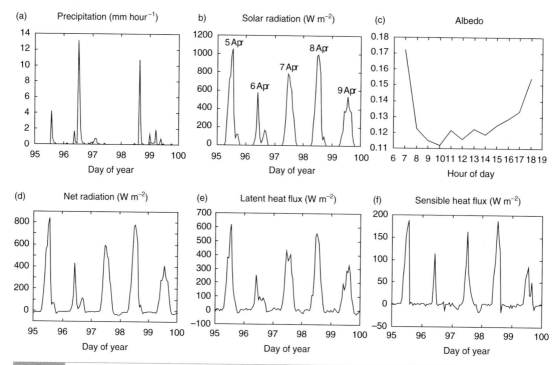

Fig. 12.5 Hourly precipitation, incident solar radiation, net radiation, latent heat flux, and sensible heat flux for a tropical rainforest in southwestern Amazonia over a 5-day period (April 5–9, 1993). Also shown is the average albedo for this period. Data are from the Anglo–Brazilian Amazonian Climate Observation Study (ABRACOS) for a forest near the town of Ji-Paraná in Rondônia. The ABRACOS study is described by Gash et al. (1996) and the tower fluxes by Grace et al. (1995, 1996).

Figure 12.5 illustrates the diurnal cycle over a 5-day period for a tropical rainforest in Amazonia. In general, most of the net radiation during the measurement period was dissipated as latent heat. For this period, the Bowen ratio was 0.20, and the evaporative fraction, defined as $\lambda E / (H + \lambda E)$, was 0.83. The occurrence of rain greatly affected surface fluxes. Cloudy and rainy conditions suppressed incoming solar radiation, net radiation, and latent heat flux on April 6 and 9. In contrast, clear skies on April 5 and 8 allowed strong solar radiation (1000 W m^{-2}) and net radiation (>750 W m^{-2}), with correspondingly high latent heat flux (>550 W m^{-2}). Overall, latent and sensible heat fluxes increased linearly with net radiation over this time period (Figure 12.6).

Figure 12.7 illustrates the diurnal cycle for an aspen forest and a jack pine forest in Canada. On an average summer day, latent heat exchange was an important means of dissipating the energy absorbed by the aspen forest. At midday, about one-half the net radiation absorbed by the forest was returned to the atmosphere as latent heat. This was over twice the sensible heat flux, indicating a well-watered site. The Bowen ratio was 0.13 and the evaporative fraction was 0.88. The jack pine forest had much less latent heat flux than the aspen forest. Midday fluxes were on the order of 100 W m^{-2} compared with 250 W m^{-2} for the aspen forest. Sensible heat was the dominant means to dissipate net radiation. The dry environment of the jack pine forest is seen in a high Bowen ratio (1.45) and low evaporative fraction (0.41).

Figure 12.8 illustrates surface energy fluxes over a 20-day period for grassland in Kansas. No rain fell from July 19 to August 2, 1987 (days 200–214). Latent heat flux was the dominant flux when soil was wet. As the soil dried, latent heat flux declined and sensible heat flux increased.

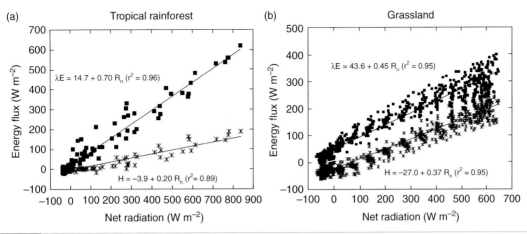

Fig. 12.6 Relationship of latent heat flux (λE) and sensible heat flux (H) with net radiation (R_n) for (a) the tropical rainforest shown in Figure 12.5 and (b) the grassland shown in Figure 12.8.

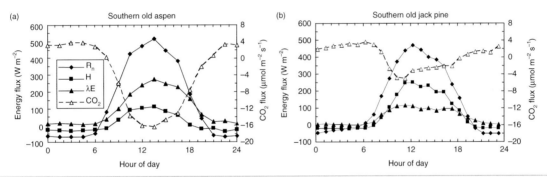

Fig. 12.7 Diurnal cycle of net radiation (R_n), sensible heat (H), latent heat (λE), and CO_2 for two boreal forests in Canada. Data are from the Boreal Ecosystem Atmosphere Study (BOREAS) for (a) a quaking aspen forest in Prince Albert National Park, Saskatchewan and (b) a jack pine forest near Nipawin, Saskatchewan. The tower fluxes are described by Blanken et al. (1997) and Baldocchi et al. (1997), respectively. Data are averaged for the period July 19–August 10, 1994 as described by Bonan et al. (1997).

Rainfall towards the end of the 20-day period wetted the soil and latent heat flux increased from the previous day. Soil heat flux was a significant portion of the energy budget throughout the period. For this period, the Bowen ratio was 0.31 and the evaporative fraction was 0.76. Latent and sensible heat fluxes increased linearly with net radiation (Figure 12.6).

Figure 12.9 illustrates surface energy fluxes for semiarid vegetation in the Sonoran Desert near Tucson, Arizona. During dry periods, latent heat flux was small (about 50 W m⁻²), and most of the net radiation was dissipated as sensible heat. Sensible heat flux peaked in the middle of the afternoon. Soil heat flux was a sizable portion of the surface energy balance and attained peak values before noon. The Bowen ratio averaged for this day was 6.92 and the evaporative fraction was 0.13, both indicative of dry soil. In late August, monsoonal rains wetted the soil and produced a different diurnal cycle. Sensible heat flux decreased substantially compared with the dry day and latent heat flux increased due to the wet soil. The lower Bowen ratio (0.63) and higher evaporative fraction (0.61) are indicative of the wetter soil.

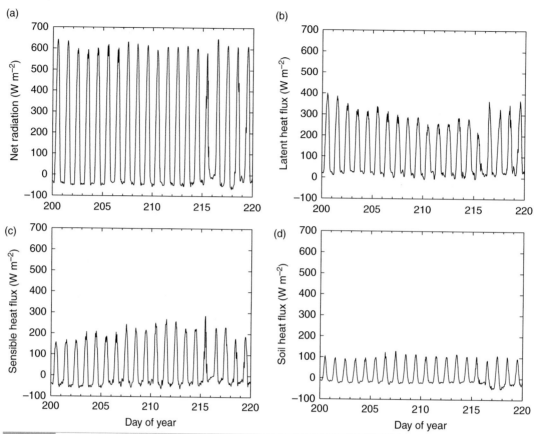

Fig. 12.8 30-minute fluxes of net radiation, latent heat, sensible heat, and soil heat for grassland during the 20-day period July 19–August 7, 1987. Data are from the First ISLSCP (International Satellite Land Surface Climatology Project) Field Experiment (FIFE) and averaged for the 15 km × 15 km study site near Manhattan, Kansas, as described by Betts and Ball (1998).

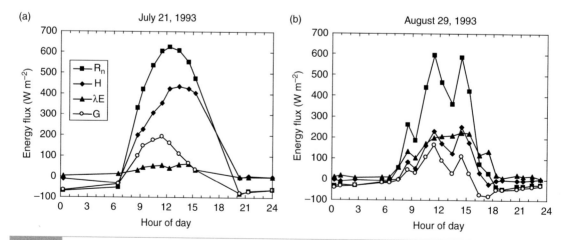

Fig. 12.9 Diurnal cycle of net radiation (R_n), sensible heat (H), latent heat (λE), and soil heat (G) for semiarid Sonoran desert near Tucson, Arizona for (a) a typical dry day with clear sky and (b) wet soil during monsoonal rain. Data from Unland et al. (1996).

12.6 | Energy Balance Model

The effect of different surfaces on energy fluxes and surface climate can be understood using a simple one-dimensional model of the surface energy balance. The principles of these models have been described elsewhere (Gates 1980; Campbell and Norman 1998; Monteith and Unsworth 2013).

12.6.1 Net Radiation

Not all of the solar radiation impinging on the ground heats the surface. Some of the incident solar radiation is reflected; the remainder is absorbed. The amount reflected is:

$$S\uparrow = rS\downarrow \tag{12.7}$$

where $S\downarrow$ is the incident radiation onto the surface and r is the albedo, defined as the fraction of $S\downarrow$ reflected by the surface. The remainder, $(1 - r)S\downarrow$, is the solar radiation absorbed by the surface. Albedo varies with wavelength. Plant canopies, for example, reflect more radiation in the near-infrared waveband (wavelengths greater than 0.7 μm) than in the visible waveband (less than 0.7 μm).

Albedo varies over the course of a day depending on solar zenith angle (Figure 12.5c), and the albedo of the land surface also varies seasonally depending on the amount of leaves, snow cover, and soil moisture. In seasonally snow-covered regions, for example, the surface albedo of cropland is much greater during winter than that of forests (Figure 12.10). Trees protrude above the snow and mask the high albedo of the underlying snow. Some generalizations are evident. Albedo typically ranges from 0.80–0.95 for fresh snow to as little as 0.03–0.10 for water at low solar zenith angle (Table 12.1). Snow, deserts, and glaciers have high albedo. Urban surfaces have low albedo. Vegetation has low albedo, typically ranging from 0.05 to 0.25, with forests absorbing more solar radiation than grasslands or croplands. Soil albedo generally decreases with coarser particle size. Coarse soil particles trap radiation through multiple reflections among adjacent particles. In contrast, fine soil particles expose a relatively uniform surface, trapping less radiation. Soil albedo also decreases with soil wetness because radiation is trapped by internal reflection.

Table 12.1 | Broadband albedo of various surfaces

Surface	Albedo
Natural	
Fresh snow	0.80–0.95
Old snow	0.45–0.70
Desert	0.20–0.45
Glacier	0.20–0.40
Soil	0.05–0.40
Cropland	0.18–0.25
Grassland	0.16–0.26
Deciduous forest	0.15–0.20
Coniferous forest	0.05–0.15
Water	0.03–0.10
Urban	
Road	0.05–0.20
Roof	0.08–0.35
Wall	0.10–0.40
White paint	0.50–0.90
Red, brown, green paint	0.20–0.35
Black paint	0.02–0.15

Source: From Oke (1987, p. 12, p. 281).

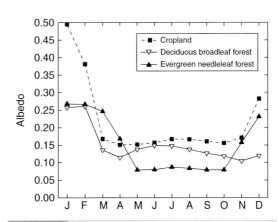

Fig. 12.10 Satellite observations of monthly surface albedo averaged for three land cover types in Northeast United States. Data from Jackson et al. (2008). See also Zhao and Jackson (2014) for a larger analysis of albedo and land cover.

Similarly, not all the longwave radiation incident on the ground heats the surface. Terrestrial objects emit electromagnetic radiation in the infrared band at long wavelengths between 3 and 100 μm. This emission is proportional to temperature raised to the fourth power. In addition, the surface reflects some of the incoming longwave radiation. The outgoing longwave radiation is:

$$L\uparrow = \varepsilon\sigma T_s^4 + (1-\varepsilon)L\downarrow \qquad (12.8)$$

where T_s is surface temperature (K) and $\sigma = 5.67 \times 10^{-8}$ W m^{-2} K^{-4} is the Stefan–Boltzmann constant. The emissivity (ε) of land surfaces generally ranges from 0.95 to 1.0. The emissivity of a surface is also its absorptivity, and $(1-\varepsilon)L\downarrow$ is the longwave radiation reflected by the surface. A surface with $\varepsilon = 1$ emits 448 W m^{-2} with $T_s = 25°C$ (298.15 K).

The net radiation at the surface is:

$$R_n = (1-r)S\downarrow + \varepsilon L\downarrow - \varepsilon\sigma T_s^4 \qquad (12.9)$$

The first two terms are the absorbed radiation, and the last term is the emitted radiation, as described also by Eq. (3.7).

12.6.2 Sensible Heat Flux

Sensible heat flux is represented by Fick's law using Eq. (3.21) and is proportional to the temperature difference between the surface and surrounding air multiplied by a conductance:

$$H = -c_p(\theta_a - T_s)g_{ah} \qquad (12.10)$$

Here, θ_a is the potential temperature of air (K) and c_p is the molar specific heat of moist air at constant pressure ($c_p = 29.2$ J mol^{-1} K^{-1} is a representative value). The conductance (g_{ah}, mol m^{-2} s^{-1}) depends on atmospheric turbulence and surface characteristics. A typical value is on the order 0.4–4 mol m^{-2} s^{-1}. A surface loses energy if its temperature is warmer than the air (a positive flux); it gains energy when it is colder than the air (a negative flux). A positive sensible heat flux warms the overlying air. For example, a surface that is 5°C warmer than the air has a sensible heat flux of 292 W m^{-2} with $g_{ah} = 2$ mol m^{-2} s^{-1}.

Potential temperature is the temperature a parcel of air at some height above the surface would have if it was brought adiabatically from its actual pressure to the surface pressure. Potential temperature is used instead of temperature in boundary layer meteorology because it is independent of pressure or volume and is conserved in vertical motion. The potential temperature at height z (m) above the surface can be approximated as $\theta_a = T_a + z\Gamma_d$, where $\Gamma_d = 0.0098$ K m^{-1} is the dry adiabatic lapse rate. If air temperature is measured close to the surface (e.g., 20–30 m), the two quantities are similar and are often interchanged.

For ground and vegetated surfaces, the conductance (g_{ah}) depends on atmospheric turbulence as influenced by surface roughness, from Eq. (13.26). Sensible heat is exchanged between land and atmosphere because of turbulent mixing of air and resultant transport of heat, generally away from the surface during the day. Turbulence occurs when wind flows over Earth's surface and the ground, trees, grasses, and other objects retard the fluid motion of air. Taller objects are rougher than shorter objects, exert more drag on air flow, and generate more turbulence, all other factors being equal.

12.6.3 Latent Heat Flux

Evapotranspiration is represented by Fick's law using Eq. (3.20) and is proportional to the vapor pressure deficit between the surface and surrounding air multiplied by a conductance. In molar notation:

$$E = -\frac{\left[e_a - e_*(T_s)\right]}{P}g_w \qquad (12.11)$$

The flux (E) and the diffusive conductance (g_w) both have units mol H$_2$O m^{-2} s^{-1} (the former is the mass flux divided by the molecular mass of water); and e_a is the vapor pressure of air, $e_*(T_s)$ is the saturation vapor pressure evaluated at the surface temperature, and here P is surface air pressure (Pa). Multiplying by the latent heat of vaporization (λ, given here as J mol^{-1}) converts to λE with units W m^{-2}. At 15°C, 2466 J are required to change 1 g of water from liquid to vapor (equivalent to 44.44 kJ mol^{-1} because the molecular mass of water is 18.02 g mol^{-1}).

A positive flux means loss of heat and water to the atmosphere.

The term $e_*(T_s) - e_a$ is the vapor pressure deficit between the evaporating surface, which is saturated with moisture, and air. Saturation vapor pressure increases exponentially with higher temperature (Figure 3.5) so that evapotranspiration increases with higher temperature, all other factors being equal. Consider, for example, a surface at sea level ($P = 101,325$ Pa) with $T_s = 25°C$. Its saturation vapor pressure is $e_*(T_s) = 3167$ Pa. If the surrounding air has vapor pressure $e_a = 1584$ Pa (50% relative humidity at 25°C), the latent heat flux is 278 W m^{-2} with $g_w = 0.40$ mol m^{-2} s^{-1}. Latent heat flux increases to 466 W m^{-2} when the same air is in contact with a surface with $T_s = 30°C$ (saturation vapor pressure is 4243 Pa).

The scientific literature describes evapotranspiration in various forms. A common expression in the biometeorology community is:

$$\lambda E = -\frac{c_p}{\gamma}\Big[e_a - e_*(T_s)\Big]g_w \tag{12.12}$$

where $\gamma = c_p P / \lambda$ is the psychrometric constant ($\gamma = 66.5$ Pa K^{-1} is representative). The boundary layer community favors specific humidity ($q = \rho_v / \rho$), given by Eq. (3.31). An equivalent form of the evaporative flux is:

$$E = -\rho\Big[q_a - q_*(T_s)\Big]g_w \tag{12.13}$$

Here, E has units kg m^{-2} s^{-1}, g_w is given in m s^{-1}, and ρ is the density of moist air (kg m^{-3}), specified by Eq. (3.29). This conductance (m s^{-1}) is related to molar conductance (mol m^{-2} s^{-1}) by Eq. (3.18).

The conductance (g_w) depends on an aerodynamic conductance (g_{aw}) that varies with atmospheric turbulence using Eq. (13.26), but also decreases as the surface becomes drier so that a dry site has a lower latent heat flux than a wet site, all other factors being equal. This latter effect is represented by a surface conductance (g_c), and the total conductance is equal to the two conductances acting in series:

$$g_w = \frac{1}{g_c^{-1} + g_{aw}^{-1}} = \frac{g_c g_{aw}}{g_c + g_{aw}} \tag{12.14}$$

Typical values for g_{aw} and g_c are on the order 0.4–4 mol m^{-2} s^{-1}.

12.6.4 Soil Heat Flux

Heat transfer by conduction between the surface and the underlying soil depends on the temperature gradient and thermal conductivity. For a surface with temperature T_s, the heat transfer to or from soil with a temperature T_g at a depth Δz (m) is:

$$G = \kappa(T_s - T_g) / \Delta z \tag{12.15}$$

where κ is thermal conductivity (W m^{-1} K^{-1}). Thermal conductivity is a measure of an object's ability to conduct heat. The rate of heat transfer by conduction increases with larger temperature gradient and with larger thermal conductivity. Sandy soils generally have a higher thermal conductivity than clay soils (Table 9.1). For a soil with $\kappa = 1.9$ W m^{-1} K^{-1}, the heat flux by conduction is 95 W m^{-2} for a temperature difference of 5°C over a distance of 10 cm.

12.6.5 Surface Temperature

The net radiation that impinges on Earth's land surface is balanced by energy lost or gained through sensible heat, latent heat, and conduction. This balance is maintained by the surface temperature. The energy balance at the surface is:

$$(1-r)S\downarrow + \varepsilon L\downarrow = \varepsilon\sigma T_s^4 + H[T_s] + \lambda E[T_s] + G[T_s] \tag{12.16}$$

The left-hand side of this equation constitutes the radiative forcing, which is the sum of absorbed solar radiation and absorbed longwave radiation, respectively. The right-hand side of the equation consists of the emitted longwave radiation, sensible heat flux, latent heat flux, and heat storage in soil by conduction, each of which is specified as a function of surface temperature (T_s). This temperature is the temperature that balances the surface energy budget and Eq. (12.16) can be solved for T_s. However, this equation is difficult to solve without numerical methods, because of the non-linear dependence of longwave radiation on T_s^4 and latent heat flux on saturation vapor pressure, $e_*(T_s)$.

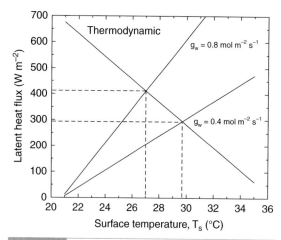

Thermodynamic, Eq. (12.17), and aerodynamic, Eq. (12.19), formulations of latent heat flux in relation to surface temperature. For this example, $R_n - G = 500$ W m^{-2} and $T_a = 25$°C. The aerodynamic equation is shown for $g_w = 0.4$ and 0.8 mol m^{-2} s^{-1}. The thermodynamic and aerodynamic equations intersect at the correct surface temperature.

The Penman–Monteith equation provides a simplified calculation for latent heat flux and T_s.

12.7 | Penman–Monteith Equation

Penman (1948) combined the thermodynamic and aerodynamic aspects of evaporation into a mathematical equation that provides a simple means to study the surface energy budget and surface temperature (see review by Monteith 1981). The evaporation of water from a saturated surface is a thermodynamic process in which energy is required to change water from liquid to vapor. From Eqs. (12.2) and (12.10), latent heat flux is:

$$\lambda E = (R_n - G) - H = (R_n - G) - c_p(T_s - T_a)g_{ah} \quad (12.17)$$

Latent heat flux increases as available energy $(R_n - G)$ increases and decreases as sensible heat flux (H) increases. For otherwise constant conditions, latent heat flux decreases as surface temperature (T_s) increases (Figure 12.11).

Evapotranspiration is also an aerodynamic process related to the turbulent transport of water vapor away from the surface. This is represented by Eq. (12.12), in which latent heat flux increases as the vapor pressure difference, given by $e_*(T_s) - e_a$, increases. The saturation vapor pressure at the surface temperature is a non-linear function of temperature, but can be approximated by:

$$e_*(T_s) = e_*(T_a) + s(T_s - T_a) \quad (12.18)$$

where $e_*(T_a)$ is the saturation vapor pressure evaluated at the air temperature (T_a) and $s = de_*(T_a)/dT$ is the slope of the saturation vapor pressure versus temperature evaluated at T_a. Substituting Eq. (12.18) into Eq. (12.12):

$$\lambda E = \frac{c_p}{\gamma}\left[e_*(T_a) + s(T_s - T_a) - e_a\right]g_w \quad (12.19)$$

Here, latent heat flux increases as surface temperature (T_s) increases (Figure 12.11).

Latent heat flux is obtained by finding the surface temperature that satisfies Eqs. (12.17) and (12.19). From Eq. (12.17):

$$T_s - T_a = \frac{(R_n - G) - \lambda E}{c_p g_{ah}} \quad (12.20)$$

Substituting this expression for $T_s - T_a$ into Eq. (12.19):

$$\lambda E = \frac{s(R_n - G) + c_p\left[e_*(T_a) - e_a\right]g_{ah}}{s + \gamma(g_{ah}/g_w)} \quad (12.21)$$

This equation shows that evapotranspiration is a weighted linear combination of available energy $(R_n - G)$ and the vapor pressure deficit of air $(e_*(T_a) - e_a)$.

Figure 12.11 illustrates the solution of these equations. In this example, $g_w = 0.4$ mol m^{-2} s^{-1} gives $T_s = 29.7$°C, and the available energy (500 W m^{-2}) is partitioned as $\lambda E = 295$ W m^{-2} and $H = 205$ W m^{-2}. A larger conductance ($g_w = 0.8$ mol m^{-2} s^{-1}) yields greater latent heat flux (412 W m^{-2}), less sensible heat flux (88 W m^{-2}), and a cooler surface (27.0°C). In general, surface temperature decreases as latent heat flux increases, all other factors being equal.

The original derivation by Penman (1948) was for evaporation over open water. Equation (12.21) has since been extended to a leaf (Chapter 15) or plant canopy (Chapter 17) by substituting the appropriate conductances and is commonly referred to as the Penman–Monteith equation (Monteith 1965). The total conductance for evapotranspiration (g_w) consists of an aerodynamic conductance (g_{aw}) and a surface conductance (g_c) acting in series, given by Eq. (12.14), and it is common to assume $g_{ah} = g_{aw}$. Then:

$$\lambda E = \frac{s(R_n - G) + c_p \left[e_*(T_a) - e_a \right] g_{ah}}{s + \gamma(1 + g_{ah} / g_c)} \tag{12.22}$$

12.8 | Soil Microclimates

The use of paper, hay, and black plastic mulches to cover soil illustrates the effect of surface properties on energy fluxes and surface temperature (Table 12.2). These mulches alter surface albedo, surface conductance for evaporation, and thermal conductivity. At midday on the warm summer day when measurements were made, the uncovered soil had a surface temperature of 40°C. Almost one-third of the net radiation was dissipated as latent heat. All mulches reduced evaporation compared with bare soil, but they differed in their effect on surface temperature. The soil covered with paper was similar in temperature to the bare soil. Although evaporation decreased compared with the bare soil, the surface absorbed much less radiation due to the high albedo of the light-colored surface. In contrast, soil covered with black plastic had a temperature of 52°C. This material was hot because of its low albedo and because the plastic barrier prevented evaporation. Instead, most of the net radiation was dissipated as sensible heat. The hay mulch also was hot (51°C). In this case, the surface absorbed radiation similar to the bare soil. However, the mulch reduced evaporation and hindered heat transfer to the underlying soil due to low thermal conductivity.

Surface albedo can be altered to change soil temperature. One study applied a white powder to reduce soil temperature (Table 12.3). The untreated soil reflected 30 percent of incoming solar radiation. Net radiation was partitioned primarily as latent heat. Whitening the soil surface increased albedo so that the soil reflected

Table 12.2 Midday summer energy balance (W m⁻²) and temperature (°C) for bare ground and soil covered with paper, hay, and black plastic mulch

	Bare ground	Mulch		
		Paper	Hay	Black plastic
Net radiation, R_n	642	433	607	712
Sensible heat, H	362	349	489	635
Latent heat, λE	195	42	84	0
Soil heat, G	85	42	35	77
Temperature, T_s	40	40	51	52

Source: From Rosenberg et al. (1983, p. 196).

Table 12.3 Surface energy budget (MJ m⁻² day⁻¹) and surface temperature (°C) for untreated and whitened soils

	Untreated soil	Whitened soil
Incoming solar radiation, $S\downarrow$	27.2	27.2
Reflected solar radiation, $S\uparrow$	8.2	16.3
Net longwave radiation, $L\downarrow - L\uparrow$	−12.8	−9.8
Net radiation, R_n	6.2	1.1
Sensible heat, H	1.9	−2.5
Latent heat, λE	4.2	3.4
Soil heat, G	0.2	0.2
Temperature, T_s	33	28

Source: From Stanhill (1965).

60 percent of incoming solar radiation. Less radiation was available to warm the soil or evaporate water. The soil cooled by 5°C and evaporation decreased by 19 percent. Sensible heat was now transferred from air to the colder surface.

12.9 | Review Questions

1. A soil has energy fluxes R_n = 450 W m^{-2}, H = 350 W m^{-2}, and λE = 45 W m^{-2}. Calculate the rate of warming. Another soil has energy fluxes R_n = 600 W m^{-2}, H = 475 W m^{-2}, and λE = 85 W m^{-2}. Which soil warms faster? Assume c_v = 2.5 MJ m^{-3} K^{-1} and Δz = 0.5 m.

2. Calculate latent and sensible heat fluxes for a site with R_n = 250 W m^{-2}, G = 0.3R_n, and Bowen ratio 0.5, 1.0, and 2.0. Which site has the highest latent heat flux?

3. Annual precipitation is 800 mm and annual net radiation is 70 W m^{-2}. Which is more likely to limit annual evapotranspiration: energy or water?

4. Use Eq. (12.6) to calculate annual evapotranspiration for a site with annual precipitation P = 1200 mm and annual net radiation R_n = 60 W m^{-2} and a second site with P = 800 mm and R_n = 100 W m^{-2}. Which site has greater runoff?

5. At a particular site, daily average R_n = 120 W m^{-2} and P = 8 mm day^{-1}. Which is more likely to limit evapotranspiration: energy or water?

6. In Figure 12.6, latent heat flux at the tropical rainforest increases at a faster rate with respect to net radiation (slope = 0.70) than at the grassland (slope = 0.45). Why might this be?

7. Use the Penman–Monteith equation to calculate latent heat flux for the following conditions: $R_n - G$ = 400 W m^{-2}, $e_*(T_a)$ = 3167 Pa and s = 189 Pa K^{-1} (values for T_a = 25°C), g_{ah} = 1.6 mol m^{-2} s^{-1}, and g_c = 1.6 mol m^{-2} s^{-1}. (a) Relative humidity is 75 percent. (b) Relative humidity is 50 percent. Use c_p = 29.2 J mol^{-1} K^{-1} and γ = 66.5 Pa K^{-1}.

8. Calculate surface temperature for 7a and 7b. Why does surface temperature vary with relative humidity?

9. Recalculate latent heat flux and surface temperature for 7a and 7b with g_c = 0.4 mol m^{-2} s^{-1}. Explain the difference.

10. Use Eq. (12.20) and Eq. (12.21) to derive an expression for T_s in relation to $R_n - G$. What is the sensitivity of surface temperature to a 1 W m^{-2} change in available energy? How do g_{ah} and g_w affect this sensitivity?

12.10 | References

Avissar, R., and Mahrer, Y. (1982). Verification study of a numerical greenhouse microclimate model. *Transactions of the American Society of Agricultural Engineers*, 25, 1711–1720.

Baldocchi, D. D., Vogel, C. A., and Hall, B. (1997). Seasonal variation of energy and water vapor exchange rates above and below a boreal jack pine forest canopy. *Journal of Geophysical Research*, 102D, 28939–28951.

Betts, A. K., and Ball, J. H. (1998). FIFE surface climate and site-average dataset 1987–89. *Journal of the Atmospheric Sciences*, 55, 1091–1108.

Blanken, P. D., Black, T. A., Yang, P. C., et al. (1997). Energy balance and canopy conductance of a boreal aspen forest: Partitioning overstory and understory components. *Journal of Geophysical Research*, 102D, 28915–28927.

Bonan, G. B., Davis, K. J., Baldocchi, D., Fitzjarrald, D., and Neumann, H. (1997). Comparison of the NCAR LSM1 land surface model with BOREAS aspen and jack pine tower fluxes. *Journal of Geophysical Research*, 102D, 29065–29075.

Budyko, M. I. (1974). *Climate and Life*. New York: Academic Press.

Budyko, M. I. (1986). *The Evolution of the Biosphere*. Dordrecht: Reidel.

Campbell, G. S., and Norman, J. M. (1998). *An Introduction to Environmental Biophysics*, 2nd ed. New York: Springer-Verlag.

Donohue, R. J., Roderick, M. L., and McVicar, T. R. (2007). On the importance of including vegetation dynamics in Budyko's hydrological model. *Hydrology and Earth System Sciences*, 11, 983–995.

Donohue, R. J., Roderick, M. L., and McVicar, T. R. (2010). Can dynamic vegetation information improve the accuracy of Budyko's hydrological model? *Journal of Hydrology*, 390, 23–34.

Gash, J. H. C., Nobre, C. A., Roberts, J. M., and Victoria, R. L. (1996). An overview of ABRACOS. In *Amazonian Deforestation and Climate*, ed. J. H. C. Gash, C. A. Nobre,

J. M. Roberts, and R. L. Victoria. New York: Wiley, pp. 1–14.

Gates, D. M. (1980). *Biophysical Ecology.* New York: Springer-Verlag.

Gentine, P., D'Odorico, P., Lintner, B. R., Sivandran, G., and Salvucci, G. (2012). Interdependence of climate, soil, and vegetation as constrained by the Budyko curve. *Geophysical Research Letters*, 39, L19404, doi:10.1029/2012GL053492.

Goulden, M. L, Winston, G. C., McMillan, A. M. S., et al. (2006). An eddy covariance mesonet to measure the effect of forest age on land–atmosphere exchange. *Global Change Biology*, 12, 2146–2162.

Grace, J., LLoyd, J., McIntyre, J., et al. (1995). Fluxes of carbon dioxide and water vapour over an undisturbed tropical forest in south-west Amazonia. *Global Change Biology*, 1, 1–12.

Grace, J., LLoyd, J., McIntyre, J., et al. (1996). Carbon dioxide flux over Amazonian rain forest in Rondônia. In *Amazonian Deforestation and Climate*, ed. J. H. C. Gash, C. A. Nobre, J. M. Roberts, and R. L. Victoria. New York: Wiley, pp. 307–318.

Jackson, R. B., Randerson, J. T., Canadell, J. G., et al. (2008). Protecting climate with forests. *Environmental Research Letters*, 3, 044006, doi:10.1088/1748-9326/3/4/044006.

Koster, R. D., and Mahanama, S. P. P. (2012). Land surface controls on hydroclimatic means and variability. *Journal of Hydrometeorology*, 13, 1604–1620.

Koster, R. D., Oki, T., and Suarez, M. J. (1999). The offline validation of land surface models: Assessing success at the annual timescale. *Journal of the Meteorological Society of Japan*, 77, 257–363.

Ma, S., Baldocchi, D. D., Xu, L., and Hehn, T. (2007). Inter-annual variability in carbon dioxide exchange of an oak/grass savanna and open grassland in California. *Agricultural and Forest Meteorology*, 147, 157–171.

Mahrer, Y., Avissar, R., Naot, O., and Katan, J. (1987). Intensified soil solarization with closed greenhouses: Numerical and experimental studies. *Agricultural and Forest Meteorology*, 41, 325–334.

Miller, S. D., Goulden, M. L., Hutyra, L. R., et al. (2011). Reduced impact logging minimally alters tropical rainforest carbon and energy exchange. *Proceedings of the National Academy of Sciences USA*, 108, 19431–19435.

Milly, P. C. D. (1994). Climate, soil water storage, and the average annual water balance. *Water Resources Research*, 30, 2143–2156.

Monteith, J. L. (1965). Evaporation and environment. In *The State and Movement of Water in Living Organisms (19th Symposia of the Society for Experimental Biology)*, ed. G. E. Fogg. New York: Academic Press, pp. 205–234.

Monteith, J. L. (1981). Evaporation and surface temperature. *Quarterly Journal of the Royal Meteorological Society*, 107, 1–27.

Monteith, J. L., and Unsworth, M. H. (2013). *Principles of Environmental Physics*, 4th ed. Amsterdam: Elsevier.

Oke, T. R. (1987). *Boundary Layer Climates*, 2nd ed. London: Routledge.

Oudin, L., Andréassian, V., Lerat, J., and Michel, C. (2008). Has land cover a significant impact on mean annual streamflow? An international assessment using 1508 catchments. *Journal of Hydrology*, 357, 303–316.

Penman, H. L. (1948). Natural evaporation from open water, bare soil and grass. *Proceedings of the Royal Society of London*, 193A, 120–145.

Rosenberg, N. J., Blad, B. L., and Verma, S. B. (1983). *Microclimate: The Biological Environment*, 2nd ed. New York: Wiley.

Schwalm, C. R., Williams, C. A., Schaefer, K., et al. (2010). A model-data intercomparison of CO_2 exchange across North America: results from the North American Carbon Program site synthesis. *Journal of Geophysical Research*, 115, G00H05, doi:10.1029/2009JG001229.

Stanhill, G. (1965). Observations on the reduction of soil temperature. *Agricultural Meteorology*, 2, 197–203.

Thomas, C. K., Law, B. E., Irvine, J., et al. (2009). Seasonal hydrology explains interannual and seasonal variation in carbon and water exchange in a semiarid mature ponderosa pine forest in central Oregon. *Journal of Geophysical Research*, 114, G04006, doi:10.1029/2009JG001010.

Trenberth, K. E., Fasullo, J. T., and Kiehl, J. (2009). Earth's global energy budget. *Bulletin of the American Meteorological Society*, 90, 311–323.

Unland, H. E., Houser, P. R., Shuttleworth, W. J., and Yang, Z.-L. (1996). Surface flux measurement and modeling at a semi-arid Sonoran Desert site. *Agricultural and Forest Meteorology*, 82, 119–153.

Urbanski, S., Barford, C., Wofsy, S., et al. (2007). Factors controlling CO_2 exchange on timescales from hourly to decadal at Harvard Forest. *Journal of Geophysical Research*, 112, G02020, doi:10.1029/2006JG000293.

Wang, K., and Dickinson, R. E. (2012). A review of global terrestrial evapotranspiration: observation, modeling, climatology, and climatic variability. *Reviews of Geophysics*, 50, RG2005, doi:10.1029/2011RG000373.

Williams, C. A., Reichstein, M., Buchmann, N., et al. (2012). Climate and vegetation controls on the surface water balance: synthesis of evapotranspiration measured across a global network of flux towers. *Water Resources Research*, 48, W06523, doi:10.1029/2011WR011586.

Zhang, L., Dawes, W. R., and Walker, G. R. (2001). Response of mean annual evapotranspiration to vegetation changes at catchment scale. *Water Resources Research*, 37, 701–708.

Zhao, K., and Jackson, R. B. (2014). Biophysical forcings of land-use changes from potential forestry activities in North America. *Ecological Monographs*, 84, 329–353.

13

Turbulent Fluxes

13.1 Chapter Summary

The fluxes of sensible and latent heat occur because turbulent mixing of air transports heat and moisture, typically away from the surface. These fluxes are described in terms of transport by mean motion (a small term) and by turbulence (the dominant term). Turbulent transport is quantified by the covariance of temperature and moisture fluctuations with vertical velocity fluctuations. Turbulent fluxes of sensible heat, latent heat, and momentum are related to logarithmic profiles of temperature, water vapor, and wind near the surface. Monin–Obukhov similarity theory describes these profiles and fluxes in the surface layer of the atmosphere and is used to derive the aerodynamic conductances (g_{ah}, g_{aw}) that regulate sensible and latent heat fluxes and an additional conductance (g_{am}) for momentum. These conductances depend on roughness length and displacement height, which vary among land cover types. Aerodynamic conductances increase with taller (rougher) vegetation, all other factors being equal.

13.2 Turbulence

The exchanges of sensible and latent heat between land and atmosphere occur because of turbulent mixing of air and resultant heat and moisture transport. The flow of air can be represented as discrete parcels of air moving vertically and horizontally. These parcels have properties such as temperature, water vapor, and momentum (mass × velocity). As the parcels of air move, they carry with them their heat, water vapor, and momentum. Turbulence creates eddies that mix air from above downward and from below upward and transports heat and water vapor in relation to the temperature and moisture of the parcels of air being mixed (Figure 13.1). For example, air near the surface is generally warmer and moister than air above. Downward movements of air, therefore, tend to decrease temperature and water vapor while upward movements of air carry heat and moisture from the surface higher into the atmosphere.

The importance of turbulence in mixing air in the atmospheric boundary layer near the surface is illustrated with two simple examples. Over the course of a summer day, productive vegetation can absorb 5–6 g C m^{-2} or more during photosynthesis. This is enough to deplete all the CO_2 in the air to a height of 30 m or so. (The molecular weight of dry air is 28.97 g mol^{-1}, while that of carbon is 12 g mol^{-1}. With a CO_2 molar mixing ratio of 380 parts per million by volume and a density of 1.225 kg m^{-3}, a cubic meter of dry air contains $380 \times 10^{-6}(12 / 28.97)1225 = 0.19$ g C m^{-3}. A net uptake of 6 g C m^{-2} would deplete carbon to a height of 31 m.) In practice, however, CO_2 is not depleted because wind mixes the air and replenishes the absorbed CO_2. Likewise, productive vegetation can transpire 5–10 kg m^{-2} (5–10 mm) of water on a hot summer day. This would increase the moisture in

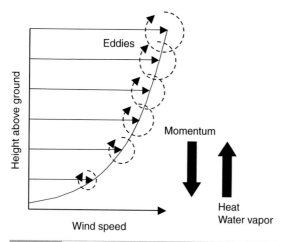

Fig. 13.1 Conceptual diagram of momentum, heat, and water vapor fluxes. Eddies transfer momentum to the surface, while heat and water are generally carried from the warm, moist surface to the cooler, drier atmosphere.

the first 100 m of the atmosphere by 50–100 g m^{-3} if it was not transported away. This is much more water than the atmosphere can hold when saturated. Instead, wind replaces the moist air with drier air.

Near the surface, turbulence generates vertical motions. Turbulence can be generated mechanically due to surface friction (often called forced convection) when wind flows over Earth's surface (Figure 13.1). The ground, trees, grasses, and other objects protruding into the atmosphere exert a retarding force on the fluid motion of air. The frictional drag imparted on air as it encounters these rough surface elements slows the flow of air near the ground. The reduction in wind speed transfers momentum from the atmosphere to the surface, creating turbulence that mixes the air and transports heat and water from the surface into the lower atmosphere. With greater height above the surface, eddies are larger so that transport of momentum, heat, and moisture is more efficient with height above the surface.

Vertical turbulent motion can also occur due to surface heating and buoyancy (often called free convection). In daytime, the surface is typically warmer than the atmosphere and strong solar heating of the land provides a source of buoyant energy. Warm air is less dense than cold air, and rising air enhances mixing and

the transport of heat and moisture away from the surface. In this unstable atmosphere, sensible heat flux is positive, temperature decreases with height, and vertical transport increases with strong surface heating. At night, longwave emission generally cools the surface more rapidly than the air above. The lowest levels of the atmosphere become stable, with cold, dense air trapped near the surface and warmer air above. Sensible heat flux is negative (i.e., toward the surface). These conditions suppress vertical motions and reduce transport.

13.3 | The Statistics of Turbulence

Turbulent motion in the atmosphere is seen in the patterns of smoke emitted from a chimney, the swirling of dust or other debris, the ripples of waves across a lake, or the fluttering of leaves and flags. Yet while turbulence is easily observed, it is difficult to describe mathematically. A deterministic description of turbulence is difficult because of its chaotic behavior. Instead, the randomness and unpredictability of turbulent motion necessitates the use of statistics to characterize turbulence.

Turbulent flow can be described mathematically by representing an instantaneous measurement as its time-mean and its fluctuating component. Consider, for example, a quantity w. Its mean over a specified time interval is $\bar{w}$. At any given time during that period, its fluctuation from the mean is $w' = w - \bar{w}$. In other words, a variable can be represented as the sum of its time-mean and fluctuating components:

$$w = \bar{w} + w' \tag{13.1}$$

The mean represents the component of w that varies slowly over time. The fluctuation from the mean is the turbulent component. The mean of w' is by definition zero. The variance of w is:

$$\sigma^2 = \frac{1}{N} \sum (w_i - \bar{w})^2 = \overline{w'w'} \tag{13.2}$$

As an example, consider the horizontal wind speed data shown in Figure 13.2 measured every 15 minutes over the course of a day. Wind

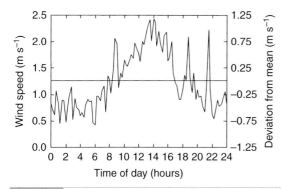

Fig. 13.2 Wind speed (left-hand axis) and deviation from mean (right-hand axis) measured over the course of a day. The thick solid line shows the diurnal mean.

speed ranged from a minimum of 0.43 m s^{-1} to a maximum of 2.42 m s^{-1} and averaged 1.25 m s^{-1}. At any given time, the instantaneous wind speed can be characterized by its deviation from the mean.

A parcel of air has scalar properties such as temperature, water vapor, or CO_2 concentration that are carried with the parcel as it is mixed horizontally and vertically in the atmosphere. (A scalar quantity has only magnitude. A vector quantity has both magnitude and direction. Temperature and gas concentration are scalars. Velocity is a vector quantity.) Indeed, the ability to efficiently mix the properties of air is an important characteristic of turbulence. Consequently, quantities co-vary over time. Consider a second quantity $c = \bar{c} + c'$. The average product $\overline{wc}$ is:

$$\overline{wc} = \bar{w}\,\bar{c} + \overline{w'c'} \qquad (13.3)$$

The term $\overline{w'c'}$ is also the covariance between the two quantities:

$$\text{cov}(w,c) = \frac{1}{N}\sum(w_i - \bar{w})(c_i - \bar{c}) = \overline{w'c'} \qquad (13.4)$$

Variables of interest in the boundary layer include: u (m s^{-1}), the velocity component in the x-direction (zonal wind); v (m s^{-1}), the velocity component in the y-direction (meridional wind); w (m s^{-1}), the vertical velocity; θ (K), the potential temperature; and c (mol mol^{-1}), the concentration of a gas. For water vapor, $c = e/P$ is the mole fraction given by Eq. (3.14). Each of these can be represented as the sum of mean and turbulent components. Of particular importance are the covariances $\overline{w'u'}$, $\overline{w'v'}$, $\overline{w'\theta'}$, and $\overline{w'c'}$, which are related to the turbulent fluxes of momentum, heat, and gases.

13.4 Turbulent Flux Definitions

The vertical flux of a scalar with concentration c is the product of its concentration times the vertical velocity (i.e., wc) so that the mean flux over a period of time is $\overline{wc} = \bar{w}\,\bar{c} + \overline{w'c'}$. The scalar flux is the sum of transport by mean motion ($\bar{w}\,\bar{c}$) and transport by turbulence ($\overline{w'c'}$). Mean transport is generally small. Instead, the latter term, which is called the turbulent flux, is the dominant term.

Scalars to calculate sensible heat flux and evapotranspiration are potential temperature and water vapor mole fraction. Sensible heat flux (H, W m^{-2}) and water vapor flux (E, mol m^{-2} s^{-1}) are:

$$H = \rho_m c_p \overline{w'\theta'}$$
$$E = \rho_m \overline{w'c'} \qquad (13.5)$$

where $\rho_m = P/\Re T$ is molar density (mol m^{-3}) from Eq. (3.11) and c_p is the specific heat of air at constant pressure (J mol^{-1} K^{-1}). The momentum flux (τ, mol m^{-1} s^{-2}) has zonal and meridional components related to u and v:

$$\tau_x = -\rho_m \overline{w'u'}$$
$$\tau_y = -\rho_m \overline{w'v'} \qquad (13.6)$$

(Momentum flux more commonly is given using the density of air (ρ, kg m^{-3}) from Eq. (3.29) with units kg m^{-1} s^{-2}. These equations are related by molecular mass (kg mol^{-1})).

One means to estimate these fluxes, known as eddy covariance, is to measure fluctuations of temperature, water vapor, wind, and vertical velocity and determine the covariance of the variable of interest with vertical velocity. Similar methodology is used to measure CO_2 flux. Eddy covariance has become a standard technique to measure turbulent fluxes at the

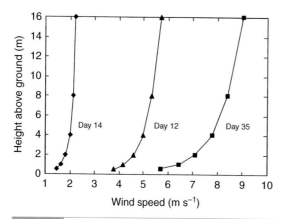

Fig. 13.3 Wind profile above sparse grassland in southeastern Australia under near-neutral conditions measured at 1630 hours on days 12, 14, and 35 of the Wangara experiment. Data from Clarke et al. (1971).

surface (Baldocchi et al. 1988; Aubinet et al. 2000, 2012; Baldocchi 2003).

13.5 | Logarithmic Wind Profiles

Under near-neutral atmospheric conditions, the wind profile in the atmosphere over flat homogenous terrain increases logarithmically with height. For example, Figure 13.3 shows winds speeds measured to a height of 16 m above sparse grassland in southeastern Australia. Wind speed is lowest near the surface and increases with greater height. The logarithmic wind profile is related to the momentum flux, which is nearly constant with height in the layer of the atmosphere near the surface (the lowest 20–50 m).

The logarithmic wind profile is derived by defining a velocity scale (u_*, m s^{-1}), also called the friction velocity, that is related to surface stress by:

$$u_*^2 = \tau / \rho_m = \left[\left(\overline{w'u'}\right)^2 + \left(\overline{w'v'}\right)^2\right]^{1/2} \qquad (13.7)$$

For convenience, the x-axis is oriented along the direction of surface wind so that $v = 0$, and:

$$u_*^2 = \tau / \rho_m = -\overline{w'u'} \qquad (13.8)$$

The momentum flux is related to the mean vertical gradient of wind by:

$$\tau = \rho_m k u_* z \left(\partial \overline{u} / \partial z\right) \qquad (13.9)$$

Here, z is height above the surface and $k = 0.4$ is the von Karman constant. Combining Eq. (13.8) and Eq. (13.9) gives:

$$\left(z / u_*\right)\partial \overline{u} / \partial z = 1 / k \qquad (13.10)$$

This equation states that when scaled by the velocity scale (u_*) the mean vertical wind gradient ($\partial \overline{u} / \partial z$) depends only on height above the surface (z). Stated another way, the mean vertical wind gradient depends solely on a characteristic velocity scale (u_*) and a characteristic length scale (z).

Integrating $\partial \overline{u} / \partial z$ between two heights ($z_2 > z_1$) gives:

$$\overline{u}_2 - \overline{u}_1 = \left(u_* / k\right)\ln\left(z_2 / z_1\right) \qquad (13.11)$$

The surface is defined as $\overline{u}_1 = 0$ at $z_1 = z_{0m}$ so that wind speed at height z is:

$$\overline{u}(z) = \left(u_* / k\right)\ln\left(z / z_{0m}\right) \qquad (13.12)$$

The term z_{0m} is known as the roughness length for momentum. It is the theoretical height at which wind speed is zero. Table 13.1 gives representative roughness lengths. Roughness length is generally less than 0.01 m for soil and increases with vegetation.

The height (z) in Eq. (13.12) is height above the ground surface. Over some surfaces, however, the protrusion of roughness elements above the surface displaces turbulent flow upward. Trees, for example, extend into the atmosphere from the ground. In this case, height must be adjusted relative to a reference height known as the zero-plane displacement height (d) and Eq. (13.12) becomes:

$$\overline{u}(z) = \left(u_* / k\right)\ln\left[(z - d) / z_{0m}\right] \qquad (13.13)$$

The displacement height is the vertical displacement caused by surface elements. This displacement height is zero for bare ground and greater than zero for vegetation. The height $z = d + z_{0m}$

Table 13.1	Roughness length of various surfaces
Surface	Roughness length (m)
Soil	0.001–0.01
Grass	
Short	0.003–0.01
Tall	0.04–0.10
Crop	0.04–0.20
Forest	1.0–6.0
Suburban	
Low density	0.4–1.2
High density	0.8–1.8
Urban	
Short building	1.5–2.5
Tall building	2.5–10

Source: From Oke (1987, pp. 57, 298).

is the height where the wind profile extrapolates to zero and is known as the apparent sink of momentum. The logarithmic wind profile is valid only for heights $z \gg d + z_{0m}$ and does not describe wind within plant canopies. Also, it only applies to an extensive uniform surface of 1 km² or more.

Vegetation increases surface roughness in relation to canopy height and density. Short grass is a smoother surface than tall grass. Forests are aerodynamically rougher than crops or grasses. Although they are in fact complex functions of plant canopies, it is often assumed that roughness length for vegetation is approximately one-tenth canopy height and displacement height is approximately seven-tenths canopy height (Campbell and Norman 1998; Shuttleworth 2012; Monteith and Unsworth 2013).

13.6 | Monin–Obukhov Similarity Theory

Equation (13.13) describes the profile of wind ($\bar{u}$) in relation to the flux of momentum ($u_*^2 = \tau / \rho_m$). It can be generalized from near-neutral conditions to all atmospheric conditions and extended to include flux–profile relationships for temperature and water vapor. The turbulent fluxes of heat, water vapor, and momentum are nearly constant with height in the layer of the atmosphere near the surface. This region of the atmosphere is known as the surface, or constant flux, layer and generally extends 20–50 m or so above the ground. The Monin–Obukhov similarity theory relates turbulent fluxes of momentum, sensible heat, and moisture in the surface layer to mean vertical gradients of wind, potential temperature, and water vapor.

The Monin–Obukhov similarity theory states that when scaled appropriately the dimensionless mean vertical gradients of wind ($\bar{u}$), potential temperature ($\bar{\theta}$), and water vapor ($\bar{c}$) are unique functions of a buoyancy parameter (ζ):

$$\left[\frac{k(z-d)}{u_*}\right]\frac{\partial \bar{u}}{\partial z} = \phi_m(\zeta)$$

$$\left[\frac{k(z-d)}{\theta_*}\right]\frac{\partial \bar{\theta}}{\partial z} = \phi_h(\zeta) \quad (13.14)$$

$$\left[\frac{k(z-d)}{c_*}\right]\frac{\partial \bar{c}}{\partial z} = \phi_w(\zeta)$$

The functions $\phi_m(\zeta)$, $\phi_h(\zeta)$, and $\phi_w(\zeta)$ are universal similarity functions that relate the constant fluxes of momentum, sensible heat, and water vapor to the mean vertical gradients of wind, temperature, and moisture in the surface layer. The characteristic velocity (u_*), temperature (θ_*), and moisture (c_*) scales are:

$$u_* u_* = -\overline{w'u'} = \tau / \rho_m$$

$$\theta_* u_* = -\overline{w'\theta'} = -H / \rho_m c_p \quad (13.15)$$

$$c_* u_* = -\overline{w'c'} = -E / \rho_m$$

Eq. (13.14) and Eq. (13.15) are extensions of Eq. (13.8) and Eq. (13.10) to account for buoyancy, with similar flux–profile equations for temperature and water vapor.

The similarity functions $\phi_m(\zeta)$, $\phi_h(\zeta)$, and $\phi_w(\zeta)$ represent the effect of buoyancy on turbulence, with:

$$\zeta = (z-d)/L \quad (13.16)$$

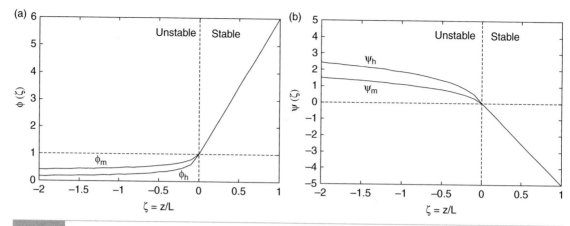

Fig. 13.4 Similarity functions (a) $\phi_m(\zeta)$ and $\phi_h(\zeta)$ and (b) $\psi_m(\zeta)$ and $\psi_h(\zeta)$ in relation to $\zeta = z / L$.

and L (m) is the Obukhov length scale. The length scale is:

$$L = -u_*^3 \theta_v / \left[kg\left(H_v / \rho_m c_p\right)\right] \qquad (13.17)$$

where $g = 9.81$ m s^{-2} is gravitational acceleration and θ_v is the potential temperature. Positive values indicate a stable atmosphere. Negative values indicate an unstable atmosphere. Moist air is less dense than dry air. To account for the effect of water vapor on buoyancy, potential temperature (θ) and sensible heat flux (H) are replaced by virtual potential temperature (θ_v) and virtual sensible heat (H_v), approximated in relation to specific humidity (q, kg kg^{-1}) by:

$$\begin{aligned} \theta_v &= \theta(1 + 0.61q) \\ H_v &= H + 0.61 c_p \theta E \end{aligned} \qquad (13.18)$$

Representative similarity functions, from Brutsaert (1982, pp. 68–71), Garratt (1992, pp. 52–54) and Shuttleworth (2012, p. 294), but see also Brutsaert (2005, pp. 46–50), are:

$$\begin{aligned} \phi_m^2(\zeta) &= \phi_h(\zeta) = \phi_w(\zeta) = (1 - 16\zeta)^{-1/2} \quad \text{for } \zeta < 0 \text{ (unstable)} \\ \phi_m(\zeta) &= \phi_h(\zeta) = \phi_w(\zeta) = 1 + 5\zeta \quad \text{for } \zeta \geq 0 \text{ (stable)} \end{aligned} \qquad (13.19)$$

These functions have values of <1 when the atmosphere is unstable and >1 when the atmosphere is stable (Figure 13.4a).

Integrating $\partial \bar{u} / \partial z$, given by Eq. (13.14), between two heights ($z_2 > z_1$) gives the profile equation:

$$\bar{u}_2 - \bar{u}_1 = \frac{u_*}{k}\left[\ln\left(\frac{z_2 - d}{z_1 - d}\right) - \psi_m\left(\frac{z_2 - d}{L}\right) + \psi_m\left(\frac{z_1 - d}{L}\right)\right] \qquad (13.20)$$

The surface is defined as $\bar{u}_1 = 0$ at $z_1 = d + z_{0m}$, and $\psi_m(z_{0m}/L)$ is small so that wind speed at height z is:

$$\bar{u}(z) = \frac{u_*}{k}\left[\ln\left(\frac{z - d}{z_{0m}}\right) - \psi_m(\zeta)\right] \qquad (13.21)$$

Equation (13.21) is analogous to Eq. (13.13), but with the function $\psi_m(\zeta)$ to account for atmospheric stability. Similarly, $\partial \bar{\theta} / \partial z$ and $\partial \bar{c} / \partial z$ integrated between height z and the surface are:

$$\begin{aligned} \bar{\theta}(z) - \bar{\theta}_s &= \frac{\theta_*}{k}\left[\ln\left(\frac{z - d}{z_{0h}}\right) - \psi_h(\zeta)\right] \\ \bar{c}(z) - \bar{c}_s &= \frac{c_*}{k}\left[\ln\left(\frac{z - d}{z_{0w}}\right) - \psi_w(\zeta)\right] \end{aligned} \qquad (13.22)$$

For temperature and water vapor, surface values are defined as $\bar{\theta}_s$ at $z = d + z_{0h}$ and $\bar{c}_s$ at $z = d + z_{0w}$, where z_{0h} and z_{0w} are the roughness lengths for heat and moisture, respectively. Similar to momentum, these are the effective height at which heat and moisture are exchanged with the atmosphere and are known as the apparent sources of heat and moisture. The surface values $\bar{\theta}_s$ and $\bar{c}_s$ are the theoretical values if the logarithmic profiles were extrapolated downward to the surface. Such extrapolation is not valid within plant canopies.

Substituting the definitions of u_*, θ_*, and c_* given by Eq. (13.15) into Eq. (13.21) and Eq. (13.22) gives profiles of wind, temperature, and water vapor in the surface layer in relation to momentum flux (τ), sensible heat flux (H), and water vapor flux (E). When sensible heat flux and evapotranspiration are positive, so that heat and water vapor are exchanged into the atmosphere, temperature and water vapor decrease with height. This typically occurs during the day. At night, when sensible heat flux is negative (i.e., toward the surface), temperature increases with height.

The functions $\psi_m(\zeta)$, $\psi_h(\zeta)$, and $\psi_w(\zeta)$ account for the influence of atmospheric stability on turbulent fluxes and are obtained from Eq. (13.19). For unstable conditions ($\zeta < 0$):

$$\psi_m(\zeta) = 2\ln\left[(1+x)/2\right] + \ln\left[(1+x^2)/2\right] \\ - 2\tan^{-1}x + \pi/2 \tag{13.23}$$

$$\psi_h(\zeta) = \psi_w(\zeta) = 2\ln\left[(1+x^2)/2\right]$$

where $x = (1 - 16\zeta)^{1/4}$. For a stable atmosphere ($\zeta \geq 0$):

$$\psi_m(\zeta) = \psi_h(\zeta) = \psi_w(\zeta) = -5\zeta \tag{13.24}$$

The functions are negative when the atmosphere is stable and positive when the atmosphere is unstable (Figure 13.4b). The vertical gradient is weaker in an unstable surface layer, when turbulence efficiently mixes the air, than when stable (Figure 13.5).

13.7 | Aerodynamic Conductances

The turbulent fluxes can be written as the bulk transfer between the atmosphere at height z with $\bar{u}$, $\bar{\theta}$, and $\bar{c}$ and the surface with $\bar{u}_s = 0$, $\bar{\theta}_s$, and $\bar{c}_s$, in a manner analogous to diffusion. The turbulent fluxes are:

$$\tau = \bar{u}\,g_{am}$$
$$H = -c_p\left(\bar{\theta} - \bar{\theta}_s\right)g_{ah} \tag{13.25}$$
$$E = -\left(\bar{c} - \bar{c}_s\right)g_{aw}$$

Here, g_{am}, g_{ah}, and g_{aw} are aerodynamic conductances (mol m^{-2} s^{-1}) for momentum, heat, and

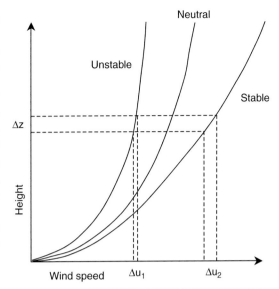

Fig. 13.5 Effect of atmospheric stability on the vertical gradient of wind speed. A given vertical increment (Δz) produces a greater difference in wind speed in stable conditions (Δu_2) than in unstable conditions (Δu_1).

water vapor, respectively, between the atmosphere at height z and the surface:

$$g_{am} = \rho_m k^2 u \left[\ln\left(\frac{z-d}{z_{0m}}\right) - \psi_m(\zeta)\right]^{-1}\left[\ln\left(\frac{z-d}{z_{0m}}\right) - \psi_m(\zeta)\right]^{-1}$$

$$g_{ah} = \rho_m k^2 u \left[\ln\left(\frac{z-d}{z_{0m}}\right) - \psi_m(\zeta)\right]^{-1}\left[\ln\left(\frac{z-d}{z_{0h}}\right) - \psi_h(\zeta)\right]^{-1}$$

$$g_{aw} = \rho_m k^2 u \left[\ln\left(\frac{z-d}{z_{0m}}\right) - \psi_m(\zeta)\right]^{-1}\left[\ln\left(\frac{z-d}{z_{0w}}\right) - \psi_w(\zeta)\right]^{-1}$$

$$\tag{13.26}$$

where u is wind speed (m s^{-1}) at height z. These conductances are obtained from Eq. (13.15) with u_* from Eq. (13.21) and θ_* and c_* from Eq. (13.22). Typical values for a neutral surface layer are 0.4–4 mol m^{-2} s^{-1} (Figure 13.6). Aerodynamic conductance increases with increasing vegetation height due to greater roughness length. These conductances are greater than the neutral value when the atmosphere is unstable, resulting in a large flux for a given vertical gradient, and less than the neutral value when the atmosphere is stable, resulting in a smaller flux for the same vertical gradient.

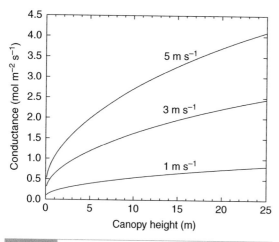

The roughness lengths for heat and moisture in a bulk aerodynamic formulation of vegetated surfaces are typically 10 percent that for momentum. That is, the apparent sources of heat and moisture occur at a lower height in the canopy than the apparent sink of momentum, and the aerodynamic conductances for sensible heat and water vapor are less than that for momentum. The derivation of this lower conductance is most obvious for neutral conditions. An additional resistance is needed to account for the smaller roughness length compared with momentum.

Fig. 13.6 Aerodynamic conductance for heat (g_{ah}) in relation to canopy height for wind speeds of 1, 3, and 5 m s^{-1}. In this example, d is 70 percent of canopy height, z_{0m} is 10 percent of canopy height, z_{0h} is 10 percent of z_{0m}, and the reference height is 10 m above the canopy. Conductance is given for neutral conditions.

13.8 | Review Questions

Use these values as needed: ρ_m = 42.3 mol m^{-3}, c_p = 29.2 J mol^{-1} K^{-1}, and λ = 44.44 kJ mol^{-1}.

1. Calculate the storage flux (μmol CO$_2$ m^{-2} s^{-1}) if for a measurement height of 3 m over grassland and an averaging period of 30 minutes the CO$_2$ mole fraction in air changes uniformly by 1 ppm. Use an air temperature of 20°C and pressure of 1013.25 hPa. Hint: What is the molar density of the air?

2. The covariance between potential temperature and vertical velocity measured by eddy covariance over a period of time is 0.2 K m s^{-1}. What is the sensible heat flux?

3. Friction velocity is u_* = 0.25 m s^{-1}, H = 200 W m^{-2}, and λE = 400 W m^{-2}. Calculate θ_* and c_*.

4. The wind speed measured at height z = 25 m is 5 m s^{-1}. Calculate the wind speed at z = 2 m above a surface with z_{0m} = 0.05 m and d = 0. Ignore the effects of atmospheric stability. What is the momentum flux?

5. Calculate surface temperature given: H = 200 W m^{-2} and θ = 30°C and u = 3 m s^{-1} at z = 20 m. The surface has the characteristics: z_{0m} = 0.05 m, z_{0h} = 0.005 m, and d = 0.35 m. Assume $\psi_m(\zeta)$ = 0.8 and $\psi_h(\zeta)$ = 1.4.

6. Calculate the aerodynamic conductance g_{ah} between height z = 25 m where u = 3 m s^{-1} and the surface for forest (z_{0m} = 1.0 m, z_{0h} = 0.1 m, and d = 7 m) and grass (z_{0m} = 0.05 m, z_{0h} = 0.005 m, and d = 0.35 m) for (a) ζ = 0 (neutral), (b) ζ = -0.5 (unstable) and (c) ζ = 0.5 (stable).

7. If the temperature at z = 25 m is θ = 20°C and H = 175 W m^{-2}, what is the surface temperature for the forest and grass in problem 6a–c? Which has the higher surface temperature (forest or grass)? How does surface temperature vary with atmospheric stability? Why?

8. The mole fraction of water vapor at z = 3 m over grassland is 0.015 mol mol^{-1} and the surface concentration is 0.03 mol mol^{-1}. Calculate the latent heat flux if wind speed is 3 m s^{-1}. The surface has: z_{0m} = 0.01 m, z_{0w} = 0.001 m, and d = 0 m. Assume neutral conditions. What is the latent heat flux if wind speed is 5 m s^{-1}?

9. In problem 8, what are the aerodynamic conductances for wind speeds 3 m s^{-1} and 5 m s^{-1}? What is the latent heat flux if these conductances decreased by a factor of two? What would the wind speeds be to give these smaller conductances?

10. Derive a formula for the additional resistance to account for the different roughness length of heat compared with momentum.

13.9 | References

Aubinet, M., Grelle, A., Ibrom, A., et al. (2000). Estimates of the annual net carbon and water exchange of forests: The EUROFLUX methodology. *Advances in Ecological Research*, 30, 113–175.

Aubinet, M., Vesala, T., and Papale, D. (2012). *Eddy Covariance: A Practical Guide to Measurement and Data Analysis*. Dordrecht: Springer.

Baldocchi, D. D. (2003). Assessing the eddy covariance technique for evaluating carbon dioxide exchange rates of ecosystems: Past, present and future. *Global Change Biology*, 9, 479–492.

Baldocchi, D. D., Hicks, B. B., and Meyers, T. P. (1988). Measuring biosphere–atmosphere exchanges of biologically related gases with micrometeorological methods. *Ecology*, 69, 1331–1340.

Brutsaert, W. (1982). *Evaporation into the Atmosphere: Theory, History, and Applications*. Dordrecht: Kluwer.

Brutsaert, W. (2005). *Hydrology: An Introduction*. Cambridge: Cambridge University Press.

Campbell, G. S., and Norman, J. M. (1998). *An Introduction to Environmental Biophysics*, 2nd ed. New York: Springer-Verlag.

Clarke, R. H., Dyer, A. J., Brook, R. R., Reid, D. G., and Troup, A. J. (1971). *The Wangara Experiment: Boundary Layer Data*, Division of Meteorological Physics Technical Paper Number 19. Melbourne: Commonwealth Scientific and Industrial Research Organization.

Garratt, J. R. (1992). *The Atmospheric Boundary Layer*. Cambridge: Cambridge University Press.

Monteith, J. L., and Unsworth, M. H. (2013). *Principles of Environmental Physics*, 4th ed. Amsterdam: Elsevier.

Oke, T. R. (1987). *Boundary Layer Climates*, 2nd ed. London: Routledge.

Shuttleworth, W. J. (2012). *Terrestrial Hydrometeorology*. Chichester: Wiley-Blackwell.

Soil Moisture and the Atmospheric Boundary Layer

14.1 | Chapter Summary

The atmospheric boundary layer is the layer of the atmosphere directly above Earth's surface. Sensible heat from the surface warms the boundary layer, and evaporated water moistens the boundary layer. The diurnal cycle of surface fluxes imparts a diurnal cycle to the boundary layer. At night, the boundary layer is typically stable with weak turbulent motion. Surface heating during the day warms the boundary layer and it becomes unstable. Soil water influences the boundary layer because of its effect on the partitioning of net radiation into sensible and latent heat. The Bowen ratio is smaller where soil water does not limit evapotranspiration, and the boundary layer is cooler, moister, and shallower than in the absence of evapotranspiration. Dry sites have lower latent heat flux and a warmer, drier, and deeper boundary layer. The changes in surface fluxes and boundary layer characteristics associated with wet soil may create conditions that favor precipitation. Surface heterogeneity in soil moisture can also generate mesoscale atmospheric circulations. A large horizontal contrast in sensible heat flux can produce a circulation similar to a sea breeze in which surface winds flow from cooler wet soil to warmer dry soil while upper winds flow in the opposite direction.

14.2 | Boundary Layer Characteristics

The atmospheric boundary layer is the layer of the atmosphere above Earth's surface that is directly affected over the course of a day by the surface through heating, cooling, friction, and the emission of atmospheric constituents such as water vapor, CO_2, dust, and pollutants. The boundary layer is the region of the atmosphere in which people live and plants grow. Processes in the boundary layer determine the climate near the ground that we experience.

The boundary layer has distinct regions. The surface layer is the layer immediately above the surface where air flow strongly depends on surface characteristics. Vertical variation in surface fluxes is negligible, and the surface layer is also referred to as the constant flux layer. Monin–Obukhov similarity theory describes flux–profile relationships in the constant flux layer (Chapter 13). However, these relationships fail in the layer of flow within and close to the plant canopy, known as the roughness sublayer. The outer layer is the region above the surface layer where flow is not as greatly influenced by the surface. It constitutes most of the boundary layer. The outer layer can have strong convective motion during periods of intense surface heating, such as during the day, and is often called the convective mixed layer. Under strong

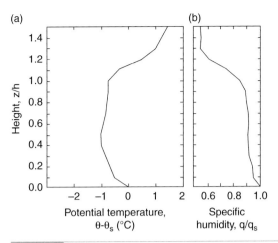

Fig. 14.1 Mean profiles of (a) potential temperature (θ) and (b) specific humidity (q) in the convective boundary layer measured above sparse grassland in southeastern Australia at 1500 local time during the Wangara experiment (Clarke et al. 1971). Height (z) is scaled by mixed layer depth (h). Potential temperature is shown as the deviation from the surface temperature (θ_s). Specific humidity is scaled by the surface humidity (q_s). The average mixed layer depth is $h = 915$ m and the mean surface humidity is $q_s = 4.1$ g kg^{-1}. Redrawn from Mahrt (1976).

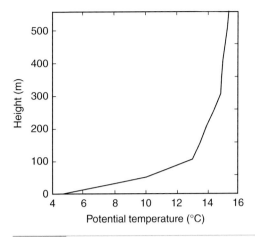

Fig. 14.2 Potential temperature profile at 0020 local time on day 8 of the Wangara experiment. Redrawn from André and Mahrt (1982).

mixing, the temperature profile shows little vertical variation and turbulent fluxes decrease with height. The top of the boundary layer is defined by a stable layer under strong convective conditions.

Figure 14.1 illustrates typical profiles of potential temperature and specific humidity in the convective boundary layer. The surface layer is well marked by decreases in temperature and humidity, above which the profiles are uniform with height through the mixed layer. The boundary layer top is marked by a sharp increase in potential temperature and a decrease in specific humidity. At night, the surface cools by longwave emission and the boundary layer is typically characterized by a temperature inversion in which the surface is cooler than the air above (Figure 14.2). This temperature inversion is most prominent on clear nights.

Surface fluxes regulate the cloudless boundary layer. In the absence of clouds, the sensible heat flux at the surface largely determines the rate of warming or cooling of the boundary layer. A convergence of sensible heat warms the boundary layer while divergence leads to cooling. Similarly, the convergence or divergence of water vapor flux increases or decreases specific humidity. Evaporated water is carried from the surface into the boundary layer, where it releases latent heat during condensation and forms clouds. The presence of clouds significantly affects the boundary layer. Clouds alter radiative transfer through the atmosphere and produce local sources of heating and cooling within the boundary layer that influence its turbulent structure. Phase change associated with condensation and evaporation of water droplets are additional complications.

14.3 | Diurnal Cycle

The diurnal heating and cooling of the surface imparts a diurnal cycle to the boundary layer (Figure 14.3). At night, the surface cools by longwave emission and the boundary layer becomes stable with weak turbulent motions. This temperature inversion typically extends to a height of 100–500 m. Above this inversion lies a residual weak mixed layer that is remnant from the previous day. Upon sunrise, surface heating by solar radiation produces upward exchange of sensible heat from the surface and

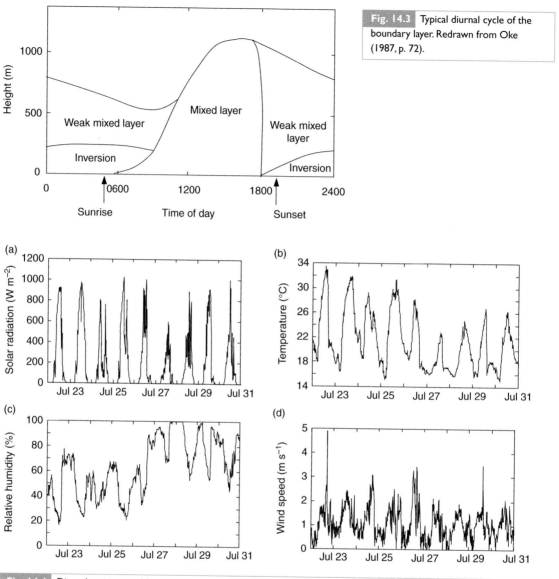

Fig. 14.3 Typical diurnal cycle of the boundary layer. Redrawn from Oke (1987, p. 72).

Fig. 14.4 Diurnal variation in (a) solar radiation, (b) air temperature, (c) relative humidity, and (d) wind speed measured in Boulder County, Colorado, July 23–31, 1997.

warms the boundary layer. The nocturnal temperature inversion erodes and is replaced with a convective mixed layer. On a hot, sunny day, strong surface heating causes the boundary layer to become strongly unstable over land, and the turbulent motions increase the depth of the boundary layer to 1–2 km or more by late afternoon. Thereafter, surface heating is no longer sufficient to maintain the mixed layer through turbulent motions. The boundary layer collapses, and after sunset the surface cools and the nocturnal inversion develops. The inversion deepens over the course of the night.

The diurnal cycle is readily evident in the surface layer. Figure 14.4 shows observed diurnal cycles of solar radiation, temperature, relative humidity, and wind for several summer days in Colorado. Solar radiation has a pronounced diurnal cycle. Peak values typically occur near noon, but there is considerable variability as clouds pass overhead (e.g., July 25). Air temperature varies in relation to solar radiation. Coldest temperatures typically occur near sunrise; warmest temperatures occur in middle to late afternoon. Clouds reduce solar radiation and reduce surface warming (compare July 23 and 28). Conversely, nighttime clouds reduce surface cooling by reradiating terrestrial longwave radiation back onto the surface. Clouds, therefore, diminish the diurnal temperature range (i.e., the difference between daily maximum and minimum temperatures). Relative humidity has a strong diurnal cycle. The amount of water vapor in the air often is lowest early in the morning, increasing to a maximum during daylight hours when evapotranspiration moistens the atmosphere. However, relative humidity is generally low during the day and increases at night when the colder air holds less water at saturation. Wind speed also has a diurnal cycle. Heating of the surface during the day provides buoyant energy that mixes the near-surface air and creates winds. At night, longwave emission of radiation cools the surface. The near-surface air becomes stable, with cold, dense air trapped near the surface. These conditions suppress turbulence and vertical mixing. Hence, winds over land normally reach maximum speeds in early afternoon and decrease at night.

Precipitation can also have a strong diurnal cycle. Air near the ground rises as it warms during the day. The rising air cools, and water vapor condenses into clouds. If conditions are right and clouds grow large enough, rain occurs in the form of afternoon thunderstorms. In the United States, precipitation has a large diurnal cycle over many areas, especially during warm seasons (Dai et al. 1999a, 2007). The strongest diurnal cycle occurs in summer in the Rocky Mountain region and in southeastern states. In these locations, it is more than twice as likely to rain during summer from 1500 to 1800 hours than any other time of day.

The diurnal heating of land varies seasonally and geographically depending on variations in solar geometry, cloudiness, and the amount of dust, pollution, and moisture in the air. Arid regions generally have a large diurnal temperature range. The dry air allows penetration of

solar radiation through the atmosphere and provides little longwave heating at night. In humid regions, the hazy sky prevents strong solar heating during the day and warms temperature at night so that diurnal temperature range is small. Coastal areas have a small diurnal temperature range because of the moderating influence of water. Clouds have a large effect on diurnal temperature range (Dai et al. 1999b). Clouds reduce daytime heating by decreasing the amount of solar radiation that reaches the ground and warm nighttime temperatures by enhancing downward longwave radiation. Wet soils also reduce diurnal temperature range by increasing daytime evaporative cooling. As soils dry, evapotranspiration decreases and daytime temperatures become hotter as a result of the decreased cooling (Durre et al. 2000; Seneviratne et al. 2010; Hirschi et al. 2011; Mueller and Seneviratne 2012).

14.4 Soil Moisture and Surface Fluxes

Soil water influences the partitioning of net radiation into sensible and latent heat (Figure 14.5). Over the course of a typical summer day, the majority of net radiation is dissipated as latent heat for a well-watered site. In contrast, a dry site exchanges much less energy via latent heat and more as sensible heat. In general, a substantial portion of net radiation is dissipated as latent heat when water does not limit evapotranspiration, and the boundary layer is cooler and moister than in the absence of evapotranspiration. As soil dries, less water is available for evapotranspiration, and more energy is dissipated as sensible heat or stored in the ground; the lower atmosphere is likely to be warm and dry (Figure 26.1).

Figure 14.6 illustrates general soil moisture–evapotranspiration regimes. When soil moisture is greater than some critical water content, the rate of evapotranspiration is limited by available energy and is independent of soil water. Below this critical value, soil water availability limits evapotranspiration, and the rate decreases as

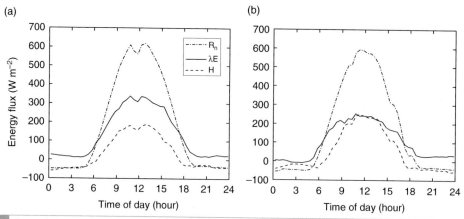

Fig. 14.5 Energy balance of (a) wet and (b) dry grassland on a typical summer day. Data are from Figure 12.8.

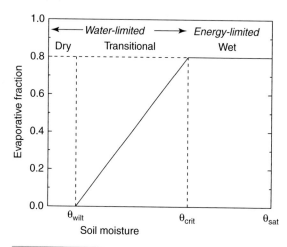

Fig. 14.6 Idealized relationship between soil water and evapotranspiration. Evapotranspiration, as a fraction of available energy, increases with wetter soil up to a maximum value at some critical water content (θ_{crit}), beyond which evapotranspiration is energy-limited. Below some minimum value (θ_{wilt}), the soil is too dry to sustain evapotranspiration. Adapted from Seneviratne et al. (2010).

the soil becomes drier. Changes in soil moisture only affect evapotranspiration in the water-limited regime, but at moisture contents below the wilting point, evapotranspiration is unaffected by soil water. Thus, there are three soil moisture–evapotranspiration regimes: wet and dry regimes, where soil moisture does not affect evapotranspiration, and a transitional

regime, where soil moisture strongly regulates evapotranspiration (Seneviratne et al. 2010).

Data collected over a 15 km × 15 km region of grassland in Kansas during the First ISLSCP (International Satellite Land Surface Climatology Project) Field Experiment (FIFE) illustrate the spatial variability in surface fluxes that can arise from surface heterogeneity in vegetation and soil moisture (Figure 14.7). On the particular day studied, the western region of the study site was warm and dry with high sensible heat flux, low latent heat flux, and low CO_2 uptake. Vegetation activity as quantified by a greenness index was low in this region. The eastern region had cooler surface temperature, low sensible heat flux, high latent heat flux, and high CO_2 uptake. Highest CO_2 uptake and latent heat flux occurred in the southeast quadrant, where vegetation activity was highest. These surface fluxes were highly correlated with one another, with CO_2 uptake increasing as latent heat flux increased and both fluxes increasing with greater vegetation greenness (Figure 14.8). Sensible heat flux increased as the surface temperature difference with air temperature increased.

The changes in surface fluxes associated with soil moisture alter the atmospheric boundary layer and may create conditions that favor precipitation. Such a feedback, where it occurs, is understood in terms of the impact of soil moisture on boundary layer stability and precipitation formation (Betts and Ball 1995, 1998; Betts

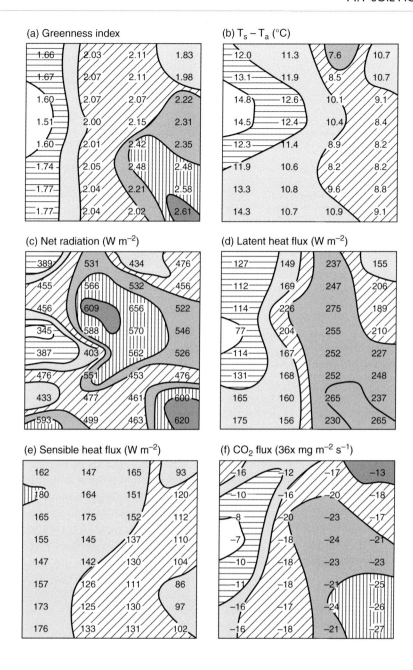

Fig. 14.7 Spatial variation of (a) greenness index, (b) surface temperature minus air temperature $(T_s - T_a)$, (c) net radiation, (d) latent heat flux, (e) sensible heat flux, and (f) CO_2 flux over the 15 km × 15 km FIFE site near Manhattan, Kansas, on August 12, 1989 between 1150 and 1355 Central Daylight Time. Data shown are spatial averages for an 8 × 4 grid (1.9 km × 3.8 km grid cell size). A negative CO_2 flux indicates uptake by vegetation. Redrawn from Desjardins et al. (1992).

et al. 1996; Eltahir 1998; Schär et al. 1999; Betts 2004, 2009; Seneviratne et al. 2010; Santanello et al. 2013). In general, wet soil has a lower Bowen ratio $(H / \lambda E)$ compared with dry soil, resulting in a shallower, moister boundary layer with lower height of cloud base. Additionally, higher soil moisture increases the net radiation at the surface by altering the balance of longwave and solar radiation. The cooler surface temperature leads to lower emission of longwave radiation at the surface, and the higher water vapor in the boundary layer leads to greater downwelling atmospheric longwave radiation. Furthermore, soil albedo generally decreases with wetter soil so that net solar radiation at the surface increases, though this is offset by greater cloud cover that reduces incoming solar radiation. In general, net radiation at the surface increases

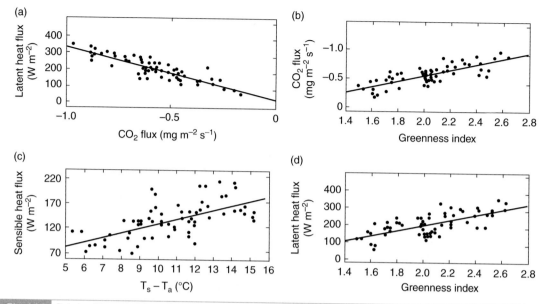

Fig. 14.8 Relationships among latent heat flux, CO_2 flux, greenness index, and sensible heat flux over the FIFE site for August 12, 1989 between 1050 and 1255 Central Daylight Time. A negative CO_2 flux indicates uptake by vegetation. Sensible heat flux is shown in relation to surface temperature minus air temperature $(T_s - T_a)$. Redrawn from Desjardins et al. (1992).

with wetter soil so that there is more energy available to heat and moisten the boundary layer. The shallower boundary layer and greater total surface heat flux (sensible heat plus latent heat) increase the convective instability of the boundary layer and create conditions that can favor convective precipitation. Thus, there may be a positive feedback between soil moisture and precipitation in which wetter soils lead to increased precipitation, which produces still wetter soils. Such a feedback has been found in atmospheric models (Chapter 26).

The feedback between soil water and precipitation is understood in terms of the release of convective instability that builds up during the development of the boundary layer over the course of a day in response to solar heating. With dry soil, the net radiative energy at the surface is converted primarily into sensible heat, with a resulting deep, well-mixed boundary layer. With wet soil, the fraction of net radiation converted to latent heat increases, the boundary layer is not as deep, temperatures are cooler, and more moisture is input into the boundary layer. Wet soils, therefore, have a comparatively large flux

of total heat into a shallow boundary layer. This increases the moist static energy per unit mass of boundary layer air. (Moist static energy is supplied by the total heat flux from the surface into the boundary layer. Moist static energy is the total energy in the boundary layer and is the sum of gravitational potential energy $[gz]$, sensible heat $[c_pT]$, and latent heat $[\lambda w]$, where g is gravitational acceleration, z is height, c_p is specific heat at constant pressure, T is temperature, λ is latent heat of vaporization, and w is water vapor mixing ratio. It is unaffected by condensation processes, which simply redistribute energy between the sensible and latent terms.) Moist static energy plays an important role in the initiation of convective storms, and moist convection redistributes this energy in the vertical.

Data collected over grassland during FIFE have given important insights to the coupling between soil moisture and the atmosphere boundary layer (Betts and Ball 1995, 1998). When partitioned based on soil water content, it is evident that summer days with dry soil had a higher Bowen ratio, were less cloudy, and had

Table 14.1 Surface Bowen ratio, available energy ($R_n - G$), surface wind speed, ΔT_{rad}, cloud cover, and lifting condensation level pressure (P_{LCL}) for 28 rainless days in July–August 1987 during FIFE partitioned by volumetric soil water content

Mean volumetric soil water	Number of days	Bowen ratio	$R_n - G$ (W m^{-2})	Wind speed (m s^{-1})	ΔT_{rad} (°C)	Cloud cover (tenths)	P_{LCL} (hPa)
0.234	10	0.36	527	7.2	1.6	4.2/2.9	127
0.157	10	0.65	528	5.9	4.9	3.2/1.4	161
0.130	8	0.85	505	5.3	7.3	1.7/0.6	231

Note: Data are averaged for a 3-hour period centered on local noon. Two estimates of cloud cover are given. ΔT_{rad} is the difference between radiative skin temperature and 2 m air temperature.

Source: From Betts and Ball (1995).

about 20 W m^{-2} less available energy at noon (Table 14.1). Reduced cloud cover on days with dry soil increased the incoming solar radiation at the surface, but this was more than offset by the greater outgoing longwave radiation from the warmer surface. Dry soils also had a higher radiative skin temperature relative to air temperature and lower mean wind speed.

Figure 14.9 shows the average diurnal cycle of net radiation, sensible heat flux, and latent heat flux for three categories of soil water content. Net radiation differed little with soil water, though the driest soil had about 20 W m^{-2} less available energy at noon, as discussed above. Differences among moisture categories in sensible heat and latent heat fluxes were small at night. However, daytime sensible heat flux increased and latent heat flux decreased with drier soil. The evaporative fraction, defined as $\lambda E / (H + \lambda E)$, had a pronounced diurnal cycle, but systematically decreased with lower soil moisture.

Plots of hourly potential temperature and mixing ratio show differences in the boundary layer related to soil moisture (Figure 14.9d). In general, mixing ratio increased over the course of a day from a morning minimum as surface evapotranspiration supplied water vapor to the boundary layer. It decreased in the afternoon as the growing boundary layer entrained drier air from above. This behavior is evident for wet soil (volumetric water content, 0.234), where

the mixing ratio increased over the course of the day until about local noon; thereafter mixing ratio decreased. The air above the drier soils had lower mixing ratios throughout the day compared with the wet soil. The mixing ratio of dry soil (volumetric water content, 0.130) had a smaller diurnal range and decreased soon after sunrise because low soil water limited evapotranspiration. Temperature increased from a morning minimum, reached a maximum in late-afternoon, and then decreased. The wet soil had a lower temperature and a smaller diurnal temperature range than the dry soil. However, equivalent potential temperature (a measure of the moist static energy) attained a higher afternoon maximum with wet soil (361 K) than with dry soil (353 K) and its diurnal temperature range increased.

The lifting condensation level is an indicator of the height of cloud base and the mixed layer depth. It can be defined as the difference in surface pressure (P_s) and saturation pressure of surface air (P^*), $P_{LCL} = P_s - P^*$. (Saturation pressure, or lifting condensation level, is the pressure at which a parcel of moist unsaturated air lifted dry adiabatically reaches saturation.) The depth of the mixed layer at the FIFE site increased over the course of the day (Figure 14.10). The afternoon maximum mixed layer depth increased with drier soil and is linked to soil moisture through evapotranspiration.

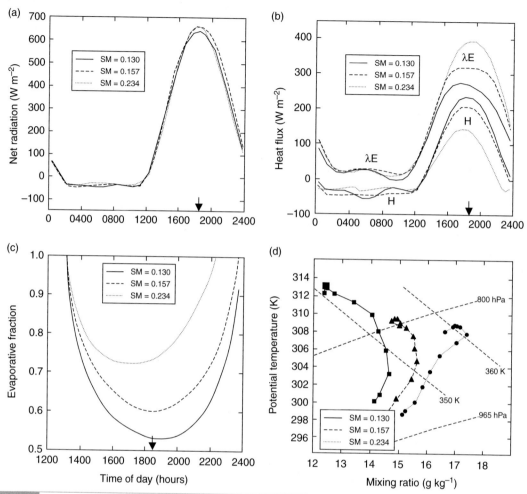

Fig. 14.9 Average diurnal cycle partitioned by volumetric soil moisture content (0.130, 0.157, and 0.234) for 28 rainless days in July–August 1987 during FIFE. The time axis is in Universal Time (UT). Local noon is about 1820 UT and is marked by an arrow. Shown are (a) net radiation, (b) sensible heat (H) and latent heat (λE) fluxes, (c) evaporative fraction, λE / (H + λE), and (d) hourly potential temperature (θ) and mixing ratio (w) measured at a height of 2 m for the daytime period (1145 to 2345 UT). Also shown are isopleths of equivalent potential temperature (θₑ) and saturation pressure (P*). Redrawn from Betts and Ball (1995).

In general, the FIFE analysis shows that increasing soil moisture leads to a higher afternoon maximum of equivalent potential temperature (and a larger diurnal temperature range of equivalent potential temperature) and lower afternoon cloud base given the same net available energy. Higher moist static energy and lower afternoon cloud base create conditions favorable for convective precipitation. Dry soil reduces evapotranspiration, and the boundary layer is warmer, drier, and deeper with lower equivalent potential temperature and higher

afternoon cloud base. A positive feedback may occur because precipitation increases soil moisture, producing a lower afternoon cloud base and higher equivalent potential temperature, which in turn favors convective precipitation.

Similar results are seen in analyses of climate model simulations (Betts 2004, 2009; Betts and Viterbo 2005). Wet soils have a lower cloud base, a cooler, moister atmospheric boundary layer, greater latent heat flux, and less sensible heat flux than dry soils. Clouds provide an important feedback on soil moisture and surface fluxes

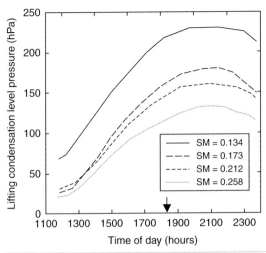

Fig. 14.10 Average diurnal cycle of lifting condensation level pressure (P_{LCL}) partitioned by volumetric soil moisture content for 94 rainless days in the period May 26– September 30, 1987 during FIFE. The time axis is in Universal Time (UT). Local noon is about 1820 UT and is marked by an arrow. Redrawn from Betts and Ball (1995).

through their effect on solar and longwave radiation.

14.5 | Surface Heterogeneity and Mesoscale Circulations

Spatially heterogeneous vegetation and soil moisture can influence boundary layer structure and generate mesoscale atmospheric circulations (Giorgi and Avissar 1997; Avissar et al. 2004; Pielke et al. 2011). The influence of surface heterogeneity arises as a result of differential surface energy fluxes and atmospheric heating. Dry surfaces, because they lack sufficient water for sustained evapotranspiration, have high sensible heat flux and low latent heat flux. The overlying air is warm and dry. In contrast, wet surfaces have high latent heat fluxes; the air is cool and moist. The contrast in surface fluxes needed to create mesoscale circulations is particularly large in semiarid climates, where wet sites are interspersed in a dry landscape (Anthes 1984; Mahfouf et al. 1987; Segal et al. 1988; Avissar and Pielke 1989; Segal and Arritt 1992; Taylor et al. 2007).

Such spatial heterogeneity occurs, for example, where patches of irrigated croplands are interspersed among dry native grasses. In such cases, the horizontal contrast in surface fluxes can produce a circulation similar to a sea breeze in which surface winds blow from the cooler, wet cropland to the warmer, dry grassland while upper winds flow in the opposite direction. Local differences in albedo can also generate mesoscale circulations (Pielke et al. 1993). Indeed, coating large areas in coastal arid climates with asphalt was proposed as a means to induce more rainfall (Black and Tarmy 1963; Black 1963).

Figure 14.11 illustrates the development of such a mesoscale circulation in an atmospheric model. In these simulations, a mesoscale atmospheric model was run once with homogenous wet forest land surface and once with a 200-km patch of dry grassland surrounded by wet forest. The difference between simulations highlights the atmospheric changes due to the dry grassland. As a result of less latent heat and more sensible heat, the near-surface air over the dry grassland is 1–4°C warmer and drier compared with the forest-only simulation. The differential surface heating between the dry grasses and wet trees on either side creates two circulations that converge over the warm grassland. The circulation on the western edge is counterclockwise, with near-surface westerly wind and easterly wind aloft. The circulation along the eastern edge is clockwise. This is accompanied by upward movement of air over the warm, dry grassland and descent on either side. These circulations transport moisture from the forest inward to the grassland and then upward so that air high in the atmosphere over the grassland is moistened.

By creating gradients in surface heating, a heterogeneous mixture of vegetation can influence precipitation (Chen and Avissar 1994; Pielke et al. 1997; Baidya Roy and Avissar 2002). For example, the landscape of northern Georgia is a mixture of forest and farmland. Simulations with a mesoscale atmospheric model contrasted the effects of an idealized homogeneous forest landscape with a landscape consisting of the observed mixture of forest and farmland (Pielke et al. 1997). In these simulations, covering a

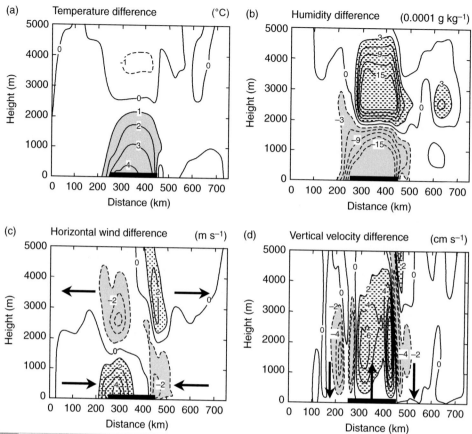

Fig. 14.11 Mesoscale circulations created by wet and dry vegetation on a summer day. Data show the difference (experiment minus control) between a simulation with a patch of dry grasses surrounded by wet trees (experiment) and a control simulation with homogenous wet trees. The dry grass extends from $x = 250$ to 450 km (thick horizontal line). Data are shown as a two-dimensional cross-section of height (left axis) and west-to-east distance (bottom axis). (a) Air temperature. Shading denotes warming greater than $1°C$. (b) Specific humidity. Stippling denotes increases and shading denotes decreases. (c) Horizontal zonal (west-to-east) wind. Positive values (stippling) indicate westerly wind. Negative values (shaded) indicate easterly wind. The four large arrows show wind direction along the surface and aloft. (d) Vertical velocity. Positive values (stippling) indicate rising motion. Negative values (shaded) indicate descending motion. The three large arrows indicate direction of motion. Adapted from Seth and Giorgi (1996).

spatial domain of 210 km by 210 km, forests are wet so that latent heat flux is high and sensible heat flux is low. Other vegetation types are water-limited. On the particular summer day simulated, the simulation with heterogeneous vegetation produces a large cloud in the early afternoon with widespread convective precipitation. The homogenous vegetation has little convective activity, and clouds form later in the afternoon. Similar simulations in a region of west Texas also found convective activity depends on the type of vegetation specified (Pielke et al.

1997). For the particular late spring day simulated, cloud development and convective precipitation are more extensive and vigorous in a simulation with the observed heterogeneous mixture of crops, trees, and short grass prairie than in a simulation with only short grass. The near-surface atmosphere is also moister than in the short grass simulation. This is a result of differential surface heating associated with the different vegetation types. The dry grassland has high sensible heat flux and low latent heat flux compared with other vegetation.

14.6 | Review Questions

1. Why is air near the ground typically much hotter than air a few meters higher on a sunny summer day with little wind? What effect does strong wind have on this temperature profile?

2. Why is the warmest temperature found in late afternoon on a cloudless day even though the Sun's radiation is greatest at noon?

3. When camping during the summer in the desert Southwest of the United States, would it be advisable to bring a heavy, warm shirt? Would a similar shirt be necessary when camping in southeastern United States?

4. The following table gives energy fluxes measured over the course of a day for two grassland sites. Determine the Bowen ratio and evaporative fraction. Describe likely differences in soil moisture and the boundary layer between these two sites.

Site	$R_n - G$	H	λE
A	527	139	388
B	505	232	273

5. In a particular locale, greenness index increases along a gradient from unproductive to productive vegetation. Describe likely changes in latent heat flux, sensible heat flux, and surface temperature along this gradient.

6. Which is likely to be hotter on a warm, sunny summer day: an urban asphalt streetscape or an urban grass park? Explain why.

7. Describe how paving desert with asphalt might induce rainfall.

8. Describe the effect that irrigated cropland has on regional sensible and latent heat flux in a native shortgrass prairie. How might the atmosphere be affected?

14.7 | References

André, J. C., and Mahrt, L. (1982). The nocturnal surface inversion and influence of clear-air radiative cooling. *Journal of the Atmospheric Sciences*, 39, 864–878.

Anthes, R. A. (1984). Enhancement of convective precipitation by mesoscale variations in vegetative covering in semiarid regions. *Journal of Climate and Applied Meteorology*, 23, 541–554.

Avissar, R., and Pielke, R. A. (1989). A parameterization of heterogeneous land surfaces for atmospheric numerical models and its impact on regional meteorology. *Monthly Weather Review*, 117, 2113–2136.

Avissar, R., Weaver, C. P., Werth, D., et al. (2004). The regional climate. In *Vegetation, Water, Humans and the Climate: A New Perspective on an Interactive System*, ed. P. Kabat, M. Claussen, P. A. Dirmeyer, et al. Berlin: Springer-Verlag, pp. 21–32.

Baidya Roy, S., and Avissar, R. (2002). Impact of land use/land cover change on regional hydrometeorology in Amazonia. *Journal of Geophysical Research*, 107, 8037, doi:10.1029/2000JD000266.

Betts, A. K. (2004). Understanding hydrometeorology using global models. *Bulletin of the American Meteorological Society*, 85, 1673–1688.

Betts, A. K. (2009). Land-surface-atmosphere coupling in observations and models. *Journal of Advances in Modeling Earth Systems*, 1, doi:10.3894/JAMES.2009.1.4.

Betts, A. K., and Ball, J. H. (1995). The FIFE surface diurnal cycle climate. *Journal of Geophysical Research*, 100D, 25679–25693.

Betts, A. K., and Ball, J. H. (1998). FIFE surface climate and site-average dataset 1987–89. *Journal of the Atmospheric Sciences*, 55, 1091–1108.

Betts, A. K., and Viterbo, P. (2005). Land-surface, boundary layer, and cloud-field coupling over the southwestern Amazon in ERA-40. *Journal of Geophysical Research*, 110, D14108, doi:10.1029/2004JD005702.

Betts, A. K., Ball, J. H., Beljaars, A. C. M., Miller, M. J., and Viterbo, P. A. (1996). The land surface–atmosphere interaction: A review based on observational and global modeling perspectives. *Journal of Geophysical Research*, 101D, 7209–7225.

Black, J. F. (1963). Weather control: Use of asphalt coatings to tap solar energy. *Science*, 139, 226–227.

Black, J. F., and Tarmy, B. L. (1963). The use of asphalt coatings to increase rainfall. *Journal of Applied Meteorology*, 2, 557–564.

Chen, F. and Avissar, R. (1994). Impact of land-surface moisture variability on local shallow convective cumulus and precipitation in large-scale models. *Journal of Applied Meteorology*, 33, 1382–1401.

Clarke, R. H., Dyer, A. J., Brook, R. R., Reid, D. G., and Troup, A. J. (1971). *The Wangara Experiment: Boundary Layer Data*, Division of Meteorological Physics Technical Paper Number 19. Melbourne, Australia: Commonwealth Scientific and Industrial Research Organization.

Dai, A., Giorgi, F., and Trenberth, K. E. (1999a). Observed and model-simulated diurnal cycles of precipitation over the contiguous United States. *Journal of Geophysical Research*, 104D, 6377–6402.

Dai, A., Trenberth, K. E., and Karl, T. R. (1999b). Effects of clouds, soil moisture, precipitation and water vapor on diurnal temperature range. *Journal of Climate*, 12, 2451–2473.

Dai, A., Lin, X., and Hsu, K.-L. (2007). The frequency, intensity, and diurnal cycle of precipitation in surface and satellite observations over low- and mid-latitudes. *Climate Dynamics*, 29, 727–744.

Desjardins, R. L., Schuepp, P. H., MacPherson, J. I., and Buckley, D. J. (1992). Spatial and temporal variations of the fluxes of carbon dioxide and sensible and latent heat over the FIFE site. *Journal of Geophysical Research*, 97D, 18467–18475.

Durre, I., Wallace, J. M., and Lettenmaier, D. P. (2000). Dependence of extreme daily maximum temperatures on antecedent soil moisture in the contiguous United States during summer. *Journal of Climate*, 13, 2641–2651.

Eltahir, E. A. B. (1998). A soil moisture–rainfall feedback mechanism, 1: Theory and observations. *Water Resources Research*, 34, 765–776.

Giorgi, F., and Avissar, R. (1997). Representation of heterogeneity effects in earth system modeling: Experience from land surface modeling. *Reviews of Geophysics*, 35, 413–437.

Hirschi, M., Seneviratne, S. I., Alexandrov, V., et al. (2011). Observational evidence for soil-moisture impact on hot extremes in southeastern Europe. *Nature Geoscience*, 4, 17–21.

Mahfouf, J.-F., Richard, E., and Mascart, P. (1987). The influence of soil and vegetation on the development of mesoscale circulations. *Journal of Climate and Applied Meteorology*, 26, 1483–1495.

Mahrt, L. (1976). Mixed layer moisture structure. *Monthly Weather Review*, 104, 1403–1407.

Mueller, B., and Seneviratne, S. I. (2012). Hot days induced by precipitation deficits at the global scale. *Proceedings of the National Academy of Sciences USA*, 109, 12398–12403.

Oke, T. R. (1987). *Boundary Layer Climates*, 2nd ed. London: Routledge.

Pielke, R. A., Rodriguez, J. H., Eastman, J. L., Walko, R. L., and Stocker, R. A. (1993). Influence of albedo variability in complex terrain on mesoscale systems. *Journal of Climate*, 6, 1798–1806.

Pielke, R. A., Lee, T. J., Copeland, J. H., et al. (1997). Use of USGS-provided data to improve weather and climate simulations. *Ecological Applications*, 7, 3–21.

Pielke, R. A., Sr., Pitman, A., Niyogi, D., et al. (2011). Land use/land cover changes and climate: Modeling analysis and observational evidence. *WIREs Climate Change*, 2, 828–850.

Santanello, J. A., Jr., Peters-Lidard, C. D., Kennedy, A., and Kumar, S. V. (2013). Diagnosing the nature of land–atmosphere coupling: A case study of dry/wet extremes in the U.S. southern Great Plains. *Journal of Hydrometeorology*, 14, 3–24.

Schär, C., Lüthi, D., and Beyerle, U. (1999). The soil–precipitation feedback: A process study with a regional climate model. *Journal of Climate*, 12, 722–741.

Segal, M., and Arritt, R. W. (1992). Nonclassical mesoscale circulations caused by surface sensible heat-flux gradients. *Bulletin of the American Meteorological Society*, 73, 1593–1604.

Segal, M., Avissar, R., McCumber, M. C., and Pielke, R. A. (1988). Evaluation of vegetation effects on the generation and modification of mesoscale circulations. *Journal of the Atmospheric Sciences*, 45, 2268–2292.

Seneviratne, S. I., Corti, T., Davin, E. L., et al. (2010). Investigating soil moisture–climate interactions in a changing climate: a review. *Earth-Science Reviews*, 99, 125–161.

Seth, A., and Giorgi, F. (1996). Three-dimensional model study of organized mesoscale circulations induced by vegetation. *Journal of Geophysical Research*, 101D, 7371–7391.

Taylor, C. M., Parker, D. J., and Harris, P. P. (2007). An observational case study of mesoscale atmospheric circulations induced by soil moisture. *Geophysical Research Letters*, 34, L15801, doi:10.1029/2007GL030572.

Part IV

Biometeorology

Leaf Temperature and Energy Fluxes

Chapter Summary

Individual leaves in the plant canopy absorb radiation and exchange sensible heat and latent heat with the surrounding air. Sensible heat flux and latent heat flux are important biophysical processes that govern leaf temperature. The efficient transfer away from the leaf surface of heat during convection and moisture during transpiration cools the leaf. Leaf boundary layer conductance and stomatal conductance regulate these fluxes. Large conductances increase fluxes, all other factors being equal. The size and shape of leaves govern boundary layer conductance, as well as the depth of the boundary layer over the leaf. A thin boundary layer, which is typical of small or deeply lobed leaves, allows strong coupling between the leaf and the surrounding air (high conductance). Larger or less lobed leaves typically have a thicker boundary layer and are more decoupled from the surrounding air (low conductance). As a result, leaf morphology closely matches the environment. Warm to hot climates with low light levels favor large leaves. Sunny environments and cold climates favor small leaves.

15.2 Leaf Energy Budget

A leaf absorbs solar and longwave radiation, emits longwave radiation, and exchanges sensible and latent heat with the surrounding air.

The net radiation (R_n) is balanced by sensible heat (H) and latent heat (λE):

$$R_n = Q_a - 2\varepsilon_l \sigma T_l^4 = H + \lambda E \qquad (15.1)$$

This equation is similar to the energy balance of the land surface, as in Eq. (12.2), but without soil heat storage. Here, Q_a is the radiative forcing, defined as the sum of absorbed solar radiation and incident longwave radiation. The leaf emits longwave radiation from its upper and lower surfaces. A common leaf emissivity is $\varepsilon_l = 0.98$. The difference between Q_a and emitted longwave radiation is the net radiation.

The fluxes of sensible and latent heat depend on leaf temperature, and models of leaf energy fluxes solve for the temperature (T_l, K) that balances the energy budget. The energy balance of a leaf is:

$$Q_a = 2\varepsilon_l \sigma T_l^4 + 2c_p \left(T_l - T_a\right) g_{bh}$$
$$+ \frac{c_p}{\gamma} \left[e_*\left(T_l\right) - e_a\right] \frac{1}{g_{bw}^{-1} + g_{sw}^{-1}} \qquad (15.2)$$

The leaf fluxes are comparable to those for the land surface, with Eq. (12.10) for H and Eq. (12.12) for λE. For a leaf, however, sensible heat flux is in relation to the leaf boundary layer conductance (g_{bh}, mol m^{-2} s^{-1}) and occurs from both sides of a leaf. Latent heat flux is in relation to the vapor pressure deficit between the leaf, assumed to be saturated with moisture, and the surrounding air ($e_*(T_l) - e_a$). The total leaf conductance is given by the boundary layer

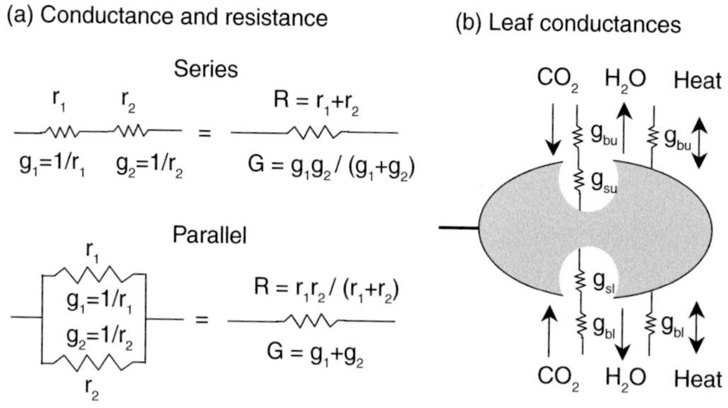

Fig. 15.1 Electrical network analogy for heat, water, and CO_2 exchange by leaves. (a) Resistance (r) and conductance (g) for two resistors in series and parallel. Also shown is the total resistance (R) and conductance (G). (b) Heat, water vapor, and CO_2 exchange as a network of leaf boundary layer (g_b) and stomatal (g_s) conductances for upper and lower leaf surfaces.

conductance for water vapor (g_{bw}, mol m^{-2} s^{-1}) and stomatal conductance for water vapor (g_{sw}, mol m^{-2} s^{-1}) acting in series. It is common to assume that transpiration occurs from only one side of the leaf.

15.3 | Leaf Conductances

The fluxes of sensible heat, latent heat, and CO_2 from a leaf can be represented as a diffusion process analogous to electrical networks (Figure 15.1). The electrical current between two points on a conducting wire equals the voltage difference divided by the electrical resistance. For an electrical circuit with two resistors connected in series, the total resistance is the sum of the individual resistances. The additive property of resistors in series is one reason why leaf fluxes are commonly expressed in terms of resistances. Plant physiologists, however, commonly use conductance (the inverse of resistance), because the flux is directly proportional to conductance and because conductance and flux have the same units (when using molar units). Conductance and molar notation are used throughout the book.

Similar to electrical networks, the diffusion of materials is related to the concentration difference divided by a resistance to diffusion, or multiplied by a conductance. For sensible heat, this diffusion conductance is the boundary layer conductance for heat. The transpiration diffusion conductance depends on two conductances connected in series: a stomatal conductance for water vapor from inside the leaf to the leaf

surface and the boundary layer conductance for water vapor from the leaf surface to the air. The total leaf conductance is $g_l = 1/(g_{bw}^{-1} + g_{sw}^{-1})$. If stomata are located on both sides of the leaf, the upper and lower conductances acting in parallel determine the overall leaf conductance.

The boundary layer conductance governs heat and moisture exchange between the leaf surface and the air around the leaf. This conductance depends on leaf size (ℓ, m) and wind speed (u, m s^{-1}) and is approximated per unit leaf area (one-sided) as:

$$g_{bh} = 0.203 (u/\ell)^{0.5}$$
$$g_{bw} = 0.223 (u/\ell)^{0.5} \tag{15.3}$$

This expression for boundary layer conductance is derived for air moving smoothly across a flat plate – a condition known as laminar forced convection (Gates 1980; Campbell and Norman 1998; Monteith and Unsworth 2013). The conductance varies with temperature and pressure in relation to the diffusivities of heat and water. The values used in Eq. (15.3) are for 1013.25 hPa and 15°C. The conductance for water vapor is greater than that for heat because of the larger diffusivity of water. Additionally, the conductance for a flat plate is multiplied by an empirical correction factor (a typical value is 1.5) to give the leaf values in Eq. (15.3). For most applications, it is common to assume that $g_{bh} = g_{bw} = g_b$. The conductance decreases with larger leaf size and increases with greater wind speed (Figure 15.2).

The boundary layer conductance represents the conductance to heat and moisture transfer between the leaf surface and ambient air above the leaf surface. Air moving across a leaf slows near the leaf surface and increases in speed with distance from the surface (Figure 15.3). Full wind flow occurs only at some distance from the leaf surface. This transition zone, in which wind speed increases with distance from the surface, is known as the leaf boundary layer. It is typically 1–10 mm thick. The boundary layer is also a region of temperature and moisture transition from a typically hot, moist leaf surface to cooler, drier air away from the surface. The boundary layer regulates heat and moisture exchange between a leaf and the air. A thin boundary layer produces a large conductance (small resistance) to heat and moisture transfer. The leaf is closely coupled to the air and has a temperature similar to that of air. A thick boundary layer produces a small conductance (large resistance) to heat and moisture transfer. Conditions at the leaf surface are decoupled from the surrounding air and the leaf is several degrees warmer than air.

Stomatal conductance acts in series with boundary layer conductance to regulate transpiration. Transpiration occurs when stomata open to allow a leaf to absorb CO_2 during photosynthesis (Figure 15.3). At the same time, water diffuses out of the saturated cavities within the foliage to the drier air surrounding the leaf. Stomata open and close in response to a variety of conditions (Chapter 16): they open with higher light levels; they close with temperatures colder or hotter than some optimum; they close as the

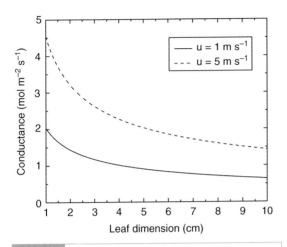

Fig. 15.2 Leaf boundary layer conductance (g_{bh}) in relation to leaf size for wind speeds of 1 and 5 m s^{-1}.

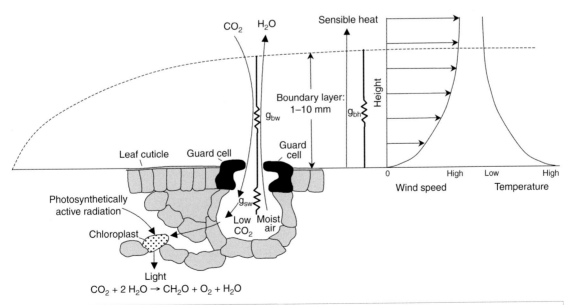

Fig. 15.3 Leaf boundary layer processes. Shown are stomata and associated CO_2 and water fluxes. These fluxes are regulated by stomatal (g_{sw}) and boundary layer (g_{bw}) conductances. Also shown are boundary layer thickness and associated wind and temperature profiles. Sensible heat flux is regulated by g_{bh}.

Table 15.1 Leaf temperature for radiative forcing of 1400, 1100, and 800 W m^{-2} with (a) longwave radiation only ($L\uparrow$), (b) longwave radiation and convection ($L\uparrow + H$), and (c) longwave radiation, convection, and transpiration ($L\uparrow + H + \lambda E$)

Q_a (W m^{-2})	$L\uparrow$	Temperature (°C)					
		$L\uparrow + H$			$L\uparrow + H + \lambda E$		
		0.1 m s^{-1}	1 m s^{-1}	5 m s^{-1}	0.1 m s^{-1}	1 m s^{-1}	5 m s^{-1}
1400	62	48	41	38	38	36	34
1100	42	38	36	36	33	33	33
800	18	28	32	33	26	29	31

Note: Leaf temperature is calculated from Eq. (15.2). Air temperature is 35°C, relative humidity is 50 percent, and wind speeds are 0.1, 1, and 5 m s^{-1}. Stomatal conductance is 0.4 mol m^{-2} s^{-1} and leaf dimension is 5 cm.

soil dries; they close if the surrounding air is too dry; and they vary with atmospheric CO_2 concentration. Stomatal conductance is a measure of how open the pores are and varies from about 0.4 mol m^{-2} s^{-1} or more when stomata are open to less 0.04 mol m^{-2} s^{-1} when stomata are closed. Stomata are typically, but not always, located on the lower leaf surface. Such leaves are termed hypostomatous. If stomata are on both sides of the leaf (amphistomatous), the upper and lower surfaces act in parallel (Figure 15.1).

15.4 | Leaf Fluxes and Temperature

Leaf temperature is the temperature that balances the leaf energy budget, given by Eq. (15.2). Analysis of the energy budget and the resulting leaf temperature under a variety of environmental conditions gives insight to the leaf microclimate. For example, Table 15.1 shows the importance of sensible and latent heat fluxes in reducing leaf temperature under a variety of radiative forcings and wind speeds for a summer day with air temperature equal to 35°C. The leaf has a radiative forcing equal to 1400, 1100, and 800 W m^{-2}, which is representative of values for clear sky at midday, cloudy sky at midday, and cloudy sky in late afternoon. If longwave radiation is the only means to dissipate this energy, the leaf has temperatures of 62°C, 42°C, and 18°C with high, moderate, and low radiative forcings.

Heat loss by convection (i.e., sensible heat) cools the leaf. Under calm conditions, with a wind speed of 0.1 m s^{-1}, sensible heat loss decreases leaf temperature by 14°C (to a temperature of 48°C) with the high radiative forcing and by 4°C (to a temperature of 38°C) with the moderate forcing. Higher wind speeds lead to lower temperatures. At 5 m s^{-1}, the temperature of the leaf exposed to the high radiative forcing is reduced from 62°C to 38°C. This example illustrates the powerful effect wind has in transporting heat away from a leaf exposed to high radiative forcing, thereby cooling the leaf. At low radiative forcing, convection warms the leaf because it is colder than the surrounding air and heat is transferred from the air to the leaf.

Latent heat exchange also decreases leaf temperature. Under calm conditions (0.1 m s^{-1}) and high radiative forcing, transpiration decreases leaf temperature by an additional 10°C, from a temperature of 48°C with longwave radiation and convection to a temperature of 38°C. Higher winds result in even lower temperatures. With a wind speed of 5 m s^{-1}, leaf temperature decreases from a lethal temperature of 62°C with longwave radiation only to a more comfortable temperature of 34°C. Cooling by transpiration is greatest with large radiative forcing and decreases as radiation decreases. It is largest for calm conditions and decreases as wind increases. The cooling effect of evaporation is why we sweat, and it is why a person may feel comfortable in dry climates, where low relative humidity results in rapid evaporation of sweat, but hot and uncomfortable in humid climates, where evaporation is not as efficient.

15.5 | Leaf–Air Coupling

The leaf environment is coupled to the surrounding air through a network of conductances that regulates sensible and latent heat fluxes. Sensible heat flux is directly proportional to the boundary layer conductance. Latent heat flux is proportional to the boundary layer and stomatal conductances acting in series. The Penman–Monteith equation, from Eq. (12.22), can be extended to a leaf (Monteith 1965; Jarvis and McNaughton 1986) and gives insights to the degree to which these conductances regulate the coupling between the leaf and air. For a hypostomatous leaf (with stomata on one side):

$$\lambda E = \frac{sR_n + 2c_p\left[e_*(T_a) - e_a\right]g_b}{s + 2\gamma(1 + g_b/g_{sw})} \qquad (15.4)$$

For convenience, it is assumed that $g_{bh} = g_{bw} = g_b$.

The role of stomata in regulating leaf transpiration can be seen in the two limiting cases when leaf boundary layer conductance is very small or very large (Jarvis and McNaughton 1986). If the leaf boundary layer conductance is very small, so that the leaf is decoupled from the surrounding air by a thick boundary layer:

$$\lambda E = sR_n/(s + 2\gamma) \qquad (15.5)$$

which is known as the equilibrium evaporation rate. In this case, transpiration is independent of stomatal conductance and depends chiefly on the net radiation available to evaporate water.

If the boundary layer conductance is very large, so that there is strong coupling between conditions at the leaf surface and outside the leaf boundary layer, transpiration is at a rate imposed by vapor pressure deficit and stomatal conductance:

$$\lambda E = \frac{c_p}{\gamma}\left[e_*(T_a) - e_a\right]g_{sw} \qquad (15.6)$$

In this case, an increase or decrease in stomatal conductance causes a proportional increase or decrease in transpiration. In between these two extremes of equilibrium and imposed transpiration, intermediate degrees of stomatal control prevail. The degree of coupling between a leaf and surrounding air depends on leaf size and wind speed (Figure 15.2). Small leaves, with high boundary layer conductance, approach strong coupling. Large leaves, with low boundary layer conductance, are weakly coupled. Leaves in still air are decoupled from the surrounding air while moving air results in strong coupling.

15.6 | Leaf Size and Shape

Principles of water-use efficiency and heat and gas exchange result in an optimal leaf form for a given environment. Heat and moisture exchange with the surrounding air regulate leaf temperature. Under sunny conditions, high sensible heat exchange decreases leaf temperature; low sensible heat exchange creates a warmer temperature. Loss of water during transpiration also decreases leaf temperature, because of the large amount of energy needed to change water from liquid to vapor. Leaf size and shape influence the ease with which a leaf exchanges heat and moisture with air.

Leaf size and shape affect boundary layer conductance. A small leaf has a higher boundary layer conductance to heat and moisture transfer than does a large leaf exposed to the same wind (Figure 15.2). This is because a small leaf has relatively little surface area relative to its perimeter length. Consequently, small leaves have a thin boundary layer and efficient heat transfer. Conditions at the leaf surface are closely coupled to the air, and leaf temperature is similar to that of the surrounding air. In contrast, a large leaf has a large surface area relative to perimeter length. Large leaves have a thick boundary layer, low boundary layer conductance, and inefficient heat transfer. They are decoupled from the surrounding air so that leaf temperature is several degrees warmer than that of air. Similarly, deep lobes on leaves decrease the surface area relative to perimeter length, resulting in higher boundary layer conductance than leaves without lobes.

Observations and theoretical studies show that the size and shape of leaves are a compromise among leaf energy exchange, leaf temperature, and photosynthesis. There is an optimal leaf size for a given environment (Parkhurst and Loucks 1972; Givnish and Vermeij 1976; Woodward 1993). Leaves growing in sunny environments are smaller and more deeply lobed than leaves growing in shaded environments. Leafy plants growing in hot, arid desert environments or cold arctic and alpine environments have small leaves. In part, this relates to the influence of leaf dimension on boundary layer conductance and the efficiency with which heat and moisture are transported away from a leaf.

Under the assumption that leaf size is determined so as to maximize water-use efficiency (the photosynthetic carbon gain for a given amount of transpirational water loss), Parkhurst and Loucks (1972) showed that warm to hot climates with low light conditions, such as might be found in the understory of temperate and tropical forests, favor large leaves (Figure 15.4). Sunny environments (e.g., the forest overstory) and cold climates favor small leaves. Givnish and Vermeij (1976) examined the influence of moisture on leaf size. They found large leaves are expected in the humid, shaded environment of the forest understory while small leaves occur in the sunny, dry conditions of the forest overstory (Figure 15.5). In a sunny environment, large leaf size increases transpiration so that large leaves are favored only in mesic conditions. Conversely, large leaves impede transpiration in a shaded environment; small leaves are favored with increasingly moist conditions.

The fossil record reveals the effect of energy exchange and stomata on leaf form. Early vascular plants were leafless or had short cylindrical leaves. Some 40–50 million years passed between the appearance of the first land plants (about 425 million years ago) and the origin of flat leaves that resemble those of modern plants. This may be related to high atmospheric CO_2 concentrations that prevailed during early plant life and a subsequent precipitous decline

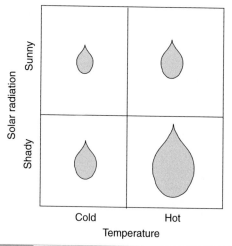

Fig. 15.4 Leaf size in relation to solar radiation and temperature. Adapted from Parkhurst and Loucks (1972).

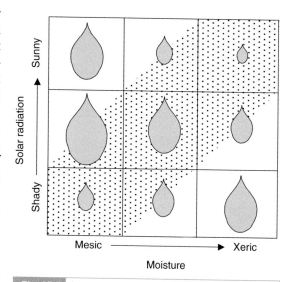

Fig. 15.5 Leaf size in relation to solar radiation and moisture. The stippled area shows the habitats likely in nature and between the forest understory and overstory. Adapted from Givnish and Vermeij (1976).

in CO_2 that facilitated the evolution of broad leaves (Beerling et al. 2001; Osborne et al. 2004; Beerling and Berner 2005). In a CO_2-enriched atmosphere, plants needed fewer stomata to absorb CO_2 for photosynthesis. However, broad flat leaves with a low density of stomata are prone to overheating because of low transpirational cooling. The appearance of broad leaves is associated with a 90 percent decline in atmospheric CO_2 between 400 and 350 million years ago, from concentrations well in excess of 1000 ppm to ~300 ppm. Plants developed more stomata as atmospheric CO_2 declined, which allowed flat leaves to stay cool. Leaf form may also have led to the extinction of many plant species 200 million years ago (McElwain et al. 1999; Beerling and Berner 2005). The fossil record shows species with large entire leaves were replaced by species with smaller, more dissected leaves. An increase in atmospheric CO_2 at this time warmed climate. The temperature of large leaves with entire margins (i.e., no lobes) reached lethal levels in this warm climate. Small or highly lobed leaves had lower temperatures and thus had an advantage over large leaves in warm climates.

These studies suggest a strong relationship between leaf morphology and environment. The close relationship among the size of leaves, leaf shape, and leaf edges (e.g., smooth, serrated, lobed) with temperature and precipitation is one means to reconstruct past climate from fossil leaves (Wolfe 1995; Wilf 1997; Wolfe et al. 1998; Peppe et al. 2011).

15.7 | Review Questions

1. A leaf with temperature 30°C is exposed to a radiative forcing Q_a = 1200 W m^{-2}. What is the net radiation at the leaf surface?

2. Calculate boundary layer conductance for a small leaf (ℓ = 3 cm) and a large leaf (ℓ = 10 cm) with calm wind (u = 0.1 m s^{-1}) and strong wind (u = 5 m s^{-1}). Which set of conditions produces strong leaf–air coupling and which produces weak coupling?

3. Calculate leaf conductance to transpiration for (a) a hypostomatous leaf and (b) an amphistomatous leaf with g_b = 2 mol m^{-2} s^{-1} and g_{sw} = 0.2 mol m^{-2} s^{-1}. If g_b increases by a factor of two, how does the leaf conductance change?

4. Use the Penman–Monteith equation to derive equations for sensible heat flux (H) and the leaf–air temperature difference ($T_l - T_a$) for a hypostomatous leaf.

5. Use the Penman–Monteith equation for a hypostomatous leaf to calculate λE, H, and $T_l - T_a$ for the following two values of stomatal conductance and two values of net radiation: g_{sw} = 0.1 and 0.4 mol m^{-2} s^{-1}; R_n = 500 and 1000 W m^{-2}. Leaf size is ℓ = 3 cm and u = 3 m s^{-1}. Use: $e_*(T_a)$ = 3167 Pa and s = 189 Pa K^{-1} (values for T_a = 25°C); relative humidity, 75 percent; c_p = 29.2 J mol^{-1} K^{-1}; and γ = 66.5 Pa K^{-1}. How does transpiration affect leaf temperature? How does this vary with net radiation?

6. Use the Penman–Monteith equation for a hypostomatous leaf to calculate $T_l - T_a$ for conditions with R_n = 750 W m^{-2} (sunny), u = 0.1 m s^{-1} (calm wind), and g_{sw} = 0.01 mol m^{-2} s^{-1} (low transpiration). Then determine which is more effective at reducing leaf temperature: (a) Shade, R_n = 250 W m^{-2}. (b) Wind, u = 1 m s^{-1}. (c) Transpiration, g_{sw} = 0.4 mol m^{-2} s^{-1}. (d) Shade, wind, and transpiration. Leaf size is ℓ = 5 cm. Other factors are the same as in question 5.

7. Use the Penman–Monteith equation for a hypostomatous leaf to calculate $T_l - T_a$ for the following two values of leaf size and two values of stomatal conductance: ℓ = 0.75 and 12 cm; g_{sw} = 0.1 and 0.4 mol m^{-2} s^{-1}. How does leaf–air coupling vary with leaf size and stomatal conductance? Use R_n = 500 W m^{-2} and u = 3 m s^{-1}. Other factors are the same as in question 5.

8. Use the Penman–Monteith equation for a hypostomatous leaf to calculate λE, H, and $T_l - T_a$ for each of the following leaf size, stomatal conductance, and net radiation: ℓ = 3 and 10 cm; g_{sw} = 0.01 and 0.4 mol m^{-2} s^{-1}; R_n = 250 and 750 W m^{-2}. In a sunny environment, which leaf is favored (based on temperature)? How does leaf temperature differ between the small and large leaf in a shaded environment? How does soil moisture (mesic or dry environment) affect these conclusions? Use u = 3 m s^{-1}, and other factors are the same as in question 5.

9. Derive the Penman–Monteith equation for an amphistomatous leaf.

10. Determine the equilibrium and imposed evaporation rates for an amphistomatous leaf. How do these compare with a hypostomatous leaf?

11. Discuss some of the principles that can be incorporated into landscape design to alleviate the high temperatures of the urban heat island.

15.8 | References

Beerling, D. J., and Berner, R. A. (2005). Feedbacks and the coevolution of plants and atmospheric CO_2. *Proceedings of the National Academy of Sciences USA*, 102, 1302–1305.

Beerling, D. J., Osborne, C. P., and Chaloner, W. G. (2001). Evolution of leaf-form in land plants linked to atmospheric CO_2 decline in the Late Palaeozoic era. *Nature*, 410, 352–354.

Campbell, G. S., and Norman, J. M. (1998). *An Introduction to Environmental Biophysics*, 2nd ed. New York: Springer-Verlag.

Gates, D. M. (1980). *Biophysical Ecology*. New York: Springer-Verlag.

Givnish, T. J., and Vermeij, G. J. (1976). Sizes and shapes of liane leaves. *American Naturalist*, 110, 743–778.

Jarvis, P. G., and McNaughton, K. G. (1986). Stomatal control of transpiration: Scaling up from leaf to region. *Advances in Ecological Research*, 15, 1–49.

McElwain, J. C., Beerling, D. J., and Woodward, F. I. (1999). Fossil plants and global warming at the Triassic–Jurassic boundary. *Science*, 285, 1386–1390.

Monteith, J. L. (1965). Evaporation and environment. In *The State and Movement of Water in Living Organisms (19th Symposia of the Society for Experimental Biology)*, ed. G. E. Fogg. New York: Academic Press, pp. 205–234.

Monteith, J. L., and Unsworth, M. H. (2013). *Principles of Environmental Physics*, 4th ed. Amsterdam: Elsevier.

Osborne, C. P., Beerling, D. J., Lomax, B. H., and Chaloner, W. G. (2004). Biophysical constraints on the origin of leaves inferred from the fossil record. *Proceedings of the National Academy of Sciences USA*, 101, 10360–10362.

Parkhurst, D. F., and Loucks, O. L. (1972). Optimal leaf size in relation to environment. *Journal of Ecology*, 60, 505–537.

Peppe, D. J., Royer, D. L., Cariglino, B., et al. (2011). Sensitivity of leaf size and shape to climate: Global patterns and paleoclimatic applications. *New Phytologist*, 190, 724–739.

Wilf, P. (1997). When are leaves good thermometers? A new case for leaf margin analysis. *Paleobiology*, 23, 373–390.

Wolfe, J. A. (1995). Paleoclimatic estimates from tertiary leaf assemblages. *Annual Review of Earth and Planetary Sciences*, 23, 119–142.

Wolfe, J. A., Forest, C. E., and Molnar, P. (1998). Paleobotanical evidence of Eocene and Oligocene paleoaltitudes in midlatitude western North America. *Geological Society of America Bulletin*, 110, 664–678.

Woodward, F. I. (1993). Leaf responses to the environment and extrapolation to larger scales. In *Vegetation Dynamics and Global Change*, ed. A. M. Solomon and H. H. Shugart. New York: Chapman and Hall, pp. 71–100.

16

Leaf Photosynthesis and Stomatal Conductance

16.1 | Chapter Summary

Photosynthesis is the process by which plants absorb light energy and produce carbohydrates from atmospheric CO_2, but in doing so they lose water through transpiration. Hence, the cycling of energy, water, and CO_2 between land and atmosphere are inextricably linked. Photosynthesis consists of three separate processes: light-dependent reactions convert light energy into chemical energy; the Calvin cycle uses this chemical energy to reduce CO_2 to carbohydrates; and diffusion supplies CO_2 through stomata. The biological benefits gained by maximizing CO_2 uptake while minimizing water loss give rise to a wide variety of plant physiologies. This is seen in the differing C_3, C_4, and CAM photosynthetic pathways. It is also seen in the close matching of photosynthetic capacity and stomatal conductance to environment. The tight relationship between photosynthesis and stomatal conductance represents a coordinated physiological response to the prevailing environment that regulates water-use efficiency.

16.2 | Overview

The overall chemical reaction for photosynthesis is:

$$nCO_2 + 2nH_2O \xrightarrow{\text{light}} (CH_2O)_n + nO_2 + nH_2O \qquad (16.1)$$

where n is the number of molecules of CO_2 that combine with water to form the carbohydrate $(CH_2O)_n$, releasing n molecules of oxygen to the atmosphere. Carbohydrates are sugars, starches, and other related compounds containing carbon combined with hydrogen and oxygen, represented generally by $(CH_2O)_n$. They are the most abundant organic compounds in nature and provide energy, structural material, and the building blocks for other molecules.

Photosynthesis occurs in the chloroplasts of leaves. These are disk-shaped structures within plant cells that are 5–10 μm in diameter. They consist of stroma, a gel-like material containing enzymes to convert CO_2 to carbohydrates during the Calvin cycle, and thylakoids, which are embedded throughout the stroma and are the site of the light-dependent reactions. The membranes of thylakoids contain the chlorophyll and carotenoid pigments essential to photosynthesis. A single leaf mesophyll cell may contain 50 chloroplasts and a 1 mm^2 of leaf area may contain 500,000 chloroplasts.

16.3 | Light-Dependent Reactions

The light-dependent reactions convert light energy into the chemical energy required for the Calvin cycle. Absorption of light oxidizes water, providing electrons to create chemical energy and releasing oxygen. The electrons pass through a series of biochemical reactions to NADP$^+$ (oxidized nicotinamide adenine dinucleotide phosphate) where they are temporarily stored in NADPH (reduced $NADP^+$) before being passed to CO_2 to form carbohydrates. In the process of electron transfer, chemical energy in the form of ATP (adenosine triphosphate) is created from adenosine diphosphate (ADP) and inorganic phosphate (e.g., H_2PO_4, which is generically represented as P_i).

The first step in photosynthesis is the absorption of light by pigment molecules contained in chloroplasts. Light energy is transferred in discrete units called photons, or quanta. For photosynthesis, the number of photons, not the total energy, is important. A photon of light with a blue wavelength (e.g., 0.450 µm) has more energy than a photon with a red wavelength (e.g., 0.680 µm), but both have the same effect on photosynthesis. However, plants do not utilize the full spectrum of solar radiation for photosynthesis. Only radiation with wavelengths between 0.4 and 0.7 µm, known as photosynthetically active radiation, contributes to photosynthesis.

Chlorophyll is the main pigment that makes leaves green, absorbing light in violet, blue, and red wavelengths while reflecting light in green wavelengths. Another class of pigments is the carotenoids. These are red, orange, and yellow pigment molecules. The more abundant chlorophyll masks their color, which is why leaves are green. Chlorophyll and carotenoid pigments occur within chloroplasts in units of several hundred pigment molecules called photosystems. Light energy absorbed by a pigment molecule transfers within the photosystem from one pigment molecule to the next until it reaches a reaction center. This is a special chlorophyll molecule that boosts one of its electrons to a higher energy level when light energy is absorbed. This electron transfers to an acceptor molecule, initiating the electron transport reactions.

Plants have two photosystems. Photosystem I (PS I) has optimal light absorption at a wavelength of 0.700 µm. Its reaction center is known as the P700 chlorophyll pigment molecule. Photosystem II (PS II) has optimal light absorption at 0.680 µm. Its reaction center is the P680 chlorophyll pigment molecule. Both are cooperatively involved in the light reactions of photosynthesis.

The light-dependent reactions begin when light energy transferred to the P680 reaction center in PS II causes it to lose an electron (Figure 16.1). This electron moves to an electron acceptor molecule. The electron-deficient P680 molecule replaces its electron by extracting an electron from water. This splits one water molecule (H_2O) into two protons ($2H^+$), two electrons ($2e^-$), and oxygen (O). Linear electron transport passes the electron from P680 to the P700 reaction center in PS I. The P700 pigment molecule cannot accept an electron unless it has lost one. As with PS II, this occurs when light energy absorbed by surrounding pigment molecules passes to the reaction center and an electron moves to an electron acceptor.

Two electrons are required to reduce $NADP^+$ to NADPH. These are provided when one water molecule is split. The transfer of one electron requires two photons (one by each photosystem), and four photons are needed to pass two electrons from one water molecule to reduce one $NADP^+$. Two NADPH are needed in the Calvin cycle for each CO_2. Consequently, eight photons are needed to split two water molecules, which provide the four electrons to produce the two NADPH required to fix one CO_2 molecule.

Carbon fixation in the Calvin cycle requires 3 ATP molecules for every 2 NADPH. The electron transport that produces NADPH also produces ATP in a process called photophosphorylation (Figure 16.1). Some of this ATP forms during non-cyclic photophosphorylation when electrons pass from PS II to PS I. In addition, light absorbed by PS I can initiate cyclic electron transport in which an electron transfers to the electron acceptor and then passes back to P700. This cyclic photophosphorylation does not split water or form NADPH, but does produce ATP. Cyclic photophosphorylation requires an additional photon absorbed only by PS I, which together with non-cyclic photophosphorylation produces the three ATP molecules required in the Calvin cycle. Therefore, a total of 8–9 photons must be absorbed to yield 2 NADPH and 3 ATP to reduce 1 CO_2.

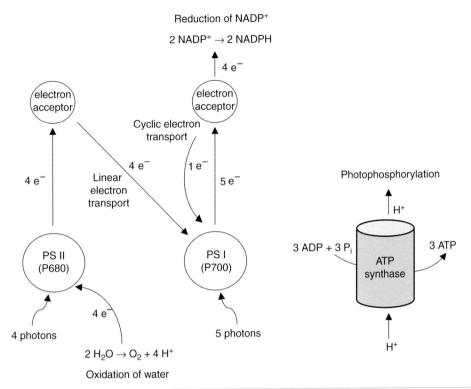

Reduction of NADP$^+$

$2\ NADP^+ \rightarrow 2\ NADPH$

$4\ e^-$

electron acceptor

electron acceptor

Cyclic electron transport

$4\ e^-$

$4\ e^-$

Linear electron transport

$1\ e^-$

$5\ e^-$

Photophosphorylation

H^+

PS II (P680)

PS I (P700)

$3\ ADP + 3\ P_i$

$3\ ATP$

ATP synthase

$4\ e^-$

4 photons

5 photons

H^+

$2\ H_2O \rightarrow O_2 + 4\ H^+$

Oxidation of water

Fig. 16.1 Light-dependent reactions of photosynthesis showing the various processes that split two H_2O yielding one O_2 and producing two NADPH and three ATP: absorption of four photons by PS II; transfer of four electrons from PS II to an acceptor molecule; oxidation of two H_2O by P680 to obtain four electrons; absorption of four photons by PS I and transfer of four electrons from PS I to an electron acceptor; linear electron transport from PS II to PS I of four electrons; and reduction of 2 NADP$^+$ to 2 NADPH using four electrons. Non-cyclic and cyclic photophosphorylation during linear electron transport and cyclic electric transport, respectively, produces 3 ATP from 3 ADP. Cyclic electron transport of one electron in PS I requires PS I absorb one additional photon.

16.4 Calvin Cycle

The NADPH and ATP produced in the light-dependent reactions are used to fix CO_2 into a carbohydrate. In many plants, the first product formed from CO_2 contains three carbon atoms. This is known as the C_3 photosynthetic pathway. The biochemical reactions that reduce CO_2 to carbohydrates are collectively known as the Calvin cycle and consist of three phases: carboxylation, reduction, and regeneration (Figure 16.2). In the carboxylation phase, the 5-carbon molecule ribulose-1,5-bisphosphate (RuBP) combines with CO_2 to form two 3-carbon molecules known as 3-phosphoglyceric acid (3-PGA). The enzyme ribulose bisphosphate carboxylase/oxygenase

(Rubisco) catalyzes this reaction. In the reduction phase, 3-PGA is reduced to the 3-carbon molecule glyceraldehyde-3-phosphate (GAP), also known as 3-phosphoglyceraldehyde or triose phosphate, when inorganic phosphate is obtained from ATP and electrons are obtained from NADPH. This also regenerates ADP and NADP$^+$ for the light-dependent reactions. Some of the GAP is exported to produce carbohydrates. The remainder is utilized in the regeneration phase, where it combines with additional ATP to regenerate RuBP. Three turns of the cycle fix three CO_2 molecules for a net production of one GAP. For each CO_2 fixed, two NADPH and three ATP are required.

In some plants (corn, sugar cane, and most tropical grasses), CO_2 is fixed into 4-carbon acids

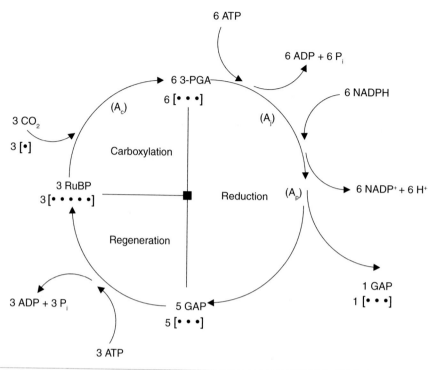

Fig. 16.2 Simplified representation of the Calvin cycle. Three turns of the cycle are depicted. The symbol [•] indicates the number of carbon atoms contained in the compound. Also shown are the general steps associated with the Rubisco (A_c), RuBP regeneration (A_j), and triose phosphate utilization (A_p) limited carboxylation rates.

(malate and aspartate) rather than 3-PGA. These species are known as C_4 plants. In this pathway, CO_2 combines with phosphoenolpyruvate (PEP) in a reaction catalyzed by the enzyme PEP carboxylase. The resulting 4-carbon molecule, oxaloacetate, is then converted to malate or aspartate. These acids are decarboxylated to yield CO_2 and pyruvate. The CO_2 enters the Calvin cycle, while the pyruvate combines with ATP to form more PEP. C_4 plants require a special leaf anatomy because the initial assimilation of CO_2 occurs in different leaf cells (mesophyll cells) than the Calvin cycle (bundle-sheath cells). It also requires two more ATP in addition to the three already used in the Calvin cycle. These synthesize PEP for continued CO_2 fixation.

The C_4 pathway seems inefficient and energetically expensive. In fact, however, C_4 plants are much more efficient at photosynthesis than C_3 plants. In the C_3 pathway, plants lose some of the CO_2 they fix in a light-enhanced process called photorespiration. This occurs because Rubisco has dual affinity for CO_2 and oxygen; it catalyzes CO_2 fixation by RuBP, but also catalyzes the oxidation of RuBP by oxygen. This reaction consumes oxygen and releases CO_2 so that the net CO_2 uptake during photosynthesis is reduced by 30–50 percent. Rubisco has a low affinity for CO_2, and the rate of photorespiration depends on the ratio of $CO_2:O_2$. In C_4 plants, the spatial separation of CO_2 assimilation into malate or aspartate (mesophyll cells) and the Calvin cycle (bundle-sheath cells) creates a high $CO_2:O_2$ ratio at the site of CO_2 fixation into 3-PGA. With relatively little O_2 compared with CO_2 in bundle-sheath cells, RuBP is not oxidized by oxygen. The C_4 plants, therefore, have little or no photorespiration and consequently have greater net photosynthetic rates than C_3 plants at high light levels and warm temperatures.

Many succulent plants such as cacti, orchids, and bromeliads use a third photosynthetic pathway called crassulacean acid metabolism (CAM). These plants grow in hot, arid climates. In this environment, they cannot open their stomata during the day to obtain CO_2 because they would quickly be desiccated by transpiration. Instead, stomata open at night, when temperatures are cooler, and CO_2 is fixed by PEP to form malate. The malate accumulates and is decarboxylated during daylight to release CO_2, which is then fixed in the Calvin cycle. Unlike C_4 plants, with their spatial separation of CO_2 fixation by PEP and carbohydrate synthesis during the Calvin cycle, CO_2 fixation and the Calvin cycle take place in the same cell. Instead, night and day temporally separate the processes.

16.5 | Net Photosynthesis

At the same time as leaves absorb CO_2 during photosynthesis, they release CO_2 in mitochondrial respiration. Mitochondrial respiration is the complement of photosynthesis. It is the process by which organic compounds are oxidized to produce the energy needed to maintain plant functions and grow new plant tissues. For glucose, the overall chemical reaction is:

$$C_6H_{12}O_6 + 6O_2 \rightarrow 6CO_2 + 6H_2O \qquad (16.2)$$

The rate of respiration depends on the biochemical quality of the tissue and increases exponentially with warmer temperatures. This respiration is different from photorespiration, which is driven by fixation of oxygen rather than CO_2 by Rubisco. The difference between CO_2 uptake during photosynthesis and CO_2 loss during mitochondrial respiration is the net CO_2 uptake by a leaf.

A variety of environmental factors influence net photosynthesis. If a leaf absorbs insufficient light, there will not be enough ATP and NADPH to fuel the Calvin cycle. When irradiance is below a certain level, typically about 20–40 µmol photon m^{-2} s^{-1}, CO_2 loss during respiration exceeds CO_2 uptake during photosynthesis. Only when light levels are above this light compensation point does a leaf gain carbon. Photosynthetic rates increase with greater irradiance until light saturation, when increased light no longer increases photosynthesis (Figure 16.3a). At these high light levels, the amount of CO_2 and Rubisco available for the Calvin cycle, not light, limit photosynthesis.

Temperature affects photosynthesis because sufficient, but not excessive, heat is a prerequisite for biochemical reactions (Figure 16.3b). The rate of photosynthesis increases with warmer temperatures up to an optimal temperature, beyond which it decreases. The optimum temperature range for most C_3 plants is 15–25°C, but the temperature range over which plants can photosynthesize is quite large.

When transpiration exceeds root uptake, plants become desiccated. Cells lose turgor. Leaves wither and become limp. The rate of photosynthesis decreases sharply as foliage water potential decreases and stomata close to prevent further desiccation (Figure 16.3c) or also with higher vapor pressure deficit (Figure 16.3d).

Higher concentration of CO_2 in the air increases photosynthetic rates (Figure 16.3e). In C_3 plants, the additional CO_2 reduces photorespiration by increasing the ratio of CO_2:O_2 reacting with Rubisco. Similar to light, the rate of photosynthesis increases with higher CO_2 concentrations up to a saturation point, beyond which photosynthesis remains constant. At this point, the supply of NADPH and ATP from the light-dependent reactions, not the amount of CO_2 available for fixation, limits photosynthesis. With low CO_2 concentrations, the supply of CO_2 limits photosynthesis. At the CO_2 compensation point, about 30–50 ppm for C_3 plants, the rate of CO_2 uptake during photosynthesis balances CO_2 loss during respiration so that there is no net gain of CO_2.

The rate of photosynthesis increases with increasing amounts of nitrogen in foliage (Figure 16.3f). Nitrogen is an essential component of chlorophyll and Rubisco. Greater amounts of nitrogen allow for more chlorophyll and Rubisco, fueling greater rates of photosynthesis.

Photosynthetic responses to environmental conditions vary among plants. Table 16.1 compares maximum photosynthetic rates under optimal conditions for several plant

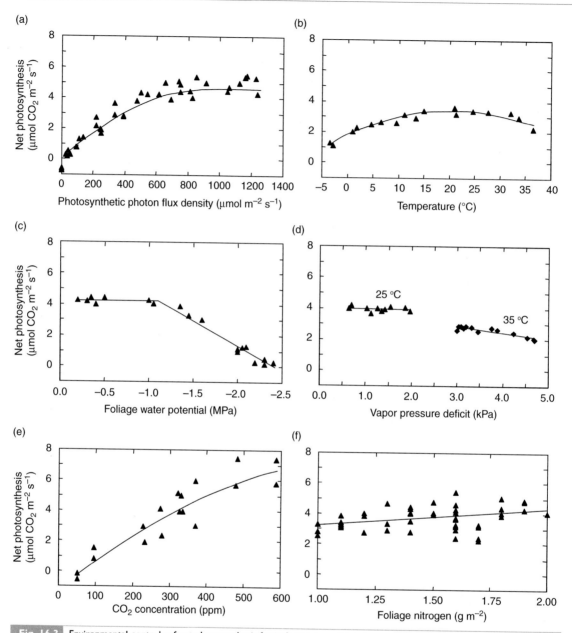

Fig. 16.3 Environmental controls of net photosynthesis for jack pine trees. Net photosynthesis is shown in response to (a) photosynthetic photon flux density, (b) temperature, (c) foliage water potential, (d) vapor pressure deficit, (e) ambient CO_2 concentration, and (f) foliage nitrogen. Data from Dang et al. (1997a,b, 1998).

types. Among herbaceous plants, those utilizing the C_4 pathway generally have the highest maximum photosynthetic rates; CAM plants generally have the lowest rates. Trees, which utilize the C_3 pathway, generally have low photosynthetic rates.

Table 16.2 compares the characteristics of the C_3, C_4, and CAM photosynthetic pathways. One difference is the presence of photorespiration in C_3 plants and absence in C_4 plants. The lower CO_2 compensation point in C_4 and CAM plants compared with C_3 plants is a result of CO_2

Table 16.1 Maximum net photosynthesis with natural CO_2 availability, saturated light intensity, optimal temperature, and adequate water

Plant type	CO_2 uptake ($\mu mol\ m^{-2}\ s^{-1}$)
Herbaceous	
C_3	
Grasses	5–15
Crops	20–40
C_4	30–60
CAM	5–10
Tree	
Tropical broadleaf evergreen	
Sunlit leaves	10–16
Shaded leaves	5–7
Broadleaf deciduous	
Sunlit leaves	10–15
Shaded leaves	3–6
Needleleaf evergreen	3–6
Needleleaf deciduous	8–10

Source: From Larcher (1995, pp. 85–86).

fixation by PEP carboxylase, which has a high affinity for CO_2, and because of low photorespiration. Plants utilizing the C_4 photosynthetic pathway show little light saturation and at full sunlight can have photosynthetic rates twice that of a C_3 plant. Because of their more efficient use of CO_2, C_4 plants attain similar or greater photosynthetic rates as C_3 plants with less water loss. Hence, they have higher water-use efficiency, defined here as the dry matter produced for a given amount of water lost in transpiration. In addition, the optimal temperature for C_4 plants is higher than that of C_3 plants. These features allow C_4 plants to grow well in warm regions with periodic drought, such as tropical savanna.

16.6 | A Photosynthesis Model

Farquhar et al. (1980) formulated a widely used model of C_3 photosynthesis (see also Farquhar and von Caemmerer 1982; von Caemmerer 2000,

2013; von Caemmerer et al. 2009; Diaz-Espejo et al. 2012; Bernacchi et al. 2013). This model represents photosynthesis based on the enzyme kinetics of Rubisco and the regeneration of RuBP in response to the supply of NADPH and ATP produced in the light-dependent reactions. In the model, photosynthesis is the lesser of two rates:

$$A_n = \min\left(A_c, A_j\right) - R_d \tag{16.3}$$

where A_c is the Rubisco-limited rate of photosynthesis, A_j is the light-limited rate allowed by RuBP regeneration, and R_d is mitochondrial respiration (all with units $\mu mol\ CO_2\ m^{-2}\ s^{-1}$). The Rubisco-limited rate is specified in relation to the maximum rate of carboxylation ($V_{c\max}$, $\mu mol\ m^{-2}\ s^{-1}$):

$$A_c = \frac{V_{c\max}\left(c_i - \Gamma_*\right)}{c_i + K_c\left(1 + o_i / K_o\right)} \tag{16.4}$$

and the RuBP-limited rate is specified in relation to the electron transport rate (J, $\mu mol\ m^{-2}\ s^{-1}$):

$$A_j = \frac{J\left(c_i - \Gamma_*\right)}{4\left(c_i + 2\Gamma_*\right)} \tag{16.5}$$

Here, c_i is the CO_2 concentration ($\mu mol\ mol^{-1}$) at the site of CO_2 fixation in the chloroplast (taken here to be the CO_2 in intercellular leaf air space), Γ_* is the CO_2 compensation point ($\mu mol\ mol^{-1}$), o_i is the O_2 concentration (equal to that of ambient air, 209 $mmol\ mol^{-1}$), and K_c ($\mu mol\ mol^{-1}$) and K_o ($mmol\ mol^{-1}$) are Michaelis–Menten constants for the carboxylation and oxygenation of Rubisco. The parameters Γ_*, K_c, and K_o depend on temperature. Typical values at 25°C are $\Gamma_* = 42.75\ \mu mol\ mol^{-1}$, $K_c = 404.9\ \mu mol\ mol^{-1}$, and $K_o = 278.4\ mmol\ mol^{-1}$ (Bernacchi et al. 2001).

The electron transport rate depends on the photosynthetically active radiation absorbed by the leaf. A common expression is the smaller of the two roots of the equation:

$$\Theta_J J^2 - \left(I_{PSII} + J_{\max}\right)J + I_{PSII}J_{\max} = 0 \tag{16.6}$$

where $J_{\max}$ is the maximum potential rate of electron transport ($\mu mol\ m^{-2}\ s^{-1}$), I_{PSII} represents the light utilized in electron transport by photosystem II ($\mu mol\ m^{-2}\ s^{-1}$), and Θ_J

Table 16.2 Photosynthetic characteristics of C_3, C_4, and CAM plants

Characteristic	C_3 plants	C_4 plants	CAM plants
Carboxylating enzyme	Rubisco	PEP carboxylase and Rubisco	Dark: PEP carboxylase Light: Rubisco
First product of photosynthesis	3-carbon acid (3-PGA)	4-carbon acids (oxaloacetate, malate, aspartate)	Dark: malate Light: 3-PGA
CO_2:ATP:NADPH	1:3:2	1:5:2	—
Location of processes	Mesophyll cells	Mesophyll cells then bundle-sheath cells	Mesophyll cells
Stomatal behavior	Open during day, close at night	Open during day, close at night	Close during day, open at night
Photorespiration	High	Low	Low
Photosynthesis inhibited by 21% O_2	Yes	No	Yes
Photosynthetic capacity	Low to high	High to very high	Medium
Light saturation	Intermediate intensity	No saturation	Intermediate to high intensity
Water-use efficiency	1–5 g kg^{-1} H_2O	3–5 g kg^{-1} H_2O	6–15 g kg^{-1} H_2O
Optimum temperature for photosynthesis	15–25°C	30–45°C	30–35°C
CO_2 compensation point	30–50 ppm	0–10 ppm	0–5 ppm

Source: From Larcher (1995, p. 64, p. 98, p. 109, p. 122).

is a curvature parameter (von Caemmerer 2000, 2013; Bernacchi et al. 2003, 2013; von Caemmerer et al. 2009). The light utilized in electron transport varies with the photosynthetically active radiation incident on a leaf ($I\downarrow$, µmol photon m^{-2} s^{-1}) as:

$$I_{PSII} = 0.5\Phi_{PSII}\alpha_l I\downarrow \qquad (16.7)$$

with α_l leaf absorptance, Φ_{PSII} the quantum yield of photosystem II (mol mol^{-1}), and the term 0.5 arises because one photon is absorbed by each of the two photosystems to move one electron. Suggested values are $\Theta_J = 0.7$ and $\Phi_{PSII} = 0.85$ (von Caemmerer 2000, 2013; von Caemmerer et al. 2009), though they vary with temperature (Bernacchi et al. 2003). Other models use $\Theta_J = 0.9$ (Medlyn et al. 2002). Leaf absorptance

for photosynthetically active radiation is commonly $\alpha_l = 0.8$.

Subsequent versions of the model introduced a third rate limited by the capacity to utilize the products of photosynthesis (triose phosphate) in the synthesis of starch and sugar (von Caemmerer 2000, 2013; von Caemmerer et al. 2009; Diaz-Espejo et al. 2012; Bernacchi et al. 2013). In its simplest form, the product-limited rate is:

$$A_p = 3T_p \qquad (16.8)$$

where T_p is the rate of triose phosphate utilization (µmol m^{-2} s^{-1}). C_4 photosynthesis can be similarly modeled, but with different expressions for A_c, A_j, and A_p (Collatz et al. 1992; von Caemmerer 2000, 2013; Diaz-Espejo et al. 2012).

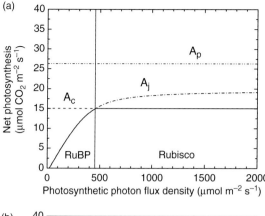

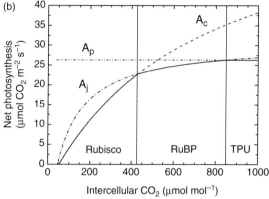

Fig. 16.4 Idealized photosynthetic response to (a) light and (b) c_i. Shown are the Rubisco-, RuBP-, and product-limited rates and the actual rate taken as the minimum of the three rates. In this example, $V_{cmax} = 70$, $J_{max} = 130$, $T_p = 9.1$, and $R_d = 1$ µmol m^{-2} s^{-1}.

Table 16.3 V_{cmax} (at 25°C) estimated for C_3 plants from leaf trait databases	
Plant type	V_{cmax25} (µmol CO_2 m^{-2} s^{-1})
Tropical trees (oxisols)	29.0 ± 7.7
Tropical trees (non-oxisols)	41.0 ± 15.1
Temperate broadleaf trees	
Evergreen	61.4 ± 27.7
Deciduous	57.7 ± 21.2
Coniferous trees	
Evergreen	62.5 ± 24.7
Deciduous	39.1 ± 11.7
Shrubs	
Evergreen	61.7 ± 24.6
Deciduous	54.0 ± 14.5
C_3 herbaceous	78.2 ± 31.1
C_3 crops	100.7 ± 36.6

Note: Shown are the mean for various plant functional types ± one standard deviation.

Source: From Kattge et al. (2009).

Figure 16.4a illustrates the components of photosynthesis using typical parameter values. Below some irradiance, light limits photosynthesis, and photosynthesis follows the RuBP-limited rate (A_j). The rate of photosynthesis increases with greater light, but beyond a certain irradiance the amount of Rubisco limits photosynthesis (A_c). Figure 16.4b similarly illustrates the A_n–c_i relationship at light saturation. At low c_i and high irradiance, the amount of Rubisco limits the rate of photosynthesis because RuBP levels are saturating, and photosynthesis follows the Rubisco-limited rate (A_c). Higher c_i increases photosynthesis until RuBP regeneration (A_j) limits photosynthesis, and dA_n / dc_i decreases. As c_i continues to increase, triose phosphate utilization (A_p) limits photosynthesis.

The parameters V_{cmax}, J_{max}, and R_d vary with temperature. Kattge et al. (2009) derived V_{cmax} at 25°C from leaf gas exchange measurements (Table 16.3). High values of V_{cmax} correlate with high leaf nitrogen, and J_{max} and R_d can be estimated from scaling relationships with V_{cmax} arising from nitrogen investment in photosynthetic capacity. The maximum rate of carboxylation (V_{cmax}) is proportional to the amount of Rubisco. The maximum potential rate of electron transport (J_{max}) varies in relation to leaf chlorophyll. Both Rubisco and chlorophyll require nitrogen, and plants allocate their resources to match photosynthetic capacity with the environment. For example, there is no need to have extra chlorophyll to trap light if the concentration of Rubisco is low. Thus, leaves with low V_{cmax} also have low J_{max}. Figure 16.5 illustrates this relationship for 109 species of C_3 plants. Estimates of V_{cmax} range from a low of 6 µmol m^{-2} s^{-1} in some trees to a high of 194 µmol m^{-2} s^{-1} for some agricultural crops. Values for J_{max} have a similar wide range, but there is a positive correlation between V_{cmax}

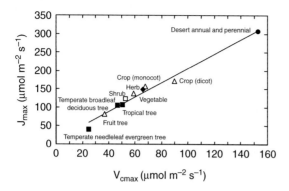

Fig. 16.5 Relationship between maximum rate of carboxylation (V_{cmax}) and maximum potential rate of electron transport (J_{max}) for 109 C_3 species. Data are shown as averages for broad groups of species. Original data for all 109 species show a similar relationship. Data from Wullschleger (1993).

and J_{max}. This reflects an optimal allocation of nitrogen to balance enzymatic (Rubisco) and light-harvesting (chlorophyll) capabilities. More precisely, $J_{max} = 1.67V_{cmax}$ at 25°C (Medlyn et al. 2002). The rates T_p and R_d similarly scale with V_{cmax}. Despite the widespread use of the photosynthesis model, V_{cmax} is a poorly constrained parameter (Rogers 2014).

16.7 | Diffusive Limitations on CO₂ Supply

For photosynthesis to occur, CO_2 must diffuse into the leaf to the chloroplasts, where it is fixed and converted to carbohydrates. Leaves have a waxy layer on the surface that restricts gas diffusion. Instead, CO_2 passes through microscopic openings in foliage known as stomata (Hetherington and Woodward 2003). Stomata are typically 10–80 μm in length with a maximum width of about 5 μm. A leaf may contain 5–1000 stomata per square millimeter of leaf area with a total pore area of less than 1–5 percent of leaf area. Stomatal conductance for CO_2 and water is directly proportional to pore width, with the maximum opening determining the upper limit to the rate of gas exchange. By varying the width of the stomatal pore, plants

control gas exchange. Stomata open to allow CO_2 uptake during photosynthesis and close to prevent desiccation during transpiration.

The rate of leaf photosynthesis can be represented as a diffusion process with:

$$A_n = \frac{g_{bw}}{1.4}(c_a - c_s) = \frac{g_{sw}}{1.6}(c_s - c_i) = g_l(c_a - c_i) \quad (16.9)$$

In these equations, c_a, c_s, and c_i are the ambient, leaf surface, and intercellular CO_2 concentration (μmol mol⁻¹), respectively, and g_{bw} and g_{sw} are leaf boundary layer and stomatal conductance to water vapor diffusion (mol H_2O m⁻² s⁻¹), respectively. The factors 1.4 and 1.6 correct for the lower diffusivity of CO_2 compared with H_2O. The factor 1.4 adjusts the boundary layer conductance for H_2O to that for CO_2 ($g_{bc} = g_{bw}/1.4$) and 1.6 similarly adjusts stomatal conductance for H_2O to that for CO_2 ($g_{sc} = g_{sw}/1.6$). The first equation represents the diffusion of CO_2 from ambient air surrounding a leaf to the leaf surface, which is directly proportional to leaf boundary layer conductance. The second equation is the diffusion of CO_2 from the leaf surface to intercellular space, which is directly proportional to stomatal conductance. The final equation is the combined CO_2 flux from ambient air to intercellular space, with the total leaf conductance equal to $g_l = 1/(1.4g_{bw}^{-1} + 1.6g_{sw}^{-1})$. Diffusion of CO_2 from ambient air through the boundary layer and stomata into intercellular air space requires a concentration gradient such that $c_a > c_s > c_i$. This is evident from the diffusion equations, which can be rewritten as:

$$c_s = c_a - \frac{1.4}{g_{bw}}A_n$$

$$c_i = c_s - \frac{1.6}{g_{sw}}A_n \quad (16.10)$$

In the photosynthesis model as presented here, the CO_2 concentration in the chloroplast is taken to be the same as the CO_2 in intercellular air space. An additional term is required to account for CO_2 diffusion from intercellular air space (with concentration c_i) through the mesophyll to the chloroplast (with concentration c_c). This conductance, termed mesophyll conductance (g_m), regulates the mesophyll diffusive

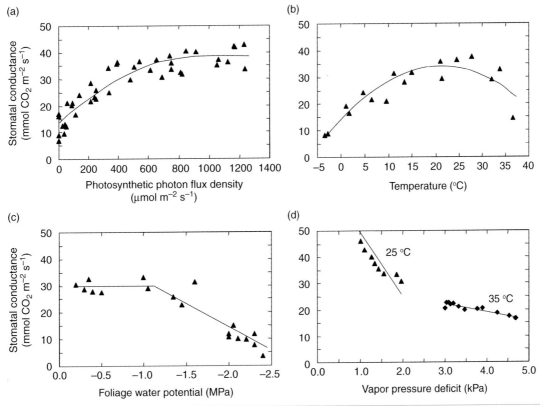

Fig. 16.6 Environmental controls of stomatal conductance for jack pine. Stomatal conductance is shown in response to (a) photosynthetic photon flux density, (b) temperature, (c) foliage water potential, and (d) vapor pressure deficit. Data from Dang et al. (1997a,b, 1998).

flux such that $c_c = c_i - A_n / g_m$ (Flexas et al. 2008, 2012; Diaz-Espejo et al. 2012; von Caemmerer 2013). The chloroplastic and intercellular CO$_2$ concentrations are equal only if the mesophyll conductance is infinitely large. In fact, however, the mesophyll conductance is finite and of similar magnitude as stomatal conductance such that $c_c < c_i$. The role of mesophyll conductance in regulating photosynthesis is the subject of considerable research.

Stomata respond to a variety of environmental factors (Figure 16.6). Except for CAM plants, stomata open with light and close in darkness. Stomata also close with temperatures warmer and colder than some optimum. Stomata close to prevent excessive water loss. This occurs in two ways. Stomatal conductance decreases as leaf water potential decreases. Low leaf water potential occurs when the loss of water by transpiration exceeds the rate of uptake from soil. Stomata close to prevent further desiccation. Additionally, the vapor pressure deficit between the leaf and air increases as the humidity of air decreases, creating a high potential for transpiration. Stomata close to prevent excessive desiccation under these conditions.

Jarvis (1976) developed an empirical model in which:

$$g_{sw} = g_{sw}(I\downarrow) f_1(T) f_2(D) f_3(\psi) f_4(c_a) \qquad (16.11)$$

Here, $g_{sw}(I\downarrow)$ is stomatal conductance as a function of photosynthetically active radiation, and $f_1(T)$, $f_2(D)$, $f_3(\psi)$, and $f_4(c_a)$ are empirical functions scaled from zero to one that adjust stomatal conductance for temperature, vapor pressure deficit, foliage water potential, and ambient CO$_2$ concentration, respectively. This approach has been used to represent stomatal conductance in the land surface models used with climate models (Dickinson et al. 1986, 1993; Sellers et al. 1986).

16.8 | Photosynthesis–Transpiration Compromise

Subsequent models of stomatal conductance were developed from a more mechanistic understanding of stomatal physiology. The physiology of stomata represents a compromise between the two conflicting goals of permitting CO_2 uptake during photosynthesis while restricting water loss during transpiration (Cowan 1977; Cowan and Farquhar 1977). When stomata open, water diffuses out of the same pathway through which CO_2 diffuses into the intercellular leaf space. The gradient of water vapor (~1000–3000 Pa, or 0.01–0.03 mol mol^{-1}) is 100 times greater than the CO_2 gradient (~30 Pa, or 300 μmol mol^{-1}). Hence, large transpiration water loss can accompany photosynthetic CO_2 uptake. C_3 plants, in particular, must allow for large amounts of CO_2 diffusion because Rubisco has a low affinity for CO_2, which leads to large water loss.

Stomata are regulated so as to maximize CO_2 gain and minimize water loss. This is evident from experiments relating stomatal conductance and photosynthesis. Plants grown under a variety of irradiances, nutrient concentrations, CO_2 concentrations, and leaf water potentials show large variation in photosynthetic rate and stomatal conductance, but photosynthesis and stomatal conductance vary in near constant proportion for a given set of conditions (Wong et al. 1978, 1979, 1985a–c). Photosynthesis and stomatal conductance measurements for jack pine trees illustrate such relationships. Over a wide range of light and foliage water potential, from full illumination to dark and from moist to desiccated, photosynthesis increases proportionally with increases in stomatal conductance (Figure 16.7). Transpiration also increases with greater conductance.

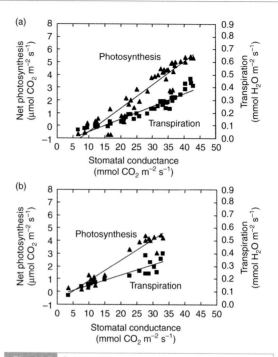

Fig. 16.7 Relationship between photosynthesis, transpiration, and stomatal conductance for jack pine. (a) Light response over a range of 0 to 1250 μmol photon m^{-2} s^{-1}. (b) Foliage water potential response over a range of −0.2 to −2.4 MPa. Data from Dang et al. (1997a,b, 1998).

Physiological measurements among a variety of plant communities also show a positive correlation between maximum stomatal conductance and maximum rate of photosynthesis (Field and Mooney 1986; Körner 1994; Hetherington and Woodward 2003). Figure 16.8 illustrates such relationships, with maximum rates of photosynthesis increasing as stomatal conductance increases. Coherent changes in photosynthetic carbon metabolism and stomatal behavior show they change in concert. Stomatal conductance varies to match the photosynthetic capacity of leaves so as to optimize gas exchange (i.e., minimize transpiration water loss while permitting photosynthetic CO_2 gain) in relation to the prevailing environment.

Plants achieve this regulation by adjusting stomatal conductance to maintain intercellular CO_2 as a nearly constant ratio of ambient CO_2 (i.e., c_i / c_a is constant) for particular

(a)

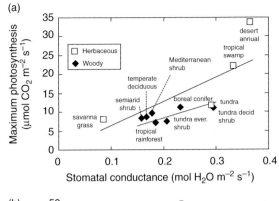

(b)

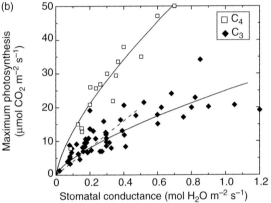

Fig. 16.8 Relationship between maximum photosynthesis and stomatal conductance. (a) Data shown are mean values for 7 types of woody vegetation and 4 types of herbaceous vegetation. The regression equations shown with these data are based on the full dataset of 55 woody plants and 18 herbaceous plants. Data from Körner (1994). (b) Data shown are for C_3 and C_4 plants (Hetherington and Woodward 2003). The dashed line for C_3 plants shows the regression through the origin for stomatal conductance less than 0.5 mol m^{-2} s^{-1}, comparable to panel (a).

environmental conditions. This is evident from the diffusion equation. Because the boundary layer conductance (~1–4 mol m^{-2} s^{-1}) is typically ten times greater than stomatal conductance (~0.1–0.4 mol m^{-2} s^{-1}), the diffusion equation can be approximated by:

$$A_n = \frac{g_{sw}}{1.6}\left(c_a - c_i\right) = \frac{g_{sw}}{1.6}c_a\left(1 - c_i/c_a\right) \quad (16.12)$$

The term $1 - c_i / c_a$ is the relative CO_2 diffusion gradient for a given ambient concentration (c_a). This equation can be rewritten to give

stomatal conductance in relation to photosynthetic rate:

$$g_{sw} = \frac{1.6A_n}{c_a\left(1 - c_i/c_a\right)} \quad (16.13)$$

A linear relationship between A_n and g_{sw}, for a given c_a, implies a constant c_i / c_a. C_4 plants have a higher rate of photosynthesis for a given stomatal conductance (Figure 16.8b), implying lower c_i / c_a. Indeed, the c_i / c_a ratio is between about 0.65 to 0.80 for C_3 plants and about 0.40 to 0.60 for C_4 plants (Hetherington and Woodward 2003). The tight relationship between photosynthesis and stomatal conductance to maintain c_i / c_a nearly constant represents a coordinated physiological response to the prevailing environment.

16.9 | A Photosynthesis–Stomatal Conductance Model

The close relationship between stomatal conductance and photosynthesis suggests that a model of photosynthesis would improve prediction of stomatal conductance. Collatz et al. (1991) formalized these concepts in an empirical model for C_3 plants that explicitly links the biochemistry of photosynthesis and the biophysics of CO_2 diffusion. Their model combines the Farquhar et al. (1980) photosynthesis model and the Ball–Berry stomatal conductance model (Ball et al. 1987). The Ball–Berry model is an extension of Eq. (16.13) and relates stomatal conductance to photosynthesis by the equation:

$$g_{sw} = g_0 + g_1 \frac{A_n}{c_s}h_s \quad (16.14)$$

where h_s is the fractional humidity at the leaf surface (dimensionless), c_s is the CO_2 concentration at the leaf surface (µmol mol^{-1}), g_1 is the slope of the relationship, and g_0 is the minimum conductance. Photosynthesis (A_n) has units µmol CO_2 m^{-2} s^{-1} while g_{sw} and g_0 have units mol H_2O m^{-2} s^{-1}. Collatz et al. (1992) described a similar model for C_4 plants. Representative parameter

3

values for C_3 plants are $g_1 = 9$ and $g_0 = 0.01$ mol H_2O m^{-2} s^{-1}, though they vary greatly among species (Medlyn et al. 2011).

Equation (16.14), with A_n from Eq. (16.3), can be used to model stomatal conductance. Solution of this set of equations describing photosynthesis and stomatal conductance requires knowledge of intercellular CO_2 (c_i) and leaf surface CO_2 (c_s). These are obtained from the diffusion equations using Eq. (16.10). A similar diffusion network is used to calculate humidity at the leaf surface (h_s). Solution of this equation set is complex and typically requires iterative solutions for A_n and g_{sw}. This approach was introduced into the land component of climate models in the mid-1990s (Bonan 1995; Sellers et al. 1996; Cox et al. 1999) and allows simulation of photosynthesis (Bonan 1995; Denning et al. 1995, 1996; Craig et al. 1998). Variants of the Ball–Berry model replace leaf surface humidity with functions of vapor pressure deficit (Leuning 1995; Medlyn et al. 2011).

The Ball–Berry stomatal model is empirical, based on correlations between stomatal conductance and photosynthesis from numerous leaf gas exchange measurements. It is appropriate for well-watered soils, where for a given relative humidity or vapor pressure deficit, stomatal conductance scales with the ratio of CO_2 assimilation to CO_2 concentration. However, how to represent stomatal closure with soil moisture stress is problematic. Some models directly impose diffusive limitations in response to soil drying by reducing the slope (g_1) between stomatal conductance and assimilation; other models impose biochemical limitations and indirectly reduce stomatal conductance by reducing assimilation (A_n) as soil moisture stress increases. Neither approach completely replicates observed stomatal response to soil moisture stress (Egea et al. 2011; De Kauwe et al. 2013). Some evidence suggests that both diffusive and biochemical limitations must be considered (Zhou et al. 2013). There is also uncertainty about the form of the soil moisture stress function (Verhoef and Egea 2014).

16.10 | Water-Use Efficiency

Although its origins are empirical, the Ball–Berry style model can be derived from the assumption that stomata maximize carbon gain while minimizing water loss (Katul et al. 2010; Medlyn et al. 2011). Water-use efficiency is defined as the ratio of carbon gain during photosynthesis to water loss during transpiration. At the leaf level, water-use efficiency is given by the ratio of photosynthesis to transpiration:

$$\frac{A_n}{E} = \frac{\frac{g_{sw}}{1.6}c_a\left(1 - c_i/c_a\right)}{Dg_{sw}} = \frac{c_a\left(1 - c_i/c_a\right)}{1.6D} \quad (16.15)$$

This equation expresses transpiration (E, mol H_2O m^{-2} s^{-1}) in relation to the vapor pressure deficit (D, mol mol^{-1}; $D = (e_i - e_a)/P$). Equation (16.15) is referred to as the instantaneous water-use efficiency. It varies with atmospheric conditions (temperature and humidity) that affect vapor pressure deficit. The biological component of water-use efficiency is expressed by the ratio of photosynthesis to stomatal conductance:

$$\frac{A_n}{g_{sw}} = \frac{c_a}{1.6}\left(1 - c_i/c_a\right) \quad (16.16)$$

Equation (16.16) is known as the intrinsic, or inherent, water-use efficiency and adjusts instantaneous water-use efficiency for the effect of vapor pressure deficit on transpiration. Intrinsic water-use efficiency is related to instantaneous water-use efficiency by $A_n/g_{sw} = DA_n/E$. Values of instantaneous water-use efficiency reported in the literature range from 0.1–7 mmol CO_2 mol^{-1} H_2O and 3–173 μmol CO_2 mol^{-1} H_2O for intrinsic water-use efficiency (Table 16.4).

Cowan (1977) and Cowan and Farquhar (1977) proposed that the physiology of stomata has evolved to constrain the rate of transpiration water loss (E) for a given unit of carbon gain (A_n). This optimization is achieved by assuming that stomatal conductance adjusts to maintain water-use efficiency constant over some time period. Formally, this means the marginal carbon gain of transpiration water loss ($\partial A_n/\partial E$) is constant; or equivalently, the marginal water

Table 16.4 Instantaneous water-use efficiency (A_n / E) and intrinsic water-use efficiency (A_n / g_{sw}) for various plant functional groups

Plants	A_n / E (mmol CO_2 mol^{-1} H_2O)			A_n / g_{sw} (μmol CO_2 mol^{-1} H_2O)		
	Mean	Min	Max	Mean	Min	Max
Crops	2.5	0.1	7.1	54	3	134
Herbs	2.9	0.3	7.1	44	3	159
Shrubs	3.1	1.0	6.1	68	6	173
Trees	1.8	0.1	5.8	60	4	157

Source: From Medrano et al. (2012).

cost of carbon gain ($\partial E / \partial A_n$) is constant, as originally proposed. Mathematical expressions for stomatal conductance can be obtained based on this principle using the Farquhar et al. (1980) photosynthesis model, but they vary among the Rubisco-limited, RuBP-limited, or co-limited rates, with different consequences for stomatal sensitivity to atmospheric CO_2 (Arneth et al. 2002; Katul et al. 2010; Medlyn et al. 2011, 2013; Vico et al. 2013). The Ball–Berry model, despite not being constructed explicitly as an optimality model, is consistent with this theory. However, water-use efficiency optimization gives a relationship in which stomatal conductance varies with $D^{-1/2}$ rather than h_s (Katul et al. 2009; Medlyn et al. 2011).

16.11 Carbon Isotopes

The process of photosynthesis discriminates against the heavier $^{13}CO_2$ relative to $^{12}CO_2$ so that the products of photosynthesis are depleted in ^{13}C (Farquhar et al. 1982; Brugnoli et al. 2012). This fractionation occurs because diffusion through stomatal pores favors $^{12}CO_2$ and discriminates against $^{13}CO_2$ (this produces a small fractionation) and because the Rubisco enzyme that catalyzes carboxylation discriminates against $^{13}CO_2$ (which gives large fractionation). Fractionation is seen in the $\delta^{13}C$ of plant material, given by the $^{13}C/^{12}C$ ratio of a sample (R_{sample}) relative to the $^{13}C/^{12}C$ ratio of the standard ($R_{standard}$):

$$\delta^{13}C = \left(\frac{R_{sample}}{R_{standard}} - 1 \right) 1000 \qquad (16.17)$$

and reported as per mil (‰). While atmospheric CO_2 has $\delta^{13}C$ of approximately –8‰, the $\delta^{13}C$ of C_3 plants ranges from about –22 to –34‰. Fractionation is smaller in C_4 plants, and $\delta^{13}C$ ranges from approximately –9 to –16‰.

The change in the relative abundance of ^{13}C between air (with $^{13}C/^{12}C$ ratio R_{air}) and plant material (with $^{13}C/^{12}C$ ratio R_{plant}) is the discrimination defined as:

$$\Delta = \left(\frac{R_{air}}{R_{plant}} - 1 \right) 1000 \qquad (16.18)$$

which is calculated using the $\delta^{13}C$ of the air and plant material:

$$\Delta = \frac{\delta^{13}C_{air} - \delta^{13}C_{plant}}{1 + \delta^{13}C_{plant}/1000} \qquad (16.19)$$

and is given in ‰ (as $\delta^{13}C$). The Δ for C_3 plants depends on the balance between CO_2 diffusion into the leaf and CO_2 fixation by Rubisco, and thus varies with the c_i / c_a ratio. When c_i / c_a is small, there is less discrimination by Rubisco, and the diffusive fractionation term ($\Delta = 4.4‰$) dominates the total fractionation. At high c_i / c_a, the fractionation by carboxylation ($\Delta = 30‰$) is the dominant term. An approximate relationship is:

$$\Delta = 4.4 + 25.6\, c_i / c_a \qquad (16.20)$$

Water-use efficiency increases as c_i / c_a decreases, and therefore Δ decreases as water-use efficiency increases. Carbon isotopes, therefore, provide a measure of water-use efficiency (Farquhar et al. 1982; Brugnoli et al. 2012; Medrano et al. 2012).

16.12 | Stomata and Atmospheric CO_2

Atmospheric CO_2 concentration has varied substantially over timescales of hundreds to millions of years (Franks et al. 2013, 2014). A prominent global trend over the twentieth century has been the increasing concentration of CO_2 in the atmosphere (Figure 3.7a), and this trend is expected to continue through the twenty-first century (Figure 2.11b). Stomata respond to elevated CO_2 through biochemical responses and through long-term anatomical changes in stomatal density and size. Common to these responses is that higher atmospheric CO_2 enhances photosynthetic rates, and stomatal conductance decreases to maintain c_i / c_a relatively constant so that water-use efficiency increases. The exact manner in which this regulation is achieved, and leaf gas exchange response to higher CO_2, is the subject of considerable research.

Studies of the physiological response of plants to short-term exposure to high CO_2 concentration routinely find greater photosynthesis and reduced stomatal conductance as CO_2 increases. This enhancement is greater for plants with the C_3 photosynthetic pathway than for C_4 plants. Figure 16.9a illustrates this response for eucalyptus, a C_3 tree (Wong et al. 1978). Conductance decreases with higher CO_2, more so at low irradiance than high irradiance. Across all the measurements, c_i / c_a remains relatively constant (~0.6–0.8).

Most such studies have been conducted in greenhouses and growth chambers and with small plants grown in pots. Free-air CO_2 enrichment (FACE) experiments allow plants to grow in the field under prevailing environmental conditions but at ambient and elevated CO_2 concentrations. Ainsworth and Long (2005) summarized results of 12 FACE studies involving more than 40 species in forest,

grassland, desert, and agricultural ecosystems exposed to CO_2 concentrations of 475–600 ppm for several years (see also Ainsworth and Rogers 2007; Wang et al. 2012; Franks et al. 2013). Light-saturated leaf photosynthetic rate increased by 31 percent on average and stomatal conductance decreased by 20 percent on average (Figure 16.10). The photosynthetic response was greater in C_3 plants (34%) than in C_4 plants (11%). Trees showed the greatest response to CO_2 enrichment with a 47 percent increase in light-saturated photosynthetic rate. C_3 plants showed signs of biochemical acclimation, or down-regulation, to elevated CO_2 concentration, with decreased photosynthetic capacity as measured by maximum carboxylation rate ($V_{c\,max}$) and maximum rate of electron transport (J_{max}). This acclimation increased with low nitrogen conditions. Similar results were found in the synthesis by Nowak et al. (2004) of 16 FACE sites in forest, grassland, desert, and bog ecosystems with CO_2 concentration elevated to 550–600 ppm and as high as 700 ppm at one site.

Changes in atmospheric CO_2 concentration also produce anatomical changes in stomatal conductance. Comparison of modern and historically preserved leaves from herbariums reveals a decline in the density of stomata on leaf surfaces (number of stomata per unit leaf area) over the past 200 years (Figure 16.9b) – a period in which atmospheric CO_2 increased greatly from preindustrial values. Other analyses reveal similar decline over the industrial era (Lammertsma et al. 2011), and the coupling between CO_2 and stomatal density has been found in fossil leaves over the past 400 million years, during which CO_2 varied from less than 300 ppm to greater than 2000 ppm (Franks and Beerling 2009; Franks et al. 2013). Such changes in stomatal density are used to reconstruct past atmospheric CO_2 concentrations (McElwain et al. 1999; Royer et al. 2001; Beerling and Royer 2002; Franks et al. 2014).

The geometry and density of stomata set a theoretical maximum stomatal conductance (Franks and Beerling 2009). The anatomical maximum stomatal conductance ($g_{s\,max}$, mol H_2O m^{-2} s^{-1}) is:

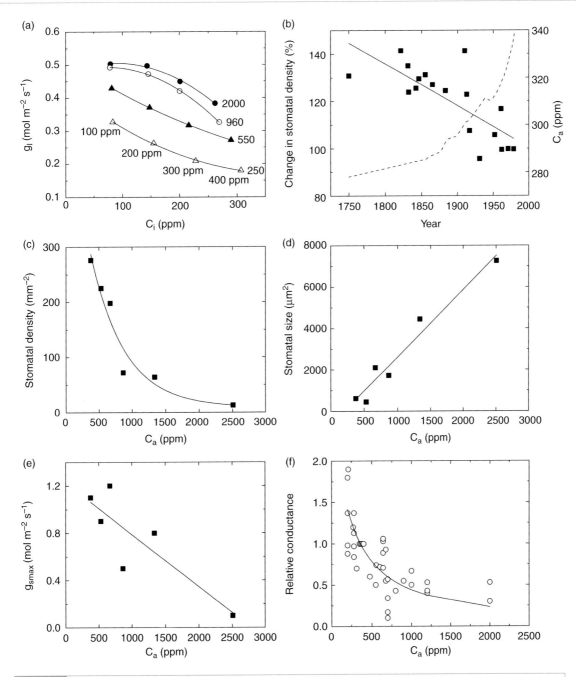

Fig. 16.9 Stomatal response to atmospheric CO$_2$ at various timescales. (a) Leaf conductance for *Eucalyptus pauciflora* in relation to short-term (< 1 hour) variation in CO$_2$ at four different irradiances (250, 550, 960, and 2000 µmol m^{-2} s^{-1}). Stomatal response is in relation to c_i. Shown also is ambient CO$_2$, which ranged from 100 to 400 ppm. Adapted from Wong et al. (1978). (b) Relative change in stomatal density in herbarium specimens of seven species of temperate trees and one species of shrub over the past 200 years. Also shown is the atmospheric CO$_2$ trend (dashed line). Adapted from Woodward (1987). (c–e) Stomatal density, size, and maximum conductance in response to atmospheric CO$_2$ concentration over the past 400 million years. Adapted from Franks et al. (2013). See also Franks and Beerling (2009). (f) Relative stomatal conductance in relation to the ambient CO$_2$ concentration. The solid line is the predicted relationship from Eq. (16.23). Data are from Franks et al. (2013).

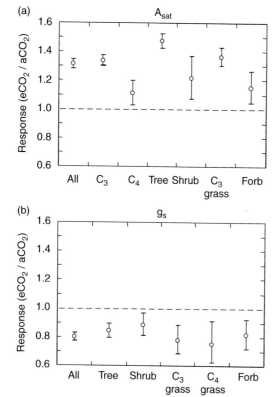

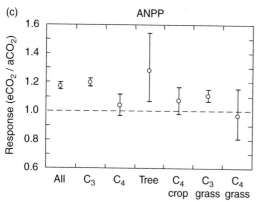

Fig. 16.10 Synthesis results from 12 FACE studies in forest, grassland, desert, and agricultural ecosystems exposed to CO_2 concentrations of 475–600 ppm. Data are the response ratio (elevated CO_2 response/ambient CO_2 response) shown as a mean (circles) and 95 percent confidence interval (bars) for all species and by plant functional type for (a) light-saturated leaf photosynthetic rate (A_{sat}), (b) stomatal conductance (g_s), and (c) aboveground production (ANPP). Data from Ainsworth and Long (2005).

$$g_{smax} = \frac{n_s a_{max} D_w \rho_m}{l + \frac{\pi}{2}(a_{max}/\pi)^{1/2}} \qquad (16.21)$$

where the number of stomata per unit area (stomatal density n_s, m^{-2}), the size of the stomata when fully opened (a_{max}, m^2), and stomatal pore depth (l, m) are anatomical traits. The term D_w is the diffusivity of water (24.0×10^{-6} m^2 s^{-1} at 15°C), and ρ_m is molar volume (42.3 mol m^{-3} at 15°C). Dow et al. (2014) give values for *Arabidopsis thaliana*, an annual flowering plant. For one particular genotype, the adaxial (upper) surface has $n_s = 80$ mm^{-2}, $a_{max} = 71.2$ μm^2, $l = 5.0$ μm, and $g_{smax} = 0.46$ mol m^{-2} s^{-1}. For the abaxial (lower) surface, $n_s = 105$ mm^{-2}, $a_{max} = 72.1$ μm^2, $l = 5.4$ μm, and $g_{smax} = 0.59$ mol m^{-2} s^{-1}. The total conductance from both surfaces is 1.05 mol m^{-2} s^{-1}. While Eq. (16.21) relates stomatal conductance to measured leaf anatomy, the anatomical g_{smax} is rarely attained and plants operate at a much lower value for the diffusive stomatal conductance (g_{sw}). However, the diffusive conductance predicted from the Ball–Berry model scales with anatomical g_{smax} (Dow et al. 2014).

Analyses of preserved leaves of nine species common in Florida show a 34 percent reduction in g_{smax} because of adaptation in stomatal density and pore size in response to a 100 ppm increase in atmosphere CO_2 over the past 150 years (Lammertsma et al. 2011). The decline in g_{smax} is consistent with optimization of carbon gain within the constraint imposed by transpiration water loss (de Boer et al. 2011). Fossilized remains from the past 400 million years also show that leaves from periods of high CO_2 concentration had relatively few but large stomata and low maximum conductance, while leaves from periods of low CO_2 concentration had many smaller stomata and higher maximum conductance (Figure 16.9c–e).

Differences among plants in g_{smax} relate to leaf anatomy. In the Florida analysis, for example, the angiosperms (e.g., maple and oak) generally have numerous small stomata and high g_{smax}, and the conifers (e.g., pine) have few large stomata and lower g_{smax} (Lammertsma et al. 2011). Angiosperms have a more elaborate hydraulic system compared with conifers. Their vein architecture and high vein length density gives a high hydraulic conductance along the soil-to-leaf pathway, which allows for high $g_{s\,max}$ (Bodribb et al. 2005, 2007; McKown et al. 2010). High leaf vein length density, and the resulting high photosynthetic capacity, was a key factor in the evolution of angiosperms ~100 million years ago in response to declining atmospheric CO_2 concentration (Brodribb and Feild 2010; Feild et al. 2011; de Boer et al. 2012).

The leaf response to elevated CO_2 is consistent across multiple timescales. The biochemical response to near-instantaneous CO_2 enrichment is a decline in stomatal conductance (Figure 16.9a). The long-term developmental response to higher CO_2 over timescales of centuries is a decline in stomatal density (Figure 16.9b), and this adaptive response is also seen in coordinated changes in stomatal density and size to regulate conductance over geologic timescales (Figure 16.9c–e). These responses are a feedback that maintains c_i / c_a nearly constant (Franks et al. 2013). Across these various timescales, the decline in stomatal conductance in response to higher CO_2 concentration is remarkably similar despite the different causal mechanisms. Indeed, the long-term response to changing CO_2 is described using an equation derived from the short-term biochemical response.

Franks et al. (2013) showed that the relative photosynthetic enhancement with elevated CO_2 is described by the equation:

$$A_{n(rel)} = \frac{A_n}{A_{n0}} = \frac{(c_a - \Gamma_*)(c_{a0} + 2\Gamma_*)}{(c_a + 2\Gamma_*)(c_{a0} - \Gamma_*)} \qquad (16.22)$$

where $c_{a0} = 360$ ppm is the reference CO_2 concentration and $\Gamma_* = 40$ ppm. This equation is an approximation of the RuBP-limited assimilation rate, given by Eq. (16.5), but using c_a instead of c_i. $A_{n(rel)}$ is the RuBP-limited assimilation rate evaluated for c_a and expressed as a ratio of the assimilation rate at c_{a0}. Similarly, the stomatal conductance at c_a relative to that at c_{a0} can be evaluated from the Ball–Berry model using Eq. (16.14):

$$g_{sw(rel)} = \frac{A_n}{c_a} \frac{c_{a0}}{A_{n0}} = \frac{A_{n(rel)}}{c_{a(rel)}} \qquad (16.23)$$

with $c_{a(rel)} = c_a / c_{a0}$. This equation is based on the assumption that the slope parameter (g_1) remains constant as c_a increases and that h_s is constant averaged over long time periods. It further requires that $g_0 \ll g_{sw}$, which is generally valid. Equation (16.23) does indeed describe the decline in stomatal conductance seen in plants grown at elevated CO_2 (Figure 16.9f). This suggests that stomata optimize for RuBP-limited carbon gain, as suggested also by water-use efficiency optimization theory (Medlyn et al. 2011, 2013).

16.13 | Review Questions

1. Why is light needed for photosynthesis?

2. Leaves reflect more solar radiation in the near-infrared waveband than in the visible waveband. Explain why this occurs.

3. What is photorespiration? Why are photorespiration rates lower in plants utilizing the C_4 photosynthetic pathway than in C_3 plants? Which type of plant shows photosynthetic inhibition at 21 percent O_2? Why?

4. Why does photosynthesis saturate at high light levels in C_3 plants? How does this compare with the light response curve of C_4 plants?

5. Why does photosynthesis saturate at high ambient CO_2 concentrations in C_3 plants? How does this compare with the CO_2 response curve of C_4 plants?

6. Some grasses utilize the C_3 photosynthetic pathway; others are C_4 plants. What type of grass is favored in a CO_2-enriched atmosphere?

7. Calculate light-saturated photosynthesis for a C_3 plant with $V_{c\,max} = 60$ µmol m^{-2} s^{-1} and $J_{max} = 100$ µmol m^{-2} s^{-1}. $K_c = 404.9$ µmol mol^{-1}, $K_o = 278.4$ mmol mol^{-1}, and $\Gamma_* = 42.75$ µmol mol^{-1}.

Use $c_i = 0.7c_a$, $o_i = 209$ mmol mol^{-1}, and $c_a = 380$ μmol mol^{-1}. $\Theta_J = 0.9$, $\Phi_{PSII} = 0.85$, and $\alpha_l = 0.8$.

8. Using the same values as question 7, calculate photosynthesis at $I \downarrow = 2000$ μmol photon m^{-2} s^{-1} for $V_{c\,max} = 30$ and 60 μmol m^{-2} s^{-1} with $J_{max} = 75$, 150, and 300 μmol m^{-2} s^{-1}. For which value of $V_{c\,max}$ is it advantageous to have high J_{max}? Is it advantageous to have high $V_{c\,max}$ in a sunny environment?

9. Repeat question 8, but with low irradiance ($I \downarrow = 100$ μmol photon m^{-2} s^{-1}). Is it advantageous to have high $V_{c\,max}$ in a shaded environment?

10. How does rising atmospheric CO_2 concentration affect leaf temperature of C_3 plants?

16.14 | References

Ainsworth, E. A., and Long, S. P. (2005). What have we learned from 15 years of free-air CO_2 enrichment (FACE)? A meta-analytical review of the responses of photosynthesis, canopy properties and plant production to rising CO_2. *New Phytologist*, 165, 351–372.

Ainsworth, E. A., and Rogers, A. (2007). The response of photosynthesis and stomatal conductance to rising [CO_2]: Mechanisms and environmental interactions. *Plant, Cell and Environment*, 30, 258–270.

Arneth, A., Lloyd, J., Šantrůčková, H., et al. (2002). Response of central Siberian Scots pine to soil water deficit and long-term trends in atmospheric CO_2 concentration. *Global Biogeochemical Cycles*, 16, 1005, 10.1029/2000GB001374.

Ball, J. T., Woodrow, I. E., and Berry, J. A. (1987). A model predicting stomatal conductance and its contribution to the control of photosynthesis under different environmental conditions. In *Progress in Photosynthesis Research*, vol. 4, ed. J. Biggins. Dordrecht: Martinus Nijhoff, pp. 221–224.

Beerling, D. J., and Royer, D. L. (2002). Reading a CO_2 signal from fossil stomata. *New Phytologist*, 153, 387–397.

Bernacchi, C. J., Singsaas, E. L., Pimentel, C., Portis, A. R. J., and Long, S. P. (2001). Improved temperature response functions for models of Rubisco-limited photosynthesis. *Plant, Cell and Environment*, 24, 253–259.

Bernacchi, C. J., Pimentel, C., and Long, S. P. (2003). *In vivo* temperature response functions of parameters required to model RuBP-limited photosynthesis. *Plant, Cell and Environment*, 26, 1419–1430.

Bernacchi, C. J., Bagley, J. E., Serbin, S. P., et al. (2013). Modelling C_3 photosynthesis from the chloroplast to the ecosystem. *Plant, Cell and Environment*, 36, 1641–1657.

Bonan, G. B. (1995). Land–atmosphere CO_2 exchange simulated by a land surface process model coupled to an atmospheric general circulation model. *Journal of Geophysical Research*, 100D, 2817–2831.

Brodribb, T. J., and Feild, T. S. (2010). Leaf hydraulic evolution led a surge in leaf photosynthetic capacity during early angiosperm diversification. *Ecology Letters*, 13, 175–183.

Brodribb, T. J., Holbrook, N. M., Zwieniecki, M. A., and Palma, B. (2005). Leaf hydraulic capacity in ferns, conifers and angiosperms: Impacts on photosynthetic maxima. *New Phytologist*, 165, 839–846.

Brodribb, T. J., Feild, T. S., and Jordan, G. J (2007). Leaf maximum photosynthetic rate and venation are linked by hydraulics. *Plant Physiology*, 144, 1890–1898.

Brugnoli, E., Loreto, F., and Ribas-Carbó, M. (2012). Stable isotopic compositions related to photosynthesis, photorespiration and respiration. In *Terrestrial Photosynthesis in a Changing Environment: A Molecular, Physiological and Ecological Approach*, ed. J. Flexas, F. Loreto, and H. Medrano. Cambridge: Cambridge University Press, pp. 152–168.

Collatz, G. J., Ball, J. T., Grivet, C., and Berry, J. A. (1991). Physiological and environmental regulation of stomatal conductance, photosynthesis and transpiration: A model that includes a laminar boundary layer. *Agricultural and Forest Meteorology*, 54, 107–136.

Collatz, G. J., Ribas-Carbo, M., and Berry, J. A. (1992). Coupled photosynthesis–stomatal conductance model for leaves of C_4 plants. *Australian Journal of Plant Physiology*, 19, 519–538.

Cowan, I. R. (1977). Stomatal behaviour and environment. *Advances in Botanical Research*, 4, 117–228.

Cowan, I. R., and Farquhar, G. D. (1977). Stomatal function in relation to leaf metabolism and environment. In *Integration of Activity in the Higher Plant*, ed.

D. H. Jennings. Cambridge: Cambridge University Press, pp. 471–505.

Cox, P. M., Betts, R. A., Bunton, C. B., et al. (1999). The impact of new land surface physics on the GCM simulation of climate and climate sensitivity. *Climate Dynamics*, 15, 183–203.

Craig, S. G., Holmén, K. J., Bonan, G. B., and Rasch, P. J. (1998). Atmospheric CO_2 simulated by the National Center for Atmospheric Research Community Climate Model, 1: Mean fields and seasonal cycles. *Journal of Geophysical Research*, 103D, 13213–13235.

Dang, Q. L., Margolis, H. A., Coyea, M. R., Sy, M., and Collatz, G. J. (1997a). Regulation of branch-level gas exchange of boreal trees: Roles of shoot water potential and vapor pressure difference. *Tree Physiology*, 17, 521–535.

Dang, Q. L., Margolis, H. A., Sy, M., et al. (1997b). Profiles of photosynthetically active radiation, nitrogen and photosynthetic capacity in the boreal forest: Implications for scaling from leaf to canopy. *Journal of Geophysical Research*, 102D, 28845–28859.

Dang, Q. L., Margolis, H. A., and Collatz, G. J. (1998). Parameterization and testing of a coupled photosynthesis–stomatal conductance model for boreal trees. *Tree Physiology* 18, 141–153.

de Boer, H. J., Lammertsma, E. I., Wagner-Cremer, F., et al. (2011). Climate forcing due to optimization of maximal leaf conductance in subtropical vegetation under rising CO_2. *Proceedings of the National Academy of Sciences USA*, 108, 4041–4046.

de Boer, H. J., Eppinga, M. B., Wassen, M. J., and Dekker, S. C. (2012). A critical transition in leaf evolution facilitated the Cretaceous angiosperm revolution. *Nature Communications*, 3, 1221, doi:10.1038/ncomms2217.

De Kauwe, M. G., Medlyn, B. E., Zaehle, S., et al. (2013). Forest water use and water use efficiency at elevated CO_2: A model–data intercomparison at two contrasting temperate forest FACE sites. *Global Change Biology*, 19, 1759–1779.

Denning, A. S., Fung, I. Y., and Randall, D. (1995). Latitudinal gradient of atmospheric CO_2 due to seasonal exchange with land biota. *Nature*, 376, 240–243.

Denning, A. S., Collatz, G. J., Zhang, C., et al. (1996). Simulations of terrestrial carbon metabolism and atmospheric CO_2 in a general circulation model. Part 1: Surface carbon fluxes. *Tellus B*, 48, 521–542.

Diaz-Espejo, A., Bernacchi, C. J., Collatz, G. J., and Sharkey, T. D. (2012). Models of photosynthesis. In *Terrestrial Photosynthesis in a Changing Environment: A Molecular, Physiological and Ecological Approach*, ed. J. Flexas, F. Loreto, and H. Medrano. Cambridge: Cambridge University Press, pp. 98–112.

Dickinson, R. E., Henderson-Sellers, A., Kennedy, P. J., and Wilson, M.F. (1986). *Biosphere–Atmosphere Transfer Scheme (BATS) for the NCAR Community Climate Model*, Technical Note NCAR/TN-275+STR. Boulder, Colorado: National Center for Atmospheric Research.

Dickinson, R. E., Henderson-Sellers, A., and Kennedy, P. J. (1993). *Biosphere–Atmosphere Transfer Scheme (BATS) Version 1e as Coupled to the NCAR Community Climate Model*, Technical Note NCAR/TN-387+STR. Boulder, Colorado: National Center for Atmospheric Research.

Dow, G. J., Bergmann, D. C., and Berry, J. A. (2014). An integrated model of stomatal development and leaf physiology. *New Phytologist*, 201, 1218–1226.

Egea, G., Verhoef, A., and Vidale, P. L. (2011). Towards an improved and more flexible representation of water stress in coupled photosynthesis–stomatal conductance models. *Agricultural and Forest Meteorology*, 151, 1370–1384.

Farquhar, G. D., and von Caemmerer, S. (1982). Modelling of photosynthetic response to environmental conditions. In *Encyclopedia of Plant Physiology New Series*, vol. 12B. *Physiological Plant Ecology. II. Water Relations and Carbon Assimilation*, ed. O. L. Lange, P. S. Nobel, C. B. Osmond, and H. Ziegler. Berlin: Springer-Verlag, pp. 549–587.

Farquhar, G. D., von Caemmerer, S., and Berry, J. A. (1980). A biochemical model of photosynthetic CO_2 assimilation in leaves of C_3 species. *Planta*, 149, 78–90.

Farquhar, G. D., O'Leary, M. H., and Berry, J. A. (1982). On the relationship between carbon isotope discrimination and the intercellular carbon dioxide concentration in leaves. *Australian Journal of Plant Physiology*, 9, 121–137.

Feild, T. S., Brodribb, T. J., Iglesias, A., et al. (2011). Fossil evidence for Cretaceous escalation in angiosperm leaf vein evolution. *Proceedings of the National Academy of Sciences USA*, 108, 8363–8366.

Field, C., and Mooney, H. A. (1986). The photosynthesis–nitrogen relationship in wild plants. In *On the Economy of Plant Form and Function*, ed. T. J. Givnish. Cambridge: Cambridge University Press, pp. 25–55.

Flexas, J., Ribas-Carbó, M., Diaz-Espejo, A., Galmés, J., and Medrano, H. (2008). Mesophyll conductance to CO_2: Current knowledge and future prospects. *Plant, Cell and Environment*, 31, 602–621.

Flexas, J., Brugnoli, E., and Warren, C. R. (2012). Mesophyll conductance to CO_2. In *Terrestrial Photosynthesis in a Changing Environment: A Molecular, Physiological and Ecological Approach*, ed. J. Flexas, F. Loreto, and H. Medrano. Cambridge: Cambridge University Press, pp. 169–185.

Franks, P. J., and Beerling, D. J. (2009). Maximum leaf conductance driven by CO_2 effects on stomatal size and density over geologic time. *Proceedings of the National Academy of Sciences USA*, 106, 10343–10347.

Franks, P. J., Adams, M. A., Amthor, J. S., et al. (2013). Sensitivity of plants to changing atmospheric CO_2 concentration: From the geological past to the next century. *New Phytologist*, 197, 1077–1094.

Franks, P. J., Royer, D. L., Beerling, D. J., et al. (2014). New constraints on atmospheric CO_2 concentration for the Phanerozoic. *Geophysical Research Letters*, 41, 4685–4694, doi:10.1002/2014GL060457.

Hetherington, A. M., and Woodward, F. I. (2003). The role of stomata in sensing and driving environmental change. *Nature*, 424, 901–908.

Jarvis, P. G. (1976). The interpretation of the variations in leaf water potential and stomatal conductance found in canopies in the field. *Philosophical Transactions of the Royal Society B*, 273, 593–610.

Kattge, J., Knorr, W., Raddatz, T., and Wirth, C. (2009). Quantifying photosynthetic capacity and its relationship to leaf nitrogen content for global-scale terrestrial biosphere models. *Global Change Biology*, 15, 976–991.

Katul, G. G., Palmroth, S., and Oren, R. (2009). Leaf stomatal responses to vapour pressure deficit under current and CO_2-enriched atmosphere explained by the economics of gas exchange. *Plant, Cell and Environment*, 32, 968–979.

Katul, G., Manzoni, S., Palmroth, S., and Oren, R. (2010). A stomatal optimization theory to describe the effects of atmospheric CO_2 on leaf photosynthesis and transpiration. *Annals of Botany*, 105, 431–442.

Körner, C. (1994). Leaf diffusive conductances in the major vegetation types of the globe. In *Ecophysiology of Photosynthesis*, ed. E.-D. Schulze and M. M. Caldwell. Berlin: Springer-Verlag, pp. 463–490.

Lammertsma, E. I., de Boer, H. J., Dekker, S. C., et al. (2011). Global CO_2 rise leads to reduced maximum stomatal conductance in Florida vegetation. *Proceedings of the National Academy of Sciences USA*, 108, 4035–4040.

Larcher, W. (1995). *Physiological Plant Ecology*, 3rd ed. New York: Springer-Verlag.

Leuning, R. (1995). A critical appraisal of a combined stomatal–photosynthesis model for C_3 plants. *Plant, Cell and Environment*, 18, 339–355.

McElwain, J. C., Beerling, D. J., and Woodward, F. I. (1999). Fossil plants and global warming at the Triassic–Jurassic boundary. *Science*, 285, 1386–1390.

McKown, A. D., Cochard, H., and Sack, L. (2010). Decoding leaf hydraulics with a spatially explicit model: Principles of venation architecture and implications for its evolution. *American Naturalist*, 175, 447–460.

Medlyn, B. E., Dreyer, E., Ellsworth, D., et al. (2002). Temperature response of parameters of a biochemically based model of photosynthesis, II: A review of experimental data. *Plant, Cell and Environment*, 25, 1167–1179.

Medlyn, B. E., Duursma, R. A., Eamus, D., et al. (2011). Reconciling the optimal and empirical approaches to modelling stomatal conductance. *Global Change Biology*, 17, 2134–2144.

Medlyn, B. E., Duursma, R. A., De Kauwe, M. G., and Prentice, I. C. (2013). The optimal stomatal response to atmospheric CO_2 concentration: Alternative solutions, alternative interpretations. *Agricultural and Forest Meteorology*, 182/183, 200–203.

Medrano, H., Gulías, J., Chaves, M. M., Galmés, J., and Flexas, J. (2012). Photosynthetic water-use efficiency. In *Terrestrial Photosynthesis in a Changing Environment: A Molecular, Physiological and Ecological Approach*, ed. J. Flexas, F. Loreto, and H. Medrano. Cambridge: Cambridge University Press, pp. 523–536.

Nowak, R. S., Ellsworth, D. S., and Smith, S. D. (2004). Functional responses of plants to elevated atmospheric CO_2: Do photosynthetic and productivity data from FACE experiments support early predictions? *New Phytologist*, 162, 253–280.

Rogers, A. (2014). The use and misuse of $V_{c,max}$ in Earth System Models. *Photosynthesis Research*, 119, 15–29.

Royer, D. L., Wing, S. L., Beerling, D. J., et al. (2001). Paleobotanical evidence for near present-day levels of atmospheric CO_2 during part of the Tertiary. *Science*, 292, 2310–2313.

Sellers, P. J., Mintz, Y., Sud, Y. C., and Dalcher, A. (1986). A simple biosphere model (SiB) for use within general circulation models. *Journal of the Atmospheric Sciences*, 43, 505–531.

Sellers, P. J., Randall, D. A., Collatz, G.J., et al. (1996). A revised land surface parameterization (SiB2) for atmospheric GCMs, Part I: Model formulation. *Journal of Climate*, 9, 676–705.

Verhoef, A., and Egea, G. (2014). Modeling plant transpiration under limited soil water: Comparison of different plant and soil hydraulic parameterizations and preliminary implications for their use in land surface models. *Agricultural and Forest Meteorology*, 191, 22–32.

Vico, G., Manzoni, S., Palmroth, S., Weih, M. and Katul, G. (2013). A perspective on optimal leaf stomatal conductance under CO_2 and light co-limitations. *Agricultural and Forest Meteorology*, 182/183, 191–199.

von Caemmerer, S. (2000). *Biochemical Models of Leaf Photosynthesis*. Collingwood, Victoria: CSIRO Publishing.

von Caemmerer, S. (2013). Steady-state models of photosynthesis. *Plant, Cell and Environment*, 36, 1617–1630.

von Caemmerer, S., Farquhar, G., and Berry, J. (2009). Biochemical model of C_3 photosynthesis. In *Photosynthesis in silico: Understanding Complexity from Molecules to Ecosystems*, ed. A. Laisk, L. Nedbal and Govindjee. Dordrecht: Springer, pp. 209–230.

Wang, D., Heckathorn, S. A., Wang, X., and Philpott, S. M. (2012). A meta-analysis of plant physiological and growth responses to temperature and elevated CO_2. *Oecologia*, 169, 1–13.

Wong, S. C., Cowan, I. R., and Farquhar, G. D. (1978). Leaf conductance in relation to assimilation in *Eucalyptus pauciflora* Sieb. ex Spreng. *Plant Physiology*, 62, 670–674.

Wong, S.-C., Cowan, I. R., and Farquhar, G. D. (1979). Stomatal conductance correlates with photosynthetic capacity. *Nature*, 282, 424–426.

Wong, S.-C., Cowan, I. R., and Farquhar, G. D. (1985a). Leaf conductance in relation to rate of CO_2 assimilation, I: Influence of nitrogen nutrition, phosphorus nutrition, photon flux density, and ambient partial pressure of CO_2 during ontogeny. *Plant Physiology*, 78, 821–825.

Wong, S.-C., Cowan, I. R., and Farquhar, G. D. (1985b). Leaf conductance in relation to rate of CO_2 assimilation, II: Effects of short-term exposures to different photon flux densities. *Plant Physiology*, 78, 826–829.

Wong, S.-C., Cowan, I. R., and Farquhar, G. D. (1985c). Leaf conductance in relation to rate of CO_2 assimilation, III: Influences of water stress and photoinhibition. *Plant Physiology*, 78, 830–834.

Woodward, F. I. (1987). Stomatal numbers are sensitive to increases in CO_2 from pre-industrial levels. *Nature*, 327, 617–618.

Wullschleger, S. D. (1993). Biochemical limitations to carbon assimilation in C_3 plants – a retrospective analysis of the A/C_i curves from 109 species. *Journal of Experimental Botany*, 44, 907–920.

Zhou, S., Duursma, R. A., Medlyn, B. E., Kelly, J. W. G., and Prentice, I. C. (2013). How should we model plant responses to drought? An analysis of stomatal and non-stomatal responses to water stress. *Agricultural and Forest Meteorology*, 182/183, 204–214.

17

Plant Canopies

17.1 | Chapter Summary

The principles that determine the temperature, energy balance, and photosynthetic rate of a leaf also determine those of plant canopies when integrated over all leaves in the canopy. These processes are related to the amount of leaf area, quantified by leaf area index. The vertical profile of leaf area in the canopy affects the distribution of radiation in the canopy and the absorption of radiation by leaves. With low leaf area index, plants absorb little solar radiation, and the overall surface albedo is largely that of soil. The absorption of radiation increases with greater leaf area index, and surface albedo responds more to the optical properties of foliage rather than soil. The integration of leaf processes over the light profile is central to the scaling of processes from a leaf to the canopy. The carbon uptake by a canopy is the integration of the photosynthetic rates of individual leaves, accounting for variations in light, microclimate, and foliage nitrogen with depth in the canopy. Similarly, canopy conductance is an aggregate measure of the conductance of individual leaves. The profile of leaf area in the canopy also affects turbulence within the canopy. The influence of vegetation on surface fluxes can be modeled by treating the soil–canopy system as an effective bulk surface, or a big leaf. The Penman–Monteith equation applied to canopies is an example of a big-leaf model. Fluxes of CO_2 and evapotranspiration measured over

forests illustrate the environmental controls of canopy fluxes and the partitioning of fluxes between the canopy and forest floor.

17.2 | Leaf Area Index

Leaf area index measures the amount of foliage in a plant canopy. Leaf area index is the projected area of leaves per unit area of ground. A square centimeter of ground covered by a leaf with an area of 1 cm² has a leaf area index of one. The same area covered by two leaves, one atop the other and each with an area of 1 cm², has a leaf area index of two. Projected, or one-sided, leaf area is different from total leaf area. For thin, flat leaves such as broadleaf trees, total leaf area is twice the projected leaf area (both sides of the leaf are included). For needleleaf trees, total leaf area is more than twice the projected leaf area. A leaf area index of two means there are 2 m² of leaf area (one-sided) covering 1 m² of ground. A typical leaf area index for a productive forest is 4–6 m² m⁻².

Cumulative leaf area index increases progressively with greater depth from the top of the canopy. Leaf area index measured below the canopy is the total leaf area. This leaf area is typically not distributed uniformly with height. Many forests have a single layer of foliage in the overstory. The oak forest shown in Figure 17.1, for example, has an overstory layer at a height between 6.5 and 3 m with a cumulative leaf area index of 4.6 m² m⁻². The foliage is densest at a

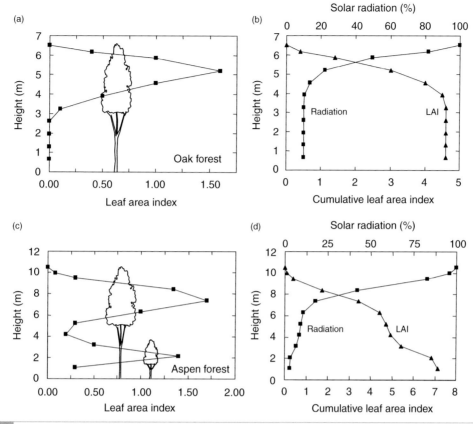

Fig. 17.1 Vertical profile of leaves in forests. The top panels show for an oak forest (a) the leaf area profile and (b) the cumulative leaf area and irradiance as a percentage of that at the top of the canopy. The bottom panels (c–d) show the same data for an aspen forest with understory. Data from Rauner (1976).

height of about 5 m. Cumulative leaf area index at this height is 3 m^2 m^{-2}, about two-thirds of the total foliage. In contrast, the aspen forest shown in Figure 17.1 has a leaf area index of 7.1 m^2 m^{-2}. Leaves are found in both an overstory situated between 10.5 and 5.5 m height and a lower understory below 3 m.

17.3 | Radiative Transfer

Plant canopies have a vertical gradient of sunlight related to the amount of leaves. An individual leaf absorbs, reflects, or transmits the sunlight that strikes it. As the canopy becomes denser with leaves, more solar radiation is absorbed or reflected and less is transmitted deeper into the canopy. A simple model of radiative transfer shows that solar radiation decreases as an exponential function of leaf area index:

$$I \downarrow (z) = I \downarrow_0 e^{-K_b L(z)} \qquad (17.1)$$

where $I \downarrow_0$ is the solar radiation at the top of the canopy, K_b is the light extinction coefficient, $L(z)$ is the cumulative leaf area index at height z, and $I \downarrow (z)$ is the irradiance at height z.

The extinction coefficient depends on leaf orientation and solar zenith angle. These determine the angle at which the solar beam strikes a leaf. A leaf receives the greatest radiation per unit surface area when the beam is perpendicular to the leaf. The angle at which solar radiation strikes a leaf depends on solar zenith angle. It also depends on leaf orientation. Some leaves are oriented horizontally while others are vertical. Many leaves are oriented randomly so that there is an equal probability of orientation in any direction (also known as a spherical leaf distribution). For these leaf orientations, the extinction coefficient is:

Horizontal: $\quad K_b = 1$

Vertical: $\qquad K_b = \dfrac{2}{\pi}\dfrac{1}{\tan B}$ $\qquad$ (17.2)

Spherical: $\quad K_b = 0.5/\sin B$

where $B = 90° - Z$ is the elevation angle above the horizon and Z is solar zenith angle.

Needles tend to have a spherical leaf distribution, evergreen and deciduous broadleaves tend to be semi-horizontal, and grasses and crops have semi-vertical foliage. Despite these complexities, a common value is $K_b = 0.5$. Using this value, a plant canopy transmits 61 percent of radiation onto the ground with $L = 1$ m² m⁻², 37 percent with $L = 2$ m² m⁻², and 22 percent with $L = 3$ m² m⁻². In a dense canopy with $L = 6$ m² m⁻², only 5 percent of solar radiation is transmitted through the canopy.

This equation for $I\downarrow(z)$ does not account for scattering of radiation within the canopy. It is derived with the assumption that leaves completely absorb light and do not reflect or transmit radiation. Scattering can be included using the equation:

$$I\downarrow(z) = I\downarrow_0 \, e^{-\sqrt{\alpha_l}\,K_b L(z)} \qquad (17.3)$$

where α_l is the absorptivity of leaves (Sellers 1985; Campbell and Norman 1998). As absorptivity decreases, the effective extinction coefficient becomes smaller and more light is transmitted through the canopy. Typical absorptivity ranges from 0.8 to 0.2–0.3 depending on solar wavelength.

Figure 17.1 shows typical vertical profiles of leaves and light within forest canopies. Solar radiation decreases rapidly in the canopy. In the oak forest, the cumulative leaf area index at a height of 5.8 m (70 cm into the canopy) is 1.4 m² m⁻² and irradiance is 50 percent of full sunlight. The total leaf area index is 4.6 m² m⁻², so that the forest floor receives only about 10 percent of full sunlight. The aspen forest has a leaf area

index of 7.1 m² m⁻². Leaves are found in both an overstory with a leaf area index of 4.9 m² m⁻² and an understory with a leaf area index of 2.2 m² m⁻². The understory receives only 8.5 percent of full sunlight. In contrast, leaves above 8.4 m in height receive more than 50 percent of full sunlight.

The exponential attenuation of light within a canopy represents the mean irradiance at any height. This irradiance is the average of some areas of the canopy in which light is unattenuated and some areas that are shaded (zero irradiance). Eq. (17.1) also describes the fraction of the canopy at cumulative leaf area index x that is sunlit:

$$f_{sun}(x) = e^{-K_b x} \qquad (17.4)$$

For a canopy with leaf area index L, the sunlit leaf area is:

$$L_{sun} = \left(1 - e^{-K_b L}\right)/K_b \qquad (17.5)$$

The remainder of the canopy is shaded. When the canopy is very dense, only a small portion is sunlit. Most of the canopy is shaded. When the Sun is high in the sky (30° zenith angle), horizontal leaves attenuate the most radiation and have the lowest sunlit leaf area (Figure 17.2). Vertically oriented leaves have the lowest extinction coefficient and have greater sunlit leaf area. Maximum sunlit leaf area is only 1 m² m⁻² for horizontal leaves and 1.7 m² m⁻² for spherically oriented leaves. When the Sun is lower in the sky (60° zenith angle), much less of the canopy is sunlit. Maximum sunlit leaf area index is less than 1 m² m⁻² for all leaf orientations.

The absorption, reflection, and transmittance of solar radiation by foliage have strong wavelength dependence. The Sun's radiation is broadly divided into two wavebands: the visible waveband at wavelengths less than 0.7 µm and the near-infrared waveband at wavelengths greater than 0.7 µm. Plants utilize light in these two wavebands differently (Table 17.1). Green leaves typically absorb more than 85 percent of the solar radiation in the visible waveband that strikes the leaf. This light is used during photosynthesis. Light in the near-infrared waveband is not utilized during photosynthesis, and rather

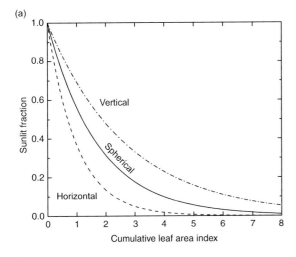

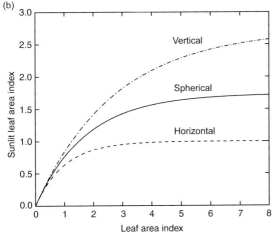

Fig. 17.2 (a) Sunlit fraction and (b) sunlit leaf area index for horizontal, spherical, and vertical leaves in relation to leaf area index with a solar zenith angle of 30°.

than absorbing this radiation, and thus possibly overheating, leaves typically absorb less than 50 percent of the radiation in the near-infrared waveband.

More complex models of radiative transfer account for the different transmission of direct beam and diffuse solar radiation and for the scattering of radiation within the canopy (Goudriaan 1977; Norman 1979; Sellers 1985; Goudriaan and van Laar 1994). Figure 17.3 illustrates results from the two-stream model of radiative transfer in plant canopies (Sellers 1985). With low leaf area index, plants absorb little solar radiation, but the absorption of radiation increases substantially with higher leaf area. With a leaf area index of 4 m² m⁻², more than 90 percent of the visible radiation is absorbed. Absorption saturates at 95 percent of the incoming radiation with a leaf area index of 6 m² m⁻². Significantly less near-infrared radiation is absorbed. At low leaf area index, more diffuse radiation is absorbed by the canopy than direct beam radiation. As leaf area index increases, this difference becomes smaller.

The canopy consists of two types of leaves (Figure 17.4). Sunlit leaves receive the unscattered direct beam radiation absorbed by the canopy and additionally a portion of the diffuse radiation (scattered direct beam radiation and atmospheric diffuse radiation) absorbed by the canopy. Shaded leaves receive only scattered direct beam and atmospheric diffuse radiation. Sunlit leaves absorb most of the direct beam radiation; shaded leaves absorb very little direct

Table 17.1 | Leaf orientation and reflection, transmission, and absorption of solar radiation by a leaf for visible and near-infrared wavebands

Vegetation	Leaf orientation	Visible			Near-infrared		
		Reflected	Transmitted	Absorbed	Reflected	Transmitted	Absorbed
Needleleaf tree	Spherical	0.07	0.05	0.88	0.35	0.10	0.55
Broadleaf tree	Semi-horizontal	0.10	0.05	0.85	0.45	0.25	0.30
Grass, crop	Semi-vertical	0.11	0.07	0.82	0.58	0.25	0.17

Source: From Dorman and Sellers (1989).

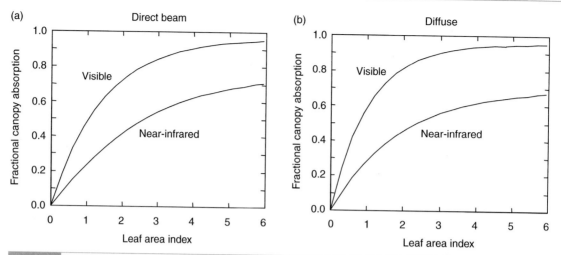

Fig. 17.3 Radiative transfer in a broadleaf forest with spherical leaf orientation in relation to leaf area index. Shown for the visible and near-infrared wavebands are (a) the fraction of direct beam solar radiation and (b) the fraction of diffuse solar radiation absorbed by the canopy using the radiative transfer model of Sellers (1985). The zenith angle is 45° and soil albedos are 0.10 (visible) and 0.20 (near-infrared). Leaf optical properties are from Table 17.1.

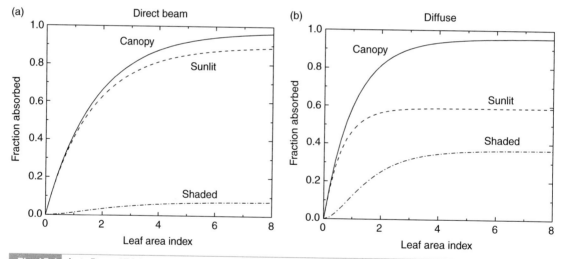

Fig. 17.4 As in Figure 17.3, but showing (a) the fraction of direct beam radiation absorbed by the total canopy and the sunlit and shaded portions of the canopy and (b) similarly for diffuse radiation. Radiative transfer uses a solution to the two-stream approximation for sunlit and shaded leaves (Dai et al. 2004) as described by Bonan et al. (2011).

beam radiation. Diffuse radiation is more equitably distributed between sunlit and shaded leaves.

The albedo of a plant canopy is the combined reflection of all plant material (leaves and stems) and the underlying ground. With low leaf area index, the albedo is largely that of soil (Figure 17.5). As leaf area index increases,

albedo responds more to the optical properties of foliage. Consequently, for ground surfaces with high albedo, canopy albedo decreases with greater leaf area index. Even when the ground is covered by snow, which has a very high albedo, plant material effectively masks the underlying snow. For leaf area index greater than 3 m^2 m^{-2}, the canopy albedo is similar to that of

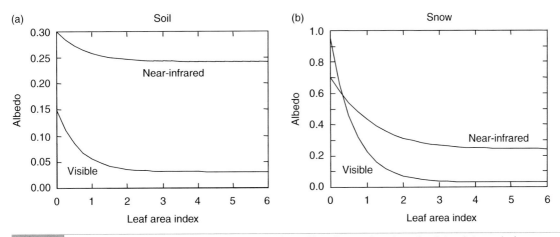

Fig. 17.5 Canopy albedo for direct beam solar radiation in the visible and near-infrared wavebands in relation to leaf area index. (a) Canopy albedo with a soil albedo of 0.15 (visible) and 0.30 (near-infrared). (b) Canopy albedo for snow-covered ground with an albedo of 0.95 (visible) and 0.70 (near-infrared). Data are for a broadleaf forest (Table 17.1) with spherical leaf orientation and for a zenith angle of 45° using the radiative transfer model of Sellers (1985).

vegetation without snow. Vegetation masking of snow albedo is evident in Figure 12.10.

Surface albedo varies for direct beam and diffuse radiation and for the visible and near-infrared wavebands. The albedo for direct beam radiation increases with greater zenith angle. This is particularly evident for spherical and semi-vertical leaf orientations. Consequently, albedo is higher early in the morning and late in the afternoon, when the Sun is near the horizon, than at midday. Figure 12.5c illustrates just such behavior in the diurnal cycle of albedo. In contrast, the albedo for diffuse radiation has no dependence on solar zenith angle and varies only slightly with leaf orientation.

Despite the complexities of radiative transfer in plant canopies, vegetation can be characterized by broad ranges of albedo (Table 12.1). The broadband albedo of vegetation averaged over all wavelengths typically ranges from 0.05 to 0.25. Forests generally have lower albedo than grasslands or croplands. Coniferous forests have lower albedo than deciduous forests.

The different optical properties of foliage in the visible and near-infrared wavebands is an important and distinguishing spectral characteristic of vegetation that facilitates the monitoring and analysis of vegetation by remote sensing. Green leaves absorb more solar radiation in the visible waveband than in the near-infrared waveband (Table 17.1). In contrast, other surfaces, especially soil, have smaller spectral differences in solar absorption. The normalized difference vegetation index (NDVI), with values ranging from –1 to +1, is a measure of the normalized difference in reflection of solar radiation in these two wavebands:

$$NDVI = (r_{nir} - r_{red})/(r_{nir} + r_{red}) \qquad (17.6)$$

where r_{nir} and r_{red} are reflectances in the near-infrared and red wavebands, respectively. Typical values for snow, lakes, and soil range from –0.2 to 0.05. Vegetated surfaces have values ranging from 0.05 to 0.70, with higher values indicating more productive vegetation. The NDVI is related to canopy photosynthetic capacity and allows monitoring of vegetation by remote sensing (Tucker et al. 1985, 1986, 1991; Myneni et al. 1997).

17.4 Canopy Photosynthesis

The photosynthetic uptake of carbon by a canopy of leaves (i.e., gross primary production; GPP) is the sum of the photosynthetic rates of the

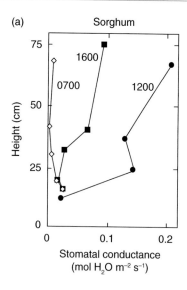

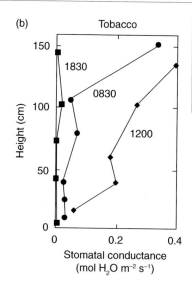

Fig. 17.6 Vertical profile of stomatal conductance in crops. (a) Sorghum at 0700, 1200, and 1600 hours. (b) Tobacco at 0830, 1200, and 1830 hours. Adapted from Jarvis and McNaughton (1986).

individual leaves, accounting for within-canopy variations in light, foliage nitrogen, and other factors. One approach to estimate canopy photosynthesis is based on the observation that biomass accumulation is proportional to the amount of radiation intercepted by the canopy (Monteith 1977). Production efficiency models simulate carbon assimilation proportional to absorbed photosynthetically active radiation times a light-use efficiency, which relates carbon gain to light absorbed (Prince and Goward 1995; Running et al. 2000, 2004; Zhao et al. 2005; Yuan et al. 2007). A typical light-use efficiency is 1.5 g C per MJ. This potential productivity is reduced by environmental constraints as:

$$GPP = E\,F\downarrow f_1(T)f_2(\theta)f_3(D) \tag{17.7}$$

where E is light-use efficiency, $F\downarrow$ is absorbed photosynthetically active radiation, and $f_1(T)$, $f_2(\theta)$, and $f_3(D)$ are empirical functions scaled from zero to one that adjust photosynthesis for temperature, soil water, and vapor pressure deficit.

Other methods explicitly scale leaf fluxes to the canopy by integrating photosynthesis and stomatal conductance over the light profile. Figure 17.6 illustrates variation of stomatal conductance with height in two canopies of crop plants. Stomatal conductance varies little with height early in the morning when low light levels limit the rate of photosynthesis. Later in the

day, stomatal conductance is greater at the top of the canopy than deeper in the canopy where leaves are shaded. The total carbon uptake during photosynthesis requires integration of leaf photosynthesis over the light profile.

Norman (1993) reviewed different methodologies for this leaf-to-canopy scaling. One method is to assume that all leaves in the canopy receive the average radiation absorbed by the canopy, but this fails to account for the non-linear extinction of light within the canopy and the non-linear dependence of photosynthesis and stomatal conductance on light. Scaling techniques need to account for the decrease in light with greater depth in the canopy. In particular, some leaves are sunlit and receive full illumination. Others are shaded and have low rates of photosynthesis.

One scaling approach analytically integrates leaf photosynthesis over the light profile using a simple model of the photosynthetic response to light combined with exponential attenuation of light in the canopy (Sellers 1985; Norman 1993). The increase in the rate of leaf photosynthesis with more absorbed photosynthetically active radiation can be represented by:

$$A_l = \frac{A_{\max}EF\downarrow}{EF\downarrow + A_{\max}} \tag{17.8}$$

where $A_{\max}$ is the maximum rate of photosynthesis at saturating light (μmol m^{-2} s^{-1}), $F\downarrow$ is

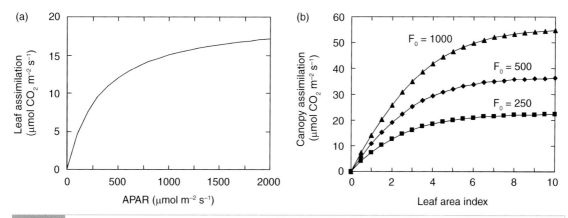

Fig. 17.7 Canopy scaling of photosynthesis. (a) Leaf photosynthesis in relation to absorbed photosynthetically active radiation (APAR). (b) Canopy photosynthesis in relation to leaf area index obtained by integrating leaf photosynthesis over an exponential light profile. Canopy photosynthesis is shown for three values of incident PAR at the top of the canopy.

absorbed photosynthetically active radiation (μmol m^{-2} s^{-1}), and E is light-use efficiency (mol mol^{-1}). If light is attenuated through the canopy exponentially as in Eq. (17.1), the integral of A_l with respect to leaf area index (L) is the total photosynthesis of the canopy, equal to:

$$GPP = \frac{A_{max}}{K_b} \ln\left[\frac{b + F_0}{b + F_0 \exp(-K_b L)}\right] \qquad (17.9)$$

where $b = A_{max} / E$ and F_0 is the photosynthetically active radiation incident on the canopy. Figure 17.7 shows this scaling for $A_{max} = 20$ μmol m^{-2} s^{-1}, E = 0.06 mol mol^{-1}, and $K_b = 0.5$. In this example, leaf photosynthesis attains a rate of 17 μmol m^{-2} s^{-1} at high irradiance. Canopy photosynthesis increases with greater leaf area index as more photosynthetically active radiation is absorbed by foliage, but saturates with leaf area index of 6–7 m^2 m^{-2} as light absorption also saturates.

Another method to integrate photosynthesis over the canopy is to consider the fraction of the canopy that is sunlit and shaded (Norman 1993). Different leaf photosynthetic rates are calculated for these two classes of leaves based on the amount of absorbed photosynthetically active radiation. Shaded leaves receive only diffuse radiation, from both the sky and direct beam radiation scattered within the canopy, and are typically in the linear portion of the photosynthetic light response curve. Sunlit leaves receive diffuse radiation and additionally the direct beam radiation intercepted in the canopy. Because they receive much more radiation than shaded leaves, sunlit leaves are typically near light saturation. Canopy photosynthesis is the sum of these two rates, each multiplied by their respective leaf area index:

$$GPP = A_{sun}L_{sun} + A_{shade}L_{shade} \qquad (17.10)$$

This approach has the advantage of accounting for the different attenuation of direct beam and diffuse radiation within the canopy.

Leaves growing in shade generally have lower photosynthetic capacity than leaves growing in sun (Table 16.1). This is a result of different investments in chlorophyll and Rubisco. These investments require much nitrogen and are energetically expensive to maintain. Plants that invest too much in photosynthetic capacity will be at a competitive disadvantage if this machinery is underutilized, such as occurs in a shaded understory. Plants that match photosynthetic capacity to local resource availability have an advantage. Based on this reasoning, the vertical profile of leaf nitrogen and photosynthetic capacity through the canopy should be related to the light profile. This concept provides another means, similar to sunlit and shaded leaves, to integrate photosynthesis and stomatal conductance for a canopy (Sellers et al. 1992).

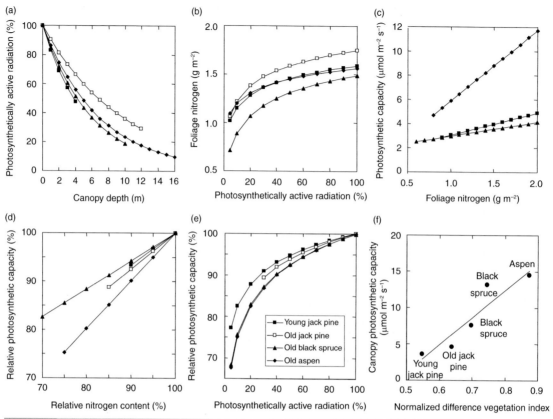

Fig. 17.8 Canopy integration of photosynthetic capacity based on relationships among light, foliage nitrogen, and photosynthetic capacity for jack pine, black spruce, and quaking aspen forests. (a) Photosynthetically active radiation as a percentage of full radiation (%PAR) in relation to depth from the top of the canopy. (b) Foliage nitrogen in relation to %PAR. (c) Leaf photosynthetic capacity in relation to foliage nitrogen. (d) Leaf photosynthetic capacity (as a percentage of that in the upper canopy) in relation to leaf nitrogen (as a percentage of that in the upper canopy). (e) Leaf relative photosynthetic capacity in relation to %PAR. (f) Canopy photosynthetic capacity integrated over all leaves in relation to the normalized difference vegetation index. Data from Dang et al. (1997).

Data collected in the boreal forests of Canada illustrate how light, foliage nitrogen, and photosynthetic capacity within tree canopies are interrelated (Figure 17.8). In jack pine, black spruce, and aspen forests, photosynthetically active radiation decreases exponentially with greater depth from the top of the canopy. Foliage nitrogen decreases with decreasing photosynthetically active radiation. Maximum photosynthetic rates under light-saturated conditions increase with greater foliage nitrogen. Jack pine and black spruce trees differ little in this relationship; aspen trees have a higher photosynthetic capacity for a given nitrogen concentration. These data show that foliage high in the canopy receives more radiation, has greater nitrogen concentration, and has greater photosynthetic capacity than foliage lower in the canopy. Indeed, throughout the canopy there is a linear relationship between photosynthetic capacity at a given height and the corresponding nitrogen content at the same height, both expressed relative to that at the top of the canopy. Relative photosynthetic capacity decreases exponentially with decreasing photosynthetically active radiation.

The original theory postulated that plants optimally allocate resources to maximize carbon gain such that area-based leaf nitrogen is distributed through the canopy in relation to the time-mean profile of photosynthetically active radiation, but it is now recognized that the nitrogen gradient is shallower than the light gradient (Hollinger 1996; Carswell et al. 2000; Meir et al. 2002; Niinemets 2007; Lloyd et al. 2010). The decline in leaf nitrogen and photosynthetic capacity is described by an exponential relationship with cumulative leaf area index, similar to the sunlit fraction:

$$f_n(x) = e^{-K_n x} \qquad (17.11)$$

Lloyd et al.'s (2010) estimates of K_n for 16 temperate broadleaf forests and 2 tropical forests range from 0.10 to 0.43 (mean, 0.19; median, 0.18).

The parameters $V_{c\max}$ and $J_{\max}$ in the Farquhar et al. (1980) photosynthetic model vary with leaf nitrogen (Rogers 2014). Many plant canopy models parameterize canopy scaling using concepts of sunlit and shaded leaves in combination with an exponential profile of foliage nitrogen (de Pury and Farquhar 1997; Wang and Leuning 1998; Dai et al. 2004; Bonan et al. 2011). The canopy is divided into sunlit and shaded fractions, and the photosynthesis–conductance model is solved using canopy-integrated parameters derived from leaf-level parameters. Canopy values for $V_{c\max}$ are found by integrating leaf nitrogen concentration, using Eq. (17.11), over the sunlit and shaded fractions of the canopy, using Eq. (17.4). Other parameters scale similarly. For example, $V_{c\max}$ integrated over the sunlit canopy is:

$$
\begin{aligned}
V_{c\max}(\text{sun}) &= \int_0^L V_{c\max 0} f_n(x) f_{sun}(x)\, dx \\
&= V_{c\max 0}\left[1 - e^{-(K_n + K_b)L}\right]\frac{1}{K_n + K_b}
\end{aligned}
\qquad (17.12)
$$

where $V_{c\max 0}$ is the value at the top of the canopy. For the shaded canopy:

$$
\begin{aligned}
V_{c\max}(\text{sha}) &= \int_0^L V_{c\max 0} f_n(x)\left[1 - f_{sun}(x)\right] dx \\
&= V_{c\max 0}\left[1 - e^{-K_n L}\right]\frac{1}{K_n} - V_{c\max}(\text{sun})
\end{aligned}
\qquad (17.13)
$$

17.5 Canopy Conductance

At the scale of an individual leaf, stomatal control of transpiration is quantified by stomatal conductance. At the scale of a canopy of leaves, an aggregate measure of surface conductance is required to account for evaporation from soil and transpiration from foliage. In particular, the physiology of individual leaves, which is directly measurable in terms of the response of stomata to light, water, temperature, and other environmental factors, must be scaled over all leaves to the canopy, where photosynthesis and transpiration can only be measured for the aggregate canopy. Whereas stomatal conductance describes the conductance of an individual leaf, canopy conductance describes the aggregate conductance of a canopy of leaves. Surface conductance additionally includes soil evaporation. In dense canopies, soil evaporation is negligible and surface and canopy conductances are effectively the same. Canopy or surface conductance can be derived from micrometeorological measurements of fluxes above the canopy.

Monteith (1965) extended the Penman equation to a canopy of leaves, and the Penman–Monteith equation is widely used for biometeorological studies. Eq. (12.22) is the Penman–Monteith equation for a canopy, and:

$$\lambda E = \frac{s(R_n - G) + c_p\left[e_*(T_a) - e_a\right]g_{ah}}{s + \gamma(1 + g_{ah}/g_c)} \qquad (17.14)$$

This equation shows that the prominent environmental controls on evapotranspiration are available energy $(R_n - G)$, atmospheric vapor pressure deficit $(e_*(T_a) - e_a)$, and aerodynamic (g_{ah}) and canopy (g_c) conductances.

The Penman–Monteith equation is a bulk surface formulation in which two conductances acting in series regulate latent heat exchange with the atmosphere. The canopy conductance (g_c) governs processes within the plant canopy, and the aerodynamic conductance (g_{ah}) governs turbulent processes above the canopy. As with an individual leaf, the canopy has degrees of coupling to the atmosphere determined by the magnitude of the aerodynamic conductance

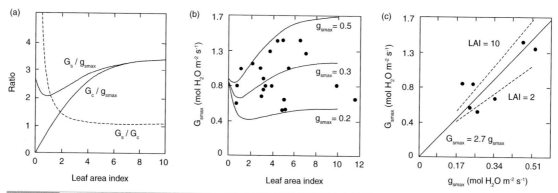

Fig. 17.9 Relationships among maximum leaf, canopy, and surface conductance when vegetation is not stressed by soil moisture and other environmental conditions. (a) Theoretical relationships among maximum leaf conductance ($g_{s\,max}$), canopy conductance (G_c), and surface conductance (G_s) with increasing leaf area index for $g_{s\,max}$ = 0.3 mol m^{-2} s^{-1}. Relationships are for particular values of net radiation, vapor pressure deficit, aerodynamic conductance, and response of stomata to light. (b) Observed maximum surface conductance ($G_{s\,max}$) in relation to leaf area index for a variety of grasslands, forests, and croplands. The solid lines show the expected relationship for three values of $g_{s\,max}$. (c) Observed $G_{s\,max}$ in relation to $g_{s\,max}$ for several grasslands, forests, and croplands. The dashed lines show the lower and upper limits of $G_{s\,max}$ expected from theory with leaf area index of 2 and 10 m^2 m^{-2}. Adapted from Kelliher et al. (1995).

(Jarvis and McNaughton 1986). With a low aerodynamic conductance, the canopy is decoupled from the atmosphere, and evapotranspiration approaches an equilibrium rate dependent on available energy:

$$\lambda E = s\left(R_n - G\right)/(s + \gamma) \tag{17.15}$$

The canopy has no influence on the rate, and the rate represents a freely evaporating surface. Where there is a high aerodynamic conductance, evapotranspiration approaches a rate imposed by vapor pressure deficit and canopy conductance:

$$\lambda E = \frac{c_p}{\gamma}\left[e_*\left(T_a\right) - e_a\right]g_c \tag{17.16}$$

The Penman–Monteith equation can be re-arranged to give the canopy conductance:

$$\frac{1}{g_c} = \frac{\varepsilon + 1}{g_{ah}}\left[\frac{\varepsilon\left(R_n - G\right)}{(\varepsilon + 1)\lambda E} - 1\right] + \frac{c_p}{\gamma}\frac{\left[e_*\left(T_a\right) - e_a\right]}{\lambda E} \tag{17.17}$$

with $\varepsilon = s/\gamma$. By measuring evapotranspiration from a canopy and if all other terms are known, the Penman–Monteith equation can be solved for g_c. This conductance captures the effects of leaf area, canopy coverage, photosynthetic capacity, and soil moisture on the partitioning of net radiation into latent heat flux.

The Penman–Monteith equation gives insight to surface and canopy conductances and their relationship with leaf area index and leaf conductance (Schulze et al. 1994; Kelliher et al. 1995). The integrated source strength of a canopy is proportional to leaf conductance times leaf area, and canopy conductance increases linearly with leaf area index at low leaf area (Figure 17.9a). Canopy conductance is relatively constant at a value largely determined by leaf conductance at high leaf area. Canopy conductance becomes constant with high leaf area because stomata close due to low light levels deep in the canopy. Surface conductance approaches canopy conductance as leaf area increases because soil evaporation becomes negligible in a dense canopy. Surface conductance significantly exceeds canopy conductance only for low leaf area index less than about 3 m^2 m^{-2}, where soil evaporation is more important. For a given leaf conductance, surface conductance is nearly independent of leaf area because the tendency for canopy conductance to decrease with low leaf area is countered by increasing proportion of soil evaporation.

Across a variety of forests, grasslands, and croplands, surface and canopy conductances derived from the Penman–Monteith equation

Table 17.2 | Maximum leaf stomatal conductance and bulk canopy conductance achieved in the field for unstressed, well-lit leaves

Plant group	Leaf (mol H_2O m^{-2} s^{-1})	Canopy (mol H_2O m^{-2} s^{-1})
Woody plants	0.23	0.83
Natural herbaceous plants	0.34	0.72
Agricultural crops	0.49	1.34

Note: Converted to mol m^{-2} s^{-1} using, $\rho_m = 42.3$ mol m^{-3}.
Source: From Kelliher et al. (1995) and Körner (1994).

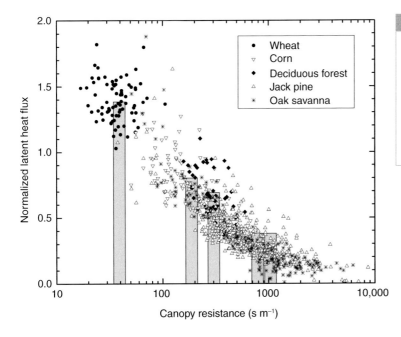

Fig. 17.10 Latent heat flux normalized by its equilibrium rate in relation to canopy resistance for wheat, corn, temperate deciduous forest, boreal jack pine conifer forest, and oak savanna. Shown are individual data points and the mean for each vegetation type. Original data from Baldocchi et al. (1997) and Baldocchi and Xu (2007).

show relationships similar to that expected from theory. Maximum surface conductance in relation to leaf area index is within the bounds determined by three values of leaf conductance representative of woody plants, natural herbaceous plants, and agricultural crops (Figure 17.9b). Maximum surface conductance increases linearly with maximum leaf conductance with a slope of ~3 (Figure 17.9c). This is consistent with theory, which predicts that for any leaf conductance surface conductance is constrained within limits set by leaf area and that surface conductance is largely determined by leaf conductance at high leaf area.

Table 17.2 compares maximum leaf stomatal conductance and canopy conductance derived from the Penman–Monteith equation for unstressed, well-lit leaves in three types of plants. A leaf conductance of 0.23 mol H_2O m^{-2} s^{-1} is typical for woody plants. Herbaceous plants have a higher conductance, and crops have the highest leaf conductance. Canopy conductance is higher because of the high leaf area index of the various plant canopies. Greater leaf area index increases latent heat exchange with the atmosphere by increasing the surface area from which moisture is lost.

Similar differences are seen among other vegetation types. Figure 17.10 compares the rate of latent heat flux normalized by the equilibrium rate in relation to canopy resistance (the inverse of canopy conductance) for several types of vegetation. Equilibrium evapotranspiration, given by Eq. (17.15), is that for a well-watered site and depends on available energy. The ratio of latent heat flux to the equilibrium flux is a measure of the partitioning of available energy to evapotranspiration. The evaporative fraction is generally high in agricultural crops, and these plants have low evaporative resistance (high canopy conductance). Forests have lower evaporative fractions compared with crops and higher resistance (lower conductance). Dry jack pine forest and oak savanna have the lowest evaporative fraction and highest resistance.

17.6 | Turbulent Transfer in Forest Canopies

Many models of turbulent fluxes in plant canopies utilize the principle of diffusion along the mean concentration gradient, similar to fluxes in the surface layer (Chapter 13). The fluxes of momentum, sensible heat, and water vapor are assumed to be proportional to the vertical gradients of wind, temperature, and specific humidity, respectively, multiplied by a turbulent diffusivity, or aerodynamic conductance. However, Monin–Obukhov stability theory fails in the layer of air immediately above the canopy, known as the roughness sublayer, and the universal similarity functions are not valid immediately above or within plant canopies. The gradient–diffusion approach also fails within plant canopies, where counter-gradient or zero-gradient fluxes and intermittent turbulence are common (Denmead and Bradley 1985; Raupach and Finnigan 1988; Baldocchi 1989; Baldocchi and Meyers 1998; Finnigan 2000). More detailed models of fluid dynamics than gradient–diffusion theory are needed to adequately represent turbulence in plant canopies. Such models include higher-order closure models based on equations for the first-order moment (e.g., mean horizontal wind velocity, mean mixing ratio) and second-order moments (e.g., vertical velocity variance, covariance of mixing ratio and vertical velocity fluctuations) associated with turbulent transfer (Wilson and Shaw 1977; Meyers and Paw U 1986, 1987; Wilson 1988; Pyles et al. 2000) or Lagrangian

models that follow the trajectories of particles (Baldocchi 1992).

Studies of turbulence in a deciduous forest illustrate some of the challenges in describing turbulent transfer in plant canopies (Baldocchi and Meyers 1988a,b, 1989; Meyers and Baldocchi 1991). The particular forest studied consists of oak and hickory trees with a canopy height of about 23 m and a leaf area index of about 5 m^2 m^{-2}. Most leaf area is in the upper 5 m of the canopy. Over 75 percent of the total leaf area is located in the upper 25 percent of the canopy. Measurements in the canopy reveal a complex turbulent structure that is highly intermittent and characterized by non-Gaussian probability distributions. Wind speed within the canopy varies with height in relation to the profile of leaf area index (Figure 17.11). Strong shear occurs in the upper 20 percent of the canopy above $0.8h$, where much of the leaf area is concentrated. In this region, wind speed decreases sharply with depth in the canopy. A reversal in the vertical gradient of wind speed occurs below this height with a local maximum in the canopy at about $0.5h$. Wind speed thereafter decreases towards the forest floor. This wind profile is not a simple exponential profile as commonly assumed, and illustrates counter-gradient momentum transport in the canopy.

Denmead and Bradley (1985) found examples of counter-gradient or zero-gradient fluxes in their measurements in a pine forest. Figure 17.12 shows profiles of potential temperature, water vapor mixing ratio, and CO_2 concentration measured in the canopy. Temperature has a maximum near the middle of the canopy. Water vapor decreases from a high near the surface, is nearly constant with height in the lower canopy space, and decreases with height in the upper canopy. The concentration of CO_2 has a minimum in the middle of the canopy. From these gradients, one would expect upward sensible heat flux above the temperature maximum and downward sensible heat flux onto the forest floor below mid-canopy. The steep gradient of water vapor near the forest floor suggests enhanced evaporation from the forest floor. The CO_2 profile implies upward transport of CO_2 from the forest floor in the lower canopy

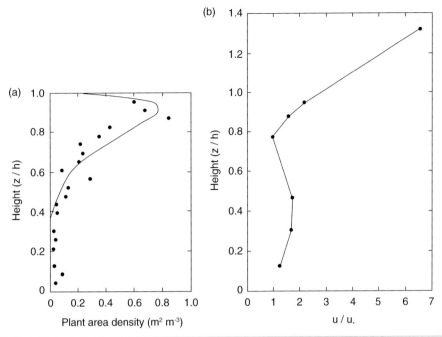

Fig. 17.11 Profiles of (a) leaf area and (b) wind speed above and in the canopy of a deciduous forest. Height (z) is given as a fraction of canopy height (h). Wind speed is normalized by friction velocity (u_*) measured above the canopy. Adapted from Baldocchi and Meyers (1988a) and Baldocchi (1989).

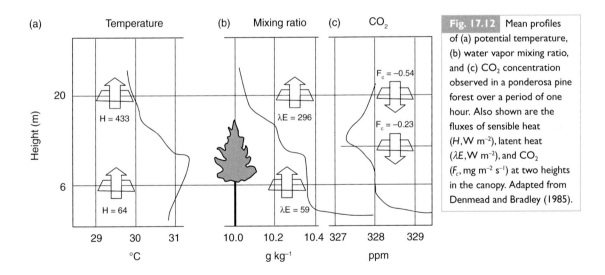

Fig. 17.12 Mean profiles of (a) potential temperature, (b) water vapor mixing ratio, and (c) CO_2 concentration observed in a ponderosa pine forest over a period of one hour. Also shown are the fluxes of sensible heat (H, W m^{-2}), latent heat (λE, W m^{-2}), and CO_2 (F_c, mg m^{-2} s^{-1}) at two heights in the canopy. Adapted from Denmead and Bradley (1985).

and downward CO_2 flux in the upper canopy. Observed fluxes show the opposite. Fluxes in the upper canopy conform to the conventional gradient–diffusion relationships. Upward fluxes of sensible heat and water vapor are accompanied by negative gradients of temperature and water vapor. The downward flux of CO_2 occurs along a positive concentration gradient. In the lower canopy, however, these fluxes are associated with counter-gradients or zero-gradients. The sensible heat flux is upward even though temperature increases with height. A large

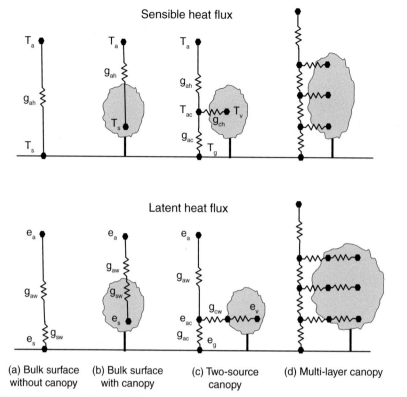

Fig. 17.13 Conductance networks for sensible heat flux (top) and latent heat flux (bottom). Networks are for a bulk surface without a canopy, vegetation with a bulk surface formulation, a two-source canopy, and multiple canopy layers. (a) The surface is the ground and fluxes are regulated by aerodynamic conductances (g_{ah}, g_{aw}). Latent heat flux also includes a soil conductance (g_{sw}). (b) The bulk canopy uses an effective surface temperature (T_s) and vapor pressure (e_s) that combines vegetation and ground. The overall conductance for latent heat is a surface conductance (g_{sw}) acting in series with an aerodynamic conductance (g_{aw}). (c) The two-source canopy partitions fluxes into vegetation and ground based on vegetation (T_v,e_v) and ground (T_g,e_g) temperature and vapor pressure. The conductance g_{ac} accounts for aerodynamic processes within the canopy. Canopy conductances account for sensible heat exchange (g_{ch}) from the vegetation to air within the canopy (T_{ac}) and latent heat exchange (g_{cw}) from the vegetation to canopy air (e_{ac}). For latent heat, this also includes stomatal conductance. (d) Multiple canopy layers.

upward flux of water vapor occurs despite zero gradient. Substantial absorption of CO_2 occurs below the middle of the canopy.

17.7 | Canopy Models

The same principles that determine the temperature and energy fluxes of a leaf also determine the temperature and energy fluxes from vegetated surfaces. Figure 17.13 illustrates some of the ways in which turbulent fluxes are modeled. One approach is to describe these fluxes simply through a bulk aerodynamic formulation with

a single aerodynamic conductance between the surface and atmosphere, as described in Chapters 12 and 13. A soil conductance can account for the extent to which the surface is not saturated with moisture.

The effects of vegetation on surface fluxes can be included by treating the soil–canopy system as an effective bulk surface, analogous to that for a non-vegetated surface but with radiative exchange integrated over the canopy and additional conductances for leaves (Figure 17.13b). This type of formulation is referred to as a big-leaf model because the canopy is treated as a single leaf scaled to represent a canopy. Canopy

flux equations in a big-leaf model are similar to those of an individual leaf, but with a leaf conductance representative for all the leaves in the canopy. The Penman–Monteith equation is a big-leaf model for latent heat flux.

Bulk surface models can be used to monitor plant canopies. If sensible heat flux and aerodynamic conductance are known from micrometeorological measurements, the effective surface temperature (T_s) is estimated from the bulk aerodynamic formulation for sensible heat using Eq. (12.10). In this case, T_s represents the aerodynamic temperature of the canopy, regulated by the aerodynamic conductance between the atmosphere and the canopy. Alternatively, estimates of T_s can be inferred from measurement of upward longwave radiation using Eq. (12.8). Here, T_s represents the temperature at which the effective surface emits longwave radiation and is known as the radiometric temperature.

In applying bulk aerodynamic formula, it is necessary to distinguish the roughness length for momentum (z_{0m}) from the roughness length for heat (z_{0h}), as discussed in Chapter 13. The latter is typically taken to be less than that of momentum (Thom 1972; Garratt and Hicks 1973; Garratt 1978; Beljaars and Holtslag 1991). If not, there can be significant difference in the surface temperature inferred from the sensible heat flux. Consider, for example, a forest canopy with a height of 20 m, displacement height d = 14 m, and roughness lengths z_{0m} = 2 m and z_{0h} = 0.2 m. The wind speed at height z = 30 m is u = 2 m s^{-1}, and for convenience the effects of atmospheric stability are neglected. From Eq. (13.26), the aerodynamic conductances evaluated using z_{0m} and z_{0h} are g_{am} = 3.1 and g_{ah} = 1.5 mol m^{-2} s^{-1}. The decreased conductance (i.e., the excess resistance) between z_{0m} and z_{0h} is 2.8 mol m^{-2} s^{-1} $(1/g_b^* = 1/g_{ah} - 1/g_{am})$. A general expression for the excess resistance between heights z_{0m} and z_{0h} is:

$$\frac{1}{g_b^*} = \frac{\ln\left(z_{0m}/z_{0h}\right)}{\rho_m k u_*} \qquad (17.18)$$

In this example, the friction velocity is u_* = 0.385 m s^{-1}, from Eq. (13.21). This excess resistance translates into a higher temperature at z_{0h} than at z_{0m}. For example, assume a sensible heat flux

H = 200 W m^{-2}, and an air temperature T_a = 30°C. The temperature at z_{0h} is found by evaluating the bulk aerodynamic formula for sensible heat with g_{ah}, using Eq. (12.10):

$$T_s = T_a + H/c_p g_{ah} \qquad (17.19)$$

so that T_s = 34.6°C (with c_p = 29.2 J mol^{-1} K^{-1}). Similarly, the temperature at z_{0m}, evaluated using g_{am} rather than g_{ah}, is T_s = 32.2°C. The temperature difference between T_s evaluated at z_{0m} and T_s evaluated at z_{0h} is:

$$\frac{T_s\left(z_{0m}\right) - T_s\left(z_{0h}\right)}{\theta_*} = \frac{1}{k}\ln\left(\frac{z_{0m}}{z_{0h}}\right) \qquad (17.20)$$

where $\theta_* = -H/\left(\rho_m c_p u_*\right)$. In this example, θ_* = –0.42 K (with ρ_m = 42.3 mol m^{-3}).

By influencing aerodynamic conductance, the height of plants has a large influence on leaf temperature. Tall forest vegetation is aerodynamically rough and has a high aerodynamic conductance (Figure 13.6). Heat is readily exchanged with the atmosphere, and the temperature of leaves is closely coupled to that of the air. In contrast, short vegetation such as grass or shrub is aerodynamically smooth, has a low aerodynamic conductance, and dissipates heat less effectively. The leaves of short vegetation are decoupled from the air and have warmer temperatures than that of the air. In cold climates, short stature may convey an advantage by warming leaf temperature (Wilson et al. 1987; Grace 1988; Grace et al. 1989). Figure 17.14 illustrates this for alpine forest and shrub vegetation. For both types of vegetation, the leaf-to-air temperature difference increases with greater net radiation. In the tall forest, the slope of this relationship is small. Leaf temperatures are similar to air temperature, with a maximum excess of less than 5°C. In the dwarf shrub vegetation, the slope is larger and the temperature excess is 15°C in bright sunshine.

As an alternative to the bulk representation of the effective surface for heat and moisture exchange, the surface can be explicitly represented by ground and vegetation (Figure 17.13c). Sensible heat is partitioned into that from foliage and that from soil, each regulated by different processes. The leaf boundary layer

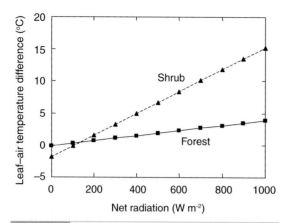

Fig. 17.14 Relationship between daytime leaf–air temperature difference and net radiation for forest (15–18 m tall) and dwarf shrub (0.1 m tall). Data are shown for wind speeds of 1–3 m s⁻¹. Data from Grace et al. (1989).

conductance integrated over all leaves in the canopy governs sensible heat flux from foliage. Turbulent processes within the canopy govern sensible heat flux from the soil. Latent heat is partitioned into soil evaporation and transpiration. Transpiration is regulated by a canopy conductance that is an integration of leaf boundary layer and stomatal conductances over all the leaves in the canopy. Soil evaporation is regulated by aerodynamic processes within the plant canopy and by soil moisture. These equations can be solved by writing the sensible and latent heat fluxes as a linear combination of atmospheric, vegetation, and ground temperatures and vapor pressure, respectively (Deardorff 1978).

The leaf and ground energy budgets represent a system of equations that can be solved for the unknown vegetation (T_v) and ground (T_g) temperatures. With reference to the one-layer canopy depicted in Figure 17.13c, three sensible heat fluxes are represented in the soil–plant–atmosphere system: the flux from vegetation to canopy air (H_v); from ground to canopy air (H_g); and from canopy air to the atmosphere (H). Assuming the canopy air has negligible capacity to store heat, the total sensible heat flux to the atmosphere is the sum of fluxes from vegetation and the ground:

$$H = c_p \left(T_{ac} - T_a \right) g_{ah} = c_p \left(T_v - T_{ac} \right) g_{ch} + c_p \left(T_g - T_{ac} \right) g_{ac}$$
$$(17.21)$$

Here, g_{ah} is the aerodynamic conductance for sensible heat, g_{ch} is the leaf boundary layer conductance scaled to the canopy, and g_{ac} is an aerodynamic conductance within the canopy. Rearranging terms in these equations, the canopy air temperature (T_{ac}), which is common to all three fluxes, is evaluated as a weighted average of the atmospheric, vegetation, and ground temperatures (T_a, T_v, and T_g):

$$T_{ac} = \frac{g_{ah} T_a + g_{ch} T_v + g_{ac} T_g}{g_{ah} + g_{ch} + g_{ac}} \quad (17.22)$$

This equation for canopy air temperature is substituted into the expression for H_v. A similar equation is derived for canopy air vapor pressure (e_{ac}) for latent heat flux, and the energy balance of the canopy is solved for the temperature (T_v) that balances net radiation, sensible heat flux, and latent heat flux. Then, with canopy fluxes known, the ground fluxes and temperature (T_g) are updated.

Alternatively, one can assume that the canopy air has some capacity to store heat (Vidale and Stöckli 2005). The storage of heat in the canopy air is:

$$\rho_m c_p \left(\partial T / \partial t \right) \partial z \quad (17.23)$$

where $\partial T / \partial t$ is the rate of change of temperature (K s⁻¹) and ∂z is canopy height (m). For a canopy with a height $z = 25$ m, a change in temperature of 1°C hr⁻¹ is associated with approximately 8 W m⁻² heat storage in the canopy. If heat storage is included, the canopy air temperature is not diagnosed as a linear combination of T_a, T_v, and T_g as above, but rather predicted from:

$$\rho_m c_p \Delta z \left(\Delta T_{ac} / \Delta t \right) = -H + H_v + H_g \quad (17.24)$$

The storage heat in the canopy air space is the difference between heat entering the air space (H_v, H_g) and sensible heat transferred to the atmosphere (H).

The preceding methodologies provide simple means to simulate fluxes from plant canopies and are commonly used in climate models (Deardorff 1978; Dickinson et al. 1986, 1993; Sellers et al. 1986, 1996a,b). They can be extended to represent multiple canopy layers

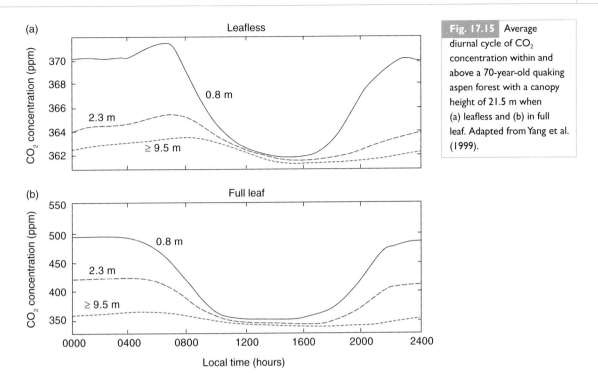

Fig. 17.15 Average diurnal cycle of CO_2 concentration within and above a 70-year-old quaking aspen forest with a canopy height of 21.5 m when (a) leafless and (b) in full leaf. Adapted from Yang et al. (1999).

(Figure 17.13d) with a simple representation of aerodynamic conductances within the canopy (Shuttleworth and Wallace 1985; Choudhury and Monteith 1988). However, two-source canopy models and multilayer models are based on gradient–diffusion theory and do not allow for counter-gradient fluxes as observed in plant canopies. More complex canopy models are required to resolve turbulent fluxes in plant canopies (Baldocchi 1992; Baldocchi and Meyers 1998; Pyles et al. 2000).

17.8 Environmental Controls of Canopy Fluxes

Surface fluxes vary over the course of a day in response to the diurnal cycle. As more radiation is received from the Sun, the land warms and more energy is dissipated as sensible and latent heat. In addition, stomata open to allow CO_2 uptake, but in doing so plants lose water during transpiration. Figure 12.7 illustrates these cycles for boreal aspen and jack pine forests on a typical summer day. In early morning, the forest

has a negative radiative balance. Sensible and latent heat fluxes are small. Stomata are closed, and trees do not absorb CO_2 during photosynthesis. Carbon dioxide is, however, lost during respiration, and there is a net flux of CO_2 from the forest to the atmosphere. As the Sun rises and solar radiation is absorbed by the forest, there is a net gain of radiation at the surface. The forest begins to warm and some of this energy is returned to the atmosphere as sensible and latent heat. Stomata open, allowing for net CO_2 uptake during photosynthesis. Fluxes are strongest in early to middle afternoon and decrease late in the afternoon when solar radiation diminishes.

The concentration of CO_2 in the near-surface atmosphere can have a strong diurnal cycle in response to CO_2 uptake during the day and loss during the night, as seen in measurements taken at the boreal aspen forest (Figure 17.15). In the dormant season, when trees are leafless, there is little diurnal variation in CO_2 concentration; nor is there a strong vertical gradient. In the growing season, however, there is a strong diurnal cycle at heights of 0.8 and 2.3 m. Nighttime concentrations range from 400 to 500 ppm while

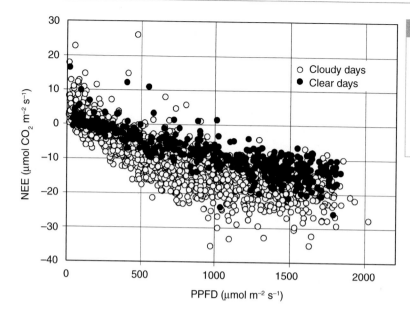

Fig. 17.16 Net ecosystem exchange of carbon (NEE) in relation to direct (clear days) and diffuse (cloudy days) photosynthetically active radiation (PPFD). Data are for a boreal aspen forest. Negative fluxes indicate carbon uptake. Adapted from Law et al. (2002).

daytime concentrations are about 350 ppm. The high nighttime concentrations arise from respiration losses and the occurrence of a strong temperature inversion that suppresses the upward turbulent transport of CO_2. The magnitude of this diurnal cycle diminishes with height so that at a height of 9.5 m and above, nighttime and daytime CO_2 concentrations are similar.

Canopy fluxes show similar functional relationships with environmental controls as do leaf fluxes. For example, the net uptake of carbon by a forest canopy increases as photosynthetically active radiation increases, saturating at high light levels (Figure 17.16). This relationship can be described by:

$$NEE = -\frac{A_{max} EF\downarrow}{EF\downarrow + A_{max}} + R_E \qquad (17.25)$$

The first term on the right-hand side of this equation is gross primary production (GPP) and is the canopy-scale equivalent of Eq. (17.8); the second term is ecosystem respiration (Lasslop et al. 2010). At the canopy scale, carbon uptake can vary depending on the amount of direct beam and diffuse radiation. Diffuse radiation penetrates more deeply into a canopy than does direct beam radiation. In the particular forest illustrated in Figure 17.16, net CO_2 uptake was greater for cloudy skies, when more of the radiation was diffuse, than for clear sky, when the radiation was primarily direct beam. Similar enhancement of CO_2 uptake when radiation is predominantly diffuse compared with when it is predominantly direct beam has been found in other forests (Hollinger et al. 1994; Goulden et al. 1997; Gu et al. 2002, 2003; Knohl and Baldocchi 2008).

Similar to leaf fluxes, GPP increases as evapotranspiration (ET) increases (Figure 17.17). At the canopy-scale, $GPP\,/\,ET$ is a measure of water-use efficiency; this quantity scaled by vapor pressure deficit ($GPP*VPD\,/\,ET$) is the canopy equivalent of intrinsic water-use efficiency. Among various forests, grasslands, and croplands, $GPP\,/\,ET$ ranges from 1 to 6 g C kg^{-1} H_2O; $GPP*VPD\,/\,ET$ ranges from 5 to 43 g C hPa kg^{-1} H_2O (Beer et al. 2009).

In deciduous forests, the emergence of leaves in spring alters the partitioning of net radiation into sensible and latent heat fluxes. Table 17.3 shows the effects of leaf emergence in a temperate deciduous forest. The majority of net radiation was dissipated as sensible heat when the canopy was leafless. After leaf emergence, latent heat was the dominant flux. The Bowen ratio above the canopy decreased from 2.1 during the dormant season to 0.5 during the growing season due to a large increase in the fraction of net radiation partitioned into latent heat

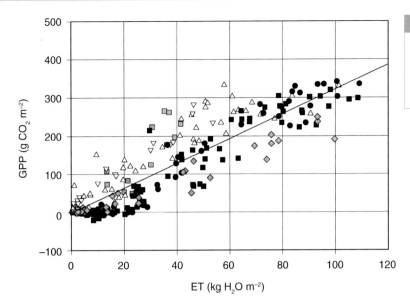

Fig. 17.17 Monthly gross primary production (GPP) in relation to monthly evapotranspiration (ET) for broadleaf deciduous forests. Adapted from Law et al. (2002).

Table 17.3 Dormant season and growing season energy fluxes above the canopy, at the forest floor, and from vegetation (canopy – floor)

	Canopy	Floor	Vegetation	Floor (%)
Dormant season				
R_n (MJ m^{-2})	720	315	405	44
H (MJ m^{-2})	419	193	226	46
λE (MJ m^{-2})	200	112	88	56
$H / \lambda E$	2.1	1.7	2.6	–
Growing season				
R_n (MJ m^{-2})	2080	287	1793	14
H (MJ m^{-2})	617	38	579	6
λE (MJ m^{-2})	1140	95	1045	8
$H / \lambda E$	0.5	0.4	0.6	–

Note: R_n, net radiation; λE, latent heat flux; H, sensible heat flux. Fluxes are summed over the dormant season (days 1–115, 305–365) and growing season (days 116–304). The forest floor contribution to canopy flux is also shown as a percentage. Source: From Wilson et al. (2000).

flux. The emergence of leaves also altered fluxes at the forest floor. In the dormant season, net radiation at the forest floor was 44 percent of that above the canopy. Sensible and latent heat fluxes at the forest floor contributed 46 percent and 56 percent, respectively, of the total canopy fluxes. After leaf emergence, net radiation at the forest floor was only 14 percent of that above the canopy, and the forest floor contributed less than 10 percent to canopy sensible and latent heat fluxes.

Leaf area index is a critical determinant of the proportion of evapotranspiration that is transpiration, and this ratio can be 0.8–0.9 or higher in closed canopies (Wang et al. 2014). The radiation profile influences the partitioning of total evapotranspiration into transpiration from foliage and evaporation from the forest floor. In

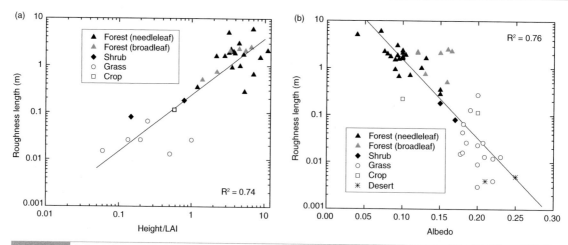

Fig. 17.18 Relationships among roughness length, canopy structure, and albedo for forest, shrub, grass, crop, and semidesert vegetation. Roughness length is shown in relation to (a) the ratio of canopy height to leaf area index and (b) albedo. Measurements were made at peak leaf area and albedo is at mid-day. Data from Cho et al. (2012).

general, the relative contribution of the forest floor and understory to total evapotranspiration is relatively little (~10%) in dense forests with high leaf area index and greater (up to 50%) in open canopy forests, where the canopy absorbs less radiation (Baldocchi and Ryu 2011). This can be seen in Figure 17.9, where the surface conductance is greater than canopy conductance at low leaf area index and approaches canopy conductance at high leaf area index and is seen also in the effects of leaf emergence in deciduous forests (Table 17.3).

The relationship between forest floor latent heat flux and available energy at the forest floor observed for a variety of forests illustrates the strong control exerted by available energy (Baldocchi et al. 2000). When the forest floor is dry, its latent heat flux increases linearly with greater available energy and accounts for about 25 percent of available energy. At high available energy beyond about 100 W m^{-2}, however, forest floor evaporation is not able to meet the potential demand and latent heat flux reaches a threshold of about 35 W m^{-2}.

Canopy structure influences canopy aerodynamics. Simplified relationships specify roughness length in relation to canopy height (h) by $z_{0m} = 0.1h$ and similarly displacement height by $d = 0.7h$ (Chapter 13). Other parameterizations additionally include a dependence on leaf area index and other factors (Raupach 1994; Sellers et al. 1996a). In an analysis of forest, grassland, cropland, and shrubland vegetation, Cho et al. (2012) found that roughness length correlates with the ratio of canopy height to leaf area index (h / L) (Figure 17.18a). Forests have large roughness length and high h / L. The shorter statured grasses, crops, and shrubs have smaller roughness length and lower h / L. The shorter vegetation also has higher surface albedo so that albedo and roughness length are negatively correlated (Figure 17.18b). This latter relationship illustrates a tradeoff between albedo and roughness length, with different consequences for net radiation (albedo) and turbulent fluxes (roughness length).

17.9 | Review Questions

1. Use Eq. (17.1) to calculate the amount of solar radiation (as a percentage of that at the top of the canopy) reaching the forest floor for a leaf area index of 4 m^2 m^{-2} and spherical, horizontal, and vertical foliage. Calculations are for a declination angle $\delta = 23°27'$, latitude $\phi = 40°$ N, and time of day is 1200 hours and 1500 hours. Zenith angle is calculated from Eq. (4.1). How does zenith angle interact with leaf orientation to affect radiative transfer?

2. What is the maximum sunlit leaf area in a canopy with spherically oriented foliage?

3. The maximum absorption of solar radiation in the visible waveband is about 95 percent. Use the production efficiency model to calculate daily canopy photosynthesis if incoming solar radiation averaged over the day is 160 W m^{-2}. Assume solar radiation is 50 percent visible waveband and 50 percent near-infrared and that environmental factors do not limit photosynthesis.

4. Contrast the production efficiency model with the GPP model given by Eq. (17.25). What is a key difference between these two models? Do they differ in their timescale?

5. Use Eq. (17.9) to calculate canopy photosynthesis for spherical, horizontal, and vertical foliage. $L = 6$ m^2 m^{-2}, $A_{max} = 20$ μmol m^{-2} s^{-1}, E = 0.06, and $F_0 = 1000$ μmol m^{-2} s^{-1}. Zenith angle is $Z = 30°$. Repeat the calculation for $Z = 60°$. How does leaf orientation interact with zenith angle to affect canopy photosynthesis?

6. Use the Penman–Monteith equation to calculate canopy conductance for the following conditions with latent heat flux equal to 300 W m^{-2}: $R_n - G = 400$ W m^{-2}, $e_*(T_a) = 3167$ Pa and $s = 189$ Pa K^{-1} (values for $T_a = 25°C$), and $g_{ah} = 2$ mol m^{-2} s^{-1}. Relative humidity is 75 percent. Use $c_p = 29.2$ J mol^{-1} K^{-1} and $\gamma = 66.5$ Pa K^{-1}.

7. Discuss differences and/or similarities among leaf conductance, surface conductance, and canopy conductance in a deciduous forest when the canopy is leafless. How does leaf emergence affect these conductances?

8. Temperature is 30.8°C at the forest floor, 31.2°C in the understory, 30.1°C in the overstory, and 28.3°C above the canopy. Sensible heat flux is 60 W m^{-2} at the forest floor and 300 W m^{-2} above the canopy. Are these indicative of gradient–diffusion or counter-gradient fluxes?

9. A surface emits longwave radiation of 500 W m^{-2}. What is the radiative temperature for an emissivity of 1.0 and 0.95?

10. Measured sensible heat flux is $H = 200$ W m^{-2}. One student evaluates surface temperature with $z_{0m} = 0.5$ m. Another student uses $z_{0m} = 0.2$ m. What is the difference in estimated surface temperature? Use $c_p = 29.2$ J mol^{-1} K^{-1}, $\rho_m = 42.3$ mol m^{-3}, and $u_* = 0.2$ m s^{-1}.

11. Calculate the change in air temperature in a 20 m tall canopy with $H_v = 85$ W m^{-2}, $H_g = 8$ W m^{-2}, and $H = 83$ W m^{-2}. How much heat is stored in the canopy? Use $\rho_m = 42.3$ mol m^{-3} and $c_p = 29.2$ J mol^{-1} K^{-1}. What is the change in temperature if $H = 93$ W m^{-2}?

17.10 | References

Baldocchi, D. D. (1989). Turbulent transfer in a deciduous forest. *Tree Physiology*, 5, 357–377.

Baldocchi, D. D. (1992). A Lagrangian random-walk model for simulating water vapor, CO_2 and sensible heat flux densities and scalar profiles over and within a soybean canopy. *Boundary-Layer Meteorology*, 61, 113–144.

Baldocchi, D. D., and Meyers, T. P. (1988a). A spectral and lag-correlation analysis of turbulence in a deciduous forest canopy. *Boundary-Layer Meteorology*, 45, 31–58.

Baldocchi, D. D., and Meyers, T. P. (1988b). Turbulence structure in a deciduous forest. *Boundary-Layer Meteorology*, 43, 345–364.

Baldocchi, D. D., and Meyers, T. P. (1989). The effects of extreme turbulent events on the estimation of aerodynamic variables in a deciduous forest canopy. *Agricultural and Forest Meteorology*, 48, 117–134.

Baldocchi, D. D., and Meyers, T. P. (1998). On using eco-physiological, micrometeorological and biogeochemical theory to evaluate carbon dioxide, water vapor and trace gas fluxes over vegetation: a perspective. *Agricultural and Forest Meteorology*, 90, 1–25.

Baldocchi, D. D., and Ryu, Y. (2011). A synthesis of forest evaporation fluxes – from days to years – as measured with eddy covariance. In *Forest Hydrology and Biogeochemistry: Synthesis of Past Research and Future Directions*, ed. D. F. Levia, D. Carlyle-Moses, and T. Tanaka. Dordrecht: Springer, pp. 101–116.

Baldocchi, D. D., and Xu, L. (2007). What limits evaporation from Mediterranean oak woodlands – the supply of moisture in the soil, physiological control by plants or the demand by the atmosphere? *Advances in Water Resources*, 30, 2113–2122.

Baldocchi, D. D., Vogel, C. A., and Hall, B. (1997). Seasonal variation of energy and water vapor exchange rates above and below a boreal jack pine forest canopy. *Journal of Geophysical Research*, 102D, 28939–28951.

Baldocchi, D. D., Law, B. E., and Anthoni, P. M. (2000). On measuring and modeling energy fluxes above the floor of a homogeneous and heterogeneous conifer forest. *Agricultural and Forest Meteorology*, 102, 187–206.

Beer, C., Ciais, P., Reichstein, M., et al. (2009). Temporal and among-site variability of inherent water use efficiency at the ecosystem level. *Global Biogeochemical Cycles*, 23, GB2018, doi:10.1029/2008GB003233.

Beljaars, A. C. M., and Holtslag, A. A. M. (1991). Flux parameterization over land surfaces for atmospheric models. *Journal of Applied Meteorology*, 30, 327–341.

Bonan, G. B., Lawrence, P. J., Oleson, K. W., et al. (2011). Improving canopy processes in the Community Land Model version 4 (CLM4) using global flux fields empirically inferred from FLUXNET data. *Journal of Geophysical Research*, 116, G02014, doi:10.1029/2010JG001593.

Campbell, G. S., and Norman, J. M. (1998). *An Introduction to Environmental Biophysics*, 2nd ed. New York: Springer-Verlag.

Carswell, F. E., Meir, P., Wandelli, E. V., et al. (2000). Photosynthetic capacity in a central Amazonian rain forest. *Tree Physiology*, 20, 179–186.

Cho, J., Miyazaki, S., Yeh, P. J.-F., et al. (2012). Testing the hypothesis on the relationship between aerodynamic roughness length and albedo using vegetation structure parameters. *International Journal of Biometeorology*, 56, 411–418.

Choudhury, B. J., and Monteith, J. L. (1988). A four-layer model for the heat budget of homogeneous land surfaces. *Quarterly Journal of the Royal Meteorological Society*, 114, 373–398.

Dai, Y., Dickinson, R. E., and Wang, Y.-P. (2004). A two-big-leaf model for canopy temperature, photosynthesis, and stomatal conductance. *Journal of Climate*, 17, 2281–2299.

Dang, Q. L., Margolis, H. A., Sy, M., et al. (1997). Profiles of photosynthetically active radiation, nitrogen and photosynthetic capacity in the boreal forest: implications for scaling from leaf to canopy. *Journal of Geophysical Research*, 102D, 28845–28859.

Deardorff, J. W. (1978). Efficient prediction of ground surface temperature and moisture, with inclusion of a layer of vegetation. *Journal of Geophysical Research*, 83C, 1889–1903.

Denmead, O. T., and Bradley, E. F. (1985). Flux–gradient relationships in a forest canopy. In *The Forest–Atmosphere Interaction*, ed. B. A. Hutchinson and B. B. Hicks. Dordrecht: Reidel, pp. 421–442.

de Pury, D. G. G., and Farquhar, G. D. (1997). Simple scaling of photosynthesis from leaves to canopies without the errors of big-leaf models. *Plant, Cell and Environment*, 20, 537–557.

Dickinson, R. E., Henderson-Sellers, A., Kennedy, P. J., and Wilson, M.F. (1986). *Biosphere–Atmosphere Transfer Scheme (BATS) for the NCAR Community Climate Model*, Technical Note NCAR/TN-275+STR. Boulder, Colorado: National Center for Atmospheric Research.

Dickinson, R. E., Henderson-Sellers, A., and Kennedy, P. J. (1993). *Biosphere–Atmosphere Transfer Scheme (BATS) Version 1e as Coupled to the NCAR Community Climate Model*, Technical Note NCAR/TN-387+STR. Boulder, Colorado: National Center for Atmospheric Research.

Dorman, J. L., and Sellers, P. J. (1989). A global climatology of albedo, roughness length and stomatal resistance for atmospheric general circulation models as represented by the simple biosphere model (SiB). *Journal of Applied Meteorology*, 28, 833–855.

Farquhar, G. D., von Caemmerer, S., and Berry, J. A. (1980). A biochemical model of photosynthetic CO_2 assimilation in leaves of C_3 species. *Planta*, 149, 78–90.

Finnigan, J. (2000). Turbulence in plant canopies. *Annual Review of Fluid Mechanics*, 32, 519–571.

Garratt, J. R. (1978). Transfer characteristics for a heterogeneous surface of large aerodynamic roughness. *Quarterly Journal of the Royal Meteorological Society*, 104, 491–502.

Garratt, J. R., and Hicks, B. B. (1973). Momentum, heat and water vapour transfer to and from natural and artificial surfaces. *Quarterly Journal of the Royal Meteorological Society*, 99, 680–687.

Goudriaan, J. (1977). *Crop Micrometeorology: A Simulation Study*. Wageningen: Center for Agricultural Publication and Documentation.

Goudriaan, J., and van Laar, H. H. (1994). *Modelling Potential Crop Growth Processes: Textbook with Exercises*. Dordrecht: Kluwer.

Goulden, M. L., Daube, B. C., Fan, S.-M., et al. (1997). Physiological responses of a black spruce forest to weather. *Journal of Geophysical Research*, 102D, 28987–28996.

Grace, J. (1988). The functional significance of short stature in montane vegetation. In *Plant Form and Vegetation Structure*, ed. M. J. A. Werger, P. J. M. van der Aart, H. J. During, and J. T. A. Verhoeven. The Hague: SPB Academic Publishing, pp. 201–209.

Grace, J., Allen, S. J., and Wilson, C. (1989). Climate and the meristem temperatures of plant communities near the tree-line. *Oecologia*, 79, 198–204.

Gu, L., Baldocchi, D., Verma, S. B., et al. (2002). Advantages of diffuse radiation for terrestrial ecosystem productivity. *Journal of Geophysical Research*, 107, 4050, 10.1029/2001JD001242.

Gu, L., Baldocchi, D. D., Wofsy, S. C., et al. (2003). Response of a deciduous forest to the Mount Pinatubo eruption: enhanced photosynthesis. *Science*, 299, 2035–2038.

Hollinger, D. Y. (1996). Optimality and nitrogen allocation in a tree canopy. *Tree Physiology*, 16, 627–634.

Hollinger, D. Y., Kelliher, F. M., Byers, J. N., et al. (1994). Carbon dioxide exchange between an undisturbed old-growth temperate forest and the atmosphere. *Ecology*, 75, 134–150.

Jarvis, P. G., and McNaughton, K. G. (1986). Stomatal control of transpiration: scaling up from leaf to region. *Advances in Ecological Research*, 15, 1–49.

Kelliher, F. M., Leuning, R., Raupach, M. R., and Schulze, E.-D. (1995). Maximum conductances for evaporation from global vegetation types. *Agricultural and Forest Meteorology*, 73, 1–16.

Knohl, A., and Baldocchi, D. D. (2008). Effects of diffuse radiation on canopy gas exchange processes in a forest ecosystem. *Journal of Geophysical Research*, 113, G02023, doi:10.1029/2007JG000663.

Körner, C. (1994). Leaf diffusive conductances in the major vegetation types of the globe. In *Ecophysiology of Photosynthesis*, ed. E.-D. Schulze and M. M. Caldwell. New York: Springer-Verlag, pp. 463–490.

Lasslop, G., Reichstein, M., Papale, D., et al. (2010). Separation of net ecosystem exchange into assimilation and respiration using a light response curve approach: Critical issues and global evaluation. *Global Change Biology*, 16, 187–208.

Law, B. E., Falge, E., Gu, L., et al. (2002). Environmental controls over carbon dioxide and water vapor exchange of terrestrial vegetation. *Agricultural and Forest Meteorology*, 113, 97–120.

Lloyd, J., Patiño, S., Paiva, R. Q., et al. (2010). Optimisation of photosynthetic carbon gain and within-canopy gradients of associated foliar traits for Amazon forest trees. *Biogeosciences*, 7, 1833–1859.

Meir, P., Kruijt, B., Broadmeadow, M., et al. (2002). Acclimation of photosynthetic capacity to irradiance in tree canopies in relation to leaf nitrogen concentration and leaf mass per unit area. *Plant, Cell and Environment*, 25, 343–357.

Meyers, T. P., and Baldocchi, D. D. (1991). The budgets of turbulent kinetic energy and Reynolds stress within and above a deciduous forest. *Agricultural and Forest Meteorology*, 53, 207–222.

Meyers, T., and Paw U, K. T. (1986). Testing of a higher-order closure model for modeling airflow within and above plant canopies. *Boundary-Layer Meteorology*, 37, 297–311.

Meyers, T., and Paw U, K. T. (1987). Modelling the plant canopy micrometeorology with higher-order closure principles. *Agricultural and Forest Meteorology*, 41, 143–163.

Monteith, J. L. (1965). Evaporation and environment. In *The State and Movement of Water in Living Organisms (19th Symposia of the Society for Experimental Biology)*, ed. G. E. Fogg. New York: Academic Press, pp. 205–234.

Monteith, J. L. (1977). Climate and the efficiency of crop production in Britain. *Philosophical Transactions of the Royal Society B*, 281, 277–294.

Myneni, R. B., Keeling, C. D., Tucker, C. J., Asrar, G., and Nemani, R. R. (1997). Increased plant growth in the northern high latitudes from 1981 to 1991. *Nature*, 386, 698–702.

Niinemets, Ü. (2007). Photosynthesis and resource distribution through plant canopies. *Plant, Cell and Environment*, 30, 1052–1071.

Norman, J. M. (1979). Modeling the complete crop canopy. In *Modification of the Aerial Environment of Plants*, ed. B. J. Barfield and J. F. Gerber. St. Joseph, Michigan: American Society of Agricultural Engineers, pp. 249–277.

Norman, J. M. (1993). Scaling processes between leaf and canopy levels. In *Scaling Physiological Processes: Leaf to Globe*, ed. J. R. Ehleringer and C. B. Field. New York: Academic Press, pp. 41–76.

Prince, S. D., and Goward, S. N. (1995). Global primary production: a remote sensing approach. *Journal of Biogeography*, 22, 815–835.

Pyles, R. D., Weare, B. C., and Paw U, K. T. (2000). The UCD advanced canopy–atmosphere–soil algorithm: Comparisons with observations from different climate and vegetation regimes. *Quarterly Journal of the Royal Meteorological Society*, 126, 2951–2980.

Rauner, J. L. (1976). Deciduous forests. In *Vegetation and the Atmosphere*: vol. 2. *Case Studies*, ed. J. L. Monteith. New York: Academic Press, pp. 241–264.

Raupach, M. R. (1994). Simplified expressions for vegetation roughness length and zero-plane displacement as functions of canopy height and area index. *Boundary-Layer Meteorology*, 71, 211–216.

Raupach, M. R., and Finnigan, J. J. (1988). Single-layer models of evaporation from plant canopies are incorrect but useful, whereas multilayer models are correct but useless: Discuss. *Australian Journal of Plant Physiology*, 15, 705–716.

Rogers, A. (2014). The use and misuse of $V_{c,max}$ in Earth System Models. *Photosynthesis Research*, 119, 15–29.

Running, S. W., Thornton, P. E., Nemani, R., and Glassy, J. M. (2000). Global terrestrial gross and net primary productivity from the Earth Observing System. In *Methods in Ecosystem Science*, ed. O. E. Sala. New York: Springer-Verlag, pp. 44–57.

Running, S. W., Nemani, R. R., Heinsch, F. A., et al. (2004). A continuous satellite-derived measure of global terrestrial primary production. *BioScience*, 54, 547–560.

Schulze, E.-D., Kelliher, F. M., Körner, C., Lloyd, J., and Leuning, R. (1994). Relationships among maximum stomatal conductance, ecosystem surface conductance, carbon assimilation rate, and plant nitrogen nutrition: A global ecology scaling exercise. *Annual Review of Ecology and Systematics*, 25, 629–660.

Sellers, P. J. (1985). Canopy reflectance, photosynthesis and transpiration. *International Journal of Remote Sensing*, 6, 1335–1372.

Sellers, P. J., Mintz, Y., Sud, Y. C., and Dalcher, A. (1986). A simple biosphere model (SiB) for use within general circulation models. *Journal of the Atmospheric Sciences*, 43, 505–531.

Sellers, P. J., Berry, J. A., Collatz, G. J., Field, C. B., and Hall, F. G. (1992). Canopy reflectance, photosynthesis, and transpiration, III: A reanalysis using improved leaf models and a new canopy integration scheme. *Remote Sensing of Environment*, 42, 187–216.

Sellers, P. J., Los, S. O., Tucker, C. J., et al. (1996a). A revised land surface parameterization (SiB2) for atmospheric GCMs, Part II: The generation of global fields of terrestrial biophysical parameters from satellite data. *Journal of Climate*, 9, 706–737.

Sellers, P. J., Randall, D. A., Collatz, G. J., et al. (1996b). A revised land surface parameterization (SiB2) for atmospheric GCMs, Part I: Model formulation. *Journal of Climate*, 9, 676–705.

Shuttleworth, W. J., and Wallace, J. S. (1985). Evaporation from sparse crops – an energy combination theory. *Quarterly Journal of the Royal Meteorological Society*, 111, 839–855.

Thom, A. S. (1972). Momentum, mass and heat exchange of vegetation. *Quarterly Journal of the Royal Meteorological Society*, 98, 124–134.

Tucker, C. J., Townshend, J. R. G., and Goff, T. E. (1985). African land-cover classification using satellite data. *Science*, 227, 369–375.

Tucker, C. J., Fung, I. Y., Keeling, C. D., and Gammon, R. H. (1986). Relationship between atmospheric CO_2 variations and a satellite-derived vegetation index. *Nature*, 319, 195–199.

Tucker, C. J., Dregne, H. E., and Newcomb, W. W. (1991). Expansion and contraction of the Sahara Desert from 1980 to 1990. *Science*, 253, 299–301.

Vidale, P. L., and Stöckli, R. (2005). Prognostic canopy air space solutions for land surface exchanges. *Theoretical and Applied Climatology*, 80, 245–257.

Wang, L., Good, S. P., and Caylor, K. K. (2014). Global synthesis of vegetation control on evapotranspiration partitioning. *Geophysical Research Letters*, 41, 6753–6757, doi:10.1002/2014GL061439.

Wang, Y.-P., and Leuning, R. (1998). A two-leaf model for canopy conductance, photosynthesis and partitioning of available energy, I: Model description and comparison with a multi-layered model. *Agricultural and Forest Meteorology*, 91, 89–111.

Wilson, C., Grace, J., Allen, S., and Slack, F. (1987). Temperature and stature: a study of temperatures in montane vegetation. *Functional Ecology*, 1, 405–414.

Wilson, J. D. (1988). A second-order closure model for flow through vegetation. *Boundary-Layer Meteorology*, 42, 371–392.

Wilson, K. B., Hanson, P. J., and Baldocchi, D. D. (2000). Factors controlling evaporation and energy partitioning beneath a deciduous forest over an annual cycle. *Agricultural and Forest Meteorology*, 102, 83–103.

Wilson, N. R., and Shaw, R. H. (1977). A higher order closure model for canopy flow. *Journal of Applied Meteorology*, 16, 1197–1205.

Yang, P. C., Black, T. A., Neumann, H. H., Novak, M. D., and Blanken, P. D. (1999). Spatial and temporal variability of CO_2 concentration and flux in a boreal aspen forest. *Journal of Geophysical Research*, 104D, 27653–27661.

Yuan, W., Liu, S., Zhou, G., et al. (2007). Deriving a light use efficiency model from eddy covariance flux data for predicting daily gross primary production across biomes. *Agricultural and Forest Meteorology*, 143, 189–207.

Zhao, M., Heinsch, F. A., Nemani, R. R., and Running, S. W. (2005). Improvements of the MODIS terrestrial gross and net primary production global data set. *Remote Sensing of Environment*, 95, 164–176.

Part V

Terrestrial Plant Ecology

18

Plant Strategies

18.1 | Chapter Summary

A plant uses the carbon gained during photosynthesis for maintenance and survival, to grow foliage, a stem, and roots, and for reproduction. It must allocate its limited available resources among growth, maintenance, and reproduction in a manner such that the species persists over time. Plants exhibit different strategies for allocating resources, collectively known as life history patterns. A successful strategy is to invest heavily in reproductive effort. The plant is small, short lived, and has many widely dispersed seeds, such as a herbaceous annual. An equally successful strategy is to be large, long lived, and have a small crop of large seeds, such as a tree. This life history favors maintenance over reproduction. There are multiple life history patterns that allow success in a given environment, but not all are successful in all environments. The environment selectively determines which strategy is successful. These life histories ensure the persistence of multiple species across the landscape in accordance with resource gradients and disturbance regimes. They give pattern to the arrangement of plant populations and communities in space and time. Three conceptualizations of plant strategies are the classifications of species into: *r*- and *K*-selected life histories; ruderal, competitor, and stress tolerator plants; and early and late successional species. More generally, plant functional types are broad classes of plants that

reduce the complexity of species diversity in ecological function to a few plant types defined by key physiological and life history characteristics. However, plant traits have a continuum of variation. These traits are understood in terms of functional tradeoffs between high metabolism and persistence.

18.2 | Carbon Balance of Plants

The net carbon gain of a plant that is available for growth is the difference between carbon uptake during photosynthesis and carbon loss during respiration (Figure 18.1). The photosynthetic uptake of an individual leaf must be summed over all foliage held by the plant to give gross primary production (GPP). Respiration loss must be summed over all tissues in the plant. The combined whole-plant respiration is termed autotrophic respiration (R_A) and is typically about 50 percent of gross primary production (Ryan 1991). Plant respiration is divided into growth respiration, which is independent of temperature, and maintenance respiration, which increases with higher temperatures. Growth respiration releases CO_2 during the synthesis of new tissues. This involves the incorporation of carbon into organic compounds and the expenditure of metabolic energy to produce the compounds. Growth respiration is typically about 25 percent of gross primary production. Maintenance respiration releases CO_2 during the carbohydrate breakdown that provides

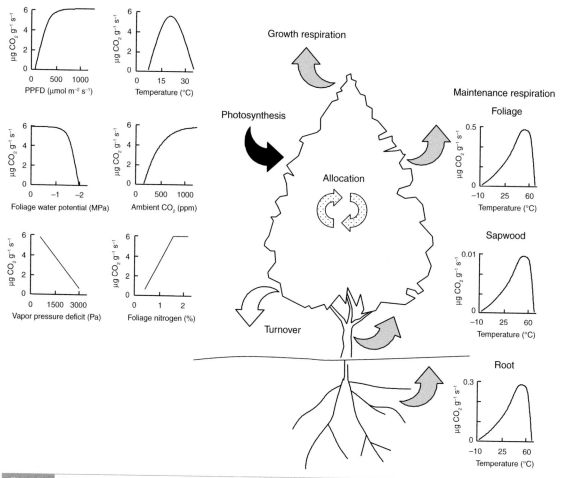

Fig. 18.1 The net carbon available for growth is the difference between carbon uptake during photosynthesis and carbon loss during growth and maintenance respiration. Photosynthesis varies with light, temperature, CO_2, water potential, and other factors. Maintenance respiration varies with temperature and differs among pools. Also shown are plant allocation of carbon and turnover loss.

the energy needed to maintain living cells. Maintenance respiration increases exponentially with higher temperature until some maximum temperature beyond which physiological activity decreases. Respiration rates vary among plant structures. Foliage and roots have higher respiration rates for the same temperature than does wood. The difference between GPP and R_A is the net primary production (NPP) available for growth.

The carbon balance of a plant varies over the course of a year due to seasonal changes in photosynthesis and respiration. This is most obvious in woody plants growing in regions with a distinct growing season (Figure 18.2). In winter, the daily carbon balance is negative; there is little or no photosynthetic uptake, but carbon is lost during autotrophic respiration. Net carbon uptake, after accounting for respiration losses, occurs during the growing season when photosynthetic uptake exceeds respiration loss.

18.3 Seasonality of Growth and Development

Phenology is the study of the onset and duration of the different phases of a plant's development

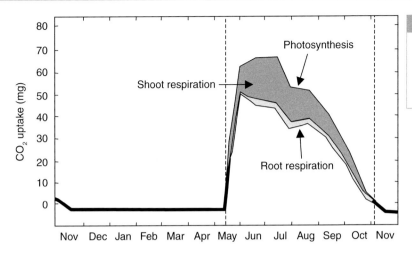

Fig. 18.2 Carbon balance (net photosynthesis, shoot respiration, root respiration) of pine seedlings growing near treeline over the course of a year. Adapted from Larcher (1995, p. 133).

during the year. Temperature, moisture, and daylength control the timing of these phases. Temperature is especially important. The rate of chemical reactions increases with higher temperatures, and temperature controls nearly all physiological processes involved in plant growth. Specific temperature regimes initiate the opening of buds, the growth of leaves, shoots and roots, the onset of flowering, seed ripening, and seed germination. These temperature requirements are measured in terms of growing degree-days, which is the accumulated daily temperature above some threshold, typically about 5°C. Measures such as growing degree-days incorporate the accumulated effect of temperature on growth and development. Many developmental processes also have a chilling requirement in which temperatures have to be below some threshold value for a certain period of time before the process can be initiated. This is especially true for budbreak and seed germination, where exposure to low temperature over several weeks or months is needed to break dormancy. Chilling requirements are expressed as the accumulated daily temperature below some threshold.

The flowering of spring wildflowers in Indiana illustrates the influence of temperature on phenology (Figure 18.3). On steep 45° slopes, the average flowering date is six days earlier on a south-facing slope than on the opposing north-facing slope despite being separated by a distance of only 46 m. The south-facing slope

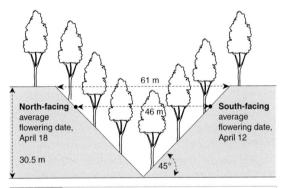

Fig. 18.3 Average flowering for nine species of spring wildflowers in an Indiana woodland growing on north- and south-facing slopes separated by 46 m. Data from Jackson (1966).

receives more radiation and is warmer than the north slope. On both slopes, flowering is triggered when the accumulated degree-hours is about 4650°C-hours, which occurs six days earlier on the south slope than on the north slope.

Length of day is also critical to many developmental processes, triggering the onset of flowering, leaf fall, and winter dormancy in many herbaceous plants, shrubs, and trees. In middle to high latitudes, the length of day varies greatly over the course of a year (Figure 4.7). At latitude 40°N, the minimum daylength is 9 hours in December; the maximum daylength is 15 hours in June. Further north, the seasonality of daylength is even more pronounced. Photoperiod, the relative duration of light and dark periods during the day, controls the timing of many physiological processes. For example, long-day plants flower in response to days longer than some maximum length; short-day plants flower when days are shorter than some maximum. Short days and exposure to low temperatures initiate dormancy and the development of frost hardiness in woody plants growing in seasonally cold climates.

In temperate and boreal forests, the most obvious phenology is the seasonal greening and senescence of deciduous trees during the growing season. In spring, as temperatures increase and days become longer, buds break and new leaves emerge. In autumn, short days, long nights, and low temperatures trigger leaf senescence that prepares deciduous trees for winter dormancy. This seasonal pattern of growth produces alternating cycles of growth and inactivity and corresponding periods of carbon utilization and storage. Deciduous trees use the carbohydrates stored in the woody tissues of branches, trunks, and roots to provide the carbon for emerging leaves in spring. These reserves are replenished during the growing season, and the surplus photosynthate is stored in branches, trunks, and roots at the end of the growing season.

The advent of satellite technology has allowed study of leaf phenology at large spatial scales. The normalized difference vegetation index (NDVI), given by Eq. (17.6), is a satellite-derived index related to leaf biomass and plant productivity. Figure 18.4 shows seasonal changes in the NDVI, illustrating the timing of leaf emergence, peak leaf area and production, and leaf senescence and dormancy. Low values indicate few leaves and low productivity; high values indicate a dense canopy and high productivity. Arid regions show little greening throughout the year. Tropical rainforests are productive year-round. Elsewhere, there are two distinct patterns to phenology represented by summergreen and raingreen plants. Cold deciduous plants in temperate and high latitudes drop leaves with the onset of cold temperature. In spring, as temperature increases and the days become longer, buds on these summergreen plants open, new foliage emerges, and plants begin to photosynthesize. Peak production typically occurs in July and August, decreasing in autumn as plants again become dormant. In tropical and subtropical latitudes, drought-deciduous plants drop leaves seasonally in relation to low precipitation and drought stress. Leaves emerge on these raingreen plants in response to sufficient precipitation.

18.4 | Allocation

Plants use the carbon absorbed by leaves during photosynthesis to maintain cellular structures and grow new tissues. Maintenance of existing tissues requires an expenditure of carbon during respiration, which reduces the carbon available for new growth. The net carbon available to a plant, along with the nutrients required for new growth, is then allocated to the growth of leaves, roots, stems, flowers, and seeds and the production of chemicals for protection from insects and herbivores. Collectively, the partitioning of resources to plant parts and functions is known as allocation. Allocation is most evident in the amount of biomass invested in various plant parts.

Allocation of available resources is a key determinant of plant growth and success. For example, high allocation to foliage ensures more leaves to capture light and absorb CO_2 for new growth. However, allocation to foliage is inefficient if there is not enough water or nutrients to support the foliage. Moreover, there usually is a limited amount of resources to spend on growth, maintenance, and reproduction, and allocation to one function is often at the expense of another function. Hence, plants must allocate resources in a way that balances conflicting needs. The resulting compromises and tradeoffs present alternative solutions that ensure survival and persistence in the landscape. Variation among plants in growth rates is determined as much by differences in allocation and how plants balance resource limitations as by different photosynthetic rates.

18.4.1 Reproduction

Allocation of resources to reproduction illustrates compromises and tradeoffs. Copious seed

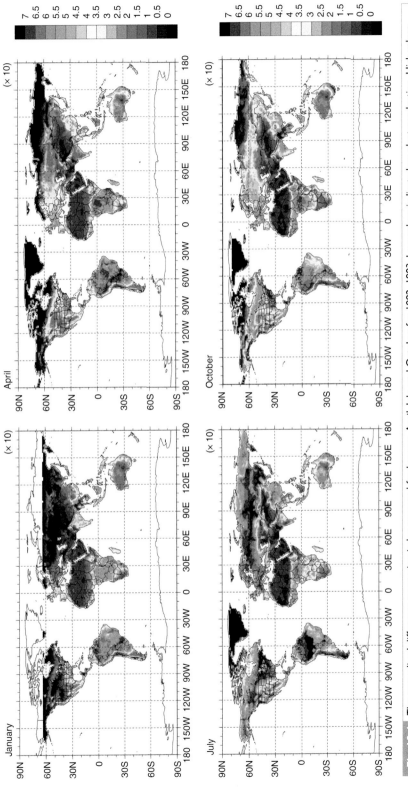

Fig. 18.4 The normalized difference vegetation index averaged for January, April, July, and October for 1982–1993. Low values indicate low plant production. High values indicate high production. See color plate section.

production increases the probability of descendants in future generations. Dispersal of a heavy blanket of seeds over a wide area ensures that at least some seeds are likely to fall on sites suitable for germination and establishment. However, high seed yield and high growth are not always compatible. Greater reproductive output diverts resources from vegetative growth and can lead to low stem or shoot growth during years of high seed production.

The number and size of seeds represent a compromise between dispersal and food reserves for germination (Grime and Jeffrey 1965; Leishman and Westoby 1994; Saverimuttu and Westoby 1996; Walters and Reich 2000; Westoby et al. 2002). The size of seeds varies from less than 10^{-6} g for the dust-like seeds of orchids to over 10^4 g for large coconuts. Tree seeds range in size from 10^{-4} g for light, wind-dispersed birch seeds to 0.1–10 g for the large nuts of beech, oak, and chestnut trees. This range in seed size is the outcome of reproductive tradeoffs. Seeds must contain sufficient carbohydrates and nutrients to support germination but must be dispersed from the parent and produced in sufficient quantities to ensure a high probability of survival. Seed number and seed size are alternatives in reproductive strategy so that plants produce many small, light seeds or fewer, larger seeds. Small seeds carry few carbohydrates to support initial growth. The seedling must depend on its own photosynthate at an early stage. However, small seeds are dispersed in large quantities and spread over large regions by wind, ensuring that some of the seeds fall on a favorable site for germination. In contrast, large, heavy seeds have sufficient initial reserves to continue growth for extended periods of time and survive in environments with low resource availability, but large seeds have a cost. They are produced in less quantity than small seeds, are not as widely dispersed, and are likely to be eaten by wildlife because they are highly nutritious.

A number of experimental studies have demonstrated the costs and advantages of seed size. For example, a study of seedling development in nine tree species growing in shade found that seedling height after 12 weeks increased with seed weight while the number of seedlings that

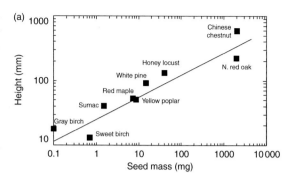

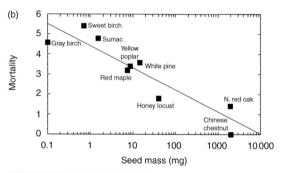

Fig. 18.5 Relationship between seed weight and seedling growth and survival for nine species of trees grown in shade. Seeds were germinated and then transplanted to grow in shade. (a) Height growth after 12 weeks. (b) Number of seedlings that died over the 12 weeks. Adapted from Grime and Jeffrey (1965).

died decreased (Figure 18.5). Gray birch, sweet birch, and sumac had the lowest seed weight, lowest height, and highest mortality. Chinese chestnut and northern red oak had the highest seed weight, greatest height, and least mortality.

18.4.2 Aboveground and Belowground Growth

The acquisition of resources for growth presents another set of tradeoffs in plants. Plants need foliage to absorb light and CO_2 and construct carbohydrates during photosynthesis. More foliage allows for more light and CO_2 acquisition and, all other factors being equal, more carbon gain. Stems provide structural support for foliage and a hydraulic pathway to leaves. Trees in particular need large woody trunks and branches to support their extensive foliage and to store and transport water, nutrients, and carbohydrates. Roots acquire water and nutrients

Table 18.1 Changes in carbon allocation in trees in relation to various stresses

Stress	Root growth	Foliage growth
Shade	Reduced	Increased
Drought	Increased	Reduced
Cold temperature	Increased	Reduced
Nitrogen deficiency	Increased	Reduced
Nitrogen surplus	Reduced	Increased

Source: From Waring (1991). See also Poorter et al. (2012).

from soil. Greater investment in aboveground foliage and stem growth comes at the expense of belowground root growth. Conversely, greater allocation to root growth allows for increased acquisition of belowground resources but at the expense of aboveground growth. Hence, resource acquisition is a compromise between aboveground foliage and shoot growth to harvest light and absorb CO_2 and belowground root growth to obtain water and nutrients.

Natural environments are seldom optimal for plant growth. Tall plants shade short neighbors. Soils may be dry or deficient in nutrients. The allocation of resources to build and maintain above- and belowground biomass is not proportioned in fixed ratios. When stressed by lack of light, water, or nutrients, plants change their pattern of resource allocation. Allocation is an integration of plant responses to multiple stresses imposed by nutrient, water, and light availability. One theory of carbon allocation holds that plants adjust their allocation so that all resources equally limit growth (Bloom et al. 1985; Chapin 1991; Gleeson and Tilman 1992). Plants typically respond to altered resource availability by allocating new biomass to the components that acquire the most limiting resource so that resource imbalances are minimized (Table 18.1). Plants in resource-rich environments, where nutrients and water are not limiting, grow best by allocating carbon to leaves. More foliage allows for greater light capture and photosynthesis. Deficiencies in belowground resources such as water and nitrogen result in increased allocation to roots and reduced allocation to foliage. A nitrogen surplus or sufficient water allows for greater leaf

production and reduced need for roots. Hence, irrigation and fertilization can increase foliage growth where these are limiting. Cold temperatures favor increased root production because low temperatures reduce the ability of roots to absorb water and nutrients.

18.5 Life History Patterns

The optimal allocation of limited resources to growth, maintenance, and reproduction is part of an overall interrelated suite of attributes that have evolved through natural selection. These functions and their necessary plant parts interact to determine fitness, defined by the number of descendants in future generations. Natural selection selectively favors those individuals that contribute the most offspring to subsequent generations. However, allocation of carbon and nutrients among the various plant parts involves compromises, and a gain in fitness from one investment may be offset by loss in another. For example, greater allocation to reproduction does not necessarily increase fitness. Less carbon is available for vegetative growth. The plant may not be able to successfully compete with neighbors. A plant can increase its fitness by reducing its reproductive effort and investing more in increased growth. Light, water, and nutrients are usurped from neighbors, ensuring the plant survives and leaves descendants.

Natural selection acts to maximize individual fitness by optimizing the form of these compromises, creating a balanced system of resource allocation to plant parts (Bloom et al. 1985; Tilman 1988; Chapin 1991; Bazzaz 1996; Grace 1997). Resources are allocated in a way that maximizes fitness. The optimal allocation of resources is evident in the morphology and life histories of plants, and the outcome of optimal resource allocation is not one specific pattern of allocation, morphology, and life strategy. Rather, light, water, and nutrients vary spatially and temporally. There are numerous alternative patterns of allocation that allow plants to successfully grow, survive, and reproduce in specific environments. Allocation must be flexible to allow for different patterns in different environments, but is constrained by the life history of a species.

A life history is an overall pattern of growth, reproduction, and longevity. A successful life history allows a species to persist through evolutionary time. One successful strategy is to remain small and allocate resources to a single episode of copious seed dispersal. Plants that reproduce once and then die are called monocarpic, or semelparous. They invest a large reproductive effort into a single flowering episode at the expense of future vegetative growth. An equally successful strategy is to grow slowly, be long lived, and have repeated reproduction during a life cycle, which may extend from several years for herbaceous plants to hundreds of years for trees. Such plants are termed polycarpic, or iteroparous.

Life histories are closely matched to environment; the environment determines which life histories are successful. Frequent, recurring disturbances that expose soil for revegetation favor plants that allocate resources to seed production and widespread dispersal. Monocarpic plants are favored in ephemeral environments with relatively low juvenile mortality and high adult mortality. However, this is a risky strategy when the environment is uncertain. A cold or dry year could destroy an entire cohort of plants, removing their descendants from future generations. In contrast, polycarpic plants are favored where juvenile mortality is high, adult mortality is low, and in uncertain environments so that the risk of a bad year is spread out over several years of reproduction.

18.5.1 | Annuals and Perennials

The distinction between annual and perennial is one example of alternative morphologies and life histories based on longevity and age of reproduction. Annual plants complete their life cycle in one year or less, germinating, growing, flowering, and setting seeds in a short period of time. Initially, they allocate resources to foliage and stems. Roots are needed to supply water and nutrients. Later, as the plant matures, resources are allocated to a single bout of seed production and dispersal before they die. *Senecio vulgaris* illustrates the ephemeral life strategy typical of herbaceous annuals (Figure 18.6). A period of

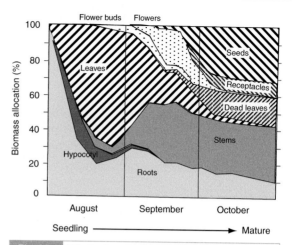

Fig. 18.6 Whole plant allocation for *Senecio vulgaris*, an annual. Adapted from Harper and Ogden (1970).

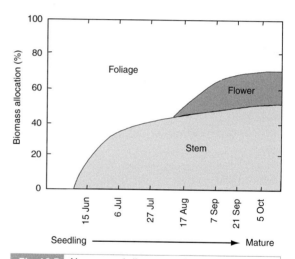

Fig. 18.7 Aboveground allocation to foliage, stems, and flowers in a herbaceous perennial (goldenrod, *Solidago speciosa*). Adapted from Abrahamson and Gadgil (1973).

vegetative growth is followed by seed production culminating in death. Three distinct growth phases are evident. The development of roots and leaves marks the first phase. During the second phase, stems grow longer and flowers emerge. The end of this phase is marked by little vegetative growth, peak flower development, and the opening of the first seed heads. In the final stage, seeds mature and leaves and roots begin to die.

Perennial plants take a longer-term allocation strategy, as illustrated by goldenrod (Figure 18.7). Foliage and stems persist

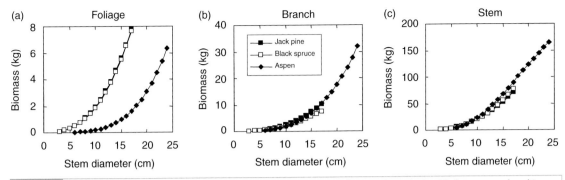

Fig. 18.8 Relationship between stem diameter and foliage, branch, and stem biomass for jack pine, black spruce, and quaking aspen trees growing in Canada. Data from Gower et al. (1997).

throughout the growing season. Moreover, perennials must allocate resources to perennial parts (e.g., buds) and storage (e.g., roots). Consequently, they produce fewer seeds than annuals, but in the subsequent growing season, while annuals start anew from seed, perennials begin growth with an established root system and carbon and nutrient stores. These stores allow more rapid initial growth than plants growing from seeds. This gives perennials an early-season height advantage, whereby they can capture light and shade neighbors.

18.5.2 Trees

Trees allocate resources in a way suited to their morphology and long lifespan. Young trees allocate much of their growth to foliage to acquire carbon and to roots to acquire water and nutrients. As trees grow larger, however, much of their biomass is woody stems and branches; relatively little biomass is foliage. The height of trees gives them a competitive advantage over smaller plants; their leaves shade the understory. However, the cost is large amounts of carbon expended in the production and maintenance of trunks and branches for mechanical support and other tissues to transport water and nutrients.

Allocation in trees is modulated by allometric constraints that maintain certain dimensional relationships between diameter and foliage, branch, and stem biomass (Figure 18.8). In particular, the structure and anatomy of trees is constrained by the need to transport water to leaves to replace that lost during transpiration

and the need to provide mechanical support to foliage. Water flows through specialized structures within wood. However, not all of the woody material in trees conducts water. Large trees have an interior core of heartwood that provides structural support but which does not conduct water. Only the outer ring of sapwood provides the material to transport water. In angiosperms, vessels are stacked end-to-end in the xylem to form a continuous tube through wood. In gymnosperms (e.g., conifers), vertically stacked, overlapping tracheids form a pathway for water flow through xylem. Because of its role in water transport, the cross-sectional area of sapwood in a tree closely relates to the amount of foliage (Waring et al. 1982). Large leaf area requires a corresponding large area of sapwood to transport water to leaves. Hence, there is a functional interdependence between foliage, which fixes carbon and transpires water, and support structures that supply water and nutrients to leaves.

Species that grow in harsh environments support less leaf area for a given sapwood cross-sectional area compared with trees growing in more favorable climates (Table 18.2). For example, Sitka spruce and Douglas fir, which grow in moderate or maritime climates along the Oregon coast, support two to six times as much leaf area for the same sapwood area as do mountain hemlock, ponderosa pine, and western juniper, which grow in harsher inland climates. A study of leaf area and sapwood area in Scots pine growing on a cool, wet site and a warm, dry site illustrates the effect of site

Table 18.2 Ratio of leaf area to sapwood area for tree species growing along a west-to-east transect in Oregon from the coast to the mountains

Species	Climate	Ratio (m² cm⁻²)
Sitka spruce	Maritime	0.44
Douglas fir	Moderate	0.32
Mountain hemlock	Subalpine	0.16
Ponderosa pine	Semiarid	0.17
Western juniper	Arid	0.07

Source: From Waring (1991). See also Waring (1983).

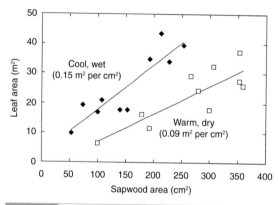

Fig. 18.9 Relationship between leaf area and sapwood area for Scots pine trees growing on a cool, wet site and a warm, dry site. Adapted from Mencuccini and Grace (1994).

conditions on allocation (Figure 18.9). The trees on both sites were grown from the same seed source, minimizing genetic variation between sites, and the stands were of similar age, density, and fertility. The trees growing on the warm, dry site had less leaf area per unit sapwood area compared with the trees on the cool, wet site. Moisture limitation on the warm, dry site allowed for less leaf area per unit sapwood area.

The relationship between sapwood area and leaf area arises because foliage must be supported by a certain area of water-transporting sapwood. Sapwood area below the required amount limits water transport, and foliage suffers water shortage. Too much sapwood, however, is an inefficient allocation of resources to non-essential tissue. Similarly, there is a broad relationship between climate and maximum leaf area index

(Grier and Running 1977; Woodward 1987, 1993; Nemani and Running 1989). Too much leaf area results in drought because evapotranspiration exceeds precipitation. Too little leaf area results in surplus soil water. Instead, there is an equilibrium between soil water, evapotranspiration, and leaf area whereby trees support a maximum leaf area for which evapotranspiration balances precipitation.

The hydraulics of water movement in trees is affected by their height. Height conveys a competitive advantage to tall trees. With increased height relative to neighbors comes greater utilization of light for photosynthesis, but this advantage is tempered by the need to move water from roots to the uppermost leaves to replenish water lost in transpiration. Hydraulic conductance to water flow decreases as a tree grows taller and water must travel a longer path. This decreased conductance may provide a physical limit to tree height (Ryan and Yoder 1997; Koch et al. 2004; Ryan et al. 2006; Niklas 2007). The rate of water transport from roots to leaves relates to the difference in water potential between leaves and roots multiplied by the hydraulic conductance. A higher tension (more negative water potential) is required to move the same amount of water through a tree with lower conductance. At high tension, air bubbles can form in the water column. These air bubbles interrupt the continuous column of water, hindering replenishment of water in the leaves and inducing stress. The decrease in conductance with greater height increases water stress in the leaves, closing stomata and reducing carbon gain during photosynthesis. However, changes in leaf area to sapwood area and other morphological traits can compensate for the decreased conductance with height.

The large investment in maintenance and support costs of trees requires compromises among growth, maintenance, and reproduction. Among broadleaf deciduous species, the typical age of maturity increases with longevity (Figure 18.10). This suggests a high early investment in growth and support at the expense of reproduction. Growth rates also decline with increased longevity. Slow-growing broadleaf deciduous species are nearly three times as

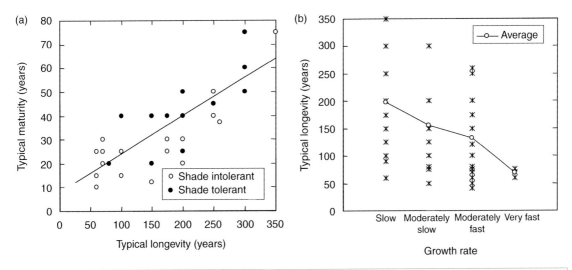

Fig. 18.10 Relationships among growth, maturity, and longevity in 87 species of broadleaf deciduous trees. (a) Age of reproduction in relation to longevity. Species are separated into shade intolerant and shade tolerant classes. (b) Longevity in relation to growth rate. Data are shown for each species and as the average longevity for each growth class. Data from Loehle (1988).

long lived as fast-growing species (190 years versus 68 years, on average). This suggests that long-lived trees invest more in maintenance than growth. Similar patterns have been found for needleleaf evergreen trees.

18.5.3 Disturbance and Competition

The roles of disturbance and competition for resources (light, water, nutrients) dominate much of our understanding of plant life histories. Recurring disturbances such as fires or windthrows create an open environment where new plants are unlikely to be shaded by neighbors. Success favors those species whose seeds fall in such sites, germinate and become established in high light environments, and grow rapidly to dominate the canopy. At the other extreme, closed canopy forests, with little sunlight on the forest floor, favor species that can tolerate shade, grow slowly in the low light of the understory, and wait for a large tree to die and create a gap in the canopy.

This distinction between disturbance and competition is embodied in the classic notion of *r*- and *K*-selected life histories (MacArthur and Wilson 1967; Gadgil and Solbrig 1972). Species that are *r*-selected maintain their abundance in the landscape through high seed production and widespread dispersal. They are short lived, with relatively little allocation to growth and large allocation to reproduction. They have numerous small seeds dispersed over large areas. They are favored in temporally varying environments where recurring disturbance creates potential for rapid colonization of open sites. In contrast, *K*-selected species maintain their abundance at or near the maximum limit for an environment. They are typically long lived and slow growing, allocating a greater proportion of resources to growth and maintenance. This allows them to survive the intense competition for resources in high-density environments, but provides less energy for reproduction. Their seeds tend to be few and large to provide sufficient resources to germinate and become established in the low light environment of a dense canopy. While the concept of *r*- and *K*-selected species has dominated much of the study of life history patterns, these better represent the endpoints of a continuum of life histories rather than a stark dichotomy. Most plants fall in between these two extremes.

An alternative classification of life history strategies recognizes the importance of resource stressed environments in shaping life histories in addition to competition and disturbance (Grime 1979). Ruderal plants live in temporary or frequently disturbed environments. They are opportunistic species adapted to disturbance. They are short lived, grow rapidly, and have a large reproductive effort at an early age. Competitor plants live in crowded environments where disturbance is infrequent, stress is low, and competition for resources favors species that compete well with others for limited resources. These plants utilize resources more efficiently than others, allocate available resources to growth, and are typically long lived, mature later, and have a small reproductive effort. Stress tolerators live in environments with limited resources where plants are physiologically stressed by lack of water, low temperatures, low nutrient availability, or low light. These plants persist under conditions of severe resource limitation by growing slowly and allocating resources to maintenance.

Plant responses to disturbance, competition, and stress produce a variety of life histories intermediate to these three extremes (Figure 18.11). Competitive ruderals are adapted to environments in which stress is low and recurring disturbances limit competition to moderate intensity. Stress tolerant competitors are adapted to relatively undisturbed environments with moderate stress. Stress tolerant ruderals are adapted to moderately disturbed, unproductive environments. A final type of plant, C-S-R strategists, is adapted to environments where moderate stress and recurring disturbances of moderate intensity combine to limit the amount of competition. Major life forms segregate along these axes of disturbance, competition, and stress. Trees and shrubs are found in environments characterized by low to moderate intensity of disturbance and tolerate a wide range of stress and competition. Annual herbs are characterized by moderate to high intensity of disturbance, low intensity of competition, and low stress. Only perennial herbs and ferns are undifferentiated with respect to disturbance, competition, and stress.

Another useful distinction is between early and late successional species (Table 18.3).

Early successional, or pioneer, species colonize recently disturbed environments. They are exposed to full sunlight. In contrast, seedlings of late successional species germinate under a forest canopy and are exposed to less sunlight. Seeds of early successional plants require light for germination and can lie dormant for many years in soil, waiting for a disturbance that opens the canopy. Seeds of late successional plants do not require full sunlight for germination and lose viability rapidly. Early successional plants are shade intolerant. They have high rates of photosynthesis at high light intensity and low rates at low light. Light saturation occurs at high light intensity. Late successional plants are shade tolerant and are photosynthetically more efficient at low light intensities than early successional plants. Shade tolerance not only influences leaf physiology, but also relates to longevity. The typical longevity of shade intolerant species of broadleaf deciduous trees averages 147 years while that of shade-tolerant species averages 191 years (Figure 18.10).

18.6 | Plant Functional Types

The classification of species into r- and K-selected plants (MacArthur and Wilson 1967; Gadgil and Solbrig 1972), ruderal, competitor, and stress tolerator plants (Grime 1979), or early and late successional plants (Bazzaz 1979, 1996; Huston and Smith 1987) represents broad classes of plant functional types that reduce the complexity of species diversity in ecological function to a few plant types. Plant functional types are defined by key physiological and life history characteristics that determine vegetation dynamics and response to changing environment. The combination of physiological and morphological traits along with climatic preferences is one basis to define functional types. The distinction between annual or perennial, evergreen or deciduous, and broadleaf or needleleaf is particularly useful because these characteristics are observable from remote sensing and are key ecological properties determining stomatal conductance, photosynthesis, and carbon allocation. Such a classification results in six plant functional

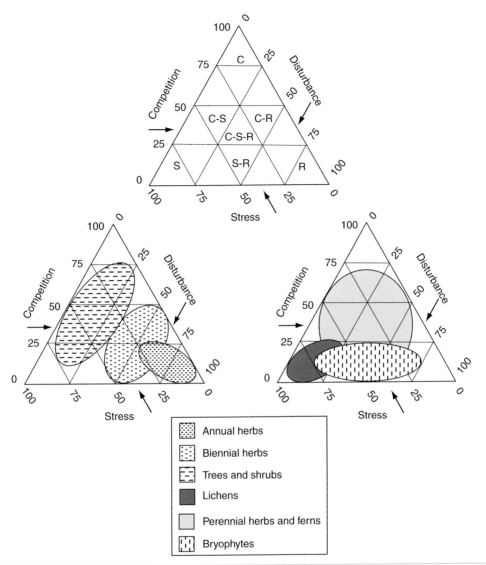

Fig. 18.11 Classification of life history strategies based on relative importance of disturbance, competition, and stress. Top: The three axes of increasing relative competition, disturbance, and stress define seven life histories – competitor (C), ruderal (R), stress tolerator (S), competitive ruderal (C-R), stress tolerant competitor (C-S), stress tolerant ruderal (S-R), and C-S-R strategist. Bottom: Distribution of annual herbs, biennial herbs, and trees and shrubs (left) and lichens, perennial herbs and ferns, and bryophytes (right) with respect to competition, disturbance, and stress. Adapted from Grime (1979, p. 57, p. 73).

types (needleleaf and broadleaf evergreen perennial, needleleaf and broadleaf deciduous perennial, broadleaf annual, grass) based on permanence of aboveground biomass, leaf longevity, and leaf type (Running et al. 1995; Ustin and Gamon 2010). Simple climate rules can then define thermal- and moisture-related varieties (Nemani and Running 1996; Bonan et al. 2002).

Among woody plants, a key functional distinction is between angiosperms (flowering plants) and gymnosperms (Table 18.4). This latter group includes conifers, cycads, and *Gingko*. Angiosperm leaves have high photosynthetic rates and stomatal conductance, are relatively inexpensive to construct (i.e., have a low mass per unit area), and are short lived. In contrast,

Table 18.3 Characteristics of early and late successional plants

	Early succession	Late succession
Seeds		
Number	Many	Few
Size	Small	Large
Dispersal	Wind, birds	Gravity, mammals
Dormancy	Long	Short
Germination	Light enhanced	Not light enhanced
Photosynthesis		
Light saturation intensity	High	Low
Light compensation point	High	Low
Efficiency at low light	Low	High
Maximum rate	High	Low
Stomatal conductance	High	Low
Morphology		
Root-to-shoot ratio	Low	High
Size at maturity	Small	Large
Structural strength	Low	High
Lifespan	Short	Long

Source: Adapted from Bazzaz (1979, 1996) and Huston and Smith (1987).

Table 18.4 Key functional differences between conifers and angiosperm trees

Trait	Conifers	Angiosperms
Leaf lifespan (years)	3–26	<1–5
Leaf mass per unit area (g m^{-2})	227	106
Maximum photosynthesis (C_3; ambient CO_2; μmol m^{-2} s^{-1})	16	30
Maximum stomatal conductance (mol m^{-2} s^{-1})	0.5	>1
Leaf venation	Single or a few parallel veins	Reticulate veins
Conduit type	Tracheids: unicellular	Vessels: multicellular
Maximum conduit diameter (μm)	~80	~500
Maximum conduit length	A few millimeters	Several meters
Maximum hydraulic conductivity (mol m^{-1} s^{-1} MPa^{-1})	260	>555

Source: From Brodribb et al. (2012).

the foliage of conifers has lower photosynthetic rates and stomatal conductance, is more expensive to construct, and persists for a longer time. These differences relate to plant hydraulics. Angiosperms have a more elaborate hydraulic system compared with conifers, with higher leaf venation, wider and longer conduits to conduct water, and higher hydraulic conductivity. This system supplies the water to sustain the high photosynthetic rates and stomatal conductance seen in angiosperms compared with conifers.

Table 18.5 | Maximum photosynthetic capacity per unit leaf mass (A_{mass}), leaf mass per unit area (LMA), leaf nitrogen per unit leaf mass (N_{mass}), and slope for photosynthesis–nitrogen relationship for 257 species of forbs, shrubs, broadleaf trees, and needleleaf trees

Functional group	A_{mass} (nmol CO_2 g^{-1} s^{-1})	LMA (g m^{-2})	N_{mass} (%)	A_{mass}–N_{mass} slope (µmol CO_2 s^{-1} g^{-1} N)
Forb	305	51	3.5	9.2
Broadleaf shrub				
Deciduous	157	71	2.1	9.7
Evergreen	62	141	1.6	2.4
Broadleaf tree				
Deciduous	139	73	2.2	4.2
Evergreen	55	112	1.5	1.5
Needleleaf tree				
Deciduous	97	100	1.9	–
Evergreen	28	263	1.2	2.8

Source: From Reich et al. (1998a).

Comparisons among plant functional groups show broad distinctions in leaf traits based on type of leaf (broadleaf, needleleaf) and leaf lifespan (deciduous, evergreen). Key traits are maximum photosynthetic capacity per unit leaf mass (A_{mass}), leaf respiration rate per unit leaf mass (R_{mass}), leaf nitrogen per unit leaf mass (N_{mass}), leaf lifespan, and the carbon cost of construction. The latter is quantified by the ratio of leaf mass to leaf surface area (LMA), or by the inverse (leaf surface area per unit leaf mass), which is termed specific leaf area (SLA). Leaf mass per unit area is a measure of carbon investment in photosynthesizing leaf area. Species with low LMA (high SLA) require little carbon to produce a unit area of foliage; they have thin leaves with large surface area per unit mass (e.g., a broadleaf). Species with high LMA (low SLA) require more carbon to produce a unit area of foliage; they have thick foliage with low surface area per unit mass (e.g., a needleleaf). Leaf lifespan describes the period over which the initial investment in carbon to produce the leaf can be recouped during photosynthesis. Deciduous leaves are shed annually, and the cost to produce the leaf must be incurred every year. That investment cost is recouped over longer periods in evergreen leaves.

In a study of 257 species, mean leaf traits differ among functional groups (Table 18.5). Forbs have the highest A_{mass} and N_{mass} and lowest LMA. Woody species have lower A_{mass} and N_{mass} and higher LMA. Within woody species, additional distinction depends on leaf type (needleleaf, broadleaf) and longevity (deciduous or less than one-year lifespan, evergreen or greater than one-year lifespan). Broadleaf trees have high A_{mass}, high N_{mass}, and low LMA compared with needleleaf trees. Deciduous species of shrubs, broadleaf trees, and needleleaf trees have higher A_{mass} and N_{mass} and lower LMA than corresponding evergreen species. The slope of the A_{mass}–N_{mass} relationship (i.e., the photosynthetic capacity per unit mass of nitrogen) also differs among functional groups. Species with long lifespan and high LMA, whether broadleaf or needleleaf, have low maximum photosynthetic rates per unit leaf nitrogen.

Similar functional differences are seen in leaf respiration. Leaf maintenance respiration relates to photosynthetic capacity and leaf traits because a high photosynthetic capacity requires a large investment in enzymes and pigments, which have high respiration costs. In a study of 69 species from four functional groups (forbs, broadleaf shrubs, broadleaf trees, needleleaf trees) ranging from alpine tundra to desert to

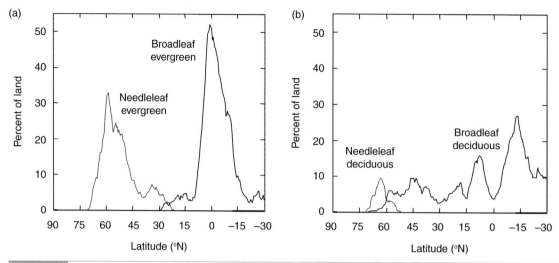

Fig. 18.12 Geographic distribution of (a) evergreen trees and (b) deciduous trees showing the percentage of land area covered by trees in relation to latitude. Trees are distinguished by needleleaf and broadleaf types. Derived from DeFries et al. (1999, 2000a,b).

tropical rainforest, R_{mass} increases with greater N_{mass} and decreases with larger LMA and longer leaf lifespan (Reich et al. 1998b). Forbs have the highest respiration rate. Needleleaf evergreen trees have the lowest respiration rate. Broadleaf shrubs and trees are intermediate in their traits.

The distinction between evergreen and deciduous represents an integrated plant response to environment (Chabot and Hicks 1982; Mooney and Gulmon 1982; Kikuzawa 1991; Reich et al. 1995b; Westoby et al. 2002; Wright et al. 2004). Evergreen trees have low photosynthetic capacity, low nitrogen concentration in foliage, and low leaf area per unit leaf mass. Their leaves have a high initial carbon cost to construct per unit photosynthesizing leaf area. However, evergreen trees, because they retain foliage throughout the year, can photosynthesize and gain carbon at all times of the year when weather is suitable. Because they retain foliage for several years, the high initial cost to construct foliage is spread out over a long period of time. Over the lifespan of their foliage, evergreen trees, therefore, recoup the high initial investment to construct foliage despite low photosynthetic capacity. In contrast, deciduous trees shed their leaves annually. The benefit is that leaf shedding reduces transpiration and desiccation during times of the year when cold temperature or

seasonal drought restrict photosynthesis. The costs are recurring annual investment of carbon to grow leaves and nutrient loss in litterfall. These costs are minimized by having a high specific leaf area so that the carbon cost to construct photosynthesizing surface area is smaller than in evergreen trees. Moreover, these trees have a high photosynthetic capacity to compensate for the short leaf lifespan.

Deciduous and evergreen leaf habits are also related to defense mechanisms to protect against herbivory. Herbivores preferentially eat leaves with high photosynthetic capacity because of their high nitrogen concentration. One successful strategy in the face of herbivory is to make foliage unpalatable through structure or chemical defenses. Evergreen leaves represent an allocation pattern to extend the payback period on the high initial carbon investment to form leaf area, maximize carbon gain despite low photosynthetic rates, conserve nutrients, and protect against herbivory. Deciduous leaves represent an alternative strategy whereby carbon gain is maximized by high photosynthetic rates in short-lived foliage that is either shed at the end of the growing season or consumed by herbivores.

These different life history patterns produce distinct geographic distributions to evergreen and deciduous trees (Figure 18.12). Evergreen trees have a bimodal geographic distribution. Needleleaf evergreen trees are most abundant in the boreal forests of Canada, northern Europe, and Russia. They also occur in the montane forests of western United States and to a lesser extent in temperate forests of mid-latitudes. Broadleaf evergreen trees dominate tropical forests along the equator. In contrast, broadleaf deciduous trees have a wide geographic distribution and are common throughout temperate forests, savannas, and tropical seasonal forests, where they lose their leaves in response to seasonal cold or drought.

Evergreen trees dominate in tropical rainforests where favorable conditions allow for photosynthesis and growth throughout the year. Evergreen trees are also abundant in mid-latitudes on dry, nutrient-poor soils. There, efficient nutrient use and increased leaf lifespan give evergreen trees a competitive advantage over deciduous trees. Evergreen trees also dominate the forests of the North American Pacific Northwest (Waring and Franklin 1979). There, the maritime climate permits photosynthesis by evergreen trees during the winter months. In contrast, photosynthesis by deciduous trees is restricted to the summer months when leaves have emerged, but during which time soil water is likely to be limiting. The dominance of evergreen trees in subarctic climates is attributed to the cold temperatures, which restrict the growing season and limit nutrient mineralization. Nutrient conservation is important. The extreme winter cold of east Siberia favors needleleaf deciduous trees to prevent winter desiccation (Gower and Richards 1990).

A physiological and morphological definition of plant functional types must be reconciled with an understanding of plant adaptations to disturbance, which is so critical to understanding vegetation responses to changing environments. Classifications such as *r*- and *K*-selection, ruderal, stress tolerant, and competitive, and early and late succession reflect the central role of disturbance in shaping community structure and composition. Other similar classifications are exploitive and conservative species (Bormann and Likens 1979) and gap and non-gap species (Shugart 1984, 1987, 1998). In many cases, morphological and physiological considerations impose correlated life history traits. For example, early succession plants, in addition to their physiological traits of shade intolerance, high rates of photosynthesis, and high photosynthetic light compensation and saturation points, are also relatively short lived, fast growing, and have small seeds that are widely dispersed. Indeed, the consistent tradeoff between high investment in photosynthesis and growth versus preferential allocation to storage, defense, and reproduction imposes correlations among vegetative and regenerative traits.

18.7 | Coordinated Functional Traits

Plant functional types are discrete characterizations of leaf functional traits, but these traits have a continuum of variation. The diversity of plant types may be better characterized by a continuum of correlated traits than by discrete plant functional types. Studies of plants growing in a variety of communities and environments and representing a diversity of life forms show coordinated relationships among leaf traits such as maximum photosynthetic capacity per unit leaf mass (A_{mass}), leaf respiration rate per unit leaf mass (R_{mass}), leaf nitrogen per unit leaf mass (N_{mass}), leaf lifespan, and the carbon cost of construction (LMA). Photosynthetic capacity, respiration rate, and leaf nitrogen are interrelated because nitrogen is integral to the enzymes and pigments necessary for photosynthesis. Consequently, there is a strong relationship between maximum photosynthetic rate and leaf nitrogen content (Field and Mooney 1986; Schulze et al. 1994). Subsequent studies have emphasized LMA and leaf lifespan in addition to nitrogen (Reich et al. 1992, 1997; Wright et al. 2004; Poorter et al. 2009).

Figure 18.13 illustrates these relationships for 22 broadleaf deciduous and 9 needleleaf evergreen temperate tree species. Photosynthetic capacity per unit leaf mass (A_{mass}) increases with higher leaf nitrogen. Broadleaf deciduous trees have a higher rate of photosynthesis for a similar range of leaf nitrogen than do needleleaf evergreen trees. The leaves of broadleaf deciduous trees have low LMA, high A_{mass}, and high

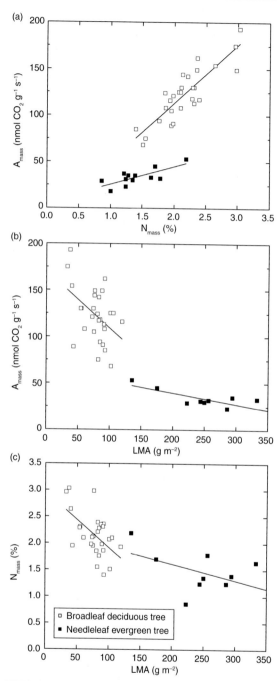

Fig. 18.13 Coordinated leaf traits for 22 broadleaf deciduous and 9 needleleaf evergreen temperate tree species. (a) Maximum rate of photosynthesis per unit leaf mass (A_{mass}) in relation to leaf nitrogen per unit leaf mass (N_{mass}). (b) A_{mass} in relation to leaf mass per unit area (LMA). (c) N_{mass} in relation to LMA. Data from Reich et al. (1995a).

N_{mass}; the long-lived needles of coniferous trees have high LMA, low A_{mass}, and low N_{mass}. Within both functional groups, A_{mass} and N_{mass} both decline with higher LMA.

Wright et al. (2004, 2005) generalized these relationships in an analysis of 2548 species from 219 families at 175 locations worldwide including arctic tundra, boreal forest, tropical rainforest, grassland, and hot and cold deserts (Figure 18.14). Annual mean temperature ranged from –16.5 to 27.5°C. Annual precipitation ranged from 133 to 5300 mm. Over this wide variety of plant communities, environments, and life forms, maximum photosynthetic capacity and leaf respiration rate increase with shorter leaf lifespan, more leaf nitrogen, and lower LMA. They called this the leaf economics spectrum, which ranges from quick to slow return on investments of carbon and nitrogen in leaves. Species with short-lived leaves generally have low LMA (i.e., low carbon cost per unit leaf area), high N_{mass}, high A_{mass}, and high R_{mass}. They reap a quick return on investment. At the other end of the spectrum are species with long leaf lifespan, high LMA, low N_{mass}, low A_{mass}, and low R_{mass}. These leaves are expensive to produce and return that investment over a long period. In between is a continuum of leaves with coordinated traits scaled between these two extremes. These functional patterns of variation among leaf structure, longevity, nutrition, and metabolism represent interdependent leaf traits that are a tradeoff between high metabolism and persistence.

Other traits relate to plant hydraulics. The hydraulic architecture of trees represents a tradeoff between high rates of photosynthesis and risk of desiccation during drought. Hydraulic failure occurs when the water column in the stem breaks (i.e., becomes air-filled, also referred to as cavitation) under high tension. This reduces stem hydraulic conductivity and water supply to foliage. The manner in which trees conduct water and resist cavitation in their stem xylem is an adaptation to tradeoffs. A hydraulic pathway with long, wide conduits transports water more efficiently than does a pathway of shorter, narrower conduits, but is also more vulnerable to drought-induced cavitation.

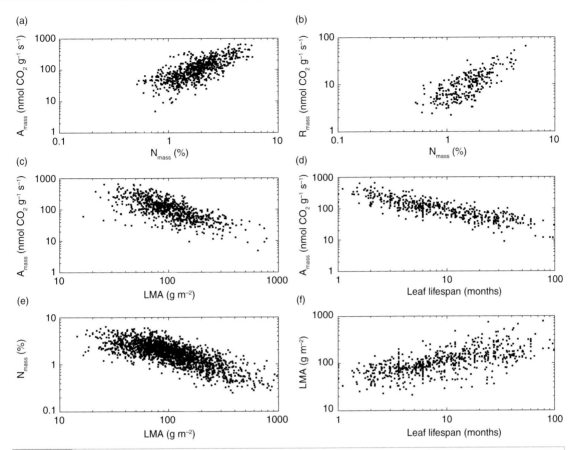

Fig. 18.14 Coordinated leaf traits for maximum rate of photosynthesis per unit leaf mass (A_{mass}), leaf respiration rate per unit leaf mass (R_{mass}), leaf nitrogen per unit leaf mass (N_{mass}), leaf mass per unit area (LMA), and leaf lifespan (LL). (a) A_{mass}–N_{mass}, 712 species. (b) R_{mass}–N_{mass}, 267 species. (c) A_{mass}–LMA, 764 species. (d) A_{mass}–LL, 512 species. (e) N_{mass}–LMA, 1958 species. (f) LMA–LL, 678 species. Data from Wright et al. (2004).

A study of 13 tropical dry forest tree species illustrates tradeoffs between hydraulic safety and efficiency (Markesteijn et al. 2011b). In these species, vulnerability to cavitation is positively correlated with stem hydraulic conductivity and negatively correlated with wood density (Figure 18.15a,b). Pioneer, early successional species are more vulnerable to cavitation than are late successional, shade-tolerant species. Early successional species have high photosynthetic rates, with high transpiration rates, and therefore require high water transport through the stem. They have high stem hydraulic conductivity and low wood density, but the cost is reduced resistance to drought-induced cavitation. They maintain a high leaf water potential that is close to the point of hydraulic failure (Figure 18.15c).

In contrast, shade-tolerant species have lower stem hydraulic conductivity, higher wood density, and operate with more hydraulic safety. Similar differences are seen between deciduous and evergreen species. A larger analysis of 40 tropical dry forest species shows similar behavior (Markesteijn et al. 2011a).

Among these tree species, wood density is a key functional trait. Wood density negatively correlates with stem hydraulic conductivity and vulnerability to cavitation. Species with high wood density have low vulnerability (are more resistant) to cavitation; species with low wood density are more vulnerable (less resistant) to cavitation. Wood density integrates various properties related to structural support, water transport, and stem water storage, and

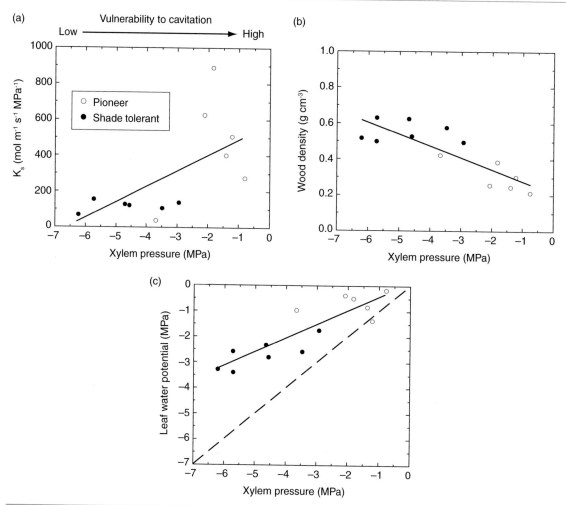

Fig. 18.15 Relationships between plant traits and vulnerability to cavitation for 13 tropical dry forest species. Vulnerability to cavitation is defined by the xylem pressure at 50 percent loss of stem hydraulic conductivity, signifying hydraulic failure and cavitation. High xylem pressure at hydraulic failure means high vulnerability to cavitation. (a) Stem hydraulic conductivity (K_s). (b) Wood density. (c) Dry season leaf water potential. The dashed line shows the point of hydraulic failure, where leaf water potential equals the xylem pressure at cavitation. The difference between that line and the regression line is the hydraulic safety margin (how close species operate to the point of hydraulic failure). Open symbols are pioneer species, and closed symbols are shade-tolerant species. Adapted from Markesteijn et al. (2011b).

in these species is a key functional trait. Wood density also relates to the tradeoff between growth rates and longevity (Kraft et al. 2010; Wright et al. 2010). Early successional, shade intolerant tree species tend to have high growth rates and short longevity and produce wood with low density. Shade-tolerant species grow slower, are longer lived, and have high wood density.

Further analysis of 226 forest tree species from 81 sites worldwide shows broad overlap in cavitation risk among species, but conifers are less vulnerable, on average, to cavitation than are angiosperms (Choat et al. 2012). Most species analyzed operate at high hydraulic risk (high vulnerability to cavitation). This represents a trade-off between growth rate and risk of mortality. Species that are hydraulically efficient can meet the high demand for water necessitated by high photosynthetic rates and fast growth. The cost, however, is less protection against drought-induced cavitation.

18.8 | Review Questions

1. Describe the likely differences in leaf phenology and growing season length for three sites located at latitude 40° N with similar elevation: 10 percent south slope, 20 percent south slope, and 20 percent northeast slope.

2. A landscape architect in Boulder, Colorado (mean January temperature, 0°C; mean July temperature, 23°C; annual precipitation, 480 mm) designs a suburban residential lawn using quaking aspen trees typically found in the mountains. Where would the trees likely grow better: on the north side of the house or on the west side?

3. Over the course of a year, a tree gains 1000 g C m^{-2} yr^{-1} in photosynthesis. Maintenance respiration and growth respiration are each 250 g C m^{-2} yr^{-1}. Fractional allocation of carbon to foliage, sapwood, and root growth is 0.40, 0.35, and 0.25, respectively. The corresponding C:N ratio of these plant parts is 25 g C g N^{-1}, 250 g C g N^{-1}, and 200 g C g N^{-1}, respectively. How much nitrogen is needed to support this growth? How is the majority of this nitrogen used?

4. The following table gives aboveground (ANPP) and belowground (BNPP) net primary production from irrigation and fertilizer application experiments in Douglas fir stands growing in New Mexico. The treatments began in 1985. What do these experiments indicate about carbon allocation in trees?

	Control	Irrigated	Fertilized
ANPP (g m^{-2} yr^{-1})			
1984	1160	900	1008
1985	1296	1125	1348

	Control	Irrigated	Fertilized
1986	1306	1410	1716
BNPP/ANPP	0.46	0.31	0.23

5. Why is the ratio of root biomass to shoot biomass low in early successional species and high in late successional species?

6. Shade intolerant tree species cannot survive in the low light environment of a closed forest canopy. Explain how shade intolerant species can coexist in a forest landscape with shade-tolerant species.

7. White spruce is a late successional tree species found in boreal forests of Alaska and Canada. Describe its leaf economics spectrum. Characterize its general life history. Where is it typically found in relation to competition, disturbance, and stress?

8. Which type of tree is expected to have larger maximum stomatal conductance: a broadleaf deciduous tree or a needleleaf evergreen tree? Why?

9. Angiosperm tree species have a wide geographic distribution and are more abundant than conifer species. Explain why.

10. Describe the tradeoff between hydraulic efficiency and hydraulic safety. Can a tree grow fast and be drought resistant? What are the characteristics that might favor a particular species in a warmer, drier planet?

18.9 | References

Abrahamson, W. G., and Gadgil, M. (1973). Growth form and reproductive effort in goldenrods (Solidago, Compositae). *American Naturalist*, 107, 651–661.

Bazzaz, F. A. (1979). The physiological ecology of plant succession. *Annual Review of Ecology and Systematics*, 10, 351–371.

Bazzaz, F. A. (1996). *Plants in Changing Environments: Linking Physiological, Population, and Community Ecology*. Cambridge: Cambridge University Press.

Bloom, A. J., Chapin, F. S., III, and Mooney, H. A. (1985). Resource limitation in plants – an economic analogy. *Annual Review of Ecology and Systematics*, 16, 363–392.

Bonan, G. B., Levis, S., Kergoat, L., and Oleson, K. W. (2002). Landscapes as patches of plant functional types: An integrating concept for climate and ecosystem models. *Global Biogeochemical Cycles*, 16, 1021, doi:10.1029/2000GB001360.

Bormann, F. H., and Likens, G. E. (1979). *Pattern and Process in a Forested Ecosystem*. New York: Springer-Verlag.

Brodribb, T. J., Pittermann, J., and Coomes, D. A. (2012). Elegance versus speed: examining the competition between conifer and angiosperm trees. *International Journal of Plant Sciences*, 173, 673–694.

Chabot, B. F., and Hicks, D. J. (1982). The ecology of leaf life spans. *Annual Review of Ecology and Systematics*, 13, 229–259.

Chapin, F. S., III (1991). Integrated responses of plants to stress. *BioScience*, 41, 29–36.

Choat, B., Jansen, S., Brodribb, T. J., et al. (2012). Global convergence in the vulnerability of forests to drought. *Nature*, 491, 752–755.

DeFries, R. S., Townshend, J. R. G., and Hansen, M. C. (1999). Continuous fields of vegetation characteristics at the global scale at 1-km resolution. *Journal of Geophysical Research*, 104D, 16911–16923.

DeFries, R. S., Hansen, M. C., and Townshend, J. R. G. (2000a). Global continuous fields of vegetation characteristics: A linear mixture model applied to multi-year 8 km AVHRR data. *International Journal of Remote Sensing*, 21, 1389–1414.

DeFries, R. S., Hansen, M. C., Townshend, J. R. G., Janetos, A. C., and Loveland, T. R. (2000b). A new global 1-km dataset of percentage tree cover derived from remote sensing. *Global Change Biology*, 6, 247–254.

Field, C., and Mooney, H. A. (1986). The photosynthesis–nitrogen relationship in wild plants. In *On the Economy of Plant Form and Function*, ed. T. J. Givnish. Cambridge: Cambridge University Press, pp. 25–55.

Gadgil, M., and Solbrig, O. T. (1972). The concept of r- and K-selection: Evidence from wild flowers and some theoretical considerations. *American Naturalist*, 106, 14–31.

Gleeson, S. K., and Tilman, D. (1992). Plant allocation and the multiple limitation hypothesis. *American Naturalist*, 139, 1322–1343.

Gower, S. T., and Richards, J. H. (1990). Larches: Deciduous conifers in an evergreen world. *BioScience*, 40, 818–826.

Gower, S. T., Vogel, J. G., Norman, J. M., et al. (1997). Carbon distribution and aboveground net primary production in aspen, jack pine, and black spruce stands in Saskatchewan and Manitoba, Canada. *Journal of Geophysical Research*, 102D, 29029–29041.

Grace, J. (1997). Toward models of resource allocation by plants. In *Plant Resource Allocation*, ed. F. A. Bazzaz and J. Grace. San Diego: Academic Press, pp. 279–291.

Grier, C. C., and Running, S. W. (1977). Leaf area of mature northwestern coniferous forests: Relation to site water balance. *Ecology*, 58, 893–899.

Grime, J. P. (1979). *Plant Strategies and Vegetation Processes*. Chichester: Wiley.

Grime, J. P., and Jeffrey, D. W. (1965). Seedling establishment in vertical gradients of sunlight. *Journal of Ecology*, 53, 621–642.

Harper, J. L., and Ogden, J. (1970). The reproductive strategy of higher plants, I: The concept of strategy with special reference to Senecio vulgaris L. *Journal of Ecology*, 58, 681–698.

Huston, M., and Smith, T. (1987). Plant succession: Life history and competition. *American Naturalist*, 130, 168–198.

Jackson, M. T. (1966). Effects of microclimate on spring flowering phenology. *Ecology*, 47, 407–415.

Kikuzawa, K. (1991). A cost-benefit analysis of leaf habit and leaf longevity of trees and their geographical pattern. *American Naturalist*, 138, 1250–1263.

Koch, G. W., Sillett, S. C., Jennings, G. M., and Davis, S. D. (2004). The limits to tree height. *Nature*, 428, 851–854.

Kraft, N. J. B., Metz, M. R., Condit, R. S., and Chave, J. (2010). The relationship between wood density and mortality in a global tropical forest data set. *New Phytologist*, 188, 1124–1136.

Larcher, W. (1995). *Physiological Plant Ecology*, 3rd ed. Berlin: Springer-Verlag.

Leishman, M. R., and Westoby, M. (1994). The role of large seed size in shaded conditions: Experimental evidence. *Functional Ecology*, 8, 205–214.

Loehle, C. (1988). Tree life history strategies: The role of defenses. *Canadian Journal of Forest Research*, 18, 209–222.

MacArthur, R. H., and Wilson, E. O. (1967). *The Theory of Island Biogeography*. Princeton: Princeton University Press.

Markesteijn, L., Poorter, L., Bongers, F., Paz, H. and Sack, L. (2011a). Hydraulics and life history of tropical dry forest tree species: Coordination of species' drought and shade tolerance. *New Phytologist*, 191, 480–495.

Markesteijn, L., Poorter, L., Paz, H., Sack, L., and Bongers, F. (2011b). Ecological differentiation in xylem cavitation resistance is associated with stem and leaf structural traits. *Plant, Cell and Environment*, 34, 137–148.

Mencuccini, M., and Grace, J. (1994). Climate influences the leaf area/sapwood area ratio in Scots pine. *Tree Physiology*, 15, 1–10.

Mooney, H. A., and Gulmon, S. L. (1982). Constraints on leaf structure and function in reference to herbivory. *BioScience*, 32, 198–206.

Nemani, R. R., and Running, S. W. (1989). Testing a theoretical climate–soil–leaf area hydrologic equilibrium of forests using satellite data and eco-system simulation. *Agricultural and Forest Meteorology*, 44, 245–260.

Nemani, R. R., and Running, S. W. (1996). Implementation of a hierarchical global vegetation classification in ecosystem function models. *Journal of Vegetation Science*, 7, 337–346.

Niklas, K. J. (2007). Maximum plant height and the biophysical factors that limit it. *Tree Physiology*, 27, 433–440.

Poorter, H., Niinemets, Ü, Poorter, L., Wright, I. J., and Villar, R. (2009). Causes and consequences of variation in leaf mass per area (LMA): A meta-analysis. *New Phytologist*, 182, 565–588.

Poorter, H., Niklas, K. J., Reich, P. B., et al. (2012). Biomass allocation to leaves, stems and roots: Meta-analyses of interspecific variation and environmental control. *New Phytologist*, 193, 30–50.

Reich, P. B., Walters, M. B., and Ellsworth, D. S. (1992). Leaf life-span in relation to leaf, plant, and stand characteristics among diverse ecosystems. *Ecological Monographs*, 62, 365–392.

Reich, P. B., Kloeppel, B. D., Ellsworth, D. S., and Walters, M. B. (1995a). Different photosynthesis–nitrogen relations in deciduous hardwood and evergreen coniferous tree species. *Oecologia*, 104, 24–30.

Reich, P. B., Koike, T., Gower, S. T., and Schoettle, A. W. (1995b). Causes and consequences of variation in conifer leaf life-span. In *Ecophysiology of Coniferous Forests*, ed. W. K. Smith and T. M. Hinckley. San Diego: Academic Press, pp. 225–254.

Reich, P. B., Walters, M. B., and Ellsworth, D. S. (1997). From tropics to tundra: global convergence in plant functioning. *Proceedings of the National Academy of Sciences USA*, 94, 13730–13734.

Reich, P. B., Ellsworth, D. S., and Walters, M. B. (1998a). Leaf structure (specific leaf area) modulates photosynthesis–nitrogen relations: Evidence from within and across species and functional groups. *Functional Ecology*, 12, 948–958.

Reich, P. B., Walters, M. B., Ellsworth, D. S., et al. (1998b). Relationships of leaf dark respiration to leaf nitrogen, specific leaf area and leaf life-span: A test across biomes and functional groups. *Oecologia*, 114, 471–482.

Running, S. W., Loveland, T. R., Pierce, L. L., Nemani, R. R., and Hunt, E. R., Jr. (1995). A remote sensing based vegetation classification logic for global land cover analysis. *Remote Sensing of Environment*, 51, 39–48.

Ryan, M. G. (1991). Effects of climate change on plant respiration. *Ecological Applications*, 1, 157–167.

Ryan, M. G., and Yoder, B. J. (1997). Hydraulic limits to tree height and tree growth. *BioScience*, 47, 235–242.

Ryan, M. G., Phillips, N., and Bond, B. J. (2006). The hydraulic limitation hypothesis revisited. *Plant, Cell and Environment*, 29, 367–381.

Saverimuttu, T., and Westoby, M. (1996). Seedling longevity under deep shade in relation to seed size. *Journal of Ecology*, 84, 681–689.

Schulze, E.-D., Kelliher, F. M., Körner, C., Lloyd, J., and Leuning, R. (1994). Relationships among maximum stomatal conductance, ecosystem surface conductance, carbon assimilation rate, and plant nitrogen nutrition: A global ecology scaling exercise. *Annual Review of Ecology and Systematics*, 25, 629–660.

Shugart, H. H. (1984). *A Theory of Forest Dynamics: The Ecological Implications of Forest Succession Models*. New York: Springer-Verlag.

Shugart, H. H. (1987). Dynamic ecosystem consequences of tree birth and death patterns. *BioScience*, 37, 596–602.

Shugart, H. H. (1998). *Terrestrial Ecosystems in Changing Environments*. Cambridge: Cambridge University Press.

Tilman, D. (1988). *Plant Strategies and the Dynamics and Structure of Plant Communities*. Princeton: Princeton University Press.

Ustin, S. L., and Gamon, J. A. (2010). Remote sensing of plant functional types. *New Phytologist*, 186, 795–816.

Walters, M. B., and Reich, P. B. (2000). Seed size, nitrogen supply, and growth rate affect tree seedling survival in deep shade. *Ecology*, 81, 1887–1901.

Waring, R. H. (1983). Estimating forest growth and efficiency in relation to canopy leaf area. *Advances in Ecological Research*, 13, 327–354.

Waring, R. H. (1991). Responses of evergreen trees to multiple stresses. In *Response of Plants to Multiple Stresses*, ed. H. A. Mooney, W. E. Winner, and E. J. Pell. San Diego: Academic Press, pp. 371–390.

Waring, R. H., and Franklin, J. F. (1979). Evergreen coniferous forests of the Pacific Northwest. *Science*, 204, 1380–1386.

Waring, R. H., Schroeder, P. E., and Oren, R. (1982). Application of the pipe model theory to predict canopy leaf area. *Canadian Journal of Forest Research*, 12, 556–560.

Westoby, M., Falster, D. S., Moles, A. T., Vesk, P. A., and Wright, I. J. (2002). Plant ecological strategies: Some leading dimensions of variation between species. *Annual Review of Ecology and Systematics*, 33, 125–159.

Woodward, F. I. (1987). *Climate and Plant Distribution*. Cambridge: Cambridge University Press.

Woodward, F. I. (1993). Leaf responses to the environment and extrapolation to larger scales. In *Vegetation*

Dynamics and Global Change, ed. A. M. Solomon and H. H. Shugart. New York: Chapman and Hall, pp. 71–100.

Wright, I. J., Reich, P. B., Westoby, M., et al. (2004). The worldwide leaf economics spectrum. *Nature*, 428, 821–827.

Wright, I. J., Reich, P. B., Cornelissen, J. H. C., et al. (2005). Assessing the generality of global leaf trait relationships. *New Phytologist*, 166, 485–496.

Wright, S. J., Kitajima, K., Kraft, N. J. B., et al. (2010). Functional traits and the growth–mortality trade-off in tropical trees. *Ecology*, 91, 3664–3674.

Populations and Communities

19.1 | Chapter Summary

This chapter continues the discussion of the preceding chapter, focusing on the arrangement of individual plants in populations and multiple species in communities. The concept of a niche is central to the understanding of the organization of species across the landscape. A niche represents the components of the environment to which a species is adapted. Evolutionary pressures have lead to niche differentiation in which species differ in preferences for resources and have different functional roles in communities. This is seen in the dispersion of species along resource gradients. Though vegetation can be classified into distinct communities of species, most communities intergrade continuously and exist within a continuum of populations. Species do not group along environmental gradients in distinct natural associations, but rather arrange individualistically according to their own physiology and life history patterns. The manner in which plant populations associate in recognizable communities is critical to understanding the response of vegetation to climate change. A plant community that exists today may have no analog under different climate. Particular species enter and dominate a given locale based on prevailing climate and other environmental conditions, the occurrence of disturbance such as fire, and their own life history patterns.

19.2 | Niche and Species Abundance

The environment is spatially heterogeneous, varying in light, temperature, soil moisture, nutrients, and other conditions. Just as physiological processes vary depending on the specific environmental conditions encountered by a plant, so too do plant species thrive over a specific range of environmental conditions. Biological performance for a particular species is typically optimal at some level of an environmental condition (e.g., soil moisture) and decreases with conditions less than or greater than optimal (Figure 19.1). In two dimensions such as soil moisture and temperature, the environmental space within which a species thrives is represented by an area formed by the intersection of moisture and temperature tolerances. Growth, for example, may be optimal over some narrow range of moisture and temperature and may decrease with conditions that are wetter or drier than optimum or soils that are warmer or colder than optimum. In three dimensions, a multidimensional cube represents the environmental conditions over which a particular species can grow.

Figure 19.2 shows actual tolerances in terms of pollen abundance. The abundance of arboreal pollen in eastern North America varies strongly with temperature and precipitation. Pollen of oak trees is most abundant with annual precipitation of 1000 mm and July temperature in

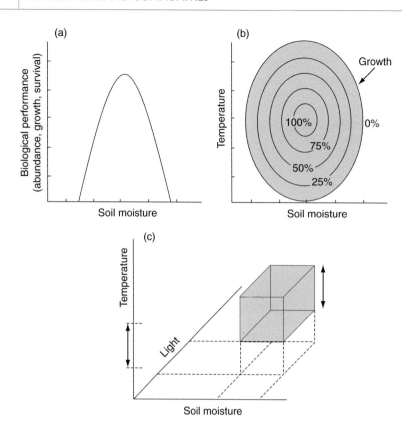

Fig. 19.1 Tolerance of a species in relation to environmental conditions. (a) Typical bell-shaped response to soil moisture. (b) Hypothetical response to soil moisture and temperature showing the range of conditions for which growth can occur and the decrease in growth away from this optimum. (c) Hypothetical range of conditions along three dimensions (soil moisture, temperature, light) over which growth is possible.

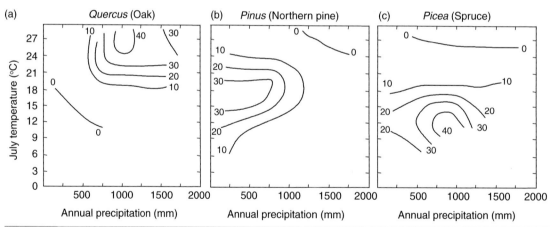

Fig. 19.2 Pollen abundance (%) in relation to July temperature and annual precipitation for (a) oak, (b) northern pine, and (c) spruce trees. Adapted from Webb et al. (1993).

excess of 24°C. Pollen of northern pine trees is most abundant in cooler, drier climates (18°C, < 750 mm). Spruce pollen is greatest in cold, moist climates (12°C, 750–1000 mm). Pollen abundance decreases as temperature and precipitation deviate from these optima.

The niche concept invokes attributes of species, specifically their tolerance of environmental conditions and the way they interact with other species, to understand the spatial distribution of species in a landscape. A niche represents all the components of the environment, both abiotic and biotic, to which a species is adapted. It includes habitat, which describes the environment where the species is found. By having different environmental preferences, multiple species can co-occur in a landscape. For example, one species may thrive on dry, nutrient-poor soil while another might prefer moist, nutrient-rich soil. The niche also includes the functional role of the species in relation to other species, which describes what it does and how it lives. The functional role of species is most obvious in animals. A common example is the size of prey. By differing in the size of preferred prey, two species can avoid direct competition for food. Indeed, the concept of niche originally was formulated for animals. Among plants, functional roles include resource utilization, the timing of biological activity, and the partitioning of light into overstory and understory environments and soil by deep- and shallow-rooted plants. Life form, phenology, and regeneration are additional functional aspects of plant niche (Grubb 1977). For example, the leaves of deciduous understory plants may emerge before those of overstory trees to allow a period of photosynthesis before the canopy closes. Early successional plants grow best in the full sunlight of exposed sites while late successional plants tolerate the low light environment of closed canopies (Table 18.3).

The concept of niche can be seen in the distribution of a species across the landscape. Light, water, nutrients, and temperature are important to plants, and different species have evolved different preferences. In this respect, niche is a multidimensional volume within which the species is likely to be found. It is closely tied to the notion of ecological limits and physiological tolerances. For example, the occurrence of *Eucalyptus pauciflora*, a common alpine eucalypt in southeastern Australia, varies in relation to type of rock (granite or sedimentary), temperature, and precipitation (Austin et al. 1990). On granite rocks, this tree species is most common where annual rainfall is greater than 1400 mm and annual mean temperature ranges from 6°C to 8°C (Figure 19.3). The probability of occurrence decreases in cooler climates and with warmer, dry climates.

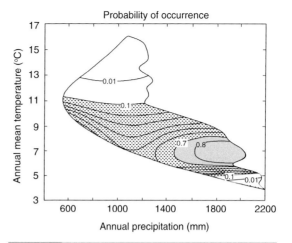

Fig. 19.3 Probability of occurrence of *Eucalyptus pauciflora* (snow gum) growing on granite rock in southeastern Australia in relation to annual mean temperature and annual precipitation. Adapted from Austin et al. (1990).

A niche represents more than just physiological tolerances and adaptations to the physical environment. A niche is shaped by the functional roles of species in communities and how these roles alter the competitive balance among species. In particular, a species that grows in isolation may tolerate a wider range of environmental conditions than when grown in a mixture of other species. Another species, perhaps better adapted to a specific portion of the environmental spectrum, may perform better and exclude the species from that portion of its potential environmental space. The fundamental niche of a species grown in isolation may, therefore, be quite different from the realized niche when grown in the presence of other species. Competition with other species may exclude a species from parts of its fundamental niche.

Computer model simulations of forest dynamics illustrate the way in which physiological tolerances, functional roles, and competition influence species distributions (Smith and Huston 1989). In their study, the authors represented the physiology and growth of trees as a tradeoff between tolerance of shade and tolerance of drought (Figure 19.4). Shade tolerant trees tend to be drought intolerant while shade intolerant trees are tolerant of low soil water. Furthermore, growth rate decreases as tolerance

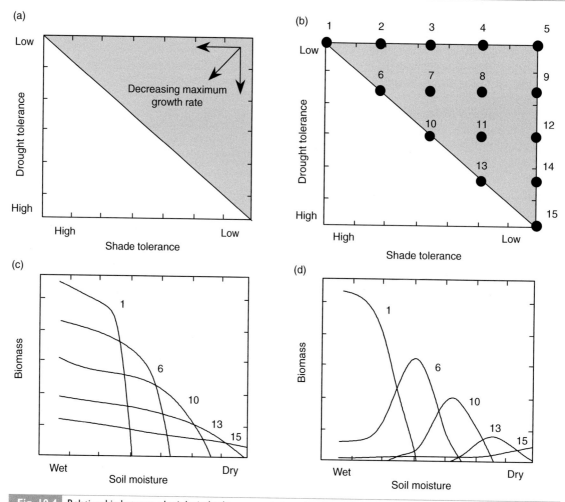

Fig. 19.4 Relationship between physiological tolerances and plant abundance along environmental gradients. Top: Possible plant strategies for light and water use represented as a continuum (a) and as 15 discrete plant functional types (b). The shaded triangle shows possible strategies based on tradeoff between light and water use. Bottom: Biomass of five plant functional types along a moisture gradient when grown in monocultures (c) and in mixed stands of all 15 plant types (d). Numbers denote plant functional types. Adapted from Smith and Huston (1989).

to low light or low soil water increases. The authors defined 15 hypothetical tree species that varied in shade and drought tolerance. When five of these species grow in monocultures (i.e., in the absence of competition with other species), their abundance along a soil moisture gradient reflects their physiological tolerances. All species grow best when soil is wet and the range of soil water over which they grow increases with drought tolerance. When grown in mixtures of all 15 species, the same five species distribute along the moisture gradient following

bell-shaped curves with distinct preferences for water availability. Each species segregates along the moisture gradient in relation to drought tolerance and shade tolerance. On wet soil, though all species can potentially grow, the species most tolerant of shade has the highest biomass. This species declines in biomass with drier soil. Drier soils selectively favor a specific combination of drought and shade tolerance for that site.

These model simulations show that tradeoffs in life history and physiology preclude a single species from always being successful

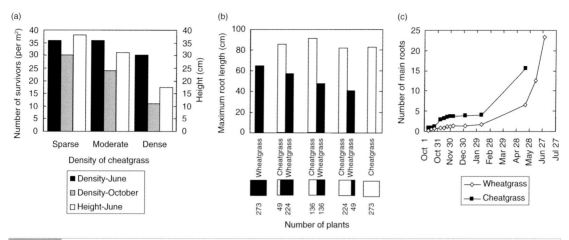

Fig. 19.5 Bluebunch wheatgrass (*Agropyron spicatum*) growth and survival in the presence of cheatgrass (*Bromus tectorum*). (a) Survival and height of wheatgrass under sparse, moderate, and dense amounts of cheatgrass. (b) Maximum root length of both grasses when grown in mixtures of 273 plants of varying amounts of wheatgrass and cheatgrass. (c) Temporal dynamics of root growth for wheatgrass and cheatgrass grown in isolation. Data from Harris (1967).

everywhere along an environmental gradient. Competition displaces each species towards the environmental conditions that it is able to tolerate but which cannot be tolerated by species that are better competitors under optimal conditions. This ecological optimum is closer to the physiological limit than to the physiological optimum. In the simulations shown in Figure 19.4, the biomass of each species is greatest closer to its physiological limit to dry soil. Competitive pressure from shade-tolerant species limits its growth on wetter soils.

A study of the introduction of non-native species provides an example of how physiological tolerances and life history patterns affect the competitive balance among species and thereby alter vegetation composition (Harris 1967). The dominant plant of pristine grasslands in the northern intermountain region of the United States, in the semiarid region between the Rocky Mountains and Cascade Mountains, used to be the native bluebunch wheatgrass (*Agropyron spicatum*). Human activities introduced cheatgrass (*Bromus tectorum*), which has since spread to dominate much of the region. Both species are winter plants. Bluebunch wheatgrass, a perennial, breaks dormancy in autumn and begins leaf growth in September or October when temperatures cool and soil moisture increases. It

grows slowly during winter, with rapid growth in spring after snowmelt. Flowers and seeds form in June and July, with summer dormancy by mid-July. Cheatgrass is a winter annual. Seeds germinate in autumn, and plants maintain slow growth throughout winter. Spring brings strong growth, followed by rapid flowering and seed production and then death by late May.

A study of the ecology of these two species showed that the presence of cheatgrass reduces the growth and survival of bluebunch wheatgrass (Figure 19.5). In one experiment, wheatgrass seeds were sown so that plants grew in different densities of cheatgrass: sparse (0–4 per m²); moderate (15–20 per m²); and dense (90–100 per m²). After sowing in October, survival of wheatgrass seedlings during the following year was inversely related to cheatgrass density. In the following June, there was little effect of cheatgrass density on survival. By October, however, only 39 percent of the plants sown in the dense cheatgrass survived while 86 percent of the plants sown in the sparse cheatgrass survived. The average height of wheatgrass decreased from 38 cm with sparse cheatgrass competition to 18 cm with dense competition. The detrimental effect of cheatgrass on wheatgrass growth was apparent when both species were grown at the same densities

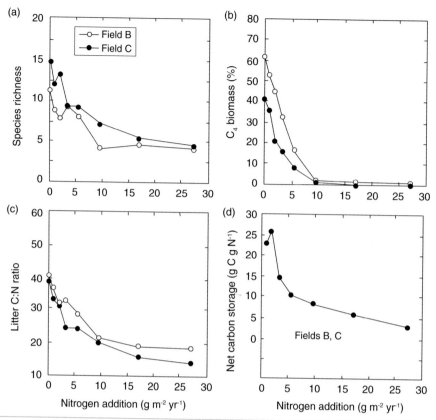

Fig. 19.6 Grassland response to 12 years of nitrogen addition. Data are mean values for several replicate plots in two old fields initially dominated by C₄ grasses. (a) Species richness (number of plants per 0.3 m²). (b) Biomass of C₄ plants as a percentage of total aboveground biomass. (c) C:N ratio of litter. (d) Net carbon storage per unit added nitrogen. Adapted from Wedin and Tilman (1996).

but in different proportions of the two species. Wheatgrass root growth declined with greater proportion of cheatgrass while cheatgrass root growth was unaffected. The cause of reduced wheatgrass growth is apparently the longer roots of cheatgrass, which provides access to soil water at the expense of wheatgrass. Cheatgrass roots grow throughout winter following germination in October. In contrast, wheatgrass roots grow little during winter. Cheatgrass grows and thrives at the expense of wheatgrass because of its winter root growth, which gives it an early growing season advantage over wheatgrass, and its earlier maturation, which depletes soil water.

Alteration of a resource can also change the competitive balance among species. For example, a 12-year study in nitrogen-limited Minnesota grasslands found that nitrogen

fertilization alters community composition, plant productivity, and carbon storage (Tilman 1987; Inouye and Tilman 1995; Wedin and Tilman 1996). In two fields dominated by native C₄ prairie grasses, 12 years of nitrogen addition reduced species richness (the number of species in a given area) by more than 50 percent with high nitrogen fertilization (Figure 19.6). The plant community shifted from dominance by native C₄ grasses to dominance by weedy C₃ grasses. While nitrogen fertilization enhanced net primary production, ecosystem carbon storage decreased with additional nitrogen. The shift in community composition from C₄ to C₃ species decreased the C:N ratio of plant litter. Detritus decomposed more rapidly because of its better chemical quality so that the overall ability to sequester carbon in the soil diminished.

19.3 Environmental Gradients and Communities

Much of the history of ecology has been dominated by a debate about the nature of plant communities and the manner in which plant populations group into distinct associations or communities (McIntosh 1981, 1985; Golley 1993). One school of thought held that communities are emergent units representing a distinct level of ecological organization (Clements 1916, 1928; Odum 1953, 1969, 1971). A second school viewed communities merely as temporally and spatially co-occurring species (Gleason 1917, 1926, 1939). They are not emergent units, but rather consist simply of species that have similar environmental tolerances and preferences.

Numerous studies of vegetation along environmental gradients, particularly in mountains, have shown that the ideas of the second school are correct. Our current understanding of plant communities is one in which floristic composition changes continuously along environmental gradients rather than discretely, resulting in a continuum of populations not a series of distinct plant associations. Species do not group along environmental gradients in distinct natural associations. They distribute individualistically according to their physiology and life history patterns. Communities with distinct floristic composition and physiognomy merely form from the overlap of species distributions,

and vegetation units such as communities are arbitrary products of classification rather than natural units clearly defined in the field. They are not emergent units, but merely comprise plant species that co-occur at a given point in space and time.

This can be seen in the dispersion of plant species along environmental gradients. The vegetation of the Great Smoky Mountains along the Tennessee–North Carolina state border illustrates the typical distribution of vegetation along environmental gradients (Whittaker 1956). In this region, elevation ranges from 460 m along the bottomlands to 2000 m at the summit of the highest peaks. Annual precipitation increases from less than 1500 mm in the lower valleys to more than 2000 mm at high elevations. The abundance of tree species changes markedly with respect to elevation (Figure 19.7). Species distributions in relation to elevation show rounded or bell-shaped curves and overlap broadly, but with distinct population centers distributed along the elevation gradient. Yellow poplar dominates low elevations on mesic sites. With higher elevation, yellow birch, mountain silverbell, sugar maple, white basswood, and yellow buckeye become progressively more dominant. High elevations above 1400 m are almost exclusively American beech forests.

The floristic composition of xeric sites is distinctly different, consisting of oak and pine forests (Figure 19.8). The dominant oaks are blackjack oak, chestnut oak, and scarlet oak.

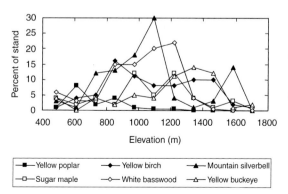
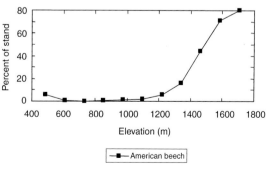

Fig. 19.7 Distribution of tree species, as a percentage of the number of trees in the stand, in relation to elevation on mesic sites in the Great Smoky Mountains circa 1940s and 1950s. Data from Whittaker (1956).

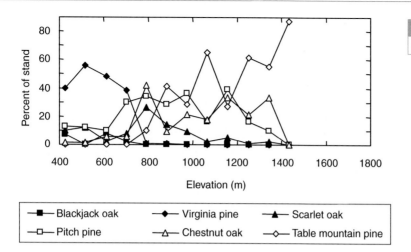

Fig. 19.8 As in Figure 19.7, but for xeric sites.

Virginia pine, pitch pine, and table mountain pine are also abundant. Each species has broad, overlapping population dispersion with respect to elevation, but distinct elevation preferences are evident. Virginia pine is the dominant tree at low elevation. It and blackjack oak grow exclusively at elevations less than 800 m. Pitch pine and table mountain pine dominate intermediate elevations in association with scarlet oak and chestnut oak. Table mountain pine dominates forests at high elevation, with lesser amounts of chestnut oak and pitch pine.

Vegetation can be classified into distinct communities or associations of species based on similar physiognomy and floristic composition. Physiognomy considers the form and structure of communities in terms of the type of plant (e.g., woody, herbaceous) and vertical structure (e.g., overstory, understory). Floristic composition considers the dominant species (e.g., oak–hickory forest). Numerous such communities occur in the Smoky Mountains, segregated along moisture and elevation gradients (Figure 19.9). Vegetation at lower and middle elevations is typically mixed deciduous cove forest and eastern hemlock forest on moist sites. Oak forests, with some hickory, grow on moderately wet slopes at low elevation. A variety of oak forests grow on moderately dry sites, giving way to oak heaths on dry sites and pine forests and heaths on xeric slopes. Red spruce and Fraser fir forests prevail at high elevations, with beech and oak forests on favorable sites.

Patches of grassy balds form on the summits of mountains. Shrub communities dominated by evergreen ericaceous shrubs form heath balds along dry ridges.

Although one or two dominant species define certain communities, several or more species comprise most communities in the Smoky Mountains. This is shown by dominance-diversity curves that graph the relative importance of each species ranked from most important to least important (Figure 19.10). For example, beech forests on sheltered south slopes above 1372 m are a mixture of American beech and lesser amounts of other species. American beech, comprising 81 percent of the trees with a diameter greater than 2.54 cm, is the most dominant. The next most abundant species is mountain silverbell (8% of trees). Yellow birch and yellow buckeye together comprise 5 percent of trees, and various other species occur in minor amounts. Cove forests between 762 and 1067 m have a more diverse community structure. Four species (mountain silverbell, white basswood, sugar maple, yellow birch) share dominance in these forests, accounting for 69 percent of trees. Six other species, with abundances of 2–6 percent, comprise 21 percent of the forest.

Most communities in the Smoky Mountains intergrade continuously and exist within a continuum of populations (Figure 19.11). Some populations are restricted to particular communities. For example, white basswood grows almost exclusively in cove forests. However,

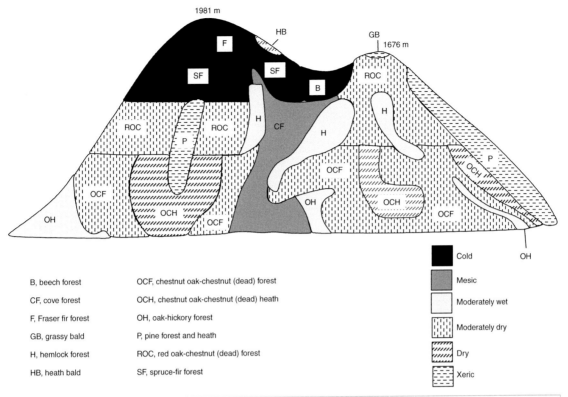

Fig. 19.9 Topographic distribution of vegetation types on a west-facing slope in the Great Smoky Mountains circa 1940s and 1950s. Adapted from Whittaker (1956).

B, beech forest   OCF, chestnut oak-chestnut (dead) forest

CF, cove forest   OCH, chestnut oak-chestnut (dead) heath

F, Fraser fir forest   OH, oak-hickory forest

GB, grassy bald   P, pine forest and heath

H, hemlock forest   ROC, red oak-chestnut (dead) forest

HB, heath bald   SF, spruce-fir forest

Cold

Mesic

Moderately wet

Moderately dry

Dry

Xeric

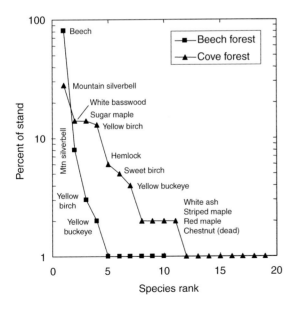

Fig. 19.10 Dominance–diversity curves for beech forests growing on sheltered south slopes above 1372 m and cove forests between 762 and 1067 m in the Great Smoky Mountains circa 1940s and 1950s. Graphs show the relative importance of species, in terms of percentage of trees in the stand, ranked from most important to least important. Data from Whittaker (1956).

chestnut oak and hemlock, while having greatest abundance in chestnut oak and hemlock communities, occur in numerous other communities. Chestnut oak can be found in abundances of 5 percent or more in oak–hickory and pine forests. Hemlock can be found in abundances of 10 percent or more in cove forests and oak–hickory forests. Other species such as red maple never form a distinct community type, but rather intermingle with other species throughout many communities.

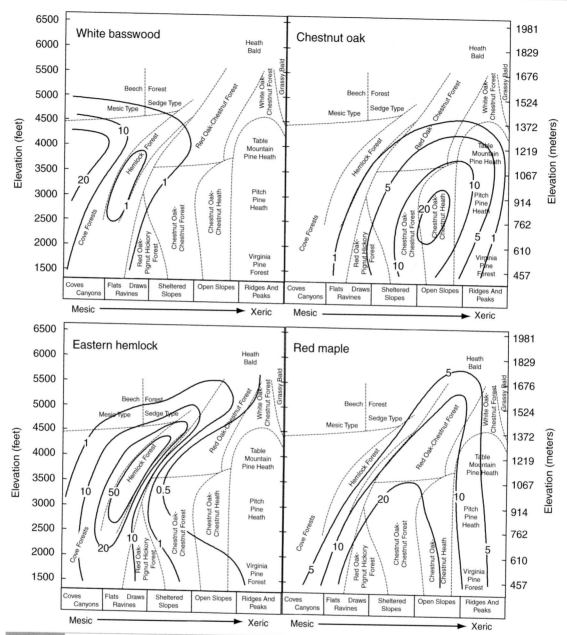

Fig. 19.11 Population distributions for white basswood, chestnut oak, eastern hemlock, and red maple in relation to elevation and moisture in the Great Smoky Mountains circa 1940s and 1950s. Contour lines show species abundance in terms of percentage of trees. Major vegetation types are delimitated. Adapted from Whittaker (1956).

19.4 | Plants in a Changing Environment

The manner in which plant populations distribute across the landscape into recognizable communities is central to understand the functioning of vegetation and its response to changing climate. A plant community that exists today may have no past analog under different climate. Rather, different species enter and dominate a particular site based on prevailing climate and other environmental conditions, the occurrence of disturbance such as fire, and their own life history patterns. At a broad scale, distinct biomes such as evergreen and deciduous forest, each with tropical, temperate, and boreal variants, can be recognized. While the biogeography of these biomes broadly correlates with climate characteristics such as temperature and precipitation (Chapter 24), understanding their response to climate change requires knowledge of how the individual species, or broad categories of plant functional types, within these biomes respond to a changing environment.

In addition to floristic composition, vegetation has other characteristics such as height, leaf area index, and biomass. Although general rules may describe the functional relationship of these characteristics with temperature and precipitation across broad climatic gradients, these, too, represent the sum of the individual plants interacting among themselves to acquire the resources necessary for growth and survival. Functional relationships derived for the prevailing climate may not be applicable under different climates if vegetation structure and composition changes.

A highly successful class of forest dynamics models simulates vegetation structure and community composition based on individual trees (Botkin et al. 1972; Shugart and West 1977; Shugart 1984; Botkin 1993). Community and ecosystem attributes are not modeled per se, but rather reflect the interactions among numerous individual trees and the interactions among species in the acquisition of light, soil water, nutrients, and other resources. Species respond individualistically to climate, disturbances that create a gap in the canopy, and site conditions such as nutrient availability, soil moisture, and soil temperature based on their specific preferences and tolerances.

19.5 | Review Questions

1. The following table shows the height of tree seedlings in relation to soil water. (a) What are the ecological limits to growth? (b) What are the preferred conditions for growth? (c) When grown with another species that is better adapted to wet soils, height growth is altered. What does this illustrate?

Volumetric soil water	Height (cm)	
	Monoculture	Mixture
0.05	0	0
0.10	0	0
0.15	3	3
0.20	7	6
0.25	10	5
0.30	16	3

Volumetric soil water	Height (cm)	
	Monoculture	Mixture
0.35	13	1
0.40	11	0
0.45	10	0

2. In an arid environment, one plant species has short roots in the upper 10 cm of soil. Another species has deeper roots that extend below 50 cm. Which species is likely to dominate? Why?

3. The following table shows stand composition (percentage of trees) for three sites in the Great Smoky Mountains. Graph the dominance–diversity curve for each stand and describe differences among stands.

Mesic, 610 meters		Xeric, 610 meters		Xeric, 1432 meters	
Species	%	Species	%	Species	%
Hemlock	27	Hemlock	1	Serviceberry	7
Mountain silverbell	1	Red maple	6	Witch hazel	3
White ash	1	White pine	10	Red maple	1
White basswood	3	Black gum	2	Black locust	1
Yellow poplar	8	Sourwood	9	Black gum	1
Yellow birch	4	Chestnut oak	2	Table mountain pine	87
Sugar maple	2	Blackjack oak	8		
Mountain magnolia	5	Scarlet oak	2		
Cucumber tree	2	Virginia pine	48		
American hornbeam	1	Pitch pine	10		
Kentucky yellowwood	1				
Dogwood	6				
Red oak	2				
Sweet birch	10				
Red maple	15				
Mountain holly	1				
Chestnut oak	1				
Black gum	2				
White oak	2				
Sourwood	4				
White pine	2				

4. The following table shows stand composition (percentage of trees) in relation to elevation in the Great Smoky Mountains. What does this data indicate about the ecological organization of plant communities?

Species	Elevation (meters)									
	488	610	731	853	975	1097	1219	1341	1463	1585
Red maple	20	25	31	26	27	13	13	10	7	3
Red oak	1	3	3	3	2	13	15	23	20	31
Sourwood	15	10	18	9	14	16	15	3	0	0
Chestnut oak	24	18	14	18	17	15	19	0	0	0

5. A particular species of woody shrub survives in an alpine environment despite slow growth rate because burial by snow protects the shrub from cold winter temperatures. Faster growing woody plant species that cannot tolerate cold temperatures are found at lower elevations. Describe what might happen to the alpine community as climate warms.

6. The boreal forest around Fairbanks, Alaska, consists of slow-growing black spruce growing on cold, nutrient-poor soils underlain with permafrost and fast-growing aspen and birch growing on warmer, nutrient-rich soils. How might the composition of this landscape change with a warmer climate?

19.6 | References

Austin, M. P., Nicholls, A. O., and Margules, C. R. (1990). Measurement of the realized qualitative niche: Environmental niches of five Eucalyptus species. *Ecological Monographs*, 60, 161–177.

Botkin, D. B. (1993). *Forest Dynamics: An Ecological Model.* Oxford: Oxford University Press.

Botkin, D. B., Janak, J. F., and Wallis, J. R. (1972). Some ecological consequences of a computer model of forest growth. *Journal of Ecology*, 60, 849–872.

Clements, F. E. (1916). *Plant Succession: An Analysis of the Development of Vegetation*, Carnegie Institution Publication Number 242. Washington, D.C.: Carnegie Institution.

Clements, F. E. (1928). *Plant Succession and Indicators.* New York: H.W. Wilson.

Gleason, H. A. (1917). The structure and development of the plant association. *Bulletin of the Torrey Botanical Club*, 44, 463–481.

Gleason, H. A. (1926). The individualistic concept of the plant association. *Bulletin of the Torrey Botanical Club*, 53, 7–26.

Gleason, H. A. (1939). The individualistic concept of the plant association. *American Midland Naturalist*, 21, 92–110.

Golley, F. B. (1993). *A History of the Ecosystem Concept in Ecology: More than the Sum of the Parts.* New Haven: Yale University Press.

Grubb, P. J. (1977). The maintenance of species richness in plant communities: The importance of the regeneration niche. *Biological Review*, 52, 107–145.

Harris, G. A. (1967). Some competitive relationships between *Agropyron spicatum* and *Bromus tectorum*. *Ecological Monographs*, 37, 89–111.

Inouye, R. S., and Tilman, D. (1995). Convergence and divergence of old-field vegetation after 11 yr of nitrogen addition. *Ecology*, 76, 1872–1887.

McIntosh, R. P. (1981). Succession and ecological theory. In *Forest Succession: Concepts and Application*, ed. D. C. West, H. H. Shugart, and D. B. Botkin. New York: Springer-Verlag, pp. 10–23.

McIntosh, R. P. (1985). *The Background of Ecology: Concept and Theory.* Cambridge: Cambridge University Press.

Odum, E. P. (1953). *Fundamentals of Ecology.* Philadelphia: Saunders.

Odum, E. P. (1969). The strategy of ecosystem development. *Science*, 164, 262–270.

Odum, E. P. (1971). *Fundamentals of Ecology*, 3rd ed. Philadelphia: Saunders.

Shugart, H. H. (1984). *A Theory of Forest Dynamics: The Ecological Implications of Forest Succession Models.* New York: Springer-Verlag.

Shugart, H. H., and West, D. C. (1977). Development of an Appalachian deciduous forest succession model and its application to assessment of the impact of the chestnut blight. *Journal of Environmental Management*, 5, 161–179.

Smith, T., and Huston, M. (1989). A theory of the spatial and temporal dynamics of plant communities. *Vegetatio*, 83, 49–69.

Tilman, D. (1987). Secondary succession and the pattern of plant dominance along experimental nitrogen gradients. *Ecological Monographs*, 57, 189–214.

Webb, T., III, Bartlein, P. J., Harrison, S. P., and Anderson, K. H. (1993). Vegetation, lake levels, and climate in eastern North America for the past 18,000 years. In *Global Climates since the Last Glacial Maximum*, ed. H. E. Wright, Jr., J. E. Kutzbach, T. Webb, III, et al. Minneapolis: University of Minnesota Press, pp. 415–467.

Wedin, D. A., and Tilman, D. (1996). Influence of nitrogen loading and species composition on the carbon balance of grasslands. *Science*, 274, 1720–1723.

Whittaker, R. H. (1956). Vegetation of the Great Smoky Mountains. *Ecological Monographs*, 26, 1–80.

Ecosystems

20.1 | Chapter Summary

This chapter extends the discussion of the preceding chapter from that of populations and communities to their functioning as ecosystems. Plants interact with one another and with soil resources as an ecosystem. The soil matrix provides water, nutrients, and other resources required for growth and survival. The availability of these resources to sustain plant growth is modulated by biological activity from plants themselves and also from microorganisms in the soil. The concept of an ecosystem embodies the interrelationships between the physical and biological environments. A terrestrial ecosystem combines living organisms and their physical environment into a functional system linked through a variety of biological, chemical, and physical processes. The structure of an ecosystem is measured by the amount of materials such as carbon and nitrogen and their distribution among living, decaying, and inorganic components. The functioning of an ecosystem is measured by processes such as photosynthesis, respiration, evapotranspiration, and elemental cycling. Plant productivity and nutrient cycling are tightly linked. High nutrient availability leads to high nutrient uptake during plant growth and high net primary production so that more nutrients return to the soil in litterfall. The good quality of the litter allows for rapid decomposition and mineralization, which reinforces the high nutrient availability. Low nutrient availability has the opposite effect. Ecosystem experiments measure responses to CO_2 enrichment, warming, and nitrogen addition and provide insight to the ecological consequences of global environmental change.

20.2 | The Ecosystem Concept

The debate within ecology of the nature of plant communities is tied to the emergence of ecosystems as a unit of study and ecosystem ecology as a field of specialization. A terrestrial ecosystem is the sum of individual organisms interacting with their neighbors to acquire the resources essential for growth and development and interacting with the physical environment to alter resource availability and the characteristics of the environment. It combines all living organisms (plants, animals, and microbes) and their physical environment into a functional system linked through biological, chemical, and physical processes. It includes climate, living and decomposing material, soil, and the circulation of energy and materials that link them.

Tansley (1935) first coined the term ecosystem, but it took many years to articulate what an ecosystem is and how to study it (McIntosh 1985; Golley 1993). Over time, ecosystem studies became organized around the cycling of carbon and nutrients (Lindeman 1942; Odum 1953, 1969, 1971; Bormann and Likens 1967; Likens et al. 1977). In this view, the structure of an ecosystem is measured by the amount of materials such as carbon and nitrogen and their distribution among living, decaying, and inorganic components. The functioning of an ecosystem is measured by its elemental cycling. These functions link the biotic and abiotic environment into a dynamic system. Like communities, much of the history of ecology has been a debate about whether ecosystems are superorganisms with emergent properties (Clements 1916, 1928; Odum 1953, 1969, 1971) or are merely the sum of its individual organisms interacting with each other and the environment (Gleason 1917, 1926, 1939).

The spatial extent of an ecosystem depends on the particular research question posed. The largest ecosystem is the planet, where the biosphere (all living organisms), hydrosphere (water), atmosphere, pedosphere (soil), and lithosphere (rock) interact to regulate temperature, precipitation, atmospheric CO_2, biodiversity, water quality, and other aspects of Earth's environment. In contrast, a decomposing log on the forest floor is a small ecosystem characterized by microbial organisms and other living creatures that inhabit the log and by flows of materials into and out of the log. Field studies of terrestrial ecosystems are typically conducted at the scale of watersheds and stands. Watershed studies allow for measurements of material flows out of a specified spatial extent (i.e., the watershed) in streamflow. A less precise scale of study is that of a stand. A stand is a relatively homogenous area of similar floristic composition, vegetation structure, soils, and microclimate such that it can be treated as a single unit of study. A typical spatial scale is 5–10 hectares (ha) (1 ha = 10,000 m²).

Vegetation data collected as part of the Hubbard Brook ecosystem study illustrate the different population, community, and ecosystem depictions of the same forest. The Hubbard Brook study has been a leading innovator in ecosystem ecology (Likens et al. 1977, 1994, 1998; Bormann and Likens 1979; Likens and Bormann 1995; Likens 2004; Fahey et al. 2005). The study site, located in the White Mountain National Forest in central New Hampshire, is part of the extensive northern hardwood forest ecosystem. The climate is cool temperate, humid continental, with long, cold winters and short, cool summers. Prior to logging between about 1909 and 1917, the vegetation is thought to have been a mature, old-age forest. Figure 20.1 shows different depictions of vegetation in the late 1950s and early 1960s in a 13.23 ha watershed with elevations ranging from 546 to 791 m. The mountains generally face towards the southeast with a slope of 21–23 percent. The soil is predominantly sandy loam.

The forest can be described by its community composition (Figure 20.1a). Three species dominate the forest: sugar maple comprises 35 percent of large trees; American beech, 27 percent; and yellow birch, 23 percent. Other less abundant species include paper birch, red spruce, balsam fir, pin cherry, and striped maple. The population size structure observed in the 1950s and 1960s indicates an all-aged forest with relatively few large, old trees and an abundance of seedlings (Figure 20.1b). The largest trees are remnants from before logging while the moderate size trees (11–20 cm diameter) regenerated during and following logging. Continued reproduction provides numerous seedlings and saplings to replace older trees as they die. The floristic composition of the forest changes with elevation (Figure 20.1c). Sugar maple, beech, and yellow birch are the most abundant trees at all elevations. Paper birch, red spruce, and balsam fir are a minor component of low and middle elevation slopes, but are more abundant on upper slopes.

Ecological studies conducted in the 1950s and 1960s found the Hubbard Brook forest ecosystem contains 16,108 g m⁻² of living plant biomass (Figure 20.1d). (Up to 80–90 percent of fresh biomass is water. Biomass is reported in terms of dry weight. Some studies report biomass in terms of carbon. The carbon content of biomass is typically 45 percent of dry weight. In this chapter, units of g m⁻² refer to dry biomass while units of g C m⁻² refer to the carbon in biomass.) Most of this biomass (82%) is contained in aboveground plant material (leaves, branches, stems); only 18 percent is belowground in roots. Annually, plants gain 1002 g m⁻² yr⁻¹ of biomass in net primary production, representing an uptake of 451 g C m⁻² yr⁻¹ from

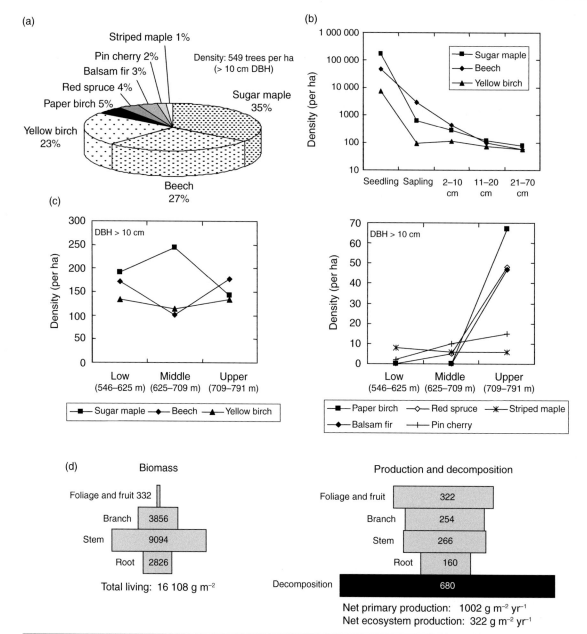

Fig. 20.1 Population structure, community composition, and ecosystem structure and function in a small watershed of the Hubbard Brook Experimental Forest circa 1956–1965. (a) Community composition in terms of the abundance of trees with a diameter at breast height (DBH) greater than 10 cm. (b) Size structure of sugar maple, beech, and yellow birch populations. Size classes are seedlings, saplings, and trees with DBH of 2–10 cm, 11–20 cm, and 21–70 cm. (c) Population density with respect to elevation (lower, middle, and upper third of slopes). Graphs show the density of trees with DBH > 10 cm for sugar maple, beech, and yellow birch (left panel) and other species (right panel). (d) Biomass distribution for foliage and fruit, branches, stems, and roots. The figure on the left shows biomass (g m^{-2}). The figure on the right shows net primary production and decomposition (g m^{-2} yr^{-1}). Boxes are proportional in size to pools and fluxes. Data from Bormann et al. (1970) and Whittaker et al. (1974). See Fahey et al. (2005) for updated carbon pools and fluxes.

the atmosphere. Not all of this carbon remains in the forest. Leaves, twigs, and woody debris fall to the ground as litter. This litter decomposes, releasing carbon back to the atmosphere. Annually, 680 g m^{-2} yr^{-1} decomposes, returning 306 g C m^{-2} yr^{-1} to the atmosphere. The difference between carbon gain during net primary production and carbon loss during decomposition, known as net ecosystem production, is 322 g m^{-2} yr^{-1} so that the ecosystem stores 145 g C m^{-2} yr^{-1}.

More recent studies of Hubbard Brook in the late 1990s show the forest gained carbon since the earlier surveys (Fahey et al. 2005). Total tree biomass in the late 1990s was 26,680 g m^{-2}, a 66 percent increase from the 1950s and 1960s. The forest floor contained 6600 g m^{-2}. These pools were near steady state, and aboveground net primary production declined by 24 percent compared with earlier surveys.

20.3 | Ecosystem Structure and Function

The emphasis in ecosystem studies on the standing stock and cycling of materials provides a means to compare diverse ecosystems such as deserts, grasslands, and forests in terms of common processes, their rates, and factors controlling these rates. Figure 20.2 depicts the flow of energy in the form of carbon among plants, litter, and soil organic matter. Some of the carbon input to the system returns to the atmosphere during plant respiration. The remainder grows new leaf, wood, and root biomass. Some of this biomass turns over and falls to the ground as litter. These litter pools decompose at various rates determined by environmental conditions (e.g., soil temperature and soil moisture) and litter type. The decomposing material becomes soil organic matter, with different turnover rates based on chemical quality and environmental conditions. A fraction of the decomposing carbon returns to the atmosphere as respiration. This representation of an ecosystem is independent of the type of ecosystem. Only the amount of carbon, transfer rates, and

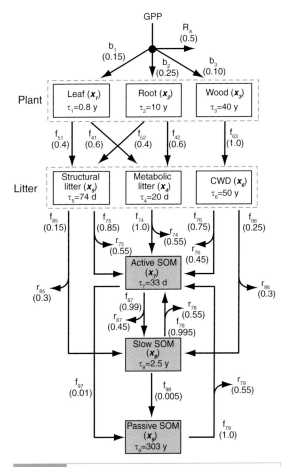

Fig. 20.2 Generalized representation of a terrestrial ecosystem as nine carbon pools and the carbon flows that connect these components. Plant carbon mass consists of leaf (x_1), root (x_2), and wood (x_3). Photosynthesis is allocated to plant material in proportion to b_i. Autotrophic respiration (R_A) is the remainder. Plant residue becomes metabolic litter (x_4), structural litter (x_5), or coarse woody debris (x_6). These pools decompose into active (x_7), slow (x_8), and passive (x_9) soil organic matter. The pools differ in turnover time (τ_i). Lines indicate carbon pathways, with f_{ij} the fraction of the total carbon loss represented by a pathway from pool j to pool i. Curved arrows denote heterotrophic respiration fluxes (R_H) for each pathway, with r_{ij} the respiration fraction. Shown are representative parameter values, with turnover time ranging from days (d) to years (y).

the factors controlling these rates vary among ecosystems.

The net carbon stored within a terrestrial ecosystem is to first order the difference between carbon uptake during photosynthesis

(a) Carbon balance

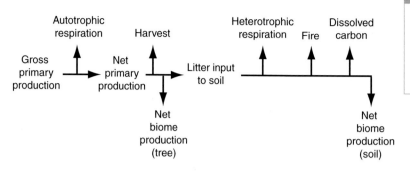

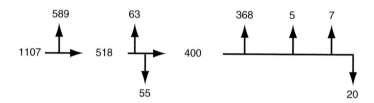

Fig. 20.3 Annual carbon balance of European forests. Fluxes are g C m^{-2} yr^{-1} and include losses from harvest, fire, and dissolved organic and inorganic carbon. Data from Schulze et al. (2009, 2010).

(b) Forests

and carbon loss during respiration. This carbon balance is termed net ecosystem production (*NEP*; Chapin et al. 2006b):

$$NEP = GPP - R_E = (GPP - R_A) - R_H = NPP - R_H \quad (20.1)$$

Here, a positive flux indicates net carbon uptake. Gross primary production (*GPP*) is the carbon uptake during photosynthesis. Some of this carbon is used in plant metabolic processes and respired back to the atmosphere. Autotrophic respiration (R_A) is the carbon loss during growth and maintenance respiration by plants. The difference is net primary production (*NPP = GPP – R_A*). This is the net carbon uptake by plants that can be used to grow new biomass (leaves, wood, roots). Measurements in grassland and forest ecosystems suggest total plant respiration is about 50 percent of gross primary production (Ryan 1991). Heterotrophic respiration (R_H) is the loss of carbon by soil microorganisms as they decompose organic debris. Together, autotrophic and heterotrophic respiration comprise total ecosystem respiration ($R_E = R_A + R_H$). The net land–atmosphere carbon flux is also frequently referred to as net ecosystem exchange (*NEE = R_E – GPP = –NEP*), positive for a source of

carbon to the atmosphere and negative for carbon uptake by ecosystems.

Carbon is also lost from an ecosystem by non-respiratory processes such as disturbance (harvesting, wildfire), herbivory, leaching, and emission of trace gases (biogenic volatile organic compounds, methane, carbon monoxide). The term net ecosystem carbon balance accounts for these additional fluxes and is the net carbon accumulation at the ecosystem scale. Net biome production is the extrapolation of net ecosystem carbon balance to large spatial scales. Carbon loss from fire and harvesting is typically minor at the scale of individual stands or ecosystems due to the low frequency of disturbance at an individual site. However, such losses can be considerably larger when aggregated across a biome and contribute greatly to net biome production. Figure 20.3 illustrates this balance for European forests. Less than 10 percent (75 g C m^{-2} yr^{-1}) of the annual gross primary production (1107 g C m^{-2} yr^{-1}) accumulates in the forests.

Luyssaert et al. (2007) estimated annual carbon fluxes in tropical, temperate, and boreal forest ecosystems based on micrometeorological flux measurements, biometric measurements,

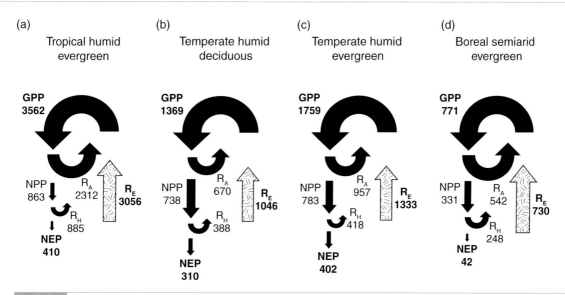

(a) Tropical humid evergreen

(b) Temperate humid deciduous

(c) Temperate humid evergreen

(d) Boreal semiarid evergreen

Fig. 20.4 Annual carbon fluxes (g C m^{-2} yr^{-1}) for (a) tropical humid evergreen, (b) temperate humid deciduous, (c) temperate humid evergreen, and (d) boreal semiarid evergreen forests. These carbon fluxes do not balance. The size of the arrows is proportional to the fluxes within each forest. Data from Luyssaert et al. (2007).

and inventory estimates (Figure 20.4). Annual gross primary production varies from more than 3500 g C m^{-2} yr^{-1} in tropical forests, to about 1400–1800 g C m^{-2} yr^{-1} in temperate forests, and about 800 g C m^{-2} yr^{-1} in boreal forests. Autotrophic respiration is 49–70 percent of gross primary production. The percentage of gross primary production returned to the atmosphere as autotrophic respiration is lowest in temperate forests and increases in warmer (tropical) and colder (boreal) climates (Piao et al. 2010). Net ecosystem production is 42–410 g C m^{-2} yr^{-1}, or 5–23 percent of gross primary production.

Figure 20.5 shows the structure of quaking aspen, black spruce, and jack pine forest ecosystems in the Prince Albert National Park in central Saskatchewan, Canada. All are even aged, with the black spruce forest considerably older. The trees are tallest and have the largest average diameter in the aspen forest, and deciduous shrubs form a continuous understory. The soil is moderately drained loam. The jack pine forest is shorter, and the trees are smaller in diameter. Reindeer lichen covers the ground in this forest. The soil is well-drained coarse sand. The black spruce forest has the smallest trees, but

their density is five to six times greater than the other forests. Feathermoss is the dominant ground cover, and there is a shrub understory. The soil is a 20–30 cm thick layer of peat over poorly drained mineral soil.

The three forests differ in total ecosystem carbon (excluding roots) and in their distribution of this carbon. The black spruce forest contains 3 times more total ecosystem carbon than the aspen forest and 6 times more than the jack pine forest. Woody material (stems and branches) is the largest living carbon pool in each forest, comprising 87 percent of living carbon in the black spruce and jack pine forests and 98 percent of living carbon in the aspen forest. The aspen forest, with the tallest and largest trees, has the most wood. The greatest difference among these forests, though, is in organic carbon on the forest floor and in soil. The black spruce forest, with its thick peat soil, contains 7 and 14 times more soil organic carbon than in the forest floor and soil of the aspen forest and jack pine forest, respectively. Soil organic carbon accounts for 88 percent of total ecosystem carbon in the black spruce forest, but only 35 percent in the aspen forest and 42 percent in the jack pine forest.

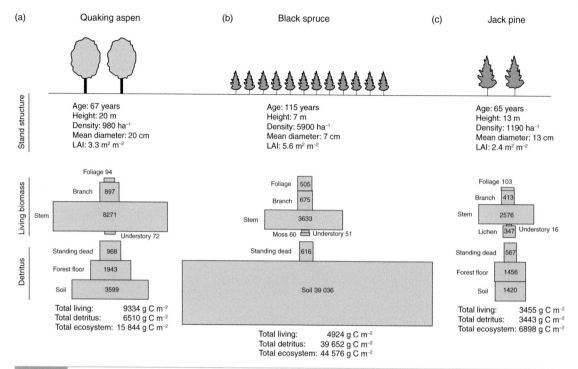

Fig. 20.5 Stand structure and carbon storage for quaking aspen, black spruce, and jack pine forests in central Saskatchewan. Living biomass is aboveground only. Standing dead, forest floor, and mineral soil carbon comprise total detritus (decaying material). Boxes are proportional in size to carbon pools and have units of g C m⁻². Data from Gower et al. (1997).

20.4 | Environmental Controls of Net Primary Production

Stand age, species composition, and site conditions such as temperature, soil moisture, and nutrient availability influence net primary production. Locally, changes in elevation and slope, by altering temperature and moisture, strongly affect growth. In general, annual tree production decreases with higher elevation due to colder temperatures (Figure 20.6). Experiments that artificially manipulate nutrients and water routinely demonstrate the importance of these resources for net primary production. In one such study, portions of a 50-year-old Douglas fir forest in the Rocky Mountains of New Mexico were for two years either irrigated weekly throughout the growing season so that precipitation was effectively doubled or fertilized once in spring with nitrogen and other nutrients (Gower et al. 1992). Production measurements prior to treatment and in untreated control plots provided baseline comparisons in the absence of water and nutrient addition. Over the two years of study, aboveground net primary production increased by 13 percent of the pretreatment value in the control untreated plots (Figure 20.7). Application of water and nutrients increased production compared with this baseline increase. Aboveground net primary production increased by 57 percent in the irrigated plots and 70 percent in the fertilized plots compared with pretreatment values. This increase was realized in more new foliage biomass. In addition, the proportion of total net primary production that was belowground in roots decreased from 46 percent in the untreated control to 31 percent in the irrigated plots and 23 percent in the fertilized plots.

Stand structure also influences net primary production. Carbon uptake during gross primary production increases with greater leaf area index. However, light absorption saturates

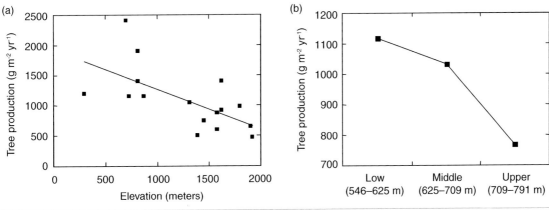

Fig. 20.6 Aboveground tree production in relation to elevation for (a) the Great Smoky Mountains circa mid-1960s and (b) the Hubbard Brook Experimental Forest circa 1956–65. Data from Whittaker (1966) and Whittaker et al. (1974).

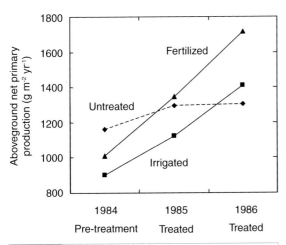

Fig. 20.7 Effect of fertilizer and irrigation on aboveground net primary production of Douglas fir over three years. Data from Gower et al. (1992).

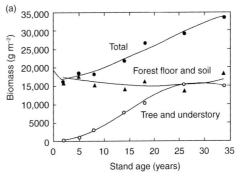

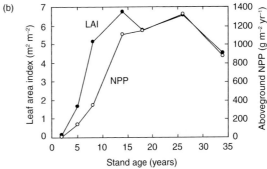

Fig. 20.8 Development following clear-cutting of commercial slash pine plantations near Gainesville, Florida. (a) Accumulation of biomass in vegetation, forest floor, and soil. (b) Aboveground net primary production (NPP) and leaf area index (LAI) for the tree stratum. Adapted from Gholz and Fisher (1982).

at high leaf area index so that gross primary production also saturates at high leaf area index. Forests typically reach peak growth relatively early in stand development, followed by age-dependent decline in productivity after canopy closure as the forest matures (He et al. 2012). Declining net primary productivity with older stand age is seen at Hubbard Brook (Fahey et al. 2005). Figure 20.8 illustrates this decline in the growth of commercial stands of slash pine in southeastern United States. Maximum leaf area index is attained in 14-year-old stands. Aboveground net primary production increases to maximum at 26 years, after which

it declines. This pattern of declining productivity is common in forests, possibly as a result of increasing respiration losses as woody biomass accumulates or decreased photosynthetic

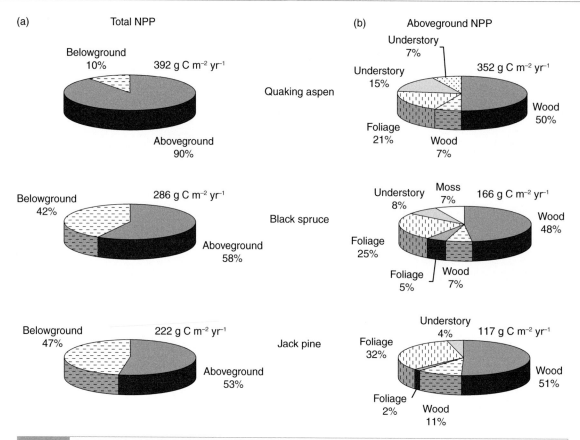

Fig. 20.9 Annual productivity for quaking aspen, black spruce, and jack pine forests in central Saskatchewan. (a) Net primary production and its allocation into aboveground and belowground production. (b) Aboveground net primary production and its allocation into biomass increment (shaded) and litterfall (dashed). Data from Gower et al. (1997) and Steele et al. (1997).

capacity, though the actual causes are uncertain (Ryan et al. 1997, 2004).

Site conditions, species composition, and stand structure can create large differences among forests in gross primary production, respiration, and carbon allocation to above- and belowground production. Figure 20.9 shows net primary production for the three boreal forests described in Figure 20.5. Total net primary production in these forests ranges from 222 to 392 g C m^{-2} yr^{-1}, with greater production in the aspen stand than for black spruce or jack pine. The aspen stand allocates less net primary production to roots (10%) than black spruce (42%) or jack pine (47%). This, combined with its greater total production, gives the aspen forest an aboveground tree production (352 g C m^{-2} yr^{-1}) that is two to three times that of the other

forests. Greater allocation of carbon to roots by needleleaf evergreen trees than by broadleaf deciduous trees is a general pattern found throughout the boreal forest.

A detailed study of carbon allocation in six needleleaf evergreen forests along a west-to-east transect in Oregon beginning at the Pacific coast and extending 225 km inland illustrates the effect of site conditions on net primary production (Figure 20.10). The Coast Range and Cascade Mountains greatly influence climate in this region. Elevation across the transect ranges from 170 to 1460 m, with annual mean temperature of 6.0–11.2°C. Annual precipitation ranges from 2510 mm in the west to 220 mm in the east. The westernmost site near the coast has a maritime climate, with cool temperatures and 2510 mm annual precipitation. Tall, productive

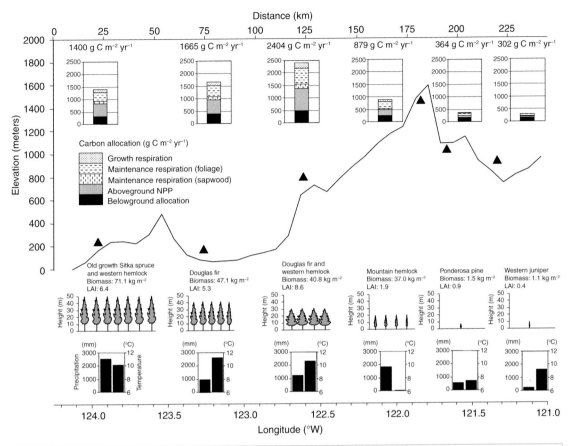

Fig. 20.10 Climate, stand structure, and carbon allocation for six forests along a west-to-east transect in Oregon between latitudes 44°N and 45°N beginning at the coast and extending 225 km inland. The elevation profile is the approximate elevation based on a 5-minute dataset at latitude 44°33.25′ N. Carbon allocation sums to gross primary production, given at the top of each bar chart. Data from Runyon et al. (1994) and Williams et al. (1997).

Sitka spruce and western hemlock trees dominate the forest. Annual precipitation decreases markedly at the inland sites, but Douglas fir and western hemlock still form tall, productive forests. Snow is rare in the coast and valley sites, but deep in the subalpine forest. Here, mountain hemlock forms an open canopy forest of moderate productivity. Drought is common at the ponderosa pine and juniper woodlands east of the Cascade Mountains. The easternmost site is located in the desert interior region created by the rain shadow of the Cascades. At this site, annual precipitation is only 220 mm, and the climate is continental, with hot, dry summers and cold winters. Both sites have widely spaced trees with low productivity. Across the transect, leaf area index ranges from 0.4 to 8.6 m^2 m^{-2}; gross

primary production ranges from 302 to 2404 g C m^{-2} yr^{-1}. The tallest forests with the densest canopy and highest production grow to the west of the Cascade Mountains. Gross primary production decreases and a greater portion of this is allocated to roots in the harsher alpine forest and arid woodlands east of the Cascades.

Not all of net primary production results in biomass increment. Foliage, twigs, branches, and bark fall to the ground as litterfall. Fine roots continually die, providing a large source of organic carbon to the soil. Trees die and create coarse woody debris as stems topple over. Figure 20.9 gives an indication of how much of net primary production is litterfall in boreal forests. Total detritus production ranges from 32 percent to 43 percent of aboveground net

Table 20.1 | Standing stocks of nutrients in a 55-year-old northern hardwood forest in the Hubbard Brook Experimental Forest

Pool	Nutrient (g m^{-2})						
	Ca	Mg	Na	K	N	S	P
Aboveground biomass	38.3	3.6	0.16	15.5	35.1	4.2	3.4
Belowground biomass	10.1	1.3	0.38	6.3	18.1	1.7	5.3
Forest floor	37.2	3.8	0.36	6.6	125.6	12.4	7.8

Source: From Likens et al. (1977, p.101).

primary production. Foliage litterfall is high in all three forests. In deciduous forests, leaf biomass turns over annually with leaf abscission in autumn. Evergreen trees also lose foliage, not all at once but slowly over several years. Needle longevity is typically a few years in pines and several years or more in spruces.

Coarse woody debris in the form of standing dead trees, downed boles, and large branches can be a large component of forest ecosystems. This debris is produced when wind uproots and snaps trees, branches break, and when fire, insects, disease, and slow growth kill trees. In the quaking aspen, black spruce, and jack pine forests of Saskatchewan (Figure 20.9), coarse woody debris production is 7–11 percent of aboveground net primary production, or about 12–25 g C m^{-2} yr^{-1}. Logs and large branches decay slowly so that the accumulated mass of coarse woody debris can be large, 1–4 kg m^{-2} in deciduous forests and 10–51 kg m^{-2} in coniferous forests (Harmon et al. 1986).

20.5 | Biogeochemical Cycles

The carbon cycle depends on parallel flows of nutrients that are absorbed by plants during growth, returned to the soil in litterfall, and released to the soil solution during decomposition. Elements such as nitrogen (N), phosphorus (P), potassium (K), calcium (Ca), magnesium (Mg), sulfur (S), and other micronutrients are essential for plant growth and development (Table 21.3). These elements reside within ecosystems in plant biomass and decaying organic material and are recycled between soil and living organisms (Table 20.1). Litterfall and coarse woody debris contribute organic matter and nutrients to the soil. This debris decays from the activities of bacteria, fungi, earthworms, and other soil microorganisms. The decomposition of organic material mineralizes nutrients bound in the organic matter to be used again in plant growth. Slow rates of decomposition result in accumulation of nutrients bound in organic debris that are unavailable for plant growth. Some nutrients are also released in the chemical weathering of rocks. Rainfall also imports many elements into an ecosystem in small quantities. Other elements, especially nitrogen, are also obtained from gases in the atmosphere.

Figure 20.11a shows the calcium cycle for a 55-year-old forest at Hubbard Brook, which is typical of elements without a prominent gas phase and where organic cycling, weathering, and cation exchange in the soil solution are the important processes. Calcium resides in living plant material (48.4 g m^{-2}), in decomposing material on the forest floor (37.2 g m^{-2}), and in the soil, where it is either available in the soil solution (51.0 g m^{-2}) or in mineral soil (960 g m^{-2}) and rock (6460 g m^{-2}). The soil complex holds most of the calcium; living plant material holds relatively little calcium. The amount of calcium available annually in the soil solution for plant use is a balance of several processes. Precipitation imports 0.22 g m^{-2} yr^{-1}. Calcium leaches from leaves and bark as water drips off leaves (throughfall) or flows down branches and stems (stemflow). A total of 0.67 g m^{-2} yr^{-1} enters the soil solution in this process. Chemical exudates from roots provide 0.35 g m^{-2} yr^{-1} to the soil solution. The two largest sources of calcium

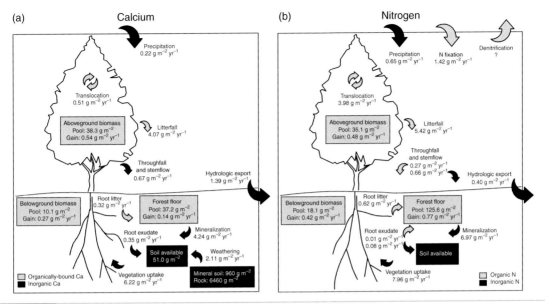

Fig. 20.11 Annual biogeochemical cycles for (a) calcium and (b) nitrogen in a 55-year-old northern hardwood forest ecosystem in the Hubbard Brook Experimental Forest. Boxes show major stores. Arrows show annual fluxes. Calcium data from Likens et al. (1977, p. 96). Nitrogen data from Bormann et al. (1977), Likens et al. (1977, p. 101), and Bormann and Likens (1979, p. 76). See Likens et al. (1998) for updated calcium pools and fluxes.

are mineralization of forest floor litter (4.24 g m^{-2} yr^{-1}) and weathering (2.11 g m^{-2} yr^{-1}). The available calcium is taken up by vegetation, held on exchange sites in the soil, or lost during runoff. Runoff carries away 18 percent (1.39 g m^{-2} yr^{-1}) of the available calcium. Plants absorb the remainder (6.22 g m^{-2} yr^{-1}) during growth. However, only 0.81 g m^{-2} yr^{-1} (13% of annual uptake) accumulates in plant biomass. Sixty-five percent of the uptake (4.07 g m^{-2} yr^{-1}) returns to the ground in aboveground litterfall. Another 0.32 g m^{-2} yr^{-1} is lost through root mortality. Leaching from leaves and bark in throughfall and stemflow and from roots results in a loss of 1.02 g m^{-2} yr^{-1}.

The nitrogen cycle is more complex because nitrogen has a gaseous phase, occurring as diatomic nitrogen (N_2), ammonia (NH_3), nitrous oxide (N_2O), nitric oxide (NO), and nitrogen dioxide (NO_2), and because the nitrogen used in plant growth is the inorganic ions of nitrate (NO_3^-) and ammonium (NH_4^+) (Table 20.2). Nitrogen is not a large part of primary or secondary minerals, and there is little release in weathering. Instead, the source of nitrogen over geologic

Table 20.2 Main forms of nitrogen in the environment

Form	Chemical symbol
Organic N	–
Inorganic ions	
Ammonium	NH_4^+
Nitrate	NO_3^-
Nitrite	NO_2^-
Gas	
Diatomic nitrogen	N_2
Ammonia	NH_3
Nitrous oxide	N_2O
Nitric oxide	NO
Nitrogen dioxide	NO_2

timescales is the atmosphere, and the major stores of nitrogen in an ecosystem are bound in plants and soil organic matter. The nitrogen cycle consists of four main processes: nitrogen fixation – conversion of atmospheric nitrogen (N_2) to a plant usable form; mineralization – release of organically bound nitrogen as NH_4^+

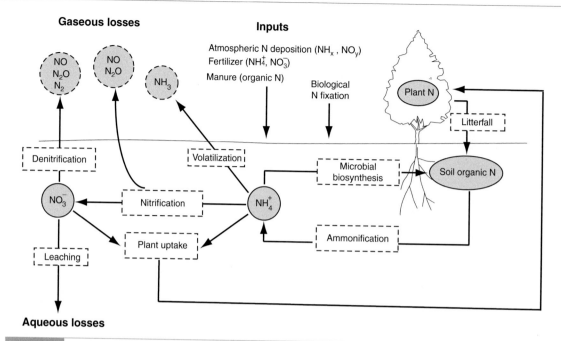

Gaseous losses

Inputs

Atmospheric N deposition (NH_x, NO_y)
Fertilizer (NH_4^+, NO_3^-)
Manure (organic N)

Biological
N fixation

NO
N_2O
N_2

NO
N_2O

NH_3

Plant N

Litterfall

Denitrification

Volatilization

Microbial
biosynthesis

Soil organic N

NO_3^-

Nitrification

NH_4^+

Plant uptake

Ammonification

Leaching

Aqueous losses

Fig. 20.12 The plant–soil nitrogen cycle. Circles indicate various pools (solid lines) or gaseous losses (dashed lines). Boxes denote processes. Also shown are natural inputs from biological nitrogen fixation and anthropogenic inputs from nitrogen deposition, fertilizer, and manure. Adapted from Bouwman et al. (2009).

during decomposition (also called ammonification); nitrification – oxidation of NH_4^+ to nitrite (NO_2^-) and of NO_2^- to NO_3^-; and denitrification – reduction of NO_3^- to N_2O and N_2 (Figure 20.12). Additionally, ammonia volatilization is the loss of NH_3 to the atmosphere. Nitrate leaching, denitrification, and ammonia volatilization are the three main nitrogen loss pathways.

Although N_2 comprises 78 percent of the atmosphere, it is unusable by plants and is converted to biologically available nitrogen by biological nitrogen fixation. Some species of plants (e.g., legumes such as alfalfa, clover, peas, and beans) have symbiotic relationships with bacteria and algae that fix N_2 into organic nitrogen for use by plants. Preindustrial biological nitrogen fixation prior to human activities is estimated to be 58 Tg N yr^{-1} (Vitousek et al. 2013). Nitrogen fixation is prevalent in agricultural land due to the cultivation of nitrogen-fixing crops, but also occurs in natural ecosystems. Biological nitrogen fixation is thought to be about 195 Tg N yr^{-1}, with a range of 100–290 Tg N yr^{-1} (Cleveland et al. 1999). It is prominent in

tropical ecosystems and less common at high latitudes (Houlton et al. 2007). Galloway et al. (2004) assessed the global nitrogen budget. They estimated biological nitrogen fixation to be 138 Tg N yr^{-1}. Lightning strikes also fix about 5 Tg N yr^{-1}, considerably less than biological nitrogen fixation. The burning of fossil fuels adds about 25 Tg N yr^{-1} when nitrogen in industrial and automotive emissions deposits onto land in precipitation or dry deposition. These inputs are smaller than the nitrogen released in the mineralization of soil organic matter.

Much of the nitrogen used in plant growth comes from internal recycling and decomposition of organic debris, which releases inorganic nitrogen in the form of ammonium (NH_4^+) and nitrate (NO_3^-). This cycle is dominated by uptake of NH_4^+ and NO_3^- from the soil solution, incorporation into plant tissues, return of nitrogen to the soil in litter, release of NH_4^+ to the soil solution as organic matter decomposes, and the oxidation of NH_4^+ to NO_3^- during nitrification. The transformation from organically bound nitrogen to inorganic nitrogen is known

Table 20.3 | Translocation of nutrients (g m⁻² yr⁻¹) in a 55-year-old northern hardwood forest in the Hubbard Brook Experimental Forest

	N	P	K	Ca	Mg
Foliage before senescence (B)	7.1	0.56	2.8	2.0	0.49
Foliage after senescence (A)	3.3	0.21	1.4	2.5	0.41
Foliage leaching during senescence (L)	0.2	0.01	1.1	0.2	0.06
Translocation (B–A–L)	3.6	0.34	0.3	−0.7	0.02

Source: From Ryan and Bormann (1982).

as mineralization. Ammonium is the first product of mineralization. Bacteria oxidize some of the NH_4^+ to nitrite (NO_2^-) and NO_2^- to NO_3^- (nitrification). Nitrate is more readily leached from the soil solution than NH_4^+ because soils have a greater capacity to exchange cations than anions.

A significant portion of the nitrogen required in plant growth is internally recycled within plants in a process known as translocation. Translocation is the withdrawal of nutrients from senescing leaves and subsequent storage within the plant. It is common in trees, where the nitrogen is stored in woody tissue and used for growth in the following year. Table 20.3 shows the magnitude of translocation in the Hubbard Brook study, where deciduous sugar maple, beech, and yellow birch trees are dominant. One-half of the nitrogen and 60 percent of the phosphorus in leaves prior to senescence are withdrawn before leaf fall. A large portion of the annual elemental requirement is met in this fashion. For example, 34 percent of the annual nitrogen and 30 percent of the annual phosphorus requirements during growth are supplied from translocation. This contributes to the tightness of these biogeochemical cycles. Nutrients withdrawn prior to leaf fall are not subject to losses.

Figure 20.11b shows the nitrogen cycle for a 55-year-old forest at Hubbard Brook. In contrast to calcium, which is almost exclusively stored in soil and rock, nitrogen is stored either in vegetation (53.2 g m⁻²) or in debris on the forest floor (125.6 g m⁻²). Precipitation adds 0.65 g m⁻² yr⁻¹ of nitrogen in the form of NO_3^- and NH_4^+.

Nitrogen fixation imports another 1.42 g m⁻² yr⁻¹. Weathering provides little nitrogen. Of the 2.07 g m⁻² yr⁻¹ entering the ecosystem, 81 percent is retained within the ecosystem. Only 0.40 g m⁻² yr⁻¹ washes out into streams. This illustrates the overall tightness of the nitrogen cycle. Leaching from foliage, bark, and roots returns both organically bound nitrogen and inorganic nitrogen to the soil (1.02 g m⁻² yr⁻¹). However, the largest source of nitrogen is aboveground litterfall, which returns 5.42 g m⁻² yr⁻¹ to the soil. Root mortality adds a smaller amount of nitrogen (0.62 g m⁻² yr⁻¹). This litter decomposes, mineralizing nitrogen for plant uptake. The available nitrogen in the soil solution that does not wash out in runoff is used by plants during growth (7.96 g m⁻² yr⁻¹), but plants retain only 11 percent (0.9 g m⁻² yr⁻¹) of this nitrogen. The rest returns to the soil in litterfall or leaching. More nitrogen accumulates in the forest floor from nitrogen fixation, litter, and organic leaching than is lost during mineralization for a net gain of 0.77 g m⁻² yr⁻¹.

Figures 20.13 and 20.14 show the aboveground biomass and aboveground biogeochemical cycles of calcium and nitrogen for three forests in the Walker Branch watershed in eastern Tennessee. Elevation in this 97.5 ha watershed ranges from 265 m along the stream channel to 350 m along ridgetops. The vegetation is primarily oak–hickory forest, consisting chiefly of pignut hickory, mockernut hickory, chestnut oak, white oak, and red maple. Predominantly chestnut oak forests occur on dry sites along ridgetops. Yellow poplar and lesser amounts of hickory and occasional oaks and maples occur

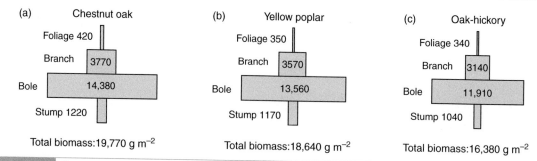

Fig. 20.13 Aboveground biomass in 70-year-old chestnut oak, yellow poplar, and oak–hickory forests in the Walker Branch watershed near Oak Ridge, Tennessee. Boxes are proportional in size to pools. Data from Edwards et al. (1989).

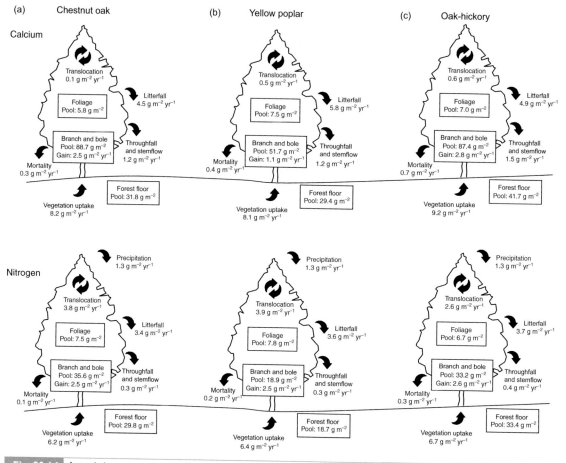

Fig. 20.14 Annual aboveground biogeochemical cycles for calcium (top) and nitrogen (bottom) in 70-year-old chestnut oak, yellow poplar, and oak–hickory forests in the Walker Branch watershed near Oak Ridge, Tennessee. Boxes show major stores. Arrows show annual fluxes. Data from Johnson and Henderson (1989).

on moist sites along streams and in valleys. Overstory trees are about 70 years old.

The three forests do not differ greatly in their cycling of calcium and nitrogen. Similar amounts of these elements return to the forest floor as litterfall and leaching in each. Uptake of calcium and nitrogen in aboveground growth is similar across the three forests, with more calcium uptake than nitrogen. Considerably more nutrients are required for foliage growth (Ca, 5.8–7.5 g m^{-2} yr^{-1}; N, 6.7–7.8 g m^{-2} yr^{-1}) than for woody growth (Ca, 1.1–2.8 g m^{-2} yr^{-1}; N, 2.5–2.6 g m^{-2} yr^{-1}). Foliage generally accounts for three-quarters of the annual elemental requirement in these deciduous forests. Translocation accounts for 28–38 percent of annual nitrogen requirement and 1–6 percent of annual calcium requirement. These forests have similar foliage mass as the Hubbard Brook forest, but considerably more woody biomass. Consequently, the forests have a greater amount of elements stored in aboveground biomass.

The forests can be compared in terms of the turnover time of calcium and nitrogen in the forest floor (estimated as the nutrient content of the forest floor divided by litterfall, throughfall, and stemflow return). Turnover times are more rapid at Walker Branch than at Hubbard Brook. The turnover time for calcium of 7.8 years at Hubbard Brook contrasts with turnover times of 5.6 years, 4.2 years, and 6.5 years for the chestnut oak, yellow poplar, and oak–hickory forests at Walker Branch, respectively. Turnover times for nitrogen are 19.8 years at Hubbard Brook versus 8.1, 4.8, and 8.1 for the chestnut oak, yellow poplar, and oak–hickory forests. Annual nitrogen return in litterfall, throughfall, and stemflow is greater at Hubbard Brook than at Walker Branch. Rather, the greater turnover time for nitrogen at Hubbard Brook arises from a considerably larger pool of nitrogen in the forest floor.

20.6 | Forest Production and Nutrient Cycling

Ecosystem production and nutrient cycling are linked (Figure 20.15). High nutrient availability leads to high nutrient uptake during plant growth. High concentrations of nutrients, especially nitrogen, in foliage allow for high photosynthetic rates. This increases net primary production so that more nutrients return to the soil in litterfall. The good chemical quality of litter allows for rapid decomposition and mineralization, which reinforces the high nutrient availability. Low nutrient availability has the

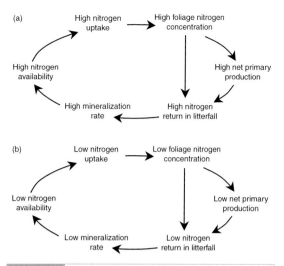

Fig. 20.15 Feedback between nutrient availability and net primary production showing that (a) high nitrogen availability reinforces high production and (b) low nitrogen availability reinforces low production.

opposite effect, reducing net primary production and nutrient return in litterfall. Litter quality is poor (low nitrogen), and litter decomposes slowly. Relatively few nutrients are mineralized, reinforcing the low nutrient availability.

A study of production and nutrient cycling in the northern hardwood forests of Wisconsin illustrates these linkages (Pastor et al. 1984). In this study area, forest composition varies with soil texture. Sugar maple forms productive stands on silty clay loam. Less productive white oak and red oak stands grow on sandy clay loam. Stands of eastern hemlock, red pine, and eastern white pine grow on sandy soils with low productivity. Aboveground net primary production, litterfall, nutrient return in litterfall, nitrogen mineralization, and species composition are highly interrelated (Figure 20.16). Annual aboveground net primary production increases with greater nitrogen mineralization. Needleleaf evergreen trees (red pine, hemlock, white pine) have the lowest production and mineralization. Sugar maple has the highest production and nitrogen mineralization. Oaks are intermediate. Nitrogen mineralization decreases as the C:N ratio of litter increases. Needleleaf evergreen trees have the poorest quality litter (high C:N ratio) and least nitrogen mineralization. Sugar maple has the

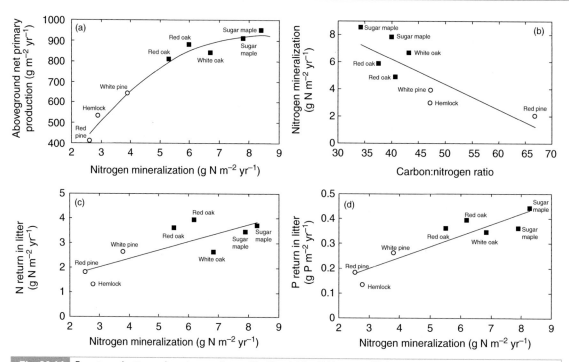

Fig. 20.16 Forest production and nutrient cycling on Blackhawk Island, Wisconsin. (a) Annual aboveground net primary production in relation to annual nitrogen mineralization. (b) Annual nitrogen mineralization in relation to litter C:N ratio. (c) Annual nitrogen return in litterfall in relation to annual nitrogen mineralization. (d) Annual phosphorus return in litterfall in relation to annual nitrogen mineralization. Data from Pastor et al. (1984).

lowest C:N ratio and highest nitrogen mineralization. Nitrogen mineralization also positively correlates with litter production and nitrogen and phosphorus return in litterfall. Needleleaf evergreen trees have the lowest return of nitrogen and phosphorus. Broadleaf deciduous trees have higher elemental return. In this region, soil texture and species composition create a nitrogen mineralization gradient. The needleleaf evergreen trees that dominate xeric sites produce low quality litter and return few nutrients in litterfall, leading to low mineralization and low production. Broadleaf deciduous forests have better quality litter, return more nutrients in litterfall, have higher rates of nitrogen mineralization, and are more productive.

Studies of production and nutrient cycling in the boreal forests of interior Alaska further illustrate interactions between the carbon and nitrogen cycles (Van Cleve et al. 1983a,b, 1986, 1991; Viereck et al. 1983; Bonan 1989, 1990; Bonan and Van Cleve 1992; Chapin et al. 2006a). The forest landscape near Fairbanks, Alaska, is a mosaic of black spruce, white spruce, quaking aspen, paper birch, and balsam poplar forests arising from topographic variation and recurring disturbances such as fires and floods (Figure 20.17). White spruce, balsam poplar, birch, and aspen form productive forests on warm, well-drained, nutrient-rich sites, along river floodplains and upland south-facing slopes. Black spruce forms unproductive forests on cold, wet, nutrient-poor soils underlain with permafrost on river terraces and upland sites.

Figure 20.18 shows the biomass, productivity, and nitrogen cycling in these forests. White spruce, birch, aspen, and balsam poplar stands have large aboveground tree biomass (11.2–16.8 kg m^{-2}) and aboveground tree production (366–562 g m^{-2} yr^{-1}). Black spruce stands accumulate less than one-half the tree biomass (5.0 kg m^{-2}) of those forests and have one-third to one-fifth the aboveground tree production (110 g m^{-2} yr^{-1}). The standing crop of

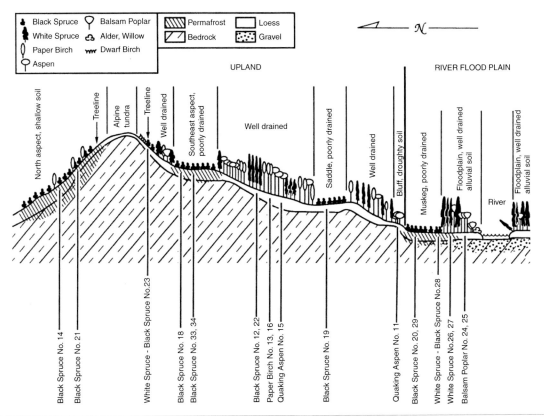

Fig. 20.17 Distribution of black spruce, white spruce, paper birch, quaking aspen, and balsam poplar forest stands near Fairbanks, Alaska, in relation to topography. Redrawn from Viereck et al. (1983).

nitrogen in tree biomass is more than twice as great in white spruce, birch, aspen, and poplar forests (26.3–28.2 g N m^{-2}) as in black spruce forests (12.3 g N m^{-2}). These productive forests have annual nitrogen requirements that are 3 times (white spruce) to 10–12 times (deciduous stands) greater than black spruce (0.53 g N m^{-2} yr^{-1}).

Black spruce has the lowest annual litterfall (43 g m^{-2} yr^{-1}) and nitrogen return in litterfall (0.3 g N m^{-2} yr^{-1}). Litterfall is 4–9 times greater in the deciduous stands, and nitrogen return in litterfall is 5–11 times greater. The deciduous stands accumulate less forest floor biomass (2.2–5.8 kg m^{-2}) than spruce (~7.5 kg m^{-2}), but have greater nitrogen concentration in the forest floor (1.2–1.5%) than spruce (0.8%).

Interactions among soil temperature, the forest floor, and litter quality control productivity and nutrient cycling in these forests (Figure 20.19). Black spruce stands have thick forest floors 12–38 cm deep. White spruce, birch, aspen, and balsam poplar have thinner forest floors (2–18 cm). With its low thermal conductivity, a thick forest floor prevents heating in summer, and soil temperature decreases as the forest floor becomes thicker. Aboveground tree production decreases with colder soils. Black spruce stands occur on the coldest soils and have the lowest production. White spruce and deciduous stands grow on warmer soils and have higher production. Rates of forest floor decomposition and nitrogen mineralization increase with warmer soil temperature. Cold soil temperatures in black spruce forests slow decomposition and nutrient mineralization. This restricts tree growth while promoting the accumulation of a thick forest floor. In contrast, warm, permafrost-free soils enhance productivity and

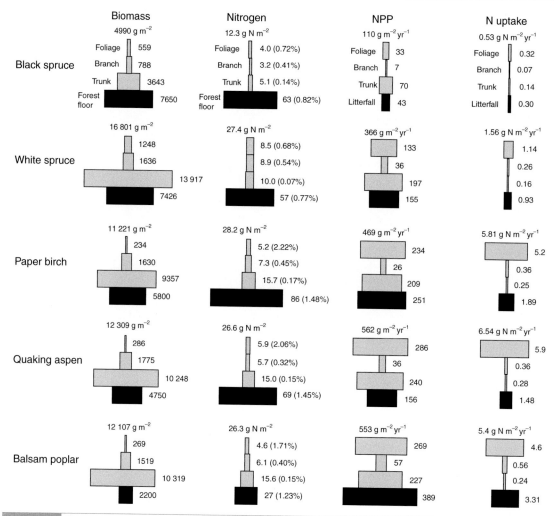

Fig. 20.18 Biomass, net primary production, and nitrogen cycling for black spruce, white spruce, paper birch, quaking aspen, and balsam poplar forests near Fairbanks, Alaska. Shown are average aboveground biomass, nitrogen content of biomass (and as a percentage of biomass, in parentheses), aboveground net primary production, and nitrogen uptake for each forest type. Also shown are forest floor biomass, forest floor nitrogen, litterfall, and nitrogen return in litterfall. Boxes are proportional in size to pools and fluxes. Data from Van Cleve et al. (1983b).

nutrient cycling through more rapid decomposition and nutrient mineralization. Forest floor chemistry interacts with soil temperature to control production and nutrient cycling. Spruce forests have forest floors with low nitrogen and high lignin concentrations, which further slows decomposition. These forests recycle material that is low in nitrogen. In contrast, more productive deciduous forests have forest floors with higher nitrogen and lower lignin concentrations and return more nitrogen in litterfall.

20.7 | Net Ecosystem Production

Eddy covariance flux measurements illustrate seasonal changes in ecosystem carbon fluxes. While only the net flux can be measured, this flux can be partitioned into gross primary production and ecosystem respiration (Reichstein et al. 2005; Lasslop et al. 2010). Figure 20.20 illustrates monthly fluxes for several ecosystems. Tundra ecosystems have small carbon

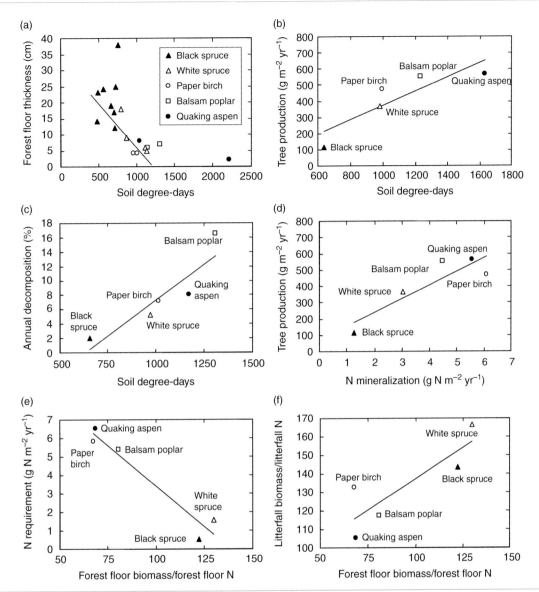

Fig. 20.19 Controls of net primary production and nutrient cycling in boreal forests near Fairbanks, Alaska. (a) Forest floor thickness in relation to soil degree-days above 0°C accumulated at a 10 cm depth from May 20 to September 10. (b) Annual aboveground tree production in relation to soil degree-days. (c) Annual forest floor decomposition in relation to soil degree-days. (d) Annual aboveground tree production in relation to annual nitrogen mineralization. (e) Annual nitrogen requirement in aboveground tree production in relation to the ratio of forest floor biomass to nitrogen. (f) Ratio of litterfall biomass to nitrogen in relation to forest floor biomass-to-nitrogen ratio. The data in panel (a) are from Viereck et al. (1983) for eight black spruce, four white spruce, two paper birch, two quaking aspen, and two balsam poplar stands. The data in panels (b)–(f) are averages for the five forest types (Van Cleve et al. 1983b; Fox and Van Cleve 1983). Van Cleve et al. (1983b) show data for individual stands. The overall patterns and conclusions are the same.

fluxes. Boreal and temperate forests have a distinct annual cycle with large carbon uptake driven by high rates of gross primary production during the growing season. The grassland has a pronounced annual cycle driven by precipitation, with greatest carbon uptake during the winter and spring wet season. The tropical rainforest has less seasonal variation.

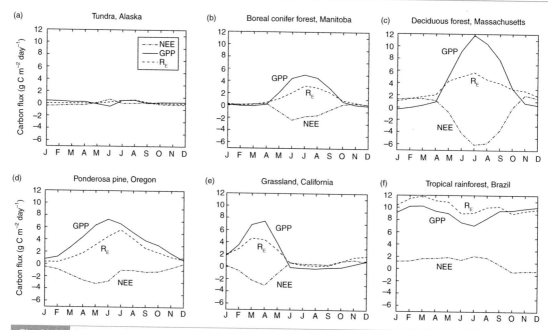

Fig. 20.20 Monthly carbon fluxes for (a) moist tundra, Alaska, (b) boreal conifer forest, Manitoba, (c) temperate deciduous forest, Massachusetts, (d) ponderosa pine, Oregon, (e) grassland, California, and (f) tropical rainforest, Brazil. Net carbon flux is given in terms of net ecosystem exchange ($NEE = -NEP = R_E - GPP$). See Figures 10.3 and 12.3 for monthly water and energy fluxes.

Long-term eddy covariance flux measurements at the Harvard Forest in central Massachusetts from 1992 through 1999 illustrate seasonal and interannual changes in net ecosystem production (Goulden et al. 1996a,b). This area of Massachusetts has a humid continental climate with warm to cool summers, cold winters, and large seasonal variation in temperature. Red oak, red maple, and scattered nearby stands of eastern white pine, red pine, and eastern hemlock dominate the site where the fluxes were measured. The forest is about 50–70 years old with a canopy height of 20–24 m. Carbon uptake by the forest varied seasonally, with carbon sequestration during the spring, summer, and autumn growing season and carbon loss during the cold dormant season (Figure 20.21a). Between 1992 and 1999, the forest typically gained 3–6 g C m^{-2} per day during the growing season and lost 1–2 g C m^{-2} per day during the non-growing season. Weather fluctuations caused large day-to-day variation in net carbon uptake during the growing season, with some days having minimal uptake or even net efflux.

Carbon uptake during gross primary production was restricted to the warm season, with peak daily uptake in excess of 10 g C m^{-2} (Figure 20.21b). Carbon loss during respiration occurred throughout the year, with relatively low values (1 g C m^{-2} per day) during the dormant season and higher values (greater than 5 g C m^{-2} per day) during the summer when plants and microbes were active. Gross primary production varied from year to year as a result of changes in growing season length. Relatively small differences in the time of leaf emergence or senescence (e.g., 6–10 days) resulted in large differences in annual gross primary production (e.g., 50 g C m^{-2}). Prolonged cloudy periods in July 1992, August 1992, and August 1994 each reduced gross primary production by 40 g C m^{-2}. Unusually warm soils in the dormant season arising from deep snow that insulated the ground resulted in high respiration losses in some years. Summer drought also increased respiration and reduced gross primary production.

Annual net carbon uptake at Harvard Forest between 1992 and 2004 ranged from 100 to 470

(a)

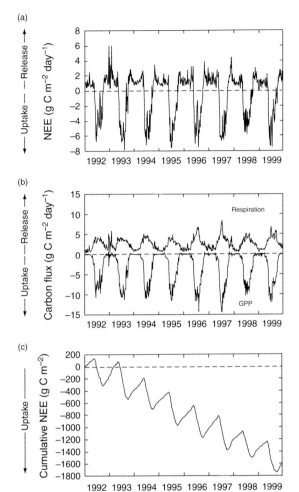

Fig. 20.21 Daily eddy covariance carbon fluxes for Harvard Forest, Massachusetts, 1992–1999 (a) Daily net ecosystem exchange. (b) Daily gross primary production and ecosystem respiration. (c) Cumulative net carbon flux. Negative values indicate carbon uptake. Positive values show carbon loss. Data from Goulden et al. (1996a,b). See also Figure 20.20c for monthly fluxes.

central Manitoba (Goulden et al. 1997, 1998). The climate is subarctic, with short, cool summers and long, cold winters. Upland areas around the tower site are dominated by dense, 10 m tall, 120-year-old black spruce trees with a feathermoss ground cover. Lowland areas are dominated by sparse, 1–6 m tall trees with sphagnum moss ground cover. Through the measurement period, carbon uptake during gross primary production ranged from 5 to 8 g C m^{-2} per day in the growing season while respiration loss ranged from 5 to 10 g C m^{-2} per day. In this forest, high rates of daily gross primary production were balanced by high respiration loss. Annually, about 600–800 g C m^{-2} were gained during gross primary production and a similar amount lost in respiration. The net carbon gain is a small residual between these two large fluxes (Figure 20.22b).

Similar to the hydrologic cycle, the metabolic activity of the terrestrial ecosystems can be seen in the isotopic composition of the carbon cycle (^{13}C/^{12}C). Atmospheric CO_2 has a δ^{13}C of approximately -8‰, but this value changes seasonally with photosynthesis and respiration and is becoming more negative over time (i.e., depleted in ^{13}C) because of fossil fuel combustion and deforestation. The δ^{13}C of C$_3$ plants ranges from about -22 to -34‰, and the isotopic composition of respired CO_2 reflects the isotopic composition of plant material. The δ^{13}C of ecosystem respiration in C$_3$ systems ranges from about -24 to -30‰ (Pataki et al. 2003; Bowling et al. 2008). Because plant material is isotopically depleted in ^{13}C relative to air, the δ^{13}C of atmospheric CO_2 is decreasing in response to anthropogenic inputs of ^{13}C-depleted CO_2 from fossil fuels and deforestation.

g C m^{-2} yr^{-1} (Figure 20.22a). This net uptake was the difference between large carbon uptake during gross primary production and similarly large carbon loss during respiration Comparably small changes in annual gross primary production or annual respiration resulted in large changes in annual net ecosystem production. Over this period, the annual net carbon uptake increased at a rate of 15 g C m^{-2} yr^{-1} per year.

Similar eddy covariance carbon fluxes have been measured at a black spruce forest in

20.8 | Ecosystem Experiments

Though short-term photosynthetic enhancement with higher CO_2 concentrations is well documented, less is known about the long-term acclimation of plants to higher CO_2, how elevated CO_2 affects carbon allocation, nitrogen availability, and nitrogen use, how physiological changes affect stand structure and composition,

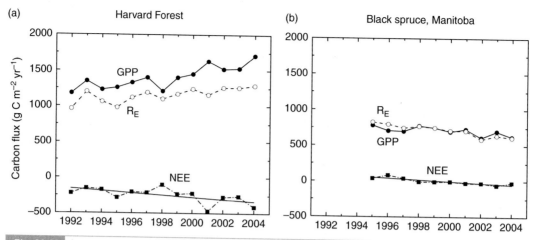

Fig. 20.22 Annual gross primary production, ecosystem respiration, and net ecosystem exchange at (a) Harvard Forest (1992–2004) and (b) an old black spruce forest, Manitoba (1995–2004). Harvard Forest data from Urbanski et al. (2007). Black spruce data from Dunn et al. (2007). See also Figure 20.20b,c for monthly fluxes.

and whether growth increases will be realized in the field where temperature, soil water, and nitrogen limit plant growth.

Much of the early research on the physiological effects of elevated CO_2 was conducted in greenhouses and growth chambers under controlled environmental conditions and with individual plants grown in pots. The applicability of these studies to plants grown in the field under prevailing meteorological conditions and with realistic ecological circumstances such as nutrient availability, stand structure, and community composition has been questioned. Subsequent studies utilized open-top chambers in the field, but these studies, too, suffered from alteration of the microclimate by the chamber and by the inability to enclose large plots.

The free-air CO_2 enrichment (FACE) methodology represents an alternative technology. Unlike greenhouses, growth chambers, or open-top chambers, free-air systems require no enclosures. A typical FACE study site is surrounded by a system of pipes that release CO_2 into the air. This allows plants to grow in the field at controlled CO_2 concentrations under fully open-air conditions. Whole ecosystems can be studied for their response to CO_2 enrichment. The technology has been applied to forest, grassland, desert, and agricultural ecosystems.

Analyses of FACE studies show increased photosynthesis, reduced stomatal conductance,

and increased plant productivity in response to CO_2 enrichment. One such FACE site was a 13-year-old loblolly pine stand at the Duke Forest, North Carolina, exposed to a 200 ppm increase in CO_2 concentration beginning in 1996 (an increase from ambient concentration of approximately 370 ppm to elevated concentration of 570 ppm). DeLucia et al. (2005) contrasted the response of that pine forest with a 10-year-old deciduous sweetgum stand near Oak Ridge, Tennessee, exposed to similar CO_2 enrichment in 1998. Net primary production increased in response to elevated CO_2 concentration by 14–26 percent over the six-year period 1997–2002 at the pine forest and by 16–38 percent over the five-year period 1998–2002 at the deciduous forest. A subsequent analysis based on data collected later in the experiments showed that water-use efficiency increased in both forests, because net primary production increased but annual transpiration remained the same or decreased (Figure 20.23).

Norby et al. (2005) compared the loblolly pine and sweetgum studies with additional FACE studies at young, mixed species forest stands dominated by aspen in Wisconsin and poplar in Italy. Across a broad range of productivity, climate, soil, and plant functional types, an increase in CO_2 concentration from ambient values of 376 ppm to elevated values of 550 ppm sustained over several years increased net

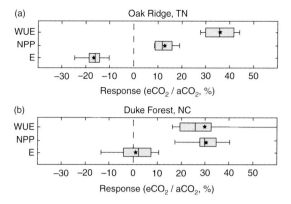

Fig. 20.23 Box and whisker plots showing the mean annual response of water-use efficiency (WUE; NPP/E), net primary production (NPP), and transpiration (E) to CO_2 enhancement at (a) Oak Ridge for the years 1999, 2004, 2007, and 2008 and (b) Duke Forest between 1996 and 2007. Shown is the elevated CO_2 response as a percentage of the ambient CO_2 measurement. The ends of the boxes show the lower (25th) and upper (75th) quartiles. The horizontal whiskers show the full range of the data. The lines in the boxes are the medians and stars are the means. Data from De Kauwe et al. (2013).

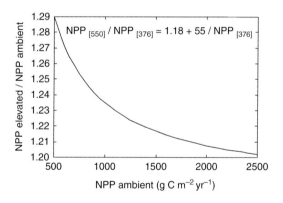

Fig. 20.24 Relationship between net primary production (g C m^{-2} yr^{-1}) at ambient CO_2 concentration (376 ppm) and elevated CO_2 concentration (550 ppm) observed for four young loblolly pine, sweetgum, aspen, and poplar temperate forests. Data from Norby et al. (2005).

primary production. The increase was greatest for low productivity stands and declined with higher productivity (Figure 20.24). Using the median net primary production at the four sites, CO_2 enrichment increased net primary production by 23 percent. Ainsworth and Long (2005) found that aboveground production increased

by 17 percent on average with elevated CO_2 concentration in a synthesis of 12 FACE studies involving more than 40 species in forest, grassland, desert, and agricultural ecosystems (Figure 16.10c). Similar results were found in the synthesis by Nowak et al. (2004) of 16 FACE sites in forest, grassland, desert, and bog ecosystems. Aboveground production was 19 percent greater on average with elevated CO_2, and total net primary production increased by 12 percent.

Whether the enhanced productivity from elevated CO_2 can be sustained is unclear. Plant productivity increases less with CO_2 enrichment when nitrogen availability is low (Oren et al. 2001; Reich et al. 2006a, 2014; Norby et al. 2010; Reich and Hobbie 2013). Nitrogen limitation reduced the stimulation of net primary productivity by elevated CO_2 over 11 years at the Oak Ridge sweetgum experiment (Norby et al. 2010). During the first 6 years of the experiment, productivity increased with CO_2 enrichment by 13–33 percent compared with the control. The increase in productivity declined in subsequent years, from 24 percent in 2001–2003 to 9 percent in 2008 (Figure 20.25a). Over long periods, enhanced tree growth leads to increased carbon storage in plant biomass and soil organic matter, which may sequester nitrogen and deplete the soil mineral nitrogen pool, a process referred to as progressive nitrogen limitation (Luo et al. 2004; Reich et al. 2006b). However, a similar decline in productivity was not seen over 8 years at the North Carolina loblolly pine experiment (Figure 20.25b). A long-term CO_2 enrichment × nitrogen fertilization experiment in perennial grassland found that low ambient soil nitrogen availability constrained the elevated CO_2 increase in plant biomass (Reich and Hobbie 2013; Reich et al. 2014).

Ecosystem responses to CO_2 enrichment are exceedingly complex (Norby and Zak 2011). The long-term fate of the additional carbon input to the ecosystems is unclear. If the increased photosynthesis is allocated to plant tissues that decompose rapidly (e.g., leaves or fine roots), the potential long-term net carbon sequestration in soils is small. Greater allocation of carbon to wood, which turns over slowly, increases

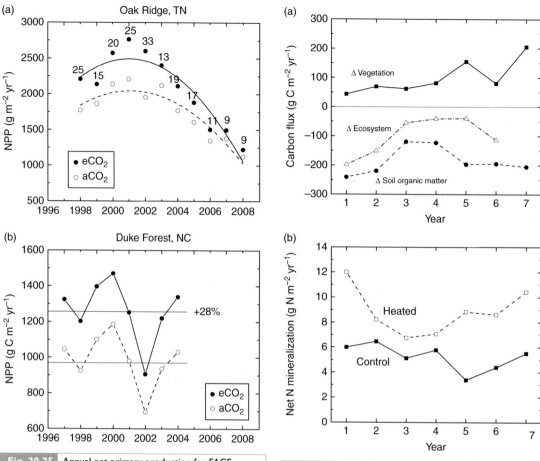

Fig. 20.25 Annual net primary production for FACE experiments at (a) Oak Ridge, Tennessee sweetgum (Norby et al. 2010) and (b) Duke Forest, North Carolina loblolly pine (McCarthy et al. 2010). Shown are data for ambient CO_2 (aCO_2) and elevated CO_2 (eCO_2). Note that the Oak Ridge data are reported as dry matter (biomass) while the Duke data are reported as the mass of carbon. Numbers for Oak Ridge show the percentage increase relative to the control. The Duke Forest did not have a similar temporal trend. Shown are the 8-year means (solid horizontal lines) and the percentage increase from the control.

Fig. 20.26 Effect of soil warming on the biogeochemistry of a temperate deciduous forest in Massachusetts. (a) Annual change in vegetation carbon, soil organic matter, and ecosystem carbon in the heated plot relative to the unheated control. (b) Net nitrogen mineralization in the control and heated areas. Adapted from Melillo et al. (2011).

carbon storage and the residence time of carbon in the system.

Climate change alters carbon storage in ecosystems by modifying carbon uptake in net primary production and carbon loss from soils. One mechanism by which this can occur is through increased soil organic matter decomposition and nitrogen availability with soil warming, which can stimulate productivity in nitrogen-limited soils (Melillo et al. 2002, 2011). A 7-year soil warming experiment in a deciduous forest in central Massachusetts illustrates these changes (Figure 20.26). The experiment elevated soil temperature by 5°C. Over the 7-year study, this warming increased carbon losses from the soil and stimulated carbon gains by trees. The cumulative soil carbon loss (1300 g C m^{-2}) exceeded the vegetation gain (700 g C m^{-2}) for a net loss of carbon (600 g C m^{-2}). The increased tree productivity occurred because the greater soil organic matter decay in the heated plots increased nitrogen mineralization by 45 percent. The trees responded with greater foliar nitrogen content and growth rates (Butler et al. 2012). Soil warming experiments in a variety of forest, shrubland, grassland, and tundra ecosystems show similar short-term enhancement of soil carbon loss, nitrogen mineralization, and net primary production (Rustad et al. 2001; Wu et al. 2011; Lu et al. 2013).

20.9 | Review Questions

1. Two scientists studying a forest ecosystem observe the temporal pattern of net primary production and biomass accumulation following recovery from fire. Initially, gross primary production exceeds respiration and the ecosystem accumulates biomass. Eventually, these two fluxes balance and biomass accumulation reaches a steady state. One scientist develops a system of mathematical equations that describe the trends in carbon pools and fluxes in terms of a box, or compartment, model. The other scientist develops a model that describes the establishment, growth, and mortality of individual trees. Both models replicate the observations. Which model better represents the ecological concept of an ecosystem?

2. Figure 20.2 represents a biogeochemical ecosystem model. Use this model to qualitatively contrast an ecosystem with high allocation of GPP to wood ($b_3 = 0.20$, $b_1 + b_2 = 0.30$) and an ecosystem with low allocation of GPP to wood ($b_3 = 0.05$, $b_1 + b_2 = 0.45$). How do the ecosystems differ in plant, litter, and soil carbon accumulation?

3. In the global carbon cycle shown in Figure 3.6b, gross primary production is 123 Pg C yr^{-1}, respiration and fire are 118.7 Pg C yr^{-1}, export to rivers is 1.7 Pg C yr^{-1}, and land use emission is 1.1 Pg C yr^{-1}. Assume autotrophic respiration is 50 percent of gross primary production and heterotrophic respiration is 55 Pg C yr^{-1}. Calculate net primary production, net ecosystem production, and net biome production. What is the total non-respiratory loss? Describe ways in which net biome production could double.

4. With reference to Figure 20.5, contrast carbon stocks in the black spruce forest with the aspen and jack pine forest. What does this suggest about the environmental conditions of these forests?

5. The following data show climate and gross primary production (GPP) for six forest stands along the Oregon transect (Figure 20.10). How does GPP vary with precipitation and temperature? Which is the better predictor of GPP?

Site	Annual precipitation (mm)	Annual mean temperature (°C)	GPP (g C m^{-2} yr^{-1})
1	2510	10.1	1400
2	980	11.2	1665
3	1180	10.6	2404
4	1810	6.0	879
5	540	7.4	364
6	220	9.1	302

6. Determine calcium budget equations for hydrologic export, vegetation gain, and forest floor gain at Hubbard Brook (Figure 20.11).

7. Determine foliage calcium and nitrogen budget equations at Walker Branch (Figure 20.14). Contrast the translocation of calcium with that of nitrogen.

8. Which of the three forests at Walker Branch has the highest percentage concentration of calcium and nitrogen in foliage?

9. A mature red oak forest in Wisconsin is harvested and replanted in white pine. Describe changes in net primary production and the nitrogen cycle between mature forests of each type.

10. The following data show carbon pools and fluxes at the Hubbard Brook forest ecosystem in the 1990s. Calculate turnover times assuming steady state (carbon input = carbon output).

Pool	Carbon (g C m^{-2})	Input (g C m^{-2} yr^{-1})
Living plant biomass	12,006	585
Woody biomass	9294	183
Forest floor	2970	319
Mineral soil	12,770	155
Coarse woody debris	1070	120

11. Calculate forest floor biomass and nitrogen turnover for black spruce, white spruce, paper birch, quaking aspen, and balsam poplar forests in Alaska (Figure 20.18). Which forests have rapid cycling of materials? Estimate forest floor turnover as forest floor mass divided by litterfall.

12. Relate nitrogen return in litterfall to aboveground net primary production in the boreal forests of Alaska (Figure 20.18). What factors contribute to the low net primary production of spruce forests?

13. How can annual net ecosystem production vary considerably from year-to-year while gross primary production and ecosystem respiration are relatively less variable?

14. The annual carbon fluxes in Figure 20.4 do not balance. Discuss reasons why.

20.10 | References

Ainsworth, E. A., and Long, S. P. (2005). What have we learned from 15 years of free-air CO_2 enrichment (FACE)? A meta-analytical review of the responses of photosynthesis, canopy properties and plant production to rising CO_2. *New Phytologist*, 165, 351–372.

Bonan, G. B. (1989). Environmental factors and ecological processes controlling vegetation patterns in boreal forests. *Landscape Ecology*, 3, 111–130.

Bonan, G. B. (1990). Carbon and nitrogen cycling in North American boreal forests, I: Litter quality and soil thermal effects in interior Alaska. *Biogeochemistry*, 10, 1–28.

Bonan, G. B., and Van Cleve, K. (1992). Soil temperature, nitrogen mineralization, and carbon source–sink relationships in boreal forests. *Canadian Journal of Forest Research*, 22, 629–639.

Bormann, F. H., and Likens, G. E. (1967). Nutrient cycling. *Science*, 155, 424–429.

Bormann, F. H., and Likens, G. E. (1979). *Pattern and Process in a Forested Ecosystem*. New York: Springer-Verlag.

Bormann, F. H., Siccama, T. G., Likens, G. E., and Whittaker, R. H. (1970). The Hubbard Brook Ecosystem Study: Composition and dynamics of the tree stratum. *Ecological Monographs*, 40, 373–388.

Bormann, F. H., Likens, G. E., and Melillo, J. M. (1977). Nitrogen budget for an aggrading northern hardwood forest ecosystem. *Science*, 196, 981–983.

Bouwman, A. F., Beusen, A. H. W., and Billen, G. (2009). Human alteration of the global nitrogen and phosphorus soil balances for the period 1970–2050. *Global Biogeochemical Cycles*, 23, GB0A04, doi:10.1029/2009GB003576.

Bowling, D. R., Pataki, D. E., and Randerson, J. T. (2008). Carbon isotopes in terrestrial ecosystem pools and CO_2 fluxes. *New Phytologist*, 178, 24–40.

Butler, S. M., Melillo, J. M., Johnson, J. E., et al. (2012). Soil warming alters nitrogen cycling in a New England forest: Implications for ecosystem function and structure. *Oecologia*, 168, 819–828.

Chapin, F. S., III, Oswood, M. W., Van Cleve, K., Viereck, L. A., and Verbyla, D. L. (2006a). *Alaska's Changing Boreal Forest*. New York: Oxford University Press.

Chapin, F. S., III, Woodwell, G. M., Randerson, J. T., et al. (2006b). Reconciling carbon-cycle concepts, terminology, and methods. *Ecosystems*, 9, 1041–1050.

Clements, F. E. (1916). *Plant Succession: An Analysis of the Development of Vegetation*, Carnegie Institution Publication Number 242. Washington, D.C.: Carnegie Institution.

Clements, F. E. (1928). *Plant Succession and Indicators*. New York: H.W. Wilson.

Cleveland, C. C., Townsend, A. R., Schimel, D. S., et al. (1999). Global patterns of terrestrial biological nitrogen (N_2) fixation in natural ecosystems. *Global Biogeochemical Cycles*, 13, 623–645.

De Kauwe, M. G., Medlyn, B. E., Zaehle, S., et al. (2013). Forest water use and water use efficiency at elevated CO_2: A model–data intercomparison at two contrasting temperate forest FACE sites. *Global Change Biology*, 19, 1759–1779.

DeLucia, E. H., Moore, D. J., and Norby, R. J. (2005). Contrasting responses of forest ecosystems to rising atmospheric CO_2: Implications for the global C cycle. *Global Biogeochemical Cycles*, 19, GB3006, doi:10.1029/2004GB002346.

Dunn, A. L., Barford, C. C., Wofsy, S. C., Goulden, M. L., and Daube, B. C. (2007). A long-term record of carbon exchange in a boreal black spruce forest: Means, responses to interannual variability, and decadal trends. *Global Change Biology*, 13, 577–590.

Edwards, N. T., Johnson, D. W., McLaughlin, S. B., and Harris, W. F. (1989). Carbon dynamics and productivity. In *Analysis of Biogeochemical Cycling Processes in Walker Branch Watershed*, ed. D. W. Johnson and R. I. Van Hook. New York: Springer-Verlag, pp. 197–232.

Fahey, T. J., Siccama, T. G., Driscoll, C. T., et al. (2005). The biogeochemistry of carbon at Hubbard Brook. *Biogeochemistry*, 75, 109–176.

Fox, J. F., and Van Cleve, K. (1983). Relationships between cellulose decomposition, Jenny's k, forest-floor nitrogen, and soil temperature in Alaskan taiga forests. *Canadian Journal of Forest Research*, 13, 789–794.

Galloway, J. N., Dentener, F. J., Capone, D. G., et al. (2004). Nitrogen cycles: Past, present, and future. *Biogeochemistry*, 70, 153–226.

Gholz, H. L., and Fisher, R. F. (1982). Organic matter production and distribution in slash pine (*Pinus elliottii*) plantations. *Ecology*, 63, 1827–1839.

Gleason, H. A. (1917). The structure and development of the plant association. *Bulletin of the Torrey Botanical Club*, 44, 463–481.

Gleason, H. A. (1926). The individualistic concept of the plant association. *Bulletin of the Torrey Botanical Club*, 53, 7–26.

Gleason, H. A. (1939). The individualistic concept of the plant association. *American Midland Naturalist*, 21, 92–110.

Golley, F. B. (1993). *A History of the Ecosystem Concept in Ecology: More than the Sum of the Parts*. New Haven: Yale University Press.

Goulden, M. L., Munger, J. W., Fan, S.-M., Daube, B. C., and Wofsy, S. C. (1996a). Measurements of carbon sequestration by long-term eddy covariance: Methods and a critical evaluation of accuracy. *Global Change Biology*, 2, 169–182.

Goulden, M. L., Munger, J. W., Fan, S.-M., Daube, B. C., and Wofsy, S. C. (1996b). Exchange of carbon dioxide by a deciduous forest: Response to interannual climate variability. *Science*, 271, 1576–1578.

Goulden, M. L., Daube, B. C., Fan, S.-M., et al. (1997). Physiological responses of a black spruce forest to weather. *Journal of Geophysical Research*, 102D, 28987–28996.

Goulden, M. L., Wofsy, S. C., Harden, J. W., et al. (1998). Sensitivity of boreal forest carbon balance to soil thaw. *Science*, 279, 214–217.

Gower, S. T., Vogt, K. A., and Grier, C. C. (1992). Carbon dynamics of Rocky Mountain Douglas-fir: Influence of water and nutrient availability. *Ecological Monographs*, 62, 43–65.

Gower, S. T., Vogel, J. G., Norman, J. M., et al. (1997). Carbon distribution and aboveground net primary production in aspen, jack pine, and black spruce stands in Saskatchewan and Manitoba, Canada. *Journal of Geophysical Research*, 102D, 29029–29041.

Harmon, M. E., Franklin, J. F., Swanson, F. J., et al. (1986). Ecology of coarse woody debris in temperate ecosystems. *Advances in Ecological Research*, 15, 133–302.

He, L., Chen, J. M., Pan, Y., Birdsey, R., and Kattge, J. (2012). Relationships between net primary productivity and forest stand age in U.S. forests. *Global Biogeochemical Cycles*, 26, GB3009, doi:10.1029/2010GB003942.

Houlton, B. Z., Wang, Y.-P., Vitousek, P. M., and Field, C. B. (2007). A unifying framework for dinitrogen fixation in the terrestrial biosphere. *Nature*, 454, 327–330.

Johnson, D. W., and Henderson, G. S. (1989). Terrestrial nutrient cycling. In *Analysis of Biogeochemical Cycling Processes in Walker Branch Watershed*, ed. D. W. Johnson and R. I. Van Hook. New York: Springer-Verlag, pp. 233–300.

Lasslop, G., Reichstein, M., Papale, P., et al. (2010). Separation of net ecosystem exchange into assimilation and respiration using a light response curve approach: Critical issues and global evaluation. *Global Change Biology*, 16, 187–208.

Likens, G. E. (2004). Some perspectives on long-term biogeochemical research from the Hubbard Brook ecosystem study. *Ecology*, 85, 2355–2362.

Likens, G. E., and Bormann, F. H. (1995). *Biogeochemistry of a Forested Ecosystem*, 2nd edn. New York: Springer-Verlag.

Likens, G. E., Bormann, F. H., Pierce, R. S., Eaton, J. S., and Johnson, N. M. (1977). *Biogeochemistry of a Forested Ecosystem*. New York: Springer-Verlag.

Likens, G. E., Driscoll, C. T., Buso, D. C., et al. (1994). The biogeochemistry of potassium at Hubbard Brook. *Biogeochemistry*, 25, 61–125.

Likens, G. E., Driscoll, C. T., Buso, D. C., et al. (1998). The biogeochemistry of calcium at Hubbard Brook. *Biogeochemistry*, 41, 89–173.

Lindeman, R. L. (1942). The trophic-dynamic aspect of ecology. *Ecology*, 23, 399–418.

Lu, M., Zhou, X., Yang, Q., et al. (2013). Responses of ecosystem carbon cycle to experimental warming: A meta-analysis. *Ecology*, 94, 726–738.

Luo, Y., Su, B., Currie, W. S., et al. (2004). Progressive nitrogen limitation of ecosystem responses to rising atmospheric carbon dioxide. *BioScience*, 54, 731–739.

Luyssaert, S., Inglima, I., Jung, M., et al. (2007). CO_2 balance of boreal, temperate, and tropical forests derived from a global database. *Global Change Biology*, 13, 2509–2537.

McCarthy, H. R., Oren, R., Johnsen, K. H., et al. (2010). Re-assessment of plant carbon dynamics at the Duke free-air CO_2 enrichment site: Interactions of atmospheric [CO_2] with nitrogen and water availability over stand development. *New Phytologist*, 185, 514–528.

McIntosh, R. P. (1985). *The Background of Ecology: Concept and Theory*. Cambridge: Cambridge University Press.

Melillo, J. M., Steudler, P. A., Aber, J. D., et al. (2002). Soil warming and carbon-cycle feedbacks to the climate system. *Science*, 298, 2173–2176.

Melillo, J. M., Butler, S., Johnson, J., et al. (2011). Soil warming, carbon–nitrogen interactions, and forest carbon budgets. *Proceedings of the National Academy of Sciences USA*, 108, 9508–9512.

Norby, R. J., and Zak, D. R. (2011). Ecological lessons from free-air CO_2 enrichment (FACE) experiments. *Annual Review of Ecology, Evolution, and Systematics*, 42, 181–203.

Norby, R. J., DeLucia, E. H., Gielen, B., et al. (2005). Forest response to elevated CO_2 is conserved across a broad range of productivity. *Proceedings of the National Academy of Sciences USA*, 102, 18052–18056.

Norby, R. J., Warren, J. M., Iversen, C. M., Medlyn, B. E., and McMurtrie, R. E. (2010). CO_2 enhancement of forest productivity constrained by limited nitrogen availability. *Proceedings of the National Academy of Sciences USA*, 107, 19368–19373.

Nowak, R. S., Ellsworth, D. S., and Smith, S. D. (2004). Functional responses of plants to elevated atmospheric CO_2 – do photosynthetic and productivity data from FACE experiments support early predictions? *New Phytologist*, 162, 253–280.

Odum, E. P. (1953). *Fundamentals of Ecology*. Philadelphia: Saunders.

Odum, E. P. (1969). The strategy of ecosystem development. *Science*, 164, 262–270.

Odum, E. P. (1971). *Fundamentals of Ecology*, 3rd edn. Philadelphia: Saunders.

Oren, R., Ellsworth, D. S., Johnsen, K. H., et al. (2001). Soil fertility limits carbon sequestration by forest ecosystems in a CO_2-enriched atmosphere. *Nature*, 411, 469–472.

Pastor, J., Aber, J. D., McClaugherty, C. A., and Melillo, J.M. (1984). Aboveground production and N and P cycling along a nitrogen mineralization gradient on Blackhawk Island, Wisconsin. *Ecology*, 65, 256–268.

Pataki, D. E., Ehleringer, J. R., Flanagan, L. B., et al. (2003). The application and interpretation of Keeling plots in terrestrial carbon cycle research. *Global Biogeochemical Cycles*, 17, 1022, doi:10.1029/2001GB001850.

Piao, S., Luyssaert, S., Ciais, P., et al. (2010). Forest annual carbon cost: A global-scale analysis of autotrophic respiration. *Ecology*, 91, 652–661.

Reich, P. B., and Hobbie, S. E. (2013). Decade-long soil nitrogen constraint on the CO_2 fertilization of plant biomass. *Nature Climate Change*, 3, 278–282.

Reich, P. B., Hobbie, S. E., Lee, T., et al. (2006a). Nitrogen limitation constrains sustainability of ecosystem response to CO_2. *Nature*, 440, 922–925.

Reich, P. B., Hungate, B. A., and Luo, Y. (2006b). Carbon–nitrogen interactions in terrestrial ecosystems in response to rising atmospheric carbon dioxide. *Annual Review of Ecology, Evolution, and Systematics*, 37, 611–636.

Reich, P. B., Hobbie, S. E., and Lee, T. D. (2014). Plant growth enhancement by elevated CO_2 eliminated by joint water and nitrogen limitation. *Nature Geoscience*, 7, 920–924.

Reichstein, M., Falge, E., Baldocchi, D., et al. (2005). On the separation of net ecosystem exchange into assimilation and ecosystem respiration: Review and improved algorithm. *Global Change Biology*, 11, 1424–1439.

Runyon, J., Waring, R. H., Goward, S. N., and Welles, J. M. (1994). Environmental limits on net primary production and light-use efficiency across the Oregon transect. *Ecological Applications*, 4, 226–237.

Rustad, L. E., Campbell, J. L., Marion, G. M., et al. (2001). A meta-analysis of the response of soil respiration, net nitrogen mineralization, and aboveground plant growth to experimental ecosystem warming. *Oecologia*, 126, 543–562.

Ryan, D. F., and Bormann, F. H. (1982). Nutrient resorption in northern hardwood forests. *BioScience*, 32, 29–32.

Ryan, M. G. (1991). Effects of climate change on plant respiration. *Ecological Applications*, 1, 157–167.

Ryan, M. G., Binkley, D., and Fownes, J. H. (1997). Age-related decline in forest productivity: Pattern and process. *Advances in Ecological Research*, 27, 213–262.

Ryan, M. G., Binkley, D., Fownes, J. H., Giardina, C. P., and Senock, R. S. (2004). An experimental test of the causes of forest growth decline with stand age. *Ecological Monographs*, 74, 393–414.

Schulze, E. D., Luyssaert, S., Ciais, P., et al. (2009). Importance of methane and nitrous oxide for Europe's terrestrial greenhouse-gas balance. *Nature Geoscience*, 2, 842–850.

Schulze, E. D., Ciais, P., Luyssaert, S., et al. (2010). The European carbon balance. Part 4: Integration of carbon and other trace-gas fluxes. *Global Change Biology*, 16, 1451–1469.

Steele, S. J., Gower, S. T., Vogel, J. G., and Norman, J. M. (1997). Root mass, net primary production and turnover in aspen, jack pine and black spruce forests in Saskatchewan and Manitoba, Canada. *Tree Physiology*, 17, 577–587.

Tansley, A. G. (1935). The use and abuse of vegetational concepts and terms. *Ecology*, 16, 284–307.

Urbanski, S., Barford, C., Wofsy, S., et al. (2007). Factors controlling CO_2 exchange on timescales from hourly to decadal at Harvard Forest. *Journal of Geophysical Research*, 112, G02020, doi:10.1029/2006JG000293.

Van Cleve, K., Dyrness, C. T., Viereck, L. A., et al. (1983a). Taiga ecosystems in interior Alaska. *BioScience*, 33, 39–44.

Van Cleve, K., Oliver, L., Schlentner, R., Viereck, L. A., and Dyrness, C. T. (1983b). Productivity and nutrient cycling in taiga forest ecosystems. *Canadian Journal of Forest Research*, 13, 747–766.

Van Cleve, K., Chapin, F. S., III, Flanagan, P. W., Viereck, L. A., and Dyrness, C. T. (1986). *Forest Ecosystems in the Alaskan Taiga: A Synthesis of Structure and Function*. New York: Springer-Verlag.

Van Cleve, K., Chapin, F. S., III, Dyrness, C. T., and Viereck, L. A. (1991). Element cycling in taiga forests: State-factor control. *BioScience*, 41, 78–88.

Viereck, L. A., Dyrness, C. T., Van Cleve, K., and Foote, M. J. (1983). Vegetation, soils, and forest productivity in selected forest types in interior Alaska. *Canadian Journal of Forest Research*, 13, 703–720.

Vitousek, P. M., Menge, D. N. L., Reed, S. C., and Cleveland, C. C. (2013). Biological nitrogen fixation: Rates, patterns and ecological controls in terrestrial ecosystems. *Philosophical Transactions of the Royal Society B*, 368, 20130119, doi:10.1098/rstb.2013.0119.

Whittaker, R. H. (1966). Forest dimensions and production in the Great Smoky Mountains. *Ecology*, 47, 103–121.

Whittaker, R. H., Bormann, F. H., Likens, G. E., and Siccama, T. G. (1974). The Hubbard Brook Ecosystem Study: Forest biomass and production. *Ecological Monographs*, 44, 233–252.

Williams, M., Rastetter, E. B., Fernandes, D. N., et al. (1997). Predicting gross primary productivity in terrestrial ecosystems. *Ecological Applications*, 7, 882–894.

Wu, Z., Dijkstra, P., Koch, G. W., Peñuelas, J., and Hungate, B. A. (2011). Responses of terrestrial ecosystems to temperature and precipitation change: A meta-analysis of experimental manipulation. *Global Change Biology*, 17, 927–942.

21

Soil Biogeochemistry

21.1 | Chapter Summary

Soils are the site of much geochemical and biological activity. Chemical weathering occurs when water, acids, and other substances react with minerals in rocks and soils. It occurs concurrent with physical weathering, which is the physical disintegration of rocks by various forces. Silicate clays and iron or aluminum oxide clays are the resistant end-products of chemical weathering. Chemical weathering releases elements into the soil solution for uptake by plants. In addition, the decomposition of plant detritus mineralizes nutrients for plant use. The rate of decomposition varies with temperature, soil water, and the chemical quality of litter. Various nitrogen trace gases diffuse out of soils during decomposition. The outcome of these processes is seen in the soil profile and its development over time. There are 12 broad classes of soil, known as soil orders, that vary in relation to degree of weathering, extent of soil development, climate, and associated vegetation. Parent material, time, topography, climate, and vegetation govern soil formation. Climate, particularly temperature and precipitation, determines the nature and rate of the weathering that occurs. Vegetation affects soil structure and fertility through the cycling of materials between plants and soil.

21.2 | Weathering

The sand, silt, and clay particles that comprise mineral soil derive from physical and chemical weathering that breaks rocks into smaller and smaller fragments until individual minerals are exposed or new minerals are created. The mineralogical composition of rocks undergoes changes during weathering and is broadly grouped into primary and secondary minerals (Table 21.1). Primary minerals are resistant to weathering and change little in chemical composition during weathering. They are prominent in the sand and silt fractions of soil. Secondary minerals are products of the chemical breakdown or alteration of less resistant minerals. Secondary minerals are small in size and dominate the clay fraction of soil. Silicate clays and iron and aluminum oxide clays are prominent secondary minerals.

Physical weathering is the physical disintegration of rocks. It occurs from the scouring of rocks by water, wind, or ice. Alternating cycles of freezing and thawing or wetting and drying mechanically disintegrate rocks because of swelling and shrinking. Plants can physically weather rocks. Roots push through small cracks in rocks, splitting them apart as the roots grow bigger. Primary minerals such as quartz and muscovite are very resistant to weathering (Figure 21.1). These eventually form sand and

Table 21.1 Primary and secondary minerals commonly found in rocks and soils

Primary mineral		Secondary minerals			
		Resistant end-product		Low resistance	
Quartz	SiO_2	Fe, Al oxide clays		Dolomite	$CaCO_3 \cdot MgCO_3$
Micas		Goethite	$FeOOH$	Calcite	$CaCO_3$
Muscovite	$KAl_3Si_3O_{10}(OH)_2$	Hematite	Fe_2O_3	Gypsum	$CaSO_4 \cdot 2H_2O$
Biotite	$KAl(Mg,Fe)_3Si_3O_{10}(OH)_2$	Gibbsite	$Al(OH)_3$		
Feldspars		Silicate clays			
Orthoclase	$KAlSi_3O_8$	Kaolinite	$Al_2Si_2O_5(OH)_4$		
Albite	$NaAlSi_3O_8$	Vermiculite	2:1 layer-silicate		
Anorthite	$CaAl_2Si_2O_8$	Montmorillonite	2:1 layer-silicate		
Hornblende[a]	$Ca_2Al_2Mg_2Fe_3Si_6O_{22}(OH)_2$	Mica	2:1 layer-silicate		
Augite[b]	$Ca_2(Al,Fe)_4(Mg,Fe)_4Si_6O_{24}$	Chlorite	2:1 layer-silicate		
Olivine	$(Mg,Fe)_2SiO_4$				

Note: Primary minerals are arranged in order of decreasing resistance to weathering. Quartz is the most resistant. Olivine is the least resistant. Secondary minerals are distinguished as resistant end-products of weathering and intermediate products with low resistance to weathering.

[a] The chemical formula for amphibole.

[b] The chemical formula for pyroxene.

Source: From Brady and Weil (1999, p. 32).

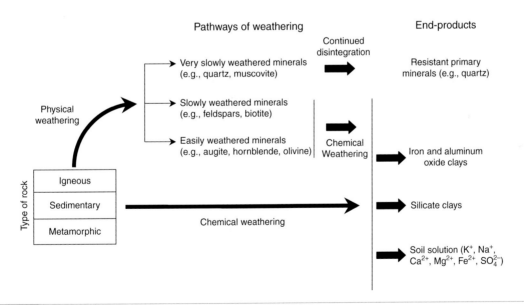

Fig. 21.1 Pathways of physical and chemical weathering. Rocks are broadly classified as igneous, sedimentary, and metamorphic, which differ in origin and chemical and mineralogical properties. Three groups of minerals remain in well-weathered soils: resistant primary minerals such as quartz; iron and aluminum oxide clays; and silicate clays. Elements in solution are taken up by plants or lost in leaching. Adapted from Brady and Weil (1999, p. 33).

Table 21.2 Chemical weathering of elements at the Hubbard Brook Experimental Forest, New Hampshire

Element	Symbol	Mass (kg per 1500 kg of bedrock)	Annual release in weathering ($g\ m^{-2}\ yr^{-1}$)	Weathering ratio	Input (%)	
					Atmosphere	Weathering
Calcium	Ca^{2+}	21.1	2.11	1.00	9	91
Sodium	Na^{+}	24.1	0.58	0.24	22	78
Magnesium	Mg^{2+}	16.5	0.35	0.21	15	85
Potassium	K^{+}	43.6	0.71	0.16	11	89
Aluminum	Al^{3+}	124.8	0.19	0.02	–	–
Silica	Si^{4+}	461.7	1.81	0.02	–	–

Note: Weathering ratio is the ratio of release to mass and indicates weathering relative to that of Ca^{2+}. –, no data.
Source: From Likens et al. (1977, p. 92) and Schlesinger (1997, p. 119).

silt particles, which are chemically inert and contribute little to soil fertility. Other primary minerals are more easily weathered, breaking up into small fragments that are then chemically weathered.

Chemical weathering is the process by which individual minerals dissolve or transform to new minerals. It occurs concurrent with physical weathering. Physical weathering increases the surface area of rocks and minerals exposed to chemical weathering; chemical weathering results in further decrease in particle size. Chemical weathering occurs when minerals in rocks and soils react with water, acids, and other substances. During chemical weathering, soluble materials are released into the soil solution for uptake by plants or export to rivers during runoff. New minerals are created, some of which are resistant to further weathering and some of which undergo additional chemical transformations (Figure 21.1).

The dominant type of chemical weathering is by carbonic acid (H_2CO_3) formed in soil from water (H_2O) and carbon dioxide (CO_2):

$$CO_2 + H_2O \rightleftharpoons H^+ + HCO_3^- \rightleftharpoons H_2CO_3 \qquad (21.1)$$

Carbonic acid decomposes rocks, releasing chemical elements into the soil solution. For example, calcite ($CaCO_3$), the dominant mineral in limestone and marble, dissolves in the presence of carbonic acid releasing calcium ions (Ca^{2+}) and HCO_3^- into solution:

$$CaCO_3 + H_2CO_3 \rightleftharpoons Ca^{2+} + 2HCO_3^- \qquad (21.2)$$

This weathering by H_2CO_3 is part of the geological carbon cycle, in which rivers carry the dissolved byproducts of weathering to oceans. Rocks weather through numerous other chemical reactions (Montgomery et al. 2000). Silicate clays and iron or aluminum oxide clays are the resistant products of chemical weathering (Table 21.1, Figure 21.1).

The chemical weathering of rocks releases elements into the soil solution for uptake by plants. The elements released vary among primary minerals (Table 21.1). Data collected at the Hubbard Brook Experimental Forest in New Hampshire give an indication of the amount of elements released annually in chemical weathering (Table 21.2). This forest is part of the extensive northern hardwood forest in northern United States and southern Canada. The climate is cool humid continental, with cold winters and cool summers. Silica and aluminum are the most abundant elements in those rocks. Calcium, sodium, magnesium, and potassium occur in lesser amounts. These elements weather differently. Calcium, sodium, magnesium, and potassium have comparatively large weathering ratios (defined as the ratio of annual element release to mass in bedrock) and are less resistant to weathering than aluminum or silica. Rock weathering is the dominant input of calcium, sodium, magnesium, and potassium to the forest.

21.3 | Decomposition and Mineralization

In addition to that released in the chemical weathering of rocks, nutrients are also contained in leaves, branches, stems, roots, and other organic matter. Over time, this material falls to the ground where microbes and other soil microorganisms consume the material, gaining energy from the carbon and breaking the residues into smaller and smaller pieces. This process is called decomposition. When the plant detritus is decomposed to the point that it is no longer recognizable, it is referred to as soil organic matter or humus. Heterotrophic respiration by microbes during decomposition returns the carbon incorporated into living organisms back to the atmosphere ($\sim$50–60 Pg C yr^{-1}). During this transformation, nutrients bound in organic material are released in an inorganic form for use by plants in a process known as mineralization.

The amount of organic material remaining at a given time can be expressed as:

$$M = M_0 e^{-kt} \tag{21.3}$$

where M_0 is the original mass, k is a decay rate (yr^{-1}), and t is time in years. The fraction of the original organic matter remaining after one year is $M / M_0 = e^{-k}$. The larger the annual decay rate, the less organic material that accumulates on the ground. For example, $k = 1$ means that 37 percent of the original litter mass remains after one year. This is representative of humid tropical climates. In cold climates, a typical value is $k = 0.1$ and 90 percent of the original litter mass remains after one year.

The rate of decomposition is a function of soil temperature and water. In general, organic material decomposes faster in warm soil than in cold soil because microbial activity increases with higher temperatures. In the example shown in Figure 21.2a, the rate of decomposition doubles for every 10°C increase in temperature. Dry soils do not have sufficient water to support microbial growth. However, extremely wet soils do not provide enough oxygen for optimal microbial activity. In Figure 21.2b,

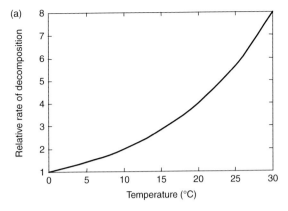

(a)

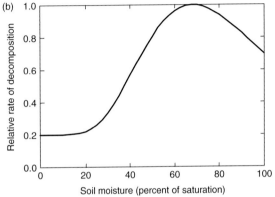

(b)

Fig. 21.2 Generalized effect of (a) temperature and (b) soil moisture on decomposition. From a terrestrial ecosystem model (Raich et al. 1991; McGuire et al. 1992).

maximum decomposition occurs with a water content of about 70 percent of saturation.

The rate of decomposition also relates to the chemical quality of litter. Litter with a high concentration of nutrients provides more nutrients to support microbial activity and therefore decomposes faster than litter with fewer nutrients. One index of litter quality is the ratio of carbon to nitrogen (C:N) in litter. Debris with a high C:N ratio decomposes slower than debris with a low C:N ratio. Another index of litter quality is the ratio of lignin to nitrogen (lignin:N). Lignin is a compound that makes plant parts structurally rigid (i.e., woody). It is common in leaves, stems, and wood and is one of the slowest plant materials to decay.

Mass loss during the first year of decomposition decreases as the lignin:N ratio in foliage

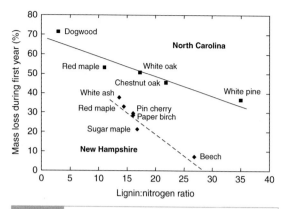

Fig. 21.3 Relationship between mass loss during the first year of decomposition and initial concentrations of lignin and nitrogen in leaf litter for six tree species in New Hampshire and five tree species in North Carolina. Data from Melillo et al. (1982).

increases. This is illustrated in Figure 21.3 for common tree species in North Carolina and New Hampshire. In the warm climate of North Carolina, more than 50–70 percent of litter mass is lost over the course of a year for red maple and dogwood trees, which have a low lignin:N ratio. In contrast, white pine litter has a high lignin:N ratio and loses less than 40 percent of its mass during one year. White oak and chestnut oak have intermediate lignin:N ratios and decay rates. In the northern hardwood forests of New Hampshire, common tree species include red maple, pin cherry, paper birch, white ash, beech, and sugar maple. Decay rates for similar lignin:N ratios are less in New Hampshire than in North Carolina, illustrating the influence of the colder climate on decomposition. Ratios of foliage lignin:N average 33 for temperate needleleaf evergreen trees and 17 for temperate broadleaf deciduous trees; C:N ratios average 64 and 46, respectively (Brovkin et al. 2012). Foliage litter from broadleaf deciduous trees, because it is of better quality, decomposes faster than that of needleleaf evergreen trees.

When plant debris falls onto the ground, nutrients (e.g., nitrogen) return to the soil as the litter decomposes and releases organically bound nutrients. However, this nitrogen is not immediately released for plant use. Microbes require nitrogen in certain concentrations to grow, and microbial growth creates demand for nutrients. This nitrogen must be available from either the decomposing material or the soil. During the initial phases of decomposition, microbial demand is greater than the supply within the decaying material. There is a net increase in the amount of nitrogen in the decomposing litter as microbes take up nitrogen from the soil and incorporate it into litter. This process, known as immobilization, reduces the amounts of nitrogen available for plant growth. As more of the material decomposes, microbial demand diminishes. When nitrogen release is greater than microbial demand, there is a net mineralization of nitrogen that is available to support plant growth. Typically, fresh litter (with high C:N ratio) immobilizes nitrogen while humus (with low C:N ratio) mineralizes nitrogen.

As a simple example, consider the carbon and nitrogen dynamics as microbes decompose 100 g C of litter. If the litter has a C:N ratio of 25:1, the decomposition mineralizes 4 g N (100 g C / 25 g C g^{-1} N); this is the gross mineralization. Assume that the C:N ratio of microbial biomass is 10:1, and that the microbial growth efficiency is 0.4. Microbial growth efficiency is the fraction of the carbon turnover that is assimilated into microbial biomass; the remainder is lost as heterotrophic respiration. If the microbes decompose 100 g C of litter, their mass increases by 40 g C and consumes 4 g N (40 g C / 10 g C g^{-1} N). With a litter C:N ratio of 25:1, enough nitrogen is mineralized to meet the microbial demand. In this example, gross mineralization is 4 g N, immobilization is 4 g N, and net mineralization (gross mineralization – immobilization) is 0 g N. With a litter C:N ratio greater than 25:1, immobilization exceeds gross mineralization and net mineralization is negative. Net mineralization is positive for litter C:N < 25.

A study by Aber and Melillo (1980) shows the dynamics of mass loss, nitrogen immobilization, and nitrogen mineralization for yellow birch leaves. Figure 21.4a illustrates the carbon and nitrogen dynamics as fresh litter decomposes. It shows the amount of nitrogen in the residue, expressed as a percentage of the original nitrogen content, plotted in relation to the

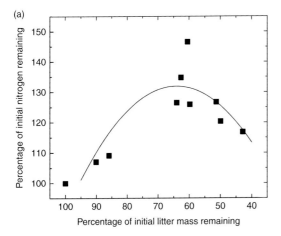

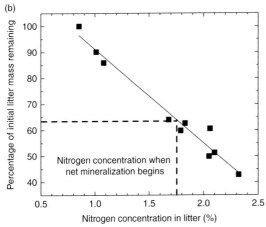

Fig. 21.4 Decomposition and nitrogen dynamics of yellow birch leaves. (a) Relationship between nitrogen content in the remaining litter (as a percentage of original nitrogen content) and remaining litter mass. (b) Relationship between litter mass (as a percentage of original mass) and nitrogen concentration in the remaining litter. Data from Aber and Melillo (1980).

remaining litter mass. As the litter decreases in mass from 100 percent to 64 percent, the amount of nitrogen in the material increases relative to the initial content because of immobilization. Thereafter, further decay results in a net release of nitrogen. Over the course of decomposition, as the mass of the leaf litter decreases, the amount of nitrogen in the decaying material increases from an initial concentration of 0.85 percent of litter mass to 2.3 percent of litter mass (Figure 21.4b). There is a linear increase in litter nitrogen as litter mass

decreases. In this example, net mineralization occurs when about 36 percent of the original litter mass has been lost, when the nitrogen concentration is about 1.75 percent (Aber and Melillo 1982). At this point, about 7.4 mg of nitrogen has been immobilized for each gram of debris decayed.

Common litter decomposition studies across various biomes illustrate the generality of these concepts. One such study was a 10-year multi-site litterbag experiment designed to investigate the effects of litter quality and climate on decomposition (Gholz et al. 2000; Parton et al. 2007; Harmon et al. 2009). The study decomposed common leaf and fine root litter at 27 sites across North America and Central America representing a range of climatic conditions from arctic to tropical and humid to arid. The litter types differed in chemical quality and represented a range of nitrogen concentration and lignin. Mass loss over the 10-year study was lowest at tundra sites, higher in deciduous forest, and greatest at tropical forest (Figure 21.5). A climatic decomposition index, calculated from monthly air temperature, precipitation, and potential evapotranspiration, accounted for geographic and seasonal variation across sites and correlated with decomposition rates (Gholz et al. 2000; Parton et al. 2007; Adair et al. 2008; Currie et al. 2010). The dynamics of nitrogen immobilization and mineralization varied based on initial litter nitrogen concentration (Figure 21.6). Across all sites, leaf litter with high initial nitrogen concentration (*Drypetes glauca*, 1.97% N) immobilized little nitrogen and instead had net mineralization. Leaf litter with low initial nitrogen (*Triticum aestivum*, 0.38% N) immobilized the most nitrogen; net mineralization did not begin until about 60 percent of the litter mass had decomposed.

As decomposition proceeds, an increasing amount of the residual material resists decay. This decay-resistant material accumulates as organic matter in the soil profile. Studies of soil development over 1000–10,000 years show soil organic matter accumulates at a rate of 0.2–12 g C m^{-2} yr^{-1}, with highest rates under cool,

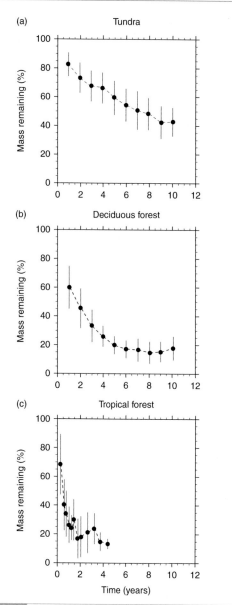

Fig. 21.5 Average litter mass remaining (as a percentage of original mass) in relation to time for leaf litter decomposed at different sites. Data are averaged across six leaf litter types for sites classified as (a) tundra (*n* = 2 sites), (b) deciduous forest (*n* = 3 sites), and (c) tropical forest (*n* = 3 sites). Shown are the observations (symbols) with ± 1 standard deviation. Data from Parton et al. (2007).

wet conditions (Schlesinger 1990). Cold, arctic soils contain in excess of 15 kg C m^{-2}; warmer climates may have less carbon, 6–9 kg C m^{-2} (Figure 24.11).

Litter decomposition and soil organic matter formation can be conceptualized in terms of several carbon pools of varying quality and decay rates and the flows among those pools. One such model is the CENTURY model (Parton et al. 1987, 1988, 1993, 1994). The model, shown in Figure 21.7, has two litter pools (structural and metabolic) and three organic matter pools (active, slow, and passive), each with different base decomposition rates. These base rates are adjusted for site conditions such as temperature, moisture, pH, texture, and other factors. The pools have a surface and a belowground, or soil, component, except the passive pool, which is only belowground. Leaf and fine root plant residue are partitioned into structural and metabolic litter based on the lignin:N ratio of the residue, with foliage assigned to surface litter and roots to belowground litter. Base turnover times are 20–46 days for metabolic litter and 74–182 days for structural litter, with more rapid turnover for belowground litter.

The five organic matter pools (two surface and three belowground) differ in base decomposition rates and abiotic controls. The active organic matter represents soil microbes and microbial products and has a base turnover of 33 days (belowground) and 61 days (surface). The slow organic matter includes resistant plant material derived from structural litter and dead wood, as well as soil-stabilized microbial products derived from the active pool. The base turnover is 2.5 years (belowground) and 12.5 years (surface). The surface slow pool additionally transfers material to the soil slow pool by physical mixing, with a base turnover time of 4–10 years. The passive pool is very resistant to decomposition, includes physically and chemically stabilized soil organic matter, and has a base turnover of 303 years.

Each carbon pool has an analogous organic nitrogen pool. Soil organic matter pools have an allowable range of C:N ratios, specified as a linear function of soil mineral nitrogen. Low mineral nitrogen results in high C:N ratios in the various organic matter pools. Mineralization or immobilization of nitrogen occurs as is necessary to maintain the C:N ratios required for the receiver pool.

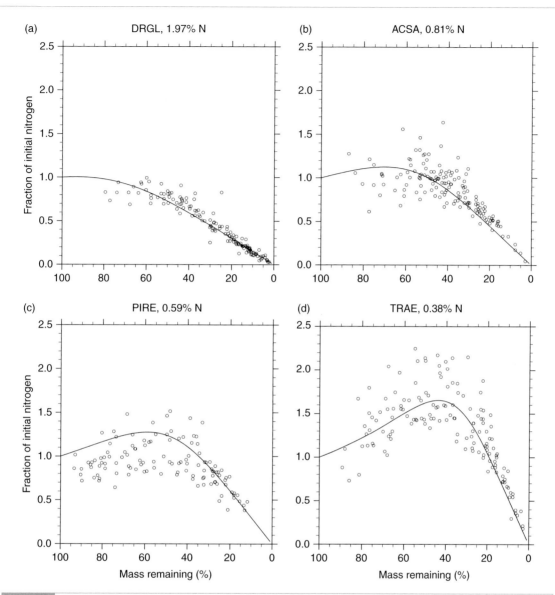

Fig. 21.6 Fraction of initial nitrogen remaining in relation to carbon mass remaining for (a) *Drypetes glauca* (DRGL), (b) *Acer saccharum* (ACSA), (c) *Pinus resinosa* (PIRE), and (d) *Triticum aestivum* (TRAE) leaf litter decomposed at tundra, forest, and humid grassland sites. These litter types varied in initial nitrogen concentration, from 0.38 percent N (TRAE) to 1.97 percent N (DRGL). Shown are the observations (open symbols) and the best-fit equation for the observations (solid lines). Data from Parton et al. (2007).

21.4 | Soil Solution

Numerous elements are essential for plant growth and development (Table 21.3). Plants require large quantities of nine elements, and these are called macronutrients. Carbon, oxygen, and hydrogen are the basic constituents of organic matter and together comprise about 94 percent of the dry mass of plants. The remaining six elements are obtained from dissolved minerals in the soil solution and serve various physiological roles. Of these nitrogen, phosphorus, and potassium are the three most

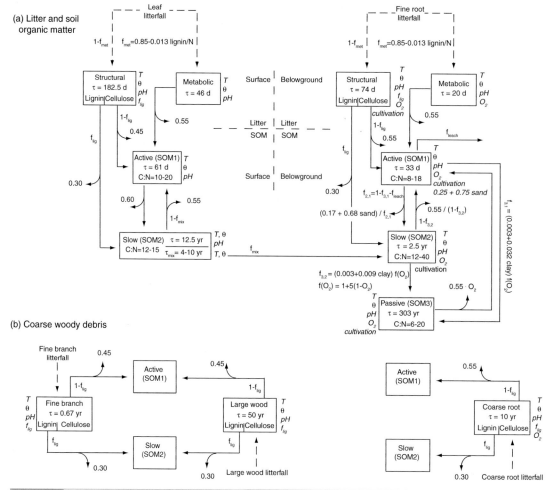

Fig. 21.7 Litter and soil organic matter pools and carbon flows represented in the CENTURY model. (a) Decomposition of leaf and fine root litter. Surface material (shown on the left) is represented by two litter pools (metabolic and structural) and active (SOM1) and slow (SOM2) organic matter pools. Belowground material (shown on the right) is represented by two litter pools (metabolic and structural) and active (SOM1), slow (SOM2), and passive (SOM3) organic matter pools. Shown is the base decomposition rate of each pool, given here as a turnover time (τ) ranging from days (d) to years (yr). The actual decomposition rate varies with soil temperature (T), soil moisture (θ), and pH. Belowground decomposition additionally varies with anaerobic conditions (O_2) and cultivation. Structural litter decomposition also depends on lignin fraction (f_{lig}). The C:N ratio of organic matter differs among pools and varies with soil mineral nitrogen. Shown is the minimum and maximum value for each pool. Solid lines indicate decomposition pathways, with curved arrows denoting heterotrophic respiration fluxes for each pathway and numbers denoting the respiration fraction. Shown also is the fraction of the total carbon flow represented by a pathway. (b) Decomposition of fine branch and large wood litter into surface pools and coarse root litter into belowground pools. Adapted from Bonan et al. (2013).

important. Nitrogen is a component of proteins, chlorophyll, and enzymes in the biochemical reactions of photosynthesis. Phosphate compounds carry energy, and phosphorus is a major component of nucleic acids. Potassium has numerous functions including production of sugars and starches, opening and closing of stomata, development of strong stems, and development of viable seeds. Plants require micronutrients in much smaller amounts and utilize these elements in various functions such as chlorophyll synthesis, photosynthesis,

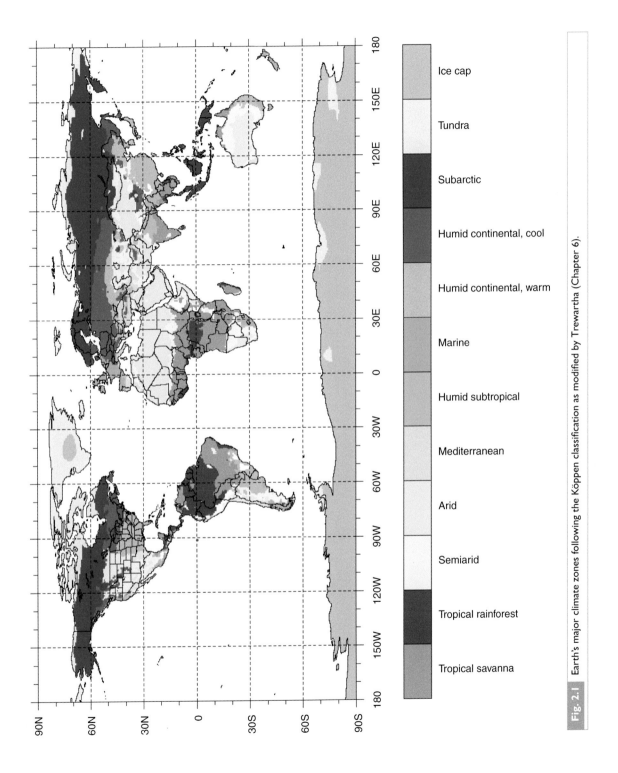

Ice cap

Tundra

Subarctic

Humid continental, cool

Humid continental, warm

Marine

Humid subtropical

Mediterranean

Arid

Semiarid

Tropical rainforest

Tropical savanna

Fig. 2.1 Earth's major climate zones following the Köppen classification as modified by Trewartha (Chapter 6).

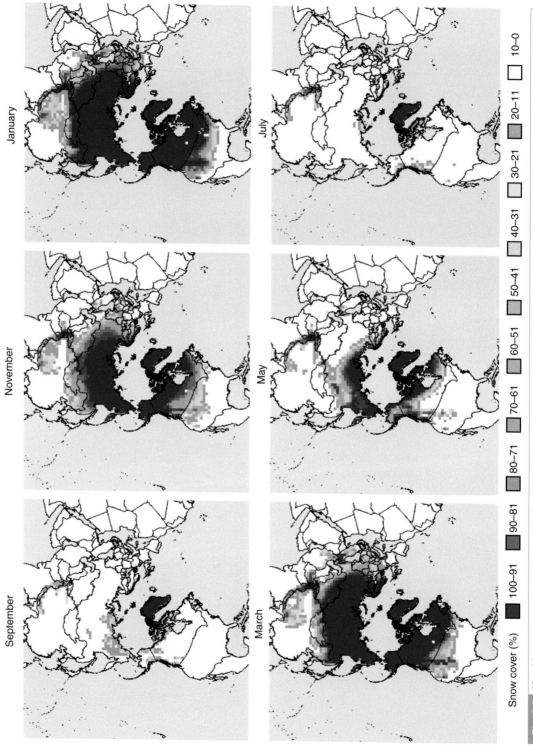

Fig. 2.2 Monthly snow cover climatology in the Northern Hemisphere for the period 1966–1999. Maps provided courtesy of David Robinson (Global Snow Lab, Rutgers University).

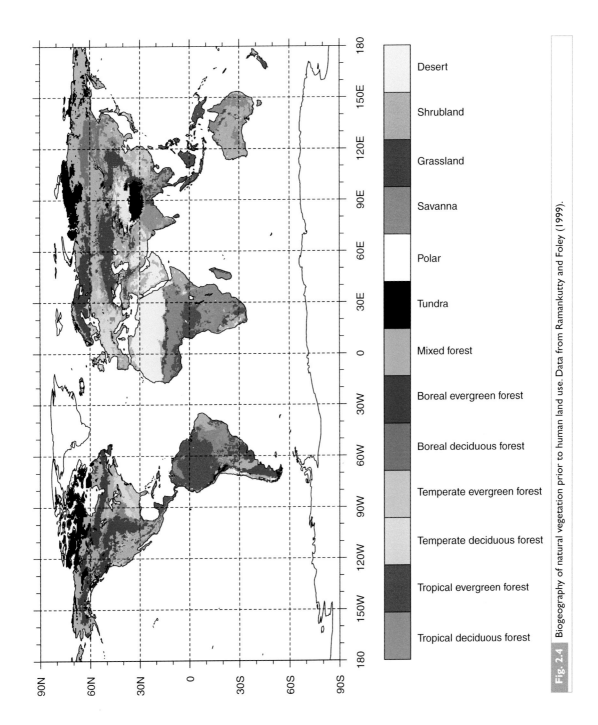

Fig. 2.4 Biogeography of natural vegetation prior to human land use. Data from Ramankutty and Foley (1999).

Desert

Shrubland

Grassland

Savanna

Polar

Tundra

Mixed forest

Boreal evergreen forest

Boreal deciduous forest

Temperate evergreen forest

Temperate deciduous forest

Tropical evergreen forest

Tropical deciduous forest

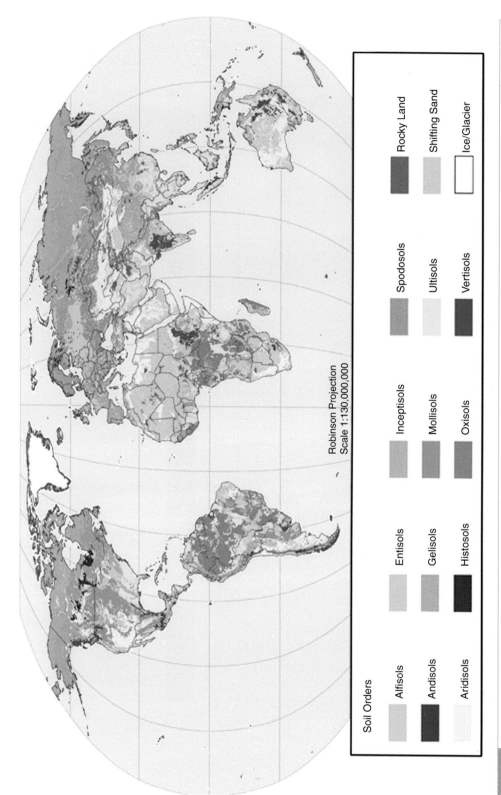

Soil Orders

Alfisols	Entisols	Inceptisols	Spodosols	Rocky Land
Andisols	Gelisols	Mollisols	Ultisols	Shifting Sand
Aridisols	Histosols	Oxisols	Vertisols	Ice/Glacier

Robinson Projection
Scale 1:130,000,000

Fig. 2.5 Geographic distribution of the twelve soil orders. Image provided courtesy of the U.S. Department of Agriculture Natural Resources Conservation Service, Soil Survey Division, World Soil Resources, Washington D.C.

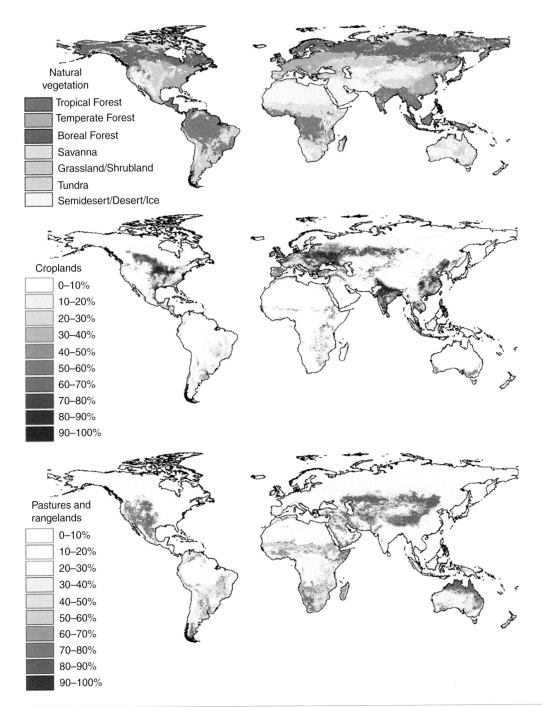

Fig. 2.8 Natural vegetation prior to human land use (top) and the extent of agricultural land during the 1990s. The panels for croplands (middle) and pastures (bottom) show the percentage of the land surface occupied by these land cover types. Reproduced from Foley et al. (2005).

Mean annual temperature

(°C)

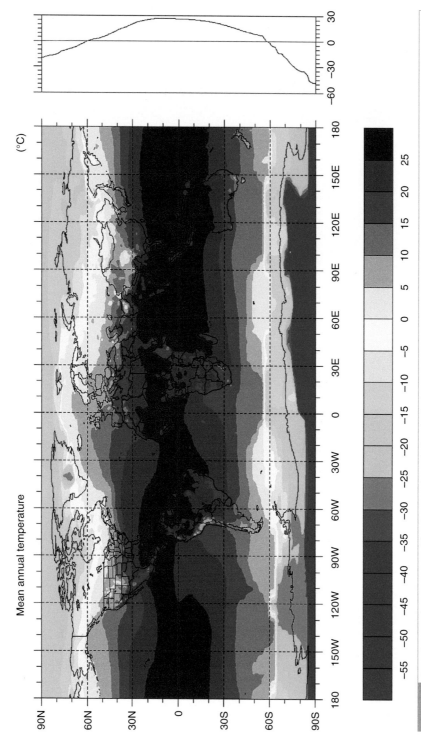

Fig. 5.6 Annual mean air temperature. The chart on the right-hand side shows temperature averaged around the world, from longitude 180° W to 180° E, as a function of latitude. Data provided by the National Center for Atmospheric Research (Boulder, Colorado).

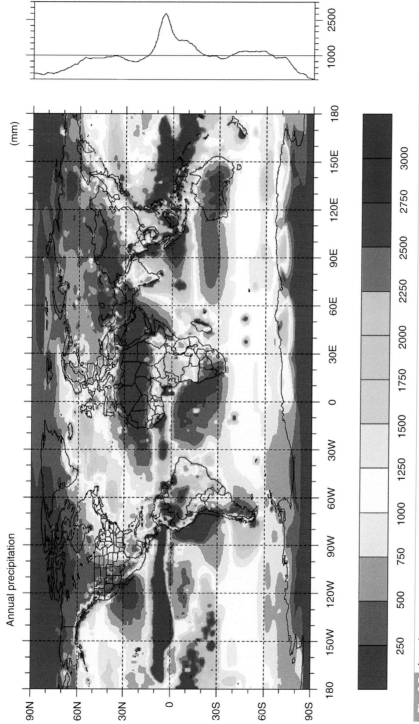

Annual precipitation

(mm)

| 250 | 500 | 750 | 1000 | 1250 | 1500 | 1750 | 2000 | 2250 | 2500 | 2750 | 3000 |

Fig. 5.7 Annual mean precipitation. The chart on the right-hand side shows precipitation averaged around the world, from longitude 180° W to 180° E, as a function of latitude. Data provided by the National Center for Atmospheric Research (Boulder, Colorado).

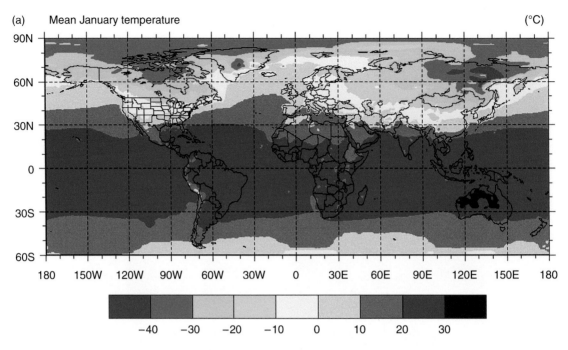

(a)   Mean January temperature   (°C)

-40   -30   -20   -10   0   10   20   30

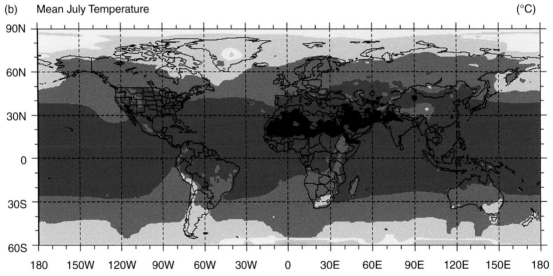

(b)   Mean July Temperature   (°C)

Fig. 5.10 Mean air temperature for (a) January and (b) July. Data provided by the National Center for Atmospheric Research (Boulder, Colorado).

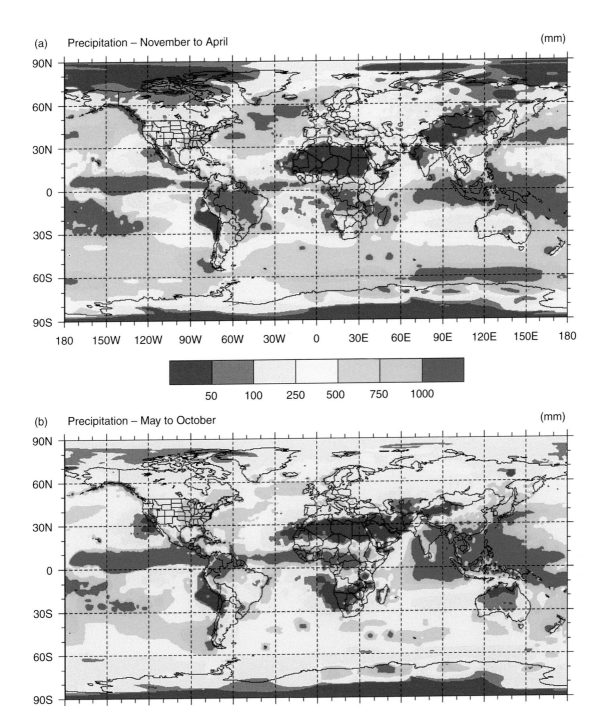

Fig. 5.12 Precipitation for (a) November–April and (b) May–October. Data provided by the National Center for Atmospheric Research (Boulder, Colorado).

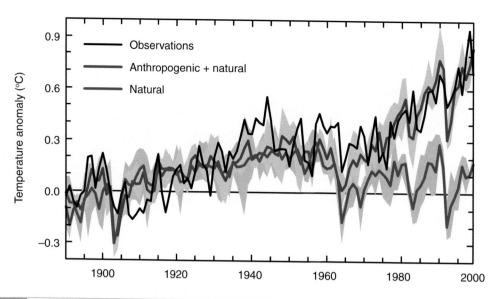

Fig. 8.15 Transient climate model simulations of the twentieth century with only natural forcings (solar irradiance, volcanic aerosols) and with natural and anthropogenic forcings (greenhouse gases, sulfate aerosols, ozone). Observed and simulated annual global mean surface air temperature is the anomaly from the 1890–1919 mean. Simulations were performed four times for each forcing. Shown are the means for each 4-member ensemble (solid line) and the ensemble range (shading). Adapted from Meehl et al. (2004).

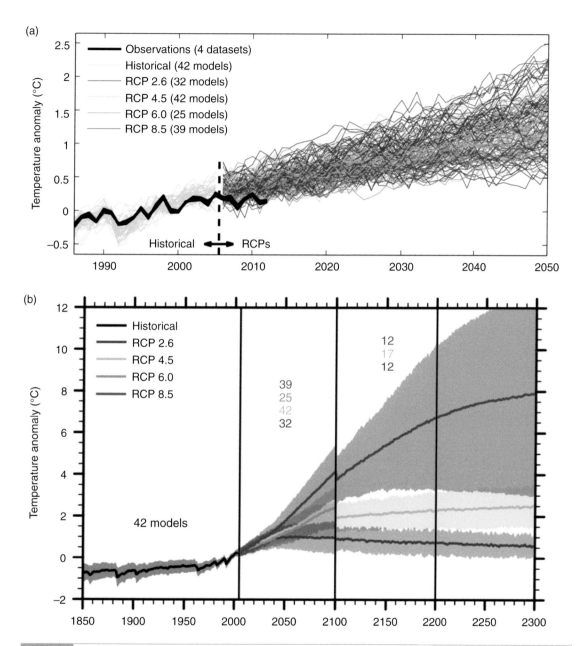

(a)

(b)

Fig. 8.16 Transient climate model simulations for the historical era and with the four representative concentration pathways (RCPs). Shown are the annual global mean surface air temperature anomalies (relative to 1986–2005) for the historical period (prior to 2005) and for the RCPs. Only one ensemble member was used from each model, and numbers in the figure indicate the number of models contributing to the different time periods. (a) Near-term climate change for 1986–2050. Each individual line is a single model realization for the historical period (gray lines) and the RCPs (colored lines). The thick black line denotes observed temperature based on four datasets. Adapted from Kirtman et al. (2013). (b) The full simulations for the historical period (1850–2005) and the RCPs (2005–2300). Solid lines show the multi-model mean and the shading denotes the spread among the individual models. The discontinuity at 2100 for RCP8.5 is an artifact of the model sample size. Adapted from Collins et al. (2013).

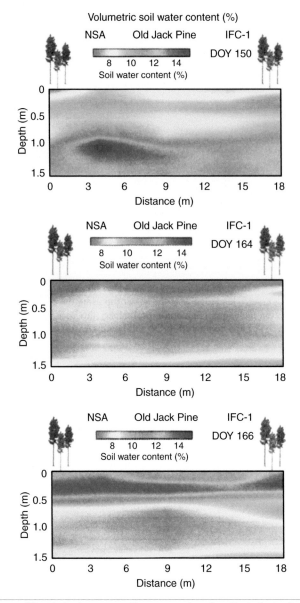

Fig. 10.7 Soil water content (%) with depth across an 18 m transect in a Canadian jack pine forest in late spring 1994. Top: moist conditions on May 30. Middle: dry down on June 13. Bottom: wetting front on June 15. Reproduced from Cuenca et al. (1997).

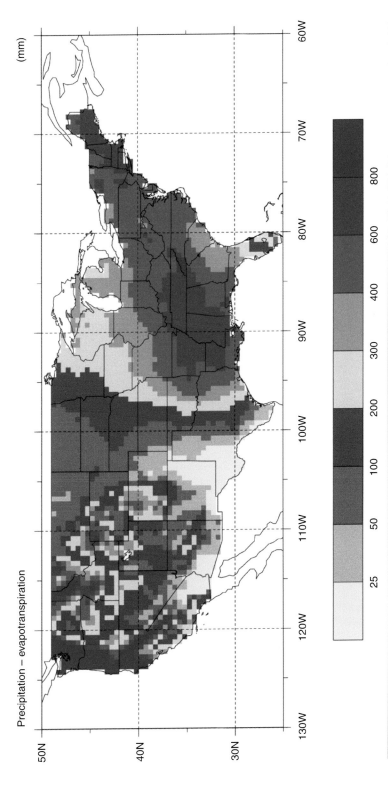

Precipitation – evapotranspiration

(mm)

25   50   100   200   300   400   600   800

Fig. 10.10 Geographic distribution of the difference between annual precipitation and evapotranspiration in the United States. Evapotranspiration is based on the water balance model of Figure 10.9 using monthly temperature and precipitation climatologies (Legates and Willmott 1990a,b) and observed soil water-holding capacity (Rosenbloom and Kittel 1996).

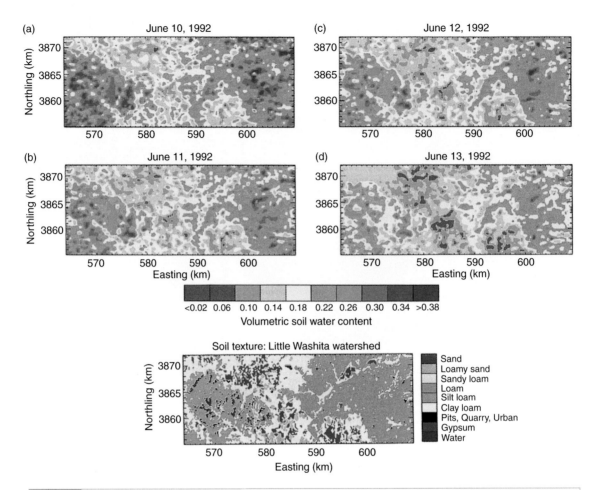

Fig. 11.7 Near-surface (0–5 cm depth) volumetric soil water content for June 10–13, 1992 in the Little Washita watershed, Oklahoma. Data have a spatial resolution of 200 m. Reproduced from Mattikalli et al. (1998).

(a)   Ringelbach catchment 3D elevation

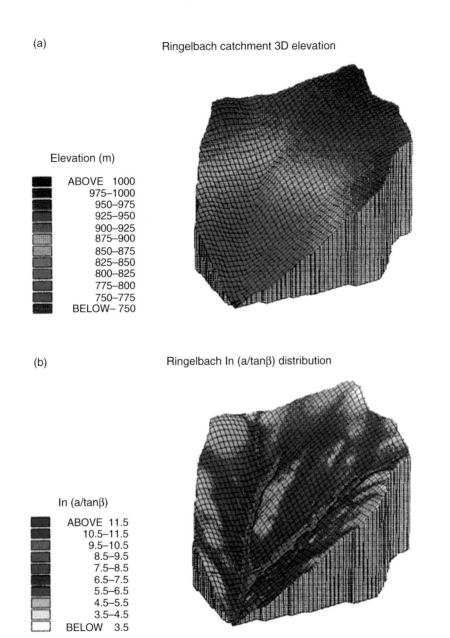

Elevation (m)

ABOVE 1000
975–1000
950–975
925–950
900–925
875–900
850–875
825–850
800–825
775–800
750–775
BELOW– 750

(b)   Ringelbach In (a/tanβ) distribution

In (a/tanβ)

ABOVE 11.5
10.5–11.5
9.5–10.5
8.5–9.5
7.5–8.5
6.5–7.5
5.5–6.5
4.5–5.5
3.5–4.5
BELOW   3.5

Fig. 11.13 Elevation (a) and topographic index (b) for the 36 ha Ringelbach watershed. Data have a spatial resolution of 5 m. Reproduced from Ambroise et al. (1996).

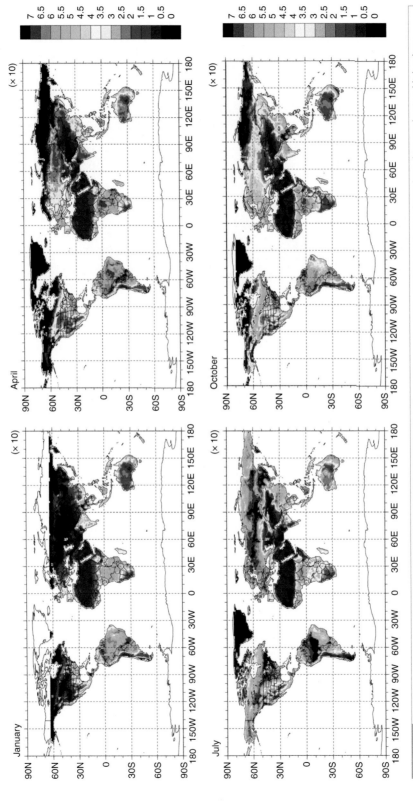

Fig. 18.4 The normalized difference vegetation index averaged for January, April, July, and October for 1982–1993. Low values indicate low plant production. High values indicate high production.

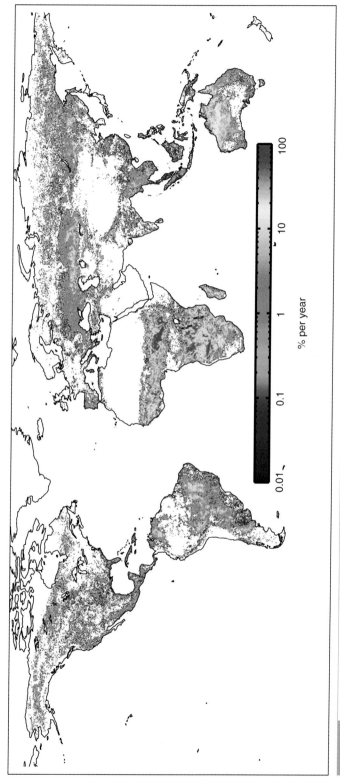

Fig. 23.5 Mean annual area burned for the period 1996–2012, expressed as the percentage of area. Reproduced from Giglio et al. (2013) and provided courtesy of Louis Giglio.

(a) CE 800

(b) CE 1400

(c) CE 1700

(d) CE 1992

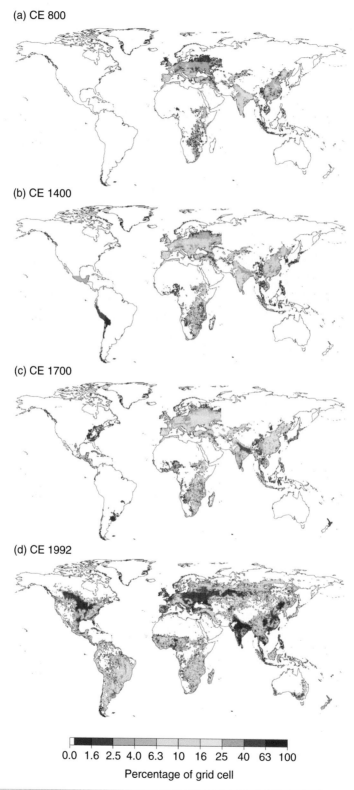

0.0 1.6 2.5 4.0 6.3 10 16 25 40 63 100
Percentage of grid cell

Fig. 23.8 Global historical cropland for the years CE 800, CE 1400, CE 1700, and CE 1992 in terms of percentage area. Reproduced from Pongratz et al. (2008) and provided courtesy of Julia Pongratz.

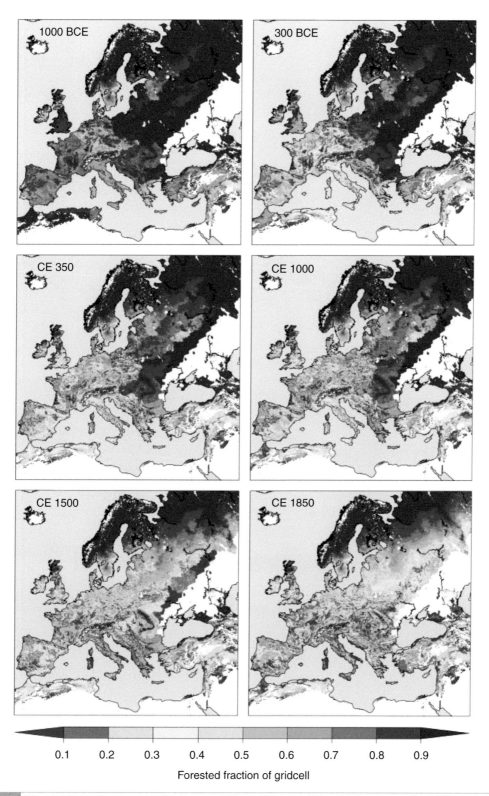

0.1   0.2   0.3   0.4   0.5   0.6   0.7   0.8   0.9

Forested fraction of gridcell

Fig. 23.9 Historical forest clearance maps for the years 1000 BCE, 300 BCE, CE 350, CE 1000, CE 1500, and CE 1850. Reprinted from Kaplan et al. (2009) with permission from Elsevier.

Fig. 23.10 Historical cropland area in the United States from 1850 to 1975 showing the fraction of the landscape covered in cropland. Data from Ramankutty and Foley (1999a,b).

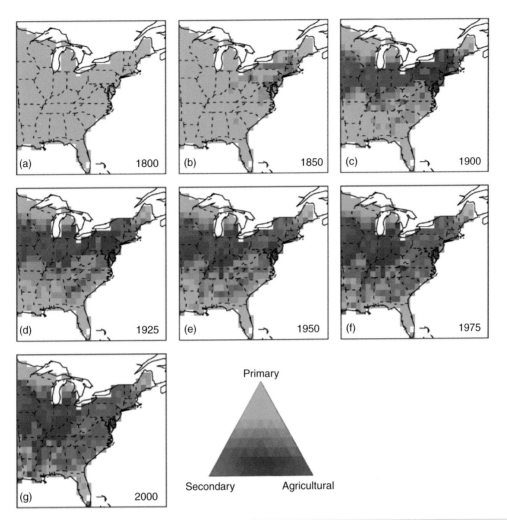

Fig. 23.11 Historical area of primary forest (green), secondary forest (blue), and agricultural land (red) in eastern United States, 1800–2000. Mixtures of land use types are indicted by color intensity and color gradients. Reproduced from Albani et al. (2006) with permission of Blackwell Publishing. See also Hurtt et al. (2006).

Normalized difference vegetation index

Annual average

(× 10)

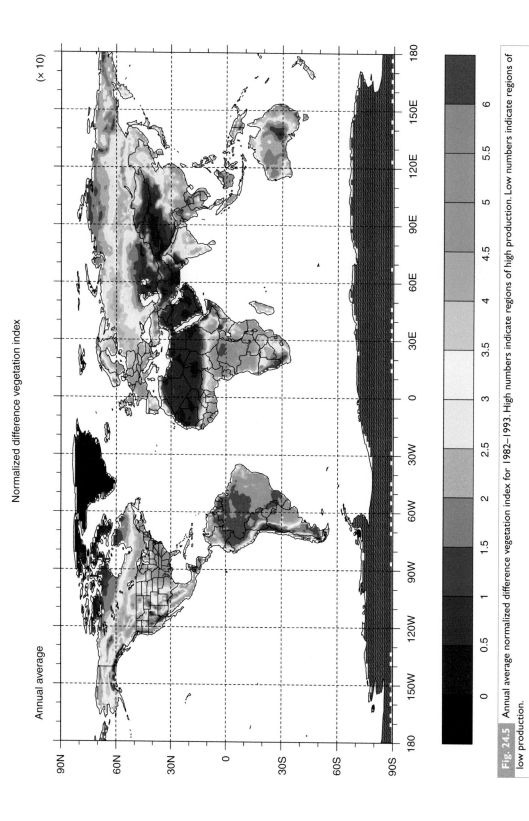

Fig. 24.5 Annual average normalized difference vegetation index for 1982–1993. High numbers indicate regions of high production. Low numbers indicate regions of low production.

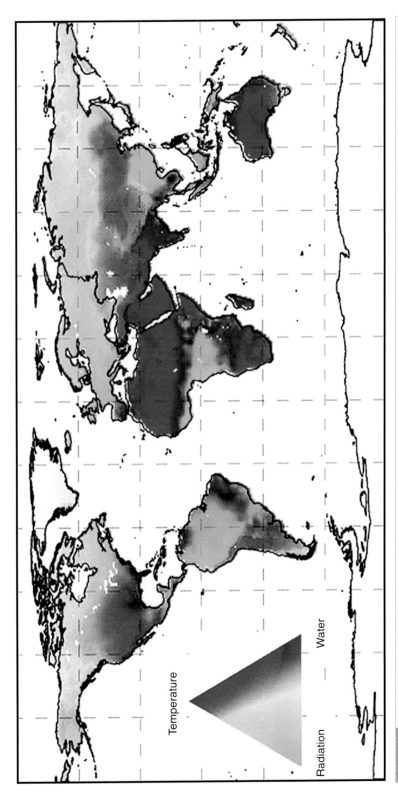

Fig. 24.6 Geographic distribution of climatic constraints to net primary production. Colors denote various limiting factors: red (water), green (radiation), and blue (temperature). Interactions among limiting factors are shown by color gradients: cyan (temperature and radiation), magenta (water and temperature), and yellow (water and radiation). Reproduced from Nemani et al. (2003) with permission of the American Association for the Advancement of Science.

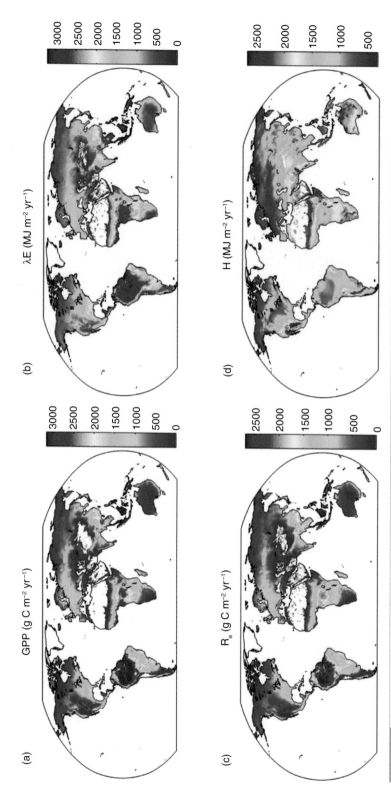

Fig. 24.7 Mean annual (a) gross primary production (GPP, g C m⁻² yr⁻¹), (b) latent heat flux (λE, MJ m⁻² yr⁻¹), (c) terrestrial ecosystem respiration (Rₑ, g C m⁻² yr⁻¹), and (d) sensible heat flux (H, MJ m⁻² yr⁻¹) for the period 1982–2008 empirically upscaled from FLUXNET eddy covariance flux towers. Reproduced from Jung et al. (2011).

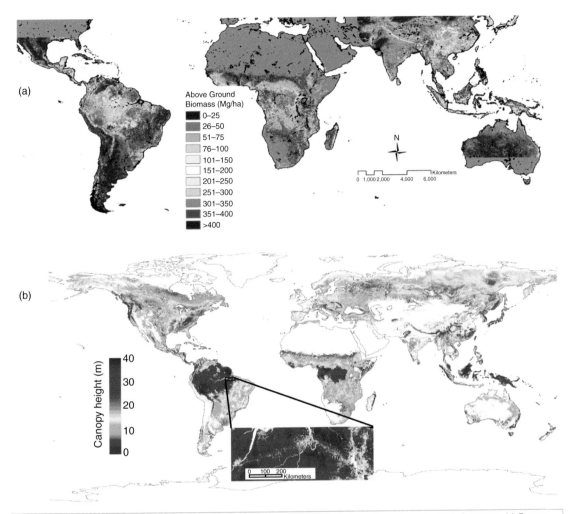

Fig. 24.9 Plant biomass and canopy height estimated from spaceborne light detection and ranging (Lidar) sensors. (a) Forest aboveground biomass in the tropics and subtropics. 10 Mg ha^{-1} = 1 kg m^{-2}. Reproduced from Saatchi et al. (2011) with permission of the National Academy of Sciences. Thurner et al. (2014) provide a similar map for boreal and temperate forests. (b) Forest canopy height. Reproduced from Simard et al. (2011). Pan et al. (2013) review forest biomass and height.

Soil carbon

(g C m^{-2})

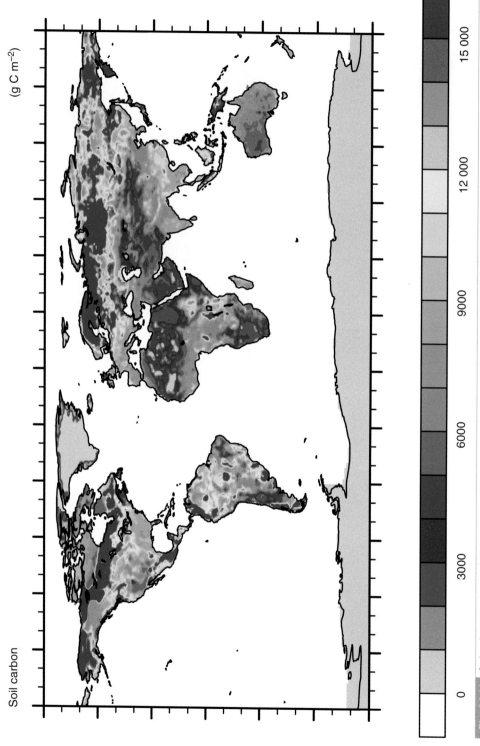

0   3000   6000   9000   12 000   15 000

Fig. 24.11 Soil carbon to a depth of 1 m from the Harmonized World Soil Database. The original data have a spatial resolution of approximately 1 km and were re-gridded to a resolution of 1° in latitude and longitude. The global total for the re-gridded data is 1259 Pg C. Data from Wieder et al. (2013). The Northern Circumpolar Soil Carbon Database (Tarnocai et al. 2009; Hugelius et al. 2013) has data specific to the Arctic.

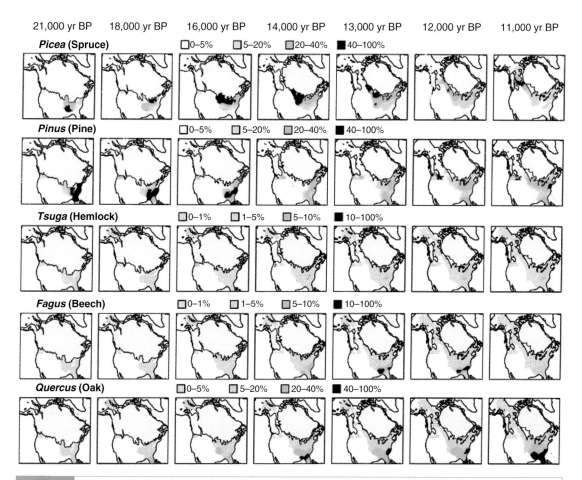

Fig. 24.15 Pollen abundance in boreal and eastern North America for spruce, pine, hemlock, beech, and oak trees between 21 kyr BP and 11 kyr BP. Adapted from Williams et al. (2004).

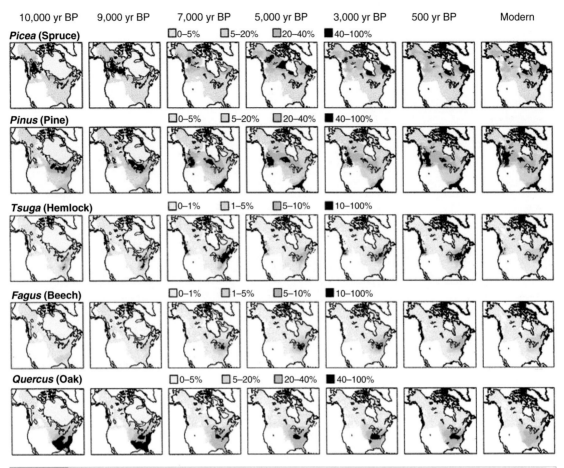

Fig. 24.16 As in Figure 24.15, but for the period 10 kyr BP to present-day (modern). Adapted from Williams et al. (2004).

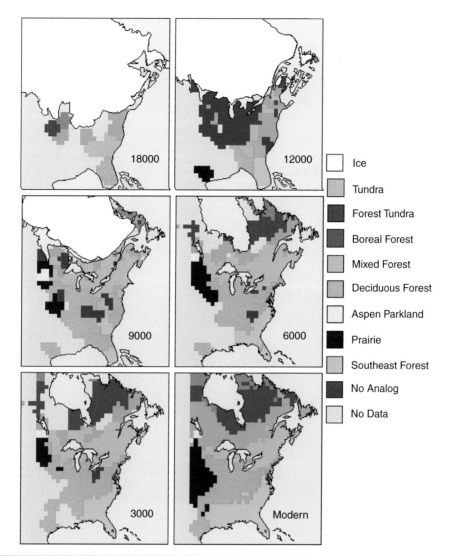

Fig. 24.17 Geographic extent of vegetation in eastern North America over the past 18 kyr BP. Vegetation types are based on pollen abundance. No analog means the pollen does not correspond to a modern vegetation type. Adapted from Overpeck et al. (1992) with graphics provided by the National Geophysical Data Center (National Oceanic and Atmospheric Administration, Boulder, Colorado).

(a) Biogeophysical

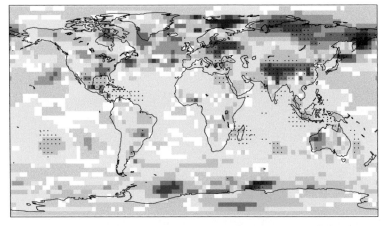

ΔT (ºC)

(b) Biogeochemical

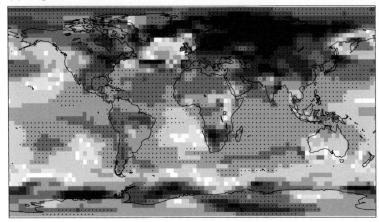

0.5
0.4
0.3
0.2
0.1
0.01
-0.01
-0.1
-0.2
-0.3
-0.4
-0.5

(c) Net

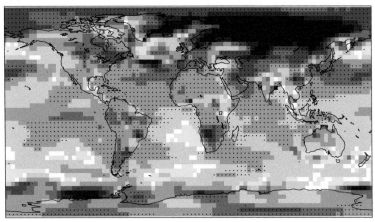

Fig. 28.21 Change in annual mean temperature due to land-cover change over the twentieth century from (a) biogeophysical effects, (b) biogeochemical effects, and (c) net effect. Areas where the change is statistically significant are dotted. Reproduced from Pongratz et al. (2010).

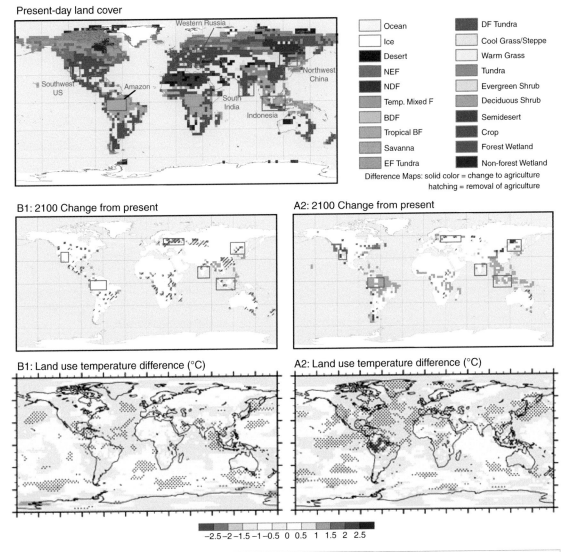

Present-day land cover

B1: 2100 Change from present

A2: 2100 Change from present

B1: Land use temperature difference (°C)

A2: Land use temperature difference (°C)

-2.5 -2 -1.5 -1 -0.5 0 0.5 1 1.5 2 2.5

Fig. 33.3 Effect of future land cover on climate in the year 2100. The top panel shows present-day land cover as represented in a climate model (Labels – B, broadleaf; N, needleleaf; E, evergreen; D, deciduous; and F, forest). The middle panels show land-cover change at 2100 for the B1 and A2 scenarios. The bottom panels show boreal summer (June–August) temperature differences due to land-cover change in the B1 and A2 scenarios. The data were calculated by subtracting the greenhouse gas forcing from a simulation including land-cover change and greenhouse gas forcings. Stippling indicates statistically significant differences. Adapted from Feddema et al. (2005).

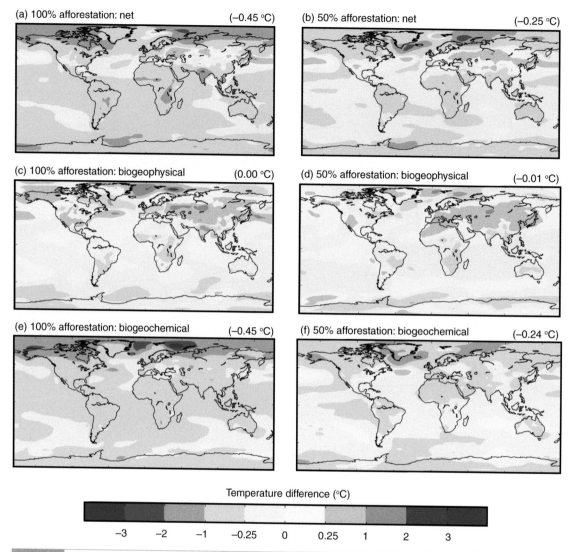

(a) 100% afforestation: net   (−0.45 °C)

(b) 50% afforestation: net   (−0.25 °C)

(c) 100% afforestation: biogeophysical   (0.00 °C)

(d) 50% afforestation: biogeophysical   (−0.01 °C)

(e) 100% afforestation: biogeochemical   (−0.45 °C)

(f) 50% afforestation: biogeochemical   (−0.24 °C)

Temperature difference (°C)

−3   −2   −1   −0.25   0   0.25   1   2   3

Fig. 33.5 Difference in annual mean temperature due to 100 percent afforestation (left) and 50 percent afforestation (right). Temperature difference is the change over the period 2081–2100 compared with a control simulation without afforestation. Shown are (a, b) the net effect, (c, d) the biogeophysical effect, and (e, f) the biogeochemical effect. The numbers in parentheses are the global average. Adapted from Arora and Montenegro (2011) and provided courtesy of Vivek Arora.

Table 21.3 Elements essential for plant growth and development, the principal form in which they are absorbed, and their usual concentration in percentage of dry weight or parts per million (mg per kg)

Element	Principal form	Concentration
Macronutrients		
Carbon	CO_2	44%
Oxygen	H_2O, O_2	44%
Hydrogen	H_2O	6%
Nitrogen	NO_3^-, NH_4^+	1–4%
Potassium	K^+	0.5–6%
Calcium	Ca^{2+}	0.2–3.5%
Phosphorus	$H_2PO_4^-$, HPO_4^{2-}	0.1–0.8%
Magnesium	Mg^{2+}	0.1–0.8%
Sulfur	SO_4^{2-}	0.05–1%
Micronutrients		
Boron	$B(OH)_3$, $B(OH)_4^-$	5–75 ppm
Chlorine	Cl^-	100–10,000 ppm
Cobalt	Co^{2+}	trace
Copper	Cu^{2+}	4–30 ppm
Iron	Fe^{2+}, Fe^{3+}	25–300 ppm
Manganese	Mn^{2+}	15–800 ppm
Molybdenum	MoO_4^{2-}	0.1–5 ppm
Sodium	Na^+	trace
Zinc	Zn^{2+}	15–100 ppm

Source: From Barbour et al. (1999, p. 335).

enzyme synthesis and activation, and to maintain ionic balance.

Mineralized elements in soil are dissolved in solution and adsorbed on clay and humus particles. Elements in solution are available for plant use but can also be leached from the soil. The negatively charged surfaces of clay and humus particles bind positively charged elements (called cations), holding them in the soil so that they do not leach. The cations bound to clay and humus particles can be replaced by other cations in a process called cation exchange. This releases cations into the soil solution, where they are available for uptake by plants. In general, cation exchange capacity increases with higher clay content. Organic particles also adsorb nutrients.

Cation exchange occurs because the mineralogical and chemical composition of clay and humus particles result in a net negative charge that must be balanced by positively charged cations. These negatively charged particles, known as micelles, attract hundreds of thousands of cations such as hydrogen (H^+), aluminum (Al^{3+}), calcium (Ca^{2+}), magnesium (Mg^{2+}), potassium (K^+), ammonium (NH_4^+), and sodium (Na^+). Cations in solution are continually exchanged with cations on clay and humus particles. For example, a calcium ion held on a clay micelle exchanges with two hydrogen ions in the soil solution:

$$[\text{micelle}]Ca^{2+} + \underset{\text{solution}}{2H^+} \rightleftharpoons [\text{micelle}]_{H^+}^{H^+} + \underset{\text{solution}}{Ca^{2+}}$$

$$(21.4)$$

Some cations are held more strongly than others. Cation adsorption decreases in the order $Al^{3+} > Ca^{2+} > Mg^{2+} > K^+ = NH_4^+ > Na^+$. Aluminum ions are strongly held while sodium ions are weakly held. Weakly held cations are most easily displaced from exchange sites, to be replaced with other cations. In addition, the relative concentration in the soil solution determines the degree to which adsorption occurs.

The total number of negatively charged exchange sites that attract positively charged cations determines cation exchange capacity. Soils with a high cation exchange capacity are potentially fertile. A high base saturation, in which calcium, magnesium, potassium, and sodium occupy most exchange sites, indicates high nutrient availability. Acidic soils have low base saturation; H^+ occupies most sites, displacing other elements to the soil solution. As such, the pH of the soil solution is an important determinant of soil fertility.

The relative concentration of hydrogen (H^+) and hydroxyl (OH^-) ions is measured by pH. An acid soil (pH < 7) has high H^+ concentration. An alkaline soil (pH > 7) has high OH^- concentration. Soil pH typically ranges from about 3 for extremely acidic peat soils to 10 or more for very alkaline soils. Most humid mineral soils have pH equal to 5–7; arid soils have a pH of 7–9. The pH of soil affects nutrient availability in two ways. First, the negative charge on clay and humus particles increases as pH increases. As a result, cation exchange capacity increases with higher pH. In addition, pH affects the availability of several essential elements. Many nutrients are most available under near neutral conditions (pH = 7) and decrease with higher or lower pH. In strongly acidic soils, the availability of macronutrients (calcium, magnesium, potassium, phosphorus, nitrogen, sulfur) as well as molybdenum and boron is curtailed. In contrast, iron, manganese, and zinc become less available as pH rises from 5 to 7–8. At low soil pH, the availability of iron, manganese, zinc, copper, and cobalt are often at such high concentrations in the soil solution to be toxic to most plants.

21.5 | Nitrogen Gaseous Losses

Emission of a variety of nitrogen gases (NH_3, NO, N_2O, and N_2) from soils to the atmosphere accompanies the mineralization of organic nitrogen (Figure 20.12). The inorganic ion ammonium (NH_4^+) and the gas ammonia (NH_3) are interchangeable by the reaction:

$$NH_4^+ + OH^- \rightleftharpoons NH_3 + H_2O \qquad (21.5)$$

The loss of NH_3 to the atmosphere is known as ammonia volatilization. This pathway of nitrogen loss is generally low in natural soils, because of low NH_4^+ concentration, but it is prevalent with fertilizer, manure, and in feedlots. Many factors affect ammonia volatilization, including temperature, soil water, pH, and wind speed. The oxidation of NH_4^+ to NO_3^- during nitrification produces NO and N_2O as byproducts. Nitrification is an aerobic process (it requires high oxygen concentrations). Denitrification is the reduction of NO_3^- to N_2 and also produces NO and N_2O. It is an anaerobic process that occurs in wet, oxygen-depleted soils.

A model of NO and N_2O production, known as the whole-in-the-pipe model (Firestone and Davidson 1989; Davidson 1991), conceptualizes the emission of these gases as the byproducts of nitrification and denitrification (Figure 21.8). Their production depends on the rate of flow of nitrogen moving through the pipe, during which a small percentage of nitrogen leaks out as NO and N_2O emissions (conceptually viewed as holes in the pipe). The flow of nitrogen through the pipe is analogous to the rates of nitrification and denitrification. Soil water content is the primary determinant of the size of the holes through which the trace gases leak. In particular, the oxidative process of nitrification dominates in moist, but not too wet, soils with water content approximately 30–60 percent of saturation; in these conditions, NO production is high. High N_2O emission occurs in wetter soils, where anaerobic conditions favor denitrification. As soils become saturated and oxygen-depleted, much of the N_2O is further reduced to N_2 by denitrifiers.

Bai et al. (2012) estimated nitrogen gas emissions from natural ecosystems using an analysis of nitrogen isotopes ($^{15}N/^{14}N$). Denitrification accounts for about 35 percent of nitrogen losses in natural systems. Global emissions of NO, N_2O, and N_2 are 16, 10, and 21 Tg N yr^{-1}, respectively (Table 21.4). Denitrification in total emits 47 Tg N yr^{-1} globally. The highest fluxes occur in tropical regions, where warm temperatures and moist soils provide favorable conditions for soil microbes. Croplands and managed grasslands are sources of additional NO and N_2O, and fertilizer application enhances gas emissions (especially N_2O). Leaching accounts for 65 percent of nitrogen losses globally, with highest percentage losses in the extra-tropics compared with the tropics.

21.6 | Soil Profile

The outcome of biological and geochemical processes, especially weathering and the leaching of secondary minerals to greater depth, is seen in the soil profile (Figure 21.9). A pit dug a few meters deep reveals several recognizable layers or horizons distinguished by color, structure,

Table 21.4 Global nitrogen gas emissions (Tg N yr^{-1})

Source	NO	N$_2$O	N$_2$
Natural	15.7 (11.2–20.3)	10.2 (7.2–13.2)	21.0 (14.9–27.1)
Agricultural	1.8	4.1	–

Note: Natural sources are from an analysis of nitrogen isotopes. Shown are the mean and range. Agricultural sources are from a literature survey.
Source: From Bai et al. (2012).

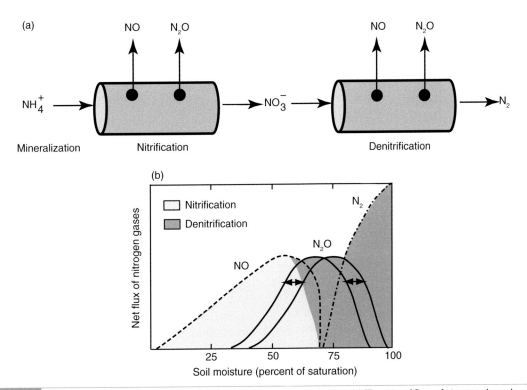

Fig. 21.8 Diagram of the hole-in-the-pipe model of nitrogen trace gas production. (a) The rate of flow of nitrogen through the pipes during nitrification and denitrification determines the gaseous losses. (b) Soil water content affects the proportion of gas losses as NO, N$_2$O, and N$_2$, conceptualized as the relative size of the holes through which NO and N$_2$O leak. Adapted from Bouwman (1998) and Davidson et al. (2000).

and aggregation. These horizons vary in thickness. In some soils, they may be well developed. In others, they may have irregular boundaries and are difficult to distinguish.

Organic debris from decomposing leaves, roots, and wood forms the O horizon on top of the mineral soil. The uppermost layer consists of undecomposed material that is easily recognizable (e.g., leaves, twigs, branches) and is called the litter layer. Immediately below this is a humus layer formed by highly decomposed organic material that is no longer distinguishable in its original form. The O horizon may be several centimeters thick in forests, where it is commonly referred to as the forest floor, but is generally absent in grasslands and unproductive sites. The A horizon, which is often referred to as topsoil, is the upper

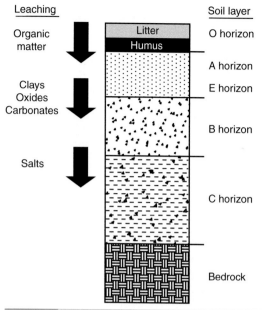

Leaching

Organic matter

Clays
Oxides
Carbonates

Salts

Soil layer

Litter
Humus

O horizon

A horizon

E horizon

B horizon

C horizon

Bedrock

Fig. 21.9 General soil profile showing the primary soil horizons. Arrows on the left depict the downward leaching of materials.

mineral soil layer. It is the zone of greatest physical, chemical, and biological activity, containing most of the soil's organic material, roots, and microorganisms. The A horizon is usually enriched in organic matter that moves downward from the surface. Percolating rainwater often carries silicate clays, iron and aluminum oxides, and other materials deeper into the soil profile. The layer where this leaching is greatest is called the E horizon. Maximum leaching leaves a concentration of resistant minerals such as quartz in sand and silt particle sizes. Below this is the B horizon. It contains much less organic material than the A horizon and is generally less weathered. The B horizon is the mineral soil zone that has been altered by the chemical deposition of materials leached from overlying horizons. It accumulates silicate clays, iron and aluminum oxides, and other secondary minerals such as calcium carbonate and gypsum from above. The C horizon is the deepest soil layer, composed of rock and mineral fragments from which the upper soil is derived. It is parent material that has not been greatly affected by soil development.

Soil carbon has a distinct vertical distribution (Figure 21.10). The percentage of soil organic carbon in the top 20 cm of soil averages 50 percent in forests and 42 percent in grasslands. Within the top meter, one-half of the soil carbon in forests and 58 percent of the soil carbon in grasslands is distributed at depths of 20–100 cm. Additional carbon is found at deeper depths. In forests, an amount of soil organic carbon equal to 56 percent of that found in the first meter is stored in the second and third meters of soil. In grasslands, this deep soil carbon is 43 percent that of the first meter.

21.7 | Soil Formation

The soil profile and its physical, chemical, and mineralogical characteristics reflect the degree of weathering and extent of soil development. Soil formation is understood in terms of five factors: parent material, time, topography, climate, and vegetation (Jenny 1980). Parent material is the underlying geologic material from which soil forms. A quartz-rich parent material such as granite or sandstone is likely to produce a sandy soil. Parent material, through its composition of primary minerals, also influences the amount and type of clay minerals found in soil. Parent material exists in several forms. It can be: residual material formed in place by underlying rock; coarse rock fragments detached and carried downslope by gravity; alluvial deposits of sediments along river floodplains and deltas; marine sediments along coastal plains; rocks and sediments left by melting glaciers; and wind-transported (eolian) materials such as dune sand, loess, and dust. Since weathering occurs over several thousand years, the length of time the material has been exposed to weathering is important. Soils forming from glacial debris have had less time to develop than soils that were not covered with glaciers during the last ice age. Alluvial deposits along floodplains are younger than upland soils. Topography (i.e., the elevation, slope, and position of a site in the landscape) can hasten or hinder soil development. Topography determines whether a

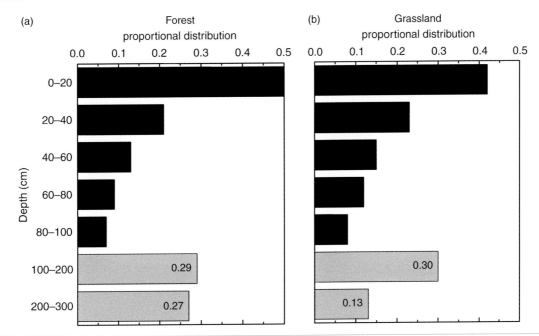

Fig. 21.10 Profiles of soil organic carbon distribution for (a) forest and (b) grassland. Black bars show the proportional distribution of soil organic carbon in the first meter of soil and sum to one. Gray bars show the additional carbon at depths of 100–200 cm and 200–300 cm, relative to the first meter. Data from Jobbágy and Jackson (2000).

site is losing soil from erosion or gaining soil from deposition. In general, uplands lose soil that washes downhill in runoff while lowlands gain soil from upslope. Shallowest soils may be found on middle slopes, where the speed and concentration of runoff is greatest. Topographic depressions can create waterlogged soils where drainage is impeded.

Climate, particularly temperature and precipitation, determines the type and rate of the weathering that occurs. Physical weathering from repeated freezing and thawing occurs in seasonally cold environments. In humid environments, repeated swelling and shrinking as soils are wetted and dried disintegrates soil particles. Chemical weathering is most rapid where climate is wet and hot. The rate of many chemical reactions increases exponentially with warmer temperatures. Heavy rains leach minerals from the surface soil layers. Hence, chemical weathering is more rapid in tropical rainforests than in temperate forests and more rapid in moist forests than in drier grasslands or deserts.

Figure 21.11 compares how forest and grassland soils typically develop from parent material. The most prominent feature of grassland soil is that it is highly enriched in organic matter added from the thick, deep root system of grasses. Consequently, grassland soils have a thick A horizon with deep distribution of soil organic matter. Because grassland soils develop under semiarid conditions, the products of chemical weathering are not rapidly leached from the soil profile. Calcium carbonate and other weathering products can accumulate in the soil. In contrast, foliage litter and woody debris accumulate on the ground in forests. The O horizon consists of easily distinguishable litter and humus layers. In cold climates, decomposition is slow and the forest floor can be several centimeters thick. In contrast, decomposition is rapid in tropical forests and little litter accumulates. Forests generally have a prominent A horizon with a pronounced zone of maximum leaching. In many temperate forests, soil water percolating through the forest floor contains organic acids derived from litter decomposition.

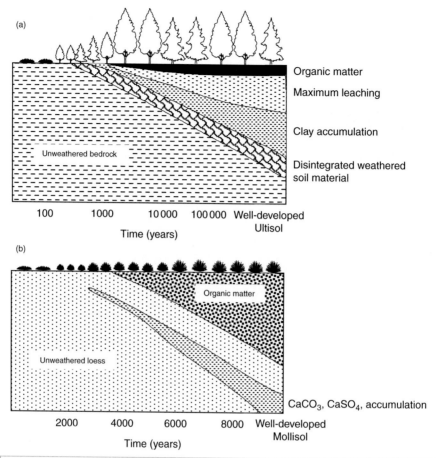

Fig. 21.11 Idealized temporal sequence of soil development from unweathered parent material. (a) Forest with ultisol soil. (b) Grassland with mollisol soil. Adapted from Brady and Weil (1999, p. 57).

These acids dominate weathering of soil minerals in the A horizon. Iron and aluminum may be leached to deeper soil layers. Where this leaching is most intense, particularly in boreal and cool temperate coniferous forests, the soil is known as a spodosol. Litter from the coniferous trees common in these forests is rich in organic acids; the soil is acidic; and materials leached from the near-surface soil accumulate in deeper soil. Podzolization is less common in warmer climates, where decomposition is more rapid and fewer acids remain to percolate through the A horizon.

Soils, then, are a product of climate and vegetation, as modified by topography, acting on the geologic parent material over time. The result

is 12 broad classes of soil, known as soil orders, that differ in color, texture, structure, and chemical and mineralogical properties. These 12 soil orders vary in relation to degree of weathering, extent of soil development, climate, and associated vegetation (Figure 21.12) and have a distinct geography that closely relates to climate and vegetation (Figure 2.5).

Entisols form in a variety of environments (e.g., rocky outcrops, deserts). Their distinguishing feature is little soil development. Inceptisols are only slightly more developed. These soils show the beginning of soil development, such as the beginning of a B horizon, but not the characteristics found in older soils. Andisols are formed on volcanic debris. They, too, have not had time

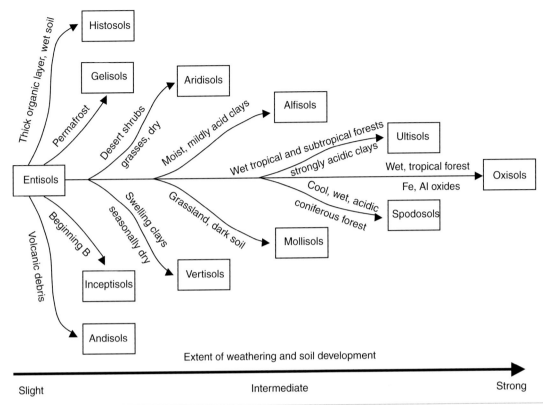

Fig. 21.12 Soil orders in relation to degree of weathering, climate, and vegetation. Adapted from Brady and Weil (1999, p. 82).

to become highly weathered. Gelisols are cold soils underlain with permafrost. The soils are often waterlogged, which in conjunction with cold temperature limits profile development. Organic matter accumulates on the surface in these cold, wet conditions. These soils are found throughout the Arctic in association with tundra vegetation. Histosols are thick organic soils that develop in wet conditions without permafrost. The saturated soil limits decomposition so that peat accumulates over time. They occur with marshes, swamps, and bogs and are most prevalent in cold climates.

Aridisols occur in deserts and other dry environments. With little precipitation and sparse, unproductive vegetation, these soils lack organic matter and there is little leaching of materials out of the profile. Vertisols typically occur in warm climates where precipitation ranges from subhumid to semiarid. Grassland is

typical native vegetation. These soils have high clay content. The clay shrinks during the dry season and swells when wetted by rainfall.

Mollisols are the thick, dark black, nutrient-rich soils with high organic matter content found in highly productive prairie vegetation. Alfisols form in cool to hot humid climates in conjunction with deciduous forests or savanna. They are characterized by a clay-enriched subsurface horizon. Ultisols are more highly weathered and acidic than alfisols. They form in warm to tropical climates. Clay mineral weathering and translocation dominate soil formation. Spodosols develop in boreal and cool temperate coniferous forests, where the acidic litter enhances leaching. Oxisols occur where climate is hot and wet year-round. They occur in association with tropical rainforest and are the most highly weathered soils, characterized by high clay content.

21.8 | Review Questions

1. Which mineral has undergone more chemical weathering: muscovite or kaolinite?

2. Which of the following minerals are common in sandy soils: kaolinite, calcite, quartz, goethite, muscovite? Which are common in clay soils?

3. In a study of litter decomposition, 70 percent of original mass remains after one year. What is the annual decay rate (k)? How much of the original mass remains after two years?

4. Leaf litter of two tree species (spruce and maple) is placed in soil at the same location. One litter sample loses 55 percent of its mass during the first year. The other loses 20 percent of its mass. Identify which sample is spruce and which is maple. Which has the highest lignin:N ratio?

5. Five tree species differ in foliage nitrogen concentration. Rank the species with respect to likely nitrogen immobilization (highest to lowest) during litter decomposition: yellow birch, 1.60 percent N; beech, 0.85 percent N; yellow poplar, 0.72 percent N; chestnut oak, 1.03 percent N; black locust, 2.45 percent N.

6. How does the decomposition of leaf, fine root, and wood litter differ in the CENTURY model?

7. With reference to Figure 21.7, the C:N ratio of leaf metabolic litter is typically > 50 while the C:N ratio of the active (SOM1) pool is 10–20. Does decomposition of leaf metabolic litter to SOM1 mineralize or immobilize nitrogen?

8. Describe a climate consequence of applying nitrogen fertilizer to enhance crop production.

9. Give two reasons why soil development is likely to be greater in a humid tropical environment than in a cold arctic environment.

10. What is the source of CO_2 in soil needed to form carbonic acid (H_2CO_3)?

21.9 | References

Aber, J. D., and Melillo, J. M. (1980). Litter decomposition: Measuring relative contributions of organic matter and nitrogen to forest soils. *Canadian Journal of Botany*, 58, 416–421.

Aber, J. D., and Melillo, J. M. (1982). Nitrogen immobilization in decaying hardwood leaf litter as a function of initial nitrogen and lignin content. *Canadian Journal of Botany*, 60, 2263–2269.

Adair E. C., Parton, W. J., Del Grosso, S. J., et al. (2008). Simple three-pool model accurately describes patterns of long-term litter decomposition in diverse climates. *Global Change Biology*, 14, 2636–2660.

Bai, E., Houlton, B. Z., and Wang, Y. P. (2012). Isotopic identification of nitrogen hotspots across natural terrestrial ecosystems. *Biogeosciences*, 9, 3287–3304.

Barbour, M. G., Burk, J. H., Pitts, W. D., Gilliam, F. S., and Schwartz, M. W. (1999). *Terrestrial Plant Ecology*, 3rd ed. Menlo Park, California: Benjamin/Cummings Publishing Company.

Bonan, G. B., Hartman, M. D., Parton, W. J., and Wieder, W. R. (2013). Evaluating litter decomposition in earth system models with long-term litter-bag experiments: An example using the Community Land Model version 4 (CLM4). *Global Change Biology*, 19, 957–974.

Bouwman, A. F. (1998). Nitrogen oxides and tropical agriculture. *Nature*, 392, 866–867.

Brady, N. C., and Weil, R. R. (1999). *The Nature and Properties of Soils*, 12th ed. Upper Saddle River, New Jersey: Prentice Hall.

Brovkin, V., van Bodegom, P. M., Kleinen, T., et al. (2012). Plant-driven variation in decomposition rates improves projections of global litter stock distribution. *Biogeosciences*, 9, 565–576.

Currie, W. S., Harmon, M. E., Burke, I. C., et al. (2010). Cross-biome transplants of plant litter show decomposition models extend to a broader climatic range but lose predictability at the decadal time scale. *Global Change Biology*, 16, 1744–1761.

Davidson, E. A. (1991). Fluxes of nitrous oxide and nitric oxide from terrestrial ecosystems. In *Microbial Production and Consumption of Greenhouse Gases: Methane, Nitrogen Oxides and Halomethanes*, ed. J. E. Rogers and W. B. Whitman. Washington, DC: American Society for Microbiology, pp. 219–235.

Davidson, E. A., Keller, M., Erickson, H. E., Verchot, L. V., and Veldkamp, E. (2000). Testing a conceptual model of soil emissions of nitrous and nitric oxides. *BioScience*, 50, 667–680.

Firestone, M. K., and Davidson, E. A. (1989). Microbiological basis of NO and N_2O production and consumption in soil. In *Exchange of Trace Gases between Terrestrial Ecosystems and the Atmosphere*, ed. M. O. Andreae and D. S. Schimel. New York: Wiley, pp. 7–21.

Gholz, H. L., Wedin, D. A., Smitherman, S. M., Harmon, M. E., and Parton, W. J. (2000). Long-term dynamics of pine and hardwood litter in contrasting environments: Toward a global model of decomposition. *Global Change Biology*, 6, 751–765.

Harmon, M. E., Silver, W. L., Fasth, B., et al. (2009). Long-term patterns of mass loss during the decomposition of leaf and fine root litter: An intersite comparison. *Global Change Biology*, 15, 1320–1338.

Jenny, H. (1980). *The Soil Resource: Origin and Behavior*. New York: Springer-Verlag.

Jobbágy, E. G., and Jackson, R. B. (2000). The vertical distribution of soil organic carbon and its relation to climate and vegetation. *Ecological Applications*, 10, 423–436.

Likens, G. E., Bormann, F. H., Pierce, R. S., Eaton, J. S., and Johnson, N. M. (1977). *Biogeochemistry of a Forested Ecosystem*. New York: Springer-Verlag.

McGuire, A. D., Melillo, J. M., Joyce, L. A., et al. (1992). Interactions between carbon and nitrogen dynamics in estimating net primary productivity for potential vegetation in North America. *Global Biogeochemical Cycles*, 6, 101–124.

Melillo, J. M., Aber, J. D., and Muratore, J. F. (1982). Nitrogen and lignin control of hardwood leaf litter decomposition dynamics. *Ecology*, 63, 621–626.

Montgomery, D. R., Zabowski, D., Ugolini, F. C., Hallberg, R. O., and Spaltenstein, H. (2000). Soils, watershed processes, and marine sediments. In *Earth System Science: From Biogeochemical Cycles to Global Change*, ed. M. C. Jacobson, R. J. Charlson, H. Rodhe, and G. H. Orians. San Diego: Academic Press, pp. 159–194.

Parton, W. J., Schimel, D. S., Cole, C. V., and Ojima, D. S. (1987). Analysis of factors controlling soil organic matter levels in Great Plains grasslands. *Soil Science Society of America Journal*, 51, 1173–1179.

Parton, W. J., Stewart, J. W. B., and Cole, C. V. (1988). Dynamics of C, N, P and S in grassland soils: A model. *Biogeochemistry*, 5, 109–131.

Parton, W. J., Scurlock, J. M. O., Ojima, D. S., et al. (1993). Observations and modeling of biomass and soil organic matter dynamics for the grassland biome worldwide. *Global Biogeochemical Cycles*, 7, 785–809.

Parton, W. J., Ojima, D. S., Cole, C. V., and Schimel, D. S. (1994). A general model for soil organic matter dynamics: Sensitivity to litter chemistry, texture and management. In *Quantitative Modeling of Soil Forming Processes*, ed. R. B. Bryant and R. W. Arnold. Madison, Wisconsin: Soil Science Society of America, pp. 147–167.

Parton, W., Silver, W. L., Burke, I. C., et al. (2007). Global-scale similarities in nitrogen release patterns during long-term decomposition. *Science*, 315, 361–364.

Raich, J. W., Rastetter, E. B., Melillo, J. M., et al. (1991). Potential net primary productivity in South America: Application of a global model. *Ecological Applications*, 1, 399–429.

Schlesinger, W. H. (1990). Evidence from chronosequence studies for a low carbon-storage potential of soils. *Nature*, 348, 232–234.

Schlesinger, W. H. (1997). *Biogeochemistry: An Analysis of Global Change*, 2nd ed. San Diego: Academic Press.

Vegetation Dynamics

22.1 | Chapter Summary

Ecosystems are not static entities, but rather are in a state of continual change. Disturbances that create clearings initiate vegetation dynamics that varies according to life history patterns and competition among plants for light, water, and nutrients. Some species grow fast and are short lived. These plants are ephemeral features in the landscape, using a life history that allows them to rapidly colonize and dominate recently disturbed patches. These denuded patches occur through processes endogenous to the landscape such as the death of a large tree that creates a gap in the canopy or through exogenous disturbances such as wildfires and hurricanes. Over time, these early dominants give way to slower growing, longer-lived species. The rise and fall of taxa is part of the life cycle of communities and ecosystems, a process ecologists call succession. It creates pattern to the arrangement of vegetation across the landscape related to disturbance history. The prevailing scientific view emphasizes succession as a population process. It is a result of the physiology, morphology, and life history of species operating in a gradient of environmental change. The differential growth, survival, and colonizing ability of species adapted to the various environments encountered during community development create shifting patterns of dominance. Individual species colonize where conditions are favorable, die out when the environment is no longer favorable, and grow in company with other species with similar environmental requirements.

22.2 | Population Dynamics

Population density changes over time through the birth, growth, and death of its members. The classic description of population growth is the logistic growth equation:

$$dN/dt = rN(1 - N/K) \qquad (22.1)$$

where N is density at time t, r is an intrinsic rate of population growth determined as the difference between birth and death rates, and K is carrying capacity. The term rN describes exponential population growth in the absence of limiting resources. However, resources are rarely unlimited, and the resources available to individuals become scarcer as population density increases. At high density, competition among individuals for available resources limits population growth. The term $(1 - N/K)$ reduces the rate of growth as population density approaches the maximum number of individuals the environment can support (i.e., the carrying capacity of the environment). This equation has an S-shape or logistic form, with exponential growth at low density declining to zero growth (i.e., constant density) at high density.

This depiction of population growth has lead to numerous theoretical insights to the study of populations and communities. For example, it

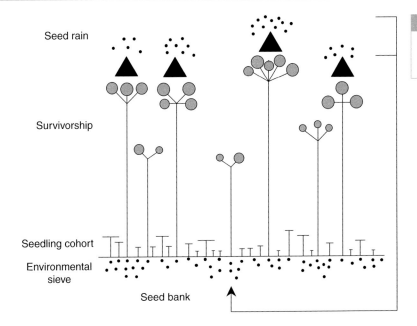

Fig. 22.1 Life cycle of a plant population from seeds to seedlings to maturity. Adapted from Harper (1977, p. 29).

Seed rain

Survivorship

Seedling cohort

Environmental sieve

Seed bank

provides the basis for *r*- and *K*-selected species (Chapter 18). Species that are *K*-selected maintain populations at or near carrying capacity. They survive and reproduce under conditions of strong competition and resource depletion. In contrast, *r*-selected species have a high intrinsic rate of population growth, colonize a site rapidly, and are favored in open environments. However, Eq. (22.1) does not accurately describe plant populations, which adjust the sizes of individuals in relation to density (Harper 1977). In addition, plants are sedentary and do not compete for space and resources with all members of a population but rather only with their immediate neighbors.

Plant population dynamics is understood in terms of demographic processes such as growth, mortality, reproduction, and dispersal and the influence of the biotic and abiotic environment on these processes (Figure 22.1). Dispersed seeds are stored in a dormant state in soil, forming a seed bank from which seedlings are recruited. The environment acts as a sieve that filters seeds from the bank. Recruitment of seedlings from the seed bank depends on safe-sites that provide the necessary conditions for germination and establishment. Only those seeds receiving suitable light, moisture, and temperature

germinate, and only some of these become established as seedlings. As seedlings grow, their roots spread out to acquire water and nutrients; leaves form to absorb light and CO_2 for photosynthesis; stems develop to support leaves. When large enough, individuals interfere with the growth of neighboring plants, competing with them for light, water, nutrients, and space. Plants that gain resources first preempt their use by others and have a competitive advantage. Success goes to the largest plant that captures space and resources at the expense of smaller neighbors. Many plants die in this process. Others can survive in the resource-depleted understory, waiting for a taller plant to die and create a gap in the canopy. Growth and survival culminates in seed production and dispersal by mature plants, which is the seed rain for the next generation. Annuals complete this life cycle over the course of a year. Woody shrubs and trees complete this cycle over many years.

This intense competition among plants for resources results in typically high rates of juvenile mortality. This is particularly true in forests, where the mortality of tree seedlings is high (Forcier 1975; Harcombe 1987). Only a few of the many thousands of seedlings produced by a tree become established in the understory.

Even fewer survive to reach the canopy. Most die as a result of failure to successfully acquire light, water, and nutrients in the presence of strong competition from neighboring individuals. Risk of mortality typically declines with increasing age, giving a concave shape to survivorship curves.

A series of experiments in the 1950s and early 1960s demonstrated the importance of crowding and competition in regulating the size of plants and the biomass of plant populations. One experiment studied temporal changes in five populations of soybean grown at different densities (Kira et al. 1953). Early in the experiment, yield increased with higher density (Figure 22.2). Mean plant mass was independent of density so that more plants yielded higher biomass. After about 45 days, during which there was no mortality, the biomass of each population was relatively constant and independent of density. Density-dependent changes in mean plant mass compensated for differences in density so that biomass was constant. This study shows that when plants are small they do not interfere with the growth of neighbors. Yield increases linearly with density. Over time, as plants grow larger, they compete for resources and interfere with the growth of neighbors. Interference occurs sooner and is greater for plants growing at high density, where plants are crowded, than at low density, where plants are widely spaced. Once competition begins, mean plant mass decreases with higher density.

A subsequent study showed that the reduction in mean plant mass is manifested in the frequency distribution of plant sizes (Koyama and Kira 1956). As plants increase in size, competition for light, water, nutrients, canopy space, and root space intensifies. Some individuals gain more resources than others and grow faster. Others are suppressed and grow little. The result is a hierarchy of plant sizes with many small individuals and fewer large individuals. The inequality of plant sizes becomes progressively greater over time as a few individuals achieve dominance and the rest are suppressed. Size inequality is greatest in dense stands, where competition and crowding is most intense (Mohler et al. 1978; Weiner 1985;

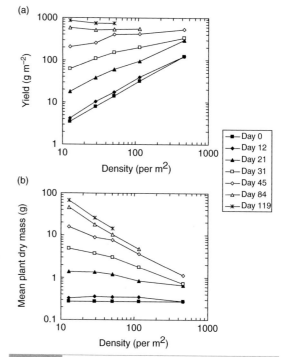

Fig. 22.2 Population dynamics of soybean over 119 days showing (a) yield and (b) mean plant mass in relation to density. Adapted from Kira et al. (1953).

Weiner and Thomas 1986; Bonan 1988, 1991; Knox et al. 1989; Miller and Weiner 1989).

The final study in this series of experiments showed that as large plants usurp the majority of resources smaller plants die and density progressively declines in a process known as self-thinning (Yoda et al. 1963). In a self-thinning population, mean plant mass ($\bar{m}$) and biomass (M) are inversely related to density (N) by the equations:

$$\bar{m} = cN^{-3/2}$$
$$M = cN^{-1/2}$$

(22.2)

where c is a constant that varies among species and with site conditions. It has been found in a wide variety of plants ranging from annual herbs to trees (White and Harper 1970; Mohler et al. 1978; Westoby 1981, 1984; Peet and Christensen 1987; Weller 1987; Lonsdale 1990; Weiner and Freckleton 2010).

Loblolly pine stands in North Carolina illustrate the dynamics of self-thinning (Peet and

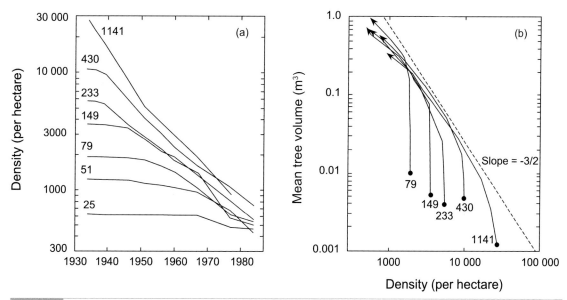

Fig. 22.3 Self-thinning in loblolly pine stands over 50 years. Numbers next to curves indicate initial stand density (per 0.1 acre plot). (a) Number of survivors in relation to stand age for seven stands of differing initial density. (b) Mean tree volume in relation to density of survivors for five stands of differing initial density. Lines show temporal trajectories of volume and density with successive points on a given line representing increasingly older stand age. The dashed line shows a slope of –3/2. Adapted from Peet and Christensen (1987).

Christensen 1980, 1987; Christensen and Peet 1981; Knox et al. 1989). One study found that despite large variation in initial density from about 600 to almost 30,000 stems per hectare, the number of surviving trees converged on a narrow range of densities of about 450–750 ha^{-1} over a 50-year period (Figure 22.3a). Mortality was greatest in the densest stand and progressively declined with lower initial density. Graphs of mean tree volume in relation to density of survivors reveal the temporal trajectory of stand dynamics (Figure 22.3b). Initially, mean tree volume increased without a corresponding decrease in density. As the canopy closed and resources became limiting, mean tree volume and density were related by a line with a slope of –3/2.

Self-thinning represents a temporal trajectory of population dynamics, and the self-thinning line describes a boundary towards which plant populations grow. It is a density-dependent upper limit to mean plant mass or stand biomass in a given environment. In the early stages of stand development or at low density, when plants do not interfere with the growth of neighbors, biomass increases toward the boundary line without a corresponding decrease in density. Once at the boundary line, further increases in biomass are achieved through decreases in density. How soon plants experience competition and mortality depends on density. High density populations reach the boundary line faster than low density populations. However, populations eventually converge on a common trajectory of biomass versus density regardless of initial density.

22.3 | Succession

Plant succession refers to temporal change in community composition and ecosystem structure following disturbance. It is a progressive change in dominance by particular groups of species, biomass accumulation, and nutrient cycling over periods of several decades to centuries. Colonization of new land not previously vegetated is known as primary succession. This occurs in response to volcanic eruptions, alluvial deposits along floodplains, formation of sand dunes, and melting of glaciers. In contrast, secondary succession is the regrowth of vegetation on previously vegetated land following a disturbance such as fire, windthrow, logging, or farm abandonment. Succession is an integral part of forest stand dynamics and the response of forests to environmental change, but also occurs in grasslands and other herbaceous communities (Bormann and Likens 1979; West et al. 1981; Shugart 1984, 1998; Botkin 1993).

Change in community composition is accompanied by changes in the environment caused by plants themselves. For example, the development of a dense canopy shades the understory, litter accumulates on the forest floor, and nutrient cycling is altered based on the amount and chemical quality of plant detritus. These changes can facilitate the establishment of other species, abetting the directional change in dominance, or inhibit the establishment of new species. Succession driven by environmental change caused by plants is known as autogenic succession. In contrast, allogenic succession is driven by external factors (e.g., climate change) that cause the environment to change.

22.3.1 Lake Michigan Sand Dunes

The growth of vegetation on sand dunes along the Indiana shoreline of Lake Michigan near Chicago is a classic example of primary succession (Cowles 1899; Olson 1958). The changing shoreline of Lake Michigan during and after glacial retreat from the last ice age left several distinct beach and dune systems exposed above lake level. Dunes are progressively older with distance from shoreline, and the vegetation of these dune systems provides a record of succession. In general, young sand dunes are covered with herbaceous annuals and perennials and woody shrubs. Pine trees and then broadleaf deciduous trees become progressively more abundant on older dunes. This vegetation change from exposed sand dune to forest occurs relatively rapidly over a few hundred years. The growth of trees results in rapid accumulation of fresh and decomposed litter accompanied by an increase in soil nitrogen and cation exchange capacity.

Successional change in community composition depends on site conditions. The most prominent pathway is a change from grasses to pine to oak (Figure 22.4). Grasses are the most abundant pioneer vegetation on newly exposed dunes. These occur in pure stands or in mixtures. Woody plants such as cottonwood trees and sand cherry shrubs may become established depending on seed availability. This primary vegetation builds up and then stabilizes the dunes, allowing for subsequent invasion of jack pine and eastern white pine trees. Various types of black oak forests that differ in understory and ground cover eventually replace the short-lived pine trees. Pioneer herbs and shrubs rapidly decline in abundance with the growth of trees several decades following dune formation. Pine declines to a minor species after the first generation or two, being replaced by black oak, which remains dominant on even the oldest dunes. Alternative successional pathways are possible depending on soil moisture. Wet depressions may develop into a tall grass prairie or red maple swamp forest. Succession from basswood to northern red oak to sugar maple without usually passing through pine and black oak stages occurs on leeward slopes and depressions, where the microclimate is more moderate and moisture accumulates.

22.3.2 Glacier Bay, Alaska

The establishment of vegetation following deglaciation at Glacier Bay in southwestern Alaska is another classic example of primary succession (Cooper 1923a,b,c, 1931, 1939; Chapin et al. 1994; Fastie 1995). Since about 1750, a large glacier that once covered the inlet has rapidly retreated at a rate averaging 400 m per year. Some 2500 km^2 of ice has melted, exposing several hundred square kilometers of glacial till and outwash along the shoreline of Glacier Bay to plant colonization. Four major successional stages characterize the growth of vegetation following deglaciation (Figure 22.5). During the first 20 years or so after glacial retreat, pioneer vegetation consisting of blue-green algae, lichens, liverworts, and forbs colonize the exposed till.

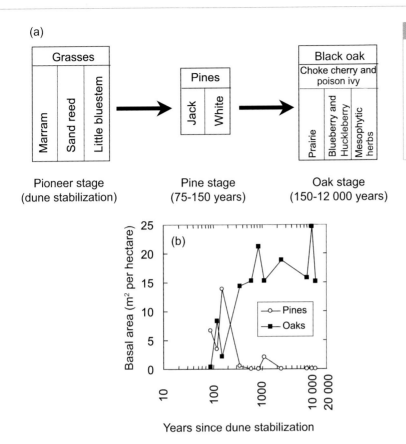

(a)

Pioneer stage (dune stabilization)

Pine stage (75-150 years)

Oak stage (150-12 000 years)

Fig. 22.4 Principal successional pathway on sand dunes along the southern shoreline of Lake Michigan. (a) Changes in community composition. (b) Tree basal area in relation to dune age. Basal area is the area of tree stems, typically at a height of about 1 m. Adapted from Olson (1958).

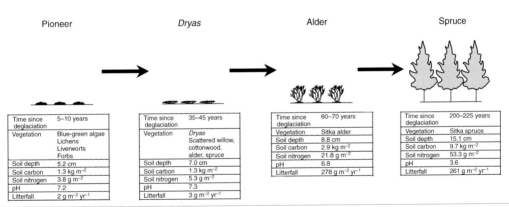

Pioneer

Dryas

Alder

Spruce

Time since deglaciation	5–10 years
Vegetation	Blue-green algae Lichens Liverworts Forbs
Soil depth	5.2 cm
Soil carbon	1.3 kg m^{-2}
Soil nitrogen	3.8 g m^{-2}
pH	7.2
Litterfall	2 g m^{-2} yr^{-1}

Time since deglaciation	35–45 years
Vegetation	*Dryas* Scattered willow, cottonwood, alder, spruce
Soil depth	7.0 cm
Soil carbon	1.3 kg m^{-2}
Soil nitrogen	5.3 g m^{-2}
pH	7.3
Litterfall	3 g m^{-2} yr^{-1}

Time since deglaciation	60–70 years
Vegetation	Sitka alder
Soil depth	8.8 cm
Soil carbon	2.9 kg m^{-2}
Soil nitrogen	21.8 g m^{-2}
pH	6.8
Litterfall	278 g m^{-2} yr^{-1}

Time since deglaciation	200–225 years
Vegetation	Sitka spruce
Soil depth	15.1 cm
Soil carbon	9.7 kg m^{-2}
Soil nitrogen	53.3 g m^{-2}
pH	3.6
Litterfall	261 g m^{-2} yr^{-1}

Fig. 22.5 Age, community composition, and average environmental characteristics of the four major successional stages following deglaciation at Glacier Bay, Alaska. Data from Chapin et al. (1994).

Scattered woody shrubs such as *Dryas drummondii*, a mat-forming shrub that fixes nitrogen, willows, and Sitka alder, another nitrogen-fixing shrub, and seedlings of black cottonwood and Sitka spruce trees may be present depending on proximity to seed sources. This pioneer community transitions to a *Dryas* stage after about 30 years following deglaciation. A continuous mat of *Dryas*, with scattered willow, alder, cottonwood, and spruce, covers the ground. About

50 years following deglaciation, alder increases in abundance to form dense thickets. The shorter *Dryas*, shaded and buried under deciduous leaf litter, does not grow well and disappears. Sitka spruce trees eventually overtop the alder, forming a needleleaf evergreen forest at about 100 years since glacial retreat.

Soils change markedly during this succession (Figure 22.5). Soil depth increases from 5 cm initially to 15 cm in mature spruce forests as a result of rapid weathering of glacial till. Soils in the pioneer stage have low soil carbon and nitrogen, high bulk density, low cation exchange capacity, and high pH. With the growth of woody vegetation and the deposition of litter, the soil becomes progressively enriched in carbon and nitrogen. Soils in spruce forests have high organic matter content, high nitrogen content, decreased bulk density, and low pH. The transition from *Dryas* to alder, an important nitrogen-fixing shrub, marks a fourfold increase in soil nitrogen and a decline in soil C:N ratio from 245:1 to 133:1. The soil C:N ratio then increases to 182:1 in the spruce forest because of the low litter quality of needles. Succession and soil development at Glacier Bay is rapid and comparable to rates observed during secondary succession. This might be because of its maritime climate, with moderate temperatures and high annual rainfall.

22.3.3 Old-Field Succession in the North Carolina Piedmont

The regrowth of forests on abandoned farmland in the North Carolina Piedmont of Southeast United States is a classic example of old-field succession (Billings 1938; Oosting 1942; Keever 1950, 1983; Bormann 1953; Christensen 1977; Peet and Christensen 1980, 1987; Christensen and Peet 1981, 1984). Oak and hickory comprise the canopy of old-growth forests, with numerous smaller broadleaf deciduous tree and shrub species in the understory. Much of these old-growth forests were cleared in the 1700s and 1800s for agriculture. As the farms were abandoned, vegetation invaded the cleared fields, passing through distinct stages of herbaceous field and pine forest before becoming a mature

Table 22.1 Old-field succession following farm abandonment on upland sites in the North Carolina Piedmont

Year	Vegetation
0	Cropland
1	Field dominated by crabgrass and horseweed
2	Field dominated by aster and ragweed
3–5	Field dominated by broomsedge with pine seedlings
10–15	Closed stand of pine trees
40	Pine forest with distinct broadleaf deciduous tree understory
70–80	Mature pine forest. Oaks and hickories replace dead pines
150–200	Oak and hickory trees dominate forest

Source: From Billings (1938) and Oosting (1942).

oak–hickory forest (Table 22.1, Figure 22.6). Annual and perennial herbaceous species cover one- and two-year-old fields. The abundance of species declines in the third year as broomsedge, a perennial grass, increases in dominance. At about this time, seedlings of loblolly pine or shortleaf pine establish. Broomsedge maintains dominance for a few years until the rapidly growing pine seedlings overtop the shorter grasses, typically by the fifth year following abandonment. Closed stands of relatively even-aged pine trees form at 10 to 15 years. Germination and survivorship of pine seedlings declines in the low light under the closed canopy, maintaining the even age-structure of the stand. Following canopy closure, stand dynamics is largely a result of growth and mortality of pine trees. By 40 years, a distinct broadleaf deciduous tree understory forms. By 70–80 years, the short-lived pine trees begin to die and the longer-lived oaks and hickories replace them in the canopy. Between about 150 and 200 years following abandonment, oak and hickory trees gain dominance and pines remain as scattered relics. The forest is uneven aged, with old, large oak and hickory trees in

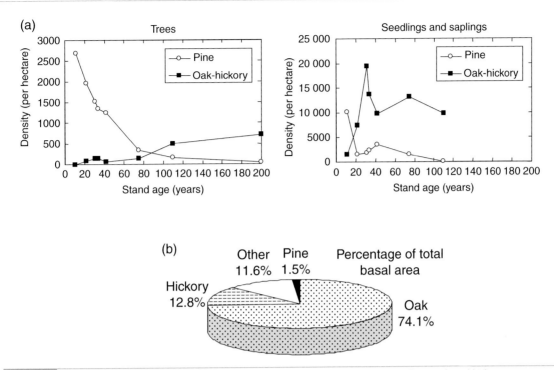

Fig. 22.6 Old-field succession on upland sites in the North Carolina Piedmont. (a) Density of pine, oak, and hickory trees (left) and seedlings/saplings (right) in relation to stand age. (b) Composition of a 200-year-old stand in terms of percentage of total basal area. Data from Billings (1938) and Oosting (1942).

the canopy, abundant oak and hickory trees in understory, and a large number of smaller subordinate trees in the understory.

22.3.4 Northern Hardwood Forests

The Hubbard Brook study provides a description of changes in community composition, biomass, and biogeochemical cycles during succession in northern hardwood forests of New England (Likens et al. 1977; Bormann and Likens 1979). During the first two to three years following clear-cutting, raspberry and blackberry (*Rubus*) flourish, complete their life cycles, and then decline in importance (Figure 22.7). These are replaced by pin cherry, a common early successional tree species restricted to recently disturbed forests. These trees grow fast and are short lived. Canopy closure occurs rapidly when large amounts of buried, viable seed form dense stands. Pin cherry dominates the canopy for the next several years. After about 25 to 35 years, sugar maple and American beech, which are tolerant of shade

and able to survive beneath the dense pin cherry canopy, become dominant. Where pin cherry is less dense, it may be co-dominant with other fast-growing species such as yellow birch and quaking aspen. These longer-lived trees establish within the first few years following disturbance and dominate the canopy for several decades before the slower growing maple and beech trees reach canopy status.

Biomass accumulation following clear-cutting generally has four phases (Figure 22.8). Reorganization is a period lasting about 10 to 20 years during which total biomass (living and detritus) decreases despite regrowth of vegetation. Loss of biomass from the forest floor and dead wood exceeds accumulation of biomass in plants. The aggradation phase is characterized by accumulation of biomass and nutrients in living plants, dead wood, and the forest floor. It begins about 15 years after clear-cutting and lasts for more than a century, culminating in peak accumulation of biomass. Accumulation

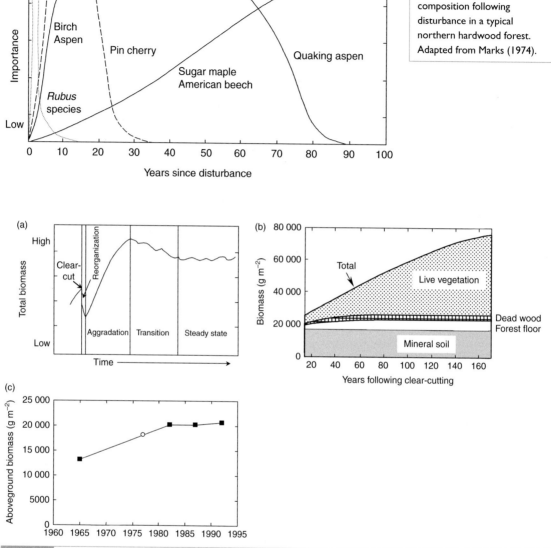

Fig. 22.7 Generalized changes in community composition following disturbance in a typical northern hardwood forest. Adapted from Marks (1974).

Fig. 22.8 Biomass accumulation following clear-cutting at the Hubbard Brook Experimental Forest. (a) General phases of ecosystem development. Adapted from Bormann and Likens (1979, p. 4). (b) General trends in biomass accumulation during aggradation. Adapted from Bormann and Likens (1979, p. 42). (c) Observed aboveground tree biomass for an undisturbed watershed in the Hubbard Brook Experimental Forest, 1965–1992. The data point for 1977 (open circle) is for trees with a diameter greater than 9.6 cm while all other years are for trees greater than 1.6 cm. Data from Likens et al. (1994). See also Fahey et al. (2005).

of organic matter on the forest floor increases available water-holding capacity and cation exchange capacity. The transition phase is a period during which total biomass declines. At steady state, total biomass fluctuates around a mean value.

The reorganization phase begins immediately following clear-cutting. It is a period during which biotic regulation of hydrologic and biogeochemical cycles is lost. Deforestation eliminates or alters transpiration, nutrient uptake by plants, decomposition, and mineralization,

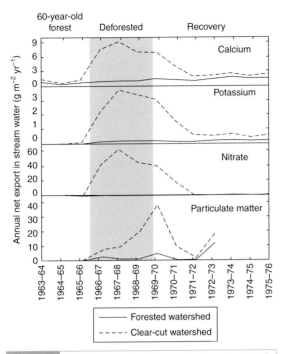

Fig. 22.9 Annual export of calcium, potassium, nitrate, and particulate matter in stream water for two watersheds in the Hubbard Brook Experimental Forest. One watershed was deforested in 1965–66 and vegetation regrowth was suppressed for three years. The other watershed was forested. Adapted from Bormann and Likens (1979, p. 149). See also Reiners (1992).

processes that regulate the hydrology and biogeochemistry of an aggrading forest ecosystem. Loss of regulation is seen in increased export of water, nutrients, and sediments from the ecosystem. Comparison of forested and deforested watersheds at Hubbard Brook demonstrates regulation of the hydrologic and biogeochemical cycles by vegetation. After clearing in 1965–66, evapotranspiration decreased and streamflow increased in the deforested watershed compared with the forested watershed (Figure 11.3). The loss of vegetation resulted in a rapid increase in the concentration of dissolved nutrients and sediments in stream water (Figure 22.9). Elimination of nutrient uptake by plants and increased rates of decomposition in the warmer, deforested site caused nutrients to flush from the system. Recovery of biotic regulation is rapid following regrowth of vegetation. Much of this is related to

the rapid colonization and growth of early successional species such as pin cherry.

Fahey et al. (2005) describe idealized patterns of net ecosystem production during forest recovery from disturbance. The forest is a source of carbon to the atmosphere of about 500–1200 g C m^{-2} yr^{-1} during the first decade as biomass debris decomposes. By about 15–20 years following disturbance, the forest becomes a carbon sink as production exceeds decomposition. During the aggradation phase, carbon accumulates in vegetation and soil, and beginning at about 30 years following disturbance the forest is a moderate carbon sink of about 200–300 g C m^{-2} yr^{-1} for the next 30–40 years. This carbon sink declines beyond about 70 years age as the forest matures and steady state is achieved.

The Hubbard Brook study illustrates the decline in net primary production typically found in old stands (Figure 22.8c). The Hubbard Brook forest accumulated biomass at a rate of 485 g m^{-2} yr^{-1} between 1965 and 1997. Accumulation declined to 46 g m^{-2} yr^{-1} between 1982 and 1992. Living tree biomass has remained relatively constant, indicating biomass accumulation is nearly balanced by mortality and is near steady state (Fahey et al. 2005). Age-dependent decline in productivity is common in many forests (He et al. 2012).

22.3.5 Boreal Forests of Interior Alaska

In the boreal forests of interior Alaska, forest communities are a successional mosaic of broadleaf deciduous and needleleaf evergreen trees that reflects recovery from recurring floods and fires (Van Cleve and Viereck 1981; Van Cleve et al. 1983a,b, 1986, 1991; Bonan 1989, 1990, 1993; Bonan and Van Cleve 1992; Chapin et al. 2006). Topographic location has a large role in determining community composition and the pathways of succession (Figure 20.17). Floodplains along rivers are a successional mix of willow and alder shrubs and highly productive balsam poplar and white spruce forests. Quaking aspen, paper birch, and white spruce form productive stands on warm, well-drained, south-facing slopes. Black spruce forms open stands with low productivity on cold, wet soils underlain with permafrost.

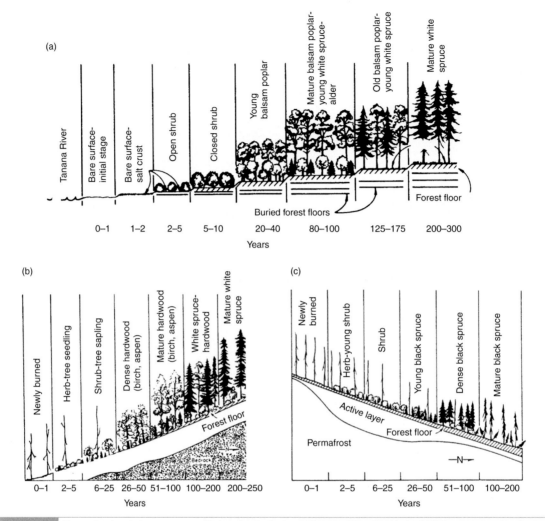

Fig. 22.10 Succession in relation to topographic location in the boreal forests of interior Alaska. (a) Primary succession along the Tanana River floodplain. (b) Post-fire secondary succession on upland south-facing slopes. (c) Post-fire secondary succession on upland north-facing slopes. Adapted from Van Cleve and Viereck (1981).

Along river floodplains, recurring floods initiate succession (Figure 22.10a). For the first one or two years, large floods deposit sediments and form terraces along river banks. As the terraces rise, flooding is less frequent and willow, alder, and balsam poplar invade the exposed sites. Shrubs increase rapidly in cover before being overtopped by trees about 20–30 years after establishment. With the formation of a closed balsam poplar canopy, the pioneer shrubs decline in abundance. Further buildup of the terrace from sediment deposits and litterfall prevents additional flooding. This allows

for the invasion of white spruce seedlings. The shift in dominance from balsam poplar to white spruce is gradual and takes about 100 years. The buildup of moss and a thick layer of organic matter on the ground is a prominent feature of this change.

On upland sites, recurring fires initiate succession (Figure 22.10b). Warm, dry sites are initially colonized by herbaceous plants, shrubs, and trees. Shrubs, primarily willow, and tree saplings dominate until about 25 years following fire, when the saplings grow into a dense stand of birch and aspen. These deciduous trees

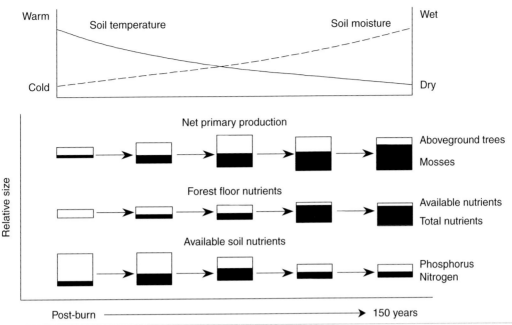

Fig. 22.11 Changes in soil moisture, soil temperature, net primary production, and nutrient cycling in a black spruce forest over 150 years following fire. Adapted from Van Cleve et al. (1983a).

dominate for the next 50 years, though white spruce seedlings and saplings grow in the understory. After about 100 years, white spruce grows into the canopy and is the dominant tree. Similar to floodplain succession, invasion of mosses and the accumulation of a thick forest floor accompany the development of a white spruce stand.

Post-fire succession on poorly drained upland sites and north slopes involves the recovery of black spruce (Figure 22.10c). During the first few years following fire, herbaceous plants and black spruce seedlings colonize the site. Many shrubs regrow from root sprouts. These shrubs grow rapidly and dominate the vegetation. A tree canopy begins to form 25–30 years after fire. Mosses invade at about this time, and a thick organic layer develops on the forest floor. This organic layer has a low thermal conductivity, insulating the soil and cooling soil temperatures. As a result, productivity declines and tree density decreases.

The general trend in community composition during succession is from deciduous forest to spruce forest with declining tree productivity and less vigorous nutrient cycling (Figure 20.19). This shift in community composition is accompanied by accumulation of organic material on the forest floor. Soil temperatures decline, slowing the rate of decomposition and further promoting the accumulation of organic material. Soil moisture increases with buildup of the forest floor. This promotes the establishment of mosses. In addition, the chemical quality of the forest floor declines with the shift from deciduous trees to spruce, further reducing decomposition rates. Nutrients accumulate in the undecomposed material on the forest floor. Tree growth declines.

Fires reset the system to earlier stages of succession. Forests older than 200–250 years are rare in the uplands around Fairbanks. Fires act as a rapid decomposer, consuming all or portions of the forest floor and replenishing the supply of nutrients. This is especially important in black spruce succession (Figure 22.11). Fires open the canopy and remove the forest floor. As a result, the soil warms. Over time, however, as mosses invade and the thick organic layer accumulates, soil temperature declines. Nutrient availability declines as mineralization rates decrease. Tree growth also declines, and in some old black spruce stands the net primary production of mosses exceeds that of trees.

22.4 | Mechanisms of Succession

Much of the debate among ecologists about succession has centered on whether it is a directional, predictable process of community change. This debate has its origins in the contrasting views of plant communities and ecosystems developed by ecologists in the early 1900s that saw either the community and ecosystem as the essential ecological unit of study or individuals and species as the essential units. The holistic school held that a community or ecosystem is an integrated entity equivalent to a superorganism. An alternative view was that plant communities and ecosystems are not superorganisms with emergent properties but rather are assemblages of individual plants and species. Attributes such as community structure and succession arise from the ecology of individuals and species interacting with one another. Much of the history of ecology has been an evaluation of these two theories (McIntosh 1981, 1985; Golley 1993).

In the holistic view, succession is orderly, predictable, and consists of discrete stages of development driven by vegetation itself (Clements 1916, 1928; Odum 1953, 1969, 1971). Succession culminates in a stable community composition, called the climax community, that is expected for a region. This view of succession is known as relay floristics or facilitation because groups of plants relay dominance to others by facilitating their entry into the community (Egler 1954; Connell and Slatyer 1977). Succession is characterized by the successive appearance and disappearance of groups of species (Figure 22.12a). Each group invades the site at a certain stage, making conditions unsuited for their own regeneration but paving the way for invasion by the next group of species. With the advent of ecosystem ecology, the superorganism concept was extended to see succession in terms of

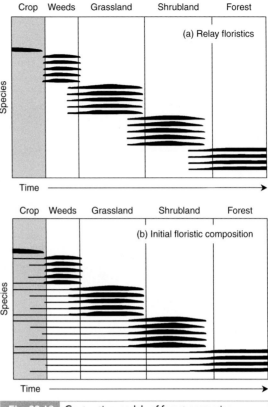

Fig. 22.12 Contrasting models of forest succession following farm abandonment illustrating (a) relay floristics and (b) initial floristic composition. Species importance at a given time increases with thicker lines. Adapted from Egler (1954).

ecosystem functions independent of the underlying plant populations. The climax ecosystem represents an equilibrium balance of functions, and succession is the process by which equilibrium is achieved.

In the alternative view, succession is seen as a population process that emphasizes individual plants and the response of species, through their various life history patterns, to environmental change (Cowles 1899; Gleason 1917, 1926, 1927, 1939). Succession occurs as a result of the differential growth, longevity, and colonizing ability of plants, expressed in the preemption of resources, canopy space, and root space. It is neither fixed nor predictable. Chance, especially the presence of seeds or sprouts to colonize a site, plays a large role in determining community composition. This view of succession is epitomized by Egler's initial floristic composition model of succession following farm abandonment (Figure 22.12b). Succession unfolds from the initial flora present in the seed bank at the time of abandonment. Seeds or vegetative parts of future dominant species are present at the time of abandonment. Progressive development relates to particular life histories. As each successive group dies out, another group of species, present from the start, assumes dominance.

These two contrasting theories have grown into a broader conceptualization of succession. No one single theory sufficiently explains all the patterns of succession found in nature (Connell and Slatyer 1977). Clementsian concepts such as facilitation can be found in various successions. Others fit an overall pattern in which community dynamics are determined by inhibition of new species until early dominants die or by tolerance of low levels of resources in later stages of succession. However, the prevailing opinion emphasizes succession as a population process (Egler 1954; Drury and Nisbet 1973; Horn 1974; Pickett 1976; Connell and Slatyer 1977; Peet and Christensen 1980, 1987; West et al. 1981; Shugart 1984, 1998; Huston and Smith 1987; Bazzaz 1996). It is a result of the different life histories of species operating in a gradient of environmental change. Succession is a consequence of differential growth, survival, and colonizing ability of species adapted to different points in an environmental gradient. Individual species colonize where conditions are favorable, die out of the community when the environment is no longer favorable, and grow in company with other species with similar environmental requirements. Seed rain and the chance that seeds or sprouts encounter a suitable safe site play a large role in vegetation development.

The prevailing view of succession emphasizes the attributes of species – their physiology, morphology, and life history – as adaptations to different environmental conditions encountered during community development. Autogenic changes during succession in the availability of light, water, and nutrients create shifting patterns of dominance as a result of tradeoffs in

the ability to effectively compete for above- and belowground resources. These tradeoffs result in a suite of traits that represent adaptation to a particular environment. Moreover, there is a tradeoff between tolerance to low resource conditions and maximum potential growth rate under high resource conditions. These tradeoffs and correlations among traits limit the ability of any one species to dominate all environmental conditions (Figure 19.4). There is an inevitable shift in the competitive ability of species over the course of succession due to these tradeoffs.

One trait that has received much attention is the different physiology of early and late successional plants (Bazzaz 1979, 1996; Huston and Smith 1987). This relates to shade tolerance. Shade tolerant trees photosynthesize at low light levels found in the understory of old-growth forests. Photosynthetic rates of shade intolerant species are not as high at low light. These species need full sunlight to have positive carbon gain. The physiological distinction between shade tolerant and intolerant species is part of a larger overall life history difference between early and late successional species (Table 18.3).

One theory of community structure and dynamics, known as the resource-ratio theory of succession, is based on resource acquisition and tradeoffs in characteristics that allow plants to effectively compete for above- and belowground resources (Tilman 1985, 1988). Species with high allocation to production of leaves and stems are effective competitors for light, an aboveground resource. Species with high allocation to root production are good competitors for belowground resources such as nitrogen. Resource supply and acquisition determine which morphologies are viable in different environments. Root specialists are favored where aboveground resources are abundant but belowground resources are limiting. Shoot specialists are favored as light becomes more limiting. Succession is a result of autogenic changes in the availability of above- and belowground resources over time that favor different allocation patterns (Figure 22.13). During the initial stages of succession, light is readily available but nitrogen availability is low. Short, leafy species with high leaf allocation, low stem allocation,

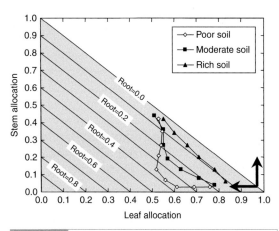

Fig. 22.13 The resource-ratio theory of succession. Allocation to leaves, stems, and roots is represented as a triangle. Each point within the triangle represents a unique plant morphology. Morphologies near the origin have a high proportion of their biomass in roots. Those near the upper left corner have a high proportion of stem while those in the lower right corner are primarily foliage. Successional changes in foliage, stem, and root allocation are illustrated for three levels of soil fertility. The two large arrows indicate the general trends in stem and foliage allocation. Adapted from Tilman (1988, p. 233).

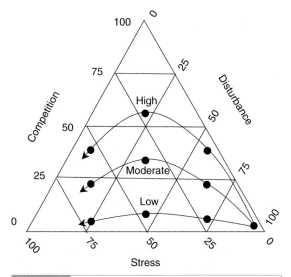

Fig. 22.14 Grime's plant strategies showing successional changes in ruderal, competitor, and stress tolerators under conditions of high, moderate, and low resource availability. Adapted from Grime (1979, p. 151).

and moderate root allocation grow best. As the community develops, the canopy closes and light availability declines. Nitrogen supply increases with the introduction of nitrogen-fixing species, production of leaf litter, and accumulation of plant detritus. This environment favors tall species with greater allocation to stem and less to foliage and roots. The actual sequence of dominant morphologies differs depending on soil fertility, though in the example shown in Figure 22.13 all three sites converge on a similar morphology because they are assumed to attain the same final nitrogen availability.

An alternative view of succession is based on Grime's (1979) classification of species into life history patterns of competitor, ruderal, and stress tolerator (Figure 18.11). This classification arises from tradeoffs among a broad range of traits including reproductive effort, dispersal, ability to capture resources, and ability to tolerate stress. It invokes disturbance, stress, and competition as opposing forces selecting for particular traits. Succession is a result of the

interplay between these different life history patterns and changing environmental conditions (Figure 22.14). Ruderal species, with their rapid ability to colonize disturbed sites, are favored in open sites with abundant resources (light, space). As vegetation develops and biomass accumulates, competition for resources increases and selects for competitive species. Depletion of resources in late stages of succession leads to replacement by stress tolerators. This view of succession is driven by autogenic decreases in resource availability as succession progresses. On less productive sites, earlier onset of resource depletion limits the appearance of highly competitive species. On unproductive sites, succession moves directly from ruderal to stress tolerator. Grace (1991) compares this model with the resource-ratio model.

Another theory of succession emphasizes vital attributes (Noble and Slatyer 1980). The concept of vital attributes combines life history patterns, resource utilization, and response to disturbance. These vital attributes are: method of arrival or persistence during and after disturbance; ability to establish and grow to maturity in the developing community; and time required

Table 22.2 Life history patterns of plants that dominate different successional stages at Glacier Bay, Alaska

Plant type	Stage	Seed mass (µg)	Height (m)	Age of reproduction (years)	Longevity (years)
Forb	Pioneer	72	0.3	1–2	20
Dwarf shrub	Dryas	97	0.1	6–8	50
Sitka alder	Alder	494	4	8	100
Sitka spruce	Spruce	2694	40	30–50	700

Source: From Chapin et al. (1994).

for an individual to reach critical life stages such as seed production. These three attributes combine to produce numerous life histories.

In general, life history patterns are related to a tradeoff between an ability to rapidly colonize and dominate a site versus slow growth and ability to compete in a resource limited environment. There is a fundamental physiological tradeoff between characteristics that allow either high photosynthesis and rapid growth under high light or the ability to grow and survive in shade. This dichotomy is seen in the broad classification of plants into categories such as *r*- and *K*-selected species (MacArthur and Wilson 1967; Gadgil and Solbrig 1972), shade intolerant and tolerant (Huston and Smith 1987), early and late successional (Bazzaz 1979, 1996; Huston and Smith 1987), ruderal and competitor (Grime 1979), exploitive and conservative (Bormann and Likens 1979), and gap and non-gap (Shugart 1984, 1987, 1998). Regardless of what they are called, species that dominate communities in the early stages of succession produce prolific crops of small, widely dispersed seeds, grow rapidly to maturity, but are short lived and unable to survive in the light-depleted understory of a dense canopy. These species have traits that take advantage of the temporary nature of disturbance. They maintain their abundance in the landscape through rapid colonization and dominance of disturbed sites enriched in light, water, and nutrients. In contrast, species that dominate during the late stages of succession have lower rates of dispersal and colonization, grow slowly, and are long lived. They maintain their presence by surviving in the intense competition for limited resources.

Life history patterns such as seed size, maximum height, age of reproduction, and longevity explain succession at Glacier Bay (Table 22.2). For example, the order in which species colonize newly exposed glacial till correlates with seed size and dispersability. Seeds of early successional species are lighter than those of later stages of succession. Early successional species also have a shorter lifespan and mature earlier than mid- and late successional species. These reproductive traits are critical to successional changes in community composition. Long-distance seed dispersal and short generations are necessary to establish and maintain plants in early successional communities far removed from seed sources. The low density of alder and spruce in pioneer communities reflects low seed rain rather than inability to germinate and establish on newly exposed sites. Alder and spruce are slower to arrive on recently deglaciated terrain than pioneer species because their heavier seeds and older age of reproduction limit seed sources.

Facilitation and inhibition among species can also be seen in the succession (Chapin et al. 1994). The change from *Dryas* to alder involves competitive displacement as the taller alder shrubs shade the shorter *Dryas*. Similarly, slow-growing, long-lived spruce trees overtop and displace alder. Hence, height and longevity, which increase throughout succession, are major factors determining community change. The progressive addition of organic matter and soil nitrogen facilitates rapid community development. Conversely, *Dryas* and alder inhibit the establishment of alder and spruce seedlings, respectively. Alder and spruce must disperse to

Table 22.3 | Percentage of stands sampled within which a species occurs in relation to successional age

Species	Pine forest				Broadleaf deciduous
	20–40 years	40–60 years	60–80 years	> 80 years	
Small, wind-dispersed seeds					
Red maple	100	100	100	100	100
Sweetgum	100	100	97	93	46
Yellow poplar	94	86	74	86	68
Large, animal-dispersed seeds					
Pignut hickory	33	55	74	86	92
Mockernut hickory	35	55	79	93	94
White oak	23	59	89	93	100
Northern red oak	55	67	86	93	94
Black oak	71	77	86	93	72

Source: From Christensen and Peet (1981).

the site before the prior colonizers modify the environment to inhibit seedling establishment. Inhibition of seedling establishment and growth by other species prevents late successional species from dominating more quickly.

In old-fields in the North Carolina Piedmont, sequential change in the dominance of herbaceous annuals and perennials to pine trees and then oaks and hickories relates to life history patterns such as time of seed dispersal and germination, environmental requirements for germination, and type of seed dispersal. Pine trees gain early dominance because their seeds are light and widely dispersed by wind and because their seedlings germinate and grow rapidly in the high light environment of old-fields. Variation among stands in the initial stocking of pine seedlings relates to the proximity of seed trees and the quality of seed. Pines lose dominance to broadleaf deciduous trees because they are relatively short-lived and because pine seedlings germinate and establish poorly in shade. Broadleaf deciduous trees are tolerant of shade and can survive and grow better in low light. Among broadleaf deciduous trees, seed size and dispersal accounts for much of the temporal variation in species abundance. Tree species with small, wind-dispersed seeds occur in stands of all ages, though with greatest frequency in

young stands (Table 22.3). In contrast, oak and hickory, with large nuts, are relatively infrequent in young pine stands and increase in frequency throughout succession. Presence of these species in young stands reflects proximity to seed trees while their abundance in late stages of succession is ensured through the growth of large canopy dominants.

Pin cherry, which contributes greatly to the rapid recovery of northern hardwood forests following disturbance, is an example of an exploitive life history (Marks 1974). Its life history is suited for efficient colonization of large gaps in the canopy. The combination of buried seeds that are viable in the soil for 15 years or more and dispersal by birds ensures a large seed bank well after disappearance from a particular site. Once established, pin cherry grows rapidly. By age 25–30, when pin cherry trees rapidly die, sufficient seeds have been produced and disseminated to ensure renewal after the next major disturbance.

22.5 | Biosphere–Atmosphere Interactions

The successional development of ecosystems alters biosphere–atmosphere coupling. The

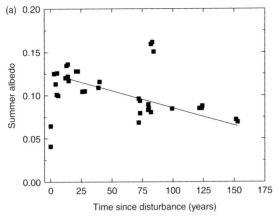

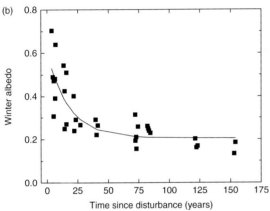

Fig. 22.15 Daily mean albedo during (a) summer and (b) winter in boreal forests of Alaska and western Canada in relation to time since disturbance. Adapted from Amiro et al. (2006).

post-fire successional development of boreal forests in North America illustrates temporal trends in surface energy fluxes. Amiro et al. (2006) presented measurements for 22 North American boreal forest sites initiated by wildfire and comprising grasses, herbs, and willow, aspen, jack pine, and black spruce trees depending on age. The 150-year forest chronosequence shows a marked change in albedo as the forests recover from fire (Figure 22.15). The general trend is a decline in albedo as the forest recovers. Summer albedo immediately following fire is low (~0.05) because of charring, but increases to about 0.12 with a deciduous canopy for a 30-year period, and then decreases to about 0.08 for mature spruce forest. Winter albedo has a pronounced

decline from about 0.7 for young forests with a sparse canopy to about 0.2 for mature forests with a dense canopy. The evaporative fraction $(\lambda E / R_n)$ is smallest (0.2) in very young stands, increases to about 0.6 in stands 10–30 years old, and declines to about 0.4 in older stands. The Bowen ratio $(H / \lambda E)$ is largest in very young sites (>2), decreases in stands 10–30 years old (0.5–1), and increases again in old stands (1–2).

Measurements at three sites in interior Alaska that differed in age since fire illustrate trends during post-fire black spruce succession (Liu et al. 2005; Liu and Randerson 2008). One site was a black spruce forest that had burned three years previously. The fire killed all vegetation, consumed much of the aboveground biomass, and removed a portion of the soil organic mat overlying the mineral soil. Three years following the fire, the boles of the dead trees remained standing, and 30 percent of the surface was covered by grasses and deciduous shrubs. The second site was a black spruce forest that had burned 15 years previously. Fire killed all the vegetation, and many of the dead trees still stood. The site was revegetated by aspen and willow with a mean canopy height of 5 m. The third site was a black spruce forest that had burned 80 years previously and had regrown. Black spruce trees with a mean canopy height of 4 m formed the overstory. Moss and soil organic matter formed a forest floor that was 11 cm thick on average.

The three sites differed in energy fluxes (Table 22.4). Annual net radiation declined by 31 percent for the 3-year site and the 15-year aspen forest compared with the 80-year black spruce forest. This difference was greatest during spring because increased snow cover at the younger sites resulted in increased surface albedo. The summer and winter seasons also had substantial decreases in net radiation at the 3-year and 15-year sites relative to the 80-year black spruce forest. Overall, more than 50 percent of the annual decrease in net radiation at both younger sites occurred in spring, 20–33 percent occurred in summer, and less than 20 percent occurred during winter. Annual sensible heat flux decreased by more than 50 percent compared with the 80-year site. Similar to net

Table 22.4 | Seasonal and annual net radiation (R_n), soil heat flux (G), sensible heat flux (H), latent heat flux (λE), and midday Bowen ratio ($H / \lambda E$) for three sites in interior Alaska following fire

Site	R_n (W m^{-2})	G (W m^{-2})	H (W m^{-2})	λE (W m^{-2})	$H/\lambda E$
Spring (March–May)					
3-year	58	4	21	15	2.2
15-year	61	5	31	18	2.5
80-year	98	2	57	23	3.5
Summer (June–August)					
3-year	124	12	44	36	1.6
15-year	117	11	35	59	0.9
80-year	138	9	51	55	1.3
Autumn (September–November)					
3-year	2	−2	−9	10	1.4
15-year	3	−3	−6	10	1.0
80-year	3	−1	−10	14	1.2
Winter (December–February)					
3-year	−31	−10	−19	3	0.5
15-year	−29	−9	−19	3	−0.5
80-year	−19	−7	−13	4	2.4
Annual					
3-year	38	1	9	16	2
15-year	38	1	10	22	2
80-year	55	1	21	24	2

Source: From Liu et al. (2005).

radiation, most of this reduction occurred in spring. Annual evapotranspiration at the 3-year site (202 mm yr^{-1}) was 33 percent less than the 80-year black spruce forest (301 mm yr^{-1}), but the 15-year aspen forest (283 mm yr^{-1}) had similar annual evapotranspiration compared with the black spruce forest.

Differences in energy fluxes among the three sites were due in part to the increased surface albedo of the younger sites. Spring differences were largely attributable to the increased surface albedo of the younger sites arising from snow cover. As the snow melted, the midday albedo of the 3-year site declined from 0.7 to 0.1, comparable to that of the 80-year black spruce forest. During summer, the 3-year site had higher midday albedo (0.12) than the 80-year site (0.08).

The deciduous canopy at the 15-year aspen forest also influenced surface fluxes. The emergence of leaves on aspen trees in spring increased latent heat flux by 50 percent in the three–week period after leaf emergence compared with the three weeks prior to leaf emergence and decreased midday Bowen ratio from 1.9 to 1.2. During summer, the 15-year aspen forest had the highest latent heat flux, lowest sensible heat flux, and lowest midday Bowen ratio because of the high leaf area and canopy conductance of the deciduous overstory. Similar increases in latent heat flux with leaf emergence are seen in other deciduous forests (Table 17.3), cooling regional climate (Chapter 26).

Goulden et al. (2011) compared the carbon cycle along a 150-year forest chronosequence in central Manitoba (Figure 22.16). The forest ecosystem accumulates carbon in live biomass and the forest floor with greater age. The ecosystem initially loses carbon (negative net ecosystem

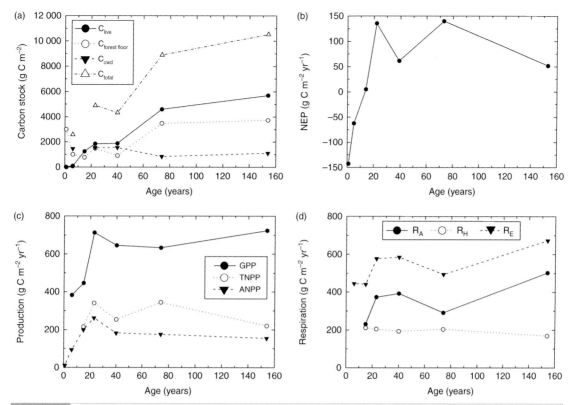

Fig. 22.16 Post-disturbance successional trends in the carbon cycle of boreal forests in central Manitoba, Canada. (a) Live biomass (C_{live}), forest floor ($C_{forest\ floor}$), coarse woody debris (C_{cwd}), and total (C_{total}) carbon stocks. (b) Net ecosystem production. (c) Gross primary production (GPP), total net primary production (TNPP), and aboveground net primary production (ANPP). (d) Autotrophic respiration (R_A), heterotrophic respiration (R_H), and ecosystem respiration (R_E). Data from Goulden et al. (2011).

production) as respiration loss from decaying detritus exceeds carbon input from gross primary production. Thereafter, as gross primary production recovers to predisturbance levels, the ecosystem gains carbon (positive net ecosystem production). Similar general patterns of substantial accumulation of carbon (and nitrogen) in forest ecosystems with age are common (Yang et al. 2011).

22.6 | Review Questions

1. Exponential population growth is described by the equation $N(t) = N_0 e^{rt}$ and logistic growth by

$$N(t) = \frac{KN_0}{N_0 + (K - N_0)e^{-rt}},$$ where $N(t)$ is density at time t, N_0 is initial density at $t = 0$, r is the rate of population growth, and K is carrying capacity. Graph and compare these two models for $N_0 = 4$, $r = 0.1$, and $K = 300$. What are the two population estimates at $t = 100$? Why is K called the carrying capacity?

2. The following table gives seedling survivorship of sugar maple and American beech in a northern hardwood forest (Forcier 1975). Characterize the shape of the survivorship curve. What does this say about the risk of mortality as the trees age?

Age class (years)	Beech	Sugar maple
0–1	1000	1000
1–2	410	583
2–3	262	317
3–4	188	225
4–5	138	166
5–6	121	118
6–7	110	84
7–8	92	54
8–9	87	48
9–10	83	42

3. For a particular plant population undergoing self-thinning $\bar{m} = 190N^{-3/2}$, where $\bar{m}$ is mean plant mass (g) and N is density per 20 cm pot. Compare mean plant mass with $N = 100$ and $N = 50$. How has biomass changed?

4. Characterize succession at Glacier Bay in terms of facilitation, inhibition, and tolerance. What vital attributes determine succession?

5. Describe which model (relay floristics or initial floristic composition) is represented by old-field succession in the North Carolina Piedmont. Characterize the vital attributes that determine old-field succession.

6. Describe the role of pin cherry in northern hardwood forest succession.

7. Describe the life history characteristics that allow species to dominate recently disturbed sites in contrast to old-growth forests.

8. Describe post-fire successional trends in albedo, energy fluxes, and carbon storage in boreal forests of North America. How might these temporal changes in biosphere–atmosphere interactions affect climate?

9. What effect would increased fire frequency have on the species composition of forests growing on permafrost-free sites in Alaska? How might this affect surface energy fluxes and climate?

22.7 | References

Amiro, B. D., Orchansky, A. L., Barr, A. G., et al. (2006). The effect of post-fire stand age on the boreal forest energy balance. *Agricultural and Forest Meteorology*, 140, 41–50.

Bazzaz, F. A. (1979). The physiological ecology of plant succession. *Annual Review of Ecology and Systematics*, 10, 351–371.

Bazzaz, F. A. (1996). *Plants in Changing Environments: Linking Physiological, Population, and Community Ecology*. Cambridge: Cambridge University Press.

Billings, W. D. (1938). The structure and development of old field shortleaf pine stands and certain associated physical properties of the soil. *Ecological Monographs*, 8, 437–499.

Bonan, G. B. (1988). The size structure of theoretical plant populations: Spatial patterns and neighborhood effects. *Ecology*, 69, 1721–1730.

Bonan, G. B. (1989). Environmental factors and ecological processes controlling vegetation patterns in boreal forests. *Landscape Ecology*, 3, 111–130.

Bonan, G. B. (1990). Carbon and nitrogen cycling in North American boreal forests, I: Litter quality and soil thermal effects in interior Alaska. *Biogeochemistry*, 10, 1–28.

Bonan, G. B. (1991). Density effects on the size structure of annual plant populations: An indication of neighbourhood competition. *Annals of Botany*, 68, 341–347.

Bonan, G. B. (1993). Physiological controls of the carbon balance of boreal forest ecosystems. *Canadian Journal of Forest Research*, 23, 1453–1471.

Bonan, G. B., and Van Cleve, K. (1992). Soil temperature, nitrogen mineralization, and carbon source–sink relationships in boreal forests. *Canadian Journal of Forest Research*, 22, 629–639.

Bormann, F. H. (1953). Factors determining the role of loblolly pine and sweetgum in early old-field succession in the Piedmont of North Carolina. *Ecological Monographs*, 23, 339–358.

Bormann, F. H., and Likens, G. E. (1979). *Pattern and Process in a Forested Ecosystem*. New York: Springer-Verlag.

Botkin, D. B. (1993). *Forest Dynamics: An Ecological Model*. Oxford: Oxford University Press.

Chapin, F. S., III, Walker, L. R., Fastie, C. L., and Sharman, L. C. (1994). Mechanisms of primary succession following deglaciation at Glacier Bay, Alaska. *Ecological Monographs*, 64, 149–175.

Chapin, F. S., III, Oswood, M. W., Van Cleve, K., Viereck, L. A., and Verbyla, D. L. (2006). *Alaska's Changing Boreal Forest*. Oxford: Oxford University Press.

Christensen, N. L. (1977). Changes in structure, pattern and diversity associated with climax forest maturation in Piedmont, North Carolina. *American Midland Naturalist*, 97, 176–188.

Christensen, N. L., and Peet, R. K. (1981). Secondary forest succession on the North Carolina Piedmont. In *Forest Succession: Concepts and Application*, ed. D. C. West, H. H. Shugart, and D. B. Botkin. New York: Springer-Verlag, pp. 230–245.

Christensen, N. L., and Peet, R. K. (1984). Convergence during secondary forest succession. *Journal of Ecology*, 72, 25–36.

Clements, F. E. (1916). *Plant Succession: An Analysis of the Development of Vegetation*, Carnegie Institution Publication Number 242. Washington, DC: Carnegie Institution.

Clements, F. E. (1928). *Plant Succession and Indicators*. New York: H.W. Wilson.

Connell, J. H., and Slatyer, R. O. (1977). Mechanisms of succession in natural communities and their role in community stability and organization. *American Naturalist*, 111, 1119–1144.

Cooper, W. S. (1923a). The recent ecological history of Glacier Bay, Alaska, I: The interglacial forests of Glacier Bay. *Ecology*, 4, 93–128.

Cooper, W. S. (1923b). The recent ecological history of Glacier Bay, Alaska, II: The present vegetation cycle. *Ecology*, 4, 223–246.

Cooper, W. S. (1923c). The recent ecological history of Glacier Bay, Alaska, III: Permanent quadrats at Glacier Bay: An initial report upon a long-period study. *Ecology*, 4, 355–365.

Cooper, W. S. (1931). A third expedition to Glacier Bay, Alaska. *Ecology*, 12, 61–95.

Cooper, W. S. (1939). A fourth expedition to Glacier Bay, Alaska. *Ecology*, 20, 130–155.

Cowles, H. C. (1899). The ecological relations of the vegetation on the sand dunes of Lake Michigan. *Botanical Gazette*, 27, 95–117, 167–202, 281–308, 361–391.

Drury, W. H., and Nisbet, I. C. T. (1973). Succession. *Journal of the Arnold Arboretum, Harvard University*, 54, 331–368.

Egler, F. E. (1954). Vegetation science concepts, I: Initial floristic composition, a factor in old-field vegetation development. *Vegetatio*, 4, 412–417.

Fahey, T. J., Siccama, T. G., Driscoll, C. T., et al. (2005). The biogeochemistry of carbon at Hubbard Brook. *Biogeochemistry*, 75, 109–176.

Fastie, C. L. (1995). Causes and ecosystem consequences of multiple pathways of primary succession at Glacier Bay, Alaska. *Ecology*, 76, 1899–1916.

Forcier, L. K. (1975). Reproductive strategies and the co-occurrence of climax tree species. *Science*, 189, 808–810.

Gadgil, M., and Solbrig, O. T. (1972). The concept of r- and K-selection: Evidence from wild flowers and some theoretical considerations. *American Naturalist*, 106, 14–31.

Gleason, H. A. (1917). The structure and development of the plant association. *Bulletin of the Torrey Botanical Club*, 44, 463–481.

Gleason, H. A. (1926). The individualistic concept of the plant association. *Bulletin of the Torrey Botanical Club*, 53, 7–26.

Gleason, H. A. (1927). Further views on the succession-concept. *Ecology*, 8, 299–326.

Gleason, H. A. (1939). The individualistic concept of the plant association. *American Midland Naturalist*, 21, 92–110.

Golley, F. B. (1993). *A History of the Ecosystem Concept in Ecology: More than the Sum of the Parts*. New Haven: Yale University Press.

Goulden, M. L., McMillan, A. M. S., Winston, G. C., et al. (2011). Patterns of NPP, GPP, respiration, and NEP during boreal forest succession. *Global Change Biology*, 17, 855–871.

Grace, J. B. (1991). A clarification of the debate between Grime and Tilman. *Functional Ecology*, 5, 583–587.

Grime, J. P. (1979). *Plant Strategies and Vegetation Processes*. Chichester: Wiley.

Harcombe, P. A. (1987). Tree life tables. *BioScience*, 37, 557–568.

Harper, J. L. (1977). *Population Biology of Plants*. London: Academic Press.

He, L., Chen, J. M., Pan, Y., Birdsey, R., and Kattge, J. (2012). Relationships between net primary productivity and forest stand age in U.S. forests. *Global Biogeochemical Cycles*, 26, GB3009, doi:10.1029/2010GB003942.

Horn, H. S. (1974). The ecology of secondary succession. *Annual Review of Ecology and Systematics*, 5, 25–37.

Huston, M., and Smith, T. (1987). Plant succession: Life history and competition. *American Naturalist*, 130, 168–198.

Keever, C. (1950). Causes of succession on old fields of the Piedmont, North Carolina. *Ecological Monographs*, 20, 229–250.

Keever, C. (1983). A retrospective view of old-field succession after 35 years. *American Midland Naturalist*, 110, 397–404.

Kira, T., Ogawa, H., and Sakazaki, N. (1953). Intraspecific competition among higher plants, I: Competition-yield-density interrelationship in regularly dispersed populations. *Journal of the Institute of Polytechnics, Osaka City University, Series D*, 4, 1–16.

Knox, R. G., Peet, R. K., and Christensen, N. L. (1989). Population dynamics in loblolly pine stands: Changes in skewness and size inequality. *Ecology*, 70, 1153–1166.

Koyama, H., and Kira, T. (1956). Intraspecific competition among higher plants, VIII: Frequency distribution of individual plant weight as affected by the interaction between plants. *Journal of the Institute of Polytechnics, Osaka City University, Series D*, 7, 73–94.

Likens, G. E., Bormann, F. H., Pierce, R. S., Eaton, J. S., and Johnson, N. M. (1977). *Biogeochemistry of a Forested Ecosystem*. New York: Springer-Verlag.

Likens, G. E., Driscoll, C. T., Buso, D. C., et al. (1994). The biogeochemistry of potassium at Hubbard Brook. *Biogeochemistry*, 25, 61–125.

Liu, H., and Randerson, J. T. (2008). Interannual variability of surface energy exchange depends on stand age in a boreal forest fire chronosequence. *Journal of Geophysical Research*, 113, G01006, doi:10.1029/2007JG000483.

Liu, H., Randerson, J. T., Lindfors, J., and Chapin, F. S., III (2005). Changes in the surface energy budget after fire in boreal ecosystems of interior Alaska: an annual perspective. *Journal of Geophysical Research*, 110, D13101, doi:10.1029/2004JD005158.

Lonsdale, W. M. (1990). The self-thinning rule: Dead or alive? *Ecology*, 71, 1373–1388.

MacArthur, R. H., and Wilson, E. O. (1967). *The Theory of Island Biogeography*. Princeton: Princeton University Press.

Marks, P. L. (1974). The role of pin cherry (Prunus pensylvanica L.) in the maintenance of stability in northern hardwood ecosystems. *Ecological Monographs*, 44, 73–88.

McIntosh, R. P. (1981). Succession and ecological theory. In *Forest Succession: Concepts and Application*, ed. D. C. West, H. H. Shugart, and D. B. Botkin. New York: Springer-Verlag, pp. 10–23.

McIntosh, R. P. (1985). *The Background of Ecology: Concept and Theory*. Cambridge: Cambridge University Press.

Miller, T. E., and Weiner, J. (1989). Local density variation may mimic effects of asymmetric competition on plant size variability. *Ecology*, 70, 1188–1191.

Mohler, C. L., Marks, P. L., and Sprugel, D. G. (1978). Stand structure and allometry of trees during self-thinning of pure stands. *Journal of Ecology*, 66, 599–614.

Noble, I. R., and Slatyer, R. O. (1980). The use of vital attributes to predict successional changes in plant communities subject to recurrent disturbances. *Vegetatio*, 43, 5–21.

Odum, E. P. (1953). *Fundamentals of Ecology*. Philadelphia: Saunders.

Odum, E. P. (1969). The strategy of ecosystem development. *Science*, 164, 262–270.

Odum, E. P. (1971). *Fundamentals of Ecology*, 3rd ed. Philadelphia: Saunders.

Olson, J. S. (1958). Rates of succession and soil changes on Southern Lake Michigan sand dunes. *Botanical Gazette*, 119, 125–170.

Oosting, H. J. (1942). An ecological analysis of the plant communities of Piedmont, North Carolina. *American Midland Naturalist*, 28, 1–126.

Peet, R. K., and Christensen, N. L. (1980). Succession: A population process. *Vegetatio*, 43, 131–140.

Peet, R. K., and Christensen, N. L. (1987). Competition and tree death. *BioScience*, 37, 586–594.

Pickett, S. T. A. (1976). Succession: An evolutionary interpretation. *American Naturalist*, 110, 107–119.

Reiners, W. A. (1992). Twenty years of ecosystem reorganization following experimental deforestation and regrowth suppression. *Ecological Monographs*, 62, 503–523.

Shugart, H. H. (1984). *A Theory of Forest Dynamics: The Ecological Implications of Forest Succession Models*. New York: Springer-Verlag.

Shugart, H. H. (1987). Dynamic ecosystem consequences of tree birth and death patterns. *BioScience*, 37, 596–602.

Shugart, H. H. (1998). *Terrestrial Ecosystems in Changing Environments*. Cambridge: Cambridge University Press.

Tilman, D. (1985). The resource-ratio hypothesis of plant succession. *American Naturalist*, 125, 827–852.

Tilman, D. (1988). *Plant Strategies and the Dynamics and Structure of Plant Communities*. Princeton: Princeton University Press.

Van Cleve, K., and Viereck, L. A. (1981). Forest succession in relation to nutrient cycling in the boreal forest of Alaska. In *Forest Succession: Concepts and Application*, ed. D. C. West, H. H. Shugart, and D. B. Botkin. New York: Springer-Verlag, pp. 185–211.

Van Cleve, K., Dyrness, C. T., Viereck, L. A., et al. (1983a). Taiga ecosystems in interior Alaska. *BioScience*, 33, 39–44.

Van Cleve, K., Oliver, L., Schlentner, R., Viereck, L. A., and Dyrness, C. T. (1983b). Productivity and nutrient cycling in taiga forest ecosystems. *Canadian Journal of Forest Research*, 13, 747–766.

Van Cleve, K., Chapin, F. S., III, Flanagan, P. W., Viereck, L. A., and Dyrness, C. T. (1986). *Forest Ecosystems in the Alaskan Taiga: A Synthesis of Structure and Function*. New York: Springer-Verlag.

Van Cleve, K., Chapin, F. S., III, Dyrness, C. T., and Viereck, L. A. (1991). Element cycling in taiga forests: State-factor control. *BioScience*, 41, 78–88.

Weiner, J. (1985). Size hierarchies in experimental populations of annual plants. *Ecology*, 66, 743–752.

Weiner, J., and Freckleton, R. P. (2010). Constant final yield. *Annual Review of Ecology, Evolution, and Systematics*, 41, 173–192.

Weiner, J., and Thomas, S. C. (1986). Size variability and competition in plant monocultures. *Oikos*, 47, 211–222.

Weller, D. E. (1987). A reevaluation of the –3/2 power rule of plant self-thinning. *Ecological Monographs*, 57, 23–43.

West, D. C., Shugart, H. H., and Botkin, D. B. (1981). *Forest Succession: Concepts and Application*. New York: Springer-Verlag.

Westoby, M. (1981). The place of the self-thinning rule in population dynamics. *American Naturalist*, 118, 581–587.

Westoby, M. (1984). The self-thinning rule. *Advances in Ecological Research*, 14, 167–225.

White, J., and Harper, J. L. (1970). Correlated changes in plant size and number in plant populations. *Journal of Ecology*, 58, 467–485.

Yang, Y., Luo, Y., and Finzi, A. C. (2011). Carbon and nitrogen dynamics during forest stand development: A global synthesis. *New Phytologist*, 190, 977–989.

Yoda, K., Kira, T., Ogawa, H., and Hozumi, K. (1963). Self-thinning in overcrowded pure stands under cultivated and natural conditions (Intraspecific competition among higher plants, XI.). *Journal of Biology, Osaka City University*, 14, 107–129.

Landscapes and Disturbances

23.1 Chapter Summary

Landscapes are another level of ecological organization, merging the concepts of populations, communities, and ecosystems. A landscape is a mosaic of communities and ecosystems formed by a gradient of environmental conditions. The pattern of vegetation in a landscape arises due to gradients in temperature, soil moisture, and other environmental factors. It also arises from disturbances such as fire, forest clearing, or farm abandonment that initiate secondary succession. Post-disturbance succession creates a mosaic of communities and ecosystems in various stages of development. These communities and ecosystems occur as distinct patches across the landscape. Patches are homogenous units of land with similar topography, soil, microclimate, and vegetation. Patches are the individual elements of the landscape and are embedded in a matrix of other patches, which forms the pattern of the landscape. Fire and human uses of land are important processes creating pattern in landscapes. Changes in the disturbance regime alter the structure and composition of a landscape and so alter exchanges of energy, water, and chemical constituents with the atmosphere.

23.2 Pattern and Process in Plant Communities

Many ecologists long viewed natural systems as being in balance and in equilibrium with the environment. The Clementsian view of plant succession held that succession is orderly, predictable change that culminates in a stable, climax ecosystem that represents an equilibrium balance of functions. However, this equilibrium view of nature is flawed. Fires, floods, hurricanes, herbivores, insects, and people regularly disturb plant communities, initiating the successional cycle. In this non-equilibrium view, the landscape is a mosaic of successional communities whose composition and structure are determined by the type and severity of disturbance. Disturbance and the resulting successional development of communities and ecosystems are the processes giving rise to the mosaic pattern of vegetation in the landscape (Watt 1947; Pickett 1976; Whittaker and Levin 1977; Bormann and Likens 1979; Shugart 1984). These disturbances can be as small as a 100 m² gap in the canopy created by the death of a large dominant tree or as large as hundreds of square kilometers following fire.

Disturbance removes existing vegetation and creates an environment with high sunlight and reduced competition for resources. These conditions favor the germination, establishment, and growth of early successional species. Early successional plants are maintained in the landscape by a fugitive life history and recurring disturbance. Their abundance is limited by the availability of a site to colonize and the availability of seed, which must be dispersed onto the site from other stands or be stored in the soil from prior colonization. In contrast, a long

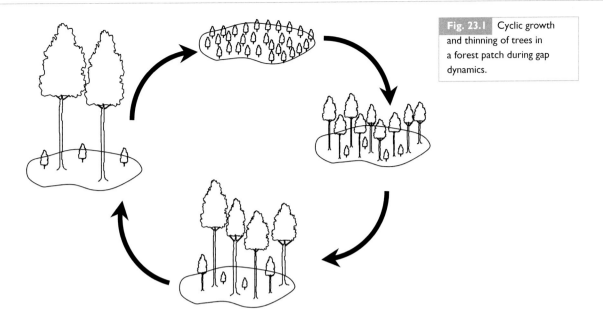

Fig. 23.1 Cyclic growth and thinning of trees in a forest patch during gap dynamics.

interval between disturbances promotes the abundance of late successional species.

23.3 | Forest Gap Dynamics

The smallest scale of disturbance is the death of a large tree, which creates small-scale cyclic dynamics within a community associated with gaps in the canopy (Figure 23.1). Treefalls are common in forests (Runkle 1981, 1982, 2000; Runkle and Yetter 1987). The death and uprooting of a tree creates a gap in the canopy related to the size of the tree. If large enough for sufficient light to reach the ground and conditions are favorable, numerous seedlings colonize the site. Many of these seedlings die. Some grow into saplings, which thin over time into a mature stand of a few large trees. Eventually these trees die, initiating a new cycle of growth. The forest landscape is a mosaic of individual patches in different stages of development (Watt 1947; Bray 1956; Bormann and Likens 1979; Shugart 1984). Within each patch, the process of establishment, thinning, and gap formation cyclically repeats. Although the state of an individual patch changes from year to year, the proportion of patches in the landscape in various developmental stages remains constant. The result is a

shifting mosaic steady state. Gap dynamics has been observed in prairie, desert, and tundra communities, but is most evident in forests.

The concept of gap dynamics and shifting mosaic steady state is illustrated with a simple model of tree growth. The accumulation of biomass in a small patch on the order of 100 m² can be described by the equation:

$$dM / dt = a - bM \qquad (23.1)$$

where M is biomass at time t. With annual net primary production $a = 1000$ g m^{-2} yr^{-1} and annual loss of biomass due to mortality $b = 0.025$ yr^{-1}, a maximum biomass of 40 kg m^{-2} accumulates in the patch. This accumulation is achieved after about 200–250 years of growth. If the trees lived forever, biomass accumulation would saturate at 40 kg m^{-2} and remain constant. However, the trees in a patch have an annual probability of mortality. If mortality is independent of age, the probability a tree will survive to age n is:

$$p = (1 - c)^n \qquad (23.2)$$

where c is the annual probability of mortality. For long-lived trees such as oak or hickory, which have a maximum age of about 300 years, $c = 0.015$ means that 1 percent of the trees survive to an age of 300 years. Thus, if mortality

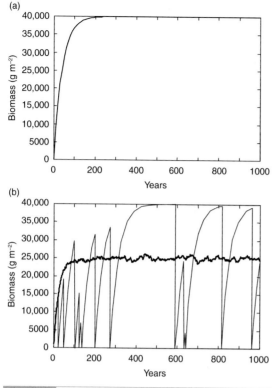

Fig. 23.2 Cyclic biomass accumulation and shifting mosaic steady state from gap dynamics. (a) Deterministic accumulation of biomass in a single patch in the absence of mortality. (b) Biomass accumulation with recurring mortality for a single patch (thin line) and averaged over a landscape of 500 patches (thick line).

is independent among patches, trees in 1.5 percent of the patches die each year, resetting biomass accumulation to zero and restarting patch dynamics. In a landscape of 500 patches, each patch undergoes cyclic dynamics related to deterministic tree growth and stochastic mortality (Figure 23.2). However, the biomass of the landscape, averaged over all patches, equilibrates after a short period. At this point, the landscape consists of a mosaic of patches, some with high biomass and some with low biomass.

The concept of a shifting mosaic steady state is seen in the model of biomass accumulation following clear-cutting in the Hubbard Brook Experimental Forest (Figure 22.8a). A large disturbance such as a clear-cut imposes an even-age structure to the regrowing forest. The dynamics of most patches is synchronized for about 100 or

150 years. These patches accumulate biomass in a few large trees as they undergo self-thinning. At the end of the aggradation phase, the majority of patches support a few even-aged trees of large size. A period of decline follows this peak biomass, when the old canopy dominants begin to die and understory trees replace them. Death is not synchronized, but varies randomly among patches. This transition and decline to steady state represents a change from an even-aged to an all-aged forest. At steady state, the landscape is a mosaic of patches of all ages ranging from recent gaps to mature trees.

Gap dynamics influences not only biomass but also community composition and can create complex patterns of species abundance in the landscape. One theory of forest dynamics holds that treefall gaps promote coexistence of species with different resource-use patterns, dispersal, and competitive abilities (Shugart 1984, 1987, 1998). For example, shade intolerant species require a canopy gap for successful regeneration, but many are small and do not achieve sufficient size to create a large opening in the canopy. Other species, generally shade tolerant trees but also some shade intolerants, are longer lived, grow to large size, and produce large gaps in the canopy. Canopy gaps created by the larger trees maintain the presence of the small, gap-requiring species.

Cyclic microsuccession from gap dynamics is seen in old-growth northern hardwood forests (Figure 23.3). Sugar maple, American beech, and yellow birch almost exclusively comprise the canopy of old-growth stands in the Hubbard Brook Experimental Forest. All have relatively long lifespans, but have different reproductive patterns that allow them to coexist. Yellow birch, with its prolific crops of small seeds, is an exploitive species; canopy gaps created by the death and uprooting of isolated large trees maintains its presence in the landscape. Yellow birch is least tolerant of shade and has the lightest seeds. Yellow birch seeds disperse widely and germinate best in high light and exposed mineral soil of a gap. Canopy closure precludes further yellow birch seedlings and favors seedlings of sugar maple and beech, both tolerant of shade. Sugar maple is initially favored because it

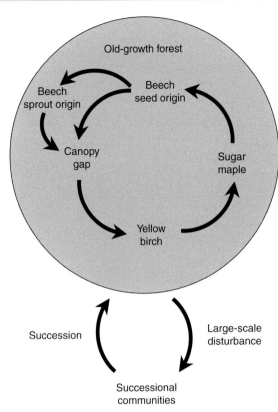

Fig. 23.3 Cyclic microsuccession among American beech, yellow birch, and sugar maple from gap dynamics in an old-growth northern hardwood forest. Adapted from Forcier (1975).

produces prolific crops of light, wind-dispersed seeds. Sugar maple saplings grow slowly under yellow birch, entering the canopy with the death of yellow birch. The dense shade of sugar maple favors beech. Its large seeds provide reserves to successfully establish in the shade of the forest floor. Beech also reproduces from root sprouts, which provides an advantage. The death of a large beech tree restarts this micro-succession. Cyclic tree replacement ensures the co-occurrence of the three species with different life histories in old-growth forests.

23.4 | Fire

Fire is a common large-scale disturbance that shapes the structure and function of many forest, scrubland, and grassland landscapes. A study of fire in the Boundary Waters Canoe Area within the Superior National Forest along the United States–Canada border in northeast Minnesota illustrates the mosaic landscape created by fire (Heinselman 1973). Fire largely determined the composition and structure of this vegetation prior to European settlement. All of the forests in a 4170 km² study area burned one to several times in a 377-year period between 1595 and 1972. Figure 23.4 illustrates the fire history for a 126 km² portion of the region. Ten fire-years occurred in this area, with the oldest dating to 1692. Fires in five years (1801, 1854, 1864, 1875, and 1910) were widespread and account for much of the area burned. For the 4170 km² region as a whole, a natural fire cycle of about 100 years prevailed prior to settlement. This is the average time required to cumulatively burn an area equal to the entire study area.

Fire is an important ecological force in the development of many plant communities. Recurring fires, both natural and human, are common to grasslands, Mediterranean vegeta-tion, savanna, and many temperate and boreal forests. Various species of grasses, shrubs, and trees have evolved a life history in response to fire. For example, the cones of some pine and spruce trees do not open to release seeds until they have been heated by a crown fire. The rapid height growth and thick bark of some pine trees provides protection from ground fires. Fires are quite common in boreal needleleaf evergreen forests, where the natural fire cycle ranges from 50 to 200 years (Heinselman 1981; Kasischke and Stocks 2000).

Satellites can be used to detect the occurrence of fires worldwide. For example, the Global Fire Emissions Database (GFED4) provides estimates of burned area since 1995 (Giglio et al. 2013). The global annual area burned between 1997 and 2011 ranged from 3.0 to 3.8 million km² and averaged 3.5 million km². Some regions of Africa and Australia burn with a frequency approach-ing one year (Figure 23.5). Africa, Southern Hemisphere South America, and Australia have the most area burned annually (Figure 23.6). Estimates of the fire return interval (biome area divided by annually burned area) for the 1990s are: boreal forest, 261 years; temperate forest,

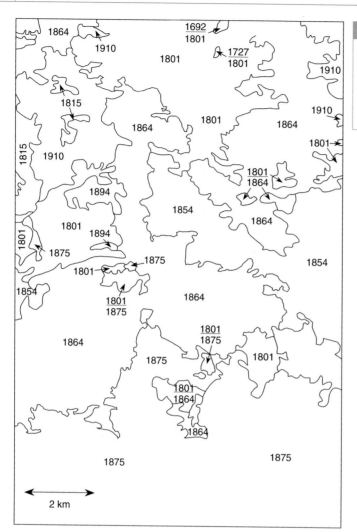

Fig. 23.4 Fire history of an approximately 9 km by 14 km region of the Boundary Waters Canoe Area, Minnesota. Stands date from after the indicated fire year. Where two years are given, stands consist of two age classes dating from separate fires. Adapted from Heinselman (1973).

138 years; temperate grassland, 88 years; tropical savanna and grassland, 5 years; and tropical forest, 34 years (Mouillot and Field 2005).

Fires alter atmospheric composition through the emission of long-lived greenhouse gases (CO_2, CH_4, and N_2O), carbon monoxide (CO), oxides of nitrogen (NO_x), nonmethane hydrocarbons, particulate matter, and aerosols during combustion (Crutzen and Andreae 1990; Andreae and Merlet 2001; Bowman et al. 2009). Estimates of the amount of carbon emitted to the atmosphere annually during fires varies depending on estimates of area burned, fuel load, combustion completeness, and emission factor. For example, Mouillot et al. (2006) estimated that 3.3 Pg C yr^{-1} was emitted by biomass burning during the 1990s. Fifty percent of this carbon was emitted during the burning of savanna, 38 percent from tropical forests, and 6 percent each from boreal and temperate forests. van der Werf et al. (2010) estimated that 2.0 Pg C yr^{-1} was emitted over the period 1997–2009. Carbon emissions varied greatly from year-to-year, ranging from a low of 1.5 Pg C yr^{-1} (2009) to a high of 2.8 Pg C yr^{-1} (1998). Slightly over one-half (52%) of the global emissions were from Africa, 15 percent from South America, 10 percent from equatorial Asia, 9 percent from boreal regions, and 7 percent from Australia.

Fires influence radiative forcing through altered atmospheric composition (e.g., CO_2, CH_4, N_2O, ozone, and aerosols) and through changes

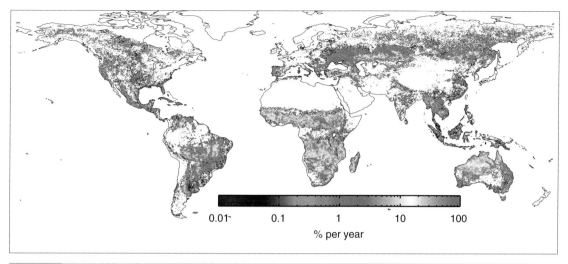

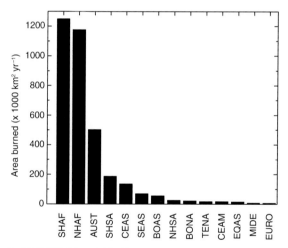

Fig. 23.6 Annual area burned for the period 1996–2011. Data from Giglio et al. (2013). Regions are: SHAF, Southern Hemisphere Africa. NHAF, Northern Hemisphere Africa. AUST, Australia. SHSA, Southern Hemisphere South America. CEAS, Central Asia. SEAS, Southeast Asia. BOAS, Boreal Asia. NHSA, Northern Hemisphere South America. BONA, Boreal North America. TENA, Temperate North America. CEAM, Central America. EQAS, Equatorial Asia. MIDE, Middle East. EURO, Europe.

carbon (soot) deposition has a lower albedo than pristine snow, which contributes to positive radiative forcing (Hansen and Nazarenko 2004; Flanner et al. 2007; Lee et al. 2013).

Randerson et al. (2006) contrasted the biogeochemical and biogeophysical consequences of fire in a black spruce forest in Alaska (Table 23.1). Long-lived greenhouse gases (CO_2, CH_4) emitted during combustion produce a positive annual radiative forcing in the first year following fire. Additional positive annual radiative forcing arises from ozone produced from trace gases emissions, black carbon deposited on snow and ice, and aerosols. The loss of forest overstory increases snow exposure, leading to higher surface albedo in spring and fall and negative annual radiative forcing. The biogeochemical warming exceeds the biogeophysical cooling in the first year following fire. However, the effects of ozone, black carbon, and aerosols are short lived. The long-term radiative forcing is a balance between post-fire increases in surface albedo and the continued positive radiative forcing from the greenhouse gas pulse emitted to the atmosphere during burning. Averaged over an 80-year fire cycle, the negative radiative forcing from surface albedo exceeds the smaller positive biogeochemical radiative forcing.

in surface albedo. In addition to direct post-fire changes in albedo (Figure 22.15), biomass burning releases black and organic carbon aerosols to the atmosphere. Dirty snow from black

Table 23.1 Radiative forcing associated with wildfire in interior Alaska

Forcing agent	Radiative forcing ($W\ m^{-2}$)	
	Year 1	Years 0–80 (mean)
CH_4 and CO_2	8	1.6
Ozone	6	0.1
Black carbon on snow	3	0.0
Black carbon on sea ice	5	0.1
Aerosols	17	0.2
Post-fire surface albedo	−5	−4.2
Total	34	−2.3

Note: Radiative forcing is per m^2 of burned area.
Source: From Randerson et al. (2006).

23.5 | Land Use

People are also an important agent of vegetation change and generate landscape pattern. Clearing of forests and grasslands for agricultural uses has led to significant loss of natural vegetation worldwide and has sculpted the present-day landscape into a mosaic of fragmented ecosystems in many regions of the world. It is estimated that approximately 18 million km^2 of land (12% of land surface) is used for cropland and another 34 million km^2 (22%) is used for pastureland and rangeland (Figure 2.8).

Wheat is the most abundant crop type, covering 22 percent of the total cultivated area of the world (Figure 23.7). It is extensive throughout the middle latitudes of North America, Europe, and Asia and also in parts of India and Australia. Maize is the second most abundant crop, covering 13 percent of the world's cropland, and is common throughout the Northern and Southern Hemispheres. Rice occupies 11 percent of cropland area, primarily in India and Southeast Asia. Barley (9% of cropland area) is grown in cold climates of Canada, northern United States, and Europe. Other crop types include soybeans (5% of crop area), pulses (dry beans and peas, 4% of cropland area), and 12 other major crop types that comprise 21 percent of crop area.

The global geography of land clearing mirrors the history of population growth (Pongratz et al. 2008; Klein Goldewijk et al. 2011). High cropland areas can be found in the Mediterranean, Middle East, and India as far back as CE 800 (Figure 23.8). Widespread deforestation in Europe occurred throughout the Middle Ages and earlier. Extensive deforestation in the United States began with European settlement in the 1600s and 1700s. The tropics are currently experiencing widespread deforestation.

23.5.1 European Deforestation

Much of the natural vegetation of Europe has been cleared for agricultural uses. Forest clearing in Europe has been documented as far back as the Neolithic culture 5000 years ago, when forests were cleared for settlement, cultivation of crops, and grazing animals (Darby, 1956; Kirby and Watkins 1998; Williams 2000, 2003; Kaplan et al. 2009; Hughes 2011). However, the pace and extent of deforestation increased greatly in the Mediterranean during the Classical period of the Greeks and Romans. Shipbuilding, use of wood for construction and fuel, forest clearing for crops, and overgrazing by livestock led to widespread loss of forests throughout the Mediterranean. Classical Greek scholars such as Homer in the ninth century BCE, Plato in the fourth century BCE, and the Roman writer Lucretius in the first century BCE each described the extensive loss of forest during those days. The large growth in population in central and western Europe during the Middle Ages led to broad loss of forest cover, so much so that conflicts between forest preservation and exploitation arose by the end of the twelfth century. One estimate is that approximately 80 percent of temperate western and central Europe was forested in CE 500, but less than one-half of those forests remained by CE 1300 and only 25 percent of the forests in central Europe remained by 1900 (Williams 2003, p. 102, p. 123). Overall, Europe experienced extensive forest loss over the past 2000 years (Figure 23.9).

A pollen sequence collected in central East Anglia between Cambridge and Norwich records human impacts on vegetation over the past 5000 years (Table 23.2). This area of the

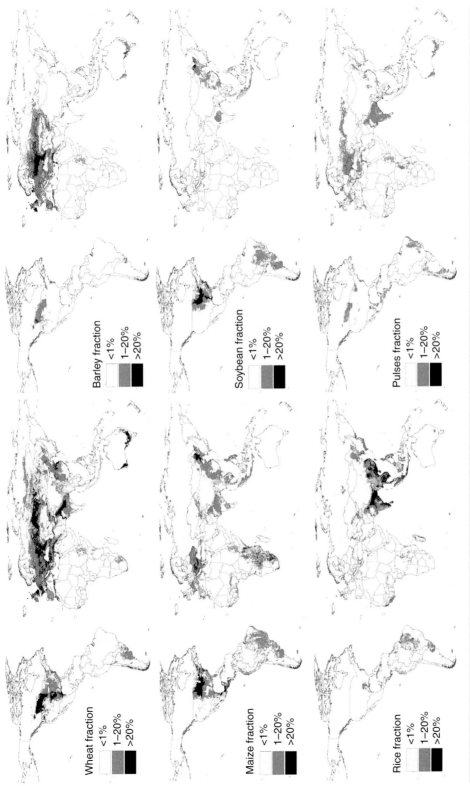

Fig. 23.7 Geographic distribution of the six most common crops in terms of percentage area. Adapted from Leff et al. (2004). See also Monfreda et al. (2008).

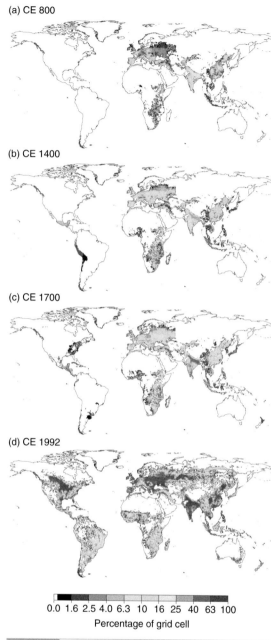

(a) CE 800

(b) CE 1400

(c) CE 1700

(d) CE 1992

0.0 1.6 2.5 4.0 6.3 10 16 25 40 63 100
Percentage of grid cell

Fig. 23.8 Global historical cropland for the years CE 800, CE 1400, CE 1700, and CE 1992 in terms of percentage area. Reproduced from Pongratz et al. (2008) and provided courtesy of Julia Pongratz. See color plate section.

important, but declined in the 1800s with the industrial era. The first 5000 years of the sequence, from 10,000 to 5000 years ago, documents the late-glacial vegetation and changing forest composition as new species migrated to the region. By 5000 years ago, the landscape was a mosaic of deciduous forests dominated by linden and oak with some elm, ash, hazel, and alder. Human influences began some 5000–3500 years ago. Elm declined noticeably as a result of Neolithic forest clearing, as it did elsewhere in Europe (Williams 2003). This caused an expansion of hazel and ruderal herbaceous plants such as plantain. The period 3500–2500 years ago saw marked declines in linden and hazel, increases in grasses and other herbaceous plants, and the first appearance of cereal-type plants, which thereafter occur throughout the sequence. This suggests an increase in cleared areas for livestock grazing and cereal cultivation. The next 1000 years saw widespread deforestation, possibly even complete clearance of forests, and major expansion of pastoral agriculture. Tree pollen declined while grasses and other herbaceous plants increased. Extensive forest clearing was made possible by the introduction of iron tools to fell trees and was necessitated by demand for wood to smelt iron. Arable agriculture expanded greatly from 1500 to 150 years ago, first in the cultivation of rye and barley and later hemp.

Deforestation occurred later in the English Lake District in Cumbria (Barker 1998). Extensive broadleaf deciduous forest dominated by oak and linden once covered the region. Forest cover declined during the period CE 900 to 1800 due to centuries of overgrazing by livestock and wood harvesting. The first period of exploitation from CE 900 to 1000 followed the Norse invasions. Sheep grazing by monks at a local abbey led to significant forest deterioration between 1100 and 1500. A period of industrial exploitation lasted from 1607 to 1800, during which time wood was harvested to produce charcoal for local iron smelting industries.

Studies of the Dutch landscape also show the history of deforestation (Dirkx 1998; Spek 1998; van Laar and den Ouden 1998; Wolf 1998). Forest clearing began in the Neolithic Period

United Kingdom is in the heart of rich farmland. Marketing, milling, and malting of grain have been important local industries since medieval times. Hemp cultivation was also locally

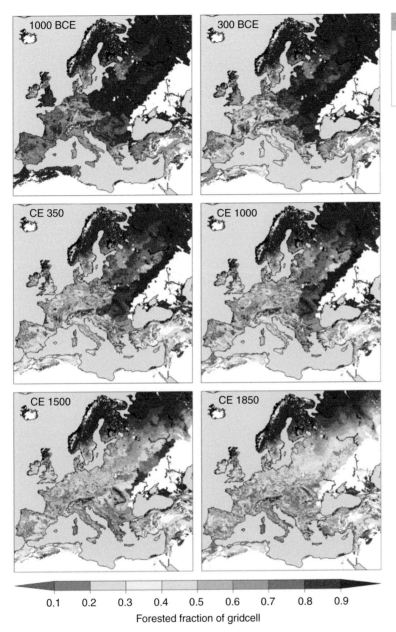

0.1   0.2   0.3   0.4   0.5   0.6   0.7   0.8   0.9

Forested fraction of gridcell

and accelerated during the Late Bronze Age and Early to Middle Iron Age (1100–250 BCE) and again in the Medieval Period (circa CE 1000). Late Medieval regulations to protect forests demonstrate significant forest decline by the 1500s. By 1800, forest covered only 4 percent of the Dutch landscape. Centuries of grazing by pigs, cattle, and sheep and collection of firewood had turned vast areas of woodland into treeless heaths. Since then there has been a general trend of

reforestation, first by private landowners and then increasing rapidly in the 1900s under government sponsorship.

Similar historical trends are seen in the region of Flanders, covering 11,000 km² of Belgium, France, and the Netherlands (Tack and Hermy 1998). Forest clearing began in earnest in the Middle Ages to create more farmland. By 1250, forests covered only 10 percent of the region, mostly as small woodlots. This

Table 23.2 | Pollen chronology in central East Anglia from present to 10,000 years before present

Age (BP)	Pollen	Land use
0–150	Decline of *Cannabis* (hemp); first appearance of *Aesculus* (horse-chestnut) and *Acer* (maple); rise of *Ulmus* (elm), *Quercus* (oak), *Tilia* (linden, lime), and *Taxus* (yew)	Abandonment of hemp cultivation; local tree planting
Expansion of arable agriculture		
150–1000	Major increase of *Cannabis* (hemp) and arable weeds	Continued expansion of arable agriculture; hemp major crop
1000–1500	Increase of *Secale* (rye) and other cereal-type pollen; first appearance of *Cannabis sativa* (hemp)	Expansion of arable agriculture (rye, barley) with introduction of ox-drawn plough
Expansion of pastoral agriculture		
1500–2500	Major increases of Gramineae (grasses) and other herbs; sharp declines of almost all trees; low abundance of cereal-type pollen	Widespread forest clearing with introduction of iron tools; pasture grazing dominant land use
Early forest clearing		
2500–3500	Sharp declines of *Tilia* (linden, lime) and *Corylus* (hazel); increase of Gramineae (grasses) and other herbs; first appearance of cereal-type pollen (oat/wheat, barley, rye)	Selective cutting of linden for fodder; forest clearing for livestock grazing; cultivation of barley and wheat
3500–5000	Sharp declines of *Ulmus* (elm), *Quercus* (oak), and *Alnus* (alder); increase of *Corylus* (hazel); first appearance of *Plantago lanceolata* (plantain) and *Taxus* (yew)	Local forest clearing
Late-glacial forest growth		
5000–7800	Increases of *Alnus* (alder) and *Tilia* (linden, lime); decrease of *Corylus* (hazel); first appearance of *Fraxinus* (ash)	
7800–9200	Sharp increase of *Corylus* (hazel) and *Ulmus* (elm); decline of *Betula* (birch); increase of *Quercus* (oak)	
9200–10,000	*Betula* (birch) dominance	
> 10,000	High *Betula* (birch) and non-aboreal pollen	

Source: From Peglar et al. (1989).

was followed by a long period of reforestation to restock land for firewood. From 1250 until 1775, forest area increased to 16 percent of the Flemish landscape. Thereafter, deforestation increased again as the population grew and more farmland was needed. By 1880, the forested area had decreased to 6 percent. Since the later part of the 1800s, forest cover has generally increased.

The Belgian Ardennes followed a similar land-use history (Petit and Lambin 2002). Forest clearing by Celts occurred as early as 700–55

BCE, but the first widespread deforestation was during the Roman era (CE 50–400), when forests were cleared for wood and food. Thereafter, forest clearing was common from circa CE 1000 to the mid-1800s as the population grew and demand for forest products and farmland increased. Forest area increased beginning in the mid-1800s as a result of government policies.

23.5.2 United States Deforestation

Much of the forests of eastern United States and the grasslands of the Great Plains of central United States have been replaced with crops (Figure 23.10) or secondary forests (Figure 23.11). The history of land use in the United States is tied to European settlement and westward expansion (Williams 1989). The New England and Virginian landscapes found by European colonists in the 1600s were predominantly primary forest, though Native Americans cleared forest for villages, to grow crops, and harvest wood and burned woodlands during hunting (Williams 1989, 2000, 2003). Over the next 200–300 years, this land was cleared for towns, croplands, and forest products at a pace set by the socioeconomic development of the country (Figure 1.2). Throughout most of northeastern United States, increasing rates of deforestation in the late 1700s lead to peak forest clearing in the mid-1800s. Since then, abandonment of farmland has led to substantial increases in secondary forest cover in many areas. By 1900, increased agriculture and forest harvesting had spread to the southeastern states, followed by reversion to secondary forest several decades later. This same pattern of clearing and farm abandonment was repeated later in the Midwest and central Plains with the westward expansion of the population. By 1992, 1.7 million km^2 of savanna and grassland and 1.4 million km^2 of forest and woodland had been cleared for cultivation since 1850 (Ramankutty and Foley 1999b).

Studies of the forest history of central New England provide a detailed history of regional deforestation (Williams 1989; Foster 1992, 1993; Foster et al. 1992, 1998; Fuller et al. 1998). The presettlement forest was a mixture of broadleaf deciduous and needleleaf evergreen trees in uneven-aged stands resulting from gap dynamics and windthrows from hurricanes. Early settlers found scattered clearings made by Native Americans. Deforestation began in the 1700s and accelerated in the early 1800s during the rise of widespread agriculture. The period from 1795 to 1860 saw large and rapid transformation of the landscape. Forests were cleared extensively for pastures, hay fields, crops, firewood, and other forest products. At the peak of clearing in the mid-1800s, many upland areas of central New England had less than 25 percent forest cover. The subsequent development of urban centers, an industrial economy, and greater reliance on coal led to farm abandonment and widespread reforestation.

Figure 23.12a illustrates this forest history for Prospect Hill, a 380 ha tract in the Harvard Forest, and the surrounding town of Petersham, Massachusetts (Foster 1992, 1993; Foster et al. 1992). The town of Petersham was settled beginning in 1733. During the initial settlement, individual farmers cleared less than 2 ha of forest annually for fuel, construction, crops, and livestock. By the late 1700s, regional population growth and improved transportation gave rise to an agriculture economy. In Petersham, forest clearing for cattle and sheep pastures led to an increase in open land from 50 percent in 1800 to nearly 85 percent in 1850. At the height of agricultural activity, the Petersham landscape was a mosaic of small fields and isolated woodlots. Forest clearing of the Prospect Hill tract closely tracked that of the town. Clearing began about 1750, proceeded slowly through 1780, and increased markedly until peak deforestation about 1840 to 1850, when forests covered only about 20 percent of the tract. Pollen collected from sediments records the changing forest composition (Foster et al. 1992). Beech, maple, pine, oak, and eastern hemlock pollen decreased in abundance while the pollen of herbaceous plants associated with hay production, grazing, and crops increased. Pollen from chestnut trees also increased, because its rapid growth and prolific root sprouts allow chestnut to dominate stands after logging. Reforestation began as early as the 1840s. In Petersham and much of central New England, reforestation of

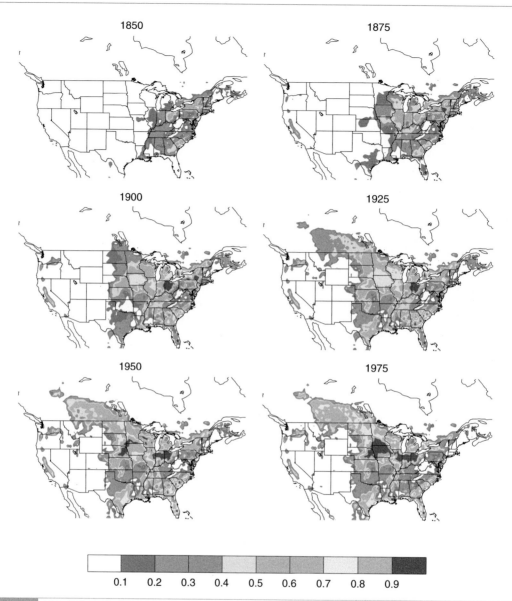

Fig. 23.10 Historical cropland area in the United States from 1850 to 1975 showing the fraction of the landscape covered in cropland. Data from Ramankutty and Foley (1999a,b). See color plate section.

agricultural fields gave rise to a thriving but short-lived timber industry in the late 1800s and early 1900s. From 1885 to 1895 virtually all of the merchantable timber on the Prospect Hill tract was logged. Abandonment and reforestation increased markedly through the late 1800s and early 1900s.

A study of Tompkins County (1250 km² area) in the central Finger Lakes region of New York

shows a similar history from forest to predominantly agricultural land in the 1800s and back to forest as farms were abandoned during the 1900s (Figure 23.12b). Forests covered the landscape prior to European settlement. Native Americans lived in the region, but cleared only a small portion of the land. Settlement and forest clearing in Tompkins County increased rapidly after 1790. Maximum area of farmland occurred

Fig. 23.11 Historical area of primary forest, secondary forest, and agricultural land in eastern United States, 1800–2000. Mixtures of land use types are indicted by color intensity and color gradients. Reproduced from Albani et al. (2006) with permission of Blackwell Publishing. See also Hurtt et al. (2006). See color plate section.

by 1900, when 81 percent of the land was open in agricultural fields or urban areas and only 19 percent was forested. Reforestation began in earnest in the early 1900s with abandonment of farmlands and continued through the 1900s. By 1980, secondary forests growing on old-fields covered 51 percent of the county. Though of similar structure and light availability, these secondary forests are dominated by red maple and eastern white pine while primary forests are dominated by sugar maple and American beech (Flinn and Marks 2007). Indeed, while much of northeastern United States is now forested (~80%), the composition of the forest has

shifted to early- and mid-successional species (e.g., red maple); late-successional species (e.g., beech, hemlock, spruce) that had been prevalent in pre-colonial forests still persist, though in much less abundance (Thompson et al. 2013).

Deforestation began later and peaked later in the Great Lakes region. Rapid settlement of land west of the Appalachian Mountains in the mid-1800s led to significant land clearing for agriculture. For example, parts of Wisconsin that were fully forested in 1831 were only 30 percent forested by 1882 and 10 percent forested by 1902 (Williams 1989, p. 367). The vegetation of the Great Lakes region has undergone significant transformation since the mid-1800s (Cole et al. 1998). The presettlement landscape of Michigan, Wisconsin, and Minnesota consisted predominantly of mesic forest, comprised of sugar maple, yellow birch, American beech, eastern hemlock, and basswood. Oak forest and savanna consisting of red oak, black oak, white oak, and bur oak and pine forest of white pine, red pine, and jack pine were also common. Other less common forest types were: boreal forest of mainly balsam fir, white spruce, and northern white cedar and some black spruce and tamarack; aspen–birch forest; and wet lowland forest of American elm, green ash, black ash, and silver maple. Since then, the extent of forest has declined by 40 percent, mostly due to conversion of mesic forest and oak forest to farmland in the southern regions (Figure 23.13). Twenty-one percent of the presettlement forest was logged, primarily pine forest in northern regions, and regrew as aspen–birch forest. The extent of pine forest declined by 78 percent, boreal forest by 62 percent, and mesic forest by 61 percent while aspen–birch forest increased by 83 percent.

A 9600-ha region of northern Wisconsin illustrates the effects of deforestation on forest structure and community composition (White and Mladenoff 1994). This region is transitional between boreal and northern hardwood forest. Old-growth stands of hemlock, white pine, sugar maple, and yellow birch dominated the pre-European settlement forest of 1860 (Table 23.3). Logging and fires were the major agents of landscape change during

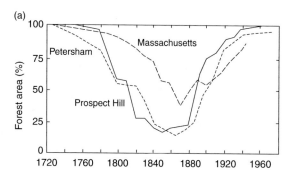

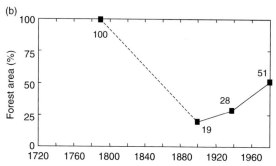

Fig. 23.12 Forest history of northeastern United States. (a) Percentage of land covered by forest in the Prospect Hill tract of the Harvard Forest in central Massachusetts, the town of Petersham, and the entire state. Adapted from Foster (1992, 1993) and Foster et al. (1992). (b) Percentage of land covered by forest in Tompkins County, New York. The value for 1900 is thought to represent the maximum extent of forest clearing. Data from Smith et al. (1993).

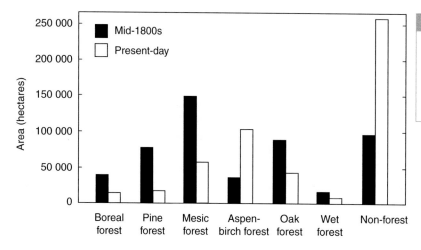

Fig. 23.13 Spatial extent of forest types in the Great Lake states of Michigan, Wisconsin, and Minnesota in the mid-1800s and present-day. Adapted from Cole et al. (1998).

Table 23.3 Percentage area of upland forest types for a 9600 ha tract in northern Wisconsin at presettlement (1860), peak human activity (1931), and current (1989) periods

	1860	1931	1989
Second-growth stands			
Pin cherry	–	37	2
Boreal mixed hardwood (quaking aspen, paper birch)	–	14	22
Mixed conifer (hemlock, white pine, white cedar, balsam fir, white spruce)	–	1	10
Hardwood/conifer (quaking aspen, paper birch, balsam fir, white spruce)	–	6	29
Northern hardwood (sugar maple, yellow birch, red maple)	–	19	30
Old-growth stands			
Old-growth hemlock (eastern hemlock, white pine, yellow birch)	46	7	4
Old-growth hardwood (sugar maple, yellow birch)	31	16	2
Old-growth hemlock/hardwood	22	–	–

Source: From White and Mladenoff (1994).

intensive settlement from 1860 to 1931, reducing the extent of old-growth forests and creating large areas of second-growth forest, mostly pin cherry, northern hardwood forests of sugar maple, yellow birch, and red maple, and boreal hardwood forests of quaking aspen and paper birch. Balsam fir, white spruce, and white cedar occurred in lesser abundance. Thereafter, logging continued but successional processes accounted for most of the landscape change and produced a diverse assemblage of boreal and northern hardwood forests.

23.5.3 Tropical Deforestation

Tropical broadleaf evergreen trees form extensive forests in the Amazon Basin of South America, the Congo Basin of Africa, and parts of Southeast Asia, Indonesia, and southeast Australia. These forests are among the most productive and biologically diverse in the world, but are being cleared, burned, and converted to agriculture. It is estimated that forest loss worldwide in the tropics during the 1990s was 55–58,000 km² per year with another 23,000 km² degraded annually (Achard et al. 2002; DeFries et al. 2002). Most of this deforestation occurred in tropical South America and tropical Asia.

The Amazonia region of Brazil has been widely studied because of the influence of these forests on climate and the carbon cycle. In the Brazilian state of Rondônia, roads, governmental colonies, and forest reserves control the spatial geography of forest clearing (Skole and Tucker 1993). Large tracts of forested land have become accessible from highways, producing a distinct pattern of land clearing. Deforestation rates in the Brazilian Amazon averaged 19,600 km² per year during the period 1996–2005, but have since declined (Figure 23.14). Governmental actions have reduced deforestation to historically low rates (Nepstad et al. 2009).

23.6 Carbon Dioxide and Other Climate Effects

The most prominent atmospheric signal of land use and land-cover change is seen in the carbon cycle. Deforestation is a source of carbon to the atmosphere whereas forest regrowth (reforestation) and the establishment of forests on non-forest land (afforestation) represent a gain in carbon storage from the atmosphere. Figure 23.15 illustrates idealized changes in carbon following forest harvesting. In this example, harvesting reduces living biomass from 70 Mg C ha⁻¹ to 10 Mg C ha⁻¹. Harvesting transfers

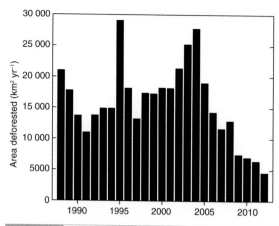

Fig. 23.14 Annual deforestation in the Amazonia region of Brazil, 1988–2012. Data provided by the Instituto Nacional de Pesquisas Espaciais, Brazil. See also Gloor et al. (2012).

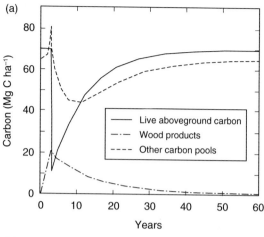

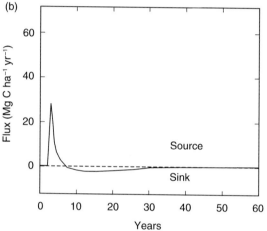

Fig. 23.15 Idealized changes in (a) ecosystem carbon pools and (b) the resulting carbon flux from harvest and regrowth in a temperate forest. 10 Mg ha^{-1} = 1 kg m^{-2}. Adapted from Houghton (2005).

the biomass to domestic and commercial wood products that decay over time. Carbon accumulates in living biomass as the forest grows, recovering to pre-harvest values by about 40 years. Dead biomass and soil organic carbon (collectively called other carbon pools) initially increase with harvesting as a result of debris left on the site. This carbon pool decreases over time as the debris decomposes and then increases as the forest ages, litterfall increases, and detritus accumulates. The net carbon flux is an immediate release of about 30 Mg C ha^{-1} yr^{-1} to the atmosphere. The forest remains a source of carbon over the next several years.

Conversion of grassland to cropland typically reduces soil organic matter by 20–40 percent during the first few decades of cultivation (Paustian et al. 2000). This results from reduced input of plant litter, increased chemical quality of crop residues, which promotes faster decomposition, and tillage, which mixes soil and breaks up aggregates. Conversely, substantial soil organic matter can accumulate with reversion of cropland to grassland (Post and Kwon 2000).

Figure 23.16 shows the estimated net carbon flux from land use between 1850 and 2005. The net carbon flux results from changes in land use (such as timber harvesting and clearing for agriculture) and accounts for the initial loss of carbon in the vegetation and also subsequent regrowth and changes in soil carbon. The net flux includes both emissions of carbon from deforestation and sinks of carbon in forests recovering from harvests or agricultural abandonment. The estimated global carbon flux from changes in land use increased from 501 Tg C (1 Tg = 10^{12} g) in 1850 to a maximum of 1712 Tg C in 1991 and has since declined slightly. Cumulatively, anthropogenic land use has added 156 Pg C to the atmosphere. Land use emissions in the United States peaked in the late-1800s and thereafter declined. Land use emissions in the tropics increased to peak rates in the late twentieth century.

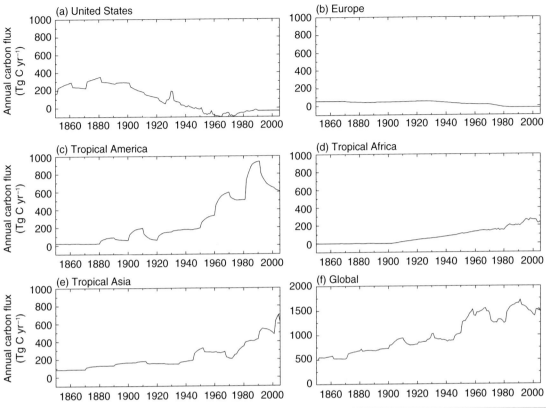

Fig. 23.16 Annual net carbon flux to the atmosphere from land use for the period 1850 to 2005. Shown are regional fluxes and the global total. Negative fluxes indicate carbon storage. Data updated from Houghton (1999, 2003) by Houghton (2008). See also Houghton et al. (2012). Data provided by the Carbon Dioxide Information Analysis Center (Oak Ridge National Laboratory, Oak Ridge, Tennessee).

Land cover change also alters the biogeophysical coupling between the land and atmosphere. Tall forests are aerodynamically rough compared with shorter grasslands and croplands (Table 13.1). Forests generally have a lower albedo compared with cropland, particularly in snow-covered regions during winter (Figure 12.10). The ratio of evapotranspiration to available energy is generally low in forest compared with some crops and lower in conifer forest than in deciduous broadleaf forest (Figure 17.10). These differences produce complex differences among vegetation types in the exchanges of energy, water, and momentum with the atmosphere (Bonan 2008). The prevailing view is that tropical deforestation warms surface climate, because the cooling resulting from the higher albedo of cropland and pastureland is offset by warming arising from reduced evapotranspiration. In contrast, temperate and boreal deforestation is thought to cool climate, primarily from higher albedo. Chapter 28 reviews these changes in greater detail.

Large-scale ecological disturbances such as wildfire and insect defoliation alter climate by disrupting ecosystem functions (O'Halloran et al. 2012). Wildfires have a particularly strong effect on surface albedo, turbulent fluxes, and biogeochemical cycles in boreal forests (Table 23.1). Insect defoliation can have a similarly strong effect on surface climate.

The mountain pine beetle epidemic in western North America forests has affected large tracts of land in British Columbia and the Rocky Mountain region of the United States (Edburg et al. 2012). In British Columbia alone, pine beetles have infested 170,000 km² of forest, about 20 percent of the total area in the province. Mountain pine beetles attack and kill pine trees with large diameters. This mortality reduces carbon uptake by forests (Kurz et al. 2008). Tree mortality also reduces evapotranspiration, which warms surface climate because energy that previously fuelled the evaporation of water becomes available to heat the land surface. Such feedbacks appear to have occurred in beetle infested forests of British Columbia (Maness et al. 2013). Over the entire region, infestation resulted in a 19 percent reduction in summertime evapotranspiration and a 1°C rise in summertime surface temperatures. Locally larger temperature increases occurred in the tree stands with the highest mortality.

23.7 | Review Questions

1. Describe how forest gap dynamics might affect the composition of a hypothetical mixture of yellow poplar and American beech in (a) a cool climate and (b) a warm climate. Both species can regenerate and grow in each climate, though growth rates are higher in the warm climate. Yellow poplar is shade intolerant, grows rapidly, attains large height, has wind-dispersed seeds, and requires large gaps for regeneration. Beech is shade tolerant, grows slowly, attains moderate to large size, is long-lived, has large seeds, can reproduce through sprouts, and does not require gaps for regeneration.

2. In the absence of canopy gaps, which species (yellow birch, sugar maple, or beech) is likely to dominate old-growth forests at Hubbard Brook?

3. With reference to Figure 22.10, describe how (a) increased fire frequency and (b) increased flooding affects white spruce forests near Fairbanks, Alaska.

4. Describe how increased logging affects the forest landscape in northern Wisconsin.

5. Old growth forests in the North Carolina Piedmont are composed primarily of oak and hickory trees in the canopy. In a particular locale, the forest is primarily pine with an oak–hickory understory. What does this indicate about historical land use?

6. Many regions of temperate forest are undergoing reforestation, while tropical regions are being deforested. How do these land-use changes affect climate?

7. How might increased fire frequency affect surface energy fluxes and climate in boreal forests?

8. Describe how defoliation by pine beetle affects surface energy fluxes and climate. Compare this with the effects of pine beetle on land–atmosphere carbon exchange. Discuss whether pine beetle defoliation is a positive or negative feedback with climate change.

23.8 | References

Achard, F., Eva, H. D., Stibig, H.-J., et al. (2002). Determination of deforestation rates of the world's humid tropical forests. *Science*, 297, 999–1002.

Albani, M., Medvigy, D., Hurtt, G. C., and Moorcroft, P. R. (2006). The contributions of land-use change, CO₂ fertilization, and climate variability to the Eastern US carbon sink. *Global Change Biology*, 12, 2370–2390.

Andreae, M. O., and Merlet P. (2001). Emissions of trace gases and aerosols from biomass burning. *Global Biogeochemical Cycles*, 15, 955–966, doi:10.1029/2000GB001382.

Barker, S. (1998). The history of the Coniston Woodlands, Cumbria, UK. In *The Ecological History of European Forests*, ed. K. J. Kirby and C. Watkins. New York: CAB International, pp. 167–183.

Bonan, G. B. (2008). Forests and climate change: Forcings, feedbacks, and the climate benefits of forests. *Science*, 320, 1444–1449.

Bormann, F. H., and Likens, G. E. (1979). *Pattern and Process in a Forested Ecosystem*. New York: Springer-Verlag.

Bowman, D. M. J. S., Balch, J. K., Artaxo, P. et al. (2009). Fire in the Earth System. *Science*, 324, 481–484.

Bray, J. R. (1956). Gap phase replacement in a maple–basswood forest. *Ecology*, 37, 598–600.

Cole, K. L., Davis, M. B., Stearns, F., Guntenspergen, G., and Walker, K. (1998). Historical landcover changes in the Great Lakes region. In *Perspectives on the Land Use History of North America: A Context for Understanding Our Changing Environment*, ed. T. D. Sisk. Biological Science Report USGS/BRD/BSR-1998-0003, U.S.

Geological Survey, Biological Resources Division, pp. 43–50.

Crutzen, P. J., and Andreae, M. O. (1990). Biomass burning in the tropics: Impact on atmospheric chemistry and biogeochemical cycles. *Science*, 250, 1669–1678.

Darby, H. C. (1956). The clearing of the woodland in Europe. In *Man's Role in Changing the Face of the Earth*, ed. W. L. Thomas, Jr. Chicago: University of Chicago Press, pp. 183–216.

DeFries, R. S., Houghton, R. A., Hansen, M. C., et al. (2002). Carbon emissions from tropical deforestation and regrowth based on satellite observations for the 1980s and 1990s. *Proceedings of the National Academy of Sciences USA*, 99, 14256–14261.

Dirkx, G. H. P. (1998). Wood-pasture in Dutch common woodlands and the deforestation of the Dutch landscape. In *The Ecological History of European Forests*, ed. K. J. Kirby and C. Watkins. New York: CAB International, pp. 53–62.

Edburg, S. L., Hicke, J. A., Brooks, P. D., et al. (2012). Cascading impacts of bark beetle-caused tree mortality on coupled biogeophysical and biogeochemical processes. *Frontiers in Ecology and the Environment*, 10, 416–424.

Flanner, M. G., Zender, C. S., Randerson, J. T., and Rasch, P. J. (2007). Present-day climate forcing and response from black carbon in snow. *Journal of Geophysical Research*, 112, D11202, doi:10.1029/2006JD008003.

Flinn, K. M., and Marks, P. L. (2007). Agricultural legacies in forest environments: tree communities, soil properties, and light availability. *Ecological Applications*, 17, 452–463.

Forcier, L. K. (1975). Reproductive strategies and the co-occurrence of climax tree species. *Science*, 189, 808–810.

Foster, D. R. (1992). Land-use history (1730–1990) and vegetation dynamics in central New England, USA. *Journal of Ecology*, 80, 753–772.

Foster, D. R. (1993). Land-use history and forest transformations in central New England. In *Humans as Components of Ecosystems*, ed. M. J. McDonnell and S. T. A. Pickett. New York: Springer-Verlag, pp. 91–110.

Foster, D. R., Zebryk, T., Schoonmaker, P., and Lezberg, A. (1992). Post-settlement history of human land-use and vegetation dynamics of a *Tsuga canadensis* (hemlock) woodlot in central New England. *Journal of Ecology*, 80, 773–786.

Foster, D. R., Motzkin, G., and Slater, B. (1998). Land-use history as long-term broad-scale disturbance:

Regional forest dynamics in central New England. *Ecosystems*, 1, 96–119.

Fuller, J. L., Foster, D. R., McLachlan, J. S., and Drake, N. (1998). Impact of human activity on regional forest composition and dynamics in central New England. *Ecosystems*, 1, 76–95.

Giglio, L., Randerson, J. T., and van der Werf, G. R. (2013). Analysis of daily, monthly, and annual burned area using the fourth-generation global fire emissions database (GFED4). *Journal of Geophysical Research: Biogeosciences*, 118, 317–328, doi:10.1002/jgrg.20042.

Gloor, M., Gatti, L., Brienen, R., et al. (2012). The carbon balance of South America: A review of the status, decadal trends and main determinants. *Biogeosciences*, 9, 5407–5430.

Hansen, J., and Nazarenko, L. (2004). Soot climate forcing via snow and ice albedos. *Proceedings of the National Academy of Sciences USA*, 101, 423–428.

Heinselman, M. L. (1973). Fire in the virgin forests of the Boundary Waters Canoe Area, Minnesota. *Quaternary Research*, 3, 329–382.

Heinselman, M. L. (1981). Fire and succession in the conifer forests of northern North America. In *Forest Succession: Concepts and Application*, ed. D. C. West, H. H. Shugart, and D. B. Botkin. New York: Springer-Verlag, pp. 374–405.

Houghton, R. A. (1999). The annual net flux of carbon to the atmosphere from changes in land use 1850–1990. *Tellus B*, 51, 298–313.

Houghton, R. A. (2003). Revised estimates of the annual net flux of carbon to the atmosphere from changes in land use and land management 1850–2000. *Tellus B*, 55, 378–390.

Houghton, R. A. (2005). Aboveground forest biomass and the global carbon balance. *Global Change Biology*, 11, 945–958.

Houghton, R.A. (2008). Carbon flux to the atmosphere from land-use changes: 1850–2005. In *TRENDS: A Compendium of Data on Global Change*. Oak Ridge, Tennessee: Carbon Dioxide Information Analysis Center, Oak Ridge National Laboratory, U.S. Department of Energy.

Houghton, R. A., House, J. I., Pongratz, J., et al. (2012). Carbon emissions from land use and land-cover change. *Biogeosciences*, 9, 5125–5142.

Hughes, J. D. (2011). Ancient deforestation revisited. *Journal of the History of Biology*, 44, 43–57.

Hurtt, G. C., Frolking, S., Fearon, M. G., et al. (2006). The underpinnings of land-use history: Three centuries of global gridded land-use transitions, wood-harvest activity, and resulting secondary lands. *Global Change Biology*, 12, 1208–1229.

Kaplan, J. O., Krumhardt, K. M., and Zimmermann, N. (2009). The prehistoric and preindustrial deforestation of Europe. *Quaternary Science Reviews*, 28, 3016–3034.

Kasischke, E. S., and Stocks, B. J. (2000). *Fire, Climate Change, and Carbon Cycling in the Boreal Forest*. New York: Springer-Verlag.

Kirby, K. J., and Watkins, C. (1998). *The Ecological History of European Forests*. New York: CAB International.

Klein Goldewijk, K., Beusen, A., van Drecht, G., and de Vos, M. (2011). The HYDE 3.1 spatially explicit database of human-induced global land-use change over the past 12,000 years. *Global Ecology and Biogeography*, 20, 73–86.

Kurz, W. A., Dymond, C. C., Stinson, G., et al. (2008). Mountain pine beetle and forest carbon feedback to climate change. *Nature*, 452, 987–990.

Lee, Y. H., Lamarque, J.-F., Flanner, M. G., et al. (2013). Evaluation of preindustrial to present-day black carbon and its albedo forcing from Atmospheric Chemistry and Climate Model Intercomparison Project (ACCMIP). *Atmospheric Chemistry and Physics*, 13, 2607–2634.

Leff, B., Ramankutty, N., and Foley, J. A. (2004). Geographic distribution of major crops across the world. *Global Biogeochemical Cycles*, 18, GB1009, doi:10.1029/2003GB002108.

Maness, H., Kushner, P. J., and Fung, I. (2013). Summertime climate response to mountain pine beetle disturbance in British Columbia. *Nature Geoscience*, 6, 65–70.

Monfreda, C., Ramankutty, N., and Foley, J. A. (2008). Farming the planet: 2. Geographic distribution of crop areas, yields, physiological types, and net primary production in the year 2000. *Global Biogeochemical Cycles*, 22, GB1022, doi:10.1029/2007GB002947.

Mouillot, F., and Field, C. B. (2005). Fire history and the global carbon budget: A $1° \times 1°$ fire history reconstruction for the 20th century. *Global Change Biology*, 11, 398–420.

Mouillot, F., Narasimha, A., Balkanski, Y., Lamarque, J.-F., and Field, C. B. (2006). Global carbon emissions from biomass burning in the 20th century. *Geophysical Research Letters*, 33, L01801, doi:10.1029/2005GL024707.

Nepstad, D., Soares-Filho, B. S., Merry, F., et al. (2009). The end of deforestation in the Brazilian Amazon. *Science*, 326, 1350–1351.

O'Halloran, T. L., Law, B. E., Goulden, M. L., et al. (2012). Radiative forcing of natural forest disturbances. *Global Change Biology*, 18, 555–565.

Paustian, K., Six, J., Elliott, E. T., and Hunt, H. W. (2000). Management options for reducing CO_2 emissions from agricultural soils. *Biogeochemistry*, 48, 147–163.

Peglar, S. M., Fritz, S. C., and Birks, H. J. B. (1989). Vegetation and land-use history at Diss, Norfolk, U.K. *Journal of Ecology*, 77, 203–222.

Petit, C. C., and Lambin, E. F. (2002). Long-term land-cover changes in the Belgian Ardennes (1775–1929): Model-based reconstruction vs. historical maps. *Global Change Biology*, 8, 616–630.

Pickett, S. T. A. (1976). Succession: An evolutionary interpretation. *American Naturalist*, 110, 107–119.

Pongratz, J., Reick, C., Raddatz, T., and Claussen, M. (2008). A reconstruction of global agricultural areas and land cover for the last millennium. *Global Biogeochemical Cycles*, 22, GB3018, doi:10.1029/2007GB003153.

Post, W. M., and Kwon, K. C. (2000). Soil carbon sequestration and land-use change: processes and potential. *Global Change Biology*, 6, 317–327.

Ramankutty, N., and Foley, J. A. (1999a). Estimating historical changes in global land cover: Croplands from 1700 to 1992. *Global Biogeochemical Cycles*, 13, 997–1027.

Ramankutty, N., and Foley, J. A. (1999b). Estimating historical changes in land cover: North American croplands from 1850 to 1992. *Global Ecology and Biogeography*, 8, 381–396.

Randerson, J. T., Liu, H., Flanner, M. G., et al. (2006). The impact of boreal forest fire on climate warming. *Science*, 314, 1130–1132.

Runkle, J. R. (1981). Gap regeneration in some old-growth forests of the eastern United States. *Ecology*, 62, 1041–1051.

Runkle, J. R. (1982). Patterns of disturbance in some old-growth mesic forests of eastern North America. *Ecology*, 63, 1533–1546.

Runkle, J. R. (2000). Canopy tree turnover in old-growth mesic forests of eastern North America. *Ecology*, 81, 554–567.

Runkle, J. R., and Yetter, T. C. (1987). Treefalls revisited: Gap dynamics in the southern Appalachians. *Ecology*, 68, 417–424.

Shugart, H. H. (1984). *A Theory of Forest Dynamics: The Ecological Implications of Forest Succession Models*. New York: Springer-Verlag.

Shugart, H. H. (1987). Dynamic ecosystem consequences of tree birth and death patterns. *BioScience*, 37, 596–602.

Shugart, H. H. (1998). *Terrestrial Ecosystems in Changing Environments*. Cambridge: Cambridge University Press.

Skole, D., and Tucker, C. (1993). Tropical deforestation and habitat fragmentation in the Amazon: Satellite data from 1978 to 1988. *Science*, 260, 1905–1910.

Smith, B. E., Marks, P. L., and Gardescu, S. (1993). Two hundred years of forest cover changes in Tompkins County, New York. *Bulletin of the Torrey Botanical Club*, 120, 229–247.

Spek, T. (1998). Interactions between humans and woodland in prehistoric and medieval Drenthe (the Netherlands): An interdisciplinary approach. In *The Ecological History of European Forests*, ed. K. J. Kirby and C. Watkins. New York: CAB International, pp. 81–93.

Tack, G., and Hermy, M. (1998). Historical ecology of woodlands in Flanders. In *The Ecological History of European Forests*, ed. K. J. Kirby and C. Watkins. New York: CAB International, pp. 283–292.

Thompson, J. R., Carpenter, D. N., Cogbill, C. V., and Foster, D. R. (2013). Four centuries of change in northeastern United States forests. *PLoS ONE*, 8(9), e72540, doi:10.1371/journal.pone.0072540.

van der Werf, G. R., Randerson, J. T., Giglio, L., et al. (2010). Global fire emissions and the contribution of deforestation, savanna, forest, agricultural, and peat fires (1997–2009). *Atmospheric Chemistry and Physics*, 10, 11707–11735.

van Laar, J. N., and den Ouden, J. B. (1998). Forest history of the Dutch province of Drenthe and its ancient woodland: A survey. In *The Ecological History of European Forests*, ed. K. J. Kirby and C. Watkins. New York: CAB International, pp. 95–106.

Watt, A. S. (1947). Pattern and process in the plant community. *Journal of Ecology*, 35, 1–22.

White, M. A., and Mladenoff, D. J. (1994). Old-growth forest landscape transitions from pre-European settlement to present. *Landscape Ecology*, 9, 191–205.

Whittaker, R. H., and Levin, S. A. (1977). The role of mosaic phenomena in natural communities. *Theoretical Population Biology*, 12, 117–139.

Williams, M. (1989). *Americans and Their Forests: A Historical Geography*. Cambridge: Cambridge University Press.

Williams, M. (2000). Dark ages and dark areas: Global deforestation in the deep past. *Journal of Historical Geography*, 26, 28–46.

Williams, M. (2003). *Deforesting the Earth: From Prehistory to Global Crisis*. Chicago: University of Chicago Press.

Wolf, R. J. A. M. (1998). Researching forest history to underpin the classification of Dutch forest ecosystems. In *The Ecological History of European Forests*, ed. K. J. Kirby and C. Watkins. New York: CAB International, pp. 265–281.

24

Global Biogeography

24.1 Chapter Summary

The structure and composition of vegetation and the functioning of terrestrial ecosystems, which at a local scale are shaped by environmental factors such as temperature and moisture, are also influenced by global climate. This is seen in the biogeography of vegetation and in the global carbon cycle. The geographic distribution of biomes closely correlates with temperature, precipitation, and evapotranspiration, and so, too, do annual net primary production and decomposition rates. Long-term changes in climate alter the biogeography and functioning of Earth's ecosystems. Global models of the terrestrial biosphere provide a quantitative framework to understand planetary ecology and the role of terrestrial ecosystems in the climate system.

24.2 Plant Geography

The broad influence of climate on macroscale ecology is evident when the complexity and diversity of terrestrial communities and ecosystems are reduced to a few biomes, or broad classes of vegetation (e.g., forest, grassland, shrubland, or desert). The natural vegetation of Earth has a distinct geographic pattern that corresponds to climate zones (Figure 2.4). The close correspondence between climate zones and biomes is readily apparent because climate zones such as tropical savanna, tropical rainforest, and tundra are named after vegetation (Table 6.1).

Tropical evergreen forests (tropical rainforests) are the dominant vegetation in hot, wet equatorial regions of South America, Africa, Southeast Asia, and Indonesia. In these regions, monthly temperatures are warm year-round, precipitation is abundant, and there is little seasonal variation in temperature or rainfall. Annual production is high. Trees are tall, often higher than 30 m, and form a thick canopy of broadleaf evergreen leaves through which little sunlight penetrates. The warm, wet conditions are optimal for decomposition so that little litter accumulates on the forest floor. Climate corresponds to the tropical rainforest zone (Figure 6.1). In tropical regions that are warm year-round but have a dry season, tropical deciduous forests are common. These drought-deciduous trees lose their leaves during the dry season. They are smaller than their rainforest counterparts and have less dense canopy coverage.

Grasses are prominent in dry tropical and temperate regions. The distinction between C_3 and C_4 grasses is important because of their different photosynthetic pathway and water-use efficiency. Grasses with the C_3 photosynthetic pathway dominate the vegetation in cool regions of the world, while C_4 plants dominant in warmer regions (Collatz et al. 1998; Still et al. 2003; Edwards et al. 2010). In dry tropical regions, forests give way to tropical savanna with widely spaced trees interspersed among tall grasses. In this tropical savanna climate, temperatures are warm throughout the year but there is a pronounced dry season (Figure 6.2). The short vegetation, low plant biomass, and deep roots are adaptations to lack of moisture. Recurring fires are common in this dry landscape. Tropical savannas occur most prominently to the north and south of the Amazon Basin in South America, to the north and south of the Congo Basin in Africa, east Africa, and northern and eastern Australia. In temperate regions, grasslands grow in dry, seasonally hot, semiarid climates with annual precipitation less than about 1000 mm (Figure 6.3). Grasslands are most prominent in the prairie of the United States Great Plains, the steppes of Central Asia, the pampa of Argentina, and the veld of South Africa.

Grasslands are transitional between forest and desert. Less precipitation results in short, widely spaced desert scrub vegetation and shrubland. The arid climate of deserts is hot and dry (Figure 6.4). Maximum daily temperatures in excess of 40°C are common. Annual precipitation is typically less than 250 mm, and some months may have less than 5 mm rainfall. The vegetation is a mix of plants that have developed different survival mechanisms. Winter annuals germinate in autumn or winter and flower in late spring. Summer annuals germinate in mid-summer and flower in late summer or autumn. Both types of annuals typically germinate after heavy rains ensure adequate moisture. Drought-deciduous plants lose leaves during dry spells. In contrast, evergreen shrubs grow throughout the year, even in periods of drought. Cacti and other succulents have specialized organs to store water and use the CAM photosynthetic pathway to minimize water loss during transpiration. Deserts occur on the eastern flanks of the subtropical high pressures near latitudes 30° N and 30° S – in southwestern United States, North Africa, southern South America, South Africa, and western Australia – and in regions far removed from sources of atmospheric moisture such as central Asia, central Australia, and the Great Basin of western United States.

Greater annual precipitation supports forest. Deciduous and evergreen trees form dense, productive temperate forests in eastern United States and the Pacific Northwest, in regions of central and southern Europe, and eastern China. Climate is typically humid subtropical, marine, or warm summer humid continental (Figures 6.6–6.8). Rainfall exceeds 1000 mm per year, summers are long and warm, and winters are cool to cold. Moderate to pronounced seasonality is a dominant feature of climate. Trees can be as tall as in tropical forests, but are composed primarily of broadleaf deciduous species, which drop their leaves in winter, and needleleaf evergreen species.

Further north, in Alaska and northern regions of Canada, Europe, and Russia, the winters are bitterly cold, the summers are cool and short, and the vegetation is a mix of evergreen and deciduous trees that form boreal forests, or taiga. The climate is cool summer humid continental or subarctic (Figures 6.9 and 6.10). Annual production and decomposition are low in this cold landscape. Trees are mostly needleleaf evergreen and are shorter and more open than their temperate counterparts. Similar forests occur in high mountain ranges such as the Rocky Mountains, Cascade Mountains, and Sierra Nevada Mountains of western United States and in the Alps of Europe. In the extreme cold of Siberia, deciduous needleleaf trees that drop their needles in winter are common.

At very high latitudes, between about 65° N and 70° N, the boreal forest gives way to the treeless tundra comprised of grass-like sedges, dwarf shrubs, lichens, and mosses. Winter temperatures are extremely cold, the growing season is short (two months or less), and soils are frozen year-round (Figure 6.11). Only the upper 50 cm or so of soil thaws during summer. Cold, wet soil restricts plant productivity and decomposition.

Chaparral, also known as Mediterranean vegetation, is a unique biome found primarily in Mediterranean regions, southern California, South Africa, and parts of Australia. Summers are hot and dry, and winters are mild and moist (Figure 6.5). As a result, plants grow during winter rather than summer. The plants are

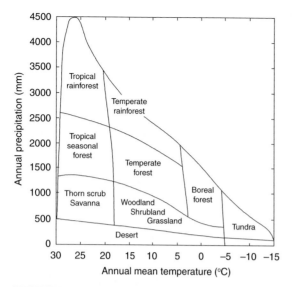

Fig. 24.1 Generalized relationships among major plant formations, annual mean temperature, and annual precipitation. Adapted from Whittaker (1975, p. 167).

characteristically short, dense woody shrubs with thick, waxy, evergreen leaves. The oily nature of these shrubs promotes wildfire.

Temperature and precipitation strongly correlate with plant geography (Figure 24.1). Temperature is important because sufficient warmth, but not excessive heat, is a prerequisite for the biochemical reactions that support life. Water is important because 80–90 percent of the mass of a plant is water. Two environmental gradients are evident in Figure 24.1. One is a latitudinal gradient from tropical forest to temperate forest to boreal forest to arctic tundra that reflects increasingly cold climates. Precipitation forms a second axis of vegetation differentiation. In the tropics, vegetation changes from rainforest to seasonal forest to savanna to desert as annual precipitation decreases. In temperate regions, forest gives way to woodland, shrubland, grassland, and desert as precipitation decreases.

Evapotranspiration also relates to plant geography. Evapotranspiration causes a typical C_3 plant to lose one liter of water for every 1–5 grams of biomass produced (Table 16.2). Evapotranspiration is an integrated measure of temperature and precipitation. Temperature indicates how much energy is available to evaporate water. Precipitation determines the amount of water that can evaporate. In his pioneering research, Charles Thornthwaite demonstrated a relationship between evapotranspiration and plant geography (Thornthwaite 1948; Thornthwaite and Mather 1955, 1957). Tropical rainforest climates, for example, are warm throughout the year, have a large positive net radiation balance, and have a high rate of potential evaporation. High annual precipitation is required to maintain a positive water balance. In contrast, subarctic or tundra climates have a short warm season, receive less radiation, and have lower evaporative potential. Less precipitation is needed to meet evaporative demand. In general, less water is evaporated in cool climates than in warm climates, and less precipitation is required to support plant growth. Conversely, more precipitation is needed to support plant growth in hot climates with high evapotranspiration loss. For example, productive forests grow in eastern United States, where annual precipitation ranges from 1000 to 1500 mm. In the tropics, a similar amount of rain creates a semiarid climate and supports savanna. Similar such changes are evident in Figure 24.1.

Budyko (1974, 1986) showed a relationship between plant geography and his radiative dryness index, defined as the ratio of net radiation to the amount of energy required to evaporate the annual precipitation (Chapter 12). Figure 24.2 illustrates this relationship. Values of the dryness index less than 0.3 correspond with tundra. Forest vegetation has values of 0.3–1. Different forest types are discriminated by net radiation. Grassland has values of 1–2, distinguished into prairie (steppe) and savanna based on net radiation. Semidesert vegetation has a value greater than 2; desert, greater than 3.

Annual temperature, precipitation, and evapotranspiration are three variables commonly used in ecological classifications of bioclimatic regions (e.g., the BIOME model; Prentice et al. 1992; Haxeltine and Prentice 1996; Kaplan et al. 2003). The Holdridge life zone classification epitomizes such bioclimatic schemes (Figure 24.3). The model depicts life zones as hexagons formed by the intersection of annual

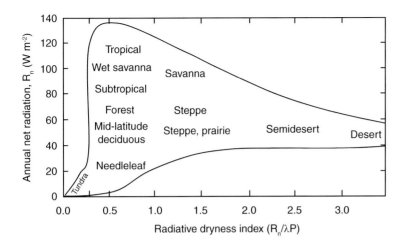

Fig. 24.2 Relationship of biomes with net radiation and radiative dryness index. This index is the ratio of net radiation to the amount of energy required to evaporate the annual precipitation ($R_n / \lambda P$ where λ is the latent heat of vaporization). After Budyko (1974, p. 348).

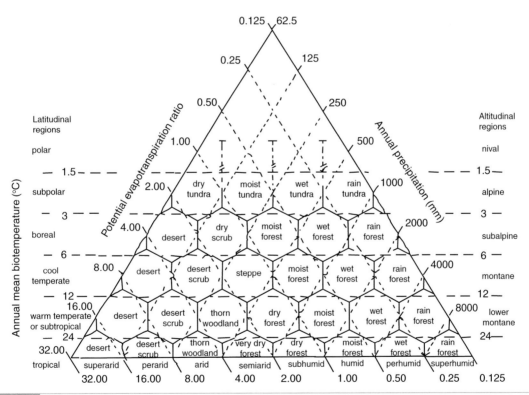

Fig. 24.3 Holdridge vegetation classification showing relationships among annual mean biotemperature, annual precipitation, annual potential evapotranspiration, and vegetation type. Annual average biotemperature is calculated from monthly temperature, setting temperature below 0°C to zero. Potential evapotranspiration ratio is annual potential evapotranspiration divided by annual precipitation. Temperature demarcations run horizontally ranging from 1.5°C to 24°C. Precipitation lines are parallel to the potential evapotranspiration ratio axis and range from 62.5 to 8000 mm. Potential evapotranspiration lines are parallel to the precipitation axis and range from 0.125 to 32. Adapted from Holdridge (1967).

mean biotemperature and annual precipitation. For example, wet tundra occurs with temperatures between 1.5°C and 3.0°C and precipitation of 250–500 mm. Wet tropical forest occurs with temperatures greater than 24°C and precipitation between 4000 and 8000 mm. Evapotranspiration, expressed as the ratio of annual potential evapotranspiration to annual precipitation, forms a third classification dimension. For example, wet tundra and wet tropical forest are perhumid despite a more than tenfold difference in precipitation because of the lower demand for water in the cold tundra environment. The Holdridge scheme has been widely used to map biogeography in response to climate change.

24.3 | Net Primary Production and Plant Biomass

Terrestrial ecosystems store large quantities of carbon. From the biome synthesis of Prentice et al. (2001), the total carbon stored in plant biomass is estimated to be 466–654 Pg C. Forests contain three-quarters of this carbon (Table 24.1). Tropical forests are the single largest store of carbon in plants, as measured by both carbon density and carbon stock. Net primary production is estimated to be about 60 Pg C yr^{-1}. Tropical forests, savannas, and grasslands together account for more than one-half of the world's annual carbon uptake by terrestrial vegetation.

Temperature, precipitation, and evapotranspiration regulate net primary production at the global scale (Figure 24.4). Across a variety of ecosystems from tundra to tropical rainforest, annual net primary production increases with warmer climate. Plant production is low in extremely cold climates. In very warm climates, excessive temperatures, often combined with low water availability, limit production. In arid climates, there is a near linear increase in production as precipitation increases. These ecosystems exhibit large sensitivity of net primary production to annual precipitation variability (Huxman et al. 2004). Productivity plateaus in

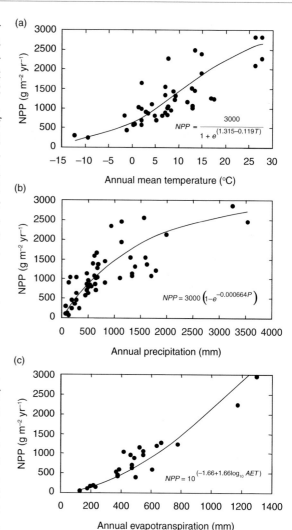

Fig. 24.4 Annual net primary production (above- and belowground) in relation to (a) annual mean temperature (Lieth 1975), (b) annual precipitation (Lieth 1975), and (c) annual evapotranspiration (Rosenzweig 1968). Del Grosso et al. (2008) provide updated relationships.

more humid climates where water does not limit production. These ecosystems exhibit low sensitivity of plant production to precipitation variability. The effect of temperature and water on plant production is evident in the relationship with evapotranspiration. Sites with low annual evapotranspiration have low productivity because temperatures are cold (e.g., tundra) or water is limiting (desert). Similar climatic dependence can be seen in eddy covariance

Table 24.1 | Estimates of terrestrial carbon density, carbon stock, and net primary production (NPP) from two different datasets

Biome	WBGU						MRS/IGBP						NPP (Pg C yr⁻¹)	
	Area	Carbon density (g C m⁻²)		Carbon stock (Pg C)			Area	Carbon density (g C m⁻²)		Carbon stock (Pg C)			WBGU	MRS
	(10¹² m²)	Plant	Soil	Plant	Soil	Total	(10¹² m²)	Plant	Soil	Plant	Soil	Total		
Forest														
Tropical	17.6	12,000	12,300	212	216	428	17.5	19,400	12,200	340	213	553	13.7	21.9
Temperate	10.4	5700	9600	59	100	159	10.4	13,400	14,700	139	153	292	6.5	8.1
Boreal	13.7	6400	34,400	88	471	559	13.7	4200	24,700	57	338	395	3.2	2.6
Tropical savanna & grassland	22.5	2900	11,700	66	264	330	27.6	2900	9000	79	247	326	17.7	14.9
Temperate grassland & shrubland	12.5	700	23,600	9	295	304	17.8	1300	9900	23	176	199	5.3	7.0
Desert & semidesert	45.5	200	4200	8	191	199	27.7	400	5700	10	159	169	1.4	3.5
Tundra	9.5	600	12,700	6	121	127	5.6	400	20,600	2	115	117	1.0	0.5
Cropland	16.0	200	8000	3	128	131	13.5	300	12,200	4	165	169	6.8	4.1
Wetland	3.5	4300	64,300	15	225	240	–	–	–	–	–	–	4.3	–
Ice	–	–	–	–	–	–	15.5	0	0	0	0	0	–	0
Total	151.2	–	–	466	2011	2477	149.3	–	–	654	1567	2221	59.9	62.6

Note: 1 Pg = 10¹⁵ g. Variations in biome classifications can lead to inconsistencies. In particular, the WBGU dataset includes ice with desert and semidesert. The MRS/IGBP dataset does not include wetlands.

Source: From Prentice et al. (2001).

Normalized difference vegetation index

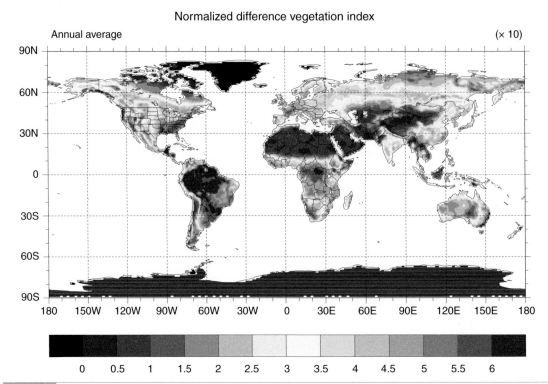

Fig. 24.5 Annual average normalized difference vegetation index for 1982–1993. High numbers indicate regions of high production. Low numbers indicate regions of low production. See color plate section.

measurements of net ecosystem exchange (Yi et al. 2010; Niu et al. 2012).

Satellite-derived measures of plant physiological activity such as the normalized difference vegetation index (NDVI, Eq. (17.6)) provide a means to monitor terrestrial ecosystems worldwide (Figure 24.5). The NDVI is related to absorption of photosynthetically active radiation in the plant canopy and can be used with production efficiency models, using Eq. (17.7), to derive net primary production (Running et al. 2004; Zhao et al. 2005). The strong relationship between climate and the NDVI is clear. Highest values occur in the tropical rainforests of South America, Africa, and Southeast Asia. The NDVI is also large in the tropical savannas of South America and Africa, in the temperate forests of eastern United States and the Pacific Northwest, and in portions of Europe, China, and eastern Australia. Arctic tundra and portions of the boreal forest have low NDVI. Lowest NDVI occurs in deserts and semiarid vegetation.

The geographic distribution of climatic constraints to net primary production can be derived from long-term climate data. Figure 24.6 shows such a derivation by Nemani et al. (2003). Water availability most strongly limits net primary production over 40 percent of Earth's vegetated area. Temperature is the primary factor limiting plant growth over 33 percent of vegetated area, and radiation limits growth over the remaining 27 percent. Multiple limiting factors occur in many regions. For example, cold winter temperatures and cloudy summers limit plant growth in high latitude regions of Eurasia and parts of Canada and Alaska. Cold winters and dry summers limit net primary production in western North America. Water limitation to plant growth is widespread throughout Africa and Australia. Net primary production in tropical South America, tropical Africa, and Indonesia is limited by high cloud cover that reduces incoming solar radiation at the surface and also by water availability in monsoon climates.

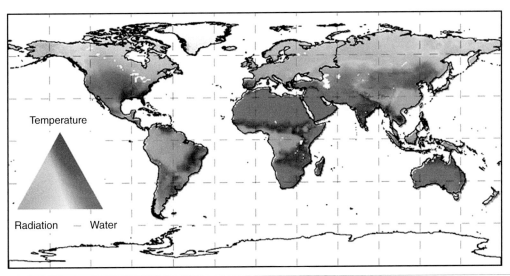

Ecosystem CO_2, latent heat, and sensible heat fluxes have similar spatial patterns as the NDVI (Figure 24.7). Annual gross primary production is estimated to be 119 ± 6 Pg C yr^{-1}, consistent with a similarly derived estimate of 123 ± 8 Pg C yr^{-1} for the period 1998–2005 (Beer et al. 2010). Ecosystem respiration is estimated to be 96 ± 6 Pg C yr^{-1}, and annual evapotranspiration is $65 \pm 3 \times 10^3$ km^3 yr^{-1}. Annual gross primary production, ecosystem respiration, and latent heat flux are largest in the equatorial tropics and the subtropics. Humid temperate regions (e.g., eastern North America, Europe, and China) also have high annual fluxes. The lowest fluxes occur in cold and dry climates. The spatial pattern of gross primary production strongly correlates with the spatial pattern of precipitation (Beer et al. 2010). Annual sensible heat flux is low in the moist tropics and highest in dry climates.

Changes in net primary production and allocation in response to climate are seen in the structure of vegetation (Figure 24.8). Vegetation stature changes from short grasses and shrubs to taller trees as annual precipitation increases. There is also a change from sparse canopies with low leaf area index to denser canopies with large leaf area. These changes are a manifestation of the dynamic balance among precipitation, soil water, and leaf area (Grier and Running 1977; Woodward 1987, 1993; Nemani and Running 1989). There is a maximum leaf area index for which evapotranspiration balances precipitation.

The geography of forest aboveground biomass and canopy height shows similar geographic patterns. Forest aboveground biomass exceeds 30 kg m^{-2} in the tropical rainforests of South America, Africa, and Southeast Asia (Figure 24.9a). Drier tropical climates support much less biomass, approximately 5–10 kg m^{-2}. Canopy height is 20–30 m or more in much of the wet tropics and the forests of eastern North America and the Pacific Northwest, Europe, and Asia (Figure 24.9b).

Some general patterns of biomass allocation in relation to environment are evident (Canadell et al. 1996; Jackson et al. 1996, 2000; Poorter et al. 2012). Rooting depth typically increases in climates with higher annual evapotranspiration, higher annual precipitation, and longer warm season (Schenk and Jackson 2002). Tundra, boreal forest, and temperate grassland have

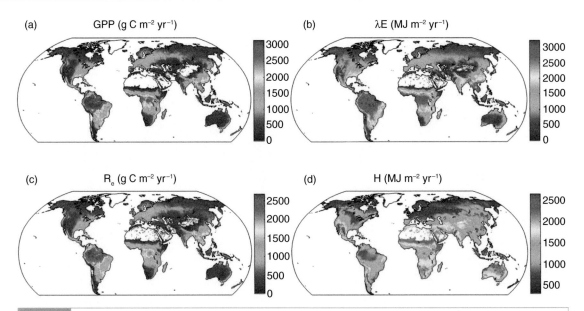

Fig. 24.7 Mean annual (a) gross primary production (GPP, g C m⁻² yr⁻¹), (b) latent heat flux (λE, MJ m⁻² yr⁻¹), (c) terrestrial ecosystem respiration (R$_e$, g C m⁻² yr⁻¹), and (d) sensible heat flux (H, MJ m⁻² yr⁻¹) for the period 1982–2008 empirically upscaled from FLUXNET eddy covariance flux towers. Reproduced from Jung et al. (2011). See color plate section.

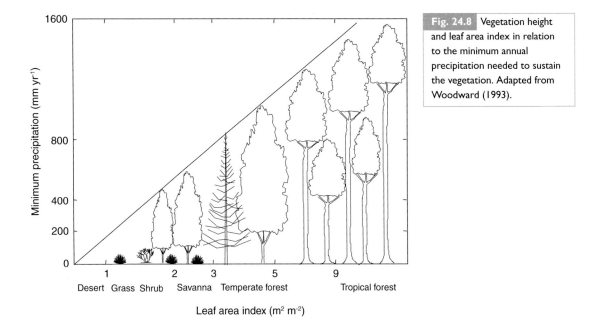

Fig. 24.8 Vegetation height and leaf area index in relation to the minimum annual precipitation needed to sustain the vegetation. Adapted from Woodward (1993).

the shallowest root profiles, with 83–93 percent of root biomass in the top 30 cm of soil (Figure 24.10). Deserts and temperate coniferous forests have the deepest root profiles, with only about 50 percent of roots in the top 30 cm. Shallow-rooted plants are most prominent in the Arctic, where permafrost and poorly drained soils restrict root growth. Plants growing in this cold environment (tundra, cold desert) have the highest ratio of root-to-shoot biomass, and in

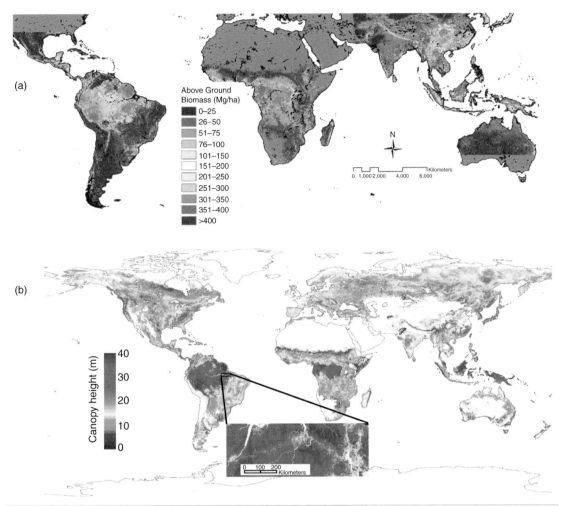

Fig. 24.9 Plant biomass and canopy height estimated from spaceborne light detection and ranging (Lidar) sensors. (a) Forest aboveground biomass in the tropics and subtropics. 10 Mg ha^{-1} = 1 kg m^{-2}. Reproduced from Saatchi et al. (2011) with permission of the National Academy of Sciences. Thurner et al. (2014) provide a similar map for boreal and temperate forests. (b) Forest canopy height. Reproduced from Simard et al. (2011). Pan et al. (2013) review forest biomass and height. See color plate section.

general cold biomes have high biomass allocation to roots. Temperate grasslands also have high root-to-shoot ratio, and in general arid biomes (grassland, desert) have high biomass allocation to roots. Woody vegetation, with large tap roots, tends to have deep root profiles. For temperate and tropical trees, 26 percent of root biomass is in the top 10 cm, 60 percent is in the top 30 cm, and 78 percent is in the top 50 cm. In contrast, grasses have 44 percent of their root biomass in the top 10 cm and 75 percent in the top 30 cm. Forests typically have

low root-to-shoot ratios because of their large aboveground woody biomass, and they have the highest biomass allocation to stems.

Although most root biomass is within the upper 50–100 cm of soil, roots can extend much deeper (Figure 24.10). Tundra plants are the shallowest rooted, extending to a depth of 50 cm on average. Desert and tropical savanna plants have roots 10–15 m deep. Deep-rooted plants are common in woody and herbaceous species across most terrestrial biomes. On average, roots of trees extend 7.0 m deep, shrubs 5.1

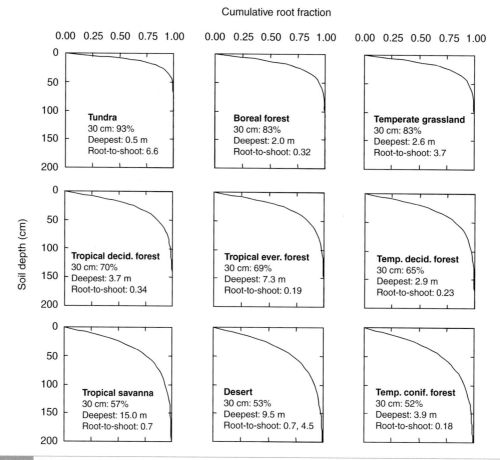

Cumulative root fraction

Fig. 24.10 Relative root abundance for tundra, boreal forest, temperate grassland, tropical deciduous forest, tropical evergreen forest, temperate deciduous forest, tropical savanna, desert, and temperate coniferous forest. Graphs show cumulative root distribution in relation to soil depth. Text boxes show the proportion of roots in the top 30 cm, average maximum rooting depth, and the average ratio of root-to-shoot biomass. Root-to-shoot ratios for desert are given separately for warm and cold deserts. Data from Jackson et al. (1996) and Canadell et al. (1996).

m, and herbaceous perennials 2.6 m (Canadell et al. 1996). Plants from arid and Mediterranean climates have the deepest rooting depths. These deep roots are a small portion of plant biomass, but are hydrologically important. They provide access to water reserves deep in the ground and help protect against intermittent droughts.

24.4 | Litterfall and Soil Carbon

Carbon storage in soil is typically comparable to or larger than plant biomass (Table 24.1). It is estimated that soils hold about 1500–2400 Pg C worldwide, depending on depth (0–100 cm depth: 1395 Pg C, Post et al. 1982; 1462–1548 Pg C, Batjes 1996; 1502 Pg C, Jobbágy and Jackson 2000; 1260 Pg C with a range of 890–1660 Pg C, Todd-Brown et al. 2013; 0–200 cm depth: 2376–2456 Pg C, Batjes 1996; 1993 Pg C, Jobbágy and Jackson 2000; and 0–300 cm depth: 2344 Pg C, Jobbágy and Jackson 2000). Figure 21.10 illustrates this vertical profile for forest and grassland. Cold or wet northern ecosystems (tundra and boreal forest) account for much of the global soil carbon (Figure 24.11).

Soil carbon (g C m^{-2})

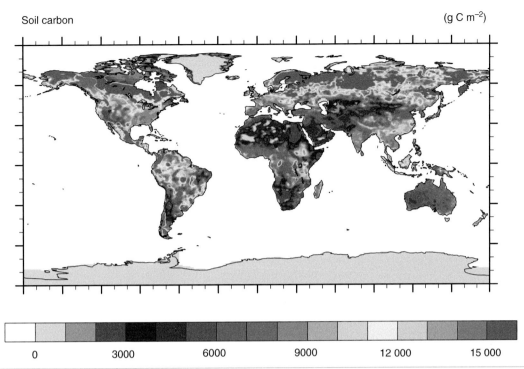

0   3000   6000   9000   12 000   15 000

Fig. 24.11 Soil carbon to a depth of 1 m from the Harmonized World Soil Database. The original data have a spatial resolution of approximately 1 km and were re-gridded to a resolution of 1° in latitude and longitude. The global total for the re-gridded data is 1259 Pg C. Data from Wieder et al. (2013). The Northern Circumpolar Soil Carbon Database (Tarnocai et al. 2009; Hugelius et al. 2013) has data specific to the Arctic. See color plate section.

Savanna and grassland also hold high amounts of soil carbon. Deserts have the least soil carbon. However, the map shown in Figure 24.11 underestimates soil organic carbon because the northern circumpolar permafrost region alone contains 1672 Pg C (0–100 cm depth: 496 Pg C; 0–300 cm depth: 1024 Pg C; with an additional 648 Pg C in deep layers; Tarnocai et al. 2009).

Soil carbon is the balance between litter input and decomposition loss. Litter production is related to net primary production. More productive ecosystems produce greater amounts of litter each year than less productive ecosystems. Across a variety of broadleaf, needleleaf, evergreen, and deciduous forests ranging from tropical to boreal, annual litter production increases with warmer and wetter climates (Figure 24.12). Tropical broadleaf forests have high litterfall (> 900 g m^{-2} yr^{-1}). Boreal needleleaf evergreen forests have low litter production (< 300 g m^{-2} yr^{-1}). Litter production in temperate forests ranges from 300 to 650 g m^{-2} yr^{-1}.

Studies of litter decomposition in a variety of climates and plant communities show that decomposition rates vary with temperature, precipitation, and evapotranspiration (Meentemeyer 1978; Gholz et al. 2000; Trofymow et al. 2002; Adair et al. 2008; Currie et al. 2010). For example, a study of pine needles placed in 26 sites from arctic to tropical and reflecting a wide variety of natural and managed ecosystems found annual decomposition increased with warmer and wetter climates (Figure 24.13). Annual evapotranspiration was a strong predictor of decomposition. Litter placed in sites with high evapotranspiration (tropical forest) lost about 40 percent of its mass over a year while litter in sites with low evapotranspiration (desert, tundra) lost only about 10 percent of its mass.

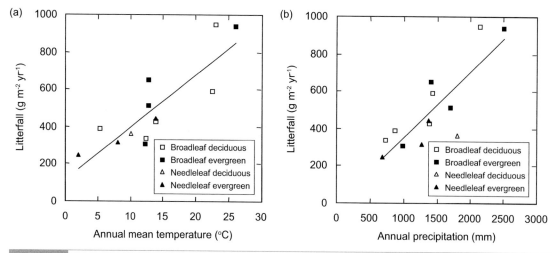

Fig. 24.12 Annual litterfall in forests in relation to (a) temperature and (b) precipitation. Data are means for 13 forest types. Data from Vogt et al. (1986). See also Matthews (1997).

24.5 | Climate Change

Climate change over periods of centuries to millennia alters the biogeography and functioning of Earth's vegetation. One of the longest timescales at which climate changes is the recurring waxing and waning of glaciers. Some 18,000 years before present (18 kyr BP) at the height of the last ice age, glaciers covered much of the Northern Hemisphere high latitudes (Figure 8.3). Over the next several thousand years, summer solar radiation in the Northern Hemisphere increased, atmospheric CO_2 concentration increased, climate warmed, and glaciers retreated (Figure 8.4).

This warming is seen in the geographic distribution of vegetation. Vegetation adapted to cold climates shifted northwards to be replaced by warm vegetation types. Pollen preserved in lake and bog sediments records this history. For example, Figure 24.14 shows the vegetation history of Anderson Pond in Tennessee over the past 16 kyr. The pollen of fir, spruce, and pine trees, species typical of boreal forests, was most abundant 16 kyr BP. However, the forest was not representative of the modern boreal forest. Oak, hickory, ash, hornbeam, and other species typical of temperate deciduous forests were also

present. Boreal species declined in abundance as climate warmed, replaced by temperate deciduous species. Between 12.5 and 9.5 kyr BP, oak and other taxa found in present-day mesophytic forests became dominant. By about 8–6 kyr BP, a forest similar to today, dominated by oak with some ash and hickory, had formed.

Vegetation throughout eastern North America had similar changes over the past 21 kyr. Figures 24.15 and 24.16 illustrate this vegetation history derived from numerous pollen diagrams. Spruce and pine trees migrated north and west, colonizing land exposed by the retreating glacier. Temperate trees such as oak and beech were restricted to southeastern United States 21 kyr BP, but increased in abundance and shifted northward as climate warmed. Hemlock expanded geographically from its glacial refuge in the Appalachian Mountains.

These pollen diagrams can be aggregated into various biomes (Figure 24.17). At the height of glaciation, spruce-dominated boreal forest and forest tundra formed a broad belt of vegetation north of latitude 35° N. In contrast to the modern eastern Canadian boreal forest, however, fir, birch, and alder were not significant in the forest. Moreover, the high presence of sedge pollen suggests these forests were more open than modern forests. Pine-dominated mixed forests grew

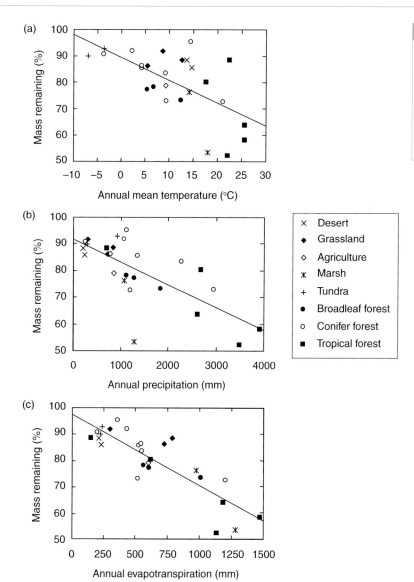

Fig. 24.13 Annual decomposition of pine litter in relation to (a) temperature, (b) precipitation, and (c) evapotranspiration. Symbols identify the 26 sites in which the litter was placed. Data from Gholz et al. (2000). Figure 21.5 shows multi-year results from the litter decomposition study.

to the south. Deciduous forests were restricted far south. The data coverage increases with time and by 12 kyr BP a large region of eastern North America can be mapped. East of the Appalachian Mountains, there was a north–south zonation to vegetation with spruce-dominated boreal forest and pine-dominated mixed forest in the north, oak-dominated deciduous forest to the south, and pine-dominated forests in Florida. However, the most significant feature is a large region in which the vegetation was not similar to any modern vegetation (Williams et al. 2001). This consisted primarily of spruce-dominated woodland with high amounts of sedge and deciduous trees such as ash and elm. Indeed, the individualistic organization of plants into communities and ecosystems makes it likely that vegetation reorganizes into novel biomes as climate changes (Williams et al. 2007).

The northern movement of vegetation is the most obvious trend over the next 12 kyr (Figure 24.17). This movement was gradual before 12 kyr BP, rapid from 12 to 9 kyr BP, and then again gradual. Modern-like vegetation did

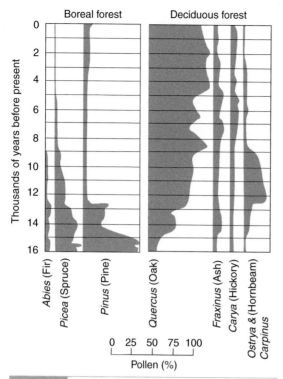

Boreal forest   Deciduous forest

Fig. 24.14 Pollen abundance at Anderson Pond, Tennessee, over the past 16 kyr BP. Adapted from Solomon et al. (1981).

not begin to develop until after 9 kyr BP. At this time, spruce-dominated boreal forest was less common than today and tundra was restricted to east of the retreating ice sheet. The deciduous forest consisted of prairie–aspen parkland in the west and northwest and mesophytic taxa in the south. Southern pine species were largely restricted to Florida until about 9 kyr BP, when southeast pine forests similar to modern forests spread northward. At the same time, prairie vegetation similar to today formed in the west and the modern deciduous forest formed in the east. Tundra and forest tundra, largely absent prior to 9 kyr BP, became more extensive. Modern-day vegetation patterns were well established by 6 kyr BP.

The correspondence of pollen abundance with temperature and precipitation (Figure 19.2) provides a means to reconstruct past climate. The observed change in vegetation reflected the overall warmer and wetter climate of eastern

North America with deglaciation. Reconstructed January and July temperatures show general warming as the ice sheet retreated (Figure 24.18). The largest increase in temperature occurred between 12 and 9 kyr BP. Warming continued until 6 kyr BP, after which July temperatures cooled somewhat in the north. January temperatures continued to increase, especially in the southeast. This reflects decreased seasonality since about 9 kyr BP brought about by changes in solar radiation in the Northern Hemisphere (Figure 8.4). At the height of glaciation, precipitation in eastern North America had a zonal pattern with high annual precipitation in the south and drier conditions to the north. The modern east-to-west decrease in precipitation developed about 9 kyr BP and strengthened by 6 kyr BP when conditions were driest in the west.

The period about 6 kyr BP is particularly noteworthy (TEMPO 1996; Prentice et al. 1996, 1998, 2000). During this time, the tilt of Earth's axis was greater than today, the eccentricity of Earth's orbit around the Sun was slightly larger, and Earth was closest to the Sun in mid-September compared with early January. This increased the amount of summertime solar radiation in the Northern Hemisphere (Figure 8.4). As a result, the climate of middle and high latitudes of the Northern Hemisphere was warmer than present. In some regions, vegetation extended northward beyond modern range limits. This is seen in the greater extent of boreal forest in eastern Canada and northern Europe.

Increased summer solar radiation 6 kyr BP created a warmer, wetter climate in North Africa compared with today (Chapter 27). Stronger than present summer solar radiation heated the African continent and increased the land–sea temperature contrast. Surface pressure dropped over land, enhancing air flow onto the continent and increasing the African summer monsoon. As a result of the wetter conditions, grasses and shrubs covered much of the modern Sahara Desert. Present climate conditions over northern Africa began to develop after 6 kyr BP, when summer solar radiation decreased to modern values, land cooled, the monsoon weakened, and subtropical deserts expanded.

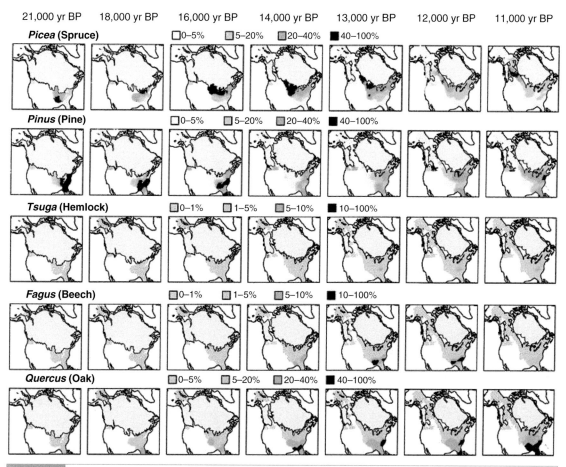

21,000 yr BP   18,000 yr BP   16,000 yr BP   14,000 yr BP   13,000 yr BP   12,000 yr BP   11,000 yr BP

Picea (Spruce)   □0–5% ▫5–20% ▪20–40% ■40–100%

Pinus (Pine)   □0–5% ▫5–20% ▪20–40% ■40–100%

Tsuga (Hemlock)   □0–1% ▫1–5% ▪5–10% ■10–100%

Fagus (Beech)   □0–1% ▫1–5% ▪5–10% ■10–100%

Quercus (Oak)   □0–5% ▫5–20% ▪20–40% ■40–100%

Fig. 24.15 Pollen abundance in boreal and eastern North America for spruce, pine, hemlock, beech, and oak trees between 21 kyr BP and 11 kyr BP. Adapted from Williams et al. (2004). See color plate section.

A key feature determining the response of vegetation to climate change is whether the vegetation is in equilibrium with climate. If so, changes in climate are matched by changes in plant geography. Alternatively, lags may be introduced if the rate of climate change exceeds the timescale at which plants migrate. This is particularly true of long-lived vegetation such as forests. Most tree species have poor dispersal, with migration rates on the order of 100–1000 m per year or less (Davis 1981; Van Minnen et al. 2000; McLachlan et al. 2005). For example, in eastern North America spruces migrated over a period of 3–4 kyr from their refuge in the Appalachian Mountains during the last glacial maximum to their present-day southern range limit (Figure 24.19). Eight

thousand years were required for oak species to migrate from their glacial refugia in southern United States to their current northern range limit. This suggests that the geographic distribution of tree species was not in equilibrium with climate at certain periods (Davis 1981; Davis et al. 1986). Instead, the rate of migration depended on the availability of seeds and the ability of seedlings to become established. Alternatively, climate change may set the rate of distribution changes, particularly northern range limits (Ordonez and Williams 2013).

In addition to climate, changes in atmospheric CO_2 concentration can alter biogeography by changing the competitive balance between C_3 and C_4 plants (Ehleringer et al. 1997; Collatz et al. 1998; Sage 2004; Beerling

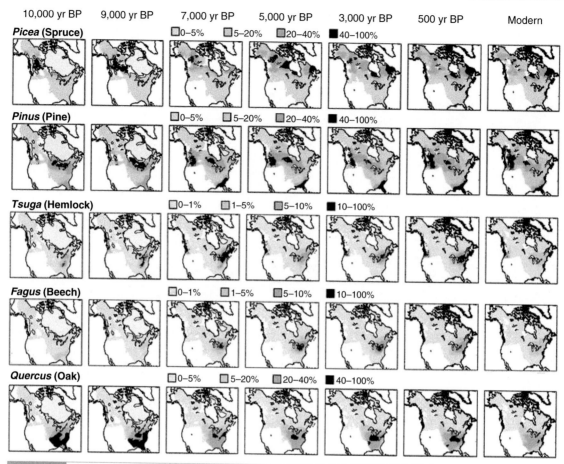

| 10,000 yr BP | 9,000 yr BP | 7,000 yr BP | 5,000 yr BP | 3,000 yr BP | 500 yr BP | Modern |

Picea (Spruce)   □0–5% ▨5–20% ▨20–40% ■40–100%

Pinus (Pine)   □0–5% ▨5–20% ▨20–40% ■40–100%

Tsuga (Hemlock)   □0–1% ▨1–5% ▨5–10% ■10–100%

Fagus (Beech)   □0–1% ▨1–5% ▨5–10% ■10–100%

Quercus (Oak)   □0–5% ▨5–20% ▨20–40% ■40–100%

Fig. 24.16 As in Figure 24.15, but for the period 10 kyr BP to present-day (modern). Adapted from Williams et al. (2004). See color plate section.

and Osborne 2006; Edwards et al. 2010; Monson and Collatz 2012). Because of their differences in photorespiration, plants utilizing the C_4 photosynthetic pathway have an advantage at high temperatures or low CO_2 concentrations, while C_3 plants have an advantage at lower temperatures or higher CO_2 concentrations. This physiological tradeoff helps to explain the present-day latitudinal distribution of C_3-dominated and C_4-dominated grasslands and why C_4 grasses rose to dominance 3–8 million years ago. However, the origin of C_4 grasslands is complex (Beerling and Osborne 2006; Edwards et al. 2010). Tropical savannas are one of the most frequently burned biomes, and recurring fire is essential for their maintenance.

24.6 Global Terrestrial Biosphere Models

Numerous numerical models have been developed to simulate net primary production, biomass accumulation, litterfall, and soil carbon in global terrestrial ecosystems. These models of planetary ecology span a range of approaches from simple empirically based models to mechanistic process models of plant physiology and biogeochemical cycles. For example, the empirical relationships shown in Figure 24.4 have been used to model vegetation productivity in relation to temperature and precipitation (Esser 1987; Friedlingstein

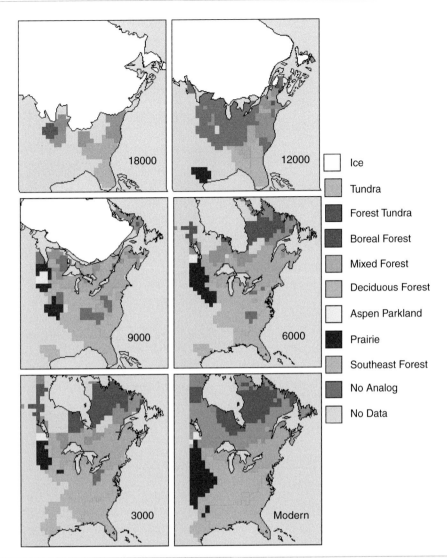

Fig. 24.17 Geographic extent of vegetation in eastern North America over the past 18 kyr BP. Vegetation types are based on pollen abundance. No analog means the pollen does not correspond to a modern vegetation type. Adapted from Overpeck et al. (1992) with graphics provided by the National Geophysical Data Center (National Oceanic and Atmospheric Administration, Boulder, Colorado). See color plate section.

et al. 1992; Dai and Fung 1993; Post et al. 1997). Process models are, however, more widely used than empirical models, and global models of terrestrial ecosystems fall into one of two general classes. Biogeochemical models simulate the carbon balance of ecosystems given a geographic distribution of vegetation, typically specified by biomes, as input to the model. The models employ algorithms for photosynthesis, respiration, allocation, and other plant physiological and microbial processes specific to the different biomes. Vegetation is represented by aggregate pools of foliage, stem, and root biomass without regard to individual plants (Figure 20.2). Dynamic global vegetation models simulate the carbon balance and also ecosystem structure and composition. These models are not developed for specific biomes, but rather use physiological algorithms for various plant functional types. Ecosystems are

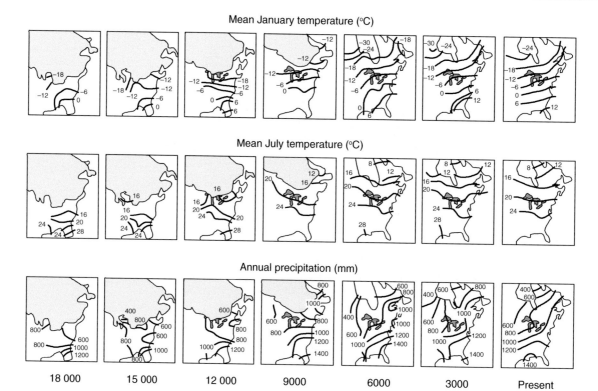

Fig. 24.18 Climate of eastern North America over the past 18 kyr BP reconstructed from pollen. The lightly shaded region shows the glacier. Adapted from Webb et al. (1993).

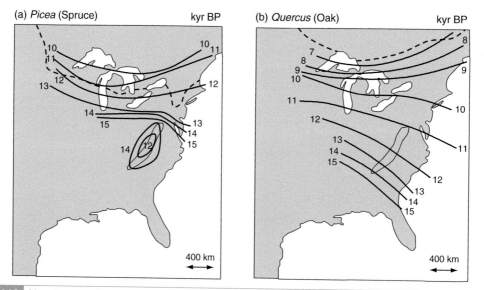

Fig. 24.19 Migration maps for (a) spruce and (b) oak. Contour lines show the time (kyr BP) of the first appearance of spruce or oak pollen in sediments. The dashed line shows the present southern range limit of spruce and northern range limit of oak. Adapted from Davis (1981).

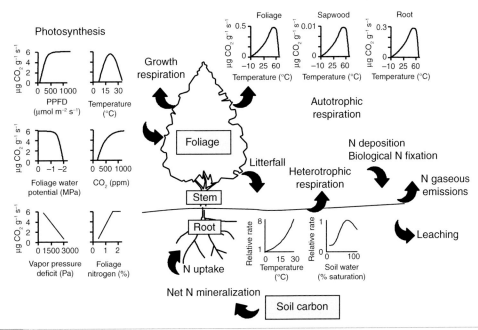

Fig. 24.20 Processes typically included in ecosystem models. Shown are the ecosystem carbon balance, environmental controls of photosynthesis and respiration, and internal carbon and nitrogen cycling.

represented as mixtures of plant functional types that change over time in response to disturbances such as fire, vegetation dynamics, and climate change.

One such biogeochemical model is the Carnegie–Ames–Stanford Approach (CASA) model (Potter et al. 1993; Randerson et al. 1996). Net primary production is related to light-use efficiency, adjusted for temperature and soil moisture stress, solar radiation, and the fraction of photosynthetically active radiation absorbed by the canopy using Eq. (17.7). This latter term is derived from satellite measurements of vegetation greenness such as the normalized difference vegetation index. Plant biomass is updated for net primary production and associated litterfall, soil carbon pools, nutrient mineralization, and nutrient allocation are calculated. The water balance is also simulated because of its control of net primary production and other ecological processes.

Another class of biogeochemical models also links carbon fluxes with the water balance and nutrient cycling, but uses more mechanistic representations of net primary production and

its environmental controls. Such global-scale ecosystem models include the CENTURY model (Parton et al. 1987, 1988, 1993), the BIOME-BGC model (Running and Coughlan 1988; Running and Gower 1991; Running and Hunt 1993; Thornton et al. 2002), the Terrestrial Ecosystem Model (Raich et al. 1991; McGuire et al. 1992; Melillo et al. 1993), and the CASACNP model (Wang et al. 2010). Physiologically based ecosystem process models simulate net primary production, decomposition, and nutrient availability in relation to leaf area index, biome type, and site conditions. These models include light, temperature, water, and nutrient limitations to net primary production, allocate carbon to grow foliage, stems, and roots, decompose litter, and mineralize nutrients (Figure 24.20). The models can simulate leaf phenology based on prevailing environmental conditions. Typical phenologies are: evergreen, in which plants maintain foliage throughout the year; summergreen, in which leaves are present during the warm season; and raingreen, in which foliage emerges during the rainy season and drops in the dry season.

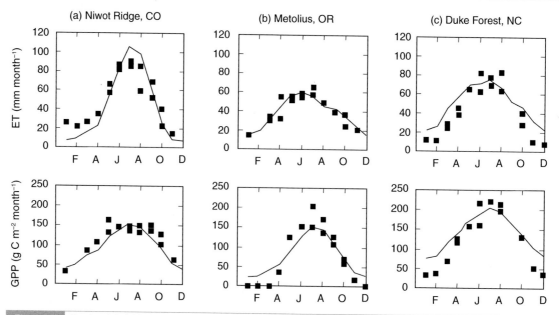

Fig. 24.21 Simulated (solid line) and observed (square) monthly evapotranspiration (top) and gross primary production (bottom) for three needleleaf evergreen forest sites in the United States. Adapted from Thornton et al. (2002).

Ecosystem models are routinely tested in their ability to simulate observed water and carbon fluxes. For example, Thornton et al. (2002) compared the BIOME-BGC model with eddy covariance measurements of evapotranspiration and carbon flux at several needleleaf evergreen forest sites in the United States. These sites included: Niwot Ridge, a 95-year-old subalpine conifer forest in Colorado with annual mean temperature of 2.1°C and annual precipitation of 808 mm; Metolius, a mixed-age old ponderosa pine forest in Oregon (8.2°C, 1251 mm); and Duke Forest, a 17-year-old loblolly pine forest in North Carolina (14.6°C, 1260 mm). Across this wide gradient in climate and stand age, the model reasonably captured monthly dynamics of evapotranspiration and gross primary production, though there are discrepancies for particular months (Figure 24.21).

Hanson et al. (2004) compared several ecosystems models with observations at the Walker Branch watershed in Tennessee for the period 1993–2000. Annual evapotranspiration was derived from two separate analyses. Mean annual evapotranspiration derived from the watershed budget of precipitation and runoff

was 613 mm per year during this period. Eddy covariance measurements provided a similar mean (601 mm yr^{-1}). Simulated mean annual evapotranspiration ranged from 463 to 801 mm yr^{-1} for nine models (Figure 24.22a). Mean annual net carbon storage derived from two biometric analyses was 218 g C m^{-2} yr^{-1}. The mean carbon uptake from eddy covariance measurements was considerably larger (648 g C m^{-2} yr^{-1}). Mean net ecosystem production simulated by six of the nine ecosystem models fell within this range of observational estimates (Figure 24.22b).

In contrast to biogeochemical models, vegetation dynamics models simulate population structure and community composition in addition to the carbon cycle and other biogeochemical cycles. The concept of gap dynamics has lead to a broad class of individual tree forest succession models called gap models (Botkin et al. 1972; Shugart and West 1977; Shugart 1984; Botkin 1993). In these models, changes in community composition, biomass, and productivity are the result of the birth, growth, and death of individual trees. Species differ in physiological tolerances of light and soil water, their utilization of these resources, and how they

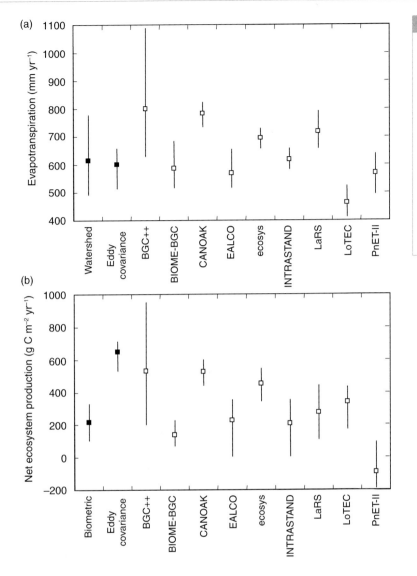

Fig. 24.22 Simulated and observed (a) annual evapotranspiration and (b) annual net ecosystem production for a broadleaf deciduous forest in the Walker Branch watershed near Oak Ridge, Tennessee. Data points are the mean (square) for the period 1993–2000. Also shown are the minimum and maximum annual fluxes (solid line). Observations are shown for two separate estimates. Model results are for nine ecosystem models. Data from Hanson et al. (2004).

alter resource availability. The result is a temporal dynamics of community composition and ecosystem development originating from the physiology and demography of individual trees. Subsequent generations of gap models included nutrient availability, other site conditions such as soil temperature and permafrost, and linkages between the biotic and abiotic environments (Pastor and Post 1986; Bonan 1990).

The strength of gap models is that they formalize the concept of succession as a population process resulting from plant demography and the different life histories of species operating in a gradient of environmental change. They have provided insights to community organization (e.g., Figure 19.4) and the response of forests to past climate change (Solomon et al. 1980, 1981; Bonan and Hayden 1990) and future climate change (Solomon 1986; Pastor and Post 1988). However, gap models require tens of thousands of individual trees to simulate a forest landscape and therefore are not applied globally.

Another type of vegetation dynamics model is known as dynamic global vegetation models (Prentice et al. 2007). These models also simulate changes in community composition, biomass, productivity, and nutrient cycling. Because these models are applied globally, they do not recognize individual species. Rather, they employ plant functional types, typically distinguished by woody or herbaceous biomass, broadleaf or needleleaf leaf form, and evergreen or deciduous leaf longevity. They do not formally recognize individual plants as in gap models. Instead, they represent cohorts of individuals based on similar size distribution or the model may represent separately an average individual plant and the density of plants.

One such dynamic global vegetation model is the Lund–Potsdam–Jena (LPJ) model (Sitch et al. 2003). This model characterizes vegetation as patches of plant functional types within a model grid cell. Each plant functional type is represented by an individual plant with the average biomass, crown area, height, and stem diameter of its population, by the number of individuals in the population, and by the fractional cover in the grid cell. Vegetation is updated in response to resource competition, allocation, mortality, biomass turnover, litterfall, establishment, and fire.

Bonan et al. (2003) used the ecological concepts of the LPJ model to develop a dynamic global vegetation model for use with climate models (Figure 25.13). The biogeography of trees in this variant of the LPJ model compares favorably with satellite observations of tree cover, and the simulated biogeography compares well with natural vegetation. The model also reproduces forest succession, as illustrated for Alaskan boreal forest (Figure 24.23). Grasses initially dominate the simulation, followed by a rapid decline in grasses and increase in deciduous trees. Deciduous trees attain peak abundance in less than 100 years, and then decline in cover as evergreen trees gain dominance. The conversion from primarily deciduous to evergreen forest occurs at 145 years, after which evergreen trees increase to 76 percent cover and deciduous trees decline to 15 percent. This simulated succession is very similar to the observed succession in interior Alaska (Figure 22.10b). Recurring fires preclude stands much older than 250 years in this region. The simulated net primary production at year 250 (386 g C m^{-2} yr^{-1}) is comparable to that observed in boreal forests of Alaska, as is vegetation biomass.

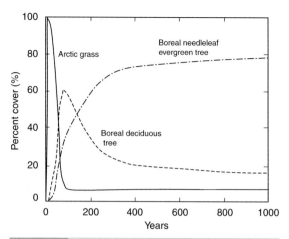

Fig. 24.23 Boreal forest stand dynamics simulated by a dynamic global vegetation model. The simulation is from initially bare ground for a single model grid cell in the boreal forest over 1000 years in the absence of disturbance. Percentage cover is the annual extent of plant functional types in the grid cell. Adapted from Bonan et al. (2003).

Many biogeochemical models and dynamic global vegetation models have been developed specifically to provide ecosystem feedback with climate change. As such, they are designed as part of a climate model and link biogeophysical, hydrologic, physiological, and biogeochemical processes into mechanistic representation of surface energy fluxes, photosynthesis, respiration, allocation, and the cycling of carbon and nutrients within ecosystems (Figure 25.2). Leaf phenology includes the timing of budburst, senescence, and leaf abscission in response to temperature and drought. Vegetation dynamics is based on net primary production, mortality, fire, and plant responses to disturbances.

Ecosystem models differ greatly in how they calculate environmental conditions such as soil water and temperature, how these conditions affect physiological processes, and how multiple resource limitations collectively affect net primary production and carbon allocation. Some models place great emphasis on physical control (e.g., light absorption, soil water, soil temperature) of ecosystem processes. Others emphasize biogeochemical controls such as nutrient availability. Plant migration presents a problem as most models limit seed availability based only on bioclimatic constraints (Van Minnen et al. 2000; Higgins and Harte 2006). As a result, simulated carbon uptake and storage can differ greatly among models. Better synthesis of observational data of net primary production (Scurlock and Olson 2002), leaf area index (Asner et al. 2003), vegetation carbon (Figure 24.9a), canopy height (Figure 24.9b), soil carbon (Figure 24.11), and carbon and energy fluxes (Figure 24.7) is needed to resolve discrepancies among global ecosystem models. Moreover, models must be comprehensively tested with a suite of data across a variety of spatial and temporal scales (Randerson et al. 2009; Luo et al. 2012).

24.7 | Review Questions

1. With reference to Figure 24.1, describe the type of natural vegetation likely to be found where annual precipitation is 750 mm and annual mean temperature is (a) 0°C, (b) 12°C, and (c) 25°C. What does this change in vegetation indicate?

2. Why does Budyko's radiative dryness index increase along a gradient from forest to grassland?

3. Based on the Holdridge classification, what type of natural vegetation is expected where annual mean biotemperature is 2°C and annual precipitation is 175 mm? What vegetation would be expected if the climate warmed by 3°C and annual precipitation increased by 200 mm?

4. Net primary production can be calculated as the minimum of the temperature and precipitation relationships in Figure 24.4. Calculate the annual precipitation below which net primary production is limited by precipitation for an annual mean air temperature of (a) 25°C, (b) 12°C, and (c) 0°C. How does this critical minimum annual precipitation vary with temperature? Why? With reference to Figure 24.1, determine the vegetation types in which net primary production is not limited by precipitation for these three values of annual mean temperature.

5. The following data from Aber et al. (1982) show mean canopy height for 13 deciduous forest stands in Wisconsin. Each stand is ranked by soil water from 1 (most xeric) to 13 (most mesic). What does this data indicate about canopy height in relation to soil water?

Soil water index	Height (m)	Soil water index	Height (m)
1	10	2	10
3	15	4	17
12	23	5	20

Soil water index	Height (m)	Soil water index	Height (m)
11	22	13	19
6	15	10	11
7	17	9	15
8	11		

6. A pollen record recovered from lake sediments contains 40 percent oak pollen. With reference to Figure 19.2, describe the climate at the time of pollen deposition. Another sample has less oak pollen and more northern pine pollen. How does this climate compare with that of the previous pollen sample?

7. With reference to Figure 24.14, contrast the climate at Anderson Pond, Tennessee, 16 kyr BP with that 10 kyr BP.

8. What is a primary distinguishing difference between biogeochemical models and dynamic global vegetation models?

9. In simulating the response of terrestrial ecosystems to climate change and possible feedbacks on climate, what timescales are appropriate for a biogeochemical model in contrast with a dynamic global vegetation model?

10. The LPJ dynamic global vegetation model simulates the average individual plant in a population as well as population density. Total plant carbon is the carbon of the average individual multiplied by population density. In contrast, the BIOME-BGC model simulates plant carbon pools without regard to population density. Why might this distinction between the LPJ and BIOME-BGC models be important for climate change research?

标记

24.8 | References

Aber, J. D., Pastor, J., and Melillo, J. M. (1982). Changes in forest canopy structure along a site quality gradient in southern Wisconsin. *American Midland Naturalist*, 108, 256–265.

Adair E. C., Parton, W. J., Del Grosso, S. J., et al. (2008). Simple three-pool model accurately describes patterns of long-term litter decomposition in diverse climates. *Global Change Biology*, 14, 2636–2660.

Asner, G. P., Scurlock, J. M. O., and Hicke, J. A. (2003). Global synthesis of leaf area index observations: Implications for ecological and remote sensing studies. *Global Ecology and Biogeography*, 12, 191–205.

Batjes, N. H. (1996). Total carbon and nitrogen in the soils of the world. *European Journal of Soil Science*, 47, 151–163.

Beer, C., Reichstein, M., Tomelleri, E., et al. (2010). Terrestrial gross carbon dioxide uptake: Global distribution and covariation with climate. *Science*, 329, 834–838.

Beerling, D. J., and Osborne, C. P. (2006). The origin of the savanna biome. *Global Change Biology*, 12, 2023–2031.

Bonan, G. B. (1990). Carbon and nitrogen cycling in North American boreal forests, I: Litter quality and soil thermal effects in interior Alaska. *Biogeochemistry*, 10, 1–28.

Bonan, G. B., and Hayden, B. P. (1990). Using a forest stand simulation model to examine the ecological and climatic significance of the late-Quaternary pine–spruce pollen zone in eastern Virginia, U.S.A. *Quaternary Research*, 33, 204–218.

Bonan, G. B., Levis, S., Sitch, S., Vertenstein, M., and Oleson, K. W. (2003). A dynamic global vegetation model for use with climate models: Concepts and description of simulated vegetation dynamics. *Global Change Biology*, 9, 1543–1566.

Botkin, D. B. (1993). *Forest Dynamics: An Ecological Model*. Oxford: Oxford University Press.

Botkin, D. B., Janak, J. F., and Wallis, J. R. (1972). Some ecological consequences of a computer model of forest growth. *Journal of Ecology*, 60, 849–872.

Budyko, M. I. (1974). *Climate and Life*. New York: Academic Press.

Budyko, M. I. (1986). *The Evolution of the Biosphere*. Dordrecht: Reidel.

Canadell, J., Jackson, R. B., Ehleringer, J. R., et al. (1996). Maximum rooting depth of vegetation types at the global scale. *Oecologia*, 108, 583–595.

Collatz, G. J., Berry, J. A., and Clark, J. S. (1998). Effects of climate and atmospheric CO_2 partial pressure on the global distribution of C_4 grasses: Present, past, and future. *Oecologia*, 114, 441–454.

Currie, W. S., Harmon, M. E., Burke, I. C., et al. (2010). Cross-biome transplants of plant litter show decomposition models extend to a broader climatic range but lose predictability at the decadal time scale. *Global Change Biology*, 16, 1744–1761.

Dai, A., and Fung, I. Y. (1993). Can climate variability contribute to the "missing" CO_2 sink? *Global Biogeochemical Cycles*, 7, 599–609.

Davis, M. B. (1981). Quaternary history and the stability of forest communities. In *Forest Succession: Concepts and Application*, ed. D. C. West, H. H. Shugart, and D. B. Botkin. New York: Springer-Verlag, pp. 132–153.

Davis, M. B., Woods, K. D., Webb, S. L., and Futyma, R. P. (1986). Dispersal versus climate: expansion of *Fagus* and *Tsuga* into the Upper Great Lakes region. *Vegetatio*, 67, 93–103.

Del Grosso, S., Parton, W., Stohlgren, T., et al. (2008). Global potential net primary production predicted from vegetation class, precipitation, and temperature. *Ecology*, 89, 2117–2126.

Edwards, E. J., Osborne, C. P., Strömberg, C. A. E., et al. (2010). The origins of C_4 grasslands: integrating evolutionary and ecosystem science. *Science*, 328, 587–591.

Ehleringer, J. R., Cerling, T. E., and Helliker, B. R. (1997). C_4 photosynthesis, atmospheric CO_2, and climate. *Oecologia*, 112, 285–299.

Esser, G. (1987). Sensitivity of global carbon pools and fluxes to human and potential climatic impacts. *Tellus B*, 39, 245–260.

Friedlingstein, P., Delire, C., Müller, J. F., and Gérard, J. C. (1992). The climate induced variation of the continental biosphere: A model simulation of the last glacial maximum. *Geophysical Research Letters*, 19, 897–900.

Gholz, H. L., Wedin, D. A., Smitherman, S. M., Harmon, M. E., and Parton, W. J. (2000). Long-term dynamics of pine and hardwood litter in contrasting environments: Toward a global model of decomposition. *Global Change Biology*, 6, 751–765.

Grier, C. C., and Running, S. W. (1977). Leaf area of mature northwestern coniferous forests: Relation to site water balance. *Ecology*, 58, 893–899.

Hanson, P. J., Amthor, J. S., Wullschleger, S. D., et al. (2004). Oak forest carbon and water simulations: Model intercomparisons and evaluations against independent data. *Ecological Monographs*, 74, 443–489.

Haxeltine, A., and Prentice, I. C. (1996). BIOME3: An equilibrium terrestrial biosphere model based on ecophysiological constraints, resource availability, and competition among plant functional types. *Global Biogeochemical Cycles*, 10, 693–709.

Higgins, P. A. T., and Harte, J. (2006). Biophysical and biogeochemical responses to climate change depend on dispersal and migration. *BioScience*, 56, 407–417.

Holdridge, L. R. (1967). *Life Zone Ecology*. San Jose, Costa Rica: Tropical Science Center.

Hugelius, G., Tarnocai, C., Broll, G., et al. (2013). The Northern Circumpolar Soil Carbon Database: Spatially distributed datasets of soil coverage and soil carbon storage in the northern permafrost regions. *Earth System Science Data*, 5, 3–13.

Huxman, T. E., Smith, M. D., Fay, P. A., et al. (2004). Convergence across biomes to a common rain-use efficiency. *Nature*, 429, 651–654.

Jackson, R. B., Canadell, J., Ehleringer, J. R., et al. (1996). A global analysis of root distributions for terrestrial biomes. *Oecologia*, 108, 389–411.

Jackson, R. B., Schenk, H. J., Jobbágy, E. G., et al. (2000). Belowground consequences of vegetation change and their treatment in models. *Ecological Applications*, 10, 470–483.

Jobbágy, E. G., and Jackson, R. B. (2000). The vertical distribution of soil organic carbon and its relation to climate and vegetation. *Ecological Applications*, 10, 423–436.

Jung, M., Reichstein, M., Margolis, H. A., et al. (2011). Global patterns of land–atmosphere fluxes of carbon dioxide, latent heat, and sensible heat derived from eddy covariance, satellite, and meteorological observations. *Journal of Geophysical Research*, 116, G00J07, doi:10.1029/2010JG001566.

Kaplan, J. O., Bigelow, N. H., Prentice, I. C., et al. (2003). Climate change and Arctic ecosystems, 2: Modeling, paleodata-model comparisons, and future projections. *Journal of Geophysical Research*, 108, 8171, doi:10.1029/2002JD002559.

Lieth, H. (1975). Modeling the primary productivity of the world. In *Primary Productivity of the Biosphere*, ed. H. Lieth and R. H. Whittaker. New York: Springer-Verlag, pp. 237–263.

Luo, Y. Q., Randerson, J. T., Abramowitz, G., et al. (2012). A framework for benchmarking land models. *Biogeosciences*, 9, 3857–3874.

Matthews, E. (1997). Global litter production, pools, and turnover times: Estimates from measurement

data and regression models. *Journal of Geophysical Research*, 102D, 18771–18800.

McGuire, A. D., Melillo, J. M., Joyce, L. A., et al. (1992). Interactions between carbon and nitrogen dynamics in estimating net primary productivity for potential vegetation in North America. *Global Biogeochemical Cycles*, 6, 101–124.

McLachlan, J. S., Clark, J. S., and Manos, P. S. (2005). Molecular indicators of tree migration capacity under rapid climate change. *Ecology*, 86, 2088–2098.

Meentemeyer, V. (1978). Macroclimate and lignin control of litter decomposition rates. *Ecology*, 59, 465–472.

Melillo, J. M., McGuire, A. D., Kicklighter, D. W., et al. (1993). Global climate change and terrestrial net primary production. *Nature*, 363, 234–240.

Monson, R. K., and Collatz, G. J. (2012). The ecophysiology and global biology of C_4 photosynthesis. In *Terrestrial Photosynthesis in a Changing Environment: A Molecular, Physiological and Ecological Approach*, ed. J. Flexas, F. Loreto, and H. Medrano. Cambridge: Cambridge University Press, pp. 54–70.

Nemani, R. R., and Running, S. W. (1989). Testing a theoretical climate–soil–leaf area hydrologic equilibrium of forests using satellite data and ecosystem simulation. *Agricultural and Forest Meteorology*, 44, 245–260.

Nemani, R. R., Keeling, C. D., Hashimoto, H., et al. (2003). Climate-driven increases in global terrestrial net primary production from 1982 to 1999. *Science*, 300, 1560–1563.

Niu, S., Luo, Y., Fei, S., et al. (2012). Thermal optimality of net ecosystem exchange of carbon dioxide and underlying mechanisms. *New Phytologist*, 194, 775–783.

Ordonez, A., and Williams, J. W. (2013). Climatic and biotic velocities for woody taxa distributions over the last 16 000 years in eastern North America. *Ecology Letters*, 16, 773–781.

Overpeck, J. T., Webb, R. S., and Webb, T., III (1992). Mapping eastern North American vegetation change of the past 18 ka: No-analogs and the future. *Geology*, 20, 1071–1074.

Pan, Y., Birdsey, R. A., Phillips, O. L., and Jackson, R. B. (2013). The structure, distribution, and biomass of the world's forests. *Annual Review of Ecology, Evolution, and Systematics*, 44, 593–622.

Parton, W. J., Schimel, D. S., Cole, C. V., and Ojima, D. S. (1987). Analysis of factors controlling soil organic matter levels in Great Plains grasslands. *Soil Science Society of America Journal*, 51, 1173–1179.

Parton, W. J., Stewart, J. W. B., and Cole, C. V. (1988). Dynamics of C, N, P and S in grassland soils: A model. *Biogeochemistry*, 5, 109–131.

Parton, W. J., Scurlock, J. M. O., Ojima, D. S., et al. (1993). Observations and modeling of biomass and soil organic matter dynamics for the grassland biome worldwide. *Global Biogeochemical Cycles*, 7, 785–809.

Pastor, J., and Post, W. M. (1986). Influence of climate, soil moisture, and succession on forest carbon and nitrogen cycles. *Biogeochemistry*, 2, 3–27.

Pastor, J., and Post, W. M. (1988). Response of northern forests to CO_2-induced climate change. *Nature*, 334, 55–58.

Poorter, H., Niklas, K. J., Reich, P. B., et al. (2012). Biomass allocation to leaves, stems and roots: Meta-analyses of interspecific variation and environmental control. *New Phytologist*, 193, 30–50.

Post, W. M., Emanuel, W. R., Zinke, P. J., and Stangenberger, A. G. (1982). Soil carbon pools and world life zones. *Nature*, 298, 156–159.

Post, W. M., King, A. W., and Wullschleger, S. D. (1997). Historical variations in terrestrial biospheric carbon storage. *Global Biogeochemical Cycles*, 11, 99–109.

Potter, C. S., Randerson, J. T., Field, C. B., et al. (1993). Terrestrial ecosystem production: A process model based on global satellite and surface data. *Global Biogeochemical Cycles*, 7, 811–841.

Prentice, I. C., Cramer, W., Harrison, S. P., et al. (1992). A global biome model based on plant physiology and dominance, soil properties and climate. *Journal of Biogeography*, 19, 117–134.

Prentice, I. C., Guiot, J., Huntley, B., Jolly, D., and Cheddadi, R. (1996). Reconstructing biomes from palaeoecological data: A general method and its application to European pollen data at 0 and 6 ka. *Climate Dynamics*, 12, 185–194.

Prentice, I. C., Harrison, S. P., Jolly, D., and Guiot, J. (1998). The climate and biomes of Europe at 6000 yr BP: Comparison of model simulations and pollen-based reconstructions. *Quaternary Science Reviews*, 17, 659–668.

Prentice, I. C., Jolly, D., and BIOME6000 (2000). Mid-Holocene and glacial-maximum vegetation geography of the northern continents and Africa. *Journal of Biogeography*, 27, 507–519.

Prentice, I. C., Farquhar, G. D., Fasham, M. J. R., et al. (2001). The carbon cycle and atmospheric carbon dioxide. In *Climate Change 2001: The Scientific Basis. Contribution of Working Group I to the Third Assessment Report of the Intergovernmental Panel on Climate Change*, ed. J. T. Houghton, Y. Ding, D. J. Griggs, et al. Cambridge: Cambridge University Press, pp. 183–237.

Prentice, I. C., Bondeau, A., Cramer, W., et al. (2007). Dynamic Global Vegetation Modeling: Quantifying terrestrial ecosystem responses to large-scale environmental change. In *Terrestrial Ecosystems in a Changing World*, ed. J. G. Canadell, D. E. Pataki, and L. F. Pitelka. Berlin: Springer, pp. 175–192.

Raich, J. W., Rastetter, E. B., Melillo, J. M., et al. (1991). Potential net primary productivity in South America: Application of a global model. *Ecological Applications*, 1, 399–429.

Randerson, J. T., Thompson, M. V., Malmstrom, C. M., Field, C. B., and Fung, I. Y. (1996). Substrate limitations for heterotrophs: Implications for models that estimate the seasonal cycle of atmospheric CO_2. *Global Biogeochemical Cycles*, 10, 585–602.

Randerson, J. T., Hoffman, F. M., Thornton, P. E., et al. (2009). Systematic assessment of terrestrial biogeochemistry in coupled climate–carbon models. *Global Change Biology*, 15, 2462–2484.

Rosenzweig, M. L. (1968). Net primary productivity of terrestrial communities: Prediction from climatological data. *American Naturalist*, 102, 67–74.

Running, S. W., and Coughlan, J. C. (1988). A general model of forest ecosystem processes for regional applications, I: Hydrological balance, canopy gas exchange and primary production processes. *Ecological Modelling*, 42, 125–154.

Running, S. W., and Gower, S. T. (1991). FOREST-BGC, a general model of forest ecosystem processes for regional applications, II: Dynamic carbon allocation and nitrogen budgets. *Tree Physiology*, 9, 147–160.

Running, S. W., and Hunt, E. R., Jr. (1993). Generalization of a forest ecosystem process model for other biomes, BIOME-BGC, and an application for global-scale models. In *Scaling Physiological Processes: Leaf to Globe*, ed. J. R. Ehleringer and C. B. Field. New York: Academic Press, pp. 141–158.

Running, S. W., Nemani, R. R., Heinsch, F. A., et al. (2004). A continuous satellite-derived measure of global terrestrial primary production. *BioScience*, 54, 547–560.

Saatchi, S. S., Harris, N. L., Brown, S., et al. (2011). Benchmark map of forest carbon stocks in tropical regions across three continents. *Proceedings of the National Academy of Sciences USA*, 108, 9899–9904.

Sage, R. F. (2004). The evolution of C_4 photosynthesis. *New Phytologist*, 161, 341–370.

Schenk, H. J., and Jackson, R. B. (2002). The global biogeography of roots. *Ecological Monographs*, 72, 311–328.

Scurlock, J. M. O., and Olson, R. J. (2002). Terrestrial net primary productivity – a brief history and a new worldwide database. *Environmental Reviews*, 10, 91–110.

Shugart, H. H. (1984). *A Theory of Forest Dynamics: The Ecological Implications of Forest Succession Models*. New York: Springer-Verlag.

Shugart, H. H., and West, D. C. (1977). Development of an Appalachian deciduous forest succession model and its application to assessment of the impact of the chestnut blight. *Journal of Environmental Management*, 5, 161–179.

Simard, M., Pinto, N., Fisher, J. B., and Baccini, A. (2011), Mapping forest canopy height globally with spaceborne lidar. *Journal of Geophysical Research*, 116, G04021, doi:10.1029/2011JG001708.

Sitch, S., Smith, B., Prentice, I. C., et al. (2003). Evaluation of ecosystem dynamics, plant geography and terrestrial carbon cycling in the LPJ dynamic global vegetation model. *Global Change Biology*, 9, 161–185.

Solomon, A. M. (1986). Transient response of forests to CO_2-induced climate change: Simulation modeling experiments in eastern North America. *Oecologia*, 68, 567–579.

Solomon, A. M., Delcourt, H. R., West, D. C., and Blasing, T. J. (1980). Testing a simulation model for reconstruction of prehistoric forest-stand dynamics. *Quaternary Research*, 14, 275–293.

Solomon, A. M., West, D. C., and Solomon, J. A. (1981). Simulating the role of climate change and species immigration in forest succession. In *Forest Succession: Concepts and Application*, ed. D. C. West, H. H. Shugart, and D. B. Botkin. New York: Springer-Verlag, pp. 154–177.

Still, C. J., Berry, J. A., Collatz, G. J., and DeFries, R. S. (2003). Global distribution of C_3 and C_4 vegetation: Carbon cycle implications. *Global Biogeochemical Cycles*, 17, 1006, doi:10.1029/2001GB001807.

Tarnocai, C., Canadell, J. G., Schuur, E. A. G., et al. (2009). Soil organic carbon pools in the northern circumpolar permafrost region. *Global Biogeochemical Cycles*, 23, GB2023, doi:10.1029/2008GB003327.

TEMPO (1996). Potential role of vegetation feedback in the climate sensitivity of high-latitude regions: A case study at 6000 years B.P. *Global Biogeochemical Cycles*, 10, 727–736.

Thornthwaite, C. W. (1948). An approach toward a rational classification of climate. *Geographical Review*, 38, 55–94.

Thornthwaite, C. W., and Mather, J. R. (1955). *The Water Balance*, Publications in Climatology Volume 8, Number 1. Centerton, New Jersey: Drexel Institute of Technology.

Thornthwaite, C. W., and Mather, J. R. (1957). *Instructions and Tables for Computing Potential Evapotranspiration and the Water Balance*, Publications in Climatology Volume 10, Number 3. Centerton, New Jersey: Drexel Institute of Technology.

Thornton, P. E., Law, B. E., Gholz, H. L., et al. (2002). Modeling and measuring the effects of disturbance history and climate on carbon and water budgets in evergreen needleleaf forests. *Agricultural and Forest Meteorology*, 113, 185–222.

Thurner, M., Beer, C., Santoro, M., et al. (2014). Carbon stock and density of northern boreal and temperate forests. *Global Ecology and Biogeography*, 23, 297–310.

Todd-Brown, K. E. O., Randerson, J. T., Post, W. M., et al. (2013). Causes of variation in soil carbon simulations from CMIP5 Earth system models and comparison with observations. *Biogeosciences*, 10, 1717–1736.

Trofymow, J. A., Moore, T. R., Titus, B., et al. (2002). Rates of litter decomposition over 6 years in Canadian forests: Influence of litter quality and climate. *Canadian Journal of Forest Research*, 32, 789–804.

Van Minnen, J. G., Leemans, R., and Ihle, F. (2000). Defining the importance of including transient ecosystem responses to simulate C-cycle dynamics in a global change model. *Global Change Biology*, 6, 595–611.

Vogt, K. A., Grier, C. C., and Vogt, D. J. (1986). Production, turnover, and nutrient dynamics of above- and belowground detritus of world forests. *Advances in Ecological Research*, 15, 303–377.

Wang, Y. P., Law, R. M., and Pak, B. (2010). A global model of carbon, nitrogen and phosphorus cycles for the terrestrial biosphere. *Biogeosciences*, 7, 2261–2282.

Webb, T., III, Bartlein, P. J., Harrison, S. P., and Anderson, K. H. (1993). Vegetation, lake levels, and climate in eastern North America for the past 18,000 years. In *Global Climates since the Last Glacial Maximum*, ed. H. E. Wright, Jr., J. E. Kutzbach, T. Webb, III, et al. Minneapolis: University of Minnesota Press, pp. 415–467.

Whittaker, R. H. (1975). *Communities and Ecosystems*, 2nd ed. New York: MacMillan.

Wieder, W. R., Bonan, G. B., and Allison, S. D. (2013). Global soil carbon projections are improved by modelling microbial processes. *Nature Climate Change*, 3, 909–912.

Williams, J. W., Shuman, B. N., and Webb, T., III (2001). Dissimilarity analyses of Late-Quaternary vegetation and climate in eastern North America. *Ecology*, 82, 3346-3362.

Williams, J. W., Shuman, B. N., Webb, T., III, Bartlein, P. J., and Leduc, P. L. (2004). Late-Quaternary vegetation dynamics in North America: Scaling from taxa to biomes. *Ecological Monographs*, 74, 309-334.

Williams, J. W., Jackson, S. T., and Kutzbach, J. E. (2007). Projected distributions of novel and disappearing climates by 2100 AD. *Proceedings of the National Academy of Sciences USA*, 104, 5738-5742.

Woodward, F. I. (1987). *Climate and Plant Distribution*. Cambridge: Cambridge University Press.

Woodward, F. I. (1993). Leaf responses to the environment and extrapolation to larger scales. In *Vegetation Dynamics and Global Change*, ed. A. M. Solomon and H. H. Shugart. New York: Chapman and Hall, pp. 71-100.

Yi, C., Ricciuto, D., Li, R., et al. (2010). Climate control of terrestrial carbon exchange across biomes and continents. *Environmental Research Letters*, 5, 034007, doi:10.1088/1748-9326/5/3/034007.

Zhao, M., Heinsch, F. A., Nemani, R. R., and Running, S. W. (2005). Improvements of the MODIS terrestrial gross and net primary production global data set. *Remote Sensing of Environment*, 95, 164-176.

Part VI

Terrestrial Forcings and Feedbacks

Terrestrial Ecosystems and Earth System Models

25.1 | Chapter Summary

Much of our understanding of how land surface processes and terrestrial ecosystems affect weather, climate, and atmospheric composition comes from numerical models of surface energy fluxes, the hydrologic cycle, and biogeochemical cycles coupled to atmospheric models. Land surface models are coupled to atmospheric models to simulate the absorption of radiation at the land surface, the exchanges of sensible and latent heat between land and atmosphere, storage of heat in soil, and the frictional drag of vegetation and other surface elements on wind. These models were initially developed to provide the surface boundary conditions of radiative and turbulent fluxes required by atmospheric models. They have since evolved to simulate the hydrologic cycle, biogeochemical cycles, and vegetation dynamics so that the land and atmosphere are represented as a coupled system. This chapter reviews the historical development of land surface models. Model evaluation is discussed, as well as application of the models in climate model experiments.

25.2 | Hydrometeorological Models

Global climate models represent a set of numerical equations that describe the large-scale circulation of the atmosphere and ocean and their physical state, including interactions among oceans, atmosphere, land, and sea ice that affect climate. The land surface fluxes of energy, moisture, and momentum and the associated hydrologic cycle that regulates them have long been represented in global climate models. In these models, absorption of radiation at the surface, the reflection of solar radiation and emission of longwave radiation, sensible and latent heat fluxes, storage of heat in soil, and frictional drag of the surface on wind influence climate. The land surface models used with climate models provide these biogeophysical boundary conditions at the land–atmosphere interface. They partition net radiation at the surface into sensible and latent heat fluxes, soil heat storage, and snow melt. They also partition precipitation into runoff, evaporation, and water storage in snow or soil. The most recent versions of these models simulate biogeochemical cycles (e.g., carbon), wildfires, land use, and land-cover change. The models update the state variables (snow cover, soil moisture, soil temperature, vegetation cover, leaf area index, and carbon pools) that regulate surface fluxes with the atmosphere.

25.2.1 First Generation Models

The first generation of land surface models used aerodynamic bulk transfer equations and simple prescriptions of albedo, surface roughness, and soil water without explicitly representing vegetation or the hydrologic cycle (e.g., Manabe et al. 1965; Williamson et al. 1987). In such

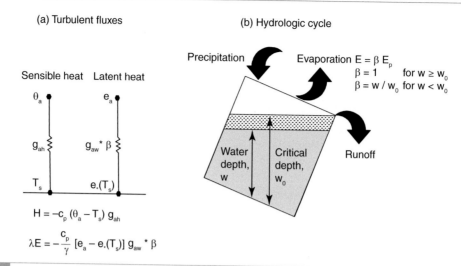

(a) Turbulent fluxes

(b) Hydrologic cycle

models, the energy balance at the land surface is represented by:

$$(1 - r)S\downarrow + \varepsilon L\downarrow = \varepsilon\sigma T_s^4 + H + \lambda E + G \qquad (25.1)$$

The left-hand side of this equation constitutes the radiative forcing, which is the sum of absorbed solar radiation and absorbed longwave radiation, respectively, where $S\downarrow$ is incoming solar radiation, $L\downarrow$ is incoming atmospheric longwave radiation, r is surface albedo, and ε is surface emissivity (Chapter 12). The right-hand side of the equation consists of emitted longwave radiation, sensible heat flux (H), latent heat flux (λE), and heat storage in soil by conduction (G). These fluxes vary with surface temperature (T_s) so that the equation is solved for the temperature that balances the energy budget.

First generation models provided simplified representations of energy exchange at the land surface. Surface albedos were specified from datasets, ignoring radiative transfer within plant canopies. Some models neglected soil heat storage, assuming zero heat capacity. A primary defining characteristic of this class of models was that turbulent fluxes were based on a bulk aerodynamic formulation. The models parameterized sensible and latent heat fluxes in terms of a single aerodynamic conductance between the surface and atmosphere (Figure 25.1a).

The models ignored the influence of vegetation on turbulent fluxes, and the complexity of latent heat flux was represented not by stomata and canopy conductance, but rather through a soil wetness factor (β). This is a dimensionless factor scaled from 0 (dry) to 1 (wet) that adjusts potential latent heat flux for the extent that soil water limits evapotranspiration (Figure 25.1a). Initially, this was a specified parameter provided by a dataset so that soil moisture and the hydrologic cycle on land were not simulated (e.g., Manabe et al. 1965; Williamson et al. 1987). In some early models, the land surface was assumed to be wet everywhere (i.e., $\beta = 1$); others specified $\beta \leq 1$ to represent arid and semiarid landscapes. Application of these models demonstrated the importance of evapotranspiration in regulating global climate (Shukla and Mintz 1982).

When the hydrologic cycle on land was included, it was treated in a simplified manner. Manabe (1969) used a bucket model to represent water storage in soil (Figure 25.1b). The soil is treated as a bucket with a maximum water-holding capacity (w_0), which is taken to be 75 percent of field capacity. Field capacity is set to 15 cm of water everywhere. Precipitation fills the bucket up to the depth w_0. Evapotranspiration depletes the bucket, and the change in soil water is the difference between precipitation and evapotranspiration. The ratio of soil water (w) to maximum water-holding capacity (w_0) determines the decrease in evapotranspiration as the soil dries out:

$$\beta = \begin{cases} 1 & \text{for} \quad w \geq w_0 \\ w/w_0 & \text{for} \quad w < w_0 \end{cases} \qquad (25.2)$$

There is no runoff for $w < w_0$. Thereafter, precipitation in excess of evapotranspiration runs off.

25.2.2 Second Generation Models

Deardorff (1978) developed the essential characteristics of the second generation of land surface models. His model included a single-layer parameterization of the plant canopy and distinguished foliage and ground fluxes (Figure 17.13c). It partitioned latent heat flux into separate fluxes of evaporation of intercepted water, evaporation from soil, and transpiration and introduced stomatal conductance, which in his model included the effects of light and soil water. The model also included a two-layer soil, using an approach known as the force–restore method to simulate diurnal and seasonal variation in ground heat flux and soil temperature. This is called force–restore because the forcing of soil temperature by soil heat flux is modified by a restoring term associated with deep soil temperature. In the absence of soil heat flux, the upper soil temperature is restored over time to that of the deep soil temperature. Two soil layers differentiate near-surface soil temperature, which responds to the diurnal cycle, from deeper soil, which responds to the annual cycle of temperature. A similar force–restore approach simulates soil water in the upper and deep soil layers. Near-surface soil water responds rapidly to the diurnal cycle and wetting events while deeper soil water responds at a longer timescale.

By the mid-1980s, development of second generation models to include the hydrologic cycle and the effects of vegetation and soil on energy and water fluxes was well underway (Dickinson 1983). The Biosphere–Atmosphere Transfer Scheme (BATS) (Dickinson et al. 1986, 1993) and the Simple Biosphere Model (SiB) (Sellers et al. 1986) epitomize this class of

models. Processes represented in these models include (Figure 25.2a): radiative transfer in the plant canopy that distinguishes the visible and near-infrared wavebands and direct beam and diffuse radiation; absorption and emission of longwave radiation; momentum transfer arising from vegetated canopies, including turbulence within the canopy; sensible heat exchange from foliage and soil; latent heat exchange from evaporation of intercepted water, soil evaporation, and transpiration; the control of transpiration by stomata; and heat transfer in multilayered soil.

One of the main characteristics of second generation models is their representation of the biological controls of evapotranspiration. Water intercepted by plant canopies evaporates at a rate regulated by the amount of intercepted water and canopy aerodynamics. Soil water and turbulent transfer at the ground surface regulate evaporation from soil. In contrast, stomata regulate transpiration so that transpiration responds to a distinctly different set of environmental factors. In models such as BATS and SiB, stomatal conductance is modeled using an approach similar to Jarvis (1976), in which stomatal conductance varies in response to photosynthetically active radiation, temperature, vapor pressure deficit, foliage water potential, and ambient CO_2 concentration through empirical relationships, given by Eq. (16.11).

The hydrologic cycle is represented in terms of interception, throughfall, stemflow, infiltration, runoff, soil water, snow, evaporation, and transpiration (Figure 25.2b). Snow accumulation and melt is modeled explicitly, typically in a multilayered snow pack. Principles of soil physics are used to simulate vertical gradients in soil water, typically using the Richards equation or some variant, given by Eq. (9.8).

Second generation land surface models require that the surface be described in terms of its vegetation and soil. The models represent the morphology of vegetation in terms of horizontal area, represented by fractional vegetation cover, and height, leaf area index, and root profile (Figure 25.3). Because surface energy fluxes and the hydrologic cycle differ between vegetation and soil, the models allow

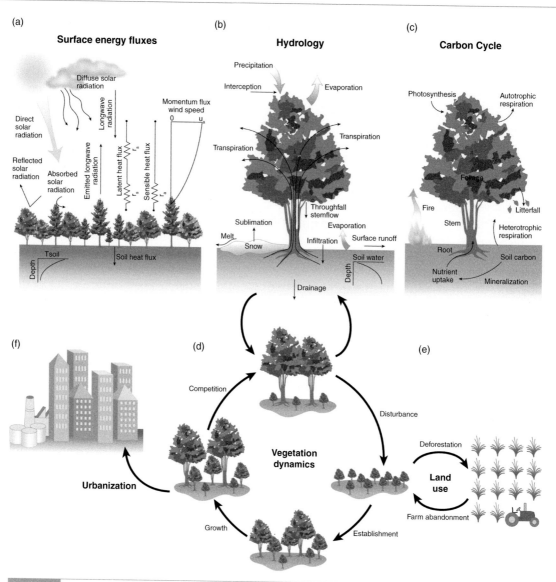

Fig. 25.2 Physical processes by which land affects climate and which are represented in the land surface models used with climate models include (a) surface energy fluxes and (b) the hydrologic cycle. The current generation of Earth system models additionally include biogeochemical and ecosystem processes governing (c) the carbon cycle and (d) vegetation dynamics. Some models also include (e) land use and (f) urbanization to represent human alteration of the biosphere. Adapted from Bonan (2008).

for vegetated and unvegetated patches within a grid cell. The land surface is characterized by vegetation types, or biomes, that vary in important morphological and physiological properties (Tables 25.1 and 25.2). The soil is characterized by texture classes that determine thermal properties such as heat capacity and thermal conductivity (Table 9.1) and hydraulic properties such as porosity, hydraulic conductivity, and suction (Table 9.2). Model experiments demonstrated biogeophysical regulation of climate by vegetation, e.g., through studies of tropical deforestation (Dickinson and Henderson–Sellers 1988).

Table 25.1 Land cover types used in the Simple Biosphere Model (SiB2) and the Biosphere-Atmosphere Transfer Scheme (BATS)

SiB2	BATS
Broadleaf evergreen forest	Broadleaf evergreen forest
Broadleaf deciduous forest	Broadleaf deciduous forest
Broad- and needleleaf forest	Mixed woodland
Needleleaf evergreen forest	Needleleaf evergreen forest
Needleleaf deciduous forest	Needleleaf deciduous forest
Broadleaf shrubland	Evergreen shrubland
Dwarf trees and shrubs	Deciduous shrubland
C_3 grassland/ agriculture	Tall grassland
C_4 grassland	Short grassland
	Tundra
	Desert
	Semidesert
	Crop/mixed farming
	Irrigated crop
	Inland water
	Bog/marsh
	Ice cap/glacier

Source: From Sellers et al. (1996c) and Dickinson et al. (1993).

25.2.3 Third Generation Models

The third generation of models directly linked photosynthesis and stomatal conductance. The Farquhar–von Caemmerer–Berry photosynthesis model, given by Eq. (16.3), and the Ball–Berry stomatal conductance model, given by Eq. (16.14), were introduced into the land component of climate models in the mid-1990s (Bonan 1995; Sellers et al. 1996c; Cox et al. 1999). The quintessential experiment with third generation models was to investigate the reduction in stomatal conductance with a doubling of atmospheric CO_2 and its effect on climate (Sellers et al. 1996a; Bounoua et al. 1999). It is now common to distinguish the direct radiative effects of

atmospheric CO_2 from the indirect physiological effects associated with stomatal conductance. The inclusion of photosynthesis also allowed climate models to simulate atmospheric CO_2 (Bonan 1995; Denning et al. 1995, 1996a,b; Craig et al. 1998) and identified feedbacks from ozone and stomata (Sitch et al. 2007) and photosynthetic enhancement by diffuse radiation (Mercado et al. 2009).

The photosynthesis–stomatal conductance model requires a more detailed understanding of plant physiology than that of second generation land surface models. Important model parameters are: $V_{c\,max}$, the maximum rate of carboxylation; J_{max}, the maximum potential rate of electron transport; g_1, an empirical coefficient that relates photosynthesis and stomatal conductance; and g_0, minimum stomatal conductance (Chapter 16). Leaf trait databases of $V_{c\,max}$ (Table 16.3) and scaling relationships such as that between $V_{c\,max}$ and J_{max} (Figure 16.5) can be used to infer parameters for various plant functional types. However, values of $V_{c\,max}$ vary greatly among models and remain poorly constrained (Rogers 2014).

Of particular importance to these models is the integration of leaf photosynthesis over the plant canopy. In contrast to the simple approach of multiplying leaf conductance by leaf area index to obtain canopy conductance, third generation models employ detailed scaling parameterizations based on vertical profiles of light and leaf nitrogen in the plant canopy (Chapter 17). One approach is to partition the canopy into sunlit and shaded leaves with decreased photosynthetic capacity with depth in the canopy based on profiles of nitrogen using Eq. (17.4) and Eq. (17.11), so that the sunlit canopy has higher $V_{c\,max}$ than the shaded canopy, from Eq. (17.12) and Eq. (17.13).

25.2.4 Land Cover Representation

Land surface models require as input a spatial depiction of vegetation on the model's gridded domain. Associated with each land cover type are specified properties such as albedo (or leaf and stem optical properties), stomatal conductance, and surface roughness. Leaf area index and fractional vegetation cover are also

Table 25.2 Vegetation and land cover parameters in the Biosphere–Atmosphere Transfer Scheme (BATS)

Land cover	FV_{max}	LAI (m² m⁻²)		z_0 (m)	Albedo		g_{smax} (mol m⁻² m⁻²)
		Max	Min		VIS	NIR	
Broadleaf evergreen forest	0.9	6	5	2.0	0.04	0.20	0.28
Broadleaf deciduous forest	0.8	6	1	0.8	0.08	0.28	0.21
Mixed woodland	0.8	6	3	0.8	0.06	0.24	0.21
Needleleaf evergreen forest	0.8	6	5	1.0	0.05	0.23	0.21
Needleleaf deciduous forest	0.8	6	1	1.0	0.05	0.23	0.21
Evergreen shrubland	0.8	6	5	0.1	0.05	0.23	0.21
Deciduous shrubland	0.8	6	1	0.1	0.08	0.28	0.21
Tall grassland	0.8	6	0.5	0.1	0.08	0.30	0.21
Short grassland	0.8	2	0.5	0.02	0.10	0.30	0.21
Tundra	0.6	6	0.5	0.04	0.10	0.30	0.21
Desert	0.0	0	0	0.05	0.20	0.40	0.21
Semidesert	0.1	6	0.5	0.10	0.17	0.34	0.21
Crop/mixed farming	0.85	6	0.5	0.06	0.10	0.30	0.35
Irrigated crop	0.8	6	0.5	0.06	0.08	0.28	0.21
Inland water	0.0	0	0	0.0024	0.07	0.20	—
Bog/marsh	0.8	6	0.5	0.03	0.06	0.18	0.21
Ice cap/glacier	0.0	0	0	0.01	0.80	0.60	—

Note: FV_{max}, maximum fractional vegetation cover; LAI, maximum and minimum leaf area index; z_0, roughness length; g_{smax}, maximum stomatal conductance converted to mol m⁻² s⁻¹ using $\rho_m = 42.3$ mol m⁻³. Albedos are for the visible (VIS) and near-infrared (NIR) wavebands.

Source: From Dickinson et al. (1993).

required. In early generation models, these vegetation parameters were obtained from syntheses of ecological literature. For example, the BATS model represents the land surface with 17 land cover types (Table 25.1), and look-up tables associate the vegetation with a suite of parameter values (Table 25.2). The SiB model utilizes twelve land cover types, later reduced to nine in SiB2 (Table 25.1).

The characterization of surface heterogeneity by land surface models has evolved with the development of the models. Early land surface models represented a model grid cell with a single homogenous vegetation type, taken as the dominant vegetation type. Mixtures of different vegetation types were accommodated by assuming vegetation is homogenously mixed over the grid cell with effective parameters for mixed vegetation.

Subsequently, land surface models explicitly recognized surface heterogeneity in vegetation and soil within a model grid cell. One approach

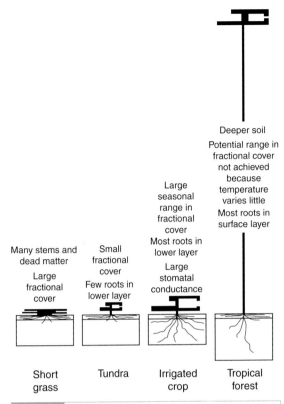

Fig. 25.3 Schematic representation of the morphology of vegetation in the Biosphere–Atmosphere Transfer Scheme. Reproduced from Dickinson et al. (1993).

25.3 | Satellite Land Data Products

The development of land surface models proceeded in tandem with that of land data products obtained from satellite data. Early generation land surface models used preexisting maps and atlases to obtain the required vegetation geography (Matthews 1983; Olson et al. 1983; Wilson and Henderson-Sellers 1985; Dorman and Sellers 1989). Now, datasets of land cover, leaf area index, and fractional vegetation cover derived from satellite data products are used to represent vegetation.

Beginning in the mid-1990s, global datasets derived from satellite remote sensing products became available for use with the models. One such dataset is land cover classification. The models require specification of land cover type (Table 25.1). The first datasets were at a coarse spatial resolution of one degree (DeFries and Townshend 1994) and later at resolutions of 8-km (DeFries et al. 1998) and 1-km (Loveland et al. 2000; Hansen et al. 2000; Friedl et al. 2002). At the same time, global datasets of leaf area index and FPAR (fraction of absorbed photosynthetically active radiation) became available at spatial resolutions of 1° (Sellers et al. 1994, 1996b), 0.5° (Nemani et al. 1996), and 8-km (Myneni et al. 1997; Buermann et al. 2002). The second version of the SiB model was specifically developed to utilize these satellite datasets (Sellers et al. 1996c), and they were introduced in climate models with good results (Chase et al. 1996; Randall et al. 1996; Bounoua et al. 2000; Buermann et al. 2001).

Land surface models specify the fractional area of the grid cell covered with vegetation to represent the different energy fluxes and hydrologic cycles of vegetation and bare soil. The concept of fractional vegetation cover was introduced by Deardorff (1978) and included in second generation models such as BATS and SiB. Initially, values were prescribed by biome (Dickinson et al. 1986, 1993; Dorman and Sellers 1989). Global datasets of fractional vegetation cover have since been developed at 1-km resolution (DeFries et al. 1999; Zeng et al. 2000), 8-km (Zeng et al. 2003), and 0.15° (Gutman and

uses a statistical technique to represent surface heterogeneity (Entekhabi and Eagleson 1989; Avissar 1992; Li and Avissar 1994). Probability density functions are derived for the variables used in the model equations. Grid average surface fluxes are calculated by applying these probability density functions in the solution of the model equations. Alternatively, the mosaic approach represents surface heterogeneity by dividing a model grid cell into a number of smaller patches of homogenous vegetation or soil (Avissar and Pielke 1989; Koster and Suarez 1992a,b; Seth et al. 1994; Essery et al. 2003; Melton and Arora 2014). Each patch is individually modeled and has its own surface climate. The grid cell average is the weighted average of the individual patches. The mosaic approach is now routinely used to represent surface heterogeneity in land surface models.

Ignatov 1998). Including these datasets in land surface models improved climate simulation (Barlage and Zeng 2004).

The representation of sparse or partial canopy cover in land surface models is problematic. Early models such as BATS and SiB allowed vegetated fraction and leaf area index to vary seasonally, though the distinction between the two is not clear. Furthermore, the concept of fractional vegetation cover is not necessarily consistent between satellite products and models. Forest canopies can have small-scale gaps on the order of 100–1000 m², but also large openings on the order of 10–100 km² or more related to large-scale disturbances such as fire. The approach used to model surface energy fluxes and the hydrologic cycle differs greatly between these two extremes. The former landscape can be adequately modeled through a big-leaf representation of a homogenous forest canopy with a fractional vegetation cover of one and a global leaf area index that is also equal to the local leaf area index. The latter landscape is better represented not as a homogenous sparse canopy, but rather as a patch of dense vegetation covering the vegetated fraction of the grid cell and another patch of bare soil covering the remainder of the grid cell. The local leaf area index of the vegetated patch is greater than the global leaf are index of the grid cell. Similarly, widely spaced shrubland or desert scrub vegetation may be best modeled as separate vegetated and bare patches. The fractional vegetation cover product derived from satellite data does not distinguish between these two different characterizations of plant canopies (Price 1992), but mostly measures canopy sparseness, as seen, for example, in open shrubland land cover.

Land surface models have expanded beyond their hydrometeorological roots to include photosynthesis, stomatal physiology, and the carbon cycle. Mechanistic modeling of leaf physiology and the carbon cycle poses a challenge in mixed life form biomes, where the various plant types can differ greatly in photosynthesis, phenology, carbon allocation, and other physiological traits. Biomes that mix physiologically and morphologically distinct plant types are inconsistent with the leaf physiological and whole-plant allocation parameterizations needed to represent ecological processes in the models. One solution used in ecological models is to represent vegetation as mixtures of plant functional types. A similar approach can be used in land surface models.

Remotely sensed vegetation continuous fields provide an alternative to discrete land cover classification, especially for mixed vegetation types (DeFries et al. 1995, 1997, 1999, 2000a,b; Hansen and DeFries 2004). This approach describes the land surface in terms of the pixel's fractional cover of different plant life forms. Some products characterize tree cover at 1-km spatial resolution, which is further distinguished by leaf type (needleleaf or broadleaf) and leaf longevity (evergreen or deciduous) (DeFries et al. 1999, 2000a,b). Other products recognize tree cover, bare ground, and non-tree (herbaceous) vegetation at 500-m resolution (Hansen et al. 2002, 2003).

Global datasets of vegetation continuous fields allow a representation of the land surface as continuous mixtures of plant functional types rather than as discrete biomes. Vegetation continuous fields facilitate subdivision of a model grid cell into patches of distinct plant functional types. One such approach uses seven primary plant functional types: needleleaf evergreen or deciduous tree, broadleaf evergreen or deciduous tree, shrub, grass, and crop (Bonan et al. 2002). Arctic, boreal, temperate, and tropical trees, C_3 and C_4 grasses, and evergreen and deciduous shrubs are physiological variants of the seven primary functional types derived from biogeography rules. However, the procedures by which to unmix subpixel mosaics of plant types and leaf area index from grid cell satellite products are problematic (Lawrence and Chase 2007). Moreover, plant functional types are themselves discrete representations of continuous variation in plant traits (Chapter 18). New techniques are needed to represent this continuous trait variation (Wang et al. 2012; Verheijen et al. 2013; Reich et al. 2014).

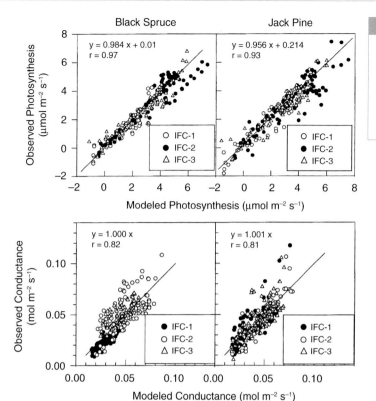

Fig. 25.4 Comparison between modeled and measured photosynthesis (top) and stomatal conductance (bottom) for branch samples of black spruce and jack pine trees. Data were collected during three intensive field campaigns (IFCs). Model parameters were derived for each IFC. Adapted from Dang et al. (1998).

25.4 | Model Evaluation

The physiological, biogeochemical, and hydro-meteorological processes simulated by land surface models can be evaluated with a variety of data across scales from leaf to canopy to global. For example, simulated photosynthesis and stomatal conductance can be compared with leaf measurements, as illustrated in a study by Dang et al. (1998). The leaf model is similar to the photosynthesis–conductance model described in Chapter 16. Data were collected for black spruce and jack pine trees during three field campaigns over the course of the growing season and were used to estimate required model parameters. Figure 25.4 compares observed and modeled leaf photosynthesis in response to photosynthetically active radiation (0–1450 μmol m^{-2} s^{-1}), leaf temperature (–5°C to 35°C), and CO_2 concentration (50–900 ppm). Modeled photosynthesis closely matches observed photosynthesis, as does stomatal conductance.

Eddy covariance flux tower measurements at specific locations provide data with which to test models at the canopy scale. Initially (during the early 1990s), such observations extended for only short (a few weeks or less) field campaigns. For example, Bonan et al. (1997) compared measured fluxes of net radiation, sensible heat, latent heat, and CO_2 for a jack pine forest, obtained during the summer of 1994, with modeled fluxes. In this case, the land surface model was run uncoupled from an atmospheric model, instead forced with the observed meteorology at the tower site. When averaged over a 23-day period, the model reasonably replicates the diurnal cycle (Figure 25.5). The most notable discrepancy with observations is that the model fails to capture the observed midday depression in latent heat flux.

The development of the Community Land Model (CLM) from version 3 (Dickinson et al. 2006) to version 3.5 (Lawrence et al. 2007) to version 4 (Lawrence et al. 2011, 2012) illustrates the evaluation of models with flux tower data.

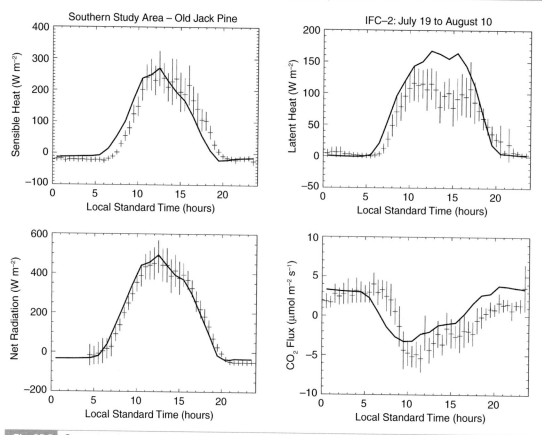

Fig. 25.5 Comparison between modeled and measured sensible heat flux, latent heat flux, net radiation, and CO_2 flux for a jack pine stand during IFC-2 (July 19–August 10, 1994) of the Boreal Ecosystem Atmosphere Study (BOREAS). See Figure 12.7 for additional information. Shown is the average diurnal cycle. Crosses show observed fluxes ± 2 standard error of the mean. The solid line is the average modeled fluxes. Reproduced from Bonan et al. (1997).

Figure 25.6 shows a model–data evaluation using observed hourly net radiation, sensible heat flux, and latent heat flux for a tropical rainforest in Amazonia collected over several weeks during 1993. The CLM3 simulations overestimate sensible heat flux and underestimate latent heat flux compared with the observations. Changes to the parameterization of latent heat flux in CLM3.5 give better agreement with the observations.

Eddy covariance flux tower measurements have become routine, with multiyear observations at many sites organized into coordinated observational networks. Long-term observations collected in a variety of biomes can critically guide model development and evaluation, illustrated by Stöckli et al. (2008) for CLM3 and CLM3.5. Measurements made in a temperate deciduous forest show that CLM3 has low soil moisture compared with observations, resulting in low latent heat flux and high sensible heat flux, especially during the summer growing season (Figure 25.7). Improvements to the parameterization of infiltration, runoff, soil evaporation, and groundwater in CLM3.5 produce wetter soil, higher latent heat flux, and lower sensible heat flux, which better matches observations. Further analyses for 15 flux tower sites from temperate, Mediterranean, tropical, boreal, and subalpine climate zones and with multiple years of data showed the generality of these improvements.

With the advent of global flux datasets derived from individual tower sites (Jung et al. 2011), the models can be confronted with data of biosphere functioning at the global scale

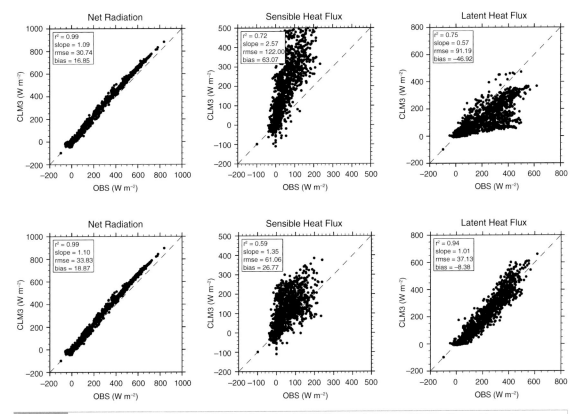

Fig. 25.6 Comparison of hourly observed (OBS) and simulated (CLM3) net radiation, sensible heat flux, and latent heat flux for tropical rainforest in southwestern Amazonia for the period April 4–July 26, 1993. Observed fluxes are from the Anglo–Brazilian Amazonian Climate Observation Study (ABRACOS). See Figure 12.5 for additional information. Modeled fluxes are for the Community Land Model version 3 (top) and a modified version of the model (bottom). Data provided courtesy of Keith Oleson and David Lawrence (National Center for Atmospheric Research, Boulder, Colorado).

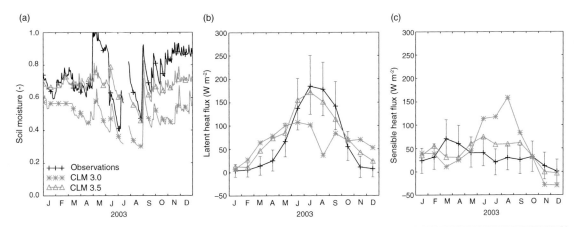

Fig. 25.7 Model simulations (gray lines) compared with observations (black lines) for Morgan Monroe State Forest, Indiana, during 2003. Shown are (a) soil moisture relative to saturation at 30 cm depth, (b) monthly latent heat flux, and (c) monthly sensible heat flux. Error bars show estimated uncertainties of observed fluxes. The gray lines show simulations using version 3.0 and version 3.5 of the Community Land Model. Adapted from Stöckli et al. (2008).

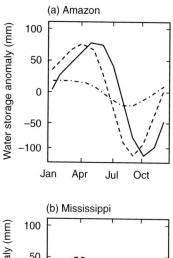

(a) Amazon

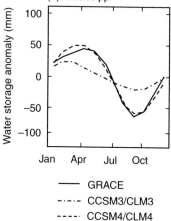

(b) Mississippi

—— GRACE
–·–·– CCSM3/CLM3
– – – – CCSM4/CLM4

Fig. 25.8 Monthly soil water storage from GRACE and from simulations with version 3 and version 4 of the Community Land Model for (a) the Amazon basin and (b) the Mississippi basin. Adapted from Lawrence et al. (2012).

(Figure 24.7). Eddy covariance measurements of sensible heat flux, latent heat flux, and CO_2 flux are routinely used to evaluate models and to inform model development at individual or multiple tower sites (Randerson et al. 2009; Blyth et al. 2011; Lawrence et al. 2011; Wang et al. 2011) and globally (Bonan et al. 2011, 2012).

The simulated hydrologic cycle can be evaluated at a variety of scales, as illustrated in the development of the Community Land Model. Flux tower analyses reveal deficiencies in simulated soil moisture and low latent heat flux for tropical forest (Figure 25.6) and temperate deciduous forest (Figure 25.7) with CLM3. Comparisons with estimates of seasonal

variation in water storage from the Gravity Recovery and Climate Experiment (GRACE) highlight model performance at the basin scale. Climate model simulations using CLM3 have weak seasonal change compared with the observations for the Mississippi and Amazon basins, while simulations using CLM4 better match the observations with a pronounced annual cycle (Figure 25.8). River flow is an integrator of the hydrologic cycle over large regions. Models can be compared with observed river flow to test how well they simulate the large-scale hydrologic cycle. Figure 25.9 shows such a comparison for the 50 largest rivers simulated using the Community Land Model. Although the model can deviate significantly from observations for particular river basins, it reproduces the general features of the observations. In these climate simulations, deviations from the observations reflect precipitation biases in the climate model and/or biases in the partitioning of precipitation into evapotranspiration and runoff.

The surface climate simulated by climate models can be compared with observations to verify how well the model reproduces the observed climatology. Figure 25.10 shows such a comparison for eastern Canada as simulated by one model in comparison with other versions of the model. The models reproduce with reasonable agreement the annual cycle of temperature, precipitation, runoff, and snow depth.

The Community Land Model development illustrates the use of observations at multiple spatial and temporal scales to guide model development. Comparison among multiple models is also an important means to assess and advance the state of knowledge. The Project for Intercomparison of Land–surface Parameterization Schemes (PILPS) compared simulated energy and water fluxes of numerous land surface models (Henderson-Sellers et al. 1996). When run uncoupled from an atmospheric model and forced with meteorological data over the course of one or more years, the various land surface models give quite different surface energy fluxes and hydrologic cycles despite identical atmospheric forcing and similar depictions of vegetation and soil texture. Initial model comparisons used synthetic

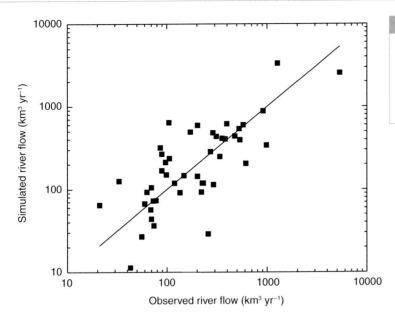

Fig. 25.9 Observed annual river flow for the 50 largest rivers and simulated annual river flow. The solid line shows the 1:1 relationship. Data provided courtesy of Keith Oleson (National Center for Atmospheric Research, Boulder, Colorado).

meteorological data for a tropical forest and a mid-latitude grassland (Pitman et al. 1999). Partitioning of net radiation between sensible and latent heat and partitioning of precipitation between evapotranspiration and runoff, key processes in the models, differed greatly among models. Interactions between evapotranspiration and runoff determine much of the differences among models, because these set the maximum attainable soil moisture (Koster and Milly 1997).

Subsequent studies used observed meteorology for particular sites and compared model simulations with observations. One such site was the Cabauw, a grassland in the Netherlands. At this location, the representation of stomata and the effects of soil moisture stress on evapotranspiration are important distinguishing characteristics determining differences among models in annual energy and water fluxes (Chen et al. 1997; Qu et al. 1998). Soil moisture data from the HAPEX-MOBILHY study in southern France revealed large differences among models in simulated soil moisture, as well as evapotranspiration, runoff, and other fluxes (Shao and Henderson-Sellers 1996a,b). Differences among models in the parameterization of soil evaporation were particularly important (Desborough et al. 1996). Simulations for Valdai,

a seasonally snow covered grassland meadow in northern Russia, showed that the models can reproduce the general features of snow hydrology, including the annual cycle and interannual variability, but different formulations of snow processes, especially ablation, produce large differences among models (Schlosser et al. 2000; Slater et al. 2001). Of particular importance in this seasonally frozen soil is the representation of phase change (Luo et al. 2003). Release of energy as soil water freezes prevents soil from becoming extremely cold. Simulated soil temperatures were colder in land surface models that did not include phase change than in those that included phase change.

Extensions of these single point simulations to catchments showed that land surface models can capture the general energy balance and hydrology, especially riverflow, of large watersheds. Such studies include the 566,000 km² Red–Arkansas River basin in the southern Great Plains region of the United States (Liang et al. 1998; Lohmann et al. 1998; Wood et al. 1998), the 58,000 km² Torne–Kalix River system in northern Scandinavia (Bowling et al. 2003a,b; Nijssen et al. 2003), and the 86,000 km² Rhône River basin in France (Boone et al. 2004). These model comparisons reveal that the land surface models used with climate models

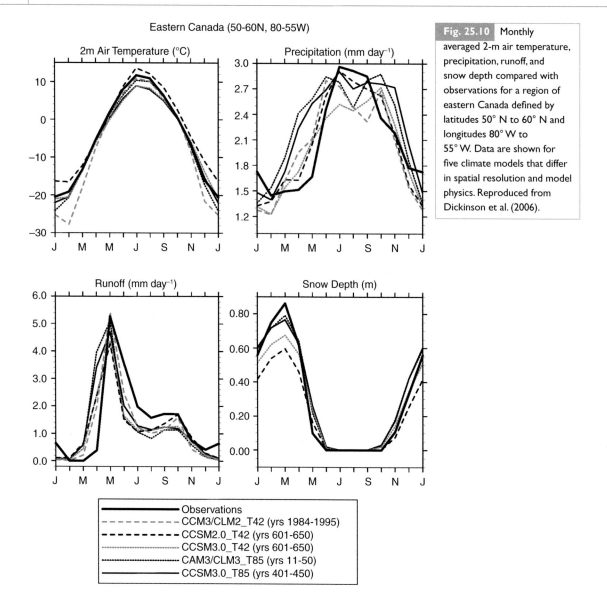

Fig. 25.10 Monthly averaged 2-m air temperature, precipitation, runoff, and snow depth compared with observations for a region of eastern Canada defined by latitudes 50° N to 60° N and longitudes 80° W to 55° W. Data are shown for five climate models that differ in spatial resolution and model physics. Reproduced from Dickinson et al. (2006).

can differ greatly in how they characterize vegetation and parameterize surface energy fluxes and the hydrologic cycle. Nonetheless, there is a distinction among first, second, and third generation models in which the simple early models perform poorly compared with later models (Henderson-Sellers et al. 2003; Pitman 2003).

There is still much to learn about how to represent hydrometeorological processes in land surface models. A comparison of seven models, for example, reveals substantial differences in the response of temperature and evapotranspiration to a commonly imposed land-cover change (Pitman et al. 2009; Boiser et al. 2012; de Noblet-Ducoudré et al. 2012). A broader comparison among 15 models also shows divergent temperature and evapotranspiration responses among models to land-cover change (Kumar et al. 2013). Models also differ in key aspects of the hydrologic cycle (Dirmeyer 2011) and produce different hydrologic cycles, even when coupled to the same atmospheric model (Wei et al. 2010).

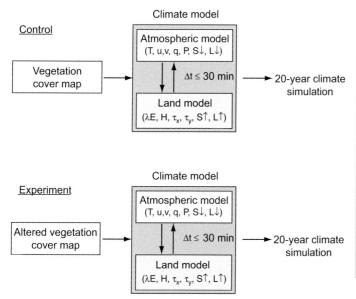

Control

Climate model

Vegetation cover map → Atmospheric model (T, u,v, q, P, $S\downarrow$, $L\downarrow$)

$\Delta t \le 30$ min

Land model (λE, H, τ_x, τ_y, $S\uparrow$, $L\uparrow$)

→ 20-year climate simulation

Experiment

Climate model

Altered vegetation cover map → Atmospheric model (T, u,v, q, P, $S\downarrow$, $L\downarrow$)

$\Delta t \le 30$ min

Land model (λE, H, τ_x, τ_y, $S\uparrow$, $L\uparrow$)

→ 20-year climate simulation

Fig. 25.11 Paired climate model simulations to examine the influence of altered vegetation cover on climate. The atmosphere model provides temperature (T), wind (u, v), humidity (q), precipitation (P), solar radiation ($S\downarrow$), and longwave radiation ($L\downarrow$) to the land model. The land model returns the surface fluxes of latent heat (λE), sensible heat (H), momentum (τ_x, τ_y), reflected solar radiation ($S\uparrow$), and emitted longwave radiation ($L\uparrow$). The coupled land–atmosphere model is integrated for many model years with a time step typically less than 30 minutes. The climate effect of altered vegetation cover is the difference between the experiment climate and the control climate (experiment – control).

25.5 | Land–Atmosphere Coupling Experiments

Paired climate model simulations are performed to examine the impact of land surface processes on climate (Figure 25.11). Such changes in the land surface may be introduced through a new dataset of, for example, leaf area index or land cover or through alternative parameterizations of a specific process such as stomatal conductance or soil water. For example, one simulation might simulate climate with the current vegetation cover. A second simulation might simulate climate with an altered vegetation cover. The difference between the two simulated climates is the effect of the altered vegetation on climate. Atmospheric models have short time steps to solve the numerical equations, typically 30 minutes or less for coarse spatial resolution global climate models and even less for high spatial resolution numerical weather prediction models. Studies of numerical weather predication typically integrate the model for a period of a several days or weeks. For climate simulations, the model is integrated for many model years (e.g., 20 years or more) to establish the impact of the land surface on the simulated climatology.

Climate model simulations have routinely demonstrated the sensitivity of climate to soil water and snow at the seasonal-to-interannual timescale (Chapter 26). Numerous studies have demonstrated the importance of vegetation, particularly leaf area index, surface roughness, rooting depth, and stomatal conductance, in determining regional and global climate. Changes in land cover through natural vegetation dynamics or through human uses of land alter these surface characteristics and have been shown to affect climate. Particular areas of research include the greening of desert landscapes in response to increased rainfall and changes in the boreal forest–tundra ecotone in response to climate change (Chapter 27). Other studies consider human influences on climate through tropical deforestation, overgrazing of dryland vegetation, and clearing of temperate forests for cropland (Chapter 28).

25.6 | Earth System Models

Models of the terrestrial biosphere have been coupled to global climate models to investigate the effects of terrestrial ecosystems on climate. One initial approach, known as asynchronous equilibrium coupling, took advantage of the

(a) Asynchronous coupling

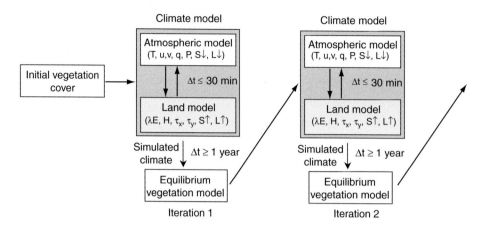

(b) Synchronous coupling

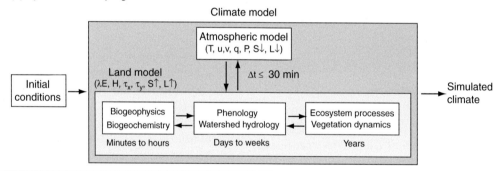

Fig. 25.12 Methods for including dynamic vegetation in climate models. (a) Asynchronous equilibrium coupling. (b) Integrated synchronous coupling. Adapted from Foley et al. (1998).

relationships between climate and biogeography to interactively change vegetation cover (Figure 25.12a). Climate is simulated with an initial vegetation cover. This climate is then used in a biogeography model to simulate the geographic distribution of vegetation. The vegetation, in turn, is used as input to the climate model to obtain a new climate. Climate and vegetation are iterated in this manner several times until a stable solution is obtained. One of the first such couplings used the Holdridge vegetation model shown in Figure 24.3 (Henderson-Sellers 1993; Henderson-Sellers and McGuffie 1995).

A better coupling integrates the long time-scale of vegetation dynamics with the short timescale physiological and hydrometeorological processes (Figure 25.12b). The development of global biogeochemical models and dynamic

global vegetation models in the late-1990s (Chapter 24) and their coupling to land surface models (Foley et al. 1996, 1998, 2000) allowed conceptual advances in biogeochemistry, vegetation dynamics, and biogeography to be directly integrated into models of the coupled biosphere–atmosphere system. This approach integrates the traditional focus on biogeophysics and hydrometeorology found in land surface models with biogeochemical models to simulate the carbon cycle in relation to climate. Dynamic global vegetation models additionally simulate plant community composition and vegetation dynamics. In these models, climate determines energy exchange, water availability, and the productivity and geography of terrestrial ecosystems. In turn, the type of plants (e.g., tree or grass, needleleaf or broadleaf, evergreen or

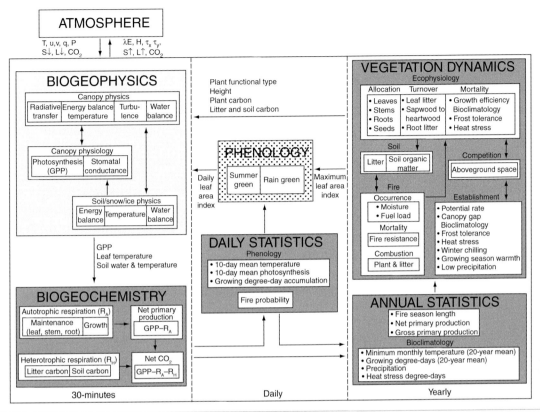

Fig. 25.13 Scope of a dynamic global vegetation model for use with climate models illustrating the linkages among biogeophysics, biogeochemistry, and vegetation dynamics. The lightly shaded biogeophysical processes represent the traditional hydrometeorological scope of land surface models. The darker boxes represent the greening of land surface models with the introduction of dynamic vegetation. Reproduced from Bonan et al. (2003).

deciduous), biomass, and leaf area influence exchanges of energy, water, momentum, and CO_2 with the atmosphere. Initial model experiments demonstrated biogeophysical feedbacks from coupled climate–vegetation dynamics in the Arctic, where expansion of trees into tundra decreases surface albedo, and in North Africa, where expansion of vegetation into desert similarly lowers albedo (Chapter 27). Other studies demonstrated biogeochemical climate feedbacks from the carbon cycle (Chapter 29).

Figure 25.13 illustrates the scope of one such model (Bonan et al. 2003), using concepts from the Lund–Potsdam–Jena (LPJ) dynamic global vegetation model (Chapter 24). Three timescales (minutes, days, and years) govern processes. Energy, water, momentum, and CO_2 are exchanged between land and atmosphere over periods of minutes to hours through short timescale biogeophysical, biogeochemical, and physiological processes. Canopy physics, soil physics, and plant physiology are interdependent and determine surface fluxes and microclimate. They also regulate biogeochemical processes such as CO_2 fluxes. The short-term biogeophysics and biogeochemistry are linked to leaf phenology, which responds to changes in temperature and soil water over periods of days to weeks. Changes in vegetation composition and structure occur over periods of years or longer in relation to gross primary production and respiration, allocation of net primary production to grow foliage, stem, and root biomass, and mortality as a result of low growth rate or fire. The growth and success of particular plant functional types are dependent on life history patterns such as evergreen and deciduous phenology, needleleaf and broadleaf foliage, C_3 and C_4 photosynthetic pathway, and temperature and precipitation preferences for biogeography. Growth and allocation are linked to soil biogeochemistry through litterfall, decomposition, and nitrogen availability. Development of the current generation of models for climate simulations continues to incorporate theoretical advances in plant demography, ecosystem dynamics, and community organization (Fisher et al. 2010; Scheiter et al. 2013). A particular challenge is to represent continuous variation in leaf traits within a plant functional type and trait variation in relation to climate (Wang et al. 2012; Verheijen et al. 2013; Reich et al. 2014).

The models are being expanded to better represent anthropogenic processes that alter the land surface. Urban land cover parameterizations simulate the effects of cities on surface energy fluxes and the hydrologic cycle and the impact of climate change on urban climates (Masson 2000; Masson et al. 2002; Best 2005; Best et al. 2006; Oleson et al. 2008, 2011; Chen et al. 2011; Grimmond et al. 2010, 2011; Demuzere et al. 2013). Crop models simulate the growth, development, and harvesting of different crop types in relation to prevailing meteorology and management practices including fertilization and irrigation (Kucharik 2003; Gervois et al. 2004, 2008; Bondeau et al. 2007; Osborne et al. 2007, 2009; Levis et al. 2012; Drewniak et al. 2013). Other model developments emphasize the effects of wood harvesting and land use on the carbon cycle (Shevliakova et al. 2009). Inclusion of these human influences on climate is part of the evolution of models of Earth's physical climate to models of the Earth system.

A current model frontier is full representation of the biogeochemical processes by which terrestrial ecosystems affect climate through various reactive gases (e.g., biogenic volatile organic compounds, nitrogen emissions, methane, ozone, and secondary organic aerosols). Global wildfire models are necessary because of the effects of fires on biogeochemical cycles and vegetation dynamics (Thonicke et al. 2001, 2010; Arora and Boer 2005; Kloster et al. 2010; Li et al. 2012, 2013), as well as chemical emissions to the atmosphere (Wiedinmyer et al. 2011). The entrainment of dust into the atmosphere is a large source of aerosols that alters the radiative balance of the atmosphere. Models of dust mobilization have been developed for inclusion in climate models (Tegen et al. 1992; Tegen and Fung 1994; Marticorena and Bergametti 1995; Zender et al. 2003; Mahowald et al. 2006). High concentrations of tropospheric ozone damage stomata and decrease plant productivity, but the exact way to parameterize this in models is uncertain (Sitch et al. 2007; Wittig et al. 2007, 2009; Ainsworth et al. 2012; Lombardozzi et al. 2013). Empirical models of biogenic volatile organic compound emissions of isoprene, monoterpenes, and other compounds (Guenther et al. 1995, 2006, 2012) have been adapted to land surface models (Levis et al. 2003; Heald et al. 2008), as have other photosynthetically-based isoprene emissions models (Arneth et al. 2011; Pacifico et al. 2011; Unger et al. 2013).

Global models of the terrestrial carbon cycle initially excluded associated biogeochemical cycles (e.g., nitrogen) that regulate the carbon cycle. An active model frontier is coupled carbon–nitrogen biogeochemistry (Thornton et al. 2007; Gerber et al. 2010; Zaehle and Friend 2010; Wang et al. 2010; Wania et al. 2012). These models of the terrestrial carbon and nitrogen cycles simulate carbon and nitrogen flows among various vegetation and soil components, nitrogen inputs for atmospheric deposition and biological nitrogen fixation, and nitrogen losses from denitrification and leaching. With the addition of nitrogen gas emissions, the models simulate the net radiative forcing of reactive nitrogen (Zaehle et al. 2011). An emerging frontier is to additionally include phosphorus (Zhang et al. 2011, 2014; Goll et al. 2012; Yang et al. 2014).

Other areas of active research include: simulation of wetlands (Wania et al. 2009, 2013; Kleinen et al. 2012; Ringeval et al. 2012) and methane emission (Gedney et al. 2004; Wania et al. 2010, 2013; Riley et al. 2011; Meng et al. 2012); better representation of permafrost and permafrost carbon (Koven et al. 2009, 2011; Swenson et al. 2012); vertically resolved profiles of soil carbon (Koven et al. 2013); and water and carbon isotopes (Riley et al. 2002, 2003; Risi et al. 2010).

Another development is to represent watershed processes by implementing the model for catchments. In the traditional approach, a rectangular grid that varies in longitude and latitude is overlain on the land surface and the model is implemented on that spatial grid. Hydrologic processes, however, may be better represented by watersheds rather than grid boxes. The catchment approach represents the land surface by a network of irregular hydrologic basins. It defines the catchment as the fundamental computational unit, typically represented at a finer spatial scale than that of atmospheric models. Only a few such land surface models using irregularly shaped watersheds have been developed (Ducharne et al. 1999, 2000; Koster et al. 2000; Chen and Kumar 2001; Tesfa et al. 2014).

The expanding multidisciplinary scientific breadth of the models is part of the growth of the atmospheric sciences towards Earth system science. Indeed, the ability to simulate biological and biogeochemical feedbacks is one of the defining aspects of the evolution of climate models to Earth system models. Moreover, although the models are designed for coupling with atmospheric models and specifically simulate terrestrial feedbacks with the atmosphere, an emerging frontier is to apply land models for climate change impacts, adaptation, and mitigation research. The models provide an integrated framework to assess physical, chemical, and biological responses to the multitude of anthropogenic perturbations in the Earth system, including climate change, CO_2, nitrogen deposition, ozone, aerosols, and land use and land-cover change. Underlying this research is the recognition that Earth's ecosystems and their coupling with the atmosphere are critical elements of global planetary change and planetary habitability.

With greater model complexity comes additional challenges and opportunities for model evaluation. The models can be tested for their simulation of carbon and water fluxes at annual timescales (e.g., Figures 24.21 and 24.22), but biogeochemical processes and ecosystem states must also be evaluated across multiple spatial

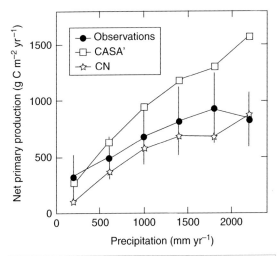

Fig. 25.14 Simulated net primary production for two biogeochemical models (CASA' and CN) coupled to the Community Land Model compared with observations. Net primary production is shown in relation to annual precipitation. Vertical bars show observational uncertainty. From Randerson et al. (2009).

and temporal scales (Randerson et al. 2009; Luo et al. 2012). Carbon cycle simulations must be assessed for long timescale (decadal to centennial) whole-plant physiological processes (e.g., net primary production, carbon allocation, litterfall), demographic processes (e.g., mortality), and biogeochemical processes (e.g., litter decomposition, soil organic matter formation). The terrestrial carbon cycle and its feedback with climate are routinely assessed in transient simulations over the twentieth century forced with reconstructed meteorology (Sitch et al. 2008; Le Quéré et al. 2009) or in coupled carbon cycle–climate simulations (Friedlingstein et al. 2006, 2014; Anav et al. 2013). The models must also be tested for their functional response to environmental gradients to ascertain process-level functioning. For example, Figure 25.14 compares net primary productivity simulated by two biogeochemical models coupled to the Community Land Model in response to precipitation. Other model evaluations compare with the annual cycle of atmospheric CO_2 (Randerson et al. 2009; Cadule et al. 2010), interannual variability (Cox et al. 2013; Piao et al. 2013; Wenzel et al. 2014), or response

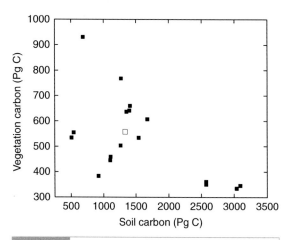

Fig. 25.15 Global vegetation and soil carbon simulated by 18 Earth system models for present-day. The open symbol shows observational estimates. Adapted from Anav et al. (2013).

to large-scale drought (Reichstein et al. 2007). Comparison with ecosystem experiments such as CO_2 enrichment (De Kauwe et al. 2013, 2014; Piao et al. 2013; Zaehle et al. 2014) and nitrogen addition (Thomas et al. 2013a,b) test model responses to perturbations in the Earth system.

Model intercomparison is an essential means to assess uncertainty in carbon cycle simulations. Earth system models differ greatly in their depiction of the present-day carbon cycle (Anav et al. 2013; Todd-Brown et al. 2013). Global soil carbon estimates vary among models by a factor of six, vegetation carbon by a factor of three, and many models greatly underestimate or overestimate global carbon pools (Figure 25.15). A variety of mathematical techniques, generally known as model–data fusion, are available to improve model predictions by estimating model parameters that best fit the observations or by assimilating observations to improve the model states; to quantify uncertainty in model output arising from parameter uncertainty; and to

identify key data deficiencies and model development needs (Zhou and Luo 2008; Wang et al. 2009; Williams et al. 2009; Dietze et al. 2014; Hararuk et al. 2014).

Scaling from individual sites to regions and the globe is a key requirement to develop and evaluate biogeochemical cycles in Earth system models, and data synthesis across scales is essential to this evaluation. For example, comparisons of models and global soil carbon data products (e.g., Figures 24.11 and 25.15) are necessary to diagnose model deficiencies. Such comparisons reveal substantial mismatches between model and observed soil carbon and highlight differences among models in soil carbon turnover (Todd-Brown et al. 2013). However, processed based studies are needed to improve the soil carbon parameterizations in Earth system models. Improvements to soil carbon parameterizations require evaluation with long-term litter decomposition experiments across large biotic and climatic gradients (e.g., Figures 21.5 and 21.6), as shown by Bonan et al. (2013). Model parameterizations must be evaluated in terms of both site-specific litter decomposition experiments and global soil carbon databases (Wieder et al. 2014). Similarly, parameterizations of gross primary production must be evaluated at the leaf-scale using leaf trait databases, at the canopy-scale using eddy covariance flux data across flux networks (e.g., Figures 12.3 and 20.20), and at the global-scale using empirically upscaled flux data products (e.g., Figure 24.7). For example, the photosynthetic parameter $V_{c\,max}$ varies greatly among models (Rogers 2014), but observations are available from leaf trait databases (Table 16.3). Comparisons across leaf, canopy, and global scales provide important constraints on Earth system models and a powerful means to assess models for their consistency with process-level knowledge across scales (Bonan et al. 2011, 2012).

25.7 | Review Questions

1. Which generation of land surface models (first generation with a bucket model or a second generation model) is more likely to capture the diurnal drying of the upper soil? Is this a necessary feature to simulate?

2. An early generation land surface model uses prescribed leaf area index obtained from satellite. A subsequent version of the model simulates leaf area. A simulation of present-day climate is not improved by the prognostic leaf model, but nor is it degraded. Discuss whether this development enhances the climate model.

3. Equation (17.14) is the Penman–Monteith equation for latent heat flux. Compare the latent heat flux for the following model grid cell when (a) the Penman–Monteith equation is evaluated directly for each individual tile to obtain the weighted average and (b) the Penman–Monteith equation is evaluated using weighted average values for g_{ah} and g_c. Why do these flux estimates differ? Use $R_n - G = 400$ W m^{-2}; $e_*(T_a) - e_a = 1000$ Pa; $s = 189$ Pa K^{-1}; $c_p = 29.2$ J mol^{-1} K^{-1}; and $\gamma = 66.5$ Pa K^{-1}.

Fractional area	g_{ah} (mol m^{-2} s^{-1})	g_c (mol m^{-2} s^{-1})
0.10	2.1	8
0.30	1.7	1
0.25	2.4	4
0.35	1.4	0.8

4. A flux tower site measured net radiation, sensible heat flux, latent heat flux, and associated meteorological forcing data over the course of a year for a tropical broadleaf evergreen forest. Soil temperature and soil water were not measured. Data are available at 30-minute time periods. Monthly leaf area index is available, as is soil texture. Describe how this data can be used to evaluate a land surface model. How is the model initialized?

5. Measurements indicate the amount of global runoff has increased over the past several decades. You hypothesize that this is related to elevated CO_2

and stomatal conductance. Describe a set of climate model simulations to test this hypothesis. Discuss the merits of the Jarvis-style model used in BATS and that used in SiB2. What would be gained by including a dynamic global vegetation model?

6. Describe two conceptual limitations to asynchronous equilibrium coupling. How does integrated synchronous coupling solve these problems?

7. How does the simulation of ecosystem feedbacks with climate change differ between biogeochemical models and dynamic global vegetation models?

8. Devise a set of simulations to test the effects of climate change, CO_2 enrichment, and nitrogen deposition on the global carbon cycle over the twentieth century. How would you account for synergistic effects among these forcings?

9. Describe an approach to evaluate the simulated terrestrial carbon cycle across spatial and temporal scales, from leaf to global and from near-instantaneous to multi-decadal. What datasets can be used in this evaluation?

10. You want to add isoprene emissions to a land surface model for air quality research. One parameterization uses an empirical model based on temperature, solar radiation, and other factors. The other parameterization mechanistically links isoprene emissions to photosynthesis. Discuss the merits of each approach.

11. The development of models of the terrestrial biosphere is marked by increasing complexity. Discuss the pros and cons of this.

12. Will models ever fully replicate the complexity of the biosphere?

25.8 | References

Ainsworth, E. A., Yendrek, C. R., Sitch, S., Collins, W. J., and Emberson, L. D. (2012). The effects of tropospheric ozone on net primary productivity and implications for climate change. *Annual Review of Plant Biology*, 63, 637–661.

Anav, A., Friedlingstein, P., Kidston, M., et al. (2013). Evaluating the land and ocean components of the global carbon cycle in the CMIP5 Earth system models. *Journal of Climate*, 26, 6801–6843.

Arneth, A., Schurgers, G., Lathiere, J., et al. (2011). Global terrestrial isoprene emission models: Sensitivity to variability in climate and vegetation. *Atmospheric Chemistry and Physics*, 11, 8037–8052.

Arora, V. K., and Boer, G. J. (2005). Fire as an interactive component of dynamic vegetation models. *Journal of Geophysical Research*, 110, G02008, doi:10.1029/2005JG000042.

Avissar, R. (1992). Conceptual aspects of a statistical-dynamical approach to represent landscape subgrid-scale heterogeneities in atmospheric models. *Journal of Geophysical Research*, 97D, 2729–2742.

Avissar, R., and Pielke, R. A. (1989). A parameterization of heterogeneous land surfaces for atmospheric numerical models and its impact on regional meteorology. *Monthly Weather Review*, 117, 2113–2136.

Barlage, M., and Zeng, X. (2004). The effects of observed fractional vegetation cover on the land surface climatology of the Community Land Model. *Journal of Hydrometeorology*, 5, 823–830.

Best, M. J. (2005). Representing urban areas within operational numerical weather prediction models. *Boundary-Layer Meteorology*, 114, 91–109.

Best, M. J., Grimmond, C. S. B., and Villani, M. G. (2006). Evaluation of the urban tile in MOSES using surface energy balance observations. *Boundary-Layer Meteorology*, 118, 503–525.

Blyth, E., Clark, D. B., Ellis, R., et al. (2011). A comprehensive set of benchmark tests for a land surface model of simultaneous fluxes of water and carbon at both the global and seasonal scale. *Geoscientific Model Development*, 4, 255–269.

Boisier, J. P., de Noblet-Ducoudré, N., Pitman, A. J., et al. (2012). Attributing the impacts of land-cover changes in temperate regions on surface temperature and heat fluxes to specific causes: Results from the first LUCID set of simulations. *Journal of Geophysical Research*, 117, D12116, doi:10.1029/2011JD017106.

Bonan, G. B. (1995). Land–atmosphere CO_2 exchange simulated by a land surface process model coupled to an atmospheric general circulation model. *Journal of Geophysical Research*, 100D, 2817–2831.

Bonan, G. B. (2008). Forests and climate change: Forcings, feedbacks, and the climate benefits of forests. *Science*, 320, 1444–1449.

Bonan, G. B., Davis, K. J., Baldocchi, D., Fitzjarrald, D., and Neumann, H. (1997). Comparison of the NCAR LSM1 land surface model with BOREAS aspen and jack pine tower fluxes. *Journal of Geophysical Research*, 102D, 29065–29075.

Bonan, G. B., Levis, S., Kergoat, L., and Oleson, K. W. (2002). Landscapes as patches of plant functional types: An integrating concept for climate and ecosystem models. *Global Biogeochemical Cycles*, 16, 1021, doi:10.1029/2000GB001360.

Bonan, G. B., Levis, S., Sitch, S., Vertenstein, M., and Oleson, K. W. (2003). A dynamic global vegetation model for use with climate models: Concepts and description of simulated vegetation dynamics. *Global Change Biology*, 9, 1543–1566.

Bonan, G. B., Lawrence, P. J., Oleson, K. W., et al. (2011). Improving canopy processes in the Community Land Model version 4 (CLM4) using global flux fields empirically inferred from FLUXNET data. *Journal of Geophysical Research*, 116, G02014, doi:10.1029/2010JG001593.

Bonan, G. B., Oleson, K. W., Fisher, R. A., Lasslop, G., and Reichstein, M. (2012). Reconciling leaf physiological traits and canopy flux data: Use of the TRY and FLUXNET databases in the Community Land Model version 4. *Journal of Geophysical Research*, 117, G02026, doi:10.1029/2011JG001913.

Bonan, G. B., Hartman, M. D., Parton, W. J., and Wieder, W. R. (2013). Evaluating litter decomposition in earth system models with long-term litterbag experiments: An example using the Community Land Model version 4 (CLM4). *Global Change Biology*, 19, 957–974.

Bondeau, A., Smith, P. C., Zaehle, S., et al. (2007). Modelling the role of agriculture for the 20th century global terrestrial carbon balance. *Global Change Biology*, 13, 679–706.

Boone, A., Habets, F., Noilhan, J., et al. (2004). The Rhône-aggregation land surface scheme intercomparison project: An overview. *Journal of Climate*, 17, 187–208.

Bounoua, L., Collatz, G. J., Sellers, P. J., et al. (1999). Interactions between vegetation and climate: Radiative and physiological effects of doubled atmospheric CO_2. *Journal of Climate*, 12, 309–324.

Bounoua, L., Collatz, G. J., Los, S. O., et al. (2000). Sensitivity of climate to changes in NDVI. *Journal of Climate*, 13, 2277–2292.

Bowling, L. C., Lettenmaier, D. P., Nijssen, B., et al. (2003a). Simulation of high-latitude hydrological processes in the Torne–Kalix basin: PILPS Phase 2(e), 1: Experiment description and summary intercomparisons. *Global and Planetary Change*, 38, 1–30.

Bowling, L. C., Lettenmaier, D. P., Nijssen, B., et al. (2003b). Simulation of high-latitude hydrological processes in the Torne–Kalix basin: PILPS Phase 2(e), 3: Equivalent model representation and sensitivity experiments. *Global and Planetary Change*, 38, 55–71.

Buermann, W., Dong, J., Zeng, X., Myneni, R. B., and Dickinson, R. E. (2001). Evaluation of the utility of satellite-based vegetation leaf area index data for climate simulations. *Journal of Climate*, 14, 3536–3551.

Buermann, W., Wang, Y., Dong, J., et al. (2002). Analysis of a multiyear global vegetation leaf area index data set. *Journal of Geophysical Research*, 107, 4646, doi:10.1029/2001JD000975.

Cadule, P., Friedlingstein, P., Bopp, L., et al. (2010). Benchmarking coupled climate–carbon models against long-term atmospheric CO_2 measurements. *Global Biogeochemical Cycles*, 24, GB2016, doi:10.1029/2009GB003556.

Chase, T. N., Pielke, R. A., Kittel, T. G. F., Nemani, R., and Running, S. W. (1996). Sensitivity of a general circulation model to global changes in leaf area index. *Journal of Geophysical Research*, 101D, 7393–7408.

Chen, F., Kusaka, H., Bornstein, R., et al. (2011). The integrated WRF/urban modelling system: Development, evaluation, and applications to urban environmental problems. *International Journal of Climatology*, 31, 273–288.

Chen, J., and Kumar, P. (2001). Topographic influence on the seasonal and interannual variation of water and energy balance of basins in North America. *Journal of Climate*, 14, 1989–2014.

Chen, T. H., Henderson-Sellers, A., Milly, P. C. D., et al. (1997). Cabauw experimental results from the Project for Intercomparison of Land-Surface Parameterization Schemes. *Journal of Climate*, 10, 1194–1215.

Cox, P. M., Betts, R. A., Bunton, C. B., et al. (1999). The impact of new land surface physics on the GCM simulation of climate and climate sensitivity. *Climate Dynamics*, 15, 183–203.

Cox, P. M., Pearson, D., Booth, B. B., et al. (2013). Sensitivity of tropical carbon to climate change constrained by carbon dioxide variability. *Nature*, 494, 341–344.

Craig, S. G., Holmén, K. J., Bonan, G. B., and Rasch, P. J. (1998). Atmospheric CO_2 simulated by the National Center for Atmospheric Research Community Climate Model, 1: Mean fields and seasonal cycles. *Journal of Geophysical Research*, 103D, 13213–13235.

Dang, Q.-L., Margolis, H. A., and Collatz, G. J. (1998). Parameterization and testing of a coupled photosynthesis–stomatal conductance model for boreal trees. *Tree Physiology*, 18, 141–153.

Deardorff, J. W. (1978). Efficient prediction of ground surface temperature and moisture, with inclusion of a layer of vegetation. *Journal of Geophysical Research*, 83C, 1889–1903.

DeFries, R. S., and Townshend, J. R. G. (1994). NDVI-derived land cover classifications at a global scale. *International Journal of Remote Sensing*, 15, 3567–3586.

DeFries, R. S., Field, C. B., Fung, I., et al. (1995). Mapping the land surface for global atmosphere–biosphere models: Toward continuous distributions of vegetation's functional properties. *Journal of Geophysical Research*, 100D, 20867–20882.

DeFries, R., Hansen, M., Steininger, M., et al. (1997). Subpixel forest cover in Central Africa from multisensor, multitemporal data. *Remote Sensing of Environment*, 60, 228–246.

DeFries, R. S., Hansen, M., Townshend, J. R. G., and Sohlberg, R. (1998). Global land cover classifications at 8 km spatial resolution: The use of training data derived from Landsat Imagery in decision tree classifiers. *International Journal of Remote Sensing*, 19, 3141–3168.

DeFries, R. S., Townshend, J. R. G., and Hansen, M. C. (1999). Continuous fields of vegetation characteristics at the global scale at 1-km resolution. *Journal of Geophysical Research*, 104D, 16911–16923.

DeFries, R. S., Hansen, M. C., and Townshend, J. R. G. (2000a). Global continuous fields of vegetation characteristics: A linear mixture model applied to multi-year 8 km AVHRR data. *International Journal of Remote Sensing*, 21, 1389–1414.

DeFries, R. S., Hansen, M. C., Townshend, J. R. G., Janetos, A. C., and Loveland, T. R. (2000b). A new global 1-km dataset of percentage tree cover derived from remote sensing. *Global Change Biology*, 6, 247–254.

De Kauwe, M. G., Medlyn, B. E., Zaehle, S., et al. (2013). Forest water use and water use efficiency at elevated CO_2: A model-data intercomparison at two contrasting temperate forest FACE sites. *Global Change Biology*, 19, 1759–1779.

De Kauwe, M. G., Medlyn, B. E., Zaehle, S., et al. (2014). Where does the carbon go? A model–data intercomparison of vegetation carbon allocation and turnover processes at two temperate forest free-air CO_2 enrichment sites. *New Phytologist*, 203, 883–899.

Demuzere, M., Oleson, K., Coutts, A. M., Pigeon, G., and van Lipzig, N. P. M. (2013). Simulating the surface energy balance over two contrasting urban environments using the Community Land Model Urban. *International Journal of Climatology*, 33, 3182–3205.

Denning, A. S., Fung, I. Y., and Randall, D. (1995). Latitudinal gradient of atmospheric CO_2 due to seasonal exchange with land biota. *Nature*, 376, 240–243.

Denning, A. S., Collatz, G. J., Zhang, C., et al. (1996a). Simulations of terrestrial carbon metabolism and atmospheric CO_2 in a general circulation model. Part 1: Surface carbon fluxes. *Tellus B*, 48, 521–542.

Denning, A. S., Randall, D. A., Collatz, G. J., and Sellers, P. J. (1996b). Simulations of terrestrial carbon metabolism and atmospheric CO_2 in a general circulation model. Part 2: Simulated CO_2 concentrations. *Tellus B*, 48, 543–567.

de Noblet-Ducoudré, N., Boisier, J.-P., Pitman, A., et al. (2012). Determining robust impacts of land-use-induced land cover changes on surface climate over North America and Eurasia: Results from the first set of LUCID experiments. *Journal of Climate*, 25, 3261–3281.

Desborough, C. E., Pitman, A. J., and Irannejad, P. (1996). Analysis of the relationship between bare soil evaporation and soil moisture simulated by 13 land surface schemes for a simple non-vegetated site. *Global and Planetary Change*, 13, 47–56.

Dickinson, R. E. (1983). Land surface processes and climate-surface albedos and energy balance. *Advances in Geophysics*, 25, 305–353.

Dickinson, R. E., and Henderson-Sellers, A. (1988). Modelling tropical deforestation: A study of GCM land-surface parameterizations. *Quarterly Journal of the Royal Meteorological Society*, 114, 439–462.

Dickinson, R. E., Henderson-Sellers, A., Kennedy, P. J., and Wilson, M. F. (1986). *Biosphere–Atmosphere Transfer Scheme (BATS) for the NCAR Community Climate Model*, Technical Note NCAR/TN-275+STR. Boulder, Colorado: National Center for Atmospheric Research.

Dickinson, R. E., Henderson-Sellers, A., and Kennedy, P. J. (1993). *Biosphere–Atmosphere Transfer Scheme (BATS) Version 1e as Coupled to the NCAR Community Climate Model*, Technical Note NCAR/TN-387+STR. Boulder, Colorado: National Center for Atmospheric Research.

Dickinson, R. E., Oleson, K. W., Bonan, G., et al. (2006). The Community Land Model and its climate statistics as a component of the Community Climate System Model. *Journal of Climate*, 19, 2302–2324.

Dietze, M. C., Serbin, S. P., Davidson, C., et al. (2014). A quantitative assessment of a terrestrial biosphere model's data needs across North American biomes. *Journal of Geophysical Research: Biogeosciences*, 119, 286–300, doi:10.1002/2013JG002392.

Dirmeyer, P. A. (2011). A history and review of the Global Soil Wetness Project (GSWP). *Journal of Hydrometeorology*, 12, 729–749.

Dorman, J. L., and Sellers, P. J. (1989). A global climatology of albedo, roughness length and stomatal resistance for atmospheric general circulation models as represented by the simple biosphere model (SiB). *Journal of Applied Meteorology*, 28, 833–855.

Drewniak, B., Song, J., Prell, J., Kotamarthi, V. R., and Jacob, R. (2013). Modeling agriculture in the Community Land Model. *Geoscientific Model Development*, 6, 495–515.

Ducharne, A., Koster, R. D., Suarez, M. J., and Kumar, P. (1999). A catchment-based land surface model for GCMs and the framework for its evaluation. *Physics and Chemistry of the Earth, Part B: Hydrology, Oceans and Atmosphere*, 24, 769–773.

Ducharne, A., Koster, R. D., Suarez, M. J., Stieglitz, M., and Kumar, P. (2000). A catchment-based approach to modeling land surface processes in a general circulation model 2: Parameter estimation and model

demonstration. *Journal of Geophysical Research*, 105D, 24823–24838.

Entekhabi, D., and Eagleson, P. S. (1989). Land surface hydrology parameterization for atmospheric general circulation models including subgrid scale spatial variability. *Journal of Climate*, 2, 816–831.

Essery, R. L. H., Best, M. J., Betts, R. A., Cox, P. M., and Taylor, C. M. (2003). Explicit representation of subgrid heterogeneity in a GCM land surface scheme. *Journal of Hydrometeorology*, 4, 530–543.

Fisher, R., McDowell, N., Purves, D., et al. (2010). Assessing uncertainties in a second-generation dynamic vegetation model caused by ecological scale limitations. *New Phytologist*, 187, 666–681.

Foley, J. A., Prentice, I. C., Ramankutty, N., et al. (1996). An integrated biosphere model of land surface processes, terrestrial carbon balance, and vegetation dynamics. *Global Biogeochemical Cycles*, 10, 603–628.

Foley, J. A., Levis, S., Prentice, I. C., Pollard, D., and Thompson, S. L. (1998). Coupling dynamic models of climate and vegetation. *Global Change Biology*, 4, 561–579.

Foley, J. A., Levis, S., Costa, M. H., Cramer, W., and Pollard, D. (2000). Incorporating dynamic vegetation cover within global climate models. *Ecological Applications*, 10, 1620–1632.

Friedl, M. A., McIver, D. K., Hodges, J. C. F., et al. (2002). Global land cover mapping from MODIS: Algorithms and early results. *Remote Sensing of Environment*, 83, 287–302.

Friedlingstein, P., Cox, P., Betts, R., et al. (2006). Climate–carbon cycle feedback analysis: Results from the C⁴MIP model intercomparison. *Journal of Climate*, 19, 3337–3353.

Friedlingstein, P., Meinshausen, M., Arora, V. K., et al. (2014). Uncertainties in CMIP5 climate projections due to carbon cycle feedbacks. *Journal of Climate*, 27, 511–526.

Gedney, N., Cox, P. M., and Huntingford, C. (2004). Climate feedback from wetland methane emissions. *Geophysical Research Letters*, 31, L20503, doi:10.1029/2004GL020919.

Gerber, S., Hedin, L. O., Oppenheimer, M., Pacala, S. W., and Shevliakova, E. (2010). Nitrogen cycling and feedbacks in a global dynamic land model. *Global Biogeochemical Cycles*, 24, GB1001, doi:10.1029/2008GB003336.

Gervois, S., de Noblet-Ducoudré, N., Viovy, N., et al. (2004). Including croplands in a global biosphere model: Methodology and evaluation at specific sites. *Earth Interactions*, 8, 1–25.

Gervois, S., Ciais, P., de Noblet-Ducoudré, N., et al. (2008). Carbon and water balance of European croplands throughout the 20th century. *Global Biogeochemical Cycles*, 22, GB2022, doi:10.1029/2007GB003018.

Goll, D. S., Brovkin, V., Parida, B. R., et al. (2012). Nutrient limitation reduces land carbon uptake in simulations with a model of combined carbon, nitrogen and phosphorus cycling. *Biogeosciences*, 9, 3547-3569.

Grimmond, C. S. B., Blackett, M., Best, M. J., et al. (2010). The international urban energy balance models comparison project: First results from phase 1. *Journal of Applied Meteorology and Climatology*, 49, 1268-1292.

Grimmond, C. S. B., Blackett, M., Best, M. J., et al. (2011). Initial results from Phase 2 of the international urban energy balance model comparison. *International Journal of Climatology*, 31, 244-272.

Guenther, A., Hewitt, C. N., Erickson, D., et al. (1995). A global model of natural volatile organic compound emissions. *Journal of Geophysical Research*, 100D, 8873-8892.

Guenther, A., Karl, T., Harley, P., et al. (2006). Estimates of global terrestrial isoprene emissions using MEGAN (Model of Emissions of Gases and Aerosols from Nature). *Atmospheric Chemistry and Physics*, 6, 3181-3210.

Guenther, A. B., Jiang, X., Heald, C. L., et al. (2012). The Model of Emissions of Gases and Aerosols from Nature version 2.1 (MEGAN2.1): An extended and updated framework for modeling biogenic emissions. *Geoscientific Model Development*, 5, 1471-1492.

Gutman, G., and Ignatov, A. (1998). The derivation of the green vegetation fraction from NOAA/AVHRR data for use in numerical weather prediction models. *International Journal of Remote Sensing*, 19, 1533-1543.

Hansen, M. C., and DeFries, R. S. (2004). Detecting long-term global forest change using continuous fields of tree-cover maps from 8-km Advanced Very High Resolution Radiometer (AVHRR) data for the years 1982-1999. *Ecosystems*, 7, 695-716.

Hansen, M. C., DeFries, R. S., Townshend, J. R. G., and Sohlberg, R. (2000). Global land cover classification at 1km spatial resolution using a classification tree approach. *International Journal of Remote Sensing*, 21, 1331-1364.

Hansen, M. C., DeFries, R. S., Townshend, J. R. G., et al. (2002). Towards an operational MODIS continuous field of percent tree cover algorithm: examples using AVHRR and MODIS data. *Remote Sensing of Environment*, 83, 303-319.

Hansen, M. C., DeFries, R. S., Townshend, J. R. G., et al. (2003). Global percent tree cover at a spatial resolution of 500 meters: First results of the MODIS vegetation continuous fields algorithm. *Earth Interactions*, 7, 1-15.

Hararuk, O., Xia, J., and Luo, Y. (2014). Evaluation and improvement of a global land model against soil carbon data using a Bayesian Markov chain Monte Carlo method. *Journal of Geophysical Research: Biogeosciences*, 119, 403-417, doi:10.1002/2013JG002535.

Heald, C. L., Henze, D. K., Horowitz, L. W., et al. (2008). Predicted change in global secondary organic aerosol concentrations in response to future climate, emissions, and land use change. *Journal of Geophysical Research*, 113, D05211, doi:10.1029/2007JD009092.

Henderson-Sellers, A. (1993). Continental vegetation as a dynamic component of a global climate model: A preliminary assessment. *Climatic Change*, 23, 337-377.

Henderson-Sellers, A., and McGuffie, K. (1995). Global climate models and "dynamic" vegetation changes. *Global Change Biology*, 1, 63-75.

Henderson-Sellers, A., McGuffie, K., and Pitman, A. J. (1996). The Project for Intercomparison of Land-surface Parameterization Schemes (PILPS): 1992 to 1995. *Climate Dynamics*, 12, 849-859.

Henderson-Sellers, A., Irannejad, P., McGuffie, K., and Pitman, A. J. (2003). Predicting land-surface climates – better skill or moving targets? *Geophysical Research Letters*, 30, 1777, doi:10.1029/2003GL017387.

Jarvis, P. G. (1976). The interpretation of the variations in leaf water potential and stomatal conductance found in canopies in the field. *Philosophical Transactions of the Royal Society B*, 273, 593-610.

Jung, M., Reichstein, M., Margolis, H. A., et al. (2011). Global patterns of land–atmosphere fluxes of carbon dioxide, latent heat, and sensible heat derived from eddy covariance, satellite, and meteorological observations. *Journal of Geophysical Research*, 116, G00J07, doi:10.1029/2010JG001566.

Kleinen, T., Brovkin, V., and Schuldt, R. J. (2012). A dynamic model of wetland extent and peat accumulation: Results for the Holocene. *Biogeosciences*, 9, 235-248.

Kloster, S., Mahowald, N. M., Randerson, J. T., et al. (2010). Fire dynamics during the 20th century simulated by the Community Land Model. *Biogeosciences*, 7, 1877-1902.

Koster, R. D., and Milly, P. C. D. (1997). The interplay between transpiration and runoff formulations in land surface schemes used with atmospheric models. *Journal of Climate*, 10, 1578-1591.

Koster, R. D., and Suarez, M. J. (1992a). Modeling the land surface boundary in climate models as a composite of independent vegetation stands. *Journal of Geophysical Research*, 97D, 2697–2715.

Koster, R. D., and Suarez, M. J. (1992b). A comparative analysis of two land surface heterogeneity representations. *Journal of Climate*, 5, 1379–1390.

Koster, R. D., Suarez, M. J., Ducharne, A., Stieglitz, M., and Kumar, P. (2000). A catchment-based approach to modeling land surface processes in a general circulation model 1: Model structure. *Journal of Geophysical Research*, 105D, 24809–24822.

Koven, C., Friedlingstein, P., Ciais, P., et al. (2009). On the formation of high-latitude soil carbon stocks: Effects of cryoturbation and insulation by organic matter in a land surface model. *Geophysical Research Letters*, 36, L21501, doi:10.1029/2009GL040150.

Koven, C. D., Ringeval, B., Friedlingstein, P., et al. (2011). Permafrost carbon–climate feedbacks accelerate global warming. *Proceedings of the National Academy of Sciences USA*, 36, 14769–14774.

Koven, C. D., Riley, W. J., Subin, Z. M., et al. (2013). The effect of vertically resolved soil biogeochemistry and alternate soil C and N models on C dynamics of CLM4. *Biogeosciences*, 10, 7109–7131.

Kucharik, C. J. (2003). Evaluation of a process-based agro-ecosystem model (Agro-IBIS) across the U.S. cornbelt: Simulations of the interannual variability in maize yield. *Earth Interactions*, 7, 1–33.

Kumar, S., Dirmeyer, P. A., Merwade, V., et al. (2013). Land use/cover change impacts in CMIP5 climate simulations: A new methodology and 21st century challenges. *Journal of Geophysical Research: Atmospheres*, 118, 6337–6353, doi:10.1002/jgrd.50463.

Lawrence, D. M., Thornton, P. E., Oleson, K. W., and Bonan, G. B. (2007). The partitioning of evapotranspiration into transpiration, soil evaporation, and canopy evaporation in a GCM: Impacts on land–atmosphere interaction. *Journal of Hydrometeorology*, 8, 862–880.

Lawrence, D. M., Oleson, K. W., Flanner, M. G., et al. (2011). Parameterization improvements and functional and structural advances in version 4 of the Community Land Model. *Journal of Advances in Modeling Earth Systems*, 3, doi:10.1029/2011 MS000045.

Lawrence, D. M., Oleson, K. W., Flanner, M. G., et al. (2012). The CCSM4 land simulation, 1850–2005: Assessment of surface climate and new capabilities. *Journal of Climate*, 25, 2240–2260.

Lawrence, P. J., and Chase, T. N. (2007). Representing a new MODIS consistent land surface in the Community

Land Model (CLM 3.0). *Journal of Geophysical Research*, 112, G01023, doi:10.1029/2006JG000168.

Le Quéré, C., Raupach, M. R., Canadell, J. G., et al. (2009). Trends in the sources and sinks of carbon dioxide. *Nature Geoscience*, 2, 831–836.

Levis, S., Wiedinmyer, C., Bonan, G. B., and Guenther, A. (2003). Simulating biogenic volatile organic compound emissions in the Community Climate System Model. *Journal of Geophysical Research*, 108, 4659, doi:10.1029/2002JD003203.

Levis, S., Bonan, G. B., Kluzek, E., et al. (2012). Interactive crop management in the Community Earth System Model (CESM1): Seasonal influences on land–atmosphere fluxes. *Journal of Climate*, 25, 4839–4859.

Li, B., and Avissar, R. (1994). The impact of spatial variability of land-surface characteristics on land-surface heat fluxes. *Journal of Climate*, 7, 527–537.

Li, F., Zeng, X. D., and Levis, S. (2012). A process-based fire parameterization of intermediate complexity in a Dynamic Global Vegetation Model. *Biogeosciences*, 9, 2761–2780.

Li, F., Levis, S., and Ward, D. S. (2013). Quantifying the role of fire in the Earth system – Part 1: Improved global fire modeling in the Community Earth System Model (CESM1). *Biogeosciences*, 10, 2293–2314.

Liang, X., Wood, E. F., Lettenmaier, D. P., et al. (1998). The Project for Intercomparison of Land-surface Parameterization Schemes (PILPS) phase 2(c) Red–Arkansas River basin experiment, 2: Spatial and temporal analysis of energy fluxes. *Global and Planetary Change*, 19, 137–159.

Lohmann, D., Lettenmaier, D. P., Liang, X., et al. (1998). The Project for Intercomparison of Land-surface Parameterization Schemes (PILPS) phase 2(c) Red–Arkansas River basin experiment, 3: Spatial and temporal analysis of water fluxes. *Global and Planetary Change*, 19, 161–179.

Lombardozzi, D., Sparks, J. P., and Bonan, G. (2013). Integrating O_3 influences on terrestrial processes: Photosynthetic and stomatal response data available for regional and global modeling. *Biogeosciences*, 10, 6815–6831.

Loveland, T. R., Reed, B. C., Brown, J. F., et al. (2000). Development of a global land cover characteristics database and IGBP DISCover from 1km AVHRR data. *International Journal of Remote Sensing*, 21, 1303–1330.

Luo, L., Robock, A., Vinnikov, K. Y., et al. (2003). Effects of frozen soil on soil temperature, spring infiltration, and runoff: Results from the PILPS 2(d) experiment at Valdai, Russia. *Journal of Hydrometeorology*, 4, 334–351.

Luo, Y. Q., Randerson, J. T., Abramowitz, G., et al. (2012). A framework for benchmarking land models. *Biogeosciences*, 9, 3857–3874.

Mahowald, N. M., Muhs, D. R., Levis, S., et al. (2006). Change in atmospheric mineral aerosols in response to climate: Last glacial period, preindustrial, modern, and doubled carbon dioxide climates. *Journal of Geophysical Research*, 111, D10202, doi:10.1029/2005JD006653.

Manabe, S. (1969). Climate and the ocean circulation, I: The atmospheric circulation and the hydrology of the Earth's surface. *Monthly Weather Review*, 97, 739–774.

Manabe, S., Smagorinsky, J., and Strickler, R. F. (1965). Simulated climatology of a general circulation model with a hydrologic cycle. *Monthly Weather Review*, 93, 769–798.

Marticorena, B., and Bergametti, G. (1995). Modeling the atmospheric dust cycle, 1: Design of a soil-derived dust emission scheme. *Journal of Geophysical Research*, 100D, 16415–16430.

Masson, V. (2000). A physically-based scheme for the urban energy budget in atmospheric models. *Boundary-Layer Meteorology*, 94, 357–397.

Masson, V., Grimmond, C. S. B., and Oke, T. R. (2002). Evaluation of the Town Energy Balance (TEB) scheme with direct measurements from dry districts in two cities. *Journal of Applied Meteorology*, 41, 1011–1026.

Matthews, E. (1983). Global vegetation and land-use: New high-resolution data bases for climate studies. *Journal of Climate and Applied Meteorology*, 22, 474–487.

Melton, J. R., and Arora, V. K. (2014). Sub-grid scale representation of vegetation in global land surface schemes: Implications for estimation of the terrestrial carbon sink. *Biogeosciences*, 11, 1021–1036.

Meng, L., Hess, P. G. M., Mahowald, N. M., et al. (2012). Sensitivity of wetland methane emissions to model assumptions: Application and model testing against site observations. *Biogeosciences*, 9, 2793–2819.

Mercado, L. M., Bellouin, N., Sitch, S., et al. (2009). Impact of changes in diffuse radiation on the global land carbon sink. *Nature*, 458, 1014–1017.

Myneni, R. B., Nemani, R. R., and Running, S. W. (1997). Estimation of global leaf area index and absorbed par using radiative transfer models. *IEEE Transactions on Geoscience and Remote Sensing*, 35, 1380–1393.

Nemani, R. R., Running, S. W., Pielke, R. A., and Chase, T. N. (1996). Global vegetation cover changes from coarse resolution satellite data. *Journal of Geophysical Research*, 101D, 7157–7162.

Nijssen, B., Bowling, L. C., Lettenmaier, D. P., et al. (2003). Simulation of high latitude hydrological processes in the Torne–Kalix basin: PILPS Phase 2(e). 2: Comparison of model results with observations. *Global and Planetary Change*, 38, 31–53.

Oleson, K. W., Bonan, G. B., Feddema, J., Vertenstein, M., and Grimmond, C. S. B. (2008). An urban parameterization for a global climate model, Part I: Formulation and evaluation for two cities. *Journal of Applied Meteorology and Climatology*, 47, 1038–1060.

Oleson, K. W., Bonan, G. B., Feddema, J., and Jackson, T. (2011). An examination of urban heat island characteristics in a global climate model. *International Journal of Climatology*, 31, 1848–1865.

Olson, J. S., Watts, J. A., and Allison, L. J. (1983). *Carbon in Live Vegetation of Major World Ecosystems*, ORNL-5862. Oak Ridge, Tennessee: Oak Ridge National Laboratory.

Osborne, T. M., Lawrence, D. M., Challinor, A. J., Slingo, J. M., and Wheeler, T. R. (2007). Development and assessment of a coupled crop–climate model. *Global Change Biology*, 13, 169–183.

Osborne, T., Slingo, J., Lawrence, D., and Wheeler, T. (2009). Examining the interaction of growing crops with local climate using a coupled crop–climate model. *Journal of Climate*, 22, 1393–1411.

Pacifico, F., Harrison, S. P., Jones, C. D., et al. (2011). Evaluation of a photosynthesis-based biogenic isoprene emission scheme in JULES and simulation of isoprene emissions under present-day climate conditions. *Atmospheric Chemistry and Physics*, 11, 4371–4389.

Piao, S., Sitch, S., Ciais, P., et al. (2013). Evaluation of terrestrial carbon cycle models for their response to climate variability and to CO_2 trends. *Global Change Biology*, 19, 2117–2132.

Pitman, A. J. (2003). The evolution of, and revolution in, land surface schemes designed for climate models. *International Journal of Climatology*, 23, 479–510.

Pitman, A. J., Henderson-Sellers, A., Desborough, C. E., et al. (1999). Key results and implications from phase 1(c) of the Project for Intercomparison of Land-surface Parametrization Schemes. *Climate Dynamics*, 15, 673–684.

Pitman, A. J., de Noblet-Ducoudré, N., Cruz, F. T., et al. (2009). Uncertainties in climate responses to past land cover change: First results from the LUCID intercomparison study. *Geophysical Research Letters*, 36, L14814, doi:10.1029/2009GL039076.

Price, J. C. (1992). Estimating vegetation amount from visible and near infrared reflectances. *Remote Sensing of Environment*, 41, 29–34.

Qu, W., Henderson-Sellers, A., Pitman, A. J., et al. (1998). Sensitivity of latent heat flux from PILPS land-surface schemes to perturbations of surface air temperature. *Journal of the Atmospheric Sciences*, 55, 1909–1927.

Randall, D. A., Dazlich, D. A., Zhang, C., et al. (1996). A revised land surface parameterization (SiB2) for GCMS, Part III: The greening of the Colorado State University General Circulation Model. *Journal of Climate*, 9, 738–763.

Randerson, J. T., Hoffman, F. M., Thornton, P. E., et al. (2009). Systematic assessment of terrestrial biogeochemistry in coupled climate–carbon models. *Global Change Biology*, 15, 2462–2484.

Reich, P. B., Rich, R. L., Lu, X., Wang, Y.-P., and Oleksyn, J. (2014). Biogeographic variation in evergreen conifer needle longevity and impacts on boreal forest carbon cycle projections. *Proceedings of the National Academy of Sciences USA*, 111, 13703–13708.

Reichstein, M., Ciais, P., Papale, D., et al. (2007). Reduction of ecosystem productivity and respiration during the European summer 2003 climate anomaly: A joint flux tower, remote sensing and modelling analysis. *Global Change Biology*, 13, 634–651.

Riley, W. J., Still, C. J., Torn, M. S., and Berry, J. A. (2002). A mechanistic model of $H_2^{18}O$ and $C^{18}OO$ fluxes between ecosystems and the atmosphere: Model description and sensitivity analyses. *Global Biogeochemical Cycles*, 16, 1095, doi:10.1029/2002GB001878.

Riley, W. J., Still, C. J., Helliker, B. R., Ribas-Carbo, M., and Berry, J. A. (2003). ^{18}O composition of CO_2 and H_2O ecosystem pools and fluxes in a tallgrass prairie: Simulations and comparisons to measurements. *Global Change Biology*, 9, 1567–1581.

Riley, W. J., Subin, Z. M., Lawrence, D. M., et al. (2011). Barriers to predicting changes in global terrestrial methane fluxes: Analyses using CLM4Me, a methane biogeochemistry model integrated in CESM. *Biogeosciences*, 8, 1925–1953.

Ringeval, B., Decharme, B., Piao, S. L., et al. (2012). Modelling sub-grid wetland in the ORCHIDEE global land surface model: Evaluation against river discharges and remotely sensed data. *Geoscientific Model Development*, 5, 941–962.

Risi, C., Bony, S., Vimeux, F., and Jouzel, J. (2010). Water-stable isotopes in the LMDZ4 general circulation model: Model evaluation for present-day and past climates and applications to climatic interpretations of tropical isotopic records. *Journal of Geophysical Research*, 115, D12118, doi:10.1029/2009JD013255.

Rogers, A. (2014). The use and misuse of $V_{c,max}$ in Earth System Models. *Photosynthesis Research*, 119, 15–29.

Scheiter, S., Langan, L., and Higgins, S. I. (2013). Next-generation dynamic global vegetation models: Learning from community ecology. *New Phytologist*, 198, 957–969.

Schlosser, C. A., Slater, A. G., Robock, A., et al. (2000). Simulations of a boreal grassland hydrology at Valdai, Russia: PILPS Phase 2(d). *Monthly Weather Review*, 128, 301–321.

Sellers, P. J., Mintz, Y., Sud, Y. C., and Dalcher, A. (1986). A simple biosphere model (SiB) for use within general circulation models. *Journal of the Atmospheric Sciences*, 43, 505–531.

Sellers, P. J., Tucker, C. J., Collatz, G. J., et al. (1994). A global 1° by 1° NDVI data set for climate studies, Part 2: The generation of global fields of terrestrial biophysical parameters from the NDVI. *International Journal of Remote Sensing*, 15, 3519–3545.

Sellers, P. J., Bounoua, L., Collatz, G. J., et al. (1996a). Comparison of radiative and physiological effects of doubled atmospheric CO_2 on climate. *Science*, 271, 1402–1406.

Sellers, P. J., Los, S. O., Tucker, C. J., et al. (1996b). A revised land surface parameterization (SiB2) for atmospheric GCMs, Part II: The generation of global fields of terrestrial biophysical parameters from satellite data. *Journal of Climate*, 9, 706–737.

Sellers, P. J., Randall, D. A., Collatz, G. J., et al. (1996c). A revised land surface parameterization (SiB2) for atmospheric GCMs, Part I: Model formulation. *Journal of Climate*, 9, 676–705.

Seth, A., Giorgi, F., and Dickinson, R. E. (1994). Simulating fluxes from heterogeneous land surfaces: Explicit subgrid method employing the biosphere–atmosphere transfer scheme (BATS). *Journal of Geophysical Research*, 99D, 18651–18667.

Shao, Y., and Henderson-Sellers, A. (1996a). Modeling soil moisture: A Project for Intercomparison of Land Surface Parameterization Schemes Phase 2(b). *Journal of Geophysical Research*, 101D, 7227–7250.

Shao, Y., and Henderson-Sellers, A. (1996b). Validation of soil moisture simulation in landsurface parameterisation schemes with HAPEX data. *Global and Planetary Change*, 13, 11–46.

Shevliakova, E., Pacala, S. W., Malyshev, S., et al. (2009). Carbon cycling under 300 years of land use change: Importance of the secondary vegetation sink. *Global Biogeochemical Cycles*, 23, GB2022, doi:10.1029/2007GB003176.

Shukla, J., and Mintz, Y. (1982). Influence of land-surface evapotranspiration on the Earth's climate. *Science*, 215, 1498–1501.

Sitch, S., Cox, P. M., Collins, W. J., and Huntingford, C. (2007). Indirect radiative forcing of climate change through ozone effects on the land-carbon sink. *Nature*, 448, 791–794.

Sitch, S., Huntingford, C., Gedney, N., et al. (2008). Evaluation of the terrestrial carbon cycle, future plant geography and climate–carbon cycle feedbacks using five Dynamic Global Vegetation Models (DGVMs). *Global Change Biology*, 14, 2015–2039.

Slater, A. G., Schlosser, C. A., Desborough, C. E., et al. (2001). The representation of snow in land surface schemes: Results from PILPS 2(d). *Journal of Hydrometeorology*, 2, 7–25.

Stöckli, R., Lawrence, D. M., Niu, G.-Y., et al. (2008). Use of FLUXNET in the Community Land Model development. *Journal of Geophysical Research*, 113, G01025, doi:10.1029/2007JG000562.

Swenson, S. C., Lawrence, D. M., and Lee, H. (2012). Improved simulation of the terrestrial hydrological cycle in permafrost regions by the Community Land Model. *Journal of Advances in Modeling Earth Systems*, 4, M08002, doi:10.1029/2012MS000165.

Tesfa, T. K., Li, H.-Y., Leung, L. R., et al. (2014). A subbasin-based framework to represent land surface processes in an Earth system model. *Geoscientific Model Development*, 7, 947–963.

Tegen, I., and Fung, I. (1994). Modeling of mineral dust in the atmosphere: sources, transport, and optical thickness. *Journal of Geophysical Research*, 99D, 22897–22914.

Tegen, I., Harrison, S. P., Kohfeld, K., et al. (2002). Impact of vegetation and preferential source areas on global dust aerosol: Results from a model study. *Journal of Geophysical Research*, 107, 4576, doi:10.1029/2001JD000963.

Thomas, R. Q., Bonan, G. B., and Goodale, C. L. (2013a). Insights into mechanisms governing forest carbon response to nitrogen deposition: A model–data comparison using observed responses to nitrogen addition. *Biogeosciences*, 10, 3869–3887.

Thomas, R. Q., Zaehle, S., Templer, P. H., and Goodale, C. L. (2013b). Global patterns of nitrogen limitation: Confronting two global biogeochemical models with observations. *Global Change Biology*, 19, 2986–2998.

Thonicke, K., Venevsky, S., Sitch, S., and Cramer, W. (2001). The role of fire disturbance for global vegetation dynamics: Coupling fire into a dynamic global vegetation model. *Global Ecology and Biogeography*, 10, 661–677.

Thonicke, K., Spessa, A., Prentice, I. C., et al. (2010). The influence of vegetation, fire spread and fire behaviour on biomass burning and trace gas emissions: results from a process-based model. *Biogeosciences*, 7, 1991–2011.

Thornton, P. E., Lamarque, J.-F., Rosenbloom, N. A., and Mahowald, N. M. (2007). Influence of carbon–nitrogen cycle coupling on land model response to CO_2 fertilization and climate variability. *Global Biogeochemical Cycles*, 21, GB4018, doi:10.1029/2006GB002868.

Todd-Brown, K. E. O., Randerson, J. T., Post, W. M., et al. (2013). Causes of variation in soil carbon simulations from CMIP5 Earth system models and comparison with observations. *Biogeosciences*, 10, 1717–1736.

Unger, N., Harper, K., Zheng, Y., et al. (2013). Photosynthesis-dependent isoprene emission from leaf to planet in a global carbon–chemistry–climate model. *Atmospheric Chemistry and Physics*, 13, 10243–10269.

Verheijen, L. M., Brovkin, V., Aerts, R., et al. (2013). Impacts of trait variation through observed trait–climate relationships on performance of an Earth system model: a conceptual analysis. *Biogeosciences*, 10, 5497–5515.

Wang, Y.-P., Trudinger, C. M., and Enting, I. G. (2009). A review of applications of model–data fusion to studies of terrestrial carbon fluxes at different scales. *Agricultural and Forest Meteorology*, 149, 1829–1842.

Wang, Y. P., Law, R. M., and Pak, B. (2010). A global model of carbon, nitrogen and phosphorus cycles for the terrestrial biosphere. *Biogeosciences*, 7, 2261–2282.

Wang, Y. P., Kowalczyk, E., Leuning, R., et al. (2011). Diagnosing errors in a land surface model (CABLE) in the time and frequency domains. *Journal of Geophysical Research*, 116, G01034, doi:10.1029/2010JG001385.

Wang, Y. P., Lu, X. J., Wright, I. J., et al. (2012). Correlations among leaf traits provide a significant constraint on the estimate of global gross primary production. *Geophysical Research Letters*, 39, L19405, doi:10.1029/2012GL053461.

Wania, R., Ross, I., and Prentice, I. C. (2009). Integrating peatlands and permafrost into a dynamic global vegetation model, 1: Evaluation and sensitivity of physical land surface processes. *Global Biogeochemical Cycles*, 23, GB3014, doi:10.1029/2008GB003412.

Wania, R., Ross, I., and Prentice, I. C. (2010). Implementation and evaluation of a new methane model within a dynamic global vegetation model: LPJ-WHyMe v1.3.1. *Geoscientific Model Development*, 3, 565–584.

Wania, R., Meissner, K. J., Eby, M., et al. (2012). Carbon–nitrogen feedbacks in the UVic ESCM. *Geoscientific Model Development*, 5, 1137–1160.

Wania, R., Melton, J. R., Hodson, E. L., et al. (2013). Present state of global wetland extent and wetland methane modelling: Methodology of a model inter-comparison project (WETCHIMP). *Geoscientific Model Development*, 6, 617–641.

Wei, J., Dirmeyer, P. A., Guo, Z., Zhang, L., and Misra, V. (2010). How much do different land models matter for climate simulation? Part I: Climatology and variability. *Journal of Climate*, 23, 3120–3134.

Wenzel, S., Cox, P. M., Eyring, V., and Friedlingstein, P. (2014). Emergent constraints on climate-carbon cycle feedbacks in the CMIP5 Earth system models. *Journal of Geophysical Research: Biogeosciences*, 119, 794–807, doi:10.1002/2013JG002591.

Wieder, W. R., Grandy, A. S., Kallenbach, C. M., and Bonan, G. B. (2014). Integrating microbial physiology and physio-chemical principles in soils with the MIcrobial-MIneral Carbon Stabilization (MIMICS) model. *Biogeosciences*, 11, 3899–3917.

Wiedinmyer, C., Akagi, S. K., Yokelson, R. J., et al. (2011). The Fire INventory from NCAR (FINN): A high resolution global model to estimate the emissions from open burning. *Geoscientific Model Development*, 4, 625–641.

Williams, M., Richardson, A. D., Reichstein, M., et al. (2009). Improving land surface models with FLUXNET data. *Biogeosciences*, 6, 1341–1359.

Williamson, D. L., Kiehl, J. T., Ramanathan, V., Dickinson, R. E., and Hack, J. J. (1987). *Description of NCAR Community Climate Model (CCM1)*, Technical Note NCAR/TN-285+STR. Boulder, Colorado: National Center for Atmospheric Research.

Wilson, M. F., and Henderson-Sellers, A. (1985). A global archive of land cover and soil data for use in general circulation climate models. *Journal of Climatology*, 5, 119–143.

Wittig, V. E., Ainsworth, E. A., and Long, S. P. (2007). To what extent do current and projected increases in surface ozone affect photosynthesis and stomatal conductance of trees? A meta-analytic review of the last 3 decades of experiments. *Plant, Cell and Environment*, 30, 1150–1162.

Wittig, V. E., Ainsworth, E. A., Naidu, S. L., Karnosky, D. F., and Long, S. P. (2009). Quantifying the impact of current and future tropospheric ozone on tree biomass, growth, physiology and biochemistry: A quantitative meta-analysis. *Global Change Biology*, 15, 396–424.

Wood, E. F., Lettenmaier, D. P., Liang, X., et al. (1998). The Project for Intercomparison of Land-surface Parameterization Schemes (PILPS) Phase 2(c) Red–Arkansas River basin experiment, 1: Experiment description and summary intercomparisons. *Global and Planetary Change*, 19, 115–135.

Yang, X., Thornton, P. E., Ricciuto, D. M., and Post, W. M. (2014). The role of phosphorus dynamics in tropical forests – a modeling study using CLM-CNP. *Biogeosciences*, 11, 1667–1681.

Zaehle, S., and Friend, A. D. (2010). Carbon and nitrogen cycle dynamics in the O-CN land surface model, 1: Model description, site-scale evaluation, and sensitivity to parameter estimates. *Global Biogeochemical Cycles*, 24, GB1005, doi:10.1029/2009GB003521.

Zaehle, S., Ciais, P., Friend, A. D., and Prieur, V. (2011). Carbon benefits of anthropogenic reactive nitrogen offset by nitrous oxide emissions. *Nature Geoscience*, 4, 601–605.

Zaehle, S., Medlyn, B. E., De Kauwe, M. G., et al. (2014). Evaluation of 11 terrestrial carbon–nitrogen cycle models against observations from two temperate Free-Air CO_2 Enrichment studies. *New Phytologist*, 202, 803–822.

Zender, C. S., Bian, H., and Newman, D. (2003). Mineral Dust Entrainment and Deposition (DEAD) model: Description and 1990s dust climatology. *Journal of Geophysical Research*, 108, 4416, doi:10.1029/2002JD002775.

Zeng, X., Dickinson, R. E., Walker, A., et al. (2000). Derivation and evaluation of global 1-km fractional vegetation cover data for land modeling. *Journal of Applied Meteorology*, 39, 826–839.

Zeng, X., Rao, P., DeFries, R. S., and Hansen, M. C. (2003). Interannual variability and decadal trend of global fractional vegetation cover from 1982 to 2000. *Journal of Applied Meteorology*, 42, 1525–1530.

Zhang, Q., Wang, Y. P., Pitman, A. J., and Dai, Y. J. (2011). Limitations of nitrogen and phosphorous on the terrestrial carbon uptake in the 20th century. *Geophysical Research Letters*, 38, L22701, doi:10.1029/2011GL049244.

Zhang, Q., Wang, Y.P., Matear, R. J., Pitman, A. J., and Dai, Y. J. (2014). Nitrogen and phosphorous limitations significantly reduce future allowable CO_2 emissions. *Geophysical Research Letters*, 41, 632–637, doi:10.1002/2013GL058352.

Zhou, T., and Luo, Y. (2008). Spatial patterns of ecosystem carbon residence time and NPP-driven carbon uptake in the conterminous United States. *Global Biogeochemical Cycles*, 22, GB3032, doi:10.1029/2007GB002939.

26

Seasonal-to-Interannual Variability

26.1 | Chapter Summary

Atmospheric and oceanic processes and their coupling dominate much of the study of seasonal-to-interannual climate variability. However, land surface processes also contribute to climate variability. Soil moisture is a key aspect of seasonal precipitation forecasts. Recycling of precipitation in evapotranspiration can lead to a positive feedback by which wet soils pump more moisture into the atmosphere, which enhances rainfall and further wets the soil. Conversely, dry soils, with low rates of evapotranspiration, can reduce rainfall. The retention of precipitation by soil and the influence of soil moisture on subsequent evapotranspiration contribute to and amplify interannual precipitation variability over tropical and middle latitudes. The presence of snow is also an important initial condition required for accurate forecasts. The high albedo of snow-covered surfaces prevents the surface from warming during the day. On warm days, a large portion of net radiation at the surface is used to melt snow. By cooling the surface and reducing the land–ocean temperature contrast, snow can influence summer precipitation in monsoon climates. The seasonal emergence of leaves in spring imparts a discernible signal to air temperature. Greater latent heat flux with leaf emergence cools air temperature.

26.2 | Soil Moisture

Soil moisture regulates boundary layer processes through the partitioning of net radiation into sensible and latent heat fluxes (Figure 26.1). Atmospheric model simulations have routinely demonstrated the importance of soil moisture, through its effect on evapotranspiration, for climate simulation (Seneviratne et al. 2010). These simulations typically alter soil moisture or more generally soil wetness (the effect of soil moisture on evapotranspiration). Such studies demonstrate a positive feedback in which wet soils pump more moisture into the atmosphere, which can enhance rainfall and further wet the soil. Conversely, dry soils, with low rates of evapotranspiration, reduce rainfall.

One experimental approach has been to artificially set soil wetness to prescribed values that do not change over time and to contrast climate simulations using wet and dry soils. Shukla and Mintz (1982) used this method to demonstrate the effect of evapotranspiration on climate. Using a global model, they compared a simulation with perpetually dry soils to that with perpetually wet soils. No evapotranspiration was allowed in the dry soil simulation. The wet simulation set evapotranspiration to the potential rate, equivalent to wet soils completely covered with vegetation. Dry soil reduces July precipitation over much of the Northern Hemisphere

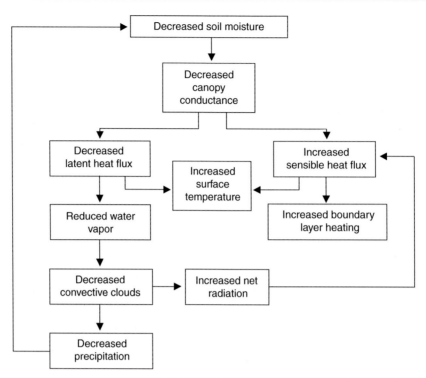

Fig. 26.1 Impacts of a decrease in soil moisture on surface climate. Adapted from Pitman (2003).

continental area compared with wet soil. Precipitation is relatively unaffected in moist tropical regions. July temperature over virtually all land areas is much warmer with dry soil compared with wet soil. This warming results from no evaporative cooling and increased solar radiation because of reduced cloudiness.

The preceding experimental approach prescribes soil wetness to a constant value. An alternative approach allows the model to calculate soil water interactively and examines the response to an initial moisture anomaly. Soil moisture is initialized to wet or dry values, and the model is integrated forward in time to see if precipitation responds to the soil moisture anomaly and if the anomaly is maintained over time. These studies show that a large initial wet or dry soil moisture perturbation can impact subsequent precipitation, which feeds back to maintain the soil moisture anomaly.

Experiments by Betts (2004) illustrate this experimental approach. Two 120-day global simulations with an atmospheric numerical model for the period May–August 1987 show

the sensitivity of precipitation to soil moisture. The two simulations differed only in initial soil moisture. A wet simulation was initialized with soils at field capacity for vegetated areas; a dry simulation was initialized with soils at 25 percent of water-holding capacity. The two simulations were analyzed in terms of precipitation (P), evapotranspiration (E), and the difference ($P - E$), which is a measure of the convergence of water vapor in the atmosphere. Where precipitation exceeds evapotranspiration ($P - E > 0$), moisture must be imported into the region to maintain precipitation (i.e., there is convergence of water vapor). Where evapotranspiration exceeds precipitation ($P - E < 0$), the excess moisture is exported (i.e., water vapor divergence). Over most regions of the Northern Hemisphere, lower soil moisture reduces precipitation and evapotranspiration by 50 percent or more during summer compared with wet soil (Table 26.1). Over the United States, central Asia, and the Amazon region of South America, the export of water vapor (divergence) increases with wet soil. In monsoonal climates (equatorial

Table 26.1 | Five-day mean summer precipitation (P), evapotranspiration (E), and water vapor convergence (P − E) for simulations with initially dry and wet soils

	Precipitation (mm per 5-day)		Evapotranspiration (mm per 5-day)		P − E (mm per 5-day)	
	Dry	Wet	Dry	Wet	Dry	Wet
Canada	5.2	12.1	5.6	11.5	−0.4	0.6
North Asia	6.0	11.2	5.6	10.8	0.4	0.4
Europe	6.0	15.5	7.0	13.9	−1.0	1.6
United States	7.3	14.8	8.6	17.6	−1.3	−2.9
Central Asia	3.2	7.5	3.5	9.2	−0.3	−1.7
Amazon	4.6	11.5	6.8	14.7	−2.2	−3.2
Equatorial Africa	12.6	18.5	8.9	13.3	3.7	5.2
African Sahel	14.8	18.2	7.1	10.0	7.6	8.1
India	28.4	32.7	9.3	14.3	19.1	18.4

Source: From Betts (2004).

Africa, the Sahel region of Africa, and India), the increase in precipitation and evapotranspiration with wet soil is relatively less. In these climates, high precipitation rates are sustained by the convergence of water vapor into the region, and the contribution of evapotranspiration to increased precipitation is less. Outside of monsoon regions, this particular atmospheric model has large coupling between soil moisture and precipitation.

Another experimental framework examines the impact of prescribed soil wetness on interannual variation in precipitation or the predictability of precipitation. In one simulation, soil wetness is calculated interactively by the climate model. In a second simulation, soil wetness is prescribed from climatological values, typically taken from the interactive simulation. This removes the interactive hydrologic coupling between land and atmosphere. Differences in simulated climate indicate the effect of interactive hydrology on precipitation. Alternatively, soil wetness is specified from simulations in which the land surface component of the climate model is forced with observed meteorology. This represents the soil wetness of the climate model if it perfectly replicated the meteorological observations. These climate model studies show that hydrologic feedbacks can influence interannual precipitation variability over land through the retention of rainfall by soil and the influence of soil moisture on subsequent evapotranspiration.

The mechanism behind the feedback between soil moisture and precipitation is understood in terms of the input of latent and sensible heat into the boundary layer (Chapter 14). When soil is dry, the net radiative energy at the surface is converted primarily into sensible heat, producing a deep, well-mixed boundary layer. Wet soil has increased evapotranspiration compared with dry soil; more of the net radiation at the surface is partitioned to latent heat rather than sensible heat. The decreased Bowen ratio results in a shallow, moist boundary layer with low height of cloud base. Wet soils also tend to have greater net radiation at the surface as a result of cooler surface temperature and decreased surface albedo. The greater total heat flux into the boundary layer combines with the shallower depth of the boundary layer to increase the convective instability of the boundary layer and create conditions that can favor convective precipitation.

Hydrometeorological land–atmosphere coupling is also understood in terms of soil moisture–evapotranspiration regimes (Figure 14.6). Radiation, not soil moisture, limits evapotranspiration in humid climates, where abundant precipitation maintains wet soils with moisture content above some maximum threshold. Below this threshold, soil moisture availability limits evapotranspiration, and evapotranspiration decreases as the soil becomes drier. In this regime, changes in precipitation (and soil moisture) affect evapotranspiration. In arid climates, precipitation is minimal, soils are extremely dry, and there is insufficient water for evapotranspiration. This conceptual framework of radiation-limited and moisture-limited regimes explains geographic patterns in evapotranspiration and provides an understanding of soil moisture influences on land–atmosphere coupling (Koster et al. 2004, 2009, 2011; Seneviratne et al. 2006, 2010; Teuling et al. 2009).

Direct comparison of soil moisture–precipitation coupling among numerous climate models in highly controlled experiments reveals that while soil moisture anomalies can induce precipitation anomalies, the response varies greatly among models. This conclusion was initially drawn from a comparison of four climate models (Koster et al. 2002). A comparison of 12 models confirmed the results of the earlier study (Koster et al. 2004, 2006a; Guo et al. 2006). Though these 12 models differ in coupling strength, regions of strong coupling between soil moisture and precipitation are common to many models. Impacts of soil moisture on rainfall are strong in regions of transition between arid and humid climates, such as the Great Plains of North America, sub-Sahara Africa, and India. This arises from high sensitivity of evapotranspiration to soil moisture in these regions and also because of large temporal variability in evapotranspiration.

The influence of soil moisture on precipitation represents the effect of soil moisture on evapotranspiration and the effect of evapotranspiration on precipitation. Soil moisture is plentiful in humid climates, and net radiation, rather than soil moisture, controls evapotranspiration. Soil moisture limits evapotranspiration in arid climates, but evapotranspiration rates are too low to affect rainfall. However, evapotranspiration is both large and controlled by soil moisture in transitional regions. Critical to this coupling strength is an enhancement of evapotranspiration by wet soil and enhancement of precipitation by increased evapotranspiration. The dependence of evapotranspiration on soil moisture is a key determinant of the coupling strength (Guo et al. 2006; Wei and Dirmeyer 2012).

Although the soil moisture–precipitation feedback has been found in climate models, it is difficult to demonstrate observationally. A study of soil moisture and rainfall over a 14-year period in Illinois found a positive correlation between soil moisture and subsequent summer rainfall (Findell and Eltahir 1997), though the results are not conclusive (Salvucci et al. 2002). Analyses of precipitation and latent heat flux show that high evaporation enhances the probability of afternoon rainfall in the United States east of the Mississippi (Findell et al. 2011).

Indirect evidence for the feedback comes from two climate model studies, which show that patterns in observed precipitation that are consistent with soil moisture–precipitation feedback can only be reproduced in climate models using interactive soil hydrology. Koster et al. (2003) analyzed a 50-year dataset of daily precipitation for the continental United States. The variance of monthly July precipitation has a distinct geographic pattern that is reproduced in a simulation with interactive soil hydrology but not when the model is run with prescribed soil wetness. A similar result is seen for sub-monthly autocorrelations of precipitation during July. These are correlations between precipitation in one 5-day period with precipitation in a subsequent 5-day period (e.g., July 1–5 and July 11–15, July 6–10 and July 16–20, etc.). Correlations between consecutive 5-day periods were not considered because these are influenced by storm duration. Subsequently, Koster and Suarez (2004) found that summer rainfall over mid-latitude continents is conditioned based on rainfall in the prior month. Observations show that above-normal precipitation in one month leads, on average, to above-normal precipitation in the following month; below-normal precipitation in one month is followed, on average, by below-normal precipitation in subsequent

Table 26.2 Percentage of hot days during summer (June–August) in relation to soil moisture in southeastern and central Europe for the period 1961–2000

Soil moisture status	Southeast Europe		Central Europe	
	Median	90th percentile	Median	90th percentile
SPI = −1.5 (moderate-to-severe drought)	19	43	16	28
SPI = −1 (mild-to-moderate drought)	16	37	14	26
SPI = 0 (normal conditions)	10	24	10	21
SPI = 1 (mild-to-moderate wetness)	4	11	7	16
SPI = 1.5 (moderate-to-severe wetness)	1	4.5	5	14

Note: The standardized precipitation index (SPI) is a proxy for drought.
Source: From Hirschi et al. (2011).

months. This can be reproduced in a climate model only when interactive soil hydrology is enabled. In both studies, the model reproduces the observations only with interactive soil hydrology. This suggests that soil moisture feedback is required to match the observations. Similar behavior is seen in other climate models, though the strength of the relationships varies considerably among models (Dirmeyer et al. 2006).

One of the more robust relationships is the correlation between soil moisture and temperature (Seneviratne et al. 2010). High rates of evapotranspiration cool the atmospheric boundary layer because more net radiation at the surface is dissipated as latent heat than as sensible heat. Conversely, high sensible heat fluxes on dry soils warm the boundary layer. Observational studies have found a negative correlation between summer temperature and precipitation over much of interior United States (Madden and Williams 1978; Karl and Quayle 1981; Namias 1983; Karl 1986; Huang and Van den Dool 1993; Durre et al. 2000; Koster et al. 2006b, 2009). In this region, hot summers are likely to also be dry; wet summers are likely to be cool. Negative correlations between summertime evapotranspiration and temperature are found in many regions of the world (Seneviratne et al. 2006, 2010).

Similar land–atmosphere feedback in which dry soils during drought amplify extreme temperatures is seen in simulations of European summer heat waves (Fischer et al. 2007; Jaeger and Seneviratne 2011; Lorenz et al. 2013). An analysis of station observations in Europe illustrates this coupling (Hirschi et al. 2011). Soil moisture deficits increase the frequency and duration of extreme summer heat in southeastern Europe, where evapotranspiration is limited by soil moisture. In this region, dry soils prior to summer produce more frequent and long-lasting hot summer temperatures. For example, over the 40-years analyzed (1961–2000), the percentage of hot days was greater in drought years compared with wet years (Table 26.2). This is seen in both the median and the 90th percentile. The increase in hot days with drought is less in central Europe, where the climate is wetter and evapotranspiration is mostly radiation-limited. This relationship between dry soils and hot extremes is found in many regions of the world (Mueller and Seneviratne 2012). Climate models predict an increase in temperature and decrease in precipitation over mid-latitude continents throughout the twenty-first century (Collins et al. 2013), and soil moisture–temperature feedbacks are especially important with a warmer, drier climate (Seneviratne et al. 2006, 2013).

The severe 1988 drought in the Mississippi River basin and the extensive flooding in the same region in 1993 may be examples of the soil moisture–precipitation feedback in which soil moisture, through its effect on evapotranspiration, accentuates floods and droughts.

Atmospheric analyses suggest that although the events originated because of anomalous atmospheric circulation patterns, surface conditions amplified these anomalies (Trenberth et al. 1988; Namias 1991; Trenberth and Branstator 1992; Trenberth and Guillemot 1996). The 1993 flood was prolonged by increased evapotranspiration that increased precipitation. The 1988 drought originated due to anomalies in sea surface temperature that affected atmospheric general circulation, but was prolonged by reduced evapotranspiration. This conclusion is supported by climate model studies (Beljaars et al. 1996; Bosilovich and Sun 1999a,b; Viterbo and Betts 1999; Pal and Eltahir 2001, 2002; Sud et al. 2003).

Dry soils, by reducing evapotranspiration, provide an important feedback during droughts. For example, a modeling study by Schubert et al. (2004a,b) indicates that dry soils accentuated the 1930s drought in the United States Great Plains. The 1930s were a period of low rainfall and high temperature throughout much of the Great Plains (Figure 7.3a) – a period aptly characterized as the Dust Bowl because of frequent dust storms arising from extremely dry and barren soil. Tropical sea surface temperatures contribute to North American droughts by altering atmospheric general circulation (Hoerling and Kumar 2003; Seager and Hoerling 2014). Simulations by Schubert et al. (2004a,b) in which the climate model was run with observed sea surface temperatures for the period 1932–1938 demonstrate that anomalous temperatures in tropical oceans caused the drought. An additional simulation disabled interactions between soil moisture and evapotranspiration by prescribing soil wetness to a seasonal climatology. Eliminating interactive soil hydrology reduces the simulated decrease in precipitation, particularly during summer, whereas with interactive hydrology the model simulates precipitation reductions comparable to observations during the warm seasons (Figure 26.2).

Increased soil moisture, with greater evapotranspiration and flux of moisture into the boundary layer, does not necessarily produce more precipitation. The effect of soil moisture on boundary layer stability is critical to

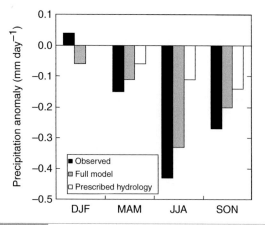

Fig. 26.2 Precipitation deficits for the period 1932–1938 averaged over the United States Great Plains. Data show the observed anomaly and anomalies simulated with and without interactive soil water. DJF, December–February; MAM, March–May; JJA, June–August; SON, September–November. Adapted from Schubert et al. (2004b).

understanding feedbacks with precipitation (Seneviratne et al. 2010). The cooler surface as a result of increased evapotranspiration may reduce atmospheric instability, thereby decreasing rather than increasing convective rainfall. Just such a process may suppress convective rainfall in the Sahel region of North Africa (Taylor and Ellis 2006). Cook et al. (2006) found that wet soils suppress rainfall in their simulations of the climate of southern Africa. Other modeling studies have found convective instability increases over dry soils (Seneviratne et al. 2010). In some regions, afternoon rainfall occurs preferentially over soils that are drier compared with the surrounding area, seen most prominently in semiarid climates (Taylor et al. 2012). These results suggest that soil moisture can act as a negative feedback to precipitation in some regions and show that our understanding of soil moisture–precipitation feedbacks is incomplete.

It is now recognized that realistic soil moisture initialization is an important determinant of temperature and precipitation forecasts (Koster et al. 2010, 2011; Seneviratne et al. 2010; van den Hurk et al. 2012; Kumar et al. 2014). Large soil moisture anomalies have a strong impact on the skill of the forecast,

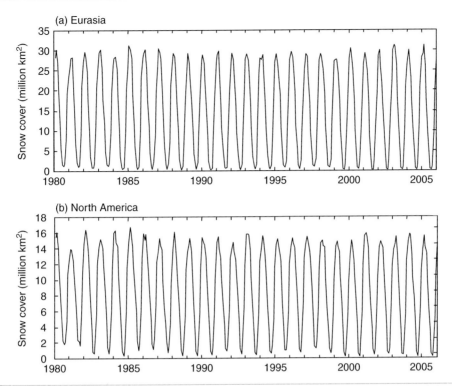

Fig. 26.3 Monthly snow cover extent during the period 1980–2005 for (a) Eurasia and (b) North America excluding Greenland. Data provided courtesy of David Robinson (Rutgers University).

and wet and dry anomalies have different impacts on skill based on radiation-limited and moisture-limited evapotranspiration regimes. Accurate representation of soil moisture feedbacks on climate requires careful consideration of soil characteristics (e.g., soil texture and water-holding capacity), hydraulic properties such as matric potential and hydraulic conductivity, and the effects of soil moisture on evapotranspiration.

26.3 | Snow

Snow cover has a pronounced annual cycle and at its annual maximum coats about 46 million km^2 of land in the Northern Hemisphere (Figure 26.3). The presence of snow on the ground affects climate through changes in surface albedo, soil insulation, and hydrology (Zhang 2005; Vavrus 2007; Xu and Dirmeyer 2013). Snow cover provides a significant negative radiative forcing, and the loss of snow cover over the past few decades has contributed to planetary warming (Flanner et al. 2011).

One of the primary climate effects arises from the high albedo of snow compared with snow-free surfaces. Albedos typically range from about 0.80–0.95 for fresh snow to as low as 0.10–0.20 for soil or vegetation (Table 12.1). The high albedo of snow is seen in satellite-derived maps of surface albedo, which show that snow-covered land has a higher albedo than snow-free land (Robinson and Kukla 1985; Jin et al. 2002; Barlage et al. 2005; Gao et al. 2005). This is particularly evident for barren land (Table 26.3). The albedo of bare surfaces increases from 0.2 without snow to more than 0.5 when covered by snow. Similar large increases in albedo with snow are seen in sparse vegetation and short stature vegetation. In contrast, the albedo of forests increases less with snow because of vegetation masking effects. Even when the ground is covered by snow,

Table 26.3 | Moderate Resolution Imaging Spectroradiometer (MODIS) mean surface albedo of Northern Hemisphere land cover types for winter (December–February) and summer (June–August) seasons

Land cover	Winter		Summer
	Snow-covered	Snow-free	
Cropland	0.55 ± 0.07	0.14 ± 0.02	0.18 ± 0.02
Grassland	0.57 ± 0.08	0.16 ± 0.02	0.18 ± 0.02
Evergreen forest	0.21 ± 0.04	0.09 ± 0.02	0.10 ± 0.01
Deciduous forest	0.24 ± 0.05	0.12 ± 0.02	0.15 ± 0.01
Bare	0.54 ± 0.11	0.21 ± 0.05	0.25 ± 0.06

Note: Mean ± one standard deviation. Data for Northern Hemisphere winter are given for snow-covered and snow-free surfaces.
Source: From Boisier et al. (2013).

foliage and wood effectively mask the underlying snow (Figure 17.5b).

A second way in which snow influences climate is during melting. When snow covers the ground, some net radiation on warm days is used to melt snow. This prevents the surface from warming above freezing until the snow melts. For example, the heat capacity of ice is approximately 2 MJ m^{-3} K^{-1}. The latent heat of fusion at 0°C is 334 MJ m^{-3} (334 J g^{-1}, Table 3.3), which is about 160 times that required to raise the temperature of ice by 1°C. Until this energy is supplied, temperature remains constant.

A third influence of snow is that its thermal conductivity is much less than that of soil. A typical value is 0.34 W m^{-1} K^{-1}, which is one-third to one-fifth that of mineral soil (Table 9.1). With low thermal conductivity, less heat is transferred by conduction. In winter, therefore, a deep snow pack on the ground acts as an insulating blanket that prevents soil from cooling. Figure 26.4 illustrates this insulating effect of snow. Prior to the onset of snow in November, air temperature and soil temperature are similar. As the air cools, the soil tracks air temperature to within 1°C. After a 10 cm snow pack covers the ground, the soil is several degrees warmer than air. Just as snow inhibits heat loss from the underlying soil in winter, it inhibits soil warming in spring. The low thermal conductivity of snow prevents heat gain by the soil. Once the snow is removed, the soil warms rapidly and again closely tracks air temperature.

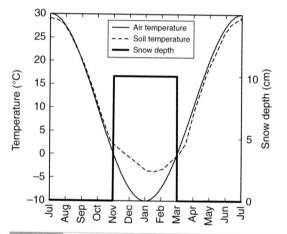

Fig. 26.4 Influence of snow cover on soil temperature from a model of soil temperature based on heat transfer and energy conservation. The left-hand axis shows air and soil temperature with no snow cover (July–November), with a 10-cm snow pack (November–March), and again with no snow (March–July). The right-hand axis shows the depth of snow on ground.

The presence of snow on the ground correlates with anomalously cold air temperatures (Walsh et al. 1982; Namias 1985; Leathers and Robinson 1993; Groisman et al. 1994; Mote 2008). Figure 26.5 illustrates this for a network of 91 stations in Northeast United States from Maine to West Virginia. Long-term mean climatological data show that both maximum and minimum temperatures are several degrees colder when snow is on the ground than when the ground is free of snow.

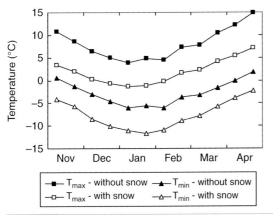

Fig. 26.5 Effect of snow cover on climatological mean daily maximum (T_{max}) and minimum (T_{min}) air temperature. Data from Leathers et al. (1995).

The correlation between snow cover and cold temperature clearly reflects variations in large-scale atmospheric circulation; cold air masses often produce snow in winter. Climate models can be used to study the extent to which snow cover feeds back to affect climate. Similar to the idealized experiments with soil moisture, climate can be simulated by models in which snow cover is prescribed to anomalously high or low values. Such simulations show that the presence of snow lowers air temperature (Walsh and Ross 1988; Cohen and Rind 1991; Walland and Simmonds 1997).

The effect of snow cover on climate is evident in the monsoon climates of India and Southeast Asia. Observations show that extensive snow cover in Eurasia leads to a weakened Indian summer monsoon with reduced precipitation (Hahn and Shukla 1976; Bamzai and Shukla 1999). Conversely, reduced snow cover leads to greater summer rainfall. The thermal contrast between land and ocean drives the monsoon. In summer, the Asian continent warms more than oceans. Low surface pressure over the continent draws cool, moist air inland from adjacent oceans, triggering heavy rainfall. Reduced snow cover in Asia results in a low land albedo, causing warmer land temperatures, greater land–sea temperature contrast, and a stronger summer monsoon. High snow cover leads to colder temperature, and the reduced land–sea temperature contrast weakens the summer monsoon.

Climate model simulations show that heavy snow cover produces a weakened Asian summer monsoon (Barnett et al. 1988, 1989; Douville and Royer 1996). Eurasian snow cover, particularly over Siberia, also affects hemispheric atmospheric circulation and plays an important role in wintertime extratropical Northern Hemisphere climate variability (Cohen and Entekhabi 1999; Cohen et al. 2001, 2012; Gong et al. 2002, 2003a,b; Saito and Cohen 2003; Fletcher et al. 2009; Smith et al. 2011). North American snow cover similarly has hemispheric climate teleconnections (Ge and Gong 2009; Sobolowski et al. 2010).

The North American monsoon may be another example of snow feedback on precipitation. The region of southwestern United States and northwestern Mexico receives most of its precipitation during the months of July–September. This is associated with a monsoonal circulation driven by summertime heating of land. Sea surface temperatures in the Pacific Ocean affect precipitation variability in this region, but interactions between land and atmosphere may also influence rainfall. Observational studies find a negative correlation between spring snow cover in the southern Rocky Mountains and summer rainfall in the United States Southwest; above-normal spring snow is followed by below-normal summer rain and vice versa (Gutzler and Preston 1997; Gutzler 2000; Zhu et al. 2005; Grantz et al. 2007).

The inverse relationship between snow and precipitation in the North American monsoon region could reflect land surface feedbacks that reduce the land–ocean temperature contrast. Delayed heating of land could arise from high spring albedo, more energy used to melt snow, and because of wetter than normal soil that increases evapotranspiration and cools surface temperature. However, the role of these feedbacks is uncertain. Observational studies do not provide unambiguous evidence of the key feedbacks among snow cover, soil moisture, and temperature (Matsui et al. 2003; Zhu et al. 2005). On the other hand, idealized climate model experiments do show such feedbacks (Small 2001; Notaro and Zarrin 2011; Feng et al. 2013), and changes in vegetation cover may also affect the monsoon (Notaro and Gutzler 2012).

26.4 | Leaf Phenology

The foliage of deciduous trees, shrubs, and grasses has a pronounced annual cycle in response to temperature and precipitation. In addition, the timing of leaf emergence and senescence and the amount of leaf area can vary from year-to-year based on prevailing temperature and precipitation, as seen in satellite datasets of leaf area (Buermann et al. 2002, 2003; Stöckli and Vidale 2004; Stöckli et al. 2011). Seasonal and interannual variability in leaf area influences atmospheric seasonality, most prominently atmospheric CO_2 concentration, but also temperature (Peñuelas et al. 2009; Richardson et al. 2013).

The seasonal emergence and senescence of leaves on deciduous trees alters sensible and latent heat fluxes (Table 17.3) and in doing so alters surface climate. In eastern United States, springtime air temperature is distinctly different after leaves emerge (Schwartz and Karl 1990; Schwartz 1992, 1996). Long-term measurements in 13 locations in north central and northeast United States show that daily maximum temperature steadily increases prior to leaf emergence, with particularly large increases in temperature during a two week period before leaves emerge (Figure 26.6). Temperature increases at a rate of 0.17°C per day well before springtime leaf emergence (46 days prior) and attains a peak rate of 0.31°C per day eight days prior to leaf emergence. After leaves emerge, the rate of temperature increase drops to less than 0.07°C per day. In this region, diurnal temperature range increases for several weeks prior to leaf emergence, but is essentially unchanged for several weeks following leafing out (Figure 26.7).

This temperature discontinuity over a period of less than a few weeks is related to increased transpiration upon leaf emergence that cools and moistens air (Fitzjarrald et al. 2001; Schwartz and Crawford 2001). The emergence of leaves in spring greatly alters the partitioning of net radiation into sensible and latent heat fluxes. Prior to leaf emergence, latent heat flux is typically a minor component of the surface energy budget, and the Bowen ratio is large. After leaf

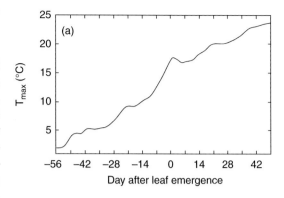

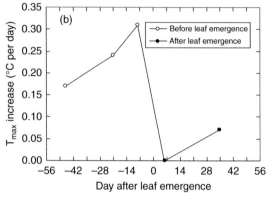

Fig. 26.6 Influence of springtime leaf emergence in eastern United States on (a) daily maximum temperature (T_{max}) and (b) rate of increase in T_{max}. Adapted from Schwartz and Karl (1990).

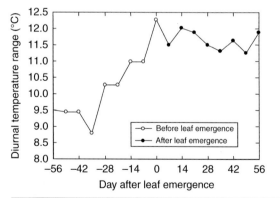

Fig. 26.7 Influence of springtime leaf emergence in eastern United States on diurnal temperature range. Adapted from Schwartz (1996).

emergence, latent heat flux increases, and the Bowen ratio decreases. The data in Table 17.3 show this response.

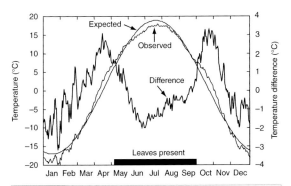

Fig. 26.8 Influence of leaf emergence on daily air temperature in west central Canada. The two thin lines show observed and expected temperature (left-hand axis). Expected temperature is from a sinusoidal curve fit to the observations. The thick line shows the difference between observed and expected temperatures (right-hand axis). The typical period in which leaves are present is also shown. Data from Hogg et al. (2000).

A distinct seasonal pattern to temperature that relates to leaf phenology is seen in west central Canada (Hogg et al. 2000). Long-term mean daily temperature for nine climate stations in the Canadian prairie provinces show that air temperature in this region is 2–3°C warmer than expected during April and October while summer temperature is up to 2°C cooler than expected (Figure 26.8). This seasonal pattern coincides with the absence or presence of leaves on deciduous trees in the forests of west central Canada. Greater sensible heat flux when leaves are absent produces the spring and autumn warming. Greater transpiration when leaves are present produces the summer cooling. A similar seasonal pattern does not occur in evergreen-dominated forests because of their low summer transpiration.

Levis and Bonan (2004) adapted a climate model to simulate leaf area based on prevailing meteorological conditions and applied the model to study the influence of leaf emergence on air temperature. Leaf area was simulated daily in response to air temperature for summergreen trees, and leaves emerged when accumulated growing degree-days exceeded some critical threshold. In one simulation, daily leaf area index was simulated interactively by the model; leaves emerge when environmental

conditions allow photosynthesis so that stomata open and transpiration commences. In a second simulation, leaf area index was prescribed according to a climatology obtained from the first simulation; leaves emerge according to calendar date rather than meteorological conditions. Surface air temperature in the prognostic leaf area simulation shows a marked increase in the days preceding leaf emergence, as in the observations (Figure 26.6). This reflects the fact that accumulated warm temperatures trigger leaf emergence in deciduous trees. This warm-up is absent with prescribed leaf area because there is no dependence of leaf emergence on temperature. The prognostic leaf area simulation also replicates the observed reduction in springtime warming trend after leaf emergence, but the prescribed leaf area simulation does not. In the former simulation, leaves emerge when conditions favor photosynthesis and transpiration. With prescribed leaf area, foliage may emerge when conditions such as cold air temperature or frozen soil preclude stomata from opening. The reduction in springtime warming only occurs when photosynthesis, stomatal conductance, and leaf emergence are synchronized with appropriate meteorological conditions.

Because of the importance of foliage in regulating surface climate, improved representations of leaf area and its phenology are included in climate models. Prognostic models of leaf area simulate the amount of foliage depending on temperature, precipitation, and plant productivity. Common predictors of leaf emergence and senescence utilize measures of accumulated springtime warmth above some temperature threshold (e.g., growing degree-days), accumulated winter temperature below some threshold (a chilling requirement), day length, and soil moisture to simulate evergreen, summergreen, and raingreen phenology. However, the phenological response to environmental drivers such as temperature, precipitation, and day length is poorly represented in models (Richardson et al. 2012, 2013).

Leaf phenology is likely to provide a feedback with climate change. In seasonally cold, extratropical climates, spring growth arrives earlier and autumn senescence is delayed with a warmer climate (Peñuelas et al. 2009; Jeong et al. 2011, 2013; Richardson et al. 2013). Interactive, rather than prescribed, vegetation and leaf area enhances variability in surface energy fluxes and precipitation (Delire et al. 2004, 2011; Crucifix et al. 2005; Wang et al. 2011). Vegetation can amplify the response to wet soil in a positive feedback because of greater leaf area that increases transpiration or provide negative feedback to a wet soil anomaly by depleting soil water (Kim and Wang 2007, 2012).

Studies of European heat waves show that delayed or weak growing season green-up amplifies extreme heat waves, such as during 2003 (Lorenz et al. 2013). The reduced leaf area during the heat wave amplified daily maximum temperature by about 0.5°C during the hottest period in August, about one-half of the warming caused by low soil moisture. In contrast, early and strong green-up contributes to enhanced evapotranspiration and surface cooling, thereby decreasing the magnitude of the warm temperature anomaly. Stéfanon et al. (2012) found similar positive and negative feedbacks in their climate simulations.

However, soil moisture–leaf area–evapotranspiration interactions can be complex, and even counterintuitive. For example, lower leaf area can dampen heat waves if the reduced evapotranspiration prevents soil moisture depletion during the summer. Moreover, long-term observations at four central and western European headwater catchments show that evapotranspiration increases during droughts (Teuling et al. 2013). Evapotranspiration in these catchments decreases only at low available soil moisture because there is sufficient water to evaporate. Instead, the reduced cloudiness during drought increases net radiation, and there is more energy available for evapotranspiration.

26.5 | Review Questions

1. Precipitation during the summer of 1956 was below normal over much of central and western United States. What impact did this probably have on soil water and how might soil water have affected rainfall? Describe a climate model experiment to test this.

2. Observations indicate that periods when snow is on the ground have air temperature colder than normal. Describe a climate model experiment to test whether this is due to the presence of snow.

3. Snow cover is anomalously high over Eurasia. Describe how summer rainfall in India is likely to be affected. Why? Describe a climate model experiment to test this hypothesis.

4. The following table shows daily average energy fluxes (W m⁻²) measured at a deciduous forest over three weekly periods during which time leaves emerge in spring. Which is the period prior to leaf emergence? Which is the period after leaves have fully emerged?

	Net radiation	Sensible heat	Latent heat
Period A	160	30	90
Period B	120	60	30
Period C	140	50	50

5. Satellite-derived monthly leaf area index is available for a 20-year period from 1981–2000. Describe a set of climate model experiments to examine the effect of interannual variation in leaf area index, as derived from satellite data, on climate variability.

26.6 | References

Bamzai, A. S., and Shukla, J. (1999). Relation between Eurasian snow cover, snow depth, and the Indian summer monsoon: An observational study. *Journal of Climate*, 12, 3117–3132.

Barlage, M., Zeng, X., Wei, H., and Mitchell, K. E. (2005). A global 0.05° maximum albedo dataset of snow-covered land based on MODIS observations. *Geophysical Research Letters*, 32, L17405, doi:10.1029/2005GL022881.

Barnett, T. P., Dümenil, L., Schlese, U., and Roeckner, E. (1988). The effect of Eurasian snow cover on global climate. *Science*, 239, 504–507.

Barnett, T. P., Dümenil, L., Schlese, U., Roeckner, E., and Latif, M. (1989). The effect of Eurasian snow cover on regional and global climate variations. *Journal of the Atmospheric Sciences*, 46, 661–685.

Beljaars, A. C. M., Viterbo, P., Miller, M. J., and Betts, A. K. (1996). The anomalous rainfall over the United States during July 1993: Sensitivity to land surface parameterization and soil moisture anomalies. *Monthly Weather Review*, 124, 362–383.

Betts, A. K. (2004). Understanding hydrometeorology using global models. *Bulletin of the American Meteorological Society*, 85, 1673–1688.

Boisier, J. P., de Noblet-Ducoudré, N., and Ciais, P. (2013). Inferring past land use-induced changes in surface albedo from satellite observations: A useful tool to evaluate model simulations. *Biogeosciences*, 10, 1501–1516.

Bosilovich, M. G., and Sun, W.-Y. (1999a). Numerical simulation of the 1993 midwestern flood: Local and remote sources of water. *Journal of Geophysical Research*, 104D, 19415–19423.

Bosilovich, M. G., and Sun, W.-Y. (1999b). Numerical simulation of the 1993 midwestern flood: Land–atmosphere interactions. *Journal of Climate*, 12, 1490–1505.

Buermann, W., Wang, Y., Dong, J., et al. (2002). Analysis of a multiyear global vegetation leaf area index data set. *Journal of Geophysical Research*, 107, 4646, doi:10.1029/2001JD000975.

Buermann, W., Anderson, B., Tucker, C. J., et al. (2003). Interannual covariability in Northern Hemisphere air temperatures and greenness associated with El Niño–Southern Oscillation and the Arctic Oscillation. *Journal of Geophysical Research*, 108, 4396, doi:10.1029/2002JD002630.

Cohen, J., and Entekhabi, D. (1999). Eurasian snow cover variability and Northern Hemisphere climate predictability. *Geophysical Research Letters*, 26, 345–348.

Cohen, J., and Rind, D. (1991). The effect of snow cover on the climate. *Journal of Climate*, 4, 689–706.

Cohen, J., Saito, K., and Entekhabi, D. (2001). The role of the Siberian high in Northern Hemisphere climate variability. *Geophysical Research Letters*, 28, 299–302.

Cohen, J. L., Furtado, J. C., Barlow, M. A., Alexeev, V. A., and Cherry, J. E. (2012). Arctic warming, increasing snow cover and widespread boreal winter cooling. *Environmental Research Letters*, 7, 014007, doi:10.1088/1748-9326/7/1/014007.

Collins, M., Knutti, R., Arblaster, J., et al. (2013). Long-term climate change: Projections, commitments and irreversibility. In *Climate Change 2013: The Physical Science Basis. Contribution of Working Group I to the Fifth Assessment Report of the Intergovernmental Panel on Climate Change*, ed. T. F. Stocker, D. Qin, G.-K. Plattner, et al. Cambridge: Cambridge University Press, pp. 1029–1136.

Cook, B. I., Bonan, G. B., and Levis, S. (2006). Soil moisture feedbacks to precipitation in southern Africa. *Journal of Climate*, 19, 4198–4206.

Crucifix, M., Betts, R. A., and Cox, P. M. (2005). Vegetation and climate variability: A GCM modelling study. *Climate Dynamics*, 24, 457–467.

Delire, C., Foley, J. A., and Thompson, S. (2004). Long-term variability in a coupled atmosphere-biosphere model. *Journal of Climate*, 17, 3947–3959.

Delire, C., de Noblet-Ducoudré, N., Sima, A., and Gouirand, I. (2011). Vegetation dynamics enhancing long-term climate variability confirmed by two models. *Journal of Climate*, 24, 2238–2257.

Dirmeyer, P. A., Koster, R. D., and Guo, Z. (2006). Do global models properly represent the feedback between land and atmosphere? *Journal of Hydrometeorology*, 7, 1177–1198.

Douville, H., and Royer, J.-F. (1996). Sensitivity of the Asian summer monsoon to an anomalous Eurasian snow cover within the Météo-France GCM. *Climate Dynamics*, 12, 449–466.

Durre, I., Wallace, J. M., and Lettenmaier, D. P. (2000). Dependence of extreme daily maximum temperatures on antecedent soil moisture in the contiguous United States during summer. *Journal of Climate*, 13, 2641–2651.

Feng, X., Bosilovich, M., Houser, P., and Chern, J.-D. (2013). Impact of land surface conditions on 2004 North American monsoon in GCM experiments. *Journal of Geophysical Research: Atmospheres*, 118, 293–305, doi:10.1029/2012JD018805.

Findell, K. L., and Eltahir, E. A. B. (1997). An analysis of the soil moisture-rainfall feedback, based on direct observations from Illinois. *Water Resources Research*, 33, 725–735.

Findell, K. L., Gentine, P., Lintner, B. R., and Kerr, C. (2011). Probability of afternoon precipitation in eastern United States and Mexico enhanced by high evaporation. *Nature Geoscience*, 4, 434–439.

Fischer, E. M., Seneviratne, S. I., Lüthi, D., and Schär, C. (2007). Contribution of land-atmosphere coupling to recent European summer heat waves. *Geophysical Research Letters*, 34, L06707, doi:10.1029/2006GL029068.

Fitzjarrald, D. R., Acevedo, O. C., and Moore, K. E. (2001). Climatic consequences of leaf presence in the eastern United States. *Journal of Climate*, 14, 598–614.

Flanner, M. G., Shell, K. M., Barlage, M., Perovich, D. K., and Tschudi, M. A. (2011). Radiative forcing and albedo feedback from the Northern Hemisphere cryosphere between 1979 and 2008. *Nature Geoscience*, 4, 151–155.

Fletcher, C. G., Hardiman, S. C., Kushner, P. J., and Cohen, J. (2009). The dynamical response to snow cover perturbations in a large ensemble of atmospheric GCM integrations. *Journal of Climate*, 22, 1208–1222.

Gao, F., Schaaf, C. B., Strahler, A. H., et al. (2005). MODIS bidirectional reflectance distribution function and albedo Climate Modeling Grid products and the variability of albedo for major global vegetation types. *Journal of Geophysical Research*, 110, D01104, doi:10.1029/2004JD005190.

Ge, Y., and Gong, G. (2009). North American snow depth and climate teleconnection patterns. *Journal of Climate*, 22, 217–233.

Gong, G., Entekhabi, D., and Cohen, J. (2002). A large-ensemble model study of the wintertime AO-NAO and the role of interannual snow perturbations. *Journal of Climate*, 15, 3488–3499.

Gong, G., Entekhabi, D., and Cohen, J. (2003a). Modeled Northern Hemisphere winter climate response to realistic Siberian snow anomalies. *Journal of Climate*, 16, 3917–3931.

Gong, G., Entekhabi, D., and Cohen, J. (2003b). Relative impacts of Siberian and North American snow anomalies on the winter Arctic Oscillation. *Geophysical Research Letters*, 30, 1848, doi:10.1029/2003GL017749.

Grantz, K., Rajagopalan, B., Clark, M., and Zagona, E. (2007). Seasonal shifts in the North American monsoon. *Journal of Climate*, 20, 1923–1935.

Groisman, P. Y., Karl, T. R., and Knight, R. W. (1994). Observed impact of snow cover on the heat balance and the rise of continental spring temperatures. *Science*, 263, 198–200.

Guo, Z., Dirmeyer, P. A., Koster, R. D., et al. (2006). GLACE: The Global Land–Atmosphere Coupling Experiment, Part II: Analysis. *Journal of Hydrometeorology*, 7, 611–625.

Gutzler, D. S. (2000). Covariability of spring snowpack and summer rainfall across the Southwest United States. *Journal of Climate*, 13, 4018–4027.

Gutzler, D. S., and Preston, J. W. (1997). Evidence for a relationship between spring snow cover in North America and summer rainfall in New Mexico. *Geophysical Research Letters*, 24, 2207–2210.

Hahn, D. G., and Shukla, J. (1976). An apparent relationship between Eurasian snow cover and Indian monsoon rainfall. *Journal of the Atmospheric Sciences*, 33, 2461–2462.

Hirschi, M., Seneviratne, S. I., Alexandrov, V., et al. (2011). Observational evidence for soil-moisture impact on hot extremes in southeastern Europe. *Nature Geoscience*, 4, 17–21.

Hoerling, M., and Kumar, A. (2003). The perfect ocean for drought. *Science*, 299, 691–694.

Hogg, E. H., Price, D. T., and Black, T. A. (2000). Postulated feedbacks of deciduous forest phenology on seasonal climate patterns in the western Canadian interior. *Journal of Climate*, 13, 4229–4243.

Huang, J., and Van den Dool, H. M. (1993). Monthly precipitation–temperature relations and temperature prediction over the United States. *Journal of Climate*, 6, 1111–1132.

Jaeger, E. B., and Seneviratne, S. I. (2011). Impact of soil moisture–atmosphere coupling on European climate extremes and trends in a regional climate model. *Climate Dynamics*, 36, 1919–1939.

Jeong, S.-J., Ho, C.-H., Gim, H.-J., and Brown, M. E. (2011). Phenology shifts at start vs. end of growing season in temperate vegetation over the Northern Hemisphere for the period 1982–2008. *Global Change Biology*, 17, 2385–2399.

Jeong, S.-J., Medvigy, D., Shevliakova, E., and Malyshev, S. (2013). Predicting changes in temperate forest budburst using continental-scale observations and models. *Geophysical Research Letters*, 40, 359–364, doi:10.1029/2012GL054431.

Jin, Y., Schaaf, C. B., Gao, F., et al. (2002). How does snow impact the albedo of vegetated land surfaces as analyzed with MODIS data? *Geophysical Research Letters*, 29, 1374, doi:10.1029/2001GL014132.

Karl, T. R. (1986). The relationship of soil moisture parameterizations to subsequent seasonal and monthly mean temperature in the United States. *Monthly Weather Review*, 114, 675–686.

Karl, T. R., and Quayle, R. G. (1981). The 1980 summer heat wave and drought in historical perspective. *Monthly Weather Review*, 109, 2055–2073.

Kim, Y., and Wang, G. (2007). Impact of vegetation feedback on the response of precipitation to antecedent soil moisture anomalies over North America. *Journal of Hydrometeorology*, 8, 534–550.

Kim, Y., and Wang, G. (2012). Soil moisture–vegetation–precipitation feedback over North America: Its sensitivity to soil moisture climatology. *Journal of Geophysical Research*, 117, D18115, doi:10.1029/2012JD017584.

Koster, R. D., and Suarez, M. J. (2004). Suggestions in the observational record of land–atmosphere feedback operating at seasonal time scales. *Journal of Hydrometeorology*, 5, 567–572.

Koster, R. D., Dirmeyer, P. A., Hahmann, A. N., et al. (2002). Comparing the degree of land–atmosphere interaction in four atmospheric general circulation models. *Journal of Hydrometeorology*, 3, 363–375.

Koster, R. D., Suarez, M. J., Higgins, R. W., and Van den Dool, H. M. (2003). Observational evidence that soil moisture variations affect precipitation. *Geophysical Research Letters*, 30, 1241, doi:10.1029/2002GL016571.

Koster, R. D., Dirmeyer, P. A., Guo, Z., et al. (2004). Regions of strong coupling between soil moisture and precipitation. *Science*, 305, 1138–1140.

Koster, R. D., Guo, Z., Dirmeyer, P. A., et al. (2006a). GLACE: The Global Land–Atmosphere Coupling Experiment, Part I: Overview. *Journal of Hydrometeorology*, 7, 590–610.

Koster, R. D., Suarez, M. J., and Schubert, S. D. (2006b). Distinct hydrological signatures in observed historical temperature fields. *Journal of Hydrometeorology*, 7, 1061–1075.

Koster, R. D., Schubert, S. D., and Suarez, M. J. (2009). Analyzing the concurrence of meteorological droughts and warm periods, with implications for the determination of evaporative regime. *Journal of Climate*, 22, 3331–3341.

Koster, R. D., Mahanama, S. P. P., Yamada, T. J., et al. (2010). Contribution of land surface initialization to subseasonal forecast skill: First results from a multi-model experiment. *Geophysical Research Letters*, 37, L02402, doi:10.1029/2009GL041677.

Koster, R. D., Mahanama, S. P. P., Yamada, T. J., et al. (2011). The second phase of the Global Land–Atmosphere Coupling Experiment: Soil moisture contributions to subseasonal forecast skill. *Journal of Hydrometeorology*, 12, 805–822.

Kumar, S., Dirmeyer, P. A., Lawrence, D. M., et al. (2014). Effects of realistic land surface initializations on subseasonal to seasonal soil moisture and temperature predictability in North America and in changing climate simulated by CCSM4. *Journal of Geophysical Research: Atmospheres*, 119, 13250–13270, doi:10.1002/2014JD022110.

Leathers, D. J., and Robinson, D. A. (1993). The association between extremes in North American snow cover extent and United States temperatures. *Journal of Climate*, 6, 1345–1355.

Leathers, D. J., Ellis, A. W., and Robinson, D. A. (1995). Characteristics of temperature depressions associated with snow cover across the Northeast United States. *Journal of Applied Meteorology*, 34, 381–390.

Levis, S., and Bonan, G. B. (2004). Simulating springtime temperature patterns in the Community Atmosphere Model coupled to the Community Land Model using prognostic leaf area. *Journal of Climate*, 17, 4531–4540.

Lorenz, R., Davin, E. L., Lawrence, D. M., Stöckli, R., and Seneviratne, S. I. (2013). How important is vegetation phenology for European climate and heat waves? *Journal of Climate*, 26, 10077–10100.

Madden, R. A., and Williams, J. (1978). The correlation between temperature and precipitation in the United States and Europe. *Monthly Weather Review*, 106, 142–147.

Matsui, T., Lakshmi, V., and Small, E. (2003). Links between snow cover, surface skin temperature, and rainfall variability in the North American monsoon system. *Journal of Climate*, 16, 1821–1829.

Mote, T. L. (2008). On the role of snow cover in depressing air temperature. *Journal of Applied Meteorology and Climatology*, 47, 2008–2022.

Mueller, B., and Seneviratne, S. I. (2012). Hot days induced by precipitation deficits at the global scale. *Proceedings of the National Academy of Sciences USA*, 109, 12398–12403.

Namias, J. (1983). Some causes of United States drought. *Journal of Climate and Applied Meteorology*, 22, 30–39.

Namias, J. (1985). Some empirical evidence for the influence of snow cover on temperature and precipitation. *Monthly Weather Review*, 113, 1542–1553.

Namias, J. (1991). Spring and summer 1988 drought over the contiguous United States – causes and prediction. *Journal of Climate*, 4, 54–65.

Notaro, M., and Gutzler, D. (2012). Simulated impact of vegetation on climate across the North American monsoon region in CCSM3.5. *Climate Dynamics*, 38, 795–814.

Notaro, M., and Zarrin, A. (2011). Sensitivity of the North American monsoon to antecedent Rocky Mountain snowpack. *Geophysical Research Letters*, 38, L17403, doi:10.1029/2011GL048803.

Pal, J. S., and Eltahir, E. A. B. (2001). Pathways relating soil moisture conditions to future summer rainfall within a model of the land–atmosphere system. *Journal of Climate*, 14, 1227–1242.

Pal, J. S., and Eltahir, E. A. B. (2002). Teleconnections of soil moisture and rainfall during the 1993 midwest summer flood. *Geophysical Research Letters*, 29, 1865, doi:10.1029/2002GL014815.

Peñuelas, J., Rutishauser, T., and Filella, I. (2009). Phenology feedbacks on climate change. *Science*, 324, 887–888.

Pitman, A. J. (2003). The evolution of, and revolution in, land surface schemes designed for climate models. *International Journal of Climatology*, 23, 479–510.

Richardson, A. D., Anderson, R. S., Arain, M. A., et al. (2012). Terrestrial biosphere models need better representation of vegetation phenology: Results from the North American Carbon Program Site Synthesis. *Global Change Biology*, 18, 566–584.

Richardson, A. D., Keenan, T. F., Migliavacca, M., et al. (2013). Climate change, phenology, and phenological control of vegetation feedbacks to the climate system. *Agricultural and Forest Meteorology*, 169, 156–173.

Robinson, D. A., and Kukla, G. (1985). Maximum surface albedo of seasonally snow-covered lands in the Northern Hemisphere. *Journal of Climate and Applied Meteorology*, 24, 402–411.

Saito, K., and Cohen, J. (2003). The potential role of snow cover in forcing interannual variability of the major Northern Hemisphere mode. *Geophysical Research Letters*, 30, 1302, doi:10.1029/2002GL016341.

Salvucci, G. D., Saleem, J. A., and Kaufmann, R. (2002). Investigating soil moisture feedbacks on precipitation with tests of Granger causality. *Advances in Water Resources*, 25, 1305–1312.

Schubert, S. D., Suarez, M. J., Pegion, P. J., Koster, R. D., and Bacmeister, J. T. (2004a). Causes of long-term drought in the U.S. Great Plains. *Journal of Climate*, 17, 485–503.

Schubert, S. D., Suarez, M. J., Pegion, P. J., Koster, R. D., and Bacmeister, J. T. (2004b). On the cause of the 1930s Dust Bowl. *Science*, 303, 1855–1859.

Schwartz, M. D. (1992). Phenology and springtime surface-layer change. *Monthly Weather Review*, 120, 2570–2578.

Schwartz, M. D. (1996). Examining the spring discontinuity in daily temperature ranges. *Journal of Climate*, 9, 803–808.

Schwartz, M. D., and Crawford, T. M. (2001). Detecting energy-balance modifications at the onset of spring. *Physical Geography*, 22, 394–409.

Schwartz, M. D., and Karl, T. R. (1990). Spring phenology: Nature's experiment to detect the effect of "green-up" on surface maximum temperatures. *Monthly Weather Review*, 118, 883–890.

Seager, R., and Hoerling, M. (2014). Atmosphere and ocean origins of North American droughts. *Journal of Climate*, 27, 4581–4606.

Seneviratne, S. I., Lüthi, D., Litschi, M., and Schär, C. (2006). Land–atmosphere coupling and climate change in Europe. *Nature*, 443, 205–209.

Seneviratne, S. I., Corti, T., Davin, E. L., et al. (2010). Investigating soil moisture–climate interactions in a changing climate: A review. *Earth-Science Reviews*, 99, 125–161.

Seneviratne, S. I., Wilhelm, M., Stanelle, T., et al. (2013). Impact of soil moisture–climate feedbacks on CMIP5 projections: First results from the GLACE-CMIP5 experiment. *Geophysical Research Letters*, 40, 5212–5217, doi:10.1002/grl.50956.

Shukla, J., and Mintz, Y. (1982). Influence of land-surface evapotranspiration on the Earth's climate. *Science*, 215, 1498–1501.

Small, E. E. (2001). The influence of soil moisture anomalies on variability of the North American monsoon system. *Geophysical Research Letters*, 28, 139–142.

Smith, K. L., Kushner, P. J., and Cohen, J. (2011). The role of linear interference in Northern Annular Mode variability associated with Eurasian snow cover extent. *Journal of Climate*, 24, 6185–6202.

Sobolowski, S., Gong, G., and Ting, M. (2010). Modeled climate state and dynamic responses to anomalous North American snow cover. *Journal of Climate*, 23, 785–799.

Stéfanon, M., Drobinski, P., D'Andrea, F., and de Noblet-Ducoudré, N. (2012). Effects of interactive vegetation phenology on the 2003 summer heat waves. *Journal of Geophysical Research*, 117, D24103, doi:10.1029/2012JD018187.

Stöckli, R., and Vidale, P. L. (2004). European plant phenology and climate as seen in a 20-year AVHRR land-surface parameter dataset. *International Journal of Remote Sensing*, 25, 3303–3330.

Stöckli, R., Rutishauser, T., Baker, I., Liniger, M. A., and Denning A. S. (2011). A global reanalysis of vegetation phenology. *Journal of Geophysical Research*, 116, G03020, doi:10.1029/2010JG001545.

Sud, Y. C., Mocko, D. M., Lau, K.-M., and Atlas, R. (2003). Simulating the midwestern U.S. drought of 1988 with a GCM. *Journal of Climate*, 16, 3946–3965.

Taylor, C. M., and Ellis, R. J. (2006). Satellite detection of soil moisture impacts on convection at the mesoscale. *Geophysical Research Letters*, 33, L03404, doi:10.1029/2005GL025252.

Taylor, C. M., de Jeu, R. A. M., Guichard, F., Harris, P. P., and Dorigo, W. A. (2012). Afternoon rain more likely over drier soils. *Nature*, 489, 423–426.

Teuling, A. J., Hirschi, M., Ohmura, A., et al. (2009). A regional perspective on trends in continental evaporation. *Geophysical Research Letters*, 36, L02404, doi:10.1029/2008GL036584.

Teuling, A. J., Van Loon, A. F., Seneviratne, S. I., et al. (2013). Evapotranspiration amplifies European summer drought. *Geophysical Research Letters*, 40, 2071–2075, doi:10.1002/grl.50495.

Trenberth, K. E., and Branstator, G. W. (1992). Issues in establishing causes of the 1988 drought over North America. *Journal of Climate*, 5, 159–172.

Trenberth, K. E., and Guillemot, C. J. (1996). Physical processes involved in the 1988 drought and 1993 floods in North America. *Journal of Climate*, 9, 1288–1298.

Trenberth, K. E., Branstator, G. W., and Arkin, P. A. (1988). Origins of the 1988 North American drought. *Science*, 242, 1640–1645.

van den Hurk, B., Doblas-Reyes, F., Balsam, G., et al. (2012). Soil moisture effects on seasonal temperature and precipitation forecast scores in Europe. *Climate Dynamics*, 38, 349–362.

Vavrus, S. (2007). The role of terrestrial snow cover in the climate system. *Climate Dynamics*, 29, 73–88.

Viterbo, P., and Betts, A. K. (1999). Impact of the ECMWF reanalysis soil water on forecasts of the July 1993 Mississippi flood. *Journal of Geophysical Research*, 104D, 19361–19366.

Walland, D. J., and Simmonds, I. (1997). Modelled atmospheric response to changes in Northern Hemisphere snow cover. *Climate Dynamics*, 13, 25–34.

Walsh, J. E., and Ross, B. (1988). Sensitivity of 30-day dynamical forecasts to continental snow cover. *Journal of Climate*, 1, 739–754.

Walsh, J. E., Tucek, D. R., and Peterson, M. R. (1982). Seasonal snow cover and short-term climatic fluctuations over the United States. *Monthly Weather Review*, 110, 1474–1485.

Wang, G., Sun, S., and Mei, R. (2011). Vegetation dynamics contributes to the multi-decadal variability of precipitation in the Amazon region. *Geophysical Research Letters*, 38, L19703, doi:10.1029/2011GL049017.

Wei, J., and Dirmeyer, P. A. (2012). Dissecting soil moisture-precipitation coupling. *Geophysical Research Letters*, 39, L19711, doi:10.1029/2012GL053038.

Xu, L., and Dirmeyer, P. (2013). Snow–atmosphere coupling strength, Part II: Albedo effect versus hydrological effect. *Journal of Hydrometeorology*, 14, 404–418.

Zhang, T. (2005). Influence of the seasonal snow cover on the ground thermal regime: an overview. *Reviews of Geophysics*, 43, RG4002, doi:10.1029/2004RG000157.

Zhu, C., Lettenmaier, D. P., and Cavazos, T. (2005). Role of antecedent land surface conditions on North American monsoon rainfall variability. *Journal of Climate*, 18, 3104–3121.

Biogeophysical Climate–Vegetation Dynamics

27.1 | Chapter Summary

This chapter examines the biogeophysical coupling between terrestrial vegetation and climate. Daisyworld is first introduced as a simple model of coupled climate–vegetation dynamics. The Daisyworld model illustrates the potential for regulation of climate by vegetation. More realistic examples of climate–vegetation interactions occur regionally in response to gradients in precipitation or temperature. The gradient from tropical rainforest to tropical deciduous forest to savanna to desert represents increasing aridity. This gradient is not only a response to soil moisture but also feeds back to affect climate, especially precipitation. An example of this is in northern and western Africa, where climate model simulations demonstrate that the expansion of vegetation into desert in response to increased rainfall feeds back to increase rainfall. Another example is the boreal forest–tundra ecotone. The transition from forest to tundra relates to cold temperature. Numerous studies show that the northward migration of trees in response to climate warming feeds back to accentuate the warming; loss of tree cover with a cold climate reinforces cold temperatures. These studies indicate widespread changes in vegetation structure and biogeography in response to climate change are likely to themselves change climate. Carbon cycle feedbacks are considered in Chapter 29.

27.2 | Biogeophysical Feedbacks

The studies outlined in Chapter 26 highlight biogeophysical processes by which Earth's land and its vegetation affect seasonal-to-interannual climate variability. Changes in vegetation structure, composition, and biogeography in response to long-term climate change similarly feed back to influence climate change. This arises from differences among vegetation in albedo, surface roughness, leaf area index, rooting depth, and canopy conductance.

The albedo of land varies with surface characteristics (Table 12.1). Vegetation generally has a lower albedo than bare soil; forests have a lower albedo than pastures or croplands. As a result, changes in vegetation such as grassland degradation or expansion and forest loss or woody encroachment alter surface albedo. An increase in surface albedo decreases net radiation at the land surface and reduces heating of the atmospheric boundary layer as well as water vapor in the boundary layer. A positive feedback develops if these changes decrease precipitation because drier soil directly increases albedo and further reduces vegetation cover (Figure 27.1a).

Surface roughness varies with land cover (Table 13.1). Vegetation has a larger roughness length than bare ground; forests have a larger roughness length than grasses. Rough surfaces generate more turbulence and have higher sensible and latent heat fluxes than smooth surfaces,

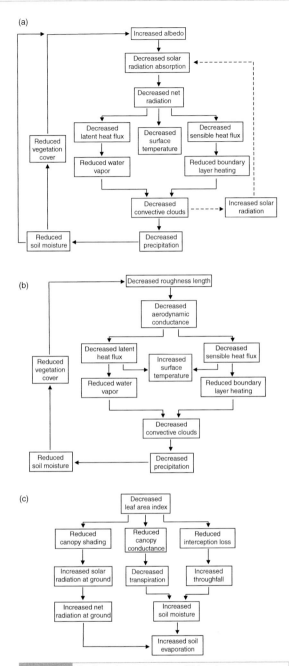

Fig. 27.1 Surface climate impacts of (a) an increase in albedo, (b) a decrease in surface roughness, and (c) a decrease in leaf area index as it affects evapotranspiration. Dashed lines represent negative feedback. Adapted from Pitman (2003).

all other factors being equal. A decrease in roughness length, by reducing aerodynamic conductance, can lead to a warmer, drier atmospheric boundary layer (Figure 27.1b).

Changes in leaf area index, rooting depth, and canopy conductance alter surface climate by modifying canopy processes (Chapter 17). Surface albedo varies with leaf area index (Figure 17.5). Canopy conductance generally decreases with lower leaf area index because there is less surface area for transpiration (Figure 17.9). Canopy conductance also varies with the photosynthetic capacity of leaves. The distinction between C_3 and C_4 plants is particularly important. The roots of trees generally extend deeper in the soil than those of herbaceous plants, providing a potentially larger pool of water to sustain transpiration (Figure 24.10). Reduced canopy conductance decreases latent heat flux, increases sensible heat flux, and can lead to a warmer, dryer, and deeper atmospheric boundary layer. A reduction in leaf area index also decreases the amount of precipitation intercepted by the canopy, with more rainfall reaching the soil and increased soil wetness (Figure 27.1c).

Figure 27.2 illustrates coupled climate–vegetation dynamics over periods of decades to millennia. Successional changes in community composition and vegetation structure in response to recurring disturbance initiate a natural cycle to ecosystem development in which, for example, a clearing is reforested. Climate change is superimposed on this successional backdrop so that changes in temperature, precipitation, and atmospheric CO_2 concentration alter ecosystem processes and might, for example, convert grassland to forest. Human activities also disturb the landscape, initiating vegetation change. The introduction of invasive or non-native species can alter resource utilization and the competitive balance among species. Abandonment of farmland initiates the successional regrowth of trees and grasses.

The nature of coupled climate–vegetation dynamics is difficult to document through direct observations. Instead, climate models

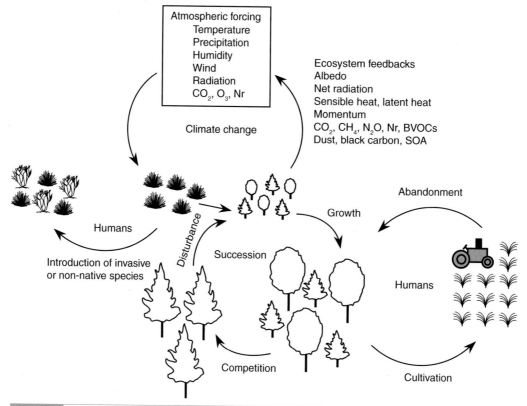

Fig. 27.2 Climate–vegetation dynamics showing the cycles of vegetation change due to climate change, succession, and human intervention.

are used. Paired climate model simulations that replace one type of vegetation with another illustrate the potential climate change arising from altered vegetation (Figure 25.11). Dynamic global vegetation models allow for direct interactive coupling of climate and vegetation (Figure 25.13). Most such studies have focused on physical climate and the effects of changing vegetation structure and community composition. In these models, vegetation growth and biogeography are influenced by temperature, precipitation, and other climatic variables. In turn, vegetation height, leaf area, rooting depth, and community composition influence albedo, radiative exchange, turbulent fluxes, and hydrology. The initial emphasis on biogeophysical feedbacks associated with coupled climate–vegetation dynamics has been expanded to include biogeochemical feedbacks,

especially the role of terrestrial ecosystems in the carbon cycle (Chapter 29).

27.3 | Daisyworld

The simplest depiction of coupled climate–vegetation dynamics is Daisyworld (Watson and Lovelock 1983; Wood et al. 2008). Daisyworld is a mathematical model of a planet with two types of daisies of different colors. One is black, has a low albedo, and reflects less solar radiation than soil. The other is white, has a high albedo, and reflects more solar radiation than soil. Black daisies, with their lower albedo, absorb more solar radiation and are locally warmer than white daisies. The growth of daisies depends on local temperature scaled to zero at temperatures of 5°C and 40°C and optimum

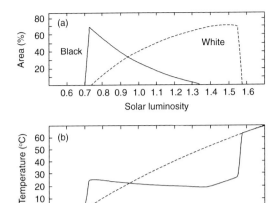

Fig. 27.3 Steady state response of Daisyworld to solar luminosity. Luminosity is the fractional change in solar constant. (a) Area of black and white daisies. (b) Planetary temperature. The dashed line shows planetary temperature without daisies. Adapted from Watson and Lovelock (1983).

at 22.5°C. This introduces a feedback in which the extent of daisy coverage affects temperature, which in turn affects daisy growth and coverage. Black daisies are warmer than white daisies and dominate in cold climates. However, an increase in the area of black daisies lowers planetary albedo, warms climate, and creates conditions in which white daisies are favored. White daisies have the opposite effect.

Figure 27.3 shows steady state values of planetary temperature and areas of black and white daisies as solar luminosity increases. The cold climates occurring at low luminosity favor black daisies because they warm the planet. White daisies are at a disadvantage because they create locally colder temperature. As black daisies spread across the planet, temperature increases. Their growth declines in the warm climate above 22.5°C while that of white daisies increases. As solar luminosity increases, therefore, the area of black daisies declines while that of white daisies, which cool the planet, increases. Across a wide range of luminosity, planetary temperature is close to optimum. Only at high (low) luminosity is temperature too hot (cold) for life. Daisyworld has generated much scientific attention because it illustrates the regulation of climate by life.

27.4 | Northern Africa

The Sahara Desert is a well studied example of the importance of vegetation to climate. About 6000 years before present (6 kyr BP), the climate of North Africa was much wetter than today (Street-Perrott and Perrott 1993; Joussaume et al. 1999; Braconnot et al. 2000, 2007). Milankovitch changes in orbital geometry increased summer solar radiation (Figure 8.4), heated the land, and strengthened the African summer monsoon. Paleobotanical data indicate grasses and shrubs covered much of North Africa, including areas that are presently desert, as a result of the wetter climate (Hoelzmann et al. 1998; Jolly et al. 1998; Prentice et al. 2000). Climate simulations show that expansion of grasses and shrubs in response to increased summer precipitation amplified the precipitation response to orbital geometry. Decreased surface albedo and increased evapotranspiration as the desert soil was vegetated strengthened the monsoonal rains.

One approach to examine vegetation feedbacks is to simulate the climate of North Africa for the period 6 kyr BP with desert replaced by vegetation. Climate model experiments by Kutzbach et al. (1996) illustrate this methodology (Table 27.1). A control simulation was performed for modern conditions. A second simulation used the orbital geometry of 6 kyr BP but with modern desert vegetation. In a third simulation for 6 kyr BP, vegetation between latitudes 15° N and 30° N was changed from desert to grassland. A fourth simulation additionally increased soil water-holding capacity and reduced soil albedo. Greater summer solar radiation due to orbital geometry increases annual precipitation between latitude 15° N and 22° N by 12 percent compared with the control simulation. Cloud cover increases, as does atmospheric moisture. Latent heat flux increases because of more net radiation at the surface and because the soils are wetter. Replacement of desert with grassland and soil enhances the summer monsoon, and the climatic response equals or exceeds that of the orbital forcing alone. Net radiation increases as a result of reduced albedo, latent heat flux increases, and the near-surface

Table 27.1 | Effect of vegetation and soil on the annual mean climate of North Africa 6 kyr BP as determined from four climate model simulations

	Control (C)	Radiative forcing (R − C)	Vegetation (RV − R)	Vegetation and soil (RVS − R)
Surface albedo (fraction)	0.25	0.00	0.00	−0.03
Net radiation (W m^{-2})	94	1	3	9
Sensible heat (W m^{-2})	56	−3	1	4
Latent heat (W m^{-2})	38	4	2	5
Precipitation (mm day^{-1})	1.28	0.15	0.08	0.20
Near-surface specific humidity (g kg^{-1})	6.2	0.6	0.3	0.6
Total cloud (fraction)	0.33	0.04	−0.01	0.00

Note: C, control simulation with modern solar radiation and vegetation. R, orbital geometry of 6 kyr BP and modern vegetation. RV, orbital geometry and vegetation of 6 kyr BP. RVS, orbital geometry, vegetation, and soil of 6 kyr BP. Radiative forcing is the difference between the radiative (R) and control (C) simulations. Vegetation and soil forcings are the climate change in addition to the radiative forcing. Data are averaged between latitude 15–22° N and longitude 0–50° E.
Source: From Kutzbach et al. (1996).

atmosphere moistens. Annual precipitation increases by 18 percent in the grassland simulation and by 28 percent in the grassland and soil simulation compared with the control simulation. The increase in precipitation reduces the area of the Sahara by 11 percent due to orbital forcing alone and by 20 percent due to feedback from vegetation and soil changes.

Climate simulations with asynchronous equilibrium vegetation coupling simulate vegetation change rather than prescribing vegetation change (Figure 25.12a). These studies also show that the geographic expansion of vegetation enhanced the summer monsoon 6 kyr BP. Figure 27.4 shows changes in vegetation cover during one such climate–vegetation simulation (Claussen and Gayler 1997). The climate model was forced with the solar radiation of 6 kyr BP, and vegetation was simulated with a biogeography model. From an initial condition of extensive desert and sparse savanna, interactive vegetation changes climate such that desert shrinks while savanna, woodland, and grassland expand. The climate–vegetation model converges on an equilibrium solution in which desert is reduced by 50 percent from its initial extent. This represents a northward shift of savanna of some 600 km in the western region of the Sahara.

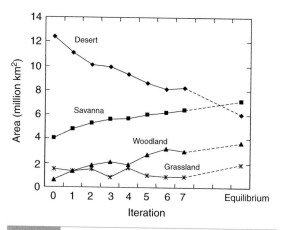

Fig. 27.4 Area of vegetation in North Africa (latitude 8.5°–36.6° N, longitude 15.5° W–52.0° E) 6 kyr BP in relation to iteration for an asynchronously coupled climate–vegetation model. Adapted from Claussen and Gayler (1997).

The northward shift of savanna amplifies the climate response to orbital forcing (Table 27.2). With interactive vegetation, surface temperature cools by 3.1°C and precipitation quadruples compared with a simulation that used a prescribed desert surface. Evapotranspiration increases because of wetter soils, but net water gain ($P − E$) increases from 8 mm per month

Table 27.2 North Africa summer climate 6 kyr BP from climate simulations with prescribed modern desert vegetation and with coupled vegetation

	Prescribed desert	Interactive vegetation
Surface temperature (°C)	34.6	31.5
Hydrologic cycle		
Precipitation, P (mm month^{-1})	30	129
Evapotranspiration, E (mm month^{-1})	22	85
$P - E$ (mm month^{-1})	8	44
Radiative fluxes		
Incident solar radiation (W m^{-2})	339	284
Reflected solar radiation (W m^{-2})	117	62
Net solar radiation (W m^{-2})	222	222
Albedo (fraction)	0.34	0.22
Net longwave radiation (W m^{-2})	−120	−90
Net radiation (W m^{-2})	102	132
Turbulent surface fluxes		
Sensible heat (W m^{-2})	62	47
Latent heat (W m^{-2})	21	82
Bowen ratio (fraction)	3.0	0.6

Note: Data are averaged for June–August and are spatially averaged between latitude 15–30° N and longitude 10° W–30° E.
Source: From Claussen and Gayler (1997).

with desert to 44 mm per month with interactive vegetation. The cooler, wetter climate alters net radiation at the surface and the partitioning of this energy into latent and sensible heat. The vegetated surface receives 30 W m^{-2} more net radiation than the desert. This energy is used to evaporate water. The Bowen ratio (i.e., the ratio of sensible to latent heat) decreases from 3 with desert to 0.6 with interactive vegetation. The simulated climate and biogeography is in better agreement with observations with interactive vegetation than without.

Other climate simulations with synchronously coupled dynamic vegetation models (Figure 25.12b) also show that desert greening enhances precipitation (Claussen 2009). For example, Levis et al. (2004) performed five climate model simulations: a control with present-day climate forcings (greenhouse gases, orbital geometry) and prescribed present-day vegetation, denoted as 0k0v; climate forcings for 6 kyr BP and prescribed present-day vegetation (6k0v); climate forcings for 6 kyr BP and dynamic vegetation (6k6v); as in the preceding simulation (6k6v) but with loam soil to increase soil water-holding capacity (6k6vt); and as in 6k6v but with loam soil and decreased soil albedo (6k6vtc). Climate forcings alone show an enhanced North Africa summer monsoon, as expected from prior studies (Table 27.3, 6k0v). Summer precipitation is 79 percent greater than in the control simulation (0k0v). With dynamic vegetation (6k6v), grasses encroach northwards and cover 28 percent of the surface (versus 11% present-day). However, this expansion of vegetation does not increase precipitation compared with that from climate forcings alone (6k0v). In part, this is because the desert soil does not increase soil water retention or decrease surface albedo. Change to loam soil does increase evapotranspiration and precipitation slightly (6k6vt), and lower soil albedo further reinforces these increases (6k6vtc). These results point to the importance of surface albedo when simulating a positive land feedback on precipitation, similar to the findings of Kutzbach et al. (1996) shown in

Table 27.3 | Vegetation and soil feedback on North Africa summer climate 6 kyr BP as determined from climate model simulations with dynamic vegetation

Simulation	Vegetation cover (fraction)		Leaf area index (m² m⁻²)	Albedo (fraction)	R_n (W m⁻²)	P (mm day⁻¹)	E (mm day⁻¹)
	Grass	Bare					
0k0v	0.11	0.89	0.4	0.25	117	2.5	1.9
6k0v	0.11	0.89	0.4	0.23	131	4.4	3.0
6k6v	0.28	0.72	1.3	0.23	129	4.4	2.9
6k6vt	0.37	0.63	1.3	0.22	130	4.7	3.1
6k6vtc	0.37	0.63	1.6	0.19	136	4.9	3.2

Note: 0k0v, present-day climate and vegetation. 6k0v, climate 6 kyr BP and present-day vegetation. 6k6v, 6 kyr BP climate and dynamic vegetation. 6k6vt, as in 6k6v but with loam soil texture and present-day soil color. 6k6vtc, as in 6k6v but with loam soil texture and decreased soil albedo. R_n, net radiation. P, precipitation. E, evapotranspiration. Data are averaged for July–September and are spatially averaged between latitude 15–25° N and longitude 12° W–34° E.
Source: From Levis et al. (2004).

Table 27.1. Other studies also indicate the magnitude of the positive precipitation feedback is sensitive to surface albedo (Bonfils et al. 2001; Knorr and Schnitzler 2006). However, vegetation does not advance far enough northward in these simulations to match paleobotanical data. While dynamic vegetation does improve simulation of the North African monsoon, additional feedbacks are necessary in this climate model to sustain a vegetated Sahara and to match paleoclimate reconstructions.

The period 6 kyr BP was part of a longer time beginning about 14.5 kyr BP when the climate of North Africa was much wetter than today (Foley et al. 2003). Then, between 6–5 kyr BP, climate abruptly became drier and the vegetation became desert. Climate model simulations suggest that this shift is related to changes in orbital geometry that weakened the summer monsoon and that vegetation feedback on precipitation amplified this change (Claussen 2009). Claussen et al. (1999, 2003) used a climate model of intermediate complexity in a transient simulation of climate from 9 kyr BP to present-day. This is coarse resolution model (10° in latitude by 51° in longitude) of the atmosphere–ocean–land–sea ice system with dynamic vegetation. It low spatial resolution

allows long climate simulations. The model was forced only with changes in orbital parameters, which leads to a gradual reduction in summer insolation in the Northern Hemisphere (Figure 27.5). Precipitation in the Sahara decreases gradually until about 5.6 kyr BP, when both precipitation and vegetation cover decrease markedly over a few hundred years. These results suggest that slow changes in solar radiation caused by Earth's orbital geometry gradually reduced precipitation. At some point, this gradual reduction in precipitation was abruptly amplified, likely by vegetation feedback, and the system switched from vegetated to desert. Other studies suggest a gradual drying of climate, implying weak climate–vegetation feedback. However, strong climate–vegetation feedback can produce a gradual decline in precipitation and transition to desert if a diversity of species with different sensitivity to soil wetness co-occur (Claussen et al. 2013).

The notion of a regime change is supported by other coupled climate–vegetation model studies that indicate the possibility of two different stable states in North Africa: a wet climate with a vegetated Sahara or a dry climate with desert. For example, climate model studies in which vegetation is simulated

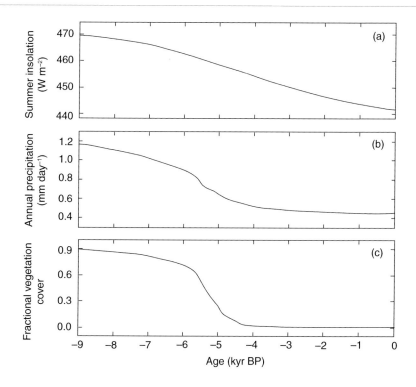

Fig. 27.5 Transient climate simulations for the past 9 kyr BP. (a) Summer (June–August) insolation averaged for the Northern Hemisphere. (b) Annual precipitation in the Sahara (spatially averaged between latitude 20°–30° N and longitude 15° W–50° E). (c) Fractional vegetation cover for the Sahara. Adapted from Claussen et al. (1999). Temporal trends in precipitation and vegetation are smoothed compared with the original figures.

asynchronously using a biogeography model show that the choice of initial vegetation cover of desert or forest can lead to different climates in the western region of the Sahara Desert (Claussen 1994, 1997, 1998; Claussen et al. 1998; Kubatzki and Claussen 1998). Under present-day orbital forcing, simulations when the model is initialized with the modern geographic extent of desert produce the present-day climate and distribution of vegetation. However, simulations in which land is initially vegetated instead of desert result in a wetter climate that supports a northward extension of savanna and shrubland from their modern distributions. A similar dichotomy of two stable climate–vegetation states is possible 21 kyr BP at the last glacial maximum but not 6 kyr BP, when a green Sahara is the only model solution. Six thousand years ago, increased summer solar radiation strengthened the monsoon and created a climate that was wet enough to maintain vegetation regardless of vegetation feedback. In contrast, the drier climate of today and at the last glacial maximum as a result of reduced summer solar radiation is sensitive to vegetation feedback.

Climate simulations by Wang and Eltahir (2000a) that used a dynamic global vegetation model also show two stable climate–vegetation regimes depending on initial conditions (Figure 27.6). An initial forest cover produces a climate–vegetation equilibrium with large annual rainfall and extensive forest vegetation over much of West Africa. Desert initial conditions give substantially less rainfall, an absence of forest, and wide distribution of grasses and desert. The existence of two stable climate–vegetation states (wet–green, dry–desert) in coupled models suggests that vegetation feedback plays an important role in the climate of this region.

Other studies with coupled climate–vegetation models also point to the importance of vegetation and soil moisture feedbacks in the Sahel region of northern Africa (Zeng et al. 1999; Wang and Eltahir 2000b,c; Wang et al. 2004; Kucharski et al. 2013). This region experienced a severe drought during the latter part of the twentieth century. While changes in sea surface temperatures drive decadal precipitation variability, vegetation and soil moisture feedbacks enhance

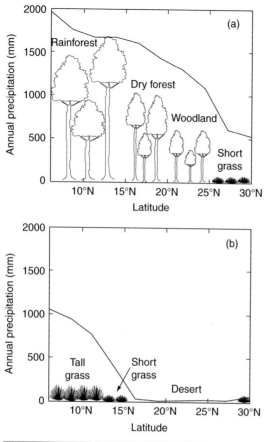

Fig. 27.6 Equilibrium climate and vegetation in West Africa in relation to latitude for (a) forest initial conditions and (b) desert initial conditions. Adapted from Wang and Eltahir (2000a).

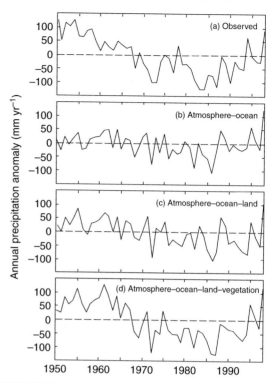

Fig. 27.7 Precipitation variability and vegetation feedback in the West African Sahel (latitude 13° N–20° N, 15° W–20° E) for 1950–1998. (a) Observed annual precipitation anomaly. (b) Simulated annual precipitation anomaly with sea surface temperatures only. (c) Simulated annual precipitation anomaly with the addition of interactive soil moisture. (d) Simulated annual precipitation anomaly with the addition of vegetation feedback. Adapted from Zeng et al. (1999).

the variability and the severity of drought. For example, Zeng et al. (1999) compared oceanic and terrestrial influences on precipitation variability for the period 1950–1998 (Figure 27.7). A simulation with interannually varying sea surface temperatures and prescribed soil moisture and vegetation cover has weak interannual precipitation variation and drying compared with the observations. Interactive soil hydrology increases the drying trend and provides a better match with observations, but the best fit with observations occurs when vegetation also responds to precipitation. Interactive vegetation influences precipitation through a positive feedback loop. Decreased rainfall leads to drier soils and reduced vegetation cover, which in

turn leads to higher surface albedo and reduced transpiration. This weakens atmospheric circulation by reducing the energy and water flux in the atmosphere, resulting in less rainfall.

In summary, a large body of literature shows that vegetation and soil moisture feedbacks are a key component of the climate of North Africa. These feedbacks enhance orbital changes in the North Africa summer monsoon, and climate models that represent these feedbacks better match paleoclimatic and paleobotanic data. The consensus is that the climate of North Africa 6 kyr BP cannot be realistically simulated without vegetation feedback on climate. Moreover, the occurrence of persistent drought in this region is triggered by forcings such as sea surface temperature but accentuated by vegetation and soil moisture feedbacks. However, there is still much to learn about vegetation–climate coupling in North Africa. Remote forcing from expanded forest cover in Eurasia may have contributed to the enhanced precipitation over North Africa 6 kyr BP. An increase in present-day extratropical forest cover decreases surface albedo and increases energy absorption in the Northern Hemisphere. The Hadley circulation moves northward to redistribute the energy, also shifting the intertropical convergence zone northward (Figure 28.20). The same mechanism may have prevailed 6 kyr BP, forced by more extensive forests in Eurasia and grassland in the Sahara that increased energy absorption and as a result produced greater precipitation over North Africa (Swann et al. 2014).

27.5 | Boreal Forests

The boreal forest is the northernmost forest, lying just south of the treeless tundra. Because of differences between boreal forest and tundra in surface albedo, surface roughness, and the partitioning of energy into latent and sensible heat, the geographic extent of these biomes is an important regulator of global climate (Bonan et al. 1995; Chapin et al. 2000, 2005; Eugster et al. 2000).

One important difference is surface albedo when snow covers the ground. Fresh snow has a high albedo, generally reflecting 80–95 percent of incident solar radiation (Table 12.1). This is also true for short tundra vegetation, which is typically buried by snow. The high albedo of snow in contrast with snow-free surfaces is an important climate feedback (Qu and Hall 2007; Fletcher et al. 2012), but the presence of trees diminishes this difference. Tall trees protrude over snow-covered ground. Foliage has a much lower albedo than snow, and dense canopies of leaves mask the high albedo of snow (Figure 17.5b). This vegetation masking of snow albedo is seen locally in comparison of summer and winter albedos for various boreal ecosystems (Table 27.4). Treeless areas have a much

Table 27.4 Daily averaged broadband albedo (fraction) during summer and with snow for boreal vegetation

	Summer	Snow-covered
Treeless		
Grassland	0.20	0.75
Wetland	0.16–0.18	0.70
Forest		
Quaking aspen	0.11	0.21
Jack pine	0.09–0.14	0.12–0.15
Spruce	0.08–0.09	0.11

Source: From Baldocchi et al. (2000).

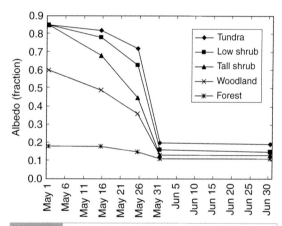

Fig. 27.8 Surface albedo of arctic vegetation during snowmelt, May 1–June 30. Data from Sturm et al. (2005a). See also Loranty et al. (2011).

higher albedo when snow is on the ground than do forests. Evergreen pine and spruce forests have low albedo in winter. Even deciduous quaking aspen forests have comparatively low winter albedo because of twigs, branches, and stems. Shrubs similarly reduce albedo compared with tundra, especially during the snow melt season (Figure 27.8). Vegetation masking of snow albedo is evident in satellite-derived maps of surface albedo during winter (Robinson and Kukla 1985; Jin et al. 2002; Barlage et al. 2005; Gao et al. 2005). Such maps show forests have a lower albedo than treeless regions, which is also evident in comparisons among biomes (Figure 12.10, Table 26.3) and seen also in a decline in albedo with greater tree cover (Loranty et al. 2014).

Climate model simulations show that the boreal forest warms climate, primarily because of its low albedo in winter and spring (Bonan et al. 1992; Thomas and Rowntree 1992; Foley et al. 1994; Betts 2000; Snyder et al. 2004; Davin and de Noblet-Ducoudré 2010). Figure 27.9 shows results from one study that compared climate simulations with the boreal forest present and with boreal forest replaced with tundra (Bonan et al. 1992). In January, temperatures for the region bounded by latitudes 40° N and 70° N are 3–7°C warmer in climate simulations with the boreal forest than without. The largest warming occurs in April. Warming persists into summer (July) and autumn (October) despite smaller differences in surface albedo between forest and tundra because warmer oceans and reduced sea ice feed back to warm climate. When compared among various global biomes, boreal forest has the greatest effect on annual mean temperature as a result of large changes in albedo (Snyder et al. 2004).

This vegetation feedback on climate has been found at other times. For example, climate was warmer than present during the late Cretaceous 66 million years ago when atmospheric CO_2 concentration was much higher than present (580 ppm). Polar deciduous forest covered the Arctic in this warm climate. Otto-Bliesner and Upchurch (1997) simulated the late Cretaceous climate once with unvegetated land and again with vegetation geography for that time. These simulations show that polar forests warm climate by reducing albedo compared with simulations without forests. Inclusion of polar forests results in a simulated climate that agrees better with reconstructions from fossil vegetation data, suggesting these forests were an important contributor to the warm climate of this period.

The forest–tundra ecotone may play a role in glaciation. At the onset of the last glaciation 115 kyr BP, Northern Hemisphere summer solar radiation was reduced by 8 percent compared with modern values and atmospheric CO_2 was 267 ppm. Climate model simulations by Gallimore and Kutzbach (1996) show that

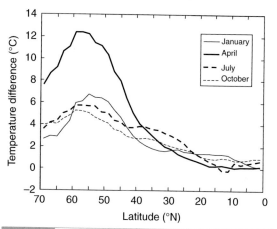

Fig. 27.9 Difference in air temperature between a climate simulation with the boreal forest present and one in which the boreal forest is replaced with tundra. Data show January, April, July, and October mean monthly temperature averaged for land as a function of latitude from the equator (0° N) to 70° N. Adapted from Bonan et al. (1992).

these changes in solar radiation and atmospheric CO_2 cool high latitude land temperature by about 5°C and increase the duration of snow cover by about one month compared with a control simulation of the present-day climate (Table 27.5). Observational evidence shows widespread changes in vegetation 115 kyr BP because of the colder climate. High latitude forests died back and were replaced with tundra and cold grassland-like vegetation. Gallimore and Kutzbach (1996) performed two additional climate simulations to examine the effects of the increased albedo associated with the expansion of tundra. In one simulation, surface albedo north of 60° N was increased to mimic reduced vegetation masking of snow albedo with modest tundra expansion. In another simulation, surface albedo was increased even more to mimic large loss of forest and extensive expansion of tundra. Increases in surface albedo with a modest expansion of tundra cools climate by an additional 2.9°C in North America and 2.8°C in Eurasia and increases the duration of snow on the ground by 18 days compared with radiative forcings. A more extensive tundra with a larger increase in albedo results in a catastrophic climate change, with summer temperatures decreasing 17–18°C and snow cover persisting almost three months longer compared with no

Table 27.5 Climate at the onset of glaciation 115 kyr BP as determined from four climate model simulations

	Control (C)	Radiative forcing (R − C)	Modest tundra expansion (RV − R)	Large tundra expansion (REV − R)
North America				
Summer temperature (°C)	12.6	−5.1	−2.9	−18.0
Snow cover (days)	237	32	18	85
Eurasia				
Summer temperature (°C)	14.3	−5.2	−2.8	−17.1
Snow cover (days)	240	24	18	84

Note: C, control simulation with present-day solar radiation and atmospheric CO_2. R, orbital geometry and atmospheric CO_2 of 115 kyr BP. RV, as in R but with modest expansion of tundra that increases surface albedo north of 60° N. REV, as in RV but with greater increase in albedo because of extensive tundra expansion. Data are averaged over land between latitudes 60°–90° N. Radiative forcing is the difference between the radiative and control simulations. Vegetation forcings are the climate change in addition to the radiative forcing.
Source: From Gallimore and Kutzbach (1996).

changes in tundra. Regions of permanent snow cover occur, indicating the onset of glaciation.

Coupled climate–vegetation models confirm that vegetation provides a positive feedback for glacial inception (de Noblet et al. 1996; Meissner et al. 2003; Calov et al. 2005; Claussen et al. 2006; Kubatzki et al. 2006). In these simulations, the cold climate as a result of reduced solar radiation and lower atmospheric CO_2 decreases the geographic extent of the boreal forest, and the expansion of tundra leads to additional cooling. Similar results are found at other time periods (Horton et al. 2010).

Coupled climate–vegetation models highlight the importance of vegetation in amplifying the climate response to orbital forcing at the last glacial maximum. For example, Levis et al. (1999) compared climate simulations of the last glacial maximum using prescribed present-day vegetation and with vegetation simulated by a dynamic global vegetation model. When vegetation is allowed to respond to the cold, dry glacial climate, forest cover decreases in the tropics and northern latitudes. Instead, tundra dominates much of the middle to high latitudes while grasslands cover the tropics and subtropics. These changes in biogeography are consistent with fossilized plant remains, which show more extensive tundra and grasses in middle to

high latitudes and forest dieback and replacement by grasses in the tropics. The simulated climate is quite different as a result of the dynamic vegetation. With dynamic vegetation, temperatures cool compared with present-day vegetation over much of Eurasia, where tree cover decreases and albedo increases in winter and spring. Temperatures warm in the tropics and subtropics where grasses replace trees and evapotranspiration is reduced in a drier climate.

The location of the treeline separating forest and tundra changed over the past 18 kyrs with the transition from glacial to interglacial (Figure 24.17). As climate warmed and the glaciers retreated northwards, the treeline migrated northwards. The period 6 kyr BP is particularly noteworthy. Changes in Earth's orbital geometry resulted in more summer solar radiation than present in the Northern Hemisphere (Figure 8.4), creating a warmer climate than present. Boreal forests extended north of the modern treeline in response to this warm climate. Foley et al. (1994) showed that the decrease in surface albedo caused by northward expansion of forest accentuated the warming. They simulated climate in response to the orbital geometry of 6 kyr BP. This orbital forcing warms high latitude land between 60° N and 90° N by 1.8°C in the annual mean. In

Table 27.6 Summer vegetation characteristics and surface energy fluxes at five sites in Alaska

	Tundra	Low shrub	Tall shrub	Woodland	Spruce forest
Leaf area index (m^2 m^{-2})	0.5	1.7	1.9	2.3	2.8
Canopy height (m)	0.10	0.25	1.5	1.7 (shrub) 7.3 (tree)	6.1
Midday albedo (fraction)	0.19	0.17	0.15	0.13	0.10
Evaporation fraction ($\lambda E/R_n$)	0.36	0.35	0.36	0.36	0.37
Bowen ratio ($H/\lambda E$)	0.94	0.98	1.06	1.15	1.22
Aerodynamic conductance (mol m^{-2} s^{-1})	0.98	2.01	1.76	3.25	3.85
Surface conductance (mol m^{-2} s^{-1})	0.24	0.25	0.25	0.29	0.25

Note: R_n, net radiation. H, sensible heat flux. λE, latent heat flux. Conductances are for midday and are converted to mol m^{-2} s^{-1} using $\rho_m = 42.3$ mol m^{-3}.
Source: From Beringer et al. (2005).

another simulation, the authors extended the northern limit of boreal forest and reduced the extent of tundra. The northward expansion of forest gives an additional warming of 1.6°C, which is comparable to that of the orbital forcing alone. This additional warming is larger in spring (4°C) than in other seasons (1°C). Other model studies also show amplification of the climate warming due to vegetation feedback (Claussen 2009).

The influence of the boreal forest on climate is more complex than just albedo. Forest and tundra ecosystems differ in their energy balance. Beringer et al. (2005) compared summer surface fluxes at five sites in Alaska representing tundra, low deciduous shrub, tall deciduous shrub, woodland treeline, and white spruce forest (Table 27.6). Albedo decreased with increased woody stature from 0.19 (tundra) to 0.10 (forest). The bulk aerodynamic conductance increased by a factor of four from tundra to forest. Evaporative fraction and bulk surface conductance were virtually identical across sites because evaporation decreased while transpiration increased along the tundra–forest transect. The Bowen ratio increased from 0.94 (tundra) to 1.22 (forest), indicative of warmer and drier sites along the transect.

The various types of boreal vegetation also differ in latent heat exchange (Table 27.7). Summertime evaporative fraction is largest over wetland and quaking aspen forests where about two-thirds to three-quarters of energy is dissipated as latent heat. Needleleaf forests such as jack pine, Scots pine, black spruce, and larch, on the other hand, have evaporative fraction ranging from one-third to one-half of available energy. Low foliage nitrogen, low photosynthetic capacity, low leaf area, high vapor pressure deficit, soil moisture deficit, and other factors combine to restrict canopy conductance in boreal needleleaf forests.

These differences in evapotranspiration affect climate. Climate model simulations show that expansion of broadleaf deciduous trees in the Arctic north of 60°N not only decreases land surface albedo, but also increases evapotranspiration on land (Swann et al. 2010). Water vapor in the atmosphere is a powerful greenhouse gas. Increased water vapor warms the Arctic climate and initiates a positive feedback whereby warmer temperature melts sea ice, which decreases ocean albedo and increases evaporation from the ocean, producing still greater warming. This feedback increases annual mean temperature in the Arctic by up to 3°C in some

Table 27.7 Mid-growing season latent heat flux (λE) normalized by net radiation (R_n) or available energy (net radiation minus soil heat flux, $R_n - G$) for various boreal vegetation

Vegetation		Normalized latent heat
Wetland		
Manitoba, Canada	$\lambda E/(R_n - G)$	0.76
Quaking aspen forest		
Saskatchewan, Canada	$\lambda E/R_n$	0.61
Jack pine forest		
Saskatchewan, Canada	$\lambda E/R_n$	0.39
Manitoba, Canada	$\lambda E/(R_n - G)$	0.34
Scots pine forest		
Russia	$\lambda E/(R_n - G)$	0.48
Sweden	$\lambda E/R_n$	0.59
Sweden	$\lambda E/(R_n - G)$	0.38
Black spruce forest		
Ontario, Canada	$\lambda E/(R_n - G)$	0.49
Saskatchewan, Canada	$\lambda E/(R_n - G)$	0.38
Quebec, Canada	$\lambda E/(R_n - G)$	0.35
Siberian larch forest		
Russia	$\lambda E/(R_n - G)$	0.44

Source: From Baldocchi et al. (2000).

regions and by 1°C over the circumpolar Arctic. The radiative forcing from higher transpiration with expanded forest cover is greater than that from surface albedo changes alone. Another study similarly found warming as a result of increased evapotranspiration and atmospheric moisture with deciduous shrub encroachment into tundra (Bonfils et al. 2012).

Changes in the disturbance regime can also affect climate. Post-fire forest succession drives differences among forests in their surface energy balance (Liu et al. 2005; Amiro et al. 2006; Liu and Randerson 2008). Surface albedo decreases as the burned forest recovers from fire (Figure 22.15). Recent burn sites have lower summer net radiation compared with mature spruce forests because of higher albedo (Table 22.4). The Bowen ratio of young aspen forest during summer (0.9) is lower than that of mature spruce forest (1.3) and its evaporative fraction (0.5) is higher (0.4). In general, there is

a clear temporal trend in fluxes related to the immediate fire, regrowth by deciduous trees, and recovery of the mature spruce forest.

Changes in the fire regime alter the age and composition of the boreal forest and thus can alter climate. Climate model simulations show that younger forests arising from increased burning cool the North American boreal climate, primarily in winter and spring (Rogers et al. 2013). A doubling of burn area cools the surface climate by 0.23°C across boreal North America during winter and spring (December–May). This is driven by increases in surface albedo; in these simulations evapotranspiration feedbacks are minor.

The mountain pine beetle epidemic in western North America forests has killed vast tracts of forests in British Columbia and the Rocky Mountain region of the United States (Edburg et al. 2012). This mortality increases surface albedo, most noticeably in winter and spring (O'Halloran et al. 2012; Bright et al. 2013; Vanderhoof et al. 2013, 2014). The mortality also decreases evapotranspiration. Such changes are evident in British Columbia (Maness et al. 2013). Over the beetle infested region, summertime evapotranspiration decreased by 19 percent and summertime surface temperatures increased by 1°C rise as a result of tree dieback. The largest decreases in evapotranspiration and warming of daytime temperature relative to nighttime temperature occurred in stands that suffered the greatest mortality (Figure 27.10). Similar changes have occurred in other beetle infested regions (Bright et al. 2013).

Wildfires and insect infestations affect climate through a variety of biogeochemical processes. Pine beetle mortality reduces carbon uptake by forests (Kurz et al. 2008). Wildfires emit CO_2, CH_4, N_2O, and aerosols to the atmosphere, but carbon accumulation increases as the forests recover (Figure 22.16). The net effect of disturbances must consider these biogeochemical changes in addition to biogeophysical forcings (Randerson et al. 2006). In boreal spruce forests, for example, the positive biogeochemical radiative forcing exceeds the negative biogeophysical radiative forcing in the first year following fire, but the opposite is true over longer time periods (Table 23.1).

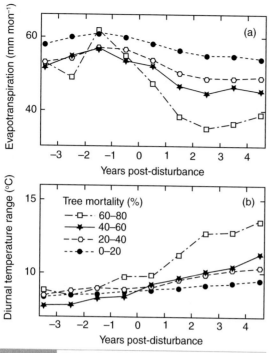

Fig. 27.10 Time series of (a) summertime evapotranspiration and (b) diurnal temperature range in British Columbia before and after pine beetle infestation in relation to degree of forest mortality. Adapted from Maness et al. (2013).

Ecologists have historically thought that climate determines the geography and functioning of the boreal forest (Larsen 1980; Bonan and Shugart 1989; Shugart et al. 1992; Hall et al. 2004). For example, the northern and southern boundaries of the boreal forest correlate with the July 13°C and 18°C isotherms, respectively. However, these correlations likely reflect coupled climate–vegetation dynamics in which the geographic extent of the boreal forest affects and is affected by climate. In addition, the forest–tundra ecotone correlates with the summer position of the Arctic front (Bryson 1966; Krebs and Berry 1970). The forest–tundra ecotone may itself control the position of this front in summer due to the large contrast in surface albedo, surface roughness, and energy exchange, which results in strong heating of the atmosphere over forest and weaker heating over tundra (Pielke and Vidale 1995). Model simulations are inconclusive (Lynch et al. 2001;

Liess et al. 2012; Snyder and Liess 2014), though the simulations differ in model resolution and domain.

27.6 | Anthropogenic Climate Change

The twentieth century has seen a prominent increase in atmospheric CO_2 concentration (Figure 3.7a), and planetary temperature has increased as a result of this and other changes in radiative forcing (Figure 8.15). The increase in atmospheric CO_2 is expected to continue throughout the twenty-first century (Figure 2.11b), producing still greater planetary warming (Figure 8.16). Various socioeconomic pathways depict a future Earth with elevated CO_2 up to ~950 ppm (Figure 2.11b) and in which the planet has warmed by an additional 2–4°C at the end of the twenty-first century (Figure 8.16). Vegetation responses to these changes in climate and atmospheric composition feed back to accentuate the changes.

In vegetated landscapes, canopy conductance controls the partitioning of net radiation into sensible and latent heat fluxes. Studies of the physiological response of plants to enhanced CO_2 concentrations routinely find reduced stomatal conductance and greater photosynthesis (Figure 16.10). Climate model simulations in which stomatal conductance decreases with a doubling of atmospheric CO_2 routinely show decreased evapotranspiration, increased sensible heat flux, and surface warming in summer. Decreased stomatal conductance may have increased continental runoff over the twentieth century (Gedney et al. 2006) and may further increase runoff over the twenty-first century (Betts et al. 2007). However, changes in leaf area and in land use also impact runoff trends (Piao et al. 2007).

The effects of increasing atmospheric CO_2 concentration on climate can be partitioned into radiative and physiological effects (Sellers et al. 1996; Bounoua et al. 1999). Radiative effects consider only climate changes associated with atmospheric radiation (i.e., the greenhouse

Table 27.8 Effect of doubled atmospheric CO_2 concentration on annual mean climate as determined from four climate model simulations

	Control (C)	Radiative forcing (R − C)	Physiological (RP − R)	Physiological and vegetation cover (RPV − R)
Ice-free northern high latitude continents (60° N–90° N)				
Temperature (°C)	−6.7	3.9	0.6	0.6
P (mm day^{-1})	0.76	0.15	0.02	0.04
E (mm day^{-1})	0.35	0.09	0.00	0.05
Leaf area index (m^2 m^{-2})	2.5	0.0	0.0	3.0
Albedo (fraction)	0.38	−0.03	0.00	−0.06
Middle latitude continents (30° N–60° N)				
Temperature (°C)	6.9	2.7	0.3	0.3
P (mm day^{-1})	1.51	0.19	−0.07	−0.02
E (mm day^{-1})	1.16	0.14	−0.03	0.04
Leaf area index (m^2 m^{-2})	5.0	0.0	0.0	2.4
Albedo (fraction)	0.17	−0.01	0.00	−0.01
Tropical continents (15° S–15° N)				
Temperature (°C)	26.6	2.1	0.1	0.0
P (mm day^{-1})	5.22	0.12	−0.17	0.03
E (mm day^{-1})	3.50	0.17	−0.02	0.15
Leaf area index (m^2 m^{-2})	5.2	0.0	0.0	2.2

Note: C, control simulation with modern CO_2 (345 ppm) and vegetation. R, doubled CO_2 (690 ppm) without changes in vegetation. RP, doubled CO_2 with changes in stomatal physiology. RPV, doubled CO_2 with changes in stomatal physiology and vegetation cover. Radiative forcing is the difference between the radiative and control simulations. Vegetation forcings are the climate change in addition to the radiative forcing. P, precipitation. E, evapotranspiration.
Source: From Levis et al. (2000).

effect). Physiological effects are changes due to reduced stomatal conductance with higher CO_2. Climate model simulations by Levis et al. (2000) illustrate this approach (Table 27.8). Three simulations contrast the radiative and physiological effects of CO_2: a control simulation with present-day CO_2 concentration; a radiative forcing simulation in which atmospheric CO_2 is doubled but stomatal conductance does not respond directly to the higher CO_2; and another doubled CO_2 simulation in which stomata respond to the higher CO_2 concentration. In general, the physiological effects of doubled CO_2 amplify the warming associated with the radiative effects of

doubled CO_2 (Table 27.8, RP − R compared with R − C). The radiative forcing generally increases evapotranspiration, warms surface temperature, and increases precipitation. The physiological forcing from reduced stomatal conductance reduces evapotranspiration and precipitation and further warms the surface.

The altered temperature and precipitation in response to elevated atmospheric CO_2 changes vegetation in addition to the physiological effects of CO_2 fertilization. Simulations with coupled climate–vegetation models show large changes in climate as a result of such vegetation changes (Betts et al. 1997, 2000; Levis et al. 2000;

Bala et al. 2006; O'ishi and Abe-Ouchi 2009; O'ishi et al. 2009; Jeong et al. 2011, 2014; Port et al. 2012). Vegetation enhancement of climate warming is particularly prominent in the Arctic, where greater tree cover reduces surface albedo.

For example, Levis et al. (2000) used a dynamic global vegetation model coupled to a climate model to study vegetation feedback with a doubling of atmospheric CO_2 concentration (Table 27.8). The radiative effect of increased atmospheric CO_2 is overall surface warming and intensification of the hydrologic cycle with increased precipitation and evapotranspiration. The reduction in stomatal conductance with higher atmospheric CO_2 reinforces the warming, but decreases precipitation and evapotranspiration in middle to low latitudes. With dynamic vegetation, leaf area index generally increases due to expansion of forests at the expense of grasses. In northern latitudes, more extensive forest enhances the spring and summer radiative warming due to a decrease in surface albedo. Without vegetation feedback, doubling atmospheric CO_2 warms temperatures over ice-free land north of latitude 45° N by 1.7–5.6°C depending on the season. Vegetation feedback enhances this warming by 1.6°C in spring, when the albedo feedback is greatest, and by 0.4°C in summer and autumn, when the albedo feedback diminishes but warming persists due to the thermal inertia of the Arctic Ocean. In middle latitudes, growth of temperate deciduous trees and grasses leads to summer cooling compared with the physiological effects as a result of greater leaf area and evapotranspiration. Spring and autumn temperatures warm due to reduced albedo as vegetation cover increases. Annual precipitation and evapotranspiration increase. In the tropics, evergreen and deciduous trees expand at the expense of grasses. This offsets the reduction in precipitation and evapotranspiration due to physiological effects.

Coupled climate–vegetation models show that vegetation feedback with a warmer climate is especially prominent in northern high latitudes, where the northward migration of trees into tundra alters climate. A more immediate feedback is the increased abundance, extent, and productivity of woody shrubs in tundra in response to warming. Such a greening of the Arctic has been observed across Eurasia and North America over the past few decades (Jia et al. 2003; Tape et al. 2006; Beck and Goetz 2011; Post et al. 2013; Xu et al. 2013). Shrubs have a lower surface albedo than tundra, particularly during the snow melt season (Figure 27.8), and shrub expansion with a warmer climate is widely expected to augment the warming (Chapin et al. 2005; Sturm et al. 2005a; Euskirchen et al. 2009; Pearson et al. 2013). Climate model simulations do indeed show that woody shrub expansion in tundra warms climate because of lower surface albedo, but also through water vapor feedback with enhanced evapotranspiration (Lawrence and Swenson 2011; Bonfils et al. 2012).

However, the climate consequences of increased shrubs are complex. The presence of shrubs also alters snow distribution and depth, the duration of snow cover, and snow thermal conductivity (Sturm et al. 2001, 2005b; Liston et al. 2002). Snow redistribution around shrubs reduces the albedo feedback by covering shrubs with snow, but also warms soil through the insulating effect of snow (Lawrence and Swenson 2011). Shrubs also shade the ground and decrease soil temperature and active layer depth compared with grassy tundra (Blok et al. 2010). This latter result suggests that shrub expansion could mitigate permafrost thaw with climate warming. Climate model simulations do indeed show a shallower active layer under shrubs, but the large-scale decrease in surface albedo with circumpolar shrub expansion warms climate and thaws the soil to deeper depths (Lawrence and Swenson 2011). Moreover, the net climate feedback in the Arctic must also include changes to carbon storage, especially permafrost thaw (Schuur et al. 2008, 2013).

Vegetation dynamics in response to climate change significantly affects the trajectory of climate change over the twentieth and twenty-first centuries, largely driven by changes in surface albedo and evapotranspiration. A much more prominent transformation of the biosphere has been brought about from anthropogenic land use and land-cover change, in particular the clearing of forests and cultivation of grass-lands to raise crops. Whereas natural vegetation dynamics amplifies greenhouse gas warming over the twentieth century, historical land use and land-cover change have an opposite effect and have cooled planetary temperature (Matthews et al. 2004; Strengers et al. 2010; Lawrence et al. 2012).

27.7 | Review Questions

1. Reconstructed sea surface temperatures and vegetation are available for 6 kyr BP. Devise a series of climate model experiments to contrast the effects of ocean and vegetation on North Africa precipitation 6 kyr BP. Account for differences in atmospheric CO_2.

2. Which climate model simulation is likely to have a stronger North Africa summer monsoon – one in which desert soil has high albedo or one in which desert soil has low albedo? In which will the strength of the monsoon be most sensitive to increased vegetation cover? Explain why.

3. Discuss the consequences of increased summer solar radiation on the present climate of North Africa. Does land clearing accentuate or mitigate this expected change?

4. Explain how boreal forest affects temperature. Why is this effect greater in spring than in winter?

5. The snow albedo feedback is an important determinant of climate response to higher atmospheric CO_2 concentration. Reduced snow cover with a warmer climate reduces surface albedo and reinforces the warming. Explain how boreal forest affects this feedback. Is this feedback greater with expansive boreal forest or with reduced extent of the boreal forest?

6. Devise a set of climate model experiments to show the effect of dynamic vegetation on the climate of the last glacial maximum. Distinguish the radiative and physiological effects of atmospheric CO_2.

7. Which type of leaf (broadleaf or cylindrical) might be favored with high atmospheric CO_2?

8. Use the following surface albedo data from Figure 27.8 to estimate how conversion of tundra to (a) low shrub and (b) tall shrub affects solar heating of land during winter (January 1), snowmelt (May 15), and summer (June 30). How might surface climate change?

Date	Tundra	Low shrub	Tall shrub
January 1	0.85	0.85	0.85
May 15	0.82	0.78	0.68
June 30	0.19	0.15	0.13

9. Describe some of the prominent changes in vegetation likely to occur over the twenty-first century in response to increased atmospheric CO_2 and climate change. How will these feed back to affect climate?

10. The original conceptualization of Daisyworld postulated that the biosphere regulates climate for its own benefit. Discuss this viewpoint. How does it differ from the notion that the biosphere affects climate such that there is coupled climate–vegetation dynamics?

27.8 | References

Amiro, B. D., Orchansky, A. L., Barr, A. G., et al. (2006). The effect of post-fire stand age on the boreal forest energy balance. *Agricultural and Forest Meteorology*, 140, 41–50.

Bala, G., Caldeira, K., Mirin, A., et al. (2006). Biogeophysical effects of CO_2 fertilization on global climate. *Tellus B*, 58, 620–627.

Baldocchi, D., Kelliher, F. M., Black, T. A., and Jarvis, P. (2000). Climate and vegetation controls on boreal zone energy exchange. *Global Change Biology*, 6(s1), 69–83.

Barlage, M., Zeng, X., Wei, H., and Mitchell, K. E. (2005). A global 0.05° maximum albedo dataset of snow-covered land based on MODIS observations. *Geophysical Research Letters*, 32, L17405, doi:10.1029/2005GL022881.

Beck, P. S. A., and Goetz, S. J. (2011). Satellite observations of high northern latitude vegetation productivity changes between 1982 and 2008: Ecological variability and regional differences. *Environmental Research Letters*, 6, 045501, doi:10.1088/1748-9326/6/4/045501.

Beringer, J., Chapin, F. S., III, Thompson, C. C., and McGuire, A. D. (2005). Surface energy exchanges along a tundra–forest transition and feedbacks to climate. *Agricultural and Forest Meteorology*, 131, 143–161.

Betts, R. A. (2000). Offset of the potential carbon sink from boreal forestation by decreases in surface albedo. *Nature*, 408, 187–190.

Betts, R. A., Cox, P. M., Lee, S. E., and Woodward, F. I. (1997). Contrasting physiological and structural vegetation feedbacks in climate change simulations. *Nature*, 387, 796–799.

Betts, R. A., Cox, P. M., and Woodward, F. I. (2000). Simulated responses of potential vegetation to doubled-CO_2 climate change and feedbacks on near-surface temperature. *Global Ecology and Biogeography*, 9, 171–180.

Betts, R. A., Boucher, O., Collins, M., et al. (2007). Projected increase in continental runoff due to plant responses to increasing carbon dioxide. *Nature*, 448, 1037–1041.

Blok, D., Heijmans, M. M. P. D., Schaepman-Strub, G., et al. (2010). Shrub expansion may reduce summer permafrost thaw in Siberian tundra. *Global Change Biology*, 16, 1296–1305.

Bonan, G. B., and Shugart, H. H. (1989). Environmental factors and ecological processes in boreal forests. *Annual Review of Ecology and Systematics*, 20, 1–28.

Bonan, G. B., Pollard, D., and Thompson, S. L. (1992). Effects of boreal forest vegetation on global climate. *Nature*, 359, 716–718.

Bonan, G. B., Chapin, F. S., III, and Thompson, S. L. (1995). Boreal forest and tundra ecosystems as components of the climate system. *Climatic Change*, 29, 145–167.

Bonfils, C., de Noblet-Ducoudré, N., Braconnot, P., and Joussaume, S. (2001). Hot desert albedo and climate change: Mid-Holocene monsoon in North Africa. *Journal of Climate*, 14, 3724–3737.

Bonfils, C. J. W., Phillips, T. J., Lawrence, D. M., et al. (2012). On the influence of shrub height and expansion on northern high latitude climate. *Environmental Research Letters*, 7, 015503, doi:10.1088/1748-9326/7/1/015503.

Bounoua, L., Collatz, G. J., Sellers, P. J., et al. (1999). Interactions between vegetation and climate: Radiative and physiological effects of doubled atmospheric CO_2. *Journal of Climate*, 12, 309–324.

Braconnot, P., Joussaume, S., de Noblet, N., and Ramstein, G. (2000). Mid-Holocene and Last Glacial Maximum African monsoon changes as simulated within the Paleoclimate Modelling Intercomparison Project. *Global and Planetary Change*, 26, 51–66.

Braconnot, P., Otto-Bliesner, B., Harrison, S., et al. (2007). Results of PMIP2 coupled simulations of the Mid-Holocene and Last Glacial Maximum – Part 1: Experiments and large-scale features. *Climate of the Past*, 3, 261–277.

Bright, B. C., Hicke, J. A., and Meddens, A. J. H. (2013). Effects of bark beetle-caused tree mortality on biogeochemical and biogeophysical MODIS products. *Journal of Geophysical Research: Biogeosciences*, 118, 974–982, doi:10.1002/jgrg.20078.

Bryson, R. A. (1966). Air masses, streamlines, and the boreal forest. *Geographical Bulletin*, 8, 228–269.

Calov, R., Ganopolski, A., Petoukhov, V., et al. (2005). Transient simulation of the last glacial inception, Part II: Sensitivity and feedback analysis. *Climate Dynamics*, 24, 563–576.

Chapin, F. S., III, McGuire, A. D., Randerson, J., et al. (2000). Arctic and boreal ecosystems of western North America as components of the climate system. *Global Change Biology*, 6(s1), 211–223.

Chapin, F. S., III, Sturm, M., Serreze, M. C., et al. (2005). Role of land-surface changes in Arctic summer warming. *Science*, 310, 657–660.

Claussen, M. (1994). On coupling global biome models with climate models. *Climate Research*, 4, 203–221.

Claussen, M. (1997). Modeling bio-geophysical feedback in the African and Indian monsoon region. *Climate Dynamics*, 13, 247–257.

Claussen, M. (1998). On multiple solutions of the atmosphere–vegetation system in present-day climate. *Global Change Biology*, 4, 549–559.

Claussen, M. (2009). Late Quaternary vegetation-climate feedbacks. *Climate of the Past*, 5, 203–216.

Claussen, M., and Gayler, V. (1997). The greening of the Sahara during the mid-Holocene: Results of an interactive atmosphere-biome model. *Global Ecology and Biogeography Letters*, 6, 369–377.

Claussen, M., Brovkin, V., Ganopolski, A., Kubatzki, C., and Petoukhov, V. (1998). Modelling global terrestrial vegetation–climate interaction. *Philosophical Transactions of the Royal Society B*, 353, 53–63.

Claussen, M., Kubatzki, C., Brovkin, V., et al. (1999). Simulation of an abrupt change in Saharan vegetation in the mid-Holocene. *Geophysical Research Letters*, 26, 2037-2040.

Claussen, M., Brovkin, V., Ganopolski, A., Kubatzki, C., and Petoukhov, V. (2003). Climate change in northern Africa: The past is not the future. *Climatic Change*, 57, 99-118.

Claussen, M., Fohlmeister, J., Ganopolski, A., and Brovkin, V. (2006). Vegetation dynamics amplifies precessional forcing. *Geophysical Research Letters*, 33, L09709, doi:10.1029/2006GL026111.

Claussen, M., Bathiany, S., Brovkin, V., and Kleinen, T. (2013). Simulated climate–vegetation interaction in semi-arid regions affected by plant diversity. *Nature Geoscience*, 6, 954-958.

Davin, E. L., and de Noblet-Ducoudré, N. (2010). Climatic impact of global-scale deforestation: Radiative versus nonradiative processes. *Journal of Climate*, 23, 97-112.

de Noblet, N. I., Prentice, I. C., Joussaume, S., et al. (1996). Possible role of atmosphere–biosphere interactions in triggering the last glaciation. *Geophysical Research Letters*, 23, 3191-3194.

Edburg, S. L., Hicke, J. A., Brooks, P. D., et al. (2012). Cascading impacts of bark beetle-caused tree mortality on coupled biogeophysical and biogeochemical processes. *Frontiers in Ecology and the Environment*, 10, 416-424.

Eugster, W., Rouse, W. R., Pielke, R. A., Sr., et al. (2000). Land–atmosphere energy exchange in Arctic tundra and boreal forest: Available data and feedbacks to climate. *Global Change Biology*, 6(s1), 84-115.

Euskirchen, E. S., McGuire, A. D., Chapin, F. S., III, Yi, S., and Thompson, C. C. (2009). Changes in vegetation in northern Alaska under scenarios of climate change, 2003-2100: Implications for climate feedbacks. *Ecological Applications*, 19, 1022-1043.

Fletcher, C. G., Zhao, H., Kushner, P. J., and Fernandes, R. (2012).Using models and satellite observations to evaluate the strength of snow albedo feedback. *Journal of Geophysical Research*, 117, D11117, doi:10.1029/2012JD017724.

Foley, J. A., Kutzbach, J. E., Coe, M. T., and Levis, S. (1994). Feedbacks between climate and boreal forests during the Holocene epoch. *Nature*, 371, 52-54.

Foley, J. A., Coe, M. T., Scheffer, M., and Wang, G. (2003). Regime shifts in the Sahara and Sahel: Interactions between ecological and climatic systems in northern Africa. *Ecosystems*, 6, 524-539.

Gallimore, R. G., and Kutzbach, J. E. (1996). Role of orbitally induced changes in tundra area in the onset of glaciation. *Nature*, 381, 503-505.

Gao, F., Schaaf, C. B., Strahler, A. H., et al. (2005). MODIS bidirectional reflectance distribution function and albedo Climate Modeling Grid products and the variability of albedo for major global vegetation types. *Journal of Geophysical Research*, 110, D01104, doi:10.1029/2004JD005190.

Gedney, N., Cox, P. M., Betts, R. A., et al. (2006). Detection of a direct carbon dioxide effect in continental river runoff records. *Nature*, 439, 835-838.

Hall, F. G., Betts, A. K., Frolking, S., et al. (2004). The boreal climate. In *Vegetation, Water, Humans and the Climate: A New Perspective on an Interactive System*, ed. P. Kabat, M. Claussen, P. A. Dirmeyer, et al. Berlin: Springer-Verlag, pp. 93-114.

Hoelzmann, P., Jolly, D., Harrison, S. P., et al. (1998). Mid-Holocene land-surface conditions in northern Africa and the Arabian peninsula: A data set for the analysis of biogeophysical feedbacks in the climate system. *Global Biogeochemical Cycles*, 12, 35-51.

Horton, D. E., Poulsen, C. J., and Pollard, D. (2010). Influence of high-latitude vegetation feedbacks on late Palaeozoic glacial cycles. *Nature Geoscience*, 3, 572-577.

Jeong, J.-H., Kug, J.-S., Linderholm, H. W., et al. (2014). Intensified Arctic warming under greenhouse warming by vegetation–atmosphere–sea ice interaction. *Environmental Research Letters*, 9, 094007, doi:10.1088/1748-9326/9/9/094007.

Jeong, S.-J., Ho, C.-H., Park, T.-W., Kim, J., and Levis, S. (2011). Impact of vegetation feedback on the temperature and its diurnal range over the Northern Hemisphere during summer in a $2 \times CO_2$ climate. *Climate Dynamics*, 37, 821-833.

Jia, G. J., Epstein, H. E., and Walker, D. A. (2003). Greening of arctic Alaska, 1981-2001. *Geophysical Research Letters*, 30, 2067, doi:10.1029/2003GL018268.

Jin, Y., Schaaf, C. B., Gao, F., et al. (2002). How does snow impact the albedo of vegetated land surfaces as analyzed with MODIS data? *Geophysical Research Letters*, 29, 1374, doi:10.1029/2001GL014132.

Jolly, D., Prentice, I. C., Bonnefille, R., et al. (1998). Biome reconstruction from pollen and plant macrofossil data for Africa and the Arabian peninsula at 0 and 6000 years. *Journal of Biogeography*, 25, 1007-1027.

Joussaume, S., Taylor, K. E., Braconnot, P., et al. (1999). Monsoon changes for 6000 years ago: Results of 18 simulations from the Paleoclimate Modeling Intercomparison Project (PMIP). *Geophysical Research Letters*, 26, 859-862.

Knorr, W., and Schnitzler, K.-G. (2006). Enhanced albedo feedback in North Africa from possible combined vegetation and soil-formation processes. *Climate Dynamics*, 26, 55–63.

Krebs, J. S., and Barry, R. G. (1970). The Arctic front and the tundra–taiga boundary in Eurasia. *Geographical Review*, 60, 548–554.

Kubatzki, C., and Claussen, M. (1998). Simulation of the global bio-geophysical interactions during the Last Glacial Maximum. *Climate Dynamics*, 14, 461–471.

Kubatzki, C., Claussen, M., Calov, R., and Ganopolski, A. (2006). Sensitivity of the last glacial inception to initial and surface conditions. *Climate Dynamics*, 27, 333–344.

Kucharski, F., Zeng, N., and Kalnay, E. (2013). A further assessment of vegetation feedback on decadal Sahel rainfall variability. *Climate Dynamics*, 40, 1453–1466.

Kurz, W. A., Dymond, C. C., Stinson, G., et al. (2008). Mountain pine beetle and forest carbon feedback to climate change. *Nature*, 452, 987–990.

Kutzbach, J., Bonan, G., Foley, J., and Harrison, S. P. (1996). Vegetation and soil feedbacks on the response of the African monsoon to orbital forcing in the early to middle Holocene. *Nature*, 384, 623–626.

Larsen, J. A. (1980). *The Boreal Ecosystem*. New York: Academic Press.

Lawrence, D. M., and Swenson, S. C. (2011). Permafrost response to increasing Arctic shrub abundance depends on the relative influence of shrubs on local soil cooling versus large-scale climate warming. *Environmental Research Letters*, 6, 045504, doi:10.1088/1748-9326/6/4/045504.

Lawrence, P. J., Feddema, J. J., Bonan, G. B., et al. (2012). Simulating the biogeochemical and biogeophysical impacts of transient land cover change and wood harvest in the Community Climate System Model (CCSM4) from 1850 to 2100. *Journal of Climate*, 25, 3071–3095.

Levis, S., Foley, J. A., and Pollard, D. (1999). CO_2, climate, and vegetation feedbacks at the Last Glacial Maximum. *Journal of Geophysical Research*, 104D, 31191–31198.

Levis, S., Foley, J. A., and Pollard, D. (2000). Large-scale vegetation feedbacks on a doubled CO_2 climate. *Journal of Climate*, 13, 1313–1325.

Levis, S., Bonan, G. B., and Bonfils, C. (2004). Soil feedback drives the mid-Holocene North African monsoon northward in fully coupled CCSM2 simulations with a dynamic vegetation model. *Climate Dynamics*, 23, 791–802.

Liess, S., Snyder, P. K., and Harding, K. J. (2012). The effects of boreal forest expansion on the summer Arctic frontal zone. *Climate Dynamics*, 38, 1805–1827.

Liston, G. E., McFadden, J. P., Sturm, M., and Pielke, R. A., Sr. (2002). Modelled changes in arctic tundra snow, energy and moisture fluxes due to increased shrubs. *Global Change Biology*, 8, 17–32.

Liu, H., and Randerson, J. T. (2008). Interannual variability of surface energy exchange depends on stand age in a boreal forest fire chronosequence. *Journal of Geophysical Research*, 113, G01006, doi:10.1029/2007JG000483.

Liu, H., Randerson, J. T., Lindfors, J., and Chapin, F. S., III (2005). Changes in the surface energy budget after fire in boreal ecosystems of interior Alaska: An annual perspective. *Journal of Geophysical Research*, 110, D13101, doi:10.1029/2004JD005158.

Loranty, M. M., Goetz, S. J., and Beck, P. S. A. (2011). Tundra vegetation effects on pan-Arctic albedo. *Environmental Research Letters*, 6, 024014, doi:10.1088/1748-9326/6/2/029601.

Loranty, M. M., Berner, L. T., Goetz, S. J., Jin, Y., and Randerson, J. T. (2014). Vegetation controls on northern high latitude snow-albedo feedback: Observations and CMIP5 model simulations. *Global Change Biology*, 20, 594–606.

Lynch, A. H., Slater, A. G., and Serreze, M. (2001). The Alaskan Arctic frontal zone: Forcing by orography, coastal contrast, and the boreal forest. *Journal of Climate*, 14, 4351–4362.

Maness, H., Kushner, P. J., and Fung, I. (2013). Summertime climate response to mountain pine beetle disturbance in British Columbia. *Nature Geoscience*, 6, 65–70.

Matthews, H. D., Weaver, A. J., Meissner, K. J., Gillett, N. P., and Eby, M. (2004). Natural and anthropogenic climate change: Incorporating historical land cover change, vegetation dynamics and the global carbon cycle. *Climate Dynamics*, 22, 461–479.

Meissner, K. J., Weaver, A. J., Matthews, H. D., and Cox, P. M. (2003). The role of land surface dynamics in glacial inception: A study with the UVic Earth System Model. *Climate Dynamics*, 21, 515–537.

O'Halloran, T. L., Law, B. E., Goulden, M. L., et al. (2012). Radiative forcing of natural forest disturbances. *Global Change Biology*, 18, 555–565.

O'ishi, R., and Abe-Ouchi, A. (2009). Influence of dynamic vegetation on climate change arising from increasing CO_2. *Climate Dynamics*, 33, 645–663.

O'ishi, R., Abe-Ouchi, A., Prentice, I. C., and Sitch, S. (2009). Vegetation dynamics and plant CO_2 responses as positive feedbacks in a greenhouse world. *Geophysical Research Letters*, 36, L11706, doi:10.1029/2009GL038217.

Otto-Bliesner, B. L., and Upchurch, G. R., Jr. (1997). Vegetation-induced warming of high-latitude regions during the Late Cretaceous period. *Nature*, 385, 804–807.

Pearson, R. G., Phillips, S. J., Loranty, M. M., et al. (2013). Shifts in Arctic vegetation and associated feedbacks under climate change. *Nature Climate Change*, 3, 673–677.

Piao, S., Friedlingstein, P., Ciais, P., et al. (2007). Changes in climate and land use have a larger direct impact than rising CO_2 on global river runoff trends. *Proceedings of the National Academy of Sciences USA*, 104, 15242–15247.

Pielke, R. A., and Vidale, P. L. (1995). The boreal forest and the polar front. *Journal of Geophysical Research*, 100D, 25755–25758.

Pitman, A. J. (2003). The evolution of, and revolution in, land surface schemes designed for climate models. *International Journal of Climatology*, 23, 479–510.

Port, U., Brovkin, V., and Claussen, M. (2012). The influence of vegetation dynamics on anthropogenic climate change. *Earth System Dynamics*, 3, 233–243.

Post, E., Bhatt, U. S., Bitz, C. M., et al. (2013). Ecological consequences of sea-ice decline. *Science*, 341, 519–524.

Prentice, I. C., Jolly, D., and BIOME6000 (2000). Mid-Holocene and glacial-maximum vegetation geography of the northern continents and Africa. *Journal of Biogeography*, 27, 507–519.

Qu, X., and Hall, A. (2007). What controls the strength of snow–albedo feedback? *Journal of Climate*, 20, 3971–3981.

Randerson, J. T., Liu, H., Flanner, M. G., et al. (2006). The impact of boreal forest fire on climate warming. *Science*, 314, 1130–1132.

Robinson, D. A., and Kukla, G. (1985). Maximum surface albedo of seasonally snow-covered lands in the Northern Hemisphere. *Journal of Climate and Applied Meteorology*, 24, 402–411.

Rogers, B. M., Randerson, J. T., and Bonan, G. B. (2013). High-latitude cooling associated with landscape changes from North American boreal forest fires. *Biogeosciences*, 10, 699–718.

Schuur, E. A. G., Bockheim, J., Canadell, J. G., et al. (2008). Vulnerability of permafrost carbon to climate change: Implications for the global carbon cycle. *BioScience*, 58, 701–714.

Schuur, E. A. G., Abbott, B. W., Bowden, W. B., et al. (2013). Expert assessment of vulnerability of permafrost carbon to climate change. *Climatic Change*, 119, 359–374.

Sellers, P. J., Bounoua, L., Collatz, G. J., et al. (1996). Comparison of radiative and physiological effects of doubled atmospheric CO_2 on climate. *Science*, 271, 1402–1406.

Shugart, H. H., Leemans, R., and Bonan, G. B. (1992). *A Systems Analysis of the Global Boreal Forest*. Cambridge: Cambridge University Press.

Snyder, P. K., and Liess, S. (2014). The simulated atmospheric response to expansion of the Arctic boreal forest biome. *Climate Dynamics*, 42, 487–503.

Snyder, P. K., Delire, C., and Foley, J. A. (2004). Evaluating the influence of different vegetation biomes on the global climate. *Climate Dynamics*, 23, 279–302.

Street-Perrott, F. A., and Perrott, R. A. (1993). Holocene vegetation, lake levels, and climate of Africa. In *Global Climates since the Last Glacial Maximum*, ed. H. E. Wright, Jr., J. E. Kutzbach, T. Webb, III, et al. Minneapolis: University of Minnesota Press, pp. 318–356.

Strengers, B. J., Müller, C., Schaeffer, M., et al. (2010). Assessing 20th century climate–vegetation feedbacks of land-use change and natural vegetation dynamics in a fully coupled vegetation–climate model. *International Journal of Climatology*, 30, 2055–2065.

Sturm, M., McFadden, J. P., Liston, G. E., et al. (2001). Snow–shrub interactions in Arctic tundra: A hypothesis with climatic implications. *Journal of Climate*, 14, 336–334.

Sturm, M., Douglas, T., Racine, C., and Liston, G. E. (2005a). Changing snow and shrub conditions affect albedo with global implications. *Journal of Geophysical Research*, 110, G01004, doi:10.1029/2005JG000013.

Sturm, M., Schimel, J., Michaelson, G., et al. (2005b). Winter biological processes could help convert arctic tundra to shrubland. *BioScience*, 55, 17–26.

Swann, A. L., Fung, I. Y., Levis, S., Bonan, G. B., and Doney, S. C. (2010). Changes in Arctic vegetation amplify high-latitude warming through the greenhouse effect. *Proceedings of the National Academy of Sciences USA*, 107, 1295–1300.

Swann, A. L. S., Fung, I. Y., Liu, Y., and Chiang, J. C. H. (2014). Remote vegetation feedbacks and the mid-Holocene green Sahara. *Journal of Climate*, 27, 4857–4870.

Tape, K., Sturm, M., and Racine, C. (2006). The evidence for shrub expansion in Northern Alaska and the Pan-Arctic. *Global Change Biology*, 12, 686–702.

Thomas, G., and Rowntree, P. R. (1992). The boreal forests and climate. *Quarterly Journal of the Royal Meteorological Society*, 118, 469–497.

Vanderhoof, M., Williams, C. A., Ghimire, B., and Rogan, J. (2013). Impact of mountain pine beetle outbreaks on forest albedo and radiative forcing, as derived from Moderate Resolution Imaging Spectroradiometer, Rocky Mountains, USA. *Journal of Geophysical Research: Biogeosciences*, 118, 1461–1471, doi:10.1002/jgrg.20120.

Vanderhoof, M., Williams, C. A., Shuai, Y., et al. (2014). Albedo-induced radiative forcing from mountain pine beetle outbreaks in forests, south-central Rocky Mountains: Magnitude, persistence, and relation to outbreak severity. *Biogeosciences*, 11, 563–575.

Wang, G., and Eltahir, E. A. B. (2000a). Biosphere–atmosphere interactions over West Africa, II: Multiple climate equilibria. *Quarterly Journal of the Royal Meteorological Society*, 126, 1261–1280.

Wang, G., and Eltahir, E. A. B. (2000b). Ecosystem dynamics and the Sahel drought. *Geophysical Research Letters*, 27, 795–798.

Wang, G., and Eltahir, E. A. B. (2000c). Role of vegetation dynamics in enhancing the low-frequency variability of the Sahel rainfall. *Water Resources Research*, 36, 1013–1021.

Wang, G., Eltahir, E. A. B., Foley, J. A., Pollard, D., and Levis, S. (2004). Decadal variability of rainfall in the Sahel: Results from the coupled GENESIS–IBIS atmosphere–biosphere model. *Climate Dynamics*, 22, 625–637.

Watson, A. J., and Lovelock, J. E. (1983). Biological homeostasis of the global environment: The parable of Daisyworld. *Tellus B*, 35, 284–289.

Wood, A. J., Ackland, G. J., Dyke, J. G., Williams, H. T. P., and Lenton, T. M. (2008). Daisyworld: A review. *Reviews of Geophysics*, 46, RG1001, doi:10.1029/2006RG000217.

Xu, L., Myneni, R. B., Chapin, F. S., III, et al. (2013). Temperature and vegetation seasonality diminishment over northern lands. *Nature Climate Change*, 3, 581–586.

Zeng, N., Neelin, J. D., Lau, K.-M., and Tucker, C. J. (1999). Enhancement of interdecadal climate variability in the Sahel by vegetation interaction. *Science*, 286, 1537–1540.

Anthropogenic Land Use and Land-Cover Change

28.1 | Chapter Summary

Changes in land cover and in human uses of land can influence climate. Conversion of forests and grasslands to agricultural land alters net radiation, the partitioning of this energy into sensible and latent heat, and the partitioning of precipitation into soil water, evapotranspiration, and runoff. Among the surface characteristics altered by land-cover change are albedo, surface roughness, leaf area index, canopy conductance, root depth, and soil texture and structure. Land degradation in arid and semiarid climates increases surface albedo, reduces evapotranspiration, and may contribute to low rainfall in these regions. Extensive deforestation and land clearing have altered the climate of vast regions of Australia. Clearing of tropical forests for pastures creates a warmer, drier climate. Clearing of temperate forests and grasslands to cultivate crops cools climate, primarily because of higher albedo. However, the climate signal associated with crops is complicated and is related to the timing of crop planting, maturation, and harvesting relative to the phenology of natural vegetation. Irrigation leads to a cooler, moister climate. The influence of historical land-cover change on climate needs to be considered as a climate forcing in addition to traditional forcings such as greenhouse gases, aerosols, solar variability, and ozone. Climate model simulations show that the biogeophysical effects of historical land-cover change have cooled climate over large regions of North America and Eurasia. This biogeophysical cooling is comparable to, but of opposite sign, greenhouse gas warming over the same period. This cooling primarily results from increases in surface albedo with deforestation. Conversions among forests, pastureland, and cropland are thought to have decreased annual evapotranspiration. The net effect of anthropogenic land use and land-cover change is the balance between these biogeophysical changes and carbon emission from land use.

28.2 | Green Planets and Brown Planets

Much of the natural vegetation of the world has been converted to cropland and rangeland (Figure 2.8). Human uses of land such as cultivation, grazing, forest clearing, and forest regrowth on abandoned farmland alter net radiation, the partitioning of this available energy between sensible and latent heat, and the partitioning of precipitation into runoff and evapotranspiration. These changes occur from modifications of albedo, surface roughness, leaf area index, rooting depth, and canopy conductance and also from changes in soil texture and structure that affect soil water. Soil texture determines how much water is available for evapotranspiration. Soil compaction with overgrazing alters hydraulic properties and changes the soil water balance. The changes in climate caused by anthropogenic land use and land cover-change have received considerable scientific interest (Bonan 2008; Pielke et al. 2007, 2011; Levis et al. 2010; Mahmood et al. 2014).

The role of vegetation in affecting global climate can be assessed in paired climate model simulations that, as an extreme, simulate a planet completely covered with well-watered vegetation and contrast this simulated climate with that of a planet completely devoid of vegetation. Shukla and Mintz (1982) performed such simulations in their pioneering study that demonstrated the importance of evapotranspiration for global climate. Other studies have since adopted similar methodology to quantify the maximum effect of vegetation on global climate.

A study by Fraedrich et al. (1999) and Kleidon et al. (2000) compared climate simulations with the extreme endpoints of land cover: a desert planet in the absence of vegetation and a green planet where all non-glaciated land is covered by forest (Table 28.1). The desert planet has no vegetation, high surface albedo, low surface roughness, and low soil water storage capacity. The green planet is covered by trees with high leaf area index and has lower albedo, increased surface roughness, and greater soil water storage. Soil water availability is further increased in the green planet by eliminating drainage and reducing runoff.

The green world has a cooler and moister climate compared with the desert world (Figure 28.1). Annual land evapotranspiration in the green planet more than triples compared with the desert planet while precipitation nearly doubles. The Bowen ratio decreases from 1.3 in the desert planet to 0.13 in the green planet. Despite the increase in precipitation, annual runoff decreases by 25 percent because a greater portion is recycled as evapotranspiration and because of the greater soil water-holding capacity. The geographic extent of Köppen climate types highlights changes in climate arising from the desert and green worlds (Table 28.2). About one-quarter of the total land area has a different climate, mostly due to reduction in the area of arid climate and increase in the area

Table 28.1 Surface boundary conditions for desert and green planet climate model simulations

	Desert	Green
Vegetation cover (fraction)	0	1.0
Albedo (fraction)	>0.28	0.12
Leaf area index (m² m⁻²)	0	10
Roughness length (m)	0.01	2
Soil depth (cm)	10	Optimized
Soil drainage	Standard model	None
Surface runoff	Standard model	Restricted

Source: From Fraedrich et al. (1999) and Kleidon et al. (2000).

of temperate climate in the green planet. More recent climate simulations contrasted a maximally forested world with a grassland world (Brovkin et al. 2009). In these simulations, the presence of trees warms temperature throughout the year and increases annual precipitation while grasses cool the climate and decrease annual precipitation.

Snyder et al. (2004) extended this approach to examine the influence of specific biomes on climate. Rather than replacing all vegetation with desert-like land cover, they systematically replaced individual biomes with bare ground. A simulation with all natural vegetation served as a control. Six experimental simulations individually replaced tropical forest, boreal forest, temperate forest, savanna, grassland, and shrubland/tundra with bare soil (Table 28.3). The presence of tropical forest creates a cooler and wetter climate compared with bare ground. Its lower surface albedo leads to warming, but this is offset by cooling from greater latent heat flux. Greater evapotranspiration contributes to high annual precipitation. Boreal forest warms the climate, primarily because trees lower surface albedo compared with snow-covered ground. Temperate forest warms temperature in winter and spring by decreasing the surface albedo

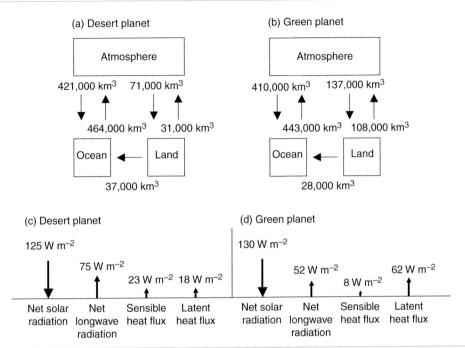

Fig. 28.1 Difference between a desert planet and a green planet. (a,b) Annual hydrologic cycle showing precipitation, evaporation, and runoff in cubic kilometers per year. (c, d) Annual surface energy fluxes over land. Data from Fraedrich et al. (1999) and Kleidon et al. (2000).

Table 28.2 | Percentage of land classified by Köppen climate types in a desert planet and a green planet

Climate type	Climate conditions	Typical vegetation	Percentage	
			Desert	Green
Tropical	$T_{min} \geq 18°C$	Tropical forest, savanna	18	19
Arid/dry	$P <$ dryness threshold	Grass, shrub, desert	28	9
Temperate	$-3°C \leq T_{min} \leq 18°C$	Temperate forest	12	29
Cold/snow	$T_{min} \leq -3°C$ and $T_{max} \geq 10°C$	Boreal forest	27	23
Ice	$T_{max} < 10°C$	Tundra	14	18

Note: T_{min}, mean temperature of coldest month. T_{max}, mean temperature of warmest month. P, annual mean precipitation.
Source: From Kleidon et al. (2000).

of snow-covered areas. It cools temperature in summer compared with bare soil because the lower albedo is offset by a higher latent heat flux compared with bare soil. The cold season warming offsets the summer cooling so that annual mean temperature is greater with forest than with bare soil. Savanna creates a cooler, wetter climate similar to tropical forest as a result of a modest decrease in surface albedo and a strong increase in latent heat flux. Grassland has similar effect. Boreal forest has the greatest effect on annual temperature as a result of strong changes in albedo. Boreal deforestation cools temperature with small effect on precipitation. Tropical forest has the greatest effect on annual precipitation as a result of strong changes in latent heat flux. Tropical deforestation warms climate and substantially decreases precipitation.

Table 28.3 | Effect of vegetation removal on annual temperature and precipitation

Biome	Area (million km²)	Surface forcing		Climate change	
		Albedo	Latent heat	ΔT (°C)	ΔP (mm day⁻¹)
Tropical forest	22.7 (16%)	Moderate	Strong	1.18	−1.34
Boreal forest	22.4 (15%)	Strong	Weak	−2.75	−0.27
Temperate forest	19.0 (13%)	Moderate	Moderate	−1.07	−0.49
Savanna	19.1 (13%)	Moderate	Strong	0.87	−1.10
Grassland	14.1 (10%)	Moderate	Moderate	0.75	−0.41
Shrubland/ tundra	24.6 (17%)	Moderate	Moderate/ weak	0.32	−0.35

Note: Percentages refer to the portion of total land area. Surface forcing refers to changes in albedo and latent heat flux ranked from strong to weak. Results are the difference (no vegetation minus control) spatially averaged over the biome where the vegetation was removed. ΔT, temperature. ΔP, precipitation.
Source: From Snyder et al. (2004).

Davin and de Noblet-Ducoudré (2010) contrasted a maximally forested world with one in which all trees are replaced with grasses. One climate model simulation depicted a world with the maximum extent of forests; another simulation replaced these forests with grasslands. Three more simulations individually considered only the albedo, surface roughness, and evapotranspiration efficiency differences between forests and grasslands. This latter process represents various parameters including rooting depth, canopy water holding capacity, and stomatal conductance. In the model, trees are more efficient at transpiring water than are grasses because of their deeper roots and larger leaf area.

Global-scale replacement of forests by grasslands increases surface albedo, and the higher albedo decreases global annual mean temperature by –1.36°C (Figure 28.2). This cooling is strongest at northern high latitudes (> 4°C) and smallest in the tropics (1°C). In contrast, the difference in evapotranspiration efficiency increases global temperature by 0.24°C. This warming is largest (> 1°C) in tropical Amazonia, tropical Africa, and Southeast Asia, occurs year-round, and is largest during the dry season (2–5°C). Temperate regions (mainly North America and Europe) have little warming in

winter, when evapotranspiration is weak, but warm by 2–5°C in summer, when the decrease in evapotranspiration is large. Conversion from forest to grassland decreases surface roughness and reduces turbulence in the boundary layer. This change in roughness increases global temperature by 0.29°C. The warming is about 1°C over most land areas and is largest in the tropics. The net biogeophysical effect of replacing forests with grassland is a cooling of –1°C. The balance among the different processes varies with latitude. The albedo effect is strongest in temperate and boreal regions of the Northern Hemisphere, where deforestation produces cooling. The net effect of deforestation in the tropics is warming, because evapotranspiration efficiency and surface roughness are the dominant influences.

28.3 | Dryland Degradation

In arid and semiarid climates, overgrazing by livestock, fuelwood collection, and other human activities can lead to land degradation that reduces vegetation cover, increases bare soil, decreases soil resources (water-holding capacity and nutrients), and increases soil salinity (Reynolds et al. 2007; D'Odorico et al.

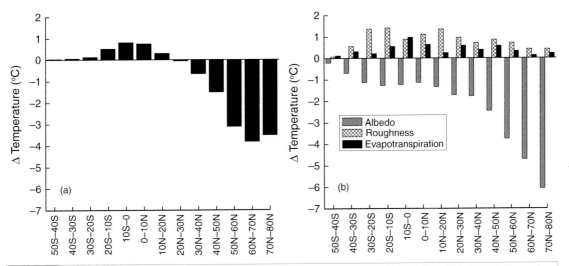

Fig. 28.2 Annual mean change in temperature zonally averaged over deforested areas. (a) Net effect. (b) Individual effects of changes in surface albedo, surface roughness, and evapotranspiration efficiency. Adapted from Davin and de Noblet-Ducoudré (2010).

2013). These changes are encompassed in the term desertification. One aspect of desertification is changes in climate resulting from land degradation.

The landscape along the United States–Mexico border in the Sonoran Desert provides a case study of the effects of overgrazing on climate (Balling 1988, 1989; Bryant et al. 1990; Balling et al. 1998). A sharp contrast in vegetation occurs along several hundred kilometers of the border. The Mexican side of the border has been heavily overgrazed. The grass is shorter, more soil is exposed, and albedo is higher than in southern Arizona, which has better grass cover. Long-term climate records show summer daily maximum air temperatures on the Mexico side of the border are 4°C warmer than nearby locations in Arizona after accounting for elevation differences (Balling 1988). Temperatures in Sonora warmed faster than stations in Arizona from 1969 to 1983 (Balling et al. 1998). The Mexico–United States temperature contrast increases with higher rainfall throughout summer (Balling 1989). This is because wet summers support dense, productive vegetation with high evapotranspiration in Arizona. In contrast, severe overgrazing restricts vegetation cover in Mexico. Although the higher albedo

Table 28.4 Energy budget along the United States–Mexico border measured at 1100 local time on September 22, 1987 after 15 days without rain

	U.S.	Mexico
Albedo, 0.3–3.0 μm (fraction)	0.13	0.17
Emitted longwave radiation (W m^{-2})	479	564
Soil heat (W m^{-2})	17	35
Sensible heat (W m^{-2})	62	234
Latent heat (W m^{-2})	476	174
Evapotranspiration (mm day^{-1})	6.6	2.4
Air temperature (°C)	24.0	26.3
Surface temperature (°C)	30.2	42.8

Source: From Bryant et al. (1990).

of the overgrazed land should cool the surface, lower vegetation cover reduces evapotranspiration and warms the surface.

The border contrast is seen in surface energy fluxes (Table 28.4). During the measurement period, the Mexico site had a higher albedo (0.17) than the site across the border (0.13). Latent heat flux after 15 days without rain was almost three times greater at the United States site than in Mexico. The lower latent heat in Mexico was balanced by higher sensible heat flux (Bowen ratio of 1.3 compared with 0.13) and more soil heat flux, which warmed the surface and air.

Other regions are experiencing woody encroachment that may impact climate (D'Odorico et al. 2013). In large regions of the southwestern United States, woody shrubs and trees have invaded grasslands and savannas. In the southern Great Plains, woody encroachment may increase precipitation and warm climate (Ge and Zou 2013). In model simulations, woody encroachment decreases surface albedo, increases leaf area index, and increases surface roughness compared with native short grasses. Net radiation increases because of the lower surface albedo. In this dry landscape, much of the additional energy is dissipated as sensible heat rather than as latent heat. The surface heating from lower albedo is greater than the evaporative cooling. Daily maximum air temperature increases by up to 0.6°C in some regions where woody plants completely replace short grass. Daily minimum temperature decreases slightly (0.1–0.2°C). Shrubs are also invading the northern Chihuahuan Desert. An increase in nighttime temperature has been observed above shrubland compared with grassland during winter (He et al. 2010, 2011). This is related to the lower vegetation fraction of shrubland in contrast with desert grassland. More energy enters the soil during the day, and this energy is released at night as longwave radiation.

Studies in and near the Negev region of southern Israel also point to the importance of vegetation in determining the climate of a semidesert landscape. Protected regions of the Negev have a much lower surface albedo than adjacent areas of the Sinai (Otterman 1974, 1977, 1981; Otterman and Tucker 1985). This albedo contrast arises from overgrazing and harvesting plants for firewood in the Sinai, but limited grazing and more abundant vegetation in the Negev. Overgrazed soils with high albedo are several degrees cooler than vegetated soil (Otterman 1974; Otterman and Tucker 1985). Daytime sensible heat flux from vegetated surfaces is up to twice that of

bare soils (Otterman 1989). This arises from the lower albedo but also from shading of soil by plants, which reduces soil heat flux. The warmer vegetated surface provides more heating of the atmosphere, greater boundary-layer growth, and greater convection – all of which increase the likelihood of rainfall (Otterman 1974, 1989). Afforestation, increased cultivation, and limited grazing since 1948 in southern Israel may have increased rainfall (Otterman et al. 1990).

Differences among vegetation in surface roughness can produce marked variation in energy fluxes and temperature in semiarid climates, where the clear sky allows strong solar heating of the land and the dry soil limits evapotranspiration. An afforestation project in southern Israel provides a case study (Rotenberg and Yakir 2010, 2011). The particular study measured the energy balance of an open canopy pine forest planted ~40 years earlier and compared this with native shrubland. The forest had a leaf area index of ~1.4 m^2 m^{-2}, a height of 11 m, and trees covered ~56 percent of the surface. The shrubland had a sparse canopy (20% vegetation cover) and short vegetation (30–50 cm tall).

During the measurement period, the forest had a lower surface albedo (by 0.1) and absorbed 24 W m^{-2} more solar radiation annually than the shrubland (Table 28.5). The forest was cooler than the shrubland, and the lower surface temperature reduced the loss of longwave radiation (25 W m^{-2}, annual mean), further increasing the net radiation gained by the forest (49 W m^{-2}, annual mean). Most of the net radiation was dissipated as sensible heat during the dry summer months. This is seen in the high Bowen ratio (4.8 on average, but >10 in summer). Despite more net radiation and little evaporative cooling, the surface temperature of the forest was 5°C cooler, on average, than the shrubland because of large sensible heat fluxes that efficiently carried the heat away from the forest canopy. The forest, with its low tree density and open canopy, had high aerodynamic coupling with the atmosphere.

The Sahel region of Africa between the Sahara desert to the north and tropical rainforest to the south is a widely studied example of land–atmosphere interactions in dryland

Table 28.5 | Forest and shrubland measurements in the semiarid climate of southern Israel

	Forest	Shrubland	Difference
Absorbed solar radiation (W m^{-2})	211	187	24
Net longwave radiation (W m^{-2})	−95	−120	25
Net radiation (W m^{-2})	116	67	49
Surface temperature (°C)	32.0	36.6	−4.6
Albedo (fraction)	0.11	0.21	−0.10
Aerodynamic conductance (mol m^{-2} s^{-1})	2.6	~0.4	−

Note: The difference is given as forest − shrubland. Surface temperature is during the day. Conductances are for midday and are converted to mol m^{-2} s^{-1} using ρ_m = 42.3 mol m^{-3}. The conductance for shrubland is estimated.

Source: From Rotenberg and Yakir (2011).

environments (Nicholson et al. 1998; Nicholson 2000; Foley et al. 2003; Xue et al. 2004; Xue 2006). Precipitation in this region of dry shrubland, grassland, and savanna is characterized by a summer monsoon driven by the thermal contrast between land and ocean. Annual precipitation increases from less than 25 mm in the desert to greater than 2000 mm along the Guinea coast (Figure 28.3). Annual precipitation in the Sahel was below normal during the latter part of the twentieth century. Between 1968 and 1997, rainfall was 25–40 percent lower than the period 1931–1960 (Figure 28.4). The anomalous low precipitation has been linked to changes in tropical sea surface temperature (Kucharski et al. 2013). However, considerable scientific research on the Sahel drought has focused on whether changes in land use and land cover reinforce drought.

Overgrazing, cultivation, and deforestation may contribute to low rainfall through a positive climate feedback mediated by increased surface albedo and decreased surface roughness. The initial concept proposed by Jule Charney and colleagues (Charney 1975; Charney et al. 1975) was based on higher surface albedo as a result of overgrazing (Figure 28.5). Overgrazing reduces vegetation cover. Surface albedo increases because the desert soil has a higher albedo than vegetation. Decreased net radiation cools the surface, which promotes subsidence of air aloft. Subsidence decreases cloud formation and convection, leading to less rainfall. In contrast, a lower albedo warms

the surface, provides more heating of the atmosphere, greater boundary-layer growth, and greater convection. Covering sandy desert areas along the Mediterranean coast of Africa with asphalt so that surface albedo is reduced has been proposed as a means to promote rainfall (Black 1963; Black and Tarmy 1963).

A more complete depiction of land–atmosphere feedback includes decreased surface roughness and soil degradation in addition to increased albedo. A study by Clark et al. (2001) typifies the simulated response of climate to land degradation. In the control climate simulation, land cover was represented with natural vegetation, predominantly sparse broadleaf trees and a shrub/ground cover mix. In a land degradation simulation, natural vegetation was replaced by broadleaf shrub/bare soil to mimic desert conditions across the Sahel. The vegetation change to sparse shrubs with bare soil increased surface albedo and decreased roughness length, vegetated fraction, and leaf area index (Table 28.6). Soil hydraulic properties were also changed to reduce water-holding capacity. Land degradation reduces precipitation in July–September (i.e., the rainy season) by more than 1 mm day^{-1}, primarily in the western region (Figure 28.6). Throughout the degraded area, precipitation (P) is 0.7 mm day^{-1} (33%) less, evapotranspiration (E) decreases by 0.5 mm day^{-1} (30%), and moisture convergence ($P-E$) decreases by 0.2 mm day^{-1} (50%) (Table 28.7). Net radiation decreases, primarily due to less absorption

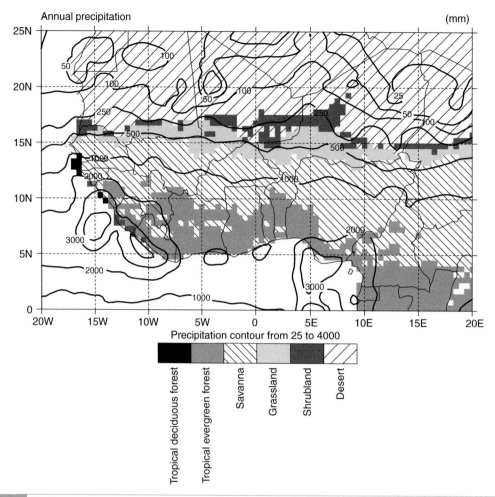

Fig. 28.3 Land cover and mean annual rainfall in northwest Africa. The Sahel refers to the region between latitudes 13° N and 20° N and longitudes 15° W and 20° E (Foley et al. 2003).

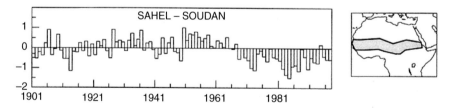

Fig. 28.4 Rainfall fluctuations over the twentieth century in the Sahel region of West Africa expressed as a standardized departure from the long-term mean. The map shows the location of the region. Reproduced from Nicholson (2000).

of solar radiation despite decreased cloud cover. Reduced net radiation and evapotranspiration leads to a warmer, drier atmospheric boundary layer.

Observations show that variation in soil moisture and vegetation affect surface energy fluxes in the West African Sahel, and this is seen also in the atmospheric boundary layer (Taylor

Table 28.6 | Surface and soil characteristics for four common vegetation types in the Sahel region of Africa

	Broadleaf evergreen tree	Broadleaf evergreen tree with ground cover	Broadleaf shrub with ground cover	Broadleaf shrub with bare soil
Vegetation				
Albedo (fraction)	0.13	0.20	0.20	0.30
Roughness length (m)	2.65	0.95	0.25	0.06
Vegetation cover (fraction)	0.98	0.30	0.10	0.10
Leaf area index ($m^2 \; m^{-2}$)	5.0	4.1	0.9	0.3
Maximum stomatal conductance ($mol \; m^{-2} \; s^{-1}$)	0.28	0.26	0.05	0.05
Root depth (cm)	100	50	50	50
Soil				
θ_{pwp} ($m^3 \; m^{-3}$)	0.12	0.13	0.05	0.04
θ_{sat} ($m^3 \; m^{-3}$)	0.42	0.42	0.44	0.44
K_{sat} ($mm \; hr^{-1}$)	72	72	634	634
ψ_{pwp} (mm)	−86	−86	−35	−35

Note: θ_{pwp}, volumetric moisture content at wilting point. θ_{sat}, volumetric moisture content at saturation. K_{sat} hydraulic conductivity at saturation. ψ_{pwp} matric potential at saturation. Albedo, roughness length, and leaf area index are averages for July–September. Stomatal conductance is converted to $mol \; m^{-2} \; s^{-1}$ using $\rho_m = 42.3 \; mol \; m^{-3}$.
Source: From Clark et al. (2001).

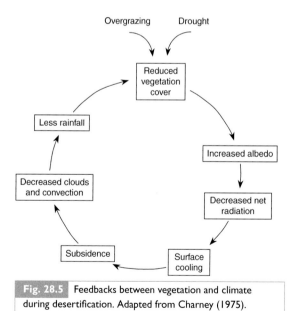

Fig. 28.5 Feedbacks between vegetation and climate during desertification. Adapted from Charney (1975).

Table 28.7 | Effect of Sahel land degradation on surface climate during the months of July–September

	Control	$D - C$
Cloud cover (fraction)	0.42	−0.06
Net radiation ($W \; m^{-2}$)	151	−29
Sensible heat ($W \; m^{-2}$)	102	−14
Latent heat ($W \; m^{-2}$)	50	−15
Surface temperature (°C)	34.0	0.2
Precipitation, P ($mm \; day^{-1}$)	2.1	−0.7
Evapotranspiration, E ($mm \; day^{-1}$)	1.7	−0.5
$P - E$ ($mm \; day^{-1}$)	0.4	−0.2

Note: Data are averaged over the degraded area. Control, control simulation. $D - C$, difference between the degraded and control simulations.
Source: From Clark et al. (2001).

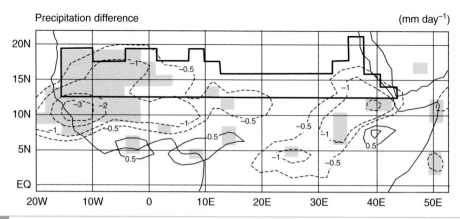

Fig. 28.6 Simulated change in precipitation (mm day⁻¹) for the months of July–September due to land degradation in the Sahel. Differences that are statistically significant at the 95 percent confidence level are shaded. The thick line shows the region of land degradation. Adapted from Clark et al. (2001).

Table 28.8 Mean annual albedo (fraction) and annual precipitation (mm) by vegetation type in West Africa

Vegetation	Albedo	Precipitation
Open shrubland	0.45	254
Closed shrubland	0.33	785
Grassland	0.30	1004
Savanna	0.25	1037
Cropland	0.23	1700

Source: From Fuller and Ottke (2002).

et al. 2011). Early afternoon measurements along a 470 km flight track (spanning 3° in latitude) found that the boundary layer above wet soil was up to 3°C cooler, 3 g kg⁻¹ moister, and half as deep as that over dry areas (Taylor et al. 2007). West Africa has a strong latitudinal gradient in surface albedo in relation to precipitation and vegetation (Samain et al. 2008). Surface albedo increases with greater aridity, from about 0.1–0.2 in savanna to >0.4 in semidesert. Similar results are seen in Table 28.8. Within the Sahel, soil moisture and vegetation control energy flux partitioning in seasonal dynamics related to the monsoon (Guichard et al. 2009; Timouk et al. 2009). The region is characterized by a short rainy season in summer and a longer dry season in the remainder of the year. Sensible heat flux is high and latent heat flux is low throughout the year at bare soil and semidesert sites. In contrast, a grassland and a woodland both show strong differences in fluxes between the dry season and the wet season. Of particular note is an increase in net radiation during the August monsoonal rains, with high latent heat flux and low sensible heat flux, in contrast to the dry season. The monsoonal rains wet the soil and trigger leaf growth at the grassland and woodland, and latent heat flux replaces sensible heat flux as the dominant flux during the rainy season. Surface albedo decreases with the green foliage, contributing to the higher net radiation, and the cooler surface temperature additionally decreases surface longwave loss.

The Sahel is a region of closely coupled interactions among ocean, atmosphere, and land (Foley et al. 2003; Kucharski et al. 2013). Initial research focused on changes in surface albedo as a cause of the drought. A more complete understanding shows drought in the Sahel is an outcome of complex interactions among changes in sea surface temperature, land use, soil water, and vegetation (Figure 27.7). The persistence of the long drought may represent a change from one climate equilibrium to another brought about by vegetation feedbacks triggered by changes in sea surface temperatures.

28.4 | Australia

Australia has experienced large changes in land cover from human activities (McAlpine et al. 2009). Extensive grazing covers approximately 43 percent of the continent, with intensive cropping and improved pastures covering another 10 percent. Much of this change occurred in the southwest and southeast regions, where 50 percent of native forests and 65 percent of native woodlands have been cleared or severely modified. The Murray–Darling basin extending over 1 million km² in southeastern Australia illustrates this change (Figure 28.7). Since European settlement in the early 1800s, the region underwent massive clearing of native woodland and forest for grazing and croplands. By the 1980s, forests, woodland, and mallee were reduced in area by 64 percent, 63 percent, and 34 percent, respectively. Tree thinning for grazing increased the area of open woodlands by 127 percent.

Narisma and Pitman (2003) used a regional climate model to study the effects of land-cover change. Compared with the land cover of 1788, when Europeans first arrived, the vegetation of 1988 has widespread replacement of trees by grasses in the southwestern area of the continent in the state of Western Australia and in southeastern regions throughout New South Wales extending south into Victoria and west into South Australia, largely as a result of livestock grazing (Figure 28.8). Shrublands have encroached into formerly sparsely vegetated land in Queensland in the northeast. The conversion from tree to grass increases surface albedo, decreases leaf area index, and decreases roughness length. The authors compared the climates simulated using the 1788 and 1988 vegetation as surface boundary conditions to the model.

The largest change in climate occurs in January, during the Australian summer, as a result of repartitioning net radiation into sensible and latent heat (Figure 28.9). Net radiation

(a) 1780

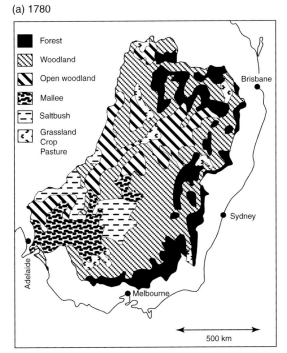

(b) 1980

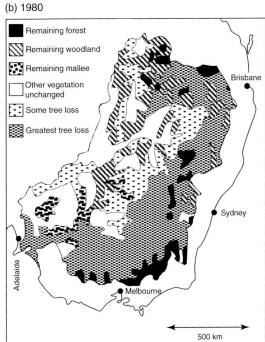

Fig. 28.7 Vegetation of the Murray–Darling Basin in southeastern Australia (a) prior to European settlement circa 1780 and (b) circa 1980. Adapted from Walker et al. (1993).

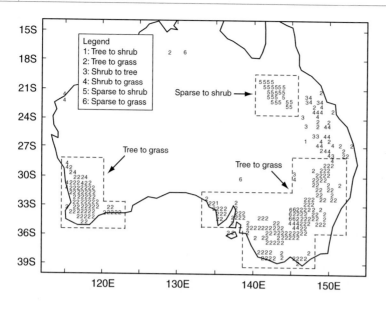

Fig. 28.8 Land-cover change in Australia (1788–1988) as represented in a climate model. Numbers denote the type of change. Land cover is unchanged in areas without numbers. Adapted from Narisma and Pitman (2003).

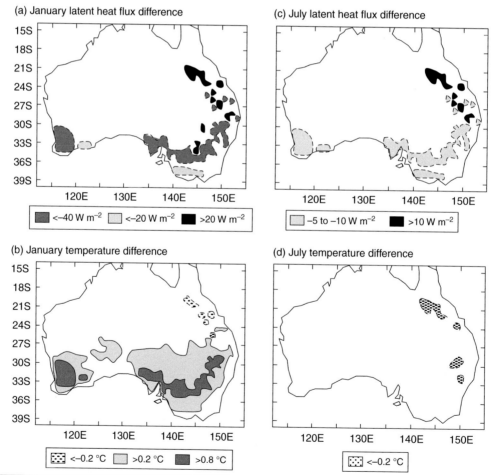

Fig. 28.9 Effect of land-cover change in Australia on latent heat flux and air temperature for (a, b) January and (c, d) July. Adapted from Narisma and Pitman (2003).

changes by less than ±20 W m⁻² with 1988 veg-etation. Latent heat flux decreases by more than 40 W m⁻² over wide regions where grasses replace trees. An increase in latent heat flux occurs with shrub encroachment, primarily due to increased leaf area index and rough-ness length. There is widespread warming in the southwest and southeast during January of 1.0–1.5°C as a result of reduced latent heat flux and corresponding increased sensible heat. Air temperature in the northeast decreases by about 0.2°C in January and 0.5°C in July as a result of increased latent heat flux and decreased sen-sible heat flux. Precipitation averaged for the three regions of large land-cover change gener-ally decreases. Other climate model studies also find summer warming in southeastern Australia and decreased summer rainfall (McAlpine et al. 2007), and the land-cover change may have increased the severity and duration of droughts (Deo et al. 2009).

The study by Narisma and Pitman (2003) suggests that land-cover change has reduced rainfall in parts of Australia. One region where this may be critical is southwestern Western Australia. The climate of this region is Mediterranean, characterized by winter rain-fall, which has decreased since the mid-1900s. The region has also had significant deforesta-tion and conversion to farmland. Pitman et al. (2004) examined the impact of land-cover change on precipitation using three differ-ent regional climate models. In all models, land-cover change decreases July rainfall by 10–20 percent along the coast and increases rainfall by 10–20 percent further inland, sim-ilar to observed trends. This is a result of reduced surface roughness, which alters advec-tion of moisture onshore from the ocean and patterns of moisture convergence and diver-gence. This suggests that land-cover change has contributed to the observed decrease in rainfall in southwest Western Australia, a con-clusion supported by other studies (Timbal and Arblaster 2006; Nair et al. 2011).

Human impact on the Australian landscape extends far back in time, to early human arrival on the continent some 45–50 kyr BP (Miller et al. 2005a; Bird et al. 2013). One such effect was burning of the landscape during hunting and for other purposes. Changes in the fire regime may have shifted the vegetation from a mosaic of drought-adapted trees and shrubs mixed with grasses to the present-day landscape of fire-adapted desert scrubland and grassland. This vegetation change may have altered cli-mate. Present-day precipitation during summer is high along the northern coast and decreases markedly inland (Figure 5.12). Climate model simulations show that the extension of the summer monsoon rainfall into the interior con-tinent is sensitive to vegetation (Miller et al. 2005b). Dense vegetation with high leaf area enhances the penetration of monsoon rain-fall inland while conversion to the present-day desert scrub vegetation reduces rainfall. Other modeling studies find that decreased vegetation cover has little effect on the monsoon (Marshall and Lynch 2008) or delays the onset of the mon-soon and decreases early season rainfall but has little effect on peak monsoonal precipitation (Notaro et al. 2011).

28.5 | Tropical Deforestation

The impact of tropical deforestation on climate has long held the interest of climate model-ers. One of the first such studies was that of Henderson-Sellers and Gornitz (1984). They used a global climate model to study the effect of replacing Amazonian rainforest with pas-ture by changing surface albedo, roughness length, and soil water-holding capacity. These surface changes decreased evapotranspiration and rainfall. A subsequent study by Dickinson and Henderson-Sellers (1988) used a global cli-mate model that included a second generation land surface model, the Biosphere–Atmosphere Transfer Scheme. All model grid cells located in South America and classified as evergreen broadleaf tree were changed to degraded grass-land (Figure 28.10a). This change in vegetation decreased roughness length, leaf area index, and vegetated fraction and altered stomatal conduc-tance. Soil texture was also changed to finer soil with more clay to reduce water-holding capac-ity, and soil color was made lighter to increase

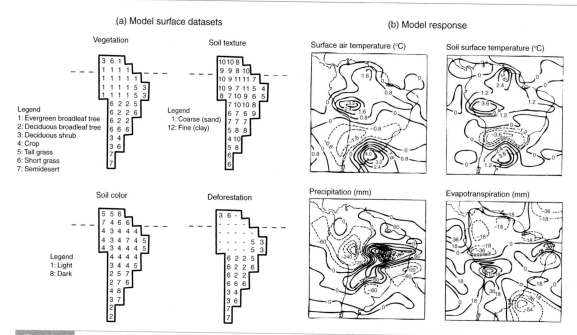

Fig. 28.10 Effect of Amazonian deforestation on simulated climate. (a) Vegetation, soil texture, and soil color in the control simulation. In the deforestation simulation, evergreen broadleaf tree was changed to degraded pasture. (b) Difference in January surface climate between the deforested and control simulations. The climate model has a spatial resolution of 4.5° latitude and 7.5° longitude. Adapted from Dickinson and Henderson-Sellers (1988).

soil albedo. The most prominent climate signal is a warming of temperature by 1–4°C and a decrease in evapotranspiration throughout the Amazon basin (Figure 28.10b). Precipitation decreases in the western Amazon, but increases in the eastern basin.

Numerous climate model studies have since examined the impact of tropical deforestation on climate. Most studies of Amazonian deforestation find that complete transformation of forest to pasture results in a warmer and drier climate (Table 28.9). Fifteen of 18 studies report an increase in annual mean air temperature ranging from 0.3°C to 3.8°C (mean, 1.7°C). Sixteen of 18 studies show decreased annual precipitation ranging from –146 to –643 mm (mean, –398 mm), and all studies find decreased evapotranspiration. Evapotranspiration decreases because of lower surface roughness, because trees have deep roots that sustain transpiration during the dry season, because interception loss decreases, and because higher surface albedo decreases net radiation. A warmer, drier climate upon

deforestation is found throughout the tropics, though the magnitude varies with region (Delire et al. 2001; Snyder et al. 2004; Voldoire and Royer 2004; Hasler et al. 2009; Lawrence and Vandecar 2015). Some studies suggest that tropical deforestation can affect extratropical climate through atmospheric teleconnections (Werth and Avissar 2002; Snyder 2010; Medvigy et al. 2013).

A key question with tropical deforestation is whether precipitation does decrease with deforestation, as the models predict. The type of land cover conversion (replacement by pasture or by cropland) affects the simulated reduction in precipitation (Costa et al. 2007). The effects of tropical deforestation may be more complex than represented in large-scale, global climate models (Pielke et al. 2007). The decrease in annual precipitation in the Amazon in response to total deforestation may be smaller in high resolution models (e.g., –62 mm yr^{-1}; Medvigy et al. 2011). Moreover, climate model studies of Amazonian deforestation are idealized and completely

Table 28.9 Annual response to Amazonian deforestation in various climate model studies

Study	Surface change		Climate change		
	Δalbedo	Δz₀	ΔT (°C)	ΔP (mm)	ΔE (mm)
Dickinson and Henderson-Sellers (1988)	+	–	3.0	0	–200
Lean and Warrilow (1989)	+	–	2.4	–490	–310
Nobre et al. (1991)	+	–	2.5	–643	–496
Dickinson and Kennedy (1992)	+	–	0.6	–511	–256
Mylne and Rowntree (1992)	+	0	–0.1	–335	–176
Henderson-Sellers et al. (1993)	+	–	0.6	–588	–232
Lean and Rowntree (1993)	+	–	2.1	–296	–201
Pitman et al. (1993)	+	–	0.7	–603	–207
Polcher and Laval (1994a)	+	0	3.8	394	–985
Polcher and Laval (1994b)	+	–	–0.1	–186	–128
McGuffie et al. (1995)	+	–	0.3	–437	–231
Sud et al. (1996)	+	–	2.0	–540	–445
Lean and Rowntree (1997)	+	–	2.3	–157	–296
Hahmann and Dickinson (1997)	+	–	1.0	–363	–149
Costa and Foley (2000)	+	–	1.4	–266	–223
Gedney and Valdes (2000)	+	–	1.3	–288	–237
Voldoire and Royer (2004)	+	–	–0.1	–146	–146
Snyder et al. (2004)	+	–	1.5	–511	–440

Note: Δalbedo and Δz₀ denote the change in surface albedo and roughness (+, increase; –, decrease; 0, no change). ΔT, ΔP, and ΔE are the simulated changes in annual temperature, precipitation, and evapotranspiration, respectively. Lawrence and Vandecar (2015) review tropical deforestation climate model simulations.

deforest the entire basin. In fact, however, deforestation is localized and patchy, and there may be a critical threshold before precipitation decreases (Lawrence and Vandecar 2015). Small scale, heterogeneous deforestation affects atmospheric processes at the mesoscale, in some cases even enhancing convection (Avissar et al. 2002; Baidya Roy and Avissar 2002; Weaver et al. 2002; Ramos da Silva and Avissar 2006; Khanna and Medvigy 2014). Deforestation produces a detectable effect on clouds in the region. Shallow clouds develop preferentially over deforested areas (Wang et al. 2009). This relates to a more unstable atmospheric boundary layer over forests that has high humidity and also to mesoscale circulations created by the contrast between interspersed patches of forests and

deforested land. Direct observational evidence for precipitation changes is lacking. However, moisture budget analyses show that air that passes over tropical forests produces more rainfall than air that moves over sparse vegetation (Spracklen et al. 2012).

Numerous observational field campaigns have been conducted in the Amazon to infer the realism of climate model simulations (Nobre et al. 2004). Data collected during the Anglo–Brazilian Amazonian Climate Observation Study (ABRACOS; Gash et al. 1996; Gash and Nobre 1997) contrast the micrometeorology of nearby forest and pasture sites in Amazonia and illustrate the effects of deforestation (Figure 28.11). One difference is the radiation balance (Culf et al. 1996). During the study, forest and pasture

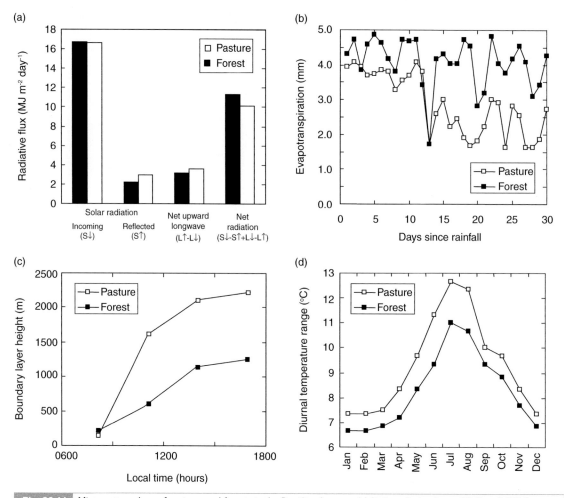

Fig. 28.11 Micrometeorology of pastures and forests in the Brazilian Amazon. (a) Daily radiative fluxes (Culf et al. 1996). (b) Daily evapotranspiration during the dry season over 30 days following heavy rainfall (Wright et al. 1992). (c) Height of the convective boundary layer (Gash and Nobre 1997). (d) Mean monthly diurnal temperature range (Culf et al. 1996).

received similar solar radiation, but the pasture had a higher albedo than the forest (0.18 versus 0.13). In addition, the pasture was warmer than the forest and emitted more longwave radiation. As a result, the net radiation balance for the pasture was 11 percent less than for the forest. Pastures are also shorter than forests and have lower aerodynamic roughness. During the wet season, pasture evapotranspiration is typically less than that of forest due to the reduced available energy and reduced roughness. During the dry season, shallow-rooted pastures tend to have low evapotranspiration because the surface soil water is depleted. In contrast, forests

have no significant reduction in evapotranspiration because the deep-rooted trees extract water from deep in the soil. This is particularly evident following heavy rain. Wright et al. (1992) found that for the first 10 days or so following heavy rainfall, pasture and forest had sustained evapotranspiration (Figure 28.11b). Thereafter, pasture evapotranspiration declined as the upper soil dried. In contrast, the forest had relatively constant evapotranspiration throughout the measurement period. Reduced evapotranspiration from pasture affects the atmospheric boundary layer. Gash and Nobre (1997) reported that the convective boundary layer over pasture

Table 28.10 Energy fluxes (W m^{-2}) and surface albedo (fraction) for forest and pasture in Rondônia, Brazil

	$S\downarrow$	$S\uparrow$	$S\downarrow-S\uparrow$	albedo	$L\downarrow$	$L\uparrow$	$L\downarrow-L\uparrow$	R_n
Forest	206	26	180	0.13	412	448	−36	144
Pasture	203	41	162	0.20	414	452	−38	124

	Wet season				Dry season			
	R_n	H	λE	$\lambda E / R_n$	R_n	H	λE	$\lambda E / R_n$
Forest	136	32	105	0.77	147	38	109	0.74
Pasture	129	46	83	0.64	113	49	64	0.56

Note: $S\downarrow$, incoming solar radiation. $S\uparrow$, reflected solar radiation. $L\downarrow$, incoming longwave radiation. $L\uparrow$, outgoing longwave radiation. R_n, net radiation. H, sensible heat flux. λE, latent heat flux. $\lambda E / R_n$, evaporative fraction.
Source: From von Randow et al. (2004).

Table 28.11 Energy fluxes (W m^{-2}) for forest in eastern Amazonia (3.0°S, 54.6°W)

	$S\downarrow$	R_n	H	$H / \lambda E$	$\lambda E / R_n$	E (mm day^{-1})	g_c (mol m^{-2} s^{-1})
Wet season	165	113	16	0.17	0.87	3.2	0.55
Dry season	204	140	21	0.17	0.83	4.0	0.52

Note: $S\downarrow$, incoming solar radiation. R_n, net radiation. H, sensible heat flux. λE, latent heat flux. $H / \lambda E$, Bowen ratio. $\lambda E / R_n$, evaporative fraction. E, evapotranspiration. g_c, surface conductance and converted to mol m^{-2} s^{-1} using $\rho_m = 42.3$ mol m^{-3}.
Source: From da Rocha et al. (2004).

was 700–1000 m higher than over forest because of stronger sensible heating (Figure 28.11c). The net result of these differences in radiation and evapotranspiration is that pastures are warmer during the day than forests (Culf et al. 1996). This is seen in the diurnal temperature range, which for the particular sites studied was larger for the pasture than for the forest, especially during the June–August dry season (Figure 28.11d).

Similar results are seen in the Large-Scale Biosphere–Atmosphere Experiment in Amazonia (LBA). von Randow et al. (2004) contrasted energy fluxes measured in forest and pasture (Table 28.10). Averaged over the measurement period, pasture had a higher albedo than forest (0.20 versus 0.13). The pasture also lost more longwave radiation than the forest so that net radiation was 14 percent lower compared with forest. At both sites, latent heat flux was the dominant turbulent flux in the wet and dry seasons. The forest had greater latent heat flux and lower sensible heat flux than did the pasture. Whereas high latent heat flux was sustained at the forest during the dry season, latent heat flux decreased at the pasture during the dry season. da Rocha et al. (2004) also found sustained latent heat flux year-round at a forest in eastern Amazonia (Table 28.11). The Bowen ratio, evaporative fraction, and surface conductance showed little sign of significant soil water stress during the dry season. Similar to earlier field programs, LBA studies show high sensible heat fluxes over pasture during the dry season create a boundary layer that is several hundred meters deeper, has warmer temperature, and has decreased moisture compared with that found over forest (Fisch et al. 2004). Both types of vegetation have similar boundary

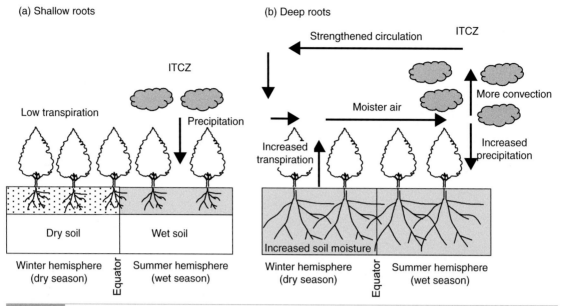

Fig. 28.12 Changes in tropical circulation and precipitation across the equator as a result of (a) shallow-rooted trees and (b) deep-rooted trees. Shaded areas denote soil water. Adapted from Kleidon and Heimann (2000).

layer depth, temperature, and humidity during the wet season.

Soil water availability for transpiration is a key control of tropical climate. Flux tower measurements reveal the importance of deep roots to sustain transpiration by trees. Studies have documented the sensitivity of climate to rooting depth (Nepstad et al. 1994; Kleidon and Heimann 2000). Deep roots sustain transpiration during the dry season, leading to a cooler, moister climate. Sustained evapotranspiration during the dry season can alter tropical atmospheric circulation and precipitation. Without deep roots, tropical vegetation in the winter hemisphere, which is experiencing a dry season, becomes water-stressed and has low transpiration. Increased transpiration as a result of deep roots increases moisture transport toward the intertropical convergence zone (Figure 28.12). More energy is available in the form of latent heat, which enhances convection and cloud cover in the summer hemisphere. An additional factor may be nocturnal redistribution of soil water during the dry season that recharges the upper soil with water from deeper in the soil profile (da Rocha et al. 2004). Such redistribution increases soil water availability and sustains evapotranspiration (Lee et al. 2005; Wang 2011).

28.6 | European Deforestation

Europe has undergone extensive land-cover change extending back to the Greek and Roman eras (Figure 23.9), and this may have changed climate. The Mediterranean region of southern Europe and northern Africa has received particular attention because of extensive deforestation over the past 2000 years. Historical records indicate the climate of the Mediterranean region was moister in Roman times (Reale and Dirmeyer 2000). At this time, forests covered much of Europe where there are now crops. Trees, shrubs, and grasses grew along the Mediterranean coast of Africa and the Middle East where there is now desert or semidesert vegetation. Climate model simulations show deforestation around the Mediterranean during the last 2000 years contributed to the dryness of the current climate (Reale and Dirmeyer 2000; Reale and Shukla 2000).

The climate model simulations of Dümenil Gates and Ließ (2001) support the notion that greater vegetation cover during the Roman era allowed more summer precipitation. Deforestation decreases summer evapotranspiration and reduces summer precipitation in some regions, particularly in the Atlas mountain range and the Iberian Peninsula. Other climate model simulations find a weaker climate response to European land-cover change (Anav et al. 2010), but that precipitation over the Iberian Peninsula is sensitive to land-cover change (Arribas et al. 2003). However, the results depend on soil moisture availability. An increase in tree cover can enhance early growing season evapotranspiration and lead to a cooler, moister climate, but depletion of soil water towards the end of the summer can diminish evapotranspiration, enhance sensible heating of the boundary layer, and produce a warmer, drier climate (Heck et al. 2001).

28.7 Land Clearing in the United States

Large regions of forests in eastern United States and grasslands in the Great Plains have been replaced with crops (Figure 23.10). Bonan (1997, 1999, 2001) and Oleson et al. (2004) performed a series of climate model simulations to study the effects of this change on the climate of the United States. In these simulations, conversion of forest to crop reduces surface roughness, decreases leaf area index, and increases stomatal conductance. Surface albedo increases in summer due to changes in leaf optical properties and in winter in snow-covered regions due to loss of trees that no longer mask the high albedo of snow. Croplands are also given loam soil texture to account for changes in soil hydraulic properties associated with agriculture.

The simulations involved three different atmospheric models, two different land surface models, and various depictions of land-cover change. Despite these differences, all simulations indicate agriculture has cooled climate, but the magnitude of the cooling depends on the particular land-cover dataset and model. The most prominent climate signal is during summer, when evapotranspiration increases, atmospheric humidity increases, and surface air temperature decreases by up to 1–2°C in the Midwest region of intensive cropland (Figure 28.13). The cooling largely arises from increased surface albedo that decreases net solar radiation at the surface. Net radiation decreases, which is balanced by lower sensible heat flux and higher latent heat flux. Feedback with the atmosphere is evident in increased cloudiness that reduces incoming solar radiation at the surface. The cooling is larger for daily maximum temperature than for daily minimum temperature so that diurnal temperature range decreases. However, the magnitude of the cooling depends on the particular model. The smaller cooling of the CLM2 relative to the NCAR LSM (Figure 28.13b,c) relates to the formulation of transpiration and within-canopy turbulent transfer that make the CLM2 a warmer, drier model compared with the NCAR LSM. Other modeling studies also indicate summer cooling associated with land-cover change due to changes in surface albedo and moisture balance (Diffenbaugh 2009).

Other studies find results differ depending on the type of land-cover change and that conversion of forest to cropland may have warmed climate. Baidya Roy et al. (2003) simulated July climate for the United States using land cover for 1700, 1910, and 1990. The model shows a warming of 0.3–0.6°C in the east where crops replaced forests between 1700 and 1910. Between 1910 and 1990, farm abandonment and reforestation in the east decreased temperatures. In contrast, agricultural expansion in the Midwest and Plains states has decreased temperature by 1°C or more. In this model, crops and grasses have similar albedo and roughness length, and the cooling where crops replace grasses is largely a result of increased evapotranspiration. Warming where crops replace forests is a balance of increased cropland albedo, reduced cropland roughness length, and deep tree roots, which sustain evapotranspiration compared with shorter-rooted crops.

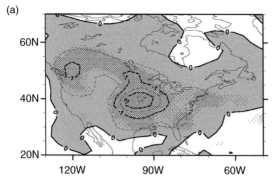

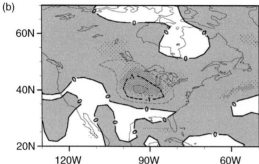

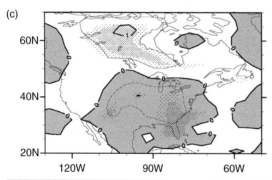

Fig. 28.13 Difference (present-day minus natural vegetation) in summer (June–August) surface air temperature. Shown are three different model simulations. All simulations used the Community Atmosphere Model (CAM2), but differed in land surface model and surface datasets. (a) National Center for Atmospheric Research land surface model (NCAR LSM) with biome datasets. (b) NCAR LSM with vegetation continuous fields datasets. (c) Community Land Model (CLM2) with vegetation continuous fields datasets. The contour interval in all figures is 0.5°C. Shading indicates cooling. Light and dense stippling indicate regions where the difference is statistically significant. Adapted from Oleson et al. (2004).

Differences among models are highlighted by several studies of reforestation and afforestation in southeastern United States. Much of this region is presently cropland or pastureland. Some modeling studies find that planting trees on this land cools summer temperature because of increased evapotranspiration (Jackson et al. 2005; Chen et al. 2012; Murphy et al. 2012). In contrast, another study reported that reforestation of croplands increases temperature throughout the year, with summertime warming up to 0.5°C in some locations (Trail et al. 2013). In this latter simulation, cropland has a higher stomatal conductance compared with forests, and high evaporative cooling from croplands has a large impact on the regional climate. A higher stomatal conductance for forests leads to cooling with reforestation.

The type of crop represented in the model is important. The studies of Bonan (1997, 1999, 2001) and Oleson et al. (2004) represented crops with a summergreen phenology typical of corn or soybean in the United States. Corn is typically sown during April and May while soybeans are sown in May and June. These crops reach silking (corn) and flowering (soybean) stages in July, mature in August, and are harvested in September and October. In contrast, other studies have represented crops by winter wheat. Winter wheat is planted in autumn. Seeds germinate prior to the onset of winter, whereupon plants become dormant until spring. Plants grow and mature throughout spring, with harvest typically in June or early July. This phenology, in which soils are not vegetated during much of summer, decreases latent heat flux, increases sensible heat flux, and warms surface air temperature upon conversion of forest to crop (Xue et al. 1996; Lamptey et al. 2005a,b). Accurate simulation of the seasonal cycle of crop growth, from planting to maturation to harvest, influences large-scale climate in midwestern United States, and the specific planting date can impact precipitation (Levis et al. 2012).

These changes occur because of differences in the seasonal cycle of leaf area and the length of time the surface is covered by growing vegetation (Twine et al. 2004; Sacks and

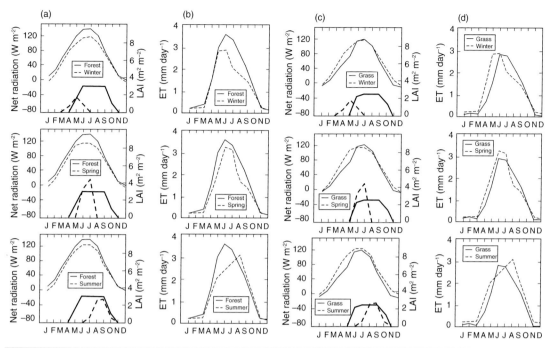

Fig. 28.14 Monthly net radiation, leaf area index (LAI), and evapotranspiration (ET) with land-cover change in the Mississippi River basin. (a) Net radiation for deciduous forest converted to winter wheat (top), spring wheat (middle), and summer crops (bottom). The thick lines overlain with net radiation show leaf area index. (b) As in (a), but for evapotranspiration. (c) Net radiation for grassland converted to winter wheat (top), spring wheat (middle), and summer crops (bottom). The thick lines overlain with net radiation show leaf area index. (d) As in (c), but for evapotranspiration. Model simulations are for a single grid cell in western Wisconsin forced with observed meteorology. Data from Twine et al. (2004).

Kucharik 2011; Levis et al. 2012). For example, Figure 28.14 compares the conversion of deciduous forest and grassland to winter wheat, spring wheat, and summer crops (corn, soybean). Net radiation decreases where forest is converted to crop. The winter decrease is due to the higher albedo of snow-covered ground. The summer decrease is due to changes in leaf optical properties and leaf area index that increase surface albedo. Monthly evapotranspiration is generally less than or equal to forest evapotranspiration, with a distinct seasonal pattern related to crop phenology. Winter wheat attains peak evapotranspiration rates in May–June when leaf area is highest and declines thereafter. Spring wheat evapotranspiration peaks in June–July with crop maturation, and summer crop evapotranspiration peaks in August. Sensible heat flux is generally less than forest during the crop growing season and greater than forest when crop fields are bare. Annual crop evapotranspiration is less than forest and has a distinct relation to phenology, which determines the duration of the evaporative season. The decrease in annual evapotranspiration is less for summer crops (15%) and greatest for winter wheat (23%).

Conversion of grassland to crop has the opposite effects (Figure 28.14c,d). Net radiation is generally greater than or equal to grassland. The greatest increase is in spring, when crop albedo is less than that of dormant grass. This difference decreases during the growing season. Crop sensible heat flux is lower than grassland during the crop growing season, but is greater in spring and autumn when crop fields are bare and evapotranspiration is low. Annual evapotranspiration is greater for crops than grassland (winter wheat, 7%; summer crops, 17%).

Observations and models indicate harvesting of winter wheat warms the surface. In Oklahoma, a large belt of winter wheat 100–150 km wide extends across the state from Texas north to Kansas. In this region, wheat typically resumes growth in early March and becomes senescent by early May with harvest during late May and early June (Figure 28.15). In contrast, adjacent vegetation is primarily summergreen grassland. While wheat is actively growing in early spring, observations show that daily maximum air temperatures are cooler than over adjacent dormant grassland (McPherson et al. 2004). This cool anomaly disappears during May as grasses grow. Once the wheat is harvested, temperatures are warmer across the wheat belt in June–August compared with grassland. The atmosphere is also moister as a result of evapotranspiration from the growing wheat, most noticeably when soil water is abundant (Haugland and Crawford 2005). Model simulations show increased latent heat flux and decreased sensible heat flux of actively growing winter wheat compared with dormant grassland, which leads to a moister, shallower atmospheric boundary layer (McPherson and Stensrud 2005).

The Canadian Prairies have undergone a significant change in agriculture over the past 30 years. More than five million hectares (ha) of land (1 ha = 10,000 m²) have been converted from summer fallow (in which the land was left bare for one year) to annual cropping. The land-use changes were largest in Saskatchewan, where 15–20 percent of the land area was converted from summer fallow to annual cropping. This change in land use has altered the summer climate, with increased evapotranspiration, decreased daily maximum temperature and diurnal temperature range, decreased incoming solar radiation, and increased precipitation (Gameda et al. 2007; Betts et al. 2013). Table 28.12 summarizes changes between the years 1953–1991 and 1991–2011. During the growing season from May 20 to August 27, relative humidity increased by 7 percent. During the first two months of the growing season (May 20 to July 18) daily maximum temperature and the diurnal range of temperature decreased by 1.2°C

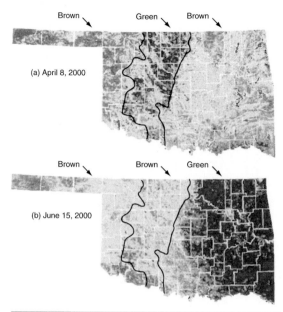

Fig. 28.15 Physiological state of vegetation in Oklahoma for the week ending on (a) April 8, 2000 and (b) June 15, 2000. A greenness index derived from the normalized difference vegetation index indicates growing vegetation. The black lines denote the winter wheat belt. The green wheat is surrounded by dormant grassland on April 8, 2000, but has been harvested by June 15, 2000 while the eastern grassland has become green. Adapted from Haugland and Crawford (2005). See also McPherson et al. (2004).

and 0.6°C, respectively, cloud cover increased by about 4 percent, reducing surface net radiation by 6 W m⁻², and precipitation increased.

28.8 | Irrigation

Irrigation is common in arid and semiarid climates, where sufficient rainfall to grow crops is lacking. Irrigation alters the local surface energy balance by decreasing surface albedo, increasing latent heat flux, and decreasing sensible heat flux. As a result, surface temperature decreases. The large contrast in sensible and latent heat fluxes between irrigated croplands and surrounding semiarid vegetation can generate mesoscale circulations akin to sea breezes between the cool, wet agricultural land and the hot, dry surrounding natural vegetation (Figure 14.11).

Table 28.12 | Changes in surface climate between 1953–1991 and 1991–2011 resulting from land-use change

	Early growing season (May 20–July 18)	Growing season (May 20–August 27)
Daily mean temperature (°C)	−0.93	−0.79
Daily maximum temperature (°C)	−1.18	−0.89
Daily minimum temperature (°C)	−0.59	−0.65
Diurnal temperature range (°C)	−0.59	−0.24
Relative humidity (%)	6.9	7.0
Cloud (tenths)	0.39	0.22
Precipitation (mm day⁻¹)	0.50	0.34

Source: From Betts et al. (2013).

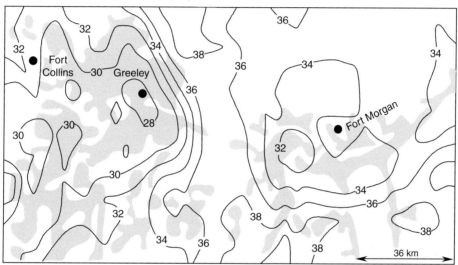

Fig. 28.16 Satellite-derived surface temperature (°C) at 1300 hours local time for the period August 1–15, 1986 for a region of northeastern Colorado near the cities of Fort Collins, Greeley, and Fort Morgan. Shading shows areas of irrigation. Adapted from Segal et al. (1988).

Climate may be changing in areas with irrigation. One such region is the Great Plains of the United States, where extensive irrigated cropland is intermixed with dry grassland (Pielke et al. 2007; Mahmood et al. 2014). In Texas, observations suggest the presence of irrigation influences precipitation (Barnston and Schickedanz 1984; Moore and Rojstaczer 2002). Other observational analyses also find enhanced rainfall in the Great Plains with irrigation (DeAngelis et al. 2010). Regional climate model simulations support the notion of increased rainfall in response to irrigation (Segal et al. 1998; Harding and Snyder 2012). Cooler surface temperatures are evident in northeastern Colorado, an area with widespread irrigated cropland among shortgrass prairie (Figure 28.16). Mesoscale atmospheric model simulations show irrigation affects the climate of northeastern Colorado (Stohlgren et al. 1998; Chase et al. 1999). Less sensible heat flux, greater latent heat flux, cooler daytime boundary layer, and more low-level moisture over irrigated areas in the plains alters atmospheric circulation between the plains and the mountains. Cooling and moistening of the boundary layer due to irrigation may have also occurred in Nebraska (Adegoke et al. 2003, 2007; Mahmood et al. 2004, 2006, 2008). Similar increases in evapotranspiration, cooling, and changes in circulation are seen with the introduction of irrigated agriculture in California (Kueppers et al. 2007; Kueppers and Snyder 2012; Sorooshian et al. 2012, 2014). Irrigation has decreased daily maximum temperature and reduced diurnal temperature range in irrigated areas of California (Lobell and Bonfils 2008).

Global climate model simulations have addressed the climatic impact of irrigation (Boucher et al. 2004; Lobell et al. 2006a,b, 2009; Sacks et al. 2009; Puma and Cook 2010; Cook et al. 2011). Such studies show regional climate influences, with cooler surface air temperature in locations of high irrigation. Irrigation is particularly prevalent in India. While global datasets of irrigated cropland are available, one of the challenges is how to model irrigation management, particularly the timing and amount of water applied (Döll and Siebert 2002).

28.9 | Land Use and Land-Cover Change as a Climate Forcing

Land-cover change in middle latitudes has likely cooled the Northern Hemisphere by increasing surface albedo. The dominant cooling occurs in northern latitudes during winter and spring, when deforestation unmasks the high albedo of snow. Croplands also have a higher albedo than forests during summer, which contributes to the cooling. For example, Brovkin et al. (2006) compared the land-cover and CO_2 forcing of climate over the past millennium simulated by six different models. All models simulate planetary warming with rising atmospheric CO_2 and cooling, particularly in the Northern Hemisphere, due to land-cover change. The cooling among models of Northern Hemisphere annual mean temperature ranges from 0.2°C to 0.4°C relative to the preindustrial era. The cooling increases throughout the nineteenth century, is greatest in the early twentieth century,

and decreases in the latter part of the twentieth century. This reflects trends of farmland expansion, abandonment, and reforestation in the Northern Hemisphere extratropics. In contrast, Northern Hemisphere annual mean temperature increases in all models by 0.4–0.7°C in response to CO_2. When both forcings are combined, the temperature increase is less than that expected from CO_2 alone. This suggests that observed climate warming over the industrial era may be smaller than that expected from rising atmospheric CO_2 alone.

A subsequent analysis of seven climate models points to a similar conclusion and highlights key uncertainties in land–atmosphere feedbacks related to land-cover change, evapotranspiration, and temperature (Pitman et al. 2009; Boisier et al. 2012; de Noblet-Ducoudré et al. 2012). Four climate simulations were performed for each model to span all combinations of present-day and preindustrial greenhouse gases and present-day and preindustrial land cover. One simulation (denoted PI) was forced with conditions representative of the preindustrial era, nominally taken as 1870 with 280 ppm atmospheric CO_2, historical sea surface temperatures, and historical land cover. Another simulation (PD) was forced with present-day (1992) atmospheric CO_2 (375 ppm), sea surface temperatures, and land cover. Two additional simulations used the preindustrial forcing with present-day land cover (PIv) and the present-day forcing with preindustrial land cover (PDv). The difference PD – PDv and PIv – PI both provide estimates of the influence of historical land-cover change on climate. Regions of large change in land cover include North America and Eurasia (Figure 28.17).

Most models (five) simulate cooling during the Northern Hemisphere summer in the regions of land-cover change, but the strength of the cooling varies considerably among models (Pitman et al. 2009). The temperature decrease varies from weak (< 0.5°C cooling) in some models to strong (> 1°C cooling) in others. The latent heat flux response also varies among models. Land-cover change decreases Northern Hemisphere summer latent heat flux in three models and increases latent heat flux in three

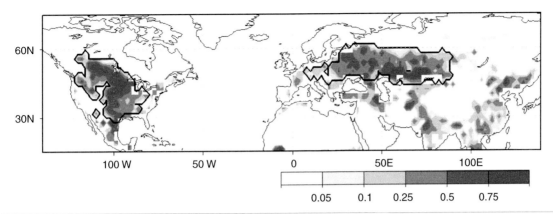

Fig. 28.17 Change in cropland and pastureland (fraction of grid cell) from the preindustrial era (1870) to present-day (1992). The two regions outlined in black denote North America and Eurasia. Reproduced from Boisier et al. (2012).

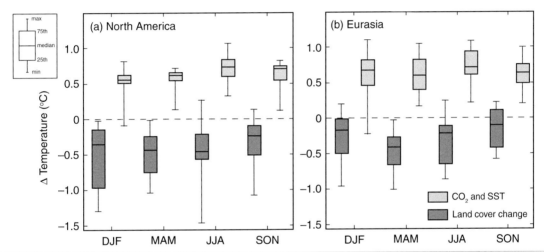

Fig. 28.18 Simulated change in surface air temperature between the preindustrial era and present-day for (a) North America and (b) Eurasia. Results are given for each of the four seasons. The top panel of each plot shows temperature changes due to greenhouse gas climate change (CO_2 and SST forcing). The bottom panel shows temperature changes due to land-cover change. Data are for seven climate models. The bottom and top of the box are the 25th and 75th percentiles, and the horizontal line within each box is the 50th percentile (the median). The whiskers (straight lines) indicate the ensemble maximum and minimum values. Adapted from de Noblet-Ducoudré et al. (2012).

models. The surface cooling is not limited to summer and occurs year-round, though with large variability among models (Figure 28.18). Historical land-cover change affects surface air temperature with a magnitude similar to (though opposite in sign) that resulting from increased greenhouse gases and a warmer ocean that occurred over the same period (de Noblet-Ducoudré et al. 2012). While these latter factors warmed temperature over North

America and Eurasia by 0.5–1°C depending on season and model, land-cover change cooled temperature by 0.5°C or more in the same regions (Figure 28.18).

Surface cooling in temperate regions is a common response to historical land-cover change simulated by climate models. This cooling occurs from increased surface albedo with deforestation, particularly during winter when snow is on the ground. Satellite measurements

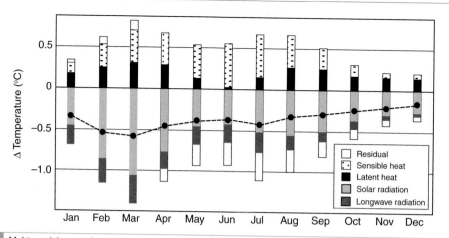

Fig. 28.19 Multi-model mean change in surface temperature due to historical land-cover change. Bars show the temperature difference arising from changes in latent heat flux, sensible heat flux, net solar radiation, downward longwave radiation, and a residual term. The dashed line shows the net surface temperature difference. Adapted from Boisier et al. (2012).

show croplands, grasslands, and bare surfaces have an albedo that is more than twice that of forests during the winter when snow is on the ground (Table 26.3). Albedo differences are smaller when the ground is snow-free or in summer, but croplands and grasslands have a higher albedo than evergreen forests (by 0.08 in summer) and deciduous forests (by 0.03).

The climate response to land-cover change is more complex than simply albedo-driven changes in surface radiation. One important difference among models is the partitioning of available energy between latent and sensible heat fluxes (Boiser et al. 2012; de Noblet-Ducoudré et al. 2012). The temperature cooling from higher surface albedo is lessened by changes in turbulent fluxes (sensible heat flux and latent heat flux) related to evapotranspiration efficiency and surface roughness that cause warming in the models (Figure 28.19). The cooling throughout the year arises from less available energy at the surface (net solar radiation plus downward longwave radiation). Less net solar radiation is the dominant term. The largest decrease in available energy is in early spring (March) because of high downwelling solar radiation combined with large changes in surface albedo due to loss of forests. The temperature cooling is 0.5°C at this time. Changes in sensible heat flux and latent heat

flux produce warming, which dampens the radiative cooling.

Deforestation may have additional climate consequences. Swann et al. (2012) found that trees in northern mid-latitudes warm the Northern Hemisphere and alter global atmospheric circulation. Their climate model simulations considered large-scale afforestation (the opposite of deforestation) by replacing C_3 grasslands and croplands between 30° and 60° N with broadleaf deciduous trees. Greater tree cover decreases surface albedo, increases the absorption of solar radiation, and increases surface air temperature. The additional energy absorbed in the Northern Hemisphere produces an imbalance between the hemispheres. The Hadley circulation shifts northward to transport more energy southward across the equator, moving the intertropical convergence zone northward (Figure 28.20). This alters precipitation in the tropics along the equator, with decreased rainfall in the Amazon basin.

Climate modeling studies generally find that temperate and boreal deforestation cools climate while tropical deforestation warms climate. The consensus is that historical land-cover change has decreased the temperature of the Northern Hemisphere and that this cooling, while small at the global scale, is regionally significant. However, there are large differences among

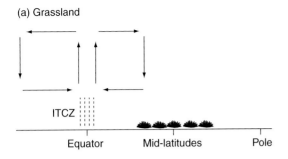

(a) Grassland

ITCZ

Equator   Mid-latitudes   Pole

(b) Afforestation

More energy transport
southward

More energy
absorbed
by forests

Equator   Mid-latitudes   Pole

Fig. 28.20 Changes in the Hadley circulation with mid-latitude afforestation. (a) Present-day grasslands. (b) Forest expansion increases energy absorption in the Northern Hemisphere. The Hadley circulation shifts northward to transport more heat across the equator, causing the intertropical convergence zone to also migrate northwards relative to present-day grasslands. Adapted from Swann et al. (2014).

models in their simulated climate response to land-cover change, even in a carefully controlled simulation protocol (Figure 28.18). The lack of consistency among models is due to the implementation of the land-cover change datasets in the models and the parameterization of crops, albedo, and evapotranspiration (Pitman et al. 2009; Boisier et al. 2012, 2013; de Noblet-Ducoudré et al. 2012). A broader comparison among 15 models also shows divergent temperature and evapotranspiration responses to land-cover change (Kumar et al. 2013).

Despite this model uncertainty, the effects of land-cover change on temperature may be emerging in the observational record. The loss of forests and increase of croplands and grasslands since preindustrial times has produced an overall cooling trend in both mean and extreme temperatures. This cooling is statistically detectable in the observed changes of warm temperature extremes and partially offsets the warming trend from greenhouse gases and other anthropogenic forcings (Christidis et al. 2013). However, the land-cover change effects are much more regional (where the land-cover change has occurred) compared with greenhouse gas warming (Pitman et al. 2012).

Another analysis compared air temperature measured over forests with that at nearby surface weather stations located in open, grassy fields for sites in North America (Lee et al. 2011) and subsequently extended for additional locations in eastern Asia and the tropical Americas (Zhang et al. 2014). In the tropics and subtropics (15° S – 20° N), the annual mean air temperature at the open sites is 0.67°C warmer than the forests; but in boreal latitudes (≥ 45° N), the opens sites are 0.95°C cooler than forests. Weaker temperature change occurs between these regions. Latitude 35° N marks the approximate transition. South of this latitude, the warming is seen in an increase in daily maximum air temperature in all months of the year. The cooling at northern locations occurs from a decrease in the daily minimum temperature throughout the year. However, climate models simulate large-scale deforestation and resulting atmospheric changes whereas the measurements in forests and clearings represent local land-cover changes.

Our prevailing understanding of the climate effects of land-cover change is based on the low surface albedo of forests compared with grassland and cropland, and additionally that forests increase evapotranspiration relative to grassland and cropland. Deforestation increases albedo, but decreases evapotranspiration. The former process cools climate, while the latter process warms climate (Figures 28.2b and 28.19). Differences among vegetation in surface albedo are well-documented in observations, particularly the low albedo of forests in seasonally snow-covered regions (Figure 12.10, Table 26.3). Differences in evapotranspiration are evident in the tropics (Figure 28.11b, Table 28.10), but are less clear in mid-latitudes and are based mostly on a conceptualization of biogeophysical processes. The greater evapotranspiration of forests is expected because trees, with their vast system of foliage, branches, and trunks, increase interception and evaporation of precipitation; because their deep roots provide a large supply of water for transpiration; and because their tall canopy provides strong aerodynamic coupling with the atmosphere.

One line of evidence comes from water budget analyses. Paired watershed studies show that forest cover decreases annual runoff and increases annual evapotranspiration compared with forest removal (Bosch and Hewlett 1982; Zhang et al. 2001; Andréassian 2004; Brown et al. 2005; Farley et al. 2005), as shown in Figure 11.3 for the Hubbard Brook watershed. Such analyses do not measure evapotranspiration directly, but instead calculate it as the difference between annual precipitation and runoff ($E = P - R$). Syntheses of water budget analyses from various catchments worldwide similarly find that forests evaporate a greater proportion of annual precipitation (E / P) compared with grasslands (Zhang et al. 2001; Farley et al. 2005). This understanding may differ for crops. Water budget analyses across Sweden show no evidence that forests evaporate more water annually than agricultural land (van der Velde et al. 2013).

Analyses of annual evapotranspiration from a variety of field measurements, models, and satellite estimates show that historical deforestation and other changes in land cover and land use have reduced global annual evapotranspiration, but they have conflicting outcomes of specific land-cover transitions (Sterling et al. 2013; Boisier et al. 2014). In the analysis of Sterling et al. (2013), non-irrigated cropland and pastureland reduce evapotranspiration compared with forests; croplands have lower evapotranspiration than grassland. Boisier et al. (2014) similarly found a robust, prominent decrease in growing season evapotranspiration where grasses replace forests, but an increase in summer evapotranspiration where crops replace grasses. The transition of forests to crops does not produce a consistent change in evapotranspiration, and what change there is is small

compared with the forest–grass and grass–crop transitions.

Other evidence for the influence of land-cover change comes from eddy covariance flux tower measurements, but such studies reveal a conflicting story. Flux tower measurements show that pine and hardwood deciduous forests in North Carolina have greater annual and growing season evapotranspiration compared with an adjacent grassland (Stoy et al. 2006), and this is seen in a cooler surface temperature of the forests (Juang et al. 2007). Other analyses of flux tower data question the conventional understanding of high rates of evapotranspiration in forests. Measurements at forest, grassland, and cropland sites in Europe show that latent heat flux is larger at the non-forest sites than at the forests when soils are moist and that the forests have higher sensible heat flux (Teuling et al. 2010). The high sensible heat flux occurs because these forests have lower albedo and absorb more energy, have stronger stomatal control of transpiration, but are well-coupled aerodynamically with the atmosphere. With dry soils, however, latent heat flux declines sharply in short-rooted grassland and cropland and sensible heat flux increases, while the forests maintain their evapotranspiration. A global synthesis of flux tower measurements across forest, grassland, cropland, and other vegetation types shows that forests do not evaporate a larger fraction of annual precipitation (E / P) compared with grassland or cropland, and in fact broadleaf deciduous and needleleaf evergreen forests have a lower evaporative fraction than grasslands (Williams et al. 2012).

Agricultural crops have high rates of leaf photosynthesis (Table 16.1) and high values of the photosynthetic parameter $V_{c\,max}$ (Table 16.3) compared with trees. Theory suggests that these leaf-scale differences should manifest at the canopy scale. Some observations do indeed show that crops have a higher canopy conductance than forests or grasses (Table 17.2). Theory also suggests that the fraction of available energy consumed in evapotranspiration scales with the product of $V_{c\,max}$ times leaf area index (Baldocchi and Meyers 1998). The evaporative fraction during the growing season is generally low in forest compared with some crops and lower in dry conifer forest than in mesic broadleaf forest (Figure 17.10). Further analyses of summer (mid-June through late August) energy fluxes measured at temperate broadleaf deciduous forests, temperate and boreal needleleaf evergreen forests, and a cropland over multiple years show key differences among sites (Wilson et al. 2002). The deciduous forests and the cropland have a lower Bowen ratio (0.25–0.50) than the coniferous forests (generally 0.5–1.0), in part because their canopy conductance is higher.

Satellite measurements of land surface temperature have been used to assess the consequences of land-cover change. Land surface temperature integrates changes in albedo, roughness, and evapotranspiration through the surface energy budget, as seen in Eq. (12.16). In the continental United States, forests have a cooler surface temperature than non-forest land, seen in the annual mean and in all seasons except winter (Wickham et al. 2012, 2013). Forests are on average 0.6°C cooler than cropland in the annual mean and 1.4°C cooler in summer (Wickham et al. 2012). This temperature difference occurs mainly in the daytime maximum; in summer, daily maximum temperature is 3.5°C cooler in forests than cropland. In winter, forests are warmer than cropland except in the southeastern region of the country (Wickham et al. 2014). In an analysis spanning North America, Zhao and Jackson (2014) also found than forests are cooler in warm seasons and warmer in cold seasons compared with adjacent grassland and cropland. The cooling occurs in daily maximum temperature, which is some 2–5°C cooler in forest compared with cropland, and a similar or larger daytime cooling occurs compared with grassland. Similar results are found in China, where the annual daytime temperature of forests is 1.1°C cooler than grassland or cropland, and seasonally up to 3°C cooler compared with croplands (Peng et al. 2014).

Conventional wisdom holds that the lower albedo of forests compared with cropland or grassland warms the surface while the higher evapotranspiration of forests cools the surface. However, changes in sensible heat flux with land-cover change may be a key driver of temperature change in temperate ecosystems and may outweigh changes in albedo. In an analysis of Eurasian and North American eddy covariance flux tower sites located in forest, grassland, and cropland, Luyssaert et al. (2014) found differences in surface temperature related to land cover, but such changes in temperature are not necessarily straightforward. Forests have the lowest annual albedo, croplands higher albedo, and grasslands highest albedo. When analyzed by land-cover transitions (e.g., forest to grassland), the change in annual mean albedo positively correlates with the change in annual mean surface temperature despite less available energy to heat the surface. Forests, grasslands, and croplands differ in surface roughness, and the amount of net radiation dissipated as sensible heat (H / R_n) increases with roughness length. At these sites, sensible heat flux is a key mechanism for surface cooling. In grasslands and croplands, the annual surface cooling from increased albedo is offset by less surface cooling from sensible heat flux because of lower roughness length.

28.10 Integrated Biogeophysical and Biogeochemical Studies

Land-cover change influences climate through biogeochemical processes that affect the carbon cycle and through biogeophysical processes that affect albedo, turbulent fluxes, and the hydrologic cycle. Greenhouse gases are well-mixed in the atmosphere, and biogeochemical feedbacks influence global climate. Biogeophysical feedbacks are more regional in scale. Climate model experiments show that the biogeochemical effects of land-cover change may offset or accentuate the biogeophysical effects.

Claussen et al. (2001) contrasted the biogeophysical and biogeochemical effects of boreal and tropical deforestation. The simulations examined changes in surface energy, moisture, and momentum fluxes (biogeophysical feedbacks) as a result of deforestation, changes in carbon pools and fluxes (biogeochemical

Table 28.13 | Contribution of biogeophysical and biogeochemical processes to changes in near-surface air temperature (annual mean, °C) as a result of boreal and tropical deforestation

	Boreal deforestation (50°–60° N)		Tropical deforestation (0°–10° S)	
	Global	Regional	Global	Regional
Biogeophysical only	−0.23	−0.82	−0.04	0.13
Biogeochemical only	0.09	0.12	0.19	0.15
Both processes	−0.11	−0.67	0.16	0.29

Note: Temperatures are the difference from a control simulation without changes in land cover.
Source: From Claussen et al. (2001).

feedbacks), and both feedbacks simultaneously (Table 28.13). Boreal deforestation cools climate as a result of higher surface albedo (trees no longer mask the high albedo of snow), but also warms climate due to carbon loss to the atmosphere. The biogeochemical warming is not enough to compensate for the biogeophysical cooling. Tropical deforestation warms climate regionally due to reduction in evapotranspiration, but produces a slight global cooling because of less moisture in the atmosphere. Deforestation releases carbon to the atmosphere, resulting in warming that exceeds that from biogeophysical processes. Other simulations similarly show boreal deforestation produces net global cooling, tropical deforestation yields net global warming, and temperate deforestation has negligible net effect on global temperature (Bala et al. 2007; Bathiany et al. 2010).

Climate model simulations have examined the importance of biogeophysical and biogeochemical feedbacks associated with land-cover change when simulating historical climate change over the industrial era. The dominant competing signals from deforestation are: an increase in surface albedo in middle to high latitudes of the Northern Hemisphere with loss of vegetation masking of snow albedo; and carbon emission to the atmosphere. Brovkin et al. (2004) found that land-cover change decreased global mean temperature by 0.26°C over the past millennium as a result of biogeophysical processes while biogeochemical processes

warmed climate by 0.18°C. The net effect of these competing processes is minor on the global scale (0.05°C cooling), but is large in temperate and high northern latitudes where the cooling due to an increase in surface albedo offsets the warming due to land-use CO_2 emission. Matthews et al. (2004) found smaller biogeophysical cooling and larger biogeochemical warming in their model simulations. They estimated biogeophysical processes cooled climate by 0.16°C over the industrial era while land-use CO_2 emission warmed climate by 0.3°C over the same period. The net effect of land-cover change over the past 150 years is to have warmed climate by 0.15°C. While the global response differs, regional cooling dominates the temperature signal in the Northern Hemisphere because of higher surface albedo, similar to Brovkin et al. (2004). Pongratz et al. (2010) simulated small biogeophysical cooling over the twentieth century (0.03°C), larger biogeochemical warming (0.16–0.18°C), and net warming (0.13–0.15°C). Although the global biogeophysical cooling is small, temperature decreases by 0.3–0.5°C in North America, Europe, India, and China, where there has been large land-cover change (Figure 28.21). The biogeochemical warming encompasses most regions of the world, and in contrast to the other studies, net warming occurs in all regions. Other simulations also find a large biogeochemical warming from historical land cover change that outweighs the biogeophysical cooling (He et al. 2014).

(a) Biogeophysical

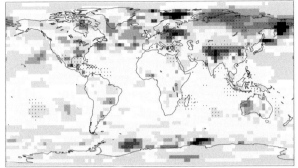

(b) Biogeochemical

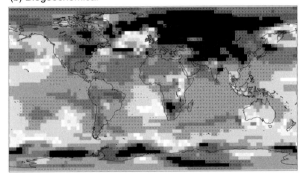

(c) Net

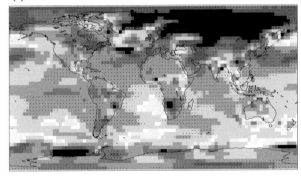

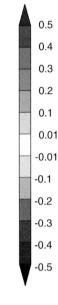

ΔT (°C)

0.5
0.4
0.3
0.2
0.1
0.01
-0.01
-0.1
-0.2
-0.3
-0.4
-0.5

Fig. 28.21 Change in annual mean temperature due to land-cover change over the twentieth century from (a) biogeophysical effects, (b) biogeochemical effects, and (c) net effect. Areas where the change is statistically significant are dotted. Reproduced from Pongratz et al. (2010). See color plate section.

28.11 | Review Questions

1. Devise a land management policy for the Sonoran desert that could be used to ameliorate the effects of climate warming.

2. Over several years, temperatures in the southern tropical Atlantic Ocean warm relative to the North Atlantic. Describe the effect this has on rainfall in the Sahel and feedbacks that accentuate or mitigate the change in rainfall.

3. Table 28.8 shows mean annual albedo and annual precipitation by vegetation type in West Africa. Does this support the theory that land degradation increases surface albedo and rainfall?

4. Describe the effect reforestation might have on the climate of Australia.

5. Using data in Table 28.10, contrast seasonal changes in evaporative fraction ($\lambda E/R_n$) and Bowen ratio ($H/\lambda E$) between pasture and forest in Amazonia. How does rooting depth affect this seasonality?

6. Each model simulation in Table 28.9 is an independent realization of climate change due to Amazonian deforestation. What is the median temperature change and the median precipitation change? How likely is it that the warming will exceed 2.3°C or that precipitation will decrease by more than 511 mm? How likely is it that the climate will become warmer and drier? What can you conclude about the effects of deforestation on climate?

7. Air temperature in a forest understory on a summer day is typically cooler than in open fields, yet the regional effect of converting temperate forest to cropland is possibly to decrease temperature during summer. Explain this difference.

8. Historical temperature records show that daily maximum air temperature increased over the past 100 years at a faster rate in the northeastern United States than in the Midwest. The predominant land use history in the Northeast is farm abandonment and reforestation while the Midwest has seen agricultural expansion. Is this confirmation that crops cool climate relative to forests?

9. Some climate modeling studies show that croplands in the United States decrease temperature.

Other studies show warming. Discuss reasons for this difference.

10. Discuss the effects of irrigation on daily maximum and daily minimum air temperature and the diurnal temperature range.

11. Contrast the three biogeophysical processes operating in the deforestation study of Davin and de Noblet-Ducoudré (2010) as discussed in the text and shown in Figure 28.2. Which is the dominant process? Compare the magnitude of the net biogeophysical effect (–1°C global mean) to the net effect estimated as the sum of the three individual processes. Do these two estimates agree? If not, why?

12. Devise a climate model experiment to examine the impact of land-cover change relative to changes in atmospheric CO_2 on climate at 2100.

13. The annual mean global temperature has increased over the latter half of the twentieth century and is likely to continue increasing during the twenty-first century because of greenhouse gases. The predominant changes in land use over these periods are farm abandonment in middle latitudes and forest clearing in tropical latitudes. Discuss the effects of this land use on the observed temperature warming and future warming in the twenty-first century.

28.12 | References

Adegoke, J. O., Pielke, R. A., Sr., Eastman, J., Mahmood, R., and Hubbard, K. G. (2003). Impact of irrigation on midsummer surface fluxes and temperature under dry synoptic conditions: A regional atmospheric model study of the U.S. High Plains. *Monthly Weather Review*, 131, 556–564.

Adegoke, J. O., Pielke, R., Sr., and Carleton, A. M. (2007). Observational and modeling studies of the impacts of agriculture-related land use change on planetary boundary layer processes in the central U.S. *Agricultural and Forest Meteorology*, 142, 203–215.

Anav, A., Ruti, P. M., Artale, V. and Valentini, R. (2010). Modelling the effects of land-cover changes on surface climate in the Mediterranean region. *Climate Research*, 41, 91–104.

Andréassian, V. (2004). Waters and forests: From historical controversy to scientific debate. *Journal of Hydrology*, 291, 1–27.

Arribas, A., Gallardo, C., Gaertner, M. A., and Castro, M. (2003). Sensitivity of the Iberian Peninsula climate to a land degradation. *Climate Dynamics*, 20, 477–489.

Avissar, R., Silva Dias, P. L., Silva Dias, M. A. F., and Nobre, C. (2002). The Large-Scale Biosphere–Atmosphere Experiment in Amazonia (LBA): Insights and future research needs. *Journal of Geophysical Research*, 107, 8086, doi:10.1029/2002JD002704.

Baidya Roy, S., and Avissar, R. (2002). Impact of land use/land cover change on regional hydrometeorology in Amazonia. *Journal of Geophysical Research*, 107, 8037, doi:10.1029/2000JD000266.

Baidya Roy, S., Hurtt, G. C., Weaver, C. P., and Pacala, S. W. (2003). Impact of historical land cover change on the July climate of the United States. *Journal of Geophysical Research*, 108, 4793, doi:10.1029/2003JD003565.

Bala, G., Caldeira, K., Wickett, M., et al. (2007). Combined climate and carbon-cycle effects of large-scale deforestation. *Proceedings of the National Academy of Sciences USA*, 104, 6550–6555.

Baldocchi, D. D., and Meyers, T. P. (1998). On using eco-physiological, micrometeorological and biogeochemical theory to evaluate carbon dioxide, water vapor and trace gas fluxes over vegetation: A perspective. *Agricultural and Forest Meteorology*, 90, 1–25.

Balling, R. C., Jr. (1988). The climatic impact of a Sonoran vegetation discontinuity. *Climatic Change*, 13, 99–109.

Balling, R. C., Jr. (1989). The impact of summer rainfall on the temperature gradient along the United States–Mexico border. *Journal of Applied Meteorology*, 28, 304–308.

Balling, R. C., Jr., Klopatek, J. M., Hildebrandt, M. L., Moritz, C. K., and Watts, C. J. (1998). Impacts of land degradation on historical temperature records from the Sonoran Desert. *Climatic Change*, 40, 669–681.

Barnston, A. G., and Schickedanz, P. T. (1984). The effect of irrigation on warm season precipitation in the southern Great Plains. *Journal of Climate and Applied Meteorology*, 23, 865–888.

Bathiany, S., Claussen, M., Brovkin, V., Raddatz, T., and Gayler, V. (2010). Combined biogeophysical and biogeochemical effects of large-scale forest cover changes in the MPI earth system model. *Biogeosciences*, 7, 1383–1399.

Betts, A. K., Desjardins, R., Worth, D., and Cerkowniak, D. (2013). Impact of land use change on the diurnal cycle climate of the Canadian Prairies. *Journal of Geophysical Research: Atmospheres*, 118, 11996–12011, doi:10.1002/2013JD020717.

Bird, M. I., Hutley, L. B., Lawes, M. J., et al. (2013). Humans, megafauna and environmental change in tropical Australia. *Journal of Quaternary Science*, 28, 439–452.

Black, J. F. (1963). Weather control: Use of asphalt coatings to tap solar energy. *Science*, 139, 226–227.

Black, J. F., and Tarmy, B. L. (1963). The use of asphalt coatings to increase rainfall. *Journal of Applied Meteorology*, 2, 557–564.

Boisier, J. P., de Noblet-Ducoudré, N., Pitman, A. J., et al. (2012). Attributing the impacts of land-cover changes in temperate regions on surface temperature and heat fluxes to specific causes: Results from the first LUCID set of simulations. *Journal of Geophysical Research*, 117, D12116, doi:10.1029/2011JD017106.

Boisier, J. P., de Noblet-Ducoudré, N., and Ciais, P. (2013). Inferring past land use-induced changes in surface albedo from satellite observations: A useful tool to evaluate model simulations. *Biogeosciences*, 10, 1501–1516.

Boisier, J. P., de Noblet-Ducoudré, N., and Ciais, P. (2014). Historical land-use-induced evapotranspiration changes estimated from present-day observations and reconstructed land-cover maps. *Hydrology and Earth System Sciences*, 18, 3571–3590.

Bonan, G. B. (1997). Effects of land use on the climate of the United States. *Climatic Change*, 37, 449–486.

Bonan, G. B. (1999). Frost followed the plow: Impacts of deforestation on the climate of the United States. *Ecological Applications*, 9, 1305–1315.

Bonan, G. B. (2001). Observational evidence for reduction of daily maximum temperature by croplands in the Midwest United States. *Journal of Climate*, 14, 2430–2442.

Bonan, G. B. (2008). Forests and climate change: Forcings, feedbacks, and the climate benefits of forests. *Science*, 320, 1444–1449.

Bosch, J. M., and Hewlett, J. D. (1982). A review of catchment experiments to determine the effect of vegetation changes on water yield and evapotranspiration. *Journal of Hydrology*, 55, 3–23.

Boucher, O., Myhre, G., and Myhre, A. (2004). Direct human influence of irrigation on atmospheric water vapour and climate. *Climate Dynamics*, 22, 597–603.

Brovkin, V., Sitch, S., von Bloh, W., et al. (2004). Role of land cover changes for atmospheric CO_2 increase and climate change during the last 150 years. *Global Change Biology*, 10, 1253–1266.

Brovkin, V., Claussen, M., Driesschaert, E., et al. (2006). Biogeophysical effects of historical land cover changes simulated by six Earth system models of intermediate complexity. *Climate Dynamics*, 26, 587–600.

Brovkin, V., Raddatz, T., Reick, C. H., Claussen, M., and Gayler, V. (2009). Global biogeophysical interactions between forest and climate. *Geophysical Research Letters*, 36, L07405, doi:10.1029/2009GL037543.

Brown, A. E., Zhang, L., McMahon, T. A., Western, A. W., and Vertessy, R. A. (2005). A review of paired catchment studies for determining changes in water yield resulting from alterations in vegetation. *Journal of Hydrology*, 310, 28–61.

Bryant, N. A., Johnson, L. F., Brazel, A. J., et al. (1990). Measuring the effect of overgrazing in the Sonoran Desert. *Climatic Change*, 17, 243–264.

Charney, J. G. (1975). Dynamics of deserts and drought in the Sahel. *Quarterly Journal of the Royal Meteorological Society*, 101, 193–202.

Charney, J., Stone, P. H., and Quirk, W. J. (1975). Drought in the Sahara: A biogeophysical feedback mechanism. *Science*, 187, 434–435.

Chase, T. N., Pielke, R. A., Sr., Kittel, T. G. F., Baron, J. S., and Stohlgren, T. J. (1999). Potential impacts on Colorado Rocky Mountain weather due to land use changes on the adjacent Great Plains. *Journal of Geophysical Research*, 104D, 16673–16690.

Chen, G.-S., Notaro, M., and Liu, Z. (2012). Simulated local and remote biophysical effects of afforestation over the Southeast United States in boreal summer. *Journal of Climate*, 25, 4511–4522.

Christidis, N., Stott, P. A., Hegerl, G. C., and Betts, R. A. (2013). The role of land use change in the recent warming of daily extreme temperatures. *Geophysical Research Letters*, 40, 589–594, doi:10.1002/grl.50159.

Clark, D. B., Xue, Y., Harding, R. J., and Valdes, P. J. (2001). Modeling the impact of land surface degradation on the climate of tropical North Africa. *Journal of Climate*, 14, 1809–1822.

Claussen, M., Brovkin, V., and Ganopolski, A. (2001). Biogeophysical versus biogeochemical feedbacks of large-scale land cover change. *Geophysical Research Letters*, 28, 1011–1014.

Cook, B. I., Puma, M. J., and Krakauer, N. Y. (2011). Irrigation induced surface cooling in the context of modern and increased greenhouse gas forcing. *Climate Dynamics*, 37, 1587–1600.

Costa, M. H., and Foley, J. A. (2000). Combined effects of deforestation and doubled atmospheric CO_2 concentrations on the climate of Amazonia. *Journal of Climate*, 13, 18–34.

Costa, M. H., Yanagi, S. N. M., Souza, P. J. O. P., Ribeiro, A., and Rocha, E. J. P. (2007). Climate change in Amazonia caused by soybean cropland expansion, as compared to caused by pastureland expansion. *Geophysical Research Letters*, 34, L07706, doi:10.1029/2007GL029271.

Culf, A. D., Esteves, J. L., de O. Marques Filho, A., and da Rocha, H. R. (1996). Radiation, temperature and humidity over forest and pasture in Amazonia. In *Amazonian Deforestation and Climate*, ed. J. H. C. Gash, C. A. Nobre, J. M. Roberts, and R. L. Victoria. New York: Wiley, pp. 175–191.

da Rocha, H. R., Goulden, M. L., Miller, S. D., et al. (2004). Seasonality of water and heat fluxes over a tropical forest in eastern Amazonia. *Ecological Applications*, 14, S22–S32.

Davin, E. L., and de Noblet-Ducoudré, N. (2010). Climatic impact of global-scale deforestation: Radiative versus nonradiative processes. *Journal of Climate*, 23, 97–112.

DeAngelis, A., Dominguez, F., Fan, Y., et al. (2010). Evidence of enhanced precipitation due to irrigation over the Great Plains of the United States. *Journal of Geophysical Research*, 115, D15115, doi:10.1029/2010JD013892.

Delire, C., Behling, P., Coe, M. T., et al. (2001). Simulated response of the atmosphere–ocean system to deforestation in the Indonesian Archipelago. *Geophysical Research Letters*, 28, 2081–2084.

de Noblet-Ducoudré, N., Boisier, J.-P., Pitman, A., et al. (2012). Determining robust impacts of land-use-induced land cover changes on surface climate over North America and Eurasia: Results from the first set of LUCID experiments. *Journal of Climate*, 25, 3261–3281.

Deo, R. C., Syktus, J. I., McAlpine, C. A., et al. (2009). Impact of historical land cover change on daily indices of climate extremes including droughts in eastern Australia. *Geophysical Research Letters*, 36, L08705, doi:10.1029/2009GL037666.

Dickinson, R. E., and Henderson-Sellers, A. (1988). Modelling tropical deforestation: A study of GCM land-surface parameterizations. *Quarterly Journal of the Royal Meteorological Society*, 114, 439–462.

Dickinson, R. E., and Kennedy, P. (1992). Impacts on regional climate of Amazon deforestation. *Geophysical Research Letters*, 19, 1947–1950.

Diffenbaugh, N. S. (2009). Influence of modern land cover on the climate of the United States. *Climate Dynamics*, 33, 945–958.

D'Odorico, P., Bhattachan, A., Davis, K. F., Ravi, S., and Runyan, C. W. (2013). Global desertification: Drivers and feedbacks. *Advances in Water Resources*, 51, 326–344.

Döll, P., and Siebert, S. (2002). Global modeling of irrigation water requirements. *Water Resources Research*, 38, 1037, doi:10.1029/2001WR000355.

Dümenil Gates, L. and Ließ, S. (2001). Impacts of deforestation and afforestation in the Mediterranean region as simulated by the MPI atmospheric GCM. *Global and Planetary Change*, 30, 309–328.

Farley, K. A., Jobbágy, E. G., and Jackson, R. B. (2005). Effects of afforestation on water yield: A global synthesis with implications for policy. *Global Change Biology*, 11, 1565–1576.

Fisch, G., Tota, J., Machado, L. A. T., et al. (2004). The convective boundary layer over pasture and forest in Amazonia. *Theoretical and Applied Climatology*, 78, 47–59.

Foley, J. A., Coe, M. T., Scheffer, M., and Wang, G. (2003). Regime shifts in the Sahara and Sahel: Interactions between ecological and climatic systems in northern Africa. *Ecosystems*, 6, 524–539.

Fraedrich, K., Kleidon, A., and Lunkeit, F. (1999). A green planet versus a desert world: Estimating the effect of vegetation extremes on the atmosphere. *Journal of Climate*, 12, 3156–3163.

Fuller, D. O., and Ottke, C. (2002). Land cover, rainfall and land-surface albedo in West Africa. *Climatic Change*, 54, 181–204.

Gameda, S., Qian, B., Campbell, C. A., and Desjardins, R. L. (2007). Climatic trends associated with summerfallow in the Canadian Prairies. *Agricultural and Forest Meteorology*, 142, 170–185.

Gash, J. H. C., and Nobre, C. A. (1997). Climatic effects of Amazonian deforestation: Some results from ABRACOS. *Bulletin of the American Meteorological Society*, 78, 823–830.

Gash, J. H. C., Nobre, C. A., Roberts, J. M., and Victoria, R. L. (1996). *Amazonian Deforestation and Climate*. New York: Wiley.

Ge, J., and Zou, C. (2013). Impacts of woody plant encroachment on regional climate in the southern Great Plains of the United States. *Journal of Geophysical Research: Atmospheres*, 118, 9093–9104, doi:10.1002/jgrd.50634.

Gedney, N., and Valdes, P. J. (2000). The effect of Amazonian deforestation on the northern hemisphere circulation and climate. *Geophysical Research Letters*, 27, 3053–3056.

Guichard, F., Kergoat, L., Mougin, E., et al. (2009). Surface thermodynamics and radiative budget in the Sahelian Gourma: Seasonal and diurnal cycles. *Journal of Hydrology*, 375, 161–177.

Hahmann, A. N., and Dickinson, R. E. (1997). RCCM2-BATS model over tropical South America: Applications to tropical deforestation. *Journal of Climate*, 10, 1944–1964.

Harding, K. J., and Snyder, P. K. (2012). Modeling the atmospheric response to irrigation in the Great Plains, Part I: General impacts on precipitation and the energy budget. *Journal of Hydrometeorology*, 13, 1667–1686.

Hasler, N., Werth, D., and Avissar, R. (2009). Effects of tropical deforestation on global hydroclimate: A multimodel ensemble analysis. *Journal of Climate*, 22, 1124–1141.

Haugland, M. J., and Crawford, K. C. (2005). The diurnal cycle of land–atmosphere interactions across Oklahoma's winter wheat belt. *Monthly Weather Review*, 133, 120–130.

He, F., Vavrus, S. J., Kutzbach, J. E., et al. (2014). Simulating global and local surface temperature changes due to Holocene anthropogenic land cover change. *Geophysical Research Letters*, 41, 623–631, doi:10.1002/2013GL058085.

He, Y., D'Odorico, P., De Wekker, S. F. J., Fuentes, J. D., and Litvak, M. (2010). On the impact of shrub encroachment on microclimate conditions in the northern Chihuahuan desert. *Journal of Geophysical Research*, 115, D21120, doi:10.1029/2009JD013529.

He, Y., De Wekker, S. F. J., Fuentes, J. D., and D'Odorico, P. (2011). Coupled land-atmosphere modeling of the effects of shrub encroachment on nighttime temperatures. *Agricultural and Forest Meteorology*, 151, 1690–1697.

Heck, P., Lüthi, D., Wernli, H., and Schär, C. (2001). Climate impacts of European-scale anthropogenic vegetation changes: A sensitivity study using a regional climate model. *Journal of Geophysical Research*, 106D, 7817–7835.

Henderson-Sellers, A., and Gornitz, V. (1984). Possible climatic impacts of land cover transformations, with particular emphasis on tropical deforestation. *Climatic Change*, 6, 231–257.

Henderson-Sellers, A., Dickinson, R. E., Durbidge, T. B., et al. (1993). Tropical deforestation: Modeling local- to regional-scale climate change. *Journal of Geophysical Research*, 98D, 7289–7315.

Jackson, R. B., Jobbágy, E. G., Avissar, R., et al. (2005). Trading water for carbon with biological carbon sequestration. *Science*, 310, 1944–1947.

Juang, J.-Y., Katul, G., Siqueira, M., Stoy, P., and Novick, K. (2007). Separating the effects of albedo from eco-physiological changes on surface temperature along a successional chronosequence in the southeastern United States. *Geophysical Research Letters*, 34, L21408, doi:10.1029/2007GL031296.

Khanna, J., and Medvigy, D. (2014). Strong control of surface roughness variations on the simulated dry season regional atmospheric response to contemporary deforestation in Rondônia, Brazil. *Journal of Geophysical Research: Atmospheres*, 119, 13067–13078, doi:10.1002/2014JD022278.

Kleidon, A., and Heimann, M. (2000). Assessing the role of deep rooted vegetation in the climate system with model simulations: Mechanism, comparison to observations and implications for Amazonian deforestation. *Climate Dynamics*, 16, 183–199.

Kleidon, A., Fraedrich, K., and Heimann, M. (2000). A green planet versus a desert world: Estimating the maximum effect of vegetation on the land surface climate. *Climatic Change*, 44, 471–493.

Kucharski, F., Zeng, N., and Kalnay, E. (2013). A further assessment of vegetation feedback on decadal Sahel rainfall variability. *Climate Dynamics*, 40, 1453–1466.

Kueppers, L. M., and Snyder, M. A. (2012). Influence of irrigated agriculture on diurnal surface energy and water fluxes, surface climate, and atmospheric circulation in California. *Climate Dynamics*, 38, 1017–1029.

Kueppers, L. M., Snyder, M. A., and Sloan, L. C. (2007). Irrigation cooling effect: Regional climate forcing by land-use change. *Geophysical Research Letters*, 34, L03703, doi:10.1029/2006GL028679.

Kumar, S., Dirmeyer, P. A., Merwade, V., et al. (2013). Land use/cover change impacts in CMIP5 climate simulations: A new methodology and 21st century challenges. *Journal of Geophysical Research: Atmospheres*, 118, 6337–6353, doi:10.1002/jgrd.50463.

Lamptey, B. L., Barron, E. J., and Pollard, D. (2005a). Impacts of agriculture and urbanization on the climate of the Northeastern United States. *Global and Planetary Change*, 49, 203–221.

Lamptey, B. L., Barron, E. J., and Pollard, D. (2005b). Simulation of the relative impact of land cover and carbon dioxide to climate change from 1700 to 2100. *Journal of Geophysical Research*, 110, D20103, doi:10.1029/2005JD005916.

Lawrence, D., and Vandecar, K. (2015). Effects of tropical deforestation on climate and agriculture. *Nature Climate Change*, 5, 27–36.

Lean, J., and Rowntree, P. R. (1993). A GCM simulation of the impact of Amazonian deforestation on climate using an improved canopy representation. *Quarterly Journal of the Royal Meteorological Society*, 119, 509–530.

Lean, J., and Rowntree, P. R. (1997). Understanding the sensitivity of a GCM simulation of Amazonian deforestation to the specification of vegetation and soil characteristics. *Journal of Climate*, 10, 1216–1235.

Lean, J., and Warrilow, D. A. (1989). Simulation of the regional climatic impact of Amazon deforestation. *Nature*, 342, 411–413.

Lee, J.-E., Oliveira, R. S., Dawson, T. E., and Fung, I. (2005). Root functioning modifies seasonal climate. *Proceedings of the National Academy of Sciences USA*, 102, 17576–17581.

Lee, X., Goulden, M. L., Hollinger, D. Y., et al. (2011). Observed increase in local cooling effect of deforestation at higher latitudes. *Nature*, 479, 384–387.

Levis, S. (2010). Modeling vegetation and land use in models of the Earth System. *WIREs Climate Change*, 1, 840–856.

Levis, S., Bonan, G. B., Kluzek, E., et al. (2012). Interactive crop management in the Community Earth System Model (CESM1): Seasonal influences on land–atmosphere fluxes. *Journal of Climate*, 25, 4839–4859.

Lobell, D. B., and Bonfils, C. (2008). The effect of irrigation on regional temperatures: A spatial and temporal analysis of trends in California, 1934–2002. *Journal of Climate*, 21, 2063–2071.

Lobell, D. B., Bala, G., and Duffy, P. B. (2006a). Biogeophysical impacts of cropland management changes on climate. *Geophysical Research Letters*, 33, L06708, doi:10.1029/2005GL025492.

Lobell, D. B., Bala, G., Bonfils, C., and Duffy, P. B. (2006b). Potential bias of model projected greenhouse warming in irrigated regions. *Geophysical Research Letters*, 33, L13709, doi:10.1029/2006GL026770.

Lobell, D., Bala, G., Mirin, A., et al. (2009). Regional differences in the influence of irrigation on climate. *Journal of Climate*, 22, 2248–2255.

Luyssaert, S., Jammet, M., Stoy, P. C., et al. (2014). Land management and land-cover change have impacts of similar magnitude on surface temperature. *Nature Climate Change*, 4, 389–393.

Mahmood, R., Hubbard, K. G., and Carlson, C. (2004). Modification of growing-season surface temperature records in the northern Great Plains due to land-use transformation: Verification of modelling results and implication for global climate change. *International Journal of Climatology*, 24, 311–327.

Mahmood, R., Foster, S. A., Keeling, T., et al. (2006). Impacts of irrigation on 20th century temperature in the northern Great Plains. *Global and Planetary Change*, 54, 1–18.

Mahmood, R., Hubbard, K. G., Leeper, R. D., and Foster, S. A. (2008). Increase in near-surface atmospheric moisture content due to land use changes: Evidence from the observed dewpoint temperature data. *Monthly Weather Review*, 136, 1554–1561.

Mahmood, R., Pielke, R. A., Sr., Hubbard, K. G., et al. (2014). Land cover changes and their biogeophysical effects on climate. *International Journal of Climatology*, 34, 929–953.

Marshall, A. G., and Lynch, A. H. (2008). The sensitivity of the Australian summer monsoon to climate forcing during the late Quaternary. *Journal of Geophysical Research*, 113, D11107, doi:10.1029/2007JD008981.

Matthews, H. D., Weaver, A. J., Meissner, K. J., Gillett, N. P., and Eby, M. (2004). Natural and anthropogenic climate change: Incorporating historical land cover change, vegetation dynamics and the global carbon cycle. *Climate Dynamics*, 22, 461–479.

McAlpine, C. A., Syktus, J., Deo, R. C., et al. (2007). Modeling the impact of historical land cover change on Australia's regional climate. *Geophysical Research Letters*, 34, L22711, doi:10.1029/2007GL031524.

McAlpine, C. A., Syktus, J., Ryan, J. G., et al. (2009). A continent under stress: Interactions, feedbacks and risks associated with impact of modified land cover on Australia's climate. *Global Change Biology*, 15, 2206–2223.

McGuffie, K., Henderson-Sellers, A., Zhang, H., Durbidge, T. B., and Pitman, A. J. (1995). Global climate sensitivity to tropical deforestation. *Global and Planetary Change*, 10, 97–128.

McPherson, R. A., and Stensrud, D. J. (2005). Influences of a winter wheat belt on the evolution of the boundary layer. *Monthly Weather Review*, 133, 2178–2199.

McPherson, R. A., Stensrud, D. J., and Crawford, K. C. (2004). The impact of Oklahoma's winter wheat belt on the mesoscale environment. *Monthly Weather Review*, 132, 405–421.

Medvigy, D., Walko, R. L., and Avissar, R. (2011). Effects of deforestation on spatiotemporal distributions of precipitation in South America. *Journal of Climate*, 24, 2147–2163.

Medvigy, D., Walko, R. L., Otte, M. J., and Avissar, R. (2013). Simulated changes in Northwest U.S. climate in response to Amazon deforestation. *Journal of Climate*, 26, 9115–9136.

Miller, G. H., Fogel, M. L., Magee, J. W., et al. (2005a). Ecosystem collapse in Pleistocene Australia and a human role in megafaunal extinction. *Science*, 309, 287–290.

Miller, G., Mangan, J., Pollard, D., et al. (2005b). Sensitivity of the Australian Monsoon to insolation and vegetation: Implications for human impact on continental moisture balance. *Geology*, 33, 65–68.

Moore, N., and Rojstaczer, S. (2002). Irrigation's influence on precipitation: Texas High Plains, U.S.A. *Geophysical Research Letters*, 29, 10.1029/2002GL014940.

Murphy, L. N., Riley, W. J., and Collins, W. D. (2012). Local and remote climate impacts from expansion of woody biomass for bioenergy feedstock in the southeastern United States. *Journal of Climate*, 25, 7643–7659.

Mylne, M. F., and Rowntree, P. R. (1992). Modelling the effects of albedo change associated with tropical deforestation. *Climatic Change*, 21, 317–343.

Nair, U. S., Wu, Y., Kala, J., et al. (2011). The role of land use change on the development and evolution of the west coast trough, convective clouds, and precipitation in southwest Australia. *Journal of Geophysical Research*, 116, D07103, doi:10.1029/2010JD014950.

Narisma, G. T., and Pitman, A. J. (2003). The impact of 200 years of land cover change on the Australian near-surface climate. *Journal of Hydrometeorology*, 4, 424–436.

Nepstad, D. C., de Carvalho, C. R., Davidson, E. A., et al. (1994). The role of deep roots in the hydrological and carbon cycles of Amazonian forests and pastures. *Nature*, 372, 666–669.

Nicholson, S. E. (2000). Land surface processes and Sahel climate. *Reviews of Geophysics*, 38, 117–140.

Nicholson, S. E., Tucker, C. J., and Ba, M. B. (1998). Desertification, drought, and surface vegetation: An example from the West African Sahel. *Bulletin of the American Meteorological Society*, 79, 815–829.

Nobre, C. A., Sellers, P. J., and Shukla, J. (1991). Amazonian deforestation and regional climate change. *Journal of Climate*, 4, 957–988.

Nobre, C. A., Silva Dias, M. A., Culf, A. D., et al. (2004). The Amazonian climate. In *Vegetation, Water, Humans and the Climate: A New Perspective on an Interactive System*, ed. P. Kabat, M. Claussen, P. A. Dirmeyer, et al. Berlin: Springer, pp. 79–92.

Notaro, M., Wyrwoll, K.-H., and Chen, G. (2011). Did aboriginal vegetation burning impact on the Australian summer monsoon? *Geophysical Research Letters*, 38, L11704, doi:10.1029/2011GL047774.

Oleson, K. W., Bonan, G. B., Levis, S., and Vertenstein, M. (2004). Effects of land use change on North American climate: Impact of surface datasets and model biogeophysics. *Climate Dynamics*, 23, 117–132.

Otterman, J. (1974). Baring high-albedo soils by overgrazing: A hypothesized desertification mechanism. *Science*, 186, 531–533.

Otterman, J. (1977). Anthropogenic impact on the albedo of the Earth. *Climatic Change*, 1, 137–155.

Otterman, J. (1981). Satellite and field studies of man's impact on the surface in arid regions. *Tellus*, 33, 68–77.

Otterman, J. (1989). Enhancement of surface–atmosphere fluxes by desert-fringe vegetation through reduction of surface albedo and of soil heat flux. *Theoretical and Applied Climatology*, 40, 67–79.

Otterman, J., and Tucker, C. J. (1985). Satellite measurements of surface albedo and temperatures in semidesert. *Journal of Climate and Applied Meteorology*, 24, 228–235.

Otterman, J., Manes, A., Rubin, S., Alpert, P., and Starr, D. O. C. (1990). An increase of early rains in southern Israel following land-use change? *Boundary-Layer Meteorology*, 53, 333–351.

Peng, S.-S., Piao, S., Zeng, Z., et al. (2014). Afforestation in China cools local land surface temperature. *Proceedings of the National Academy of Sciences USA*, 111, 2915–2919.

Pielke, R. A., Sr., Adegoke, J., Beltrán-Przekurat, A., et al. (2007). An overview of regional land-use and land-cover impacts on rainfall. *Tellus B*, 59, 587–601.

Pielke, R. A., Sr., Pitman, A., Niyogi, D., et al. (2011). Land use/land cover changes and climate: Modeling analysis and observational evidence. *WIREs Climate Change*, 2, 828–850.

Pitman, A. J., Durbidge, T. B., Henderson-Sellers, A., and McGuffie, K. (1993). Assessing climate model sensitivity to prescribed deforested landscapes. *International Journal of Climatology*, 13, 879–898.

Pitman, A. J., Narisma, G. T., Pielke, R. A., Sr., and Holbrook, N. J. (2004). Impact of land cover change on the climate of southwest Western Australia. *Journal of Geophysical Research*, 109, D18109, doi:10.1029/2003JD004347.

Pitman, A. J., de Noblet-Ducoudré, N., Cruz, F. T., et al. (2009). Uncertainties in climate responses to past land cover change: First results from the LUCID intercomparison study. *Geophysical Research Letters*, 36, L14814, doi:10.1029/2009GL039076.

Pitman, A. J., de Noblet-Ducoudré, N., Avila, F. B., et al. (2012). Effects of land cover change on temperature and rainfall extremes in multi-model ensemble simulations. *Earth System Dynamics*, 3, 213–231.

Polcher, J., and Laval, K. (1994a). The impact of African and Amazonian deforestation on tropical climate. *Journal of Hydrology*, 155, 389–405.

Polcher, J., and Laval, K. (1994b). A statistical study of the regional impact of deforestation on climate in the LMD GCM. *Climate Dynamics*, 10, 205–219.

Pongratz, J., Reick, C. H., Raddatz, T., and Claussen, M. (2010). Biogeophysical versus biogeochemical climate response to historical anthropogenic land cover change. *Geophysical Research Letters*, 37, L08702, doi:10.1029/2010GL043010.

Puma, M. J., and Cook, B. I. (2010). Effects of irrigation on global climate during the 20th century. *Journal of Geophysical Research*, 115, D16120, doi:10.1029/2010JD014122.

Ramos da Silva, R., and Avissar, R. (2006). The hydrometeorology of a deforested region of the Amazon basin. *Journal of Hydrometeorology*, 7, 1028–1042.

Reale, O., and Dirmeyer, P. (2000). Modeling the effects of vegetation on Mediterranean climate during the Roman Classical Period, Part I: Climate history and model sensitivity. *Global and Planetary Change*, 25, 163–184.

Reale, O., and Shukla, J. (2000). Modeling the effects of vegetation on Mediterranean climate during the Roman Classical Period, Part II: Model simulation. *Global and Planetary Change*, 25, 185–214.

Reynolds, J. F., Stafford Smith, D. M., Lambin, E. F., et al. (2007). Global desertification: Building a science for dryland development. *Science*, 316, 847–851.

Rotenberg, E., and Yakir, D. (2010). Contribution of semi-arid forests to the climate system. *Science*, 327, 451–454.

Rotenberg, E., and Yakir, D. (2011). Distinct patterns of changes in surface energy budget associated with forestation in the semiarid region. *Global Change Biology*, 17, 1536–1548.

Sacks, W. J., and Kucharik, C. J. (2011). Crop management and phenology trends in the U.S. Corn Belt: Impacts on yields, evapotranspiration and energy balance. *Agricultural and Forest Meteorology*, 151, 882–894.

Sacks, W. J., Cook, B. I., Buenning, N., Levis, S., and Helkowski, J. H. (2009). Effects of global irrigation on the near-surface climate. *Climate Dynamics*, 33, 159–175.

Samain, O., Kergoat, L., Hiernaux, P., et al. (2008). Analysis of the in situ and MODIS albedo variability at multiple timescales in the Sahel. *Journal of Geophysical Research*, 113, D14119, doi:10.1029/2007JD009174.

Segal, M., Avissar, R., McCumber, M. C., and Pielke, R. A. (1988). Evaluation of vegetation effects on the generation and modification of mesoscale circulations. *Journal of the Atmospheric Sciences*, 45, 2268–2292.

Segal, M., Pan, Z., Turner, R. W., and Takle, E. S. (1998). On the potential impact of irrigated areas in North America on summer rainfall caused by large-scale systems. *Journal of Applied Meteorology*, 37, 325–331.

Shukla, J., and Mintz, Y. (1982). Influence of land-surface evapotranspiration on the Earth's climate. *Science*, 215, 1498–1501.

Snyder, P. K. (2010). The influence of tropical deforestation on the Northern Hemisphere climate by atmospheric teleconnections. *Earth Interactions*, 14(4), 1–34.

Snyder, P. K., Delire, C., and Foley, J. A. (2004). Evaluating the influence of different vegetation biomes on the global climate. *Climate Dynamics*, 23, 279–302.

Sorooshian, S., Li, J., Hsu, K., and Gao, X. (2012). Influence of irrigation schemes used in regional climate models on evapotranspiration estimation: Results and comparative studies from California's Central Valley agricultural regions. *Journal of Geophysical Research*, 117, D06107, doi:10.1029/2011JD016978.

Sorooshian, S., AghaKouchak, A., and Li, J. (2014). Influence of irrigation on land hydrological processes over California. *Journal of Geophysical Research: Atmospheres*, 119, 13137–13152, doi:10.1002/2014JD022232.

Stohlgren, T. J., Chase, T. N., Pielke, R. A., Sr., Kittel, T. G. F., and Baron, J. S. (1998). Evidence that local land use practices influence regional climate, vegetation, and stream flow patterns in adjacent natural areas. *Global Change Biology*, 4, 495–504.

Stoy, P. C., Katul, G. G., Siqueira, M. B., et al. (2006). Separating the effects of climate and vegetation on evapotranspiration along a successional chronosequence in the southeastern US. *Global Change Biology*, 12, 2115–2135.

Spracklen, D. V., Arnold, S. R., and Taylor, C. M. (2012). Observations of increased tropical rainfall preceded by air passage over forests. *Nature*, 489, 282–285.

Sterling, S. M., Ducharne, A., and Polcher, J. (2013). The impact of global land-cover change on the terrestrial water cycle. *Nature Climate Change*, 3, 385–390.

Sud, Y. C., Walker, G. K., Kim, J.-H., et al. (1996). Biogeophysical consequences of a tropical deforestation scenario: A GCM simulation study. *Journal of Climate*, 9, 3225–3247.

Swann, A. L. S., Fung, I. Y., and Chiang, J. C. H. (2012). Mid-latitude afforestation shifts general circulation and tropical precipitation. *Proceedings of the National Academy of Sciences USA*, 109, 712–716.

Swann, A. L. S., Fung, I. Y., Liu, Y., and Chiang, J. C. H. (2014). Remote vegetation feedbacks and the mid-Holocene green Sahara. *Journal of Climate*, 27, 4857–4870.

Taylor, C. M., Parker, D. J., and Harris, P. P. (2007). An observational case study of mesoscale atmospheric circulations induced by soil moisture. *Geophysical Research Letters*, 34, L15801, doi:10.1029/2007GL030572.

Taylor, C. M., Parker, D. J., Kalthoff, N., et al. (2011). New perspectives on land–atmosphere feedbacks from the African Monsoon Multidisciplinary Analysis. *Atmospheric Science Letters*, 12, 38–44.

Teuling, A. J., Seneviratne, S. I., Stöckli, R., et al. (2010). Contrasting response of European forest and grassland energy exchange to heatwaves. *Nature Geoscience*, 3, 722–727.

Timbal, B., and Arblaster, J. M. (2006). Land cover change as an additional forcing to explain the rainfall decline in the south west of Australia. *Geophysical Research Letters*, 33, L07717, doi:10.1029/2005GL025361.

Timouk, F., Kergoat, L., Mougin, E., et al. (2009). Response of surface energy balance to water regime and vegetation development in a Sahelian landscape. *Journal of Hydrology*, 375, 178–189.

Trail, M., Tsimpidi, A. P., Liu, P., et al. (2013). Potential impact of land use change on future regional climate in the Southeastern U.S.: Reforestation and crop land conversion. *Journal of Geophysical Research: Atmospheres*, 118, 11577–11588, doi:10.1002/2013JD020356.

Twine, T. E., Kucharik, C. J., and Foley, J. A. (2004). Effects of land cover change on the energy and water balance of the Mississippi River basin. *Journal of Hydrometeorology*, 5, 640–655.

van der Velde, Y., Lyon, S. W., and Destouni, G. (2013). Data-driven regionalization of river discharges and emergent land cover–evapotranspiration relationships across Sweden. *Journal of Geophysical Research: Atmospheres*, 118, 2576–2587, doi:10.1002/jgrd.50224.

Voldoire, A., and Royer, J.-F. (2004). Tropical deforestation and climate variability. *Climate Dynamics*, 22, 857–874.

von Randow, C., Manzi, A. O., Kruijt, B., et al. (2004). Comparative measurements and seasonal variations in energy and carbon exchange over forest and pasture in South West Amazonia. *Theoretical and Applied Climatology*, 78, 5–26.

Walker, J., Bullen, F., and Williams, B. G. (1993). Ecohydrological changes in the Murray-Darling Basin, I: The number of trees cleared over two centuries. *Journal of Applied Ecology*, 30, 265–273.

Wang, G. (2011). Assessing the potential hydrological impacts of hydraulic redistribution in Amazonia using a numerical modeling approach. *Water Resources Research*, 47, W02528, doi:10.1029/2010WR009601.

Wang, J., Chagnon, F. J. F., Williams, E. R., et al. (2009). Impact of deforestation in the Amazon basin on cloud climatology. *Proceedings of the National Academy of Sciences USA*, 106, 3670–3674.

Weaver, C. P., Baidya Roy, S., and Avissar, R. (2002). Sensitivity of simulated mesoscale atmospheric circulations resulting from landscape heterogeneity to aspects of model configuration. *Journal of Geophysical Research*, 107, 8041, 10.1029/2001JD000376.

Werth, D., and Avissar, R. (2002). The local and global effects of Amazon deforestation. *Journal of Geophysical Research*, 107, 8087, doi:10.1029/2001JD000717.

Wickham, J. D., Wade, T. G., and Riitters, K. H. (2012). Comparison of cropland and forest surface temperatures across the conterminous United States. *Agricultural and Forest Meteorology*, 166–167, 137–143.

Wickham, J. D., Wade, T. G., and Riitters, K. H. (2013). Empirical analysis of the influence of forest extent on annual and seasonal surface temperatures for the continental United States. *Global Ecology and Biogeography*, 22, 620–629.

Wickham, J., Wade, T. G., and Riitters, K. H. (2014). An isoline separating relatively warm from relatively cool wintertime forest surface temperatures for the southeastern United States. *Global and Planetary Change*, 120, 46–53.

Williams, C. A., Reichstein, M., Buchmann, N., et al. (2012). Climate and vegetation controls on the surface water balance: Synthesis of evapotranspiration measured across a global network of flux towers. *Water Resources Research*, 48, W06523, doi:10.1029/2011WR011586.

Wilson, K. B., Baldocchi, D. D., Aubinet, M., et al. (2002). Energy partitioning between latent and sensible heat flux during the warm season at FLUXNET sites. *Water Resources Research*, 38, 1294, doi:10.1029/2001WR000989.

Wright, I. R., Gash, J. H. C., da Rocha, H. R., et al. (1992). Dry season micrometeorology of central Amazonian ranchland. *Quarterly Journal of the Royal Meteorological Society*, 118, 1083–1099.

Xue, Y. (2006). Interactions and feedbacks between climate and dryland vegetations. In *Dryland Ecohydrology*, ed. P. D'Odorico and A. Porporato. Dordrecht: Springer, pp. 85–105.

Xue, Y., Fennessy, M. J., and Sellers, P. J. (1996). Impact of vegetation properties on U.S. summer weather prediction. *Journal of Geophysical Research*, 101D, 7419–7430.

Xue, Y., Hutjes, R. W. A., Harding, R. J., et al. (2004). The Sahelian climate. In *Vegetation, Water, Humans and the Climate: A New Perspective on an Interactive System*, ed. P. Kabat, M. Claussen, P. A. Dirmeyer, et al. Berlin: Springer, pp. 59–77.

Zhang, L., Dawes, W. R., and Walker, G. R. (2001). Response of mean annual evapotranspiration to vegetation changes at catchment scale. *Water Resources Research*, 37, 701–708.

Zhang, M., Lee, X., Yu, G., et al. (2014). Response of surface air temperature to small-scale land clearing across latitudes. *Environmental Research Letters*, 9, 034002, doi:10.1088/1748-9326/9/3/034002.

Zhao, K., and Jackson, R. B. (2014). Biophysical forcings of land-use changes from potential forestry activities in North America. *Ecological Monographs*, 84, 329–353.

Carbon Cycle–Climate Feedbacks

29.1 | Chapter Summary

As shown in previous chapters, numerous climate model experiments demonstrate that vegetation exerts an important feedback on climate through energy and water cycles. In addition to these biogeophysical feedbacks, terrestrial ecosystems are coupled to climate through the carbon cycle. Terrestrial ecosystems absorb a significant portion of the annual emission of CO_2 to the atmosphere by human activities. This arises from enhanced photosynthesis as a result of climate change, increasing concentration of CO_2 in the atmosphere, or by increasing deposition of nitrogen on land. It is also caused by regrowth of forests following abandonment of farmland. At longer timescales, changes in the biogeography of ecosystems alter carbon storage on land. Climate model simulations show that the terrestrial carbon cycle is a positive climate feedback whereby a warmer climate decreases the capacity of the terrestrial biosphere to storage anthropogenic carbon emissions. The uncertainty in the carbon cycle feedback is of comparable magnitude to the uncertainty arising from physical climate processes and relates in part to plant and microbial physiological responses to temperature and plant demographic processes in response to disturbance and climate change. In addition, nitrogen limits the productivity of many terrestrial ecosystems. This control of the carbon cycle by nitrogen results in prominent carbon–nitrogen interactions including constraints on productivity increases with CO_2 enrichment, enhanced productivity with nitrogen deposition, and additional nitrogen mineralization with soil warming.

29.2 | Glacial–Interglacial Cycles

In the absence of human influences, the concentration of CO_2 in the atmosphere is the balance between oceanic and terrestrial processes. Over hundreds of thousands of years, atmospheric CO_2 has varied by about 100 ppm during glacial–interglacial cycles (Figure 8.2). Atmospheric CO_2 was lower during glacial periods (~180–200 ppm) than during interglacial periods (~270–290 ppm). The causes of this variation are unclear, but likely reflect oceanic processes (Ciais et al. 2013). In particular, the colder glacial ocean held more carbon than during warmer interglacial periods because the solubility of CO_2 increases with colder temperature. Changes in ocean circulation also contributed to lower atmospheric CO_2, while the lower sea level and increased salinity of the glacial ocean decreased oceanic carbon stocks.

In contrast with the ocean, carbon storage in the terrestrial biosphere decreased during glacial periods. Studies of the last glacial maximum (18–21 kyr before present) show a much different biosphere compared with today. Desert and desert scrub biomes were more extensive, while boreal and temperate forests decreased in area. Dry woodlands and savanna encroached into

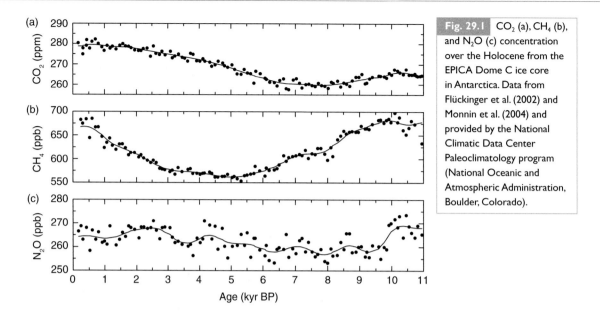

Fig. 29.1 CO_2 (a), CH_4 (b), and N_2O (c) concentration over the Holocene from the EPICA Dome C ice core in Antarctica. Data from Flückinger et al. (2002) and Monnin et al. (2004) and provided by the National Climatic Data Center Paleoclimatology program (National Oceanic and Atmospheric Administration, Boulder, Colorado).

present-day tropical rainforests. Biomes that prevailed under the cooler, drier climate of the last glacial maximum have low carbon storage per unit area, suggesting that the terrestrial biosphere stored less carbon than at present. Estimates of the carbon deficit are on the order of 300–1000 Pg C (Prentice et al. 2011; Ciais et al. 2012, 2013).

Atmospheric CO_2 was much more stable during the present interglacial, prior to the industrial era. Its concentration over the past 11 kyr varied from 260 to 280 ppm, increasing by 20 ppm over the past 7 kyr up to the industrial era (Figure 29.1). The causes of this variation are unclear, but likely reflect changes in ocean chemistry (leading to an outgassing of CO_2) and to lesser extent changes in the terrestrial biosphere (Ciais et al. 2013). High resolution ice cores show relatively minor variability in CO_2 during the last millennium until the start of the industrial era in 1750 (Figure 29.2). A prominent exception is the ~10 ppm decrease in atmospheric CO_2 around 1600.

29.3 | Present-Day Carbon Cycle

The present-day concentration of CO_2 in the atmosphere is a balance among anthropogenic emissions from fossil fuel combustion and other industrial processes, emissions from forest clearing and other land-use changes, and net uptake of CO_2 by oceans and terrestrial ecosystems. Atmospheric CO_2 increased by 40 percent over the industrial era, from 278 ppm in 1750 to 390 ppm in 2011 (Table 2.1, Figure 3.7a). Human activities emitted 555 ± 85 Pg C from fossil fuel combustion, cement production, and land-use change (Table 29.1). However, less than half of this carbon (43%) accumulated in the atmosphere (240 ± 10 Pg C). The ocean and terrestrial biosphere took up the remaining carbon in approximately equal amounts. Analyses for the past 50 years (1960–2010) similarly indicate that about 45 percent of carbon emissions remained in the atmosphere (Ballantyne et al. 2012; Le Quéré et al. 2013).

The 10-year period 2002–2011 illustrates the present-day global carbon cycle (Table 29.1). The annual input of carbon into the atmosphere from fossil fuel emissions and cement production was 8.3 ± 0.7 Pg C yr^{-1}. Land-use change emitted another 0.9 ± 0.8 Pg C yr^{-1}. Measurements of atmospheric CO_2 concentration show that the carbon in the atmosphere increased at a rate of 4.3 ± 0.2 Pg C yr^{-1}, almost one-half of total anthropogenic emissions. The remaining 4.9 Pg C yr^{-1} was taken up and stored

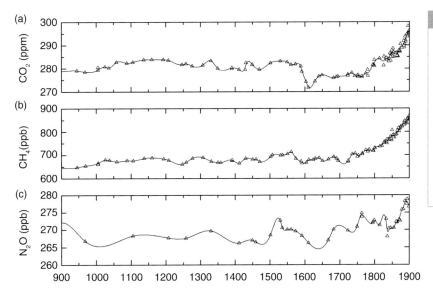

Fig. 29.2 CO_2 (a), CH_4 (b), and N_2O (c) concentration over the period 900–1900 from the Law Dome ice core in Antarctica. Data from MacFarling Meure et al. (2006) and provided by the National Climatic Data Center Paleoclimatology program (National Oceanic and Atmospheric Administration, Boulder, Colorado).

Table 29.1 Global carbon budget over the industrial era (1750–2011) and for the 10-year period 2002–2011

	1750–2011 (Pg C)	2002–2011 (Pg C yr^{-1})
Emissions		
Fossil fuel + cement production	375 ± 30	8.3 ± 0.7
Land-use change	180 ± 80	0.9 ± 0.8
Accumulations		
Atmospheric increase	240 ± 10	4.3 ± 0.2
Ocean sink	−155 ± 30	−2.4 ± 0.7
Residual terrestrial sink	−160 ± 90	−2.5 ± 1.3

Note: Data for the industrial era are cumulative fluxes. Data for 2002–2011 are annual fluxes averaged over the ten years. Positive values denote carbon sources. Negative values denote carbon sinks.
Source: From Ciais et al. (2013).

in the ocean and terrestrial biosphere. The net oceanic carbon flux is estimated to be −2.4 ± 0.7 Pg C yr^{-1} (the negative flux denotes carbon gain by oceans). Carbon balance requires a residual terrestrial flux of −2.5 ± 1.3 Pg C yr^{-1} (i.e., undisturbed terrestrial ecosystems gained carbon). The net land–atmosphere flux (−1.6 ± 1.0 Pg C yr^{-1}) is the balance between the land-use emission flux arising from human management of ecosystems and the residual land sink arising from global environmental change (e.g., higher atmospheric CO_2, climate change, increased nitrogen deposition).

Figure 29.3 illustrates decadal variability in the global carbon cycle over the last 50 years.

Global CO_2 emissions from fossil fuel combustion and cement production increased by a factor of 2.5, from 3.1 ± 0.2 Pg C yr^{-1} in the 1960s to 7.8 ± 0.4 Pg C yr^{-1} in the 2000s. Land-use emissions were nearly constant at approximately 1.5 ± 0.5 Pg C yr^{-1} during 1960–1999, but decreased to 1.0 ± 0.5 Pg C yr^{-1} during the 2000s. The atmospheric CO_2 growth rate increased from 1.7 ± 0.1 Pg C yr^{-1} in the 1960s to 4.0 ± 0.1 Pg C yr^{-1} in the 2000s, but with considerable decadal variability. The ocean and land sinks also increased over this period, from 1.2 ± 0.5 Pg C yr^{-1} in the 1960s to 2.4 ± 0.5 Pg C yr^{-1} in the 2000s (ocean) and 1.7 ± 0.7 Pg C yr^{-1} to 2.4 ± 0.8 Pg C yr^{-1} (land); both had large decadal variability.

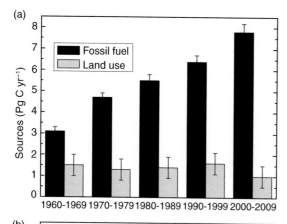

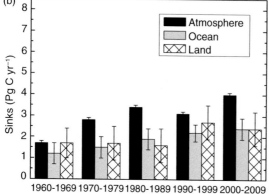

Fig. 29.3 Decadal mean global carbon budget, 1960–2009. (a) Anthropogenic emissions. (b) Atmospheric growth rate and oceanic and terrestrial sinks. Data from Le Quéré et al. (2013).

There is considerable uncertainty in the global terrestrial carbon budget. Annual carbon fluxes on land cannot be measured directly at a global scale, and various means to infer global carbon fluxes differ in assumptions, ecological processes represented, and spatial and temporal scales. The residual terrestrial sink is often calculated from fossil fuel emissions, land-use emissions, the airborne fraction, and the ocean sink to ensure carbon balance (Le Quéré et al. 2013; Ciais et al. 2013). Amospheric CO_2 inversions estimate regional ocean and land carbon fluxes to be optimally consistent with the observed atmospheric CO_2 measurements (Gurney et al. 2004; Ciais et al. 2010; Peylin et al. 2013). An important source of uncertainty in atmospheric inversions is the atmospheric transport, which is needed to infer how atmospheric CO_2 changes at a specified location in response to surface fluxes. Forest stand inventories provide direct measurements of carbon stocks, and temporal changes in stocks are used to infer carbon fluxes (Pan et al. 2011). However, these inventories are of limited spatial extent and extrapolation to regional fluxes is uncertain. Eddy covariance flux towers provide measurements of net ecosystem production, but over a small footprint around the tower. Techniques are available to empirically upscale fluxes from individual eddy covariance towers, but global representativeness is not assured (Beer et al. 2010; Jung et al. 2011). Global terrestrial biosphere models can bridge the gap from the site scale to the global scale. These models provide estimates of gross primary production, ecosystem respiration, and also the land-use change flux and provide an independent estimate of the residual terrestrial sink (Le Quéré et al. 2009, 2013; Ciais et al. 2013; Piao et al. 2013).

29.4 | Seasonal-to-Interannual Variability

Metabolic activity by terrestrial ecosystems is seen in atmospheric CO_2 concentration. At Mauna Loa, Hawaii, for example, CO_2 concentration has a pronounced annual cycle and varies by several parts per million over the course of a year, with high concentration in winter and low concentration in summer (Figure 29.4). This is in response to the seasonal growth of terrestrial ecosystems worldwide, which absorb CO_2 during the growing season and respire CO_2 during the dormant season. The seasonal pulse of CO_2 occurs throughout the Northern Hemisphere (Figure 29.5). The annual amplitude of the atmospheric CO_2 cycle (i.e., the difference between maximum and minimum concentrations) is greatest in high latitudes of the Northern Hemisphere and declines with latitudes closer to the equator. In the Southern Hemisphere, there is less land, oceans dominate atmospheric CO_2 exchange, and there is little seasonal variation in atmospheric CO_2.

Satellite and modeling analyses indicate that the annual amplitude in the Northern Hemisphere primarily reflects the seasonality

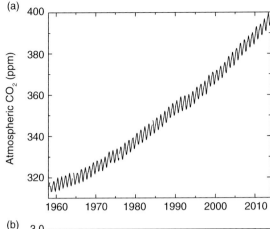

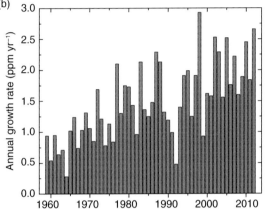

Fig. 29.4 Atmospheric CO_2 concentration measured at Mauna Loa, Hawaii, from 1959 to 2012. (a) Monthly mean concentration. (b) Annual growth rate. Data provided courtesy of Pieter Tans (National Oceanic and Atmospheric Administration, Boulder, Colorado) and Ralph Keeling (Scripps Institution of Oceanography, La Jolla, California).

Temporal changes in the amplitude and seasonality of the atmospheric CO_2 cycle are important indicators of changes in the activity of terrestrial ecosystems. For example, the annual amplitude of atmospheric CO_2 concentration at Barrow, Alaska, has increased since the early 1960s, as has that at Mauna Loa, though to a lesser extent (Figure 29.6). A large increase in amplitude is seen throughout the Northern Hemisphere north of 45° N (Graven et al. 2013). This reflects a stimulation of metabolic activity by temperate and boreal forests with greater CO_2 drawdown during the growing season (Randerson et al. 1999; Buermann et al. 2007; Barichivich et al. 2013; Graven et al. 2013). Additionally, intensification of agriculture and increased crop productivity has likely increased the amplitude of atmospheric CO_2 over the past several decades (Gray et al. 2014; Zeng et al. 2014).

By affecting photosynthesis, respiration, and fire, climate variability exerts a discernible signal in the global carbon cycle. This is seen in the effects of the El Niño–Southern Oscillation (ENSO) on terrestrial carbon fluxes. There is an enhanced source of carbon from the biosphere to the atmosphere during warm, dry El Niño years (Figure 29.7). This is also seen in annual atmospheric CO_2 growth rates, which are generally high during El Niño and low during La Niña. ENSO-driven variability in the terrestrial biosphere is seen also in satellite vegetation indices (Buermann et al. 2003). Regional drought and large wildfires contribute to the high CO_2 growth rates during El Niño episodes. During the 1997 to 1998 El Niño, wildfires emitted 2.1 Pg C, or 66 percent of the atmospheric CO_2 growth rate anomaly (van der Werf et al. 2004).

Production efficiency models provide a means to derive historical net primary production using satellite and meteorological data. The normalized difference vegetation index gives a measure of the fraction of photosynthetically active radiation absorbed by plants. Global meteorological analyses of temperature, vapor pressure, and solar radiation provide the meteorological constraints to production efficiency. Nemani et al. (2003) and Hashimoto et al. (2004) used a production efficiency model to derive net

of terrestrial net ecosystem production (Tucker et al. 1986; Randerson et al. 1997; Graven et al. 2013). Boreal forests contribute greatly to the annual amplitude at high latitudes. In these ecosystems, the timing of photosynthesis and respiration is asynchronous, with photosynthesis restricted to a short growing season while respiration occurs year-round. Grasslands and croplands also have large influence, becoming increasingly important in middle and tropical latitudes. Tropical forests contribute minimally to the annual amplitude throughout much of the Northern Hemisphere, in part because of little seasonality of ecosystem processes.

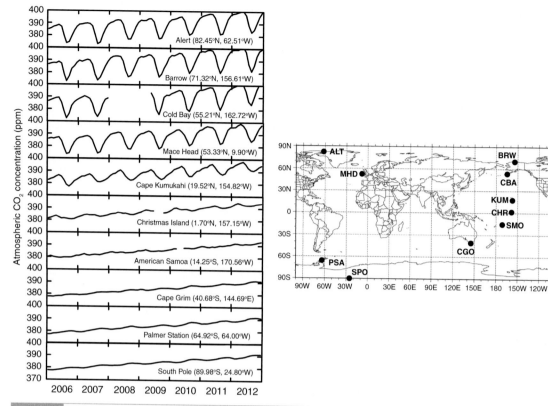

Fig. 29.5 Monthly mean atmospheric CO_2 concentration measured at 10 locations for seven years spanning 2006 to 2012. The map shows station locations. Data are from the Earth System Research Laboratory Global Monitoring Division (National Oceanic and Atmospheric Administration, Boulder, Colorado) air sampling network.

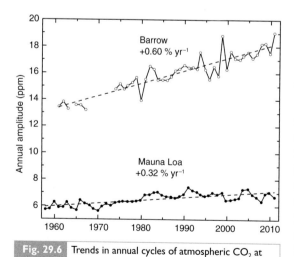

Fig. 29.6 Trends in annual cycles of atmospheric CO_2 at Barrow and Mauna Loa. Data from Graven et al. (2013).

primary production for the 1980s and 1990s and its relationship with ENSO. Their results show that global net primary production on land

decreased during El Niño events with corresponding increases in atmospheric CO_2 growth rate (Figure 29.8). This is particularly evident in 1982–1983, 1986–1987, and 1997–1998. Much of this variability originates in tropical ecosystems.

The period 1991–1993 is an exception to the general relationship between ENSO and atmospheric CO_2. The eruption of Mount Pinatubo in June 1991 resulted in a lower than expected atmospheric CO_2 growth rate because of enhanced net uptake by the land, and atmospheric CO_2 exhibits variability on interannual to decadal time scales due to large volcanic eruptions (Jones and Cox 2001; Brovkin et al. 2010; Sarmiento et al. 2010; Frölicher et al. 2013). Atmospheric CO_2 is generally lower than expected following large volcanic eruptions. Colder temperatures following eruptions decrease net primary production. However, the injection of volcanic aerosols into the

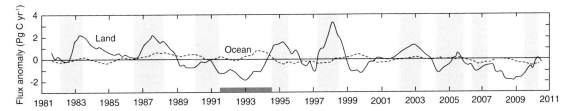

Fig. 29.7 Global carbon flux anomalies for the period 1981–2010 for net land–atmosphere (solid line) and ocean–atmosphere (dashed line) fluxes. Light gray shading indicates El Niño events, and the dark shading denotes the anomalously cool period following the Mount Pinatubo eruption. A positive flux indicates a larger than normal source of carbon to the atmosphere, or a smaller sink. Adapted from Ciais et al. (2013).

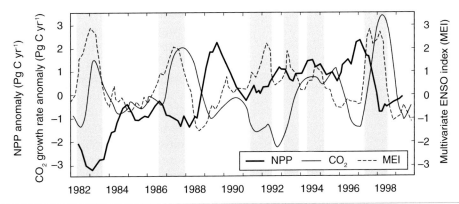

Fig. 29.8 Global net primary production (NPP) and CO_2 growth rate during the period 1982–1999 in relation to the multivariate ENSO index (MEI). NPP and CO_2 are shown as anomalies. High MEI indicates the warm phase of ENSO. Highlighted in gray are El Niños of 1982–1983, 1986–1987, 1991–1992, 1993, 1994–1995, and 1997–1998. Adapted from Hashimoto et al. (2004). See also Nemani et al. (2003).

atmosphere also increases the proportion of diffuse solar radiation. Diffuse radiation penetrates more deeply into a canopy than does direct beam radiation, which can enhance photosynthetic rates. Such an increase in carbon uptake in the months following the Mount Pinatubo eruption was observed at Harvard Forest, Massachusetts (Gu et al. 2003). Carbon loss from heterotrophic respiration, too, can be suppressed with colder temperature. An index that combines ENSO and volcanic activity explains 75 percent of the interannual variability in the land carbon flux (Raupach et al. 2008).

29.5 | Residual Terrestrial Sink

The residual land sink is the carbon accumulation by the terrestrial biosphere after accounting

for land-use emissions. The location of the residual terrestrial carbon sink is uncertain, but it likely occurs in boreal and temperate forests of the Northern Hemisphere (Gurney and Eckels 2011; Pan et al. 2011; Ciais et al. 2013). Other analyses show a greater role for tropical forests (Stephens et al. 2007; Schimel et al. 2015). Observational data indicate a large carbon sink in mature Amazonian and African tropical forests (Phillips et al. 1998, 2009; Lewis et al. 2009), but this sink is offset by deforestation emissions so that tropical forests are near neutral with respect to carbon flux.

Pan et al. (2011) used forest inventory data and changes in forest carbon stocks to estimate fluxes for the period 1990–2007 (Table 29.2). They calculated an annual global forest sink of 1.2 Pg C yr⁻¹ in boreal (0.5 ± 0.1) and temperate (0.7 ± 0.1) forests. Undisturbed tropical forests

Table 29.2 Global forest carbon budget for 1990–2007

Biome	Carbon flux (Pg C yr^{-1})
Boreal forest	−0.5 ± 0.1
Temperate forest	−0.7 ± 0.1
Tropical intact forest	−1.2 ± 0.4
Tropical net land-use change	1.3 ± 0.7
Tropical gross deforestation	2.9 ± 0.5
Tropical regrowth forest	−1.6 ± 0.5
Gross forest sink	−4.0 ± 0.7
Net forest sink	−1.1 ± 0.8

Note: Positive values denote sources. Negative values denote sinks. Boreal and temperate forests are the net flux, including land-use change and disturbance. Tropical forest fluxes are estimated separately for intact forests (unaffected by human activities) and regrowth forests (recovering from past deforestation and logging). Tropical gross deforestation is the total emission; net land-use change is the balance between gross deforestation emission and regrowth uptake.
Source: From Pan et al. (2011).

gained 1.2 ± 0.4 Pg C yr^{-1}, but lost a similar amount through land-use change (1.3 ± 0.7 Pg C yr^{-1}). The net tropical land-use change source consisted of gross deforestation emission (2.9 ± 0.5 Pg C yr^{-1}) and carbon uptake during forest regrowth following disturbance (1.6 ± 0.5 Pg C yr^{-1}). This analysis highlights the importance of forests for the global carbon cycle. Boreal, temperate, and tropical forests were a gross carbon sink of 4.0 ± 0.7 Pg C yr^{-1} during this period and a net sink of 1.1 ± 0.8 Pg C yr^{-1}. Because tropical forests were nearly carbon neutral, this carbon sink occurred mainly in temperate and boreal forests.

Several mechanisms can account for this increase in carbon storage. It may be a result of growth enhancement due to stimulation of photosynthesis by increasing atmospheric CO_2 concentrations, or it may reflect climate effects on plant productivity and ecosystem respiration. It may also arise from increasing deposition of nitrogen on land from industrial pollution and agricultural activities.

29.5.1 CO$_2$ Fertilization

Global terrestrial biosphere models forced with higher CO_2 concentrations show increased net primary production and carbon storage in response to CO_2 fertilization (Cramer et al. 2001; Sitch et al. 2008; Piao et al. 2013). This storage accounts for a large portion of the required residual terrestrial carbon uptake. In a comparison of 10 global models, Piao et al. (2013) found that higher CO_2 concentrations increased net primary production by 0.16 percent per ppm (multi-model mean) over the past three decades, in agreement with free-air CO_2 enrichment (FACE) studies (mean, 0.13% ppm^{-1}). However, it is not clear if FACE studies are representative of the global biosphere. Nor do ecosystem models fully simulate the changes in net primary production with elevated CO_2 observed at particular FACE experiments (Zaehle et al. 2014).

There is much uncertainty in the extent to which nitrogen availability restricts the CO_2 fertilization response. Experimental studies show that limited nitrogen availability can constrain the productivity enhancement from CO_2 enrichment (Chapter 20). Global terrestrial biosphere models with coupled carbon–nitrogen biogeochemistry have a lower contribution of CO_2 fertilization to the terrestrial carbon sink compared with carbon-only models (Thornton et al. 2007, 2009; Jain et al. 2009; Bonan and Levis 2010; Zaehle et al. 2010b; Zhang et al. 2011; Gerber et al. 2013; Zaehle 2013). However, there is substantial variability among models in terms of this nitrogen downregulation.

The FACE studies show much of the general patterns expected from photosynthetic theory. Yet they also raise questions about our understanding of CO_2 fertilization, how these physiological effects propagate through whole ecosystems, and our ability to represent these in models. There is large variability among sites. Most forest FACE studies have been conducted in young, fast-growing stands; whether these sites are representative of mature forests is unclear. Nor is it clear whether increased growth rates can be sustained over many years, particularly if nitrogen availability limits growth. Nitrogen cycling also plays an important role in the response of forests to increased atmospheric CO_2 through complex interactions among leaf nitrogen concentration, plant growth, carbon allocation, decomposition, and nitrogen mineralization. The long-term effect of CO_2 enrichment on net ecosystem production is also unclear. If the increased photosynthesis is allocated to wood, which decomposes slowly, the long-term net carbon gain is large; but if the carbon is allocated to plant tissues that decompose rapidly, the long-term net carbon sequestration in soils is limited.

Detecting enhanced productivity from rising CO_2 concentration in the natural world is difficult. Continuous, long-term, high-quality measurements are lacking, and the expected CO_2 fertilization response is confounded by other environmental factors such as temperature, precipitation, nitrogen deposition, and stand dynamics. However, the expected increase in water-use efficiency resulting from higher CO_2 concentration may be merging. Eddy covariance flux tower measurements in temperate and boreal forests of the Northern Hemisphere reveal that water-use efficiency has increased over the past two decades, as expected from rising atmospheric CO_2 concentration (Keenan et al. 2013), and isotopic analyses show water-use efficiency similarly has increased in tropical forests (van der Sleen et al. 2015). Higher water-use efficiency may be contributing to a greening of vegetation in warm, arid environments (Donohue et al. 2013).

29.5.2 Climate Change

Secular changes in climate also affect the global carbon cycle. A prominent trend seen in satellite measurements of the normalized difference vegetation index is longer growing season length (about 1–2 weeks) and increased summer greenness in northern high latitudes since the late twentieth century (Myneni et al. 1997; Slayback et al. 2003; Barichivich et al. 2013). Field studies of plant phenology also indicate changes in springtime leaf emergence and autumn senescence (Menzel et al. 2006; Schwartz et al. 2006; Jeong et al. 2011; Wolkovich et al. 2012). Changes in growing season length are thought to be a result of temperature increases over this period. Autumn phenology regulates

interannual variability of annual net ecosystem production (Wu et al. 2013), and greater growing season length in spring or autumn can increase plant productivity, but it is unclear if this gain is reduced by respiration losses with a warmer autumn (Piao et al. 2008; Richardson et al. 2010; Dragoni et al. 2011; Keenan et al. 2014).

Observations suggest climate change over the past several decades has increased net primary production on sites where water availability is not strongly limiting (Boisvenue and Running 2006). Nemani et al. (2003) also found that climatic constraints on net primary production eased during the 1980s and 1990s and that net primary production in many regions of the world increased during this period. However, large-scale droughts subsequently reduced net primary production during the first decade of the twenty-first century (Zhao and Running 2010).

The net effect of climate changes on the residual terrestrial sink is a complex balance of multiple ecological processes and their sensitivity to climate change. Warmer temperatures and longer growing season length can increase net primary production in temperate, boreal, and arctic ecosystems, but warmer temperatures also accelerate soil carbon loss through decomposition. Increased evaporative demand may produce soil water stress unless accompanied by greater precipitation. Most model attribution studies find only a small contribution of climate change to the terrestrial sink (Jain et al. 2009; Bonan and Levis 2010; Zaehle et al. 2010b).

The influence of climate change on the carbon cycle is broader than the effects of temperature and precipitation on ecosystem functions. The changing atmospheric composition also affects the carbon cycle. High concentrations of tropospheric ozone can induce stomatal closure and decrease leaf photosynthesis and plant productivity (Wittig et al. 2007, 2009; Ainsworth et al. 2012; Lombardozzi et al. 2013), seen, for example, in field study responses to elevated ozone for forest (Karnosky et al. 2005) and soybean (Morgan et al. 2006; Betzelberger et al. 2012). Decreased plant productivity is a positive radiative forcing by reducing the land carbon sink (Ollinger et al. 2002; Felzer et al.

2005; Sitch et al. 2007; Lombardozzi et al. 2015). Greater amounts of aerosols in the atmosphere increase the diffuse fraction of solar radiation, which enhances plant production and may contribute to the residual terrestrial sink (Mercado et al. 2009).

Extreme events such as drought and heatwaves can increase tree mortality, decrease net primary production, and increase soil respiration, thereby reducing, or even offsetting, carbon sinks (van der Molen et al. 2011; Reichstein et al. 2013). A few extreme events occurring in just a small portion of the world control much of the interannual variability in the global carbon cycle (Zscheischler et al. 2014). One such notable drought that altered the carbon cycle occurred in Europe in 2003 (Ciais et al. 2005; Reichstein et al. 2007). A strong, widespread drought in 2005 in Amazonia increased tree mortality, decreased net primary production, and caused loss of carbon (Phillips et al. 2009). A second such drought occurred in 2010, with similar decline in vegetation productivity (Lewis et al. 2011). These events had a carbon impact of 1.6 Pg C and 2.2 Pg C, respectively, from drought-induced tree death. Numerous other examples of drought and heat stress tree mortality have occurred worldwide (Allen et al. 2010).

Other types of disturbances also impact the carbon cycle. Strong winds during hurricanes or storms cause extensive windthrows and loss of branches. This tree mortality and canopy damage can be seen in the carbon cycle (Zeng et al. 2009; Negrón-Juárez et al. 2010; Fisk et al. 2013). For example, Hurricane Katrina killed over 160 million trees along the Gulf Coast of the United States in 2005 (Negrón-Juárez et al. 2010). Large-scale mountain pine beetle outbreaks in western North America result in widespread tree mortality and reduced carbon sequestration in forest ecosystems (Kurz et al. 2008a; Hicke et al. 2012). Increased disturbance from wildfire and insect outbreaks alter the carbon balance of the boreal forest (Kurz et al. 2008b; Metsaranta et al. 2010). Climate anomalies during El Niño events increase the occurrence of fires and carbon emissions in tropical forests worldwide (Page et al. 2002; van der Werf et al. 2004, 2008).

29.5.3 Nitrogen Deposition

Industrial, automotive, and agricultural activities emit nitrogen to the atmosphere (NO_x + NH_3) that is subsequently deposited onto land and ocean through precipitation (wet deposition) and dry deposition (Figure 30.1). The present-day deposition of nitrogen (NO_y + NH_x) onto land is ~63 Tg N yr^{-1}, primarily from anthropogenic sources (Figure 2.12b; Lamarque et al. 2010, 2011, 2013; Ciais et al. 2013). Greatest deposition occurs in industrialized regions of eastern United States, Europe, India, and China (Galloway et al. 2004; Holland et al. 2005; Lamarque et al. 2011; Zhang et al. 2012). Nitrogen deposition can be in excess of 1 g N m^{-2} yr^{-1} in eastern United States and near 5 g N m^{-2} yr^{-1} in portions of Europe. In China, nitrogen deposition increased from an average of about 1.3 g N m^{-2} yr^{-1} in the 1980s to 2.1 g N m^{-2} yr^{-1} in the 2000s (Liu et al. 2013). Nitrogen deposition on land is expected to increase in the future (Lamarque et al. 2011, 2013).

Aboveground net primary production of many tundra, temperate forests and grasslands, and tropical forests and grasslands increases with nitrogen addition, and this is taken as evidence that extant nitrogen availability limits plant production (LeBauer and Treseder 2008). Consequently, nitrogen inputs from atmospheric deposition can increase plant growth and litter production, resulting in enhanced terrestrial carbon storage. This storage is realized primarily in forest ecosystems as accumulation of either wood in living trees or soil organic matter from increased litterfall. However, the increase in woody biomass is restricted because trees take up only a small portion of the added nitrogen. Estimates of the nitrogen-induced terrestrial carbon gain are highly variable, from 0.25 Pg C yr^{-1} in temperate forests (Nadelhoffer et al. 1999) to 1.5–2.0 Pg C yr^{-1}, mostly in temperate forests (Holland et al. 1997). More recent model based estimates are low, about 0.2–0.3 Pg C yr^{-1} (Thornton et al. 2007; Jain et al. 2009; Bonan and Levis 2010; Zaehle et al. 2010b; Zaehle 2013).

Estimates of the total (plant and soil) carbon gain in forests are highly variable. Reported values average 24.5 g C sequestered per g N (Liu and

Table 29.3 Contribution of nitrogen deposition to carbon storage in forest ecosystems

Carbon pool	N deposition to pool (%)	C:N	N deposition to pool (Tg yr^{-1})	C uptake (Pg yr^{-1})	dC/dN$_{dep}$
Non-woody plant biomass	15	25	0.77	0.019	4
Woody biomass	5	500	0.25	0.125	25
Forest floor & soil	70	30	3.57	0.107	21
Leaching & gaseous losses	10	0	0.51	0	0
Total	100	–	5.1	0.251	49

Source: From Nadelhoffer et al. (1999).

Greaver 2009) and 41 g C g^{-1} N (Butterbach-Bahl et al. 2011); or have a likely range of 20–40 g C g^{-1} N (de Vries et al. 2009), 20–70 g C g^{-1} N (Pinder et al. 2013), 30–70 g C g^{-1} N (de Vries et al. 2008), and 50-75 g C g^{-1} N (Sutton et al. 2008). Part of the discrepancy in values depends on how nitrogen deposition is calculated, especially the form and manner (wet and dry deposition) of the nitrogen input (Högberg 2007; Magnani et al. 2007; de Vries et al. 2008; Sutton et al. 2008).

The carbon gain depends on how the additional nitrogen is distributed within the forest ecosystem and the stoichiometry of vegetation and soil organic matter pools. The C:N ratio of wood is high (200–500 g C g^{-1} N) and supports large carbon gain in forests, if the nitrogen inputs are available for plant growth. In contrast, soil organic matter has a lower C:N ratio (10–30 g C g^{-1} N). Moreover, little nitrogen is available to support carbon sequestration if the nitrogen inputs are lost from the system by gaseous losses during denitrification or by aqueous exports. Table 29.3 provides an example calculation. In this case, the forest floor and soil pool receives 70 percent of the nitrogen deposition, but only about one-half of the total carbon gain occurs in this pool because of its low C:N ratio (30 g C g^{-1} N). Instead, wood accumulates the most carbon, despite receiving only 5 percent of the nitrogen input, because of its high C:N ratio (500 g C g^{-1} N).

The tree response to nitrogen enrichment is governed by increased foliar nitrogen (Xia and Wan 2008), which can stimulate photosynthetic rates, but also by decreased allocation of carbon belowground in roots (Xia and Wan 2008). These processes increase aboveground tree production. For example, Thomas et al. (2010) estimate aboveground tree production in forests of eastern United States has increased by 61 g C g^{-1} N deposition (subsequently revised to 50 g C g^{-1} N; Thomas et al. 2013). Butterbach-Bahl et al. (2011) reported an average response equal to 25 g C g^{-1} N, and de Vries et al. (2009) concluded that aboveground tree production in forests is generally within the range 15–40 g C g^{-1} N.

The soil carbon response to nitrogen enrichment depends on many factors including plant production, allocation, litterfall, litter chemistry, and microbial activity. Nitrogen addition generally enhances aboveground tree production and litterfall, but also decreases the allocation of carbon belowground resulting in less root litter. Nitrogen enrichment can improve litter quality by decreasing the lignin:N ratio of plant detritus and through greater labile carbon. However, nitrogen fertilization experiments show widely varying responses of litter decomposition to nitrogen addition. Rates of litter decomposition generally decrease with nitrogen addition (Fog 1988; Knorr et al. 2005; Janssens et al. 2010), but studies report increased mass loss, no response, or reduced decomposition in response to nitrogen enrichment depending on litter quality, the amount of enrichment, and stage of decomposition (Fog 1988; Knorr et al. 2005; Hobbie 2008; Janssens et al. 2010; Hobbie et al. 2012). In particular, nitrogen addition

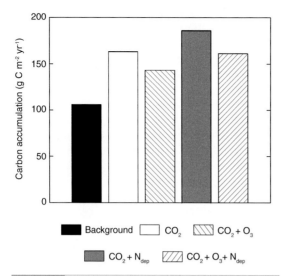

Fig. 29.9 Net carbon accumulation for the year 2000 simulated with the PnET-CN ecosystem model for forests in Northeast United States with CO_2, tropospheric ozone (O_3), and nitrogen deposition (N_{dep}) forcings. Adapted from Ollinger et al. (2002).

can increase the decomposition rate of labile litter (low lignin) and decrease the decomposition rate of low quality (high lignin) litter. Additionally, long-term increases in nitrogen can result in microbial community shifts (Frey et al. 2004; Allison and Martiny 2008). As a result of these factors, soil carbon response to nitrogen addition is less well-understood, but likely values for forests are in the range 5–35 g C g^{-1} N (de Vries et al. 2009) and average 15 g C g^{-1} N (Butterbach-Bahl et al. 2011).

29.5.4 Model Attribution Analyses

Model attribution studies simulate historical trends in the carbon cycle over the twentieth century and attribute those trends to particular forcings. The models are used in simulations forced by atmospheric CO_2 concentrations, climate, nitrogen deposition, and land use and land-cover change individually and in multi-factor combinations. These studies point to the importance of CO_2 fertilization to model historical trends in the carbon cycle.

Figure 29.9 shows carbon accumulation by forests in Northeast United States simulated by the PnET-CN ecosystem model (Ollinger et al.

2002). A simulation with historical land use, but keeping atmospheric forcings at preindustrial conditions, showed a background trend independent of atmospheric changes. Without any changes in atmospheric forcings, these forests accumulated 106 g C m^{-2} yr^{-1} at the end of the twentieth century as a result of past agricultural uses of the land. When forced with historical CO_2 concentrations, the rate of carbon accumulation increased by 54 percent above the background rate. Including the historical increase in tropospheric ozone (in addition to CO_2) reduced carbon accumulation by 12 percent. In contrast, the combination of CO_2 and nitrogen deposition increased carbon accumulation by 76 percent over the background rate and by 14 percent compared with CO_2 alone. The combination of CO_2, ozone, and nitrogen deposition increased carbon accumulation by 53 percent from the background rate. This is similar to the CO_2-only response, indicating the carbon gain from higher nitrogen deposition was offset by ozone.

An analysis with the same model but for temperate forests in the mid-Atlantic region of eastern United States found that changes in atmospheric composition over the twentieth century altered the regional carbon cycle, but climate change had lesser effect (Pan et al. 2009). Changes in atmospheric composition increased net primary production by 29 percent (14% CO_2; 17% nitrogen deposition; 6% interaction of CO_2 and nitrogen; –8% tropospheric ozone). Climate change over the same period increased productivity by 4 percent.

A similar analysis with the ORCHIDEE terrestrial biosphere model found that concurrent changes in atmospheric CO_2 and climate are the likely causes of Northern Hemisphere greening over the latter part of the twentieth century (Piao et al. 2006). Atmospheric CO_2, temperature, and precipitation account for 49 percent, 31 percent, and 13 percent, respectively, of the increase in Northern Hemisphere growing season leaf area index over the period 1980–2000, though their importance varies regionally. Greening in Siberia is associated with warming, as are trends throughout the boreal region, while trends in central North America are mostly explained by precipitation.

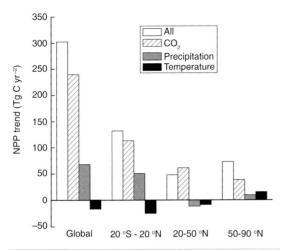

Fig. 29.10 Global and regional trends of annual net primary production (NPP) for 1980 to 2002 from ORCHIDEE simulations with all forcings and individually with only CO_2, precipitation, and temperature change. Adapted from Piao et al. (2009).

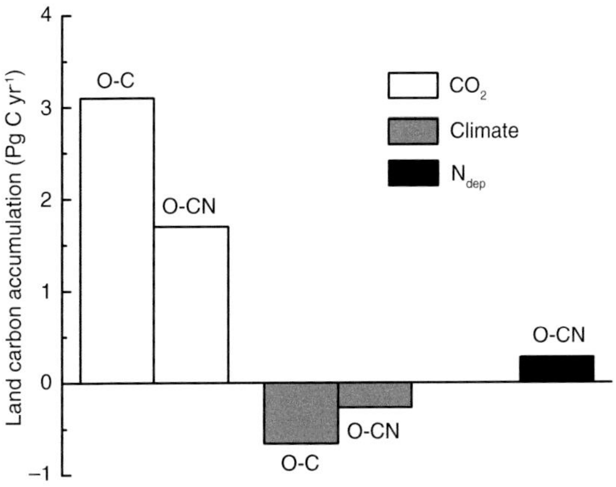

Fig. 29.11 Individual contribution of CO_2, climate change, and nitrogen deposition (N_{dep}) on net land carbon uptake in the 1990s simulated by the O-CN terrestrial biosphere model. Simulations are for carbon-only (O-C) and carbon–nitrogen (O-CN) implementations of the model. Adapted from Zaehle et al. (2010b).

A subsequent analysis with the ORCHIDEE model found similar results for net primary production (Piao et al. 2009). Global net primary production increased by 0.4 percent per year (about 300 Tg C yr^{-1} per year) over the period 1980 to 2002, mostly due to higher CO_2 (Figure 29.10). In the tropics (between 20° S and 20° N), CO_2 was the major determinant of the productivity trend, with nearly offsetting effects of precipitation and temperature changes. In northern temperate latitudes (20–50° N), climate change slightly offset the higher productivity from CO_2. Climate change enhanced the CO_2 fertilization response in boreal regions (50–90° N).

A large uncertainty is the extent to which nitrogen availability affects carbon cycle responses to CO_2 and climate change. Global terrestrial biosphere models simulate a smaller CO_2 fertilization response when nitrogen is included in the model compared with carbon-only simulations, and the carbon loss with climate change decreases because enhanced nitrogen mineralization with warmer soils stimulates plant production (Thornton et al. 2007; Jain et al. 2009; Bonan and Levis 2010; Zaehle et al. 2010b; Zhang et al. 2011; Gerber et al. 2013). For example, Figure 29.11 shows results of the O-CN terrestrial biosphere model run in carbon–nitrogen

and carbon-only implementations (Zaehle et al. 2010b). Increasing atmospheric CO_2 predominantly determines the terrestrial carbon balance in the 1990s, with minor effects from climate change and nitrogen deposition. The nitrogen cycle substantially reduces the CO_2 fertilization carbon gain and also reduces the climate-related carbon loss compared with the carbon-only simulation. While other models show similar results, there is substantial variability among models in these carbon–nitrogen interactions.

29.6 | Land-Use Emissions

Human uses of land such as forest clearing for timber and agriculture (deforestation), forest regrowth (reforestation), and the establishment of forests on non-forest land (afforestation) have an important role in the global carbon cycle as sources (deforestation) or sinks (reforestation and afforestation) of carbon. The term land use specifically refers to management practices within a particular land cover type (e.g., harvesting of wood products or selective logging in forests; fire suppression in forests; tillage and nutrient enrichment in croplands). Land-cover change refers to the conversion of one land cover type to another (e.g., deforestation that converts forest to cropland; reforestation following farm abandonment). Collectively, the sources and sinks of carbon that result from these activities are called the land use and land-cover change flux, or more generally the land-use flux (Houghton 2010).

The global land-use flux cannot be measured directly. Instead, various methodologies are used to infer fluxes. Uncertainty in these estimates arises from: different techniques to estimate the area affected by changes in land use and land cover; uncertainty in the carbon content of ecosystems and in the decay of carbon products and debris following land-cover change; and the types of management activities considered (Houghton et al. 2012). Land-use practices such as conversion to cropland and pasture, agricultural management, conservation tillage, abandonment of farmland, timber harvesting, and fire management influence the rate of carbon accumulation by terrestrial ecosystems. Moreover, the distinction between managed and unmanaged land inherent in the calculation of the land-use flux is itself imprecise, as is the separation of management effects from concurrent changes in atmospheric forcings (Gasser and Ciais 2013; Houghton 2013; Pongratz et al. 2014).

One method uses satellite estimates of changes in forest area and fires, typically for the tropics (DeFries et al. 2002; Achard et al. 2004; van der Werf et al. 2010). Alternatively, global terrestrial biosphere models calculate the land-use flux based on simulated vegetation and soil carbon in response to historical climate, atmospheric CO_2 concentrations, and other drivers of environmental change (Piao et al. 2009, 2013; Pongratz et al. 2009; Shevliakova et al. 2009). In these models, anthropogenic land-cover change is prescribed from spatially explicit datasets of land cover conversion over the industrial era (Ramankutty and Foley 1999; Klein Goldewijk 2001; Hurtt et al. 2006) and earlier in the past (Pongratz et al. 2008; Klein Goldewijk et al. 2011). A third approach to assess land-use emissions is by a bookkeeping method (Houghton et al. 1983, 2012; Houghton 2003). This combines inventory-based estimates of vegetation

and soil carbon stocks, changes in land use and land cover, and the rates at which these carbon stocks decrease or recover from land cover conversion such as deforestation or reforestation and changes in land use such as harvesting or agricultural management (e.g., Figure 23.15).

Figure 23.16 shows bookkeeping-based land-use flux estimates over the period 1850–2005. Between 1850 and 2005, changes in land use added 156 Pg C to the atmosphere. Tropical deforestation accounts for more than half of all the emission of carbon from land-use change since 1850 and almost all of the land-use emission of carbon during the 1990s. The historical legacy of farm abandonment, forest harvest, and reforestation drive the carbon balance in mid-latitude regions of North America, Europe, and Asia (Shevliakova et al. 2009; Williams et al. 2012).

The land-use flux for a particular year is reported as the net flux (the balance between sources and sinks) and reflects current and past disturbances (Houghton et al. 2012). The net flux includes instantaneous emissions that occur in the year of disturbance (e.g., enhanced decomposition, combustion during fire). It also includes historical legacy emissions from past land use and land-cover change, because woody debris left on site and wood products removed from site decay over long periods of time. These gross emissions are countered by reforestation and afforestation, which contribute to carbon sinks. Regrowing forests gain carbon over many years, and the net land-use flux includes this historical legacy sink from past disturbances. Gross land-use sources and sinks of carbon are about three times greater than the net land-use emission. The gross sink arises from the legacy of forest regrowth. Instantaneous and legacy effects contribute about equally to the gross sources.

29.7 | Coupled Carbon Cycle–Climate Models

The influence of the carbon cycle on climate can be understood in terms of two broadly defined feedbacks (Figure 29.12). In the absence of climate change, higher atmospheric CO_2

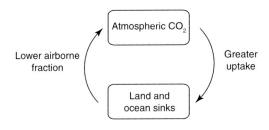

(a) Carbon–concentration feedback (–)

Lower airborne fraction

Atmospheric CO₂

Greater uptake

Land and ocean sinks

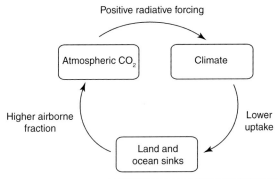

(b) Carbon–climate feedback (+)

Positive radiative forcing

Atmospheric CO₂

Climate

Higher airborne fraction

Lower uptake

Land and ocean sinks

Fig. 29.12 Schematic representation of (a) the carbon–concentration feedback and (b) the carbon–climate feedback.

concentrations drive an accumulation of carbon in the land and ocean (termed the carbon–concentration feedback). This is a negative feedback because the larger sinks reduce atmospheric CO_2. Higher atmospheric CO_2 is a positive radiative forcing that changes climate. Land and ocean carbon accumulation decrease, on land because of lower net primary production and increased heterotrophic respiration and in the ocean because of changes in ocean circulation, chemistry, and biology. This carbon–climate feedback is a positive feedback, because the decreased sinks increase atmospheric CO_2. In this framework, the total change in land carbon (ΔC_L, Pg C) is:

$$\Delta C_L = \beta_L \Delta C_A + \gamma_L \Delta T \qquad (29.1)$$

where β_L (Pg C ppm⁻¹) and γ_L (Pg C K⁻¹) are the sensitivities of land carbon storage to CO_2 and climate, respectively, and ΔC_A (ppm) and ΔT

(K) are the change in atmospheric CO_2 and planetary temperature, respectively (Friedlingstein et al. 2006). A similar relationship applies to ocean carbon, with ΔC_O, β_O and γ_O.

For example, Cramer et al. (2001) compared six global terrestrial biosphere models in simulations with increasing atmospheric CO_2 and climate change. The models were run for the period 1861 to 2100 with (a) increasing atmospheric CO_2 and a constant, preindustrial climate and (b) increasing CO_2 and climate change obtained from a global climate model. With CO_2-only forcing, the models predicted a terrestrial carbon sink of 1.4–3.8 Pg C yr⁻¹ in the 1990s (multi-model mean, 2.4 Pg C yr⁻¹) and 3.7–8.6 Pg C yr⁻¹ at the end of the twenty-first century (multi-model mean, 6.2 Pg C yr⁻¹) as a result of CO_2 fertilization. In the simulations that also included climate change, the terrestrial carbon sink decreased to 0.6–3.0 Pg C yr⁻¹ in the 1990s (multi-model mean, 1.6 Pg C yr⁻¹) and 0.3–6.6 Pg C yr⁻¹ by 2100 (multi-model mean, 3.4 Pg C yr⁻¹). Although CO_2 fertilization increases terrestrial carbon storage, climate change reduces the carbon sink because of changes in temperature and precipitation that alter the balance between net primary production and decomposition. This shows a positive feedback between climate and the carbon cycle in which climate change in response to increasing atmospheric CO_2 leads to reduced terrestrial carbon uptake and greater accumulation of carbon in the atmosphere. Friedlingstein et al. (2001) used a similar experimental protocol to deduce such a positive carbon–climate feedback.

To represent these feedbacks in climate change simulations, the scientific scope of global climate models has expanded from their original emphasis on physical processes to include the carbon cycle. Such models are generally referred to as Earth system models. The typical climate change study is a transient climate simulation from 1850 (preindustrial) to 2100 in which the model is forced with prescribed atmospheric CO_2 concentration and other forcings (Figure 8.16). Observed atmospheric CO_2 concentrations are used for the historical period and specified concentrations, derived from assumed trends in fossil fuel emissions and land use, are used for the twenty-first century. In Earth system models, the models are instead forced with carbon emissions from fossil fuel combustion and land use, and atmospheric CO_2 concentration is simulated as the balance of these emissions and ocean and terrestrial carbon fluxes.

The terrestrial biosphere responds directly to rising atmospheric CO_2 through photosynthetic enhancement, and this CO_2 fertilization is a negative feedback to higher atmospheric CO_2 concentration. Heterotrophic respiration increases with warmer soil temperature, which releases soil carbon to the atmosphere in a positive feedback. Climate change can enhance carbon uptake by net primary production (negative feedback) in cold climates where temperature increases or in dry climates where precipitation increases, and decrease net primary production (positive feedback) where soil moisture is limiting.

Models of the ocean carbon cycle are included to provide a full depiction of the global carbon cycle. Carbon dioxide dissolves in water as a function of its concentration in the air. As atmospheric CO_2 increases, the oceans respond by dissolving more CO_2. Carbon dioxide is more soluble in cold water than warm water so that the concentration of dissolved CO_2 decreases as temperature increases. Higher wind speeds increase the rate of air–sea exchange. In addition, phytoplankton and zooplankton remove CO_2 from the atmosphere. Carbon is buried in sediments as the organisms die and settle on the ocean floor. This biological pump is an important regulator of atmospheric CO_2. Differences in vertical mixing and oceanic circulation among models affect the rate at which carbon is transported from the surface to deeper waters.

The first coupled carbon cycle–climate simulation was that of Cox et al. (2000). Transient climate simulations covered the period 1860–2100 in which the model was forced with anthropogenic CO_2 emissions, and atmospheric CO_2 was calculated from these emissions and the simulated oceanic and terrestrial carbon fluxes. Two simulations isolated the carbon feedback on climate. In a biogeochemically coupled simulation, the carbon cycle responds to the increasing atmospheric CO_2 but the radiative forcing remains at preindustrial values ($CO_2 = 290$ ppm)

so that climate is unchanged (i.e., without carbon–climate feedback). The change in land carbon is $\Delta C_L = \beta_L \Delta C_A$. This is comparable to CO_2-only simulations in attribution studies. In a second biogeochemically and radiatively coupled simulation, the increasing atmospheric CO_2 additionally alters radiative forcings so that climate changes (i.e., with carbon–climate feedback). In this simulation, the change in land carbon is $\Delta C_L = \beta_L \Delta C_A + \gamma_L \Delta T$. This is comparable to CO_2 and climate attribution simulations. The difference in these two simulations is the effect of climate on the carbon cycle and quantifies the carbon–climate feedback.

By 2100, the simulated atmospheric CO_2 concentration is about 980 ppm in the simulation with climate change, about 250 ppm more than in the simulation without climate change. That is, there is a positive feedback between the carbon cycle and climate that acts to increase the portion of anthropogenic CO_2 emissions that remains in the atmosphere and amplifies the climate change. Decreased terrestrial carbon uptake contributes to the strong positive carbon–climate feedback (Figure 29.13). Without climate change, terrestrial ecosystems accumulate carbon over the simulation. With climate change, terrestrial ecosystems become a source of carbon after the middle of the twenty-first century. This is due to widespread decline in soil carbon as a result of warmer temperatures and to dieback of Amazonian rainforests as a result of a warmer and drier climate.

Friedlingstein et al. (2006) compared the carbon cycle feedback in 11 models of varying complexity for the Intergovernmental Panel on Climate Change (IPCC) fourth assessment report. Oceanic carbon uptake increases in the twenty-first century in all models, with a range of 4–10 Pg C yr^{-1} at 2100. The terrestrial carbon flux is much more variable among models, ranging from a 6 Pg C yr^{-1} source to a 11 Pg C yr^{-1} sink at 2100. All models have a positive carbon–climate feedback that amplifies global warming, but the magnitude of the feedback, quantified by the additional atmospheric CO_2, ranges from 20 ppm to 200 ppm. Much of this uncertainty arises from the biosphere (Figure 29.14). The carbon-concentration

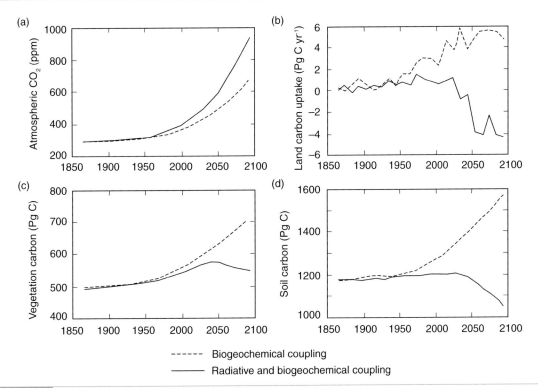

Fig. 29.13 Simulated (a) atmospheric CO_2 concentration, (b) global land carbon uptake, (c) global vegetation carbon, and (d) global soil carbon for 1860–2100 using the Hadley Centre HadCM3LC model with the TRIFFID dynamic global vegetation model. Shown are simulations with biogeochemical coupling (without carbon–climate feedback) and with radiative and biogeochemical coupling (with carbon–climate feedback). Adapted from Cox et al. (2004).

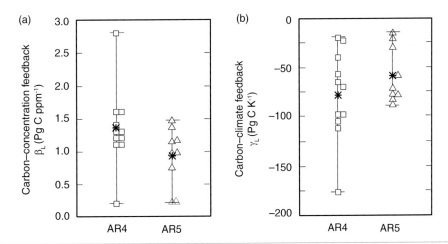

Fig. 29.14 Land carbon–concentration (a) and carbon–climate (b) feedback parameters for 11 models used in the Intergovernmental Panel on Climate Change (IPCC) fourth assessment report (AR4) (Friedlingstein et al. 2006) and 9 models used in the fifth assessment report (AR5) (Arora et al. 2013). Shown are the individual models (symbols) and the multi-model mean (star).

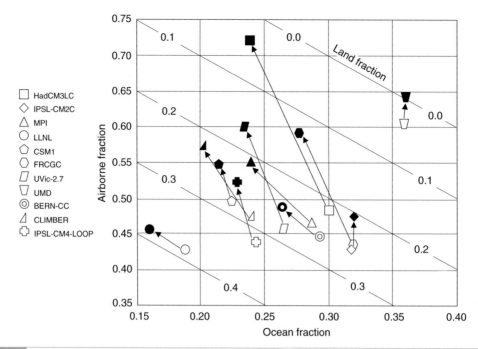

Fig. 29.15 Fraction of cumulative anthropogenic CO_2 emission in air, ocean, and land up to 2000 (open symbols) and to 2100 (closed symbols) for eleven Earth system models. Adapted from Denman et al. (2007).

feedback (β_L) over land ranges from 0.2 to 2.8 Pg C ppm^{-1} (multi-model mean, 1.4 Pg C ppm^{-1}). That is, across all models, the terrestrial biosphere gains 1.4 Pg C per ppm increase in atmospheric CO_2. The carbon–climate feedback (γ_L) over land ranges from –20 to –177 Pg C K^{-1} (multi-model mean, –79 Pg C K^{-1}). The terrestrial biosphere loses 79 Pg C per degree increase in planetary temperature.

Change in the airborne fraction (the fraction of anthropogenic emissions that remain in the atmosphere) is a measure of whether or not the global sinks of carbon on land and ocean continue to increase in proportion to emissions. All models show that the capacity of the land and ocean to store anthropogenic CO_2 emissions decreases over the twenty-first century, providing a positive feedback whereby warming further increases atmospheric CO_2 concentration (Figure 29.15). Whether such trends can be detected in the carbon cycle is a subject of considerable research (Ballantyne et al. 2012; Le Quéré et al. 2013; Lombardozzi et al. 2014).

Carbon cycle feedbacks vary regionally. For example, the evaporative demand for soil water increases with warmer climates and can exceed available soil moisture even with an increase in rainfall. Net primary production may decrease with climate warming in warm regions where greater evaporative demand dries soil. Net primary production in high latitudes may increase with warming because of more favorable temperatures.

This understanding of the carbon cycle is largely unchanged for the IPCC fifth assessment report (Figure 29.14). For this class of models and simulations, β_L is 0.2–1.5 Pg C ppm^{-1} (multi-model mean, 0.9 Pg C ppm^{-1}) and γ_L is –16 to –89 Pg C K^{-1} (multi-model mean, –58 Pg C K^{-1}). The multi-model mean β_L is about one-third weaker and the mean γ_L is about one-quarter weaker for the fifth assessment report (AR5) compared with the fourth assessment report (AR4). The methodology used to derive the feedback parameters differs between the two reports, so they are not directly comparable. The AR5 shows reduced spread across models compared with AR4, largely due to one AR4 model with high β_L and another model with high negative γ_L. In AR4, β_L varies across models by a factor of 14 and γ_L by a factor of 9. In AR5, β_L varies by a factor of 7 and γ_L by a factor of 6. Nonetheless, there is still large uncertainty in the response of terrestrial biosphere models to rising CO_2 concentrations and climate change.

Similar carbon cycle feedbacks have occurred over the historical era. Rising atmospheric CO_2 since the preindustrial era is widely accepted to have lead to carbon accumulation in the terrestrial biosphere. This enhanced vegetation growth from CO_2 fertilization may have reduced atmospheric CO_2 by 85 ppm and avoided 0.3°C in planetary warming since the preindustrial era (Shevliakova et al. 2013).

The overall carbon cycle–climate coupling can be understood in terms of five metrics (Friedlingstein et al. 2006). The first two are the carbon–concentration feedback parameters for land and ocean (β_L and β_O). Both land and ocean sinks increase with higher atmospheric CO_2. Increases in land and ocean carbon uptake in response to rising atmospheric CO_2 are a negative feedback that reduce atmospheric CO_2 (Figure 29.12a). Two other metrics are the carbon–climate feedback parameters for land and ocean (γ_L and γ_O), which characterize a positive feedback with CO_2-induced climate change (Figure 29.12b). Greater heterotrophic respiration with warming increases atmospheric CO_2, as does decreased net primary production. Climate change affects oceanic carbon uptake by altering the solubility of CO_2 and ocean circulation. The final aspect of the carbon cycle–climate system is the sensitivity of the simulated climate to atmospheric CO_2. High climate sensitivity to CO_2 means large warming for a given increase in atmospheric CO_2. This, in turn, affects the land and ocean carbon fluxes. In general, greater climate sensitivity increases the carbon–climate feedback.

There is considerable spread among models in their simulated terrestrial carbon cycle in response to twenty-first century forcings, and much of the uncertainty in carbon cycle–climate feedbacks is thought to arise from terrestrial processes (Friedlingstein et al. 2006,

2014; Jones et al. 2013; Friend et al. 2014). The uncertainty in carbon cycle feedbacks is of comparable magnitude to the uncertainty arising from physical climate processes (Huntingford et al. 2009). Multi-member ensembles that perturb key model parameter values show that a wide range of carbon cycle responses can be produced from the same model structure given a plausible range in parameter values (Booth et al. 2012, 2013; Lambert et al. 2013). Much of the variability among models in their simulation of atmospheric CO_2 over the twenty-first century relates to biases in their simulation of CO_2 concentrations over the observational era (Hoffman et al. 2014). Constraining the models to match the observed time series of atmospheric CO_2 (e.g., at Mauna Loa) may provide one path to reduce uncertainties in model simulations of the twenty-first century.

One of the important uncertainties relates to the nitrogen cycle. Global terrestrial biosphere models that include carbon–nitrogen biogeochemistry find that nitrogen availability limits the capacity of many ecosystems to sequester anthropogenic carbon emissions, thereby accelerating the accumulation of anthropogenic CO_2 in the atmosphere (Sokolov et al. 2008; Jain et al. 2009; Thornton et al. 2009; Zaehle et al. 2010a; Zhang et al. 2011; Gerber et al. 2013; Zaehle 2013). This occurs because of a smaller CO_2 fertilization response, but countered by greater nitrogen mineralization with warmer soils that stimulates carbon gain from plant production. For example, Figure 29.16 shows results of the O-CN terrestrial biosphere model run in carbon–nitrogen and carbon-only implementations from 1860–2100 (Zaehle et al. 2010a). The CO_2 fertilization carbon gain is substantially smaller with nitrogen and the loss of carbon from climate change is somewhat smaller.

The metrics by which simulations of the carbon cycle–climate system are evaluated is a key research topic (Randerson et al. 2009; Luo et al. 2012). Some common measures are vegetation carbon, soil carbon, plant productivity, and carbon turnover (Anav et al. 2013). Interannual variability in the carbon cycle is another means to assess the models (Cox et al. 2013; Piao et al. 2013). For example, Cox et al. (2013) correlated

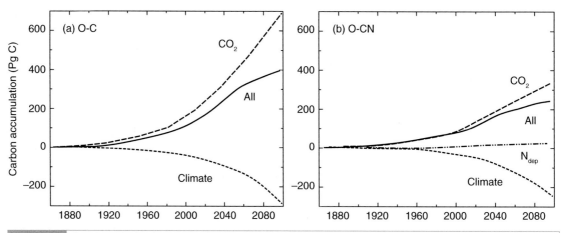

Fig. 29.16 Cumulative land carbon storage (1860–2100) in response to CO_2, climate change, nitrogen deposition (N_{dep}), and all forcings simulated by the O-CN terrestrial biosphere model. Simulations are for (a) carbon-only (O-C) and (b) carbon–nitrogen (O-CN) implementations of the model. Adapted from Zaehle et al. (2010a).

interannual anomalies in the atmospheric CO_2 growth rate (with units Pg C yr^{-1}) and anomalies in tropical temperature (with units K). This provides a measure of the observed sensitivity of the carbon cycle to temperature. They used this relationship as a constraint on γ_L simulated by models. In the models, they found that γ_{LT} (for tropical land) correlates closely with the modeled interannual sensitivity of the atmospheric CO_2 growth rate to tropical temperature anomalies (γ_{IAV}, Pg C yr^{-1} K^{-1}). They used the observed interannual sensitivity of CO_2 to temperature anomalies to reduce the uncertainty across models in γ_{LT}. They estimated that tropical land releases 53 Pg C per kelvin compared with the unconstrained model estimate of 69 Pg C K^{-1} ($\gamma_{LT} = -53$ Pg C K^{-1} compared with –69 Pg C K^{-1}). Moreover, the estimated probability of γ_{LT} more negative than –100 Pg C K^{-1} decreased from 21 percent to 0.2 percent. Similar results are found in other Earth system models (Wenzel et al. 2014), suggesting that the short-term sensitivity of the carbon cycle to temperature anomalies may be a key constraint on the long-term sensitivity of tropical carbon storage to warming.

29.8 | Compatible CO_2 Emissions

Stabilization of atmospheric CO_2 at some specified concentration in the future is a key aspect of climate policy. The amount of CO_2 in the atmosphere is the balance between anthropogenic emissions from fossil fuel combustion and land use and uptake by the terrestrial biosphere and ocean. The allowable anthropogenic CO_2 emissions to achieve a desired atmospheric CO_2 concentration can be obtained from Earth system model simulations with an interactive carbon cycle (Ciais et al. 2013; Jones et al. 2013). In emissions-driven simulations, fossil fuel emissions are prescribed input to the model, and atmospheric CO_2 (C_A) is calculated as the outcome of these fossil fuel emissions (E_{FF}), emissions from land use (E_{LU}), the flux between the atmosphere and land (F_{AL}), and the atmosphere–ocean flux (F_{AO}) calculated by the model. The resulting change in atmospheric CO_2 is:

$$dC_A/dt = E_{FF} + E_{LU} - F_{AL} - F_{AO} \qquad (29.2)$$

Earth system models can also be used in concentration-driven simulations. In these simulations, atmospheric CO_2 concentration is prescribed as input to the model. The land and ocean respond to the atmospheric CO_2 and climate change, but do not feed back to alter the CO_2 concentration. In this simulation, C_A is known, and the total anthropogenic emission allowed to obtain that C_A is obtained from F_{AL} and F_{AO} by:

$$E_{FF} + E_{LU} = dC_A/dt + F_{AL} + F_{AO} \qquad (29.3)$$

The compatible fossil fuel emissions vary among atmospheric CO_2 scenarios, as given by the various representative concentration pathways (RCPs) (Figure 2.11b), and also the resulting climate change (as it affects F_{AL} and F_{AO}). Compatible fossil fuel emissions estimates for the period 2012–2100 from various Earth system model simulations are (Ciais et al. 2013): RCP 2.6, multi-model mean 270 Pg C (multi-model range, 140–410); RCP4.5, 780 Pg C (595–1005); RCP6.0, 1060 Pg C (840–1250); and RCP8.5, 1685 Pg C (1415–1910).

The compatible emissions are greater in the RCPs that allow higher CO_2 concentration, but they also depend on the land and ocean carbon fluxes and carbon cycle feedbacks. One such feedback is the positive feedback whereby warming decreases land carbon uptake. This reduces the ability of the terrestrial biosphere to store anthropogenic carbon emissions and necessitates a reduction in anthropogenic CO_2 emissions to achieve targeted atmospheric CO_2 concentrations (Jones et al. 2006; Matthews 2005, 2006). The carbon–concentration feedback is also important because this increases the terrestrial carbon sink and allows high anthropogenic emissions. Reduced CO_2 fertilization from nitrogen or phosphorus limitation of plant productivity necessitates lower anthropogenic emissions to achieve stabilization goals.

29.9 | The Carbon Cycle and Global Change

Carbon cycle feedbacks with warming and elevated CO_2 are exceedingly complex and many processes that mediate these feedbacks are missing or poorly represented in models. The positive carbon–climate feedback is understood by the sensitivity of photosynthesis and respiration to temperature, while neglecting more complex ecosystem responses to warming such as longer growing seasons, increased nutrient availability, altered plant and microbial community composition, and changes in water usage (Luo 2007). The negative carbon–concentration feedback is based on simple conceptualizations

of photosynthetic response to elevated CO_2; the more complex response of plant carbon allocation to CO_2 is less well-known (Norby and Zak 2011).

Different parameterizations of carbon–nitrogen biogeochemistry result in divergent simulations of carbon cycle responses to global change. Models differ in the magnitude of carbon–nitrogen interactions depending on how nitrogen availability restricts photosynthesis; plant nitrogen uptake; plant and soil C:N stoichiometry, especially whether C:N ratios for various pools are constant or vary in response to nitrogen availability; nitrogen gas losses; nitrogen fixation; and competition between plants and microbes for mineral nitrogen (Zaehle and Dalmonech 2011). Phosphorus can also limit plant productivity and its response to elevated CO_2, and this is a key research frontier in Earth system models (Vitousek et al. 2010; Zhang et al. 2011, 2014; Goll et al. 2012; Yang et al. 2014).

Permafrost soils contain 1672 Pg of organic carbon (Chapter 24). As Arctic temperatures increase, this permafrost will degrade, and the soil carbon is vulnerable to decomposition and release to the atmosphere (Schuur et al. 2008, 2013). However, the magnitude of this permafrost carbon feedback is unclear because of uncertainty in simulation of soil carbon response to thawing and in the simulation of permafrost geography and its vulnerability to warming (Lawrence and Slater 2005; Koven et al. 2011, 2013; Burke et al. 2013; Slater and Lawrence 2013).

Plant physiological responses to warmer temperature or drought stress are poorly represented in models. Most models incorporate the instantaneous enzymatic response to temperature, but do not account for photosynthetic or respiratory temperature acclimation (Smith and Dukes 2013). Physiological mechanisms controlling drought-induced forest mortality include hydraulic failure of roots and branches and carbon starvation due to depletion of carbohydrate reserves (McDowell et al. 2008, 2013; McDowell and Sevanto 2010; Anderegg et al. 2012; Choat et al. 2012). Hydraulic failure occurs when the water column in xylem conduits breaks (i.e., becomes air-filled, also referred to as cavitation) because of high evaporative demand and low soil water supply. Cavitation impairs water flow and leads to tissue dehydration. Carbon starvation occurs when stomata close to prevent water stress, thereby decreasing photosynthetic carbon uptake. Continued metabolic demand for carbohydrates depletes reserves.

Demographic processes that control community composition are absent in biogeochemical models and difficult to parameterize in dynamic global vegetation models. The next generation of dynamic global vegetation models incorporates theoretical advances in plant demography, ecosystem dynamics, and community organization (Fisher et al. 2010; Scheiter et al. 2013). How to represent plant migration as climate changes remains problematic (Higgins and Harte 2006; Levis 2010).

Land use is expected to change in the future (Figure 2.12). Possible socioeconomic scenarios for the twenty-first century range from substantial global deforestation with expansion of agriculture to regional reforestation with farm abandonment. This future land use and land-cover change affects the carbon cycle and climate (Brovkin et al. 2013).

Earth system models that couple the carbon cycle and climate are abstractions of complex physical, chemical, biological, and socioeconomic processes that influence climate. Many ecological processes are represented in detail, but there is a high level of uncertainty in these processes. It will be many years, if ever, before the full complexity of nature and all the possible feedbacks are included in Earth system models. However, the picture emerging now is one of terrestrial ecosystems as an important regulator of the carbon cycle, and through this climate.

29.10 | Review Questions

1. Figure 3.6b shows one depiction of the terrestrial carbon cycle. Terrestrial ecosystems store 2.6 Pg C yr^{-1} in the absence of land use. This small carbon gain is the difference between large carbon uptake during gross primary production (123 Pg C yr^{-1}) and large carbon losses from respiration, fire, and export to rivers (120.4 Pg C yr^{-1}). What is the effect of +10 percent uncertainty in gross primary production on carbon storage on land?

2. Net primary production can be represented mathematically by the equation $NPP = 3000 / [1 + \exp(1.315 - 0.119T)]$ (Figure 24.4). Soil respiration can be represented by $R_H = aQ_{10}^{T/10}$ where $Q_{10} = 2$. For $T = 13.8°C$, calculate the relative change in NPP and R_H with a 2°C increase in temperature. Which is more sensitive to a temperature increase? Perform similar calculations for $T = -2.8°C$.

3. In a particular year, sea surface temperatures across the eastern equatorial Pacific are abnormally warm. How does this affect atmospheric CO_2? How does eruption of a large volcano affect this response?

4. Present-day global net primary production (NPP) is estimated to be 60 Pg C yr^{-1}. Experimental FACE studies indicate about a 20 percent increase in NPP on average with an increase in CO_2 concentration to about 550 ppm. If global heterotrophic (soil) respiration is 50 Pg C yr^{-1}, what is the maximum increase in net ecosystem production with an increase in CO_2 concentration to 550 ppm? How will heterotrophic respiration respond?

5. Calculate the carbon flux to the atmosphere in the first year following deforestation in the following example: carbon in vegetation of undisturbed forest, 13,500 g m^{-2}; carbon in live vegetation left on site at time of harvest, 0 g m^{-2}; carbon in wood harvested, 3000 g m^{-2}; carbon in dead plant material left in soil at time of harvest, 10,500 g m^{-2}; carbon in soils of undisturbed forests, 13,400 g m^{-2}. Assume 4 percent of the harvested wood decays in one year; 38 percent decays in 10 years; 58 percent decays in 100 years; and that this decay is linear over time. In calculating soil carbon loss, assume the minimum carbon content in soil following harvest is 6700 g m^{-2} and that this minimum is achieved 10 years following harvest (decay is linear over time).

6. For the forest in question 5, annual net primary production of undisturbed forest is 1000 g m^{-2} yr^{-1} and decomposition is 700 g m^{-2} yr^{-1}. Calculate net biome production assuming 10 percent of the forest is harvested. How does this compare with an undisturbed landscape?

7. Table 24.1 shows the amount of carbon stored in plant and soil biomass for various biomes. Use the WBGU data to calculate the change in terrestrial carbon storage if the global area of tundra is reduced by one-half and boreal forest expands to cover this area.

8. About 2000 Pg C is expected to be released to the atmosphere between 1860 and 2100 from anthropogenic activity. In a carbon cycle–climate model, 520 Pg of the anthropogenic carbon is stored on land and 420 Pg in oceans. What is the airborne fraction? How does CO_2 fertilization affect the airborne fraction? How does warmer soil affect the airborne fraction?

9. Two climate models differ in climate sensitivity to a doubling of atmospheric CO_2. One predicts a 2.0°C increase in annual mean global temperature; the other, 2.5°C. The same terrestrial and ocean carbon models are coupled to the two climate models. Which carbon cycle–climate model likely has a larger carbon–climate feedback?

10. Two terrestrial carbon models differ in the sensitivity of heterotrophic respiration to temperature. One predicts a doubling of respiration with a 10°C increase in temperature ($Q_{10} = 2.0$); the other, a 50 percent increase ($Q_{10} = 1.5$). Which model has a larger carbon–climate feedback, all other factors equal?

11. Is the climate in a biogeochemically coupled Earth system model simulation (without carbon–climate feedback) constant?

12. In a biogeochemically coupled simulation, the carbon cycle responds to the increasing atmospheric CO_2 but the radiative forcing remains at preindustrial values. In a biogeochemically and radiatively coupled simulation, CO_2 and climate both change. Describe how to perform the same set of simulations using a terrestrial biosphere model with a nitrogen cycle and forced with nitrogen deposition. What happens if nitrogen deposition is held constant at preindustrial values?

29.11 | References

Achard, F., Eva, H. D., Mayaux, P., Stibig, H.-J., and Belward, A. (2004). Improved estimates of net carbon emissions from land cover change in the tropics for the 1990s. *Global Biogeochemical Cycles*, 18, GB2008, doi:10.1029/2003GB002142.

Ainsworth, E. A., Yendrek, C. R., Sitch, S., Collins, W. J., and Emberson, L. D. (2012). The effects of tropospheric ozone on net primary productivity and implications for climate change. *Annual Review of Plant Biology*, 63, 637–661.

Allen, C. D., Macalady, A. K., Chenchouni, H., et al. (2010). A global overview of drought and heat-induced tree mortality reveals emerging climate change risks for forests. *Forest Ecology and Management*, 259, 660–684.

Allison, S. D., and Martiny, J. B. H. (2008). Resistance, resilience, and redundancy in microbial communities. *Proceedings of the National Academy of Sciences USA*, 105, 11512–11519.

Anav, A., Friedlingstein, P., Kidston, M., et al. (2013). Evaluating the land and ocean components of the global carbon cycle in the CMIP5 Earth system models. *Journal of Climate*, 26, 6801–6843.

Anderegg, W. R. L., Berry, J. A., Smith, D. D., et al. (2012). The roles of hydraulic and carbon stress in a widespread climate-induced forest die-off. *Proceedings of the National Academy of Sciences USA*, 109, 233–237.

Arora, V. K., Boer, G. J., Friedlingstein, P., et al. (2013). Carbon-concentration and carbon–climate feedbacks in CMIP5 Earth system models. *Journal of Climate*, 26, 5289–5314.

Ballantyne, A. P., Alden, C. B., Miller, J. B., Tans, P. P., and White, J. W. C. (2012). Increase in observed net carbon dioxide uptake by land and oceans during the past 50 years. *Nature*, 488, 70–72.

Barichivich, J., Briffa, K. R., Myneni, R. B., et al. (2013). Large-scale variations in the vegetation growing season and annual cycle of atmospheric CO_2 at high northern latitudes from 1950 to 2011. *Global Change Biology*, 19, 3167–3183.

Beer, C., Reichstein, M., Tomelleri, E., et al. (2010). Terrestrial gross carbon dioxide uptake: Global distribution and covariation with climate. *Science*, 329, 834–838.

Betzelberger, A. M., Yendrek, C. R., Sun, J., et al. (2012). Ozone exposure response for U.S. soybean cultivars: Linear reductions in photosynthetic potential, biomass, and yield. *Plant Physiology*, 160, 1827–1839.

Boisvenue, C., and Running, S. W. (2006). Impacts of climate change on natural forest productivity – evidence since the middle of the 20th century. *Global Change Biology*, 12, 862–882.

Bonan, G. B., and Levis, S. (2010). Quantifying carbon-nitrogen feedbacks in the Community

Land Model (CLM4). *Geophysical Research Letters*, 37, L07401, doi:10.1029/2010GL042430.

Booth, B. B. B., Jones, C. D., Collins, M., et al. (2012). High sensitivity of future global warming to land carbon cycle processes. *Environmental Research Letters*, 7, 024002, doi:10.1088/1748-9326/7/2/024002.

Booth, B. B. B., Bernie, D., McNeall, D., et al. (2013). Scenario and modelling uncertainty in global mean temperature change derived from emission-driven global climate models. *Earth System Dynamics*, 4, 95–108.

Brovkin, V., Lorenz, S. J., Jungclaus, J., et al. (2010). Sensitivity of a coupled climate–carbon cycle model to large volcanic eruptions during the last millennium. *Tellus B*, 62, 674–681.

Brovkin, V., Boysen, L., Arora, V. K., et al. (2013). Effect of anthropogenic land-use and land-cover changes on climate and land carbon storage in CMIP5 projections for the twenty-first century. *Journal of Climate*, 26, 6859–6881.

Buermann, W., Anderson, B., Tucker, C. J., et al. (2003). Interannual covariability in Northern Hemisphere air temperatures and greenness associated with El Niño–Southern Oscillation and the Arctic Oscillation. *Journal of Geophysical Research*, 108, 4396, doi:10.1029/2002JD002630.

Buermann, W., Lintner, B. R., Koven, C. D., et al. (2007). The changing carbon cycle at Mauna Loa Observatory. *Proceedings of the National Academy of Sciences USA*, 104, 4249–4254.

Burke, E. J., Jones, C. D., and Koven, C. D. (2013). Estimating the permafrost-carbon climate response in the CMIP5 climate models using a simplified approach. *Journal of Climate*, 26, 4897–4909.

Butterbach-Bahl, K., Nemitz, E., Zaehle, S., et al. (2011). Nitrogen as a threat to the European greenhouse balance. In *The European Nitrogen Assessment*, ed. M. A. Sutton, C. M. Howard, J. W. Erisman, et al. Cambridge: Cambridge University Press, pp. 434–462.

Choat, B., Jansen, S., Brodribb, T. J., et al. (2012). Global convergence in the vulnerability of forests to drought. *Nature*, 491, 752–755.

Ciais, P., Reichstein, M., Viovy, N., et al. (2005). Europe-wide reduction in primary productivity caused by the heat and drought in 2003. *Nature*, 437, 529–533.

Ciais, P., Rayner, P., Chevallier, F., et al. (2010). Atmospheric inversions for estimating CO_2 fluxes: Methods and perspectives. *Climatic Change*, 103, 69–92.

Ciais, P., Tagliabue, A., Cuntz, M., et al. (2012). Large inert carbon pool in the terrestrial biosphere during the Last Glacial Maximum. *Nature Geoscience*, 5, 74–79.

Ciais, P., Sabine, C., Bala, G., et al. (2013). Carbon and other biogeochemical cycles. In *Climate Change 2013: The Physical Science Basis. Contribution of Working Group I to the Fifth Assessment Report of the Intergovernmental Panel on Climate Change*, ed. T. F. Stocker, D. Qin, G.-K. Plattner, et al. Cambridge: Cambridge University Press, pp. 465–570.

Cox, P. M., Betts, R. A., Jones, C. D., Spall, S. A., and Totterdell, I. J. (2000). Acceleration of global warming due to carbon-cycle feedbacks in a coupled climate model. *Nature*, 408, 184–187.

Cox, P. M., Betts, R. A., Collins, M., et al. (2004). Amazonian forest dieback under climate–carbon cycle projections for the 21st century. *Theoretical and Applied Climatology*, 78, 137–156.

Cox, P. M., Pearson, D., Booth, B. B., et al. (2013). Sensitivity of tropical carbon to climate change constrained by carbon dioxide variability. *Nature*, 494, 341–344.

Cramer, W., Bondeau, A., Woodward, F. I., et al. (2001). Global response of terrestrial ecosystem structure and function to CO_2 and climate change: Results from six dynamic global vegetation models. *Global Change Biology*, 7, 357–373.

DeFries, R. S., Houghton, R. A., Hansen, M. C., et al. (2002). Carbon emissions from tropical deforestation and regrowth based on satellite observations for the 1980s and 1990s. *Proceedings of the National Academy of Sciences USA*, 99, 14256–14261.

Denman, K. L., Brasseur, G., Chidthaisong, A., et al. (2007). Couplings between changes in the climate system and biogeochemistry. In *Climate Change 2007: The Physical Science Basis. Contribution of Working Group I to the Fourth Assessment Report of the Intergovernmental Panel on Climate Change*, ed. S. Solomon, D. Qin, M. Manning, et al. Cambridge: Cambridge University Press, 499–587.

de Vries, W., Solberg, S., Dobbertin, M., et al. (2008). Ecologically implausible carbon response? *Nature*, 451, E1–E3.

de Vries, W., Solberg, S., Dobbertin, M., et al. (2009). The impact of nitrogen deposition on carbon sequestration by European forests and heathlands. *Forest Ecology and Management*, 258, 1814–1823.

Donohue, R. J., Roderick, M. L., McVicar, T. R., and Farquhar, G. D. (2013). Impact of CO_2 fertilization on maximum foliage cover across the globe's warm, arid environments. *Geophysical Research Letters*, 40, 3031–3035, doi:10.1002/grl.50563.

Dragoni, D., Schmid, H. P., Wayson, C. A., et al. (2011). Evidence of increased net ecosystem productivity associated with a longer vegetated season in a deciduous forest in south-central Indiana, USA. *Global Change Biology*, 17, 886–897.

Felzer, B., Reilly, J., Melillo, J., et al. (2005). Future effects of ozone on carbon sequestration and climate change policy using a global biogeochemical model. *Climatic Change*, 73, 345–373.

Fisher, R., McDowell, N., Purves, D., et al. (2010). Assessing uncertainties in a second-generation dynamic vegetation model caused by ecological scale limitations. *New Phytologist*, 187, 666–681.

Fisk, J. P., Hurtt, G. C., Chambers, J. Q., et al. (2013). The impacts of tropical cyclones on the net carbon balance of eastern US forests (1851–2000). *Environmental Research Letters*, 8, 045017, doi:10.1088/1748-9326/8/4/045017.

Flückiger, J., Monnin, E., Stauffer, B., et al. (2002). High-resolution Holocene N_2O ice core record and its relationship with CH_4 and CO_2. *Global Biogeochemical Cycles*, 16, 1010, doi:10.1029/2001GB001417.

Fog, K. (1988). The effect of added nitrogen on the rate of decomposition of organic matter. *Biological Reviews*, 63, 433–462.

Frey, S. D., Knorr, M., Parrent, J. L., and Simpson, R. T. (2004). Chronic nitrogen enrichment affects the structure and function of the soil microbial community in temperate hardwood and pine forests. *Forest Ecology and Management*, 196, 159–171.

Friedlingstein, P., Bopp, L., Ciais, P., et al. (2001). Positive feedback between future climate change and the carbon cycle. *Geophysical Research Letters*, 28, 1543–1546.

Friedlingstein, P., Cox, P., Betts, R., et al. (2006). Climate–carbon cycle feedback analysis: Results from the C^{4}MIP model intercomparison. *Journal of Climate*, 19, 3337–3353.

Friedlingstein, P., Meinshausen, M., Arora, V. K., et al. (2014). Uncertainties in CMIP5 climate projections due to carbon cycle feedbacks. *Journal of Climate*, 27, 511–526.

Friend, A. D., Lucht, W., Rademacher, T. T., et al. (2014). Carbon residence time dominates uncertainty in terrestrial vegetation responses to future climate and atmospheric CO_2. *Proceedings of the National Academy of Sciences USA*, 111, 3280–3285.

Frölicher, T. L., Joos, F., Raible, C. C., and Sarmiento, J. L. (2013). Atmospheric CO_2 response to volcanic eruptions: The role of ENSO, season, and variability. *Global Biogeochemical Cycles*, 27, 239–251, doi:10.1002/gbc.20028.

Galloway, J. N., Dentener, F. J., Capone, D. G., et al. (2004). Nitrogen cycles: Past, present, and future. *Biogeochemistry*, 70, 153–226.

Gasser, T., and Ciais, P. (2013). A theoretical framework for the net land-to-atmosphere CO_2 flux and its implications in the definition of "emissions from land-use change". *Earth System Dynamics*, 4, 171–186.

Gerber, S., Hedin, L. O., Keel, S. G., Pacala, S. W., and Shevliakova, E. (2013). Land use change and nitrogen feedbacks constrain the trajectory of the land carbon sink. *Geophysical Research Letters*, 40, 5218–5222, doi:10.1002/grl.50957.

Goll, D. S., Brovkin, V., Parida, B. R., et al. (2012). Nutrient limitation reduces land carbon uptake in simulations with a model of combined carbon, nitrogen and phosphorus cycling. *Biogeosciences*, 9, 3547–3569.

Graven, H. D., Keeling, R. F., Piper, S. C., et al. (2013). Enhanced seasonal exchange of CO_2 by northern ecosystems since 1960. *Science*, 341, 1085–1089.

Gray, J. M., Frolking, S., Kort, E. A., et al. (2014). Direct human influence on atmospheric CO_2 seasonality from increased cropland productivity. *Nature*, 515, 398–401.

Gu, L., Baldocchi, D. D., Wofsy, S. C., et al. (2003). Response of a deciduous forest to the Mount Pinatubo eruption: Enhanced photosynthesis. *Science*, 299, 2035–2038.

Gurney, K. R., and Eckels, W. J. (2011). Regional trends in terrestrial carbon exchange and their seasonal signatures. *Tellus B*, 63, 328–339.

Gurney, K. R., Law, R. M., Denning, A. S., et al. (2004). Transcom 3 inversion intercomparison: Model mean results for the estimation of seasonal carbon sources and sinks. *Global Biogeochemical Cycles*, 18, GB1010, doi:10.1029/2003GB002111.

Hashimoto, H., Nemani, R. R., White, M. A., et al. (2004). El Niño–Southern Oscillation-induced variability in terrestrial carbon cycling. *Journal of Geophysical Research*, 109, D23110, doi:10.1029/2004JD004959.

Hicke, J. A., Allen, C. D., Desai, A. R., et al. (2012). Effects of biotic disturbances on forest carbon cycling in the United States and Canada. *Global Change Biology*, 18, 7–34.

Higgins, P. A. T., and Harte, J. (2006). Biophysical and biogeochemical responses to climate change depend on dispersal and migration. *BioScience*, 56, 407–417.

Hobbie, S. E. (2008). Nitrogen effects on decomposition: A five-year experiment in eight temperate sites. *Ecology*, 89, 2633–2644.

Hobbie, S. E., Eddy, W. C., Buyarski, C. R., et al. (2012). Response of decomposing litter and its microbial community to multiple forms of nitrogen enrichment. *Ecological Monographs*, 82, 389–405.

Hoffman, F. M., Randerson, J. T., Arora, V. K., et al. (2014). Causes and implications of persistent atmospheric carbon dioxide biases in Earth System Models. *Journal of Geophysical Research: Biogeosciences*, 119, 141–162, doi:10.1002/2013JG002381.

Högberg, P. (2007). Nitrogen impacts on forest carbon. *Nature*, 447, 781–782.

Holland, E. A., Braswell, B. H., Lamarque, J.-F., et al. (1997). Variations in the predicted spatial distribution of atmospheric nitrogen deposition and their impact on carbon uptake by terrestrial ecosystems. *Journal of Geophysical Research*, 102D, 15849–15866.

Holland, E. A., Braswell, B. H., Sulzman, J., and Lamarque, J.-F. (2005). Nitrogen deposition onto the United States and western Europe: Synthesis of observations and models. *Ecological Applications*, 15, 38–57.

Houghton, R. A. (2003). Revised estimates of the annual net flux of carbon to the atmosphere from changes in land use and land management 1850–2000. *Tellus B*, 55, 378–390.

Houghton, R. A. (2010). How well do we know the flux of CO_2 from land-use change? *Tellus B*, 62, 337–351.

Houghton, R. A. (2013). Keeping management effects separate from environmental effects in terrestrial carbon accounting. *Global Change Biology*, 19, 2609–2612.

Houghton, R. A., Hobbie, J. E., Melillo, J. M., et al. (1983). Changes in the carbon content of terrestrial biota and soils between 1860 and 1980: A net release of CO_2 to the atmosphere. *Ecological Monographs*, 53, 235–262.

Houghton, R. A., House, J. I., Pongratz, J., et al. (2012). Carbon emissions from land use and land-cover change. *Biogeosciences*, 9, 5125–5142.

Huntingford, C., Lowe, J. A., Booth, B. B. B., et al. (2009). Contributions of carbon cycle uncertainty to future climate projection spread. *Tellus B*, 61, 355–360.

Hurtt, G. C., Frolking, S., Fearon, M. G., et al. (2006). The underpinnings of land-use history: Three centuries of global gridded land-use transitions, wood-harvest activity, and resulting secondary lands. *Global Change Biology*, 12, 1208–1229.

Jain, A., Yang, X., Kheshgi, H., et al. (2009). Nitrogen attenuation of terrestrial carbon cycle response to global environmental factors. *Global Biogeochemical Cycles*, 23, GB4028, doi:10.1029/2009GB003519.

Janssens, I. A., Dieleman, W., Luyssaert, S., et al. (2010). Reduction of forest soil respiration in response to nitrogen deposition. *Nature Geoscience*, 3, 315–322.

Jeong, S.-J., Ho, C.-H., Gim, H.-J., and Brown, M. E. (2011). Phenology shifts at start vs. end of growing season in temperate vegetation over the Northern Hemisphere for the period 1982–2008. *Global Change Biology*, 17, 2385–2399.

Jones, C. D., and Cox, P. M. (2001). Modeling the volcanic signal in the atmospheric CO_2 record. *Global Biogeochemical Cycles*, 15, 453–465.

Jones, C. D., Cox, P. M., and Huntingford, C. (2006). Climate–carbon cycle feedbacks under stabilization: Uncertainty and observational constraints. *Tellus B*, 58, 603–613.

Jones, C., Robertson, E., Arora, V., et al. (2013). Twenty-first-century compatible CO_2 emissions and airborne fraction simulated by CMIP5 Earth system models under four Representative Concentration Pathways. *Journal of Climate*, 26, 4398–4413.

Jung, M., Reichstein, M., Margolis, H. A., et al. (2011). Global patterns of land–atmosphere fluxes of carbon dioxide, latent heat, and sensible heat derived from eddy covariance, satellite, and meteorological observations. *Journal of Geophysical Research*, 116, G00J07, doi:10.1029/2010JG001566.

Karnosky, D. F., Pregitzer, K. S., Zak, D. R., et al. (2005). Scaling ozone responses of forest trees to the ecosystem level in a changing climate. *Plant, Cell and Environment*, 28, 965–981.

Keenan, T. F., Hollinger, D. Y., Bohrer, G., et al. (2013). Increase in forest water-use efficiency as atmospheric carbon dioxide concentrations rise. *Nature*, 499, 324–327.

Keenan, T. F., Gray, J., Friedl, M. A., et al. (2014). Net carbon uptake has increased through warming-induced changes in temperate forest phenology. *Nature Climate Change*, 4, 598–604.

Klein Goldewijk, K. (2001). Estimating global land use change over the past 300 years: The HYDE Database. *Global Biogeochemical Cycles*, 15, 417–433.

Klein Goldewijk, K., Beusen, A., van Drecht, G., and de Vos, M. (2011). The HYDE 3.1 spatially explicit database of human-induced global land-use change over the past 12,000 years. *Global Ecology and Biogeography*, 20, 73–86.

Knorr, M., Frey, S. D., and Curtis, P. S. (2005). Nitrogen additions and litter decomposition: A meta-analysis. *Ecology*, 86, 3252–3257.

Koven, C. D., Ringeval, B., Friedlingstein, P., et al. (2011). Permafrost carbon-climate feedbacks accelerate global warming. *Proceedings of the National Academy of Sciences USA*, 108, 14769–14774.

Koven, C. D., Riley, W. J., and Stern, A. (2013). Analysis of permafrost thermal dynamics and response to climate change in the CMIP5 Earth System Models. *Journal of Climate*, 26, 1877–1900.

Kurz, W. A., Dymond, C. C., Stinson, G., et al. (2008a). Mountain pine beetle and forest carbon feedback to climate change. *Nature*, 452, 987–990.

Kurz, W. A., Stinson, G., Rampley, G. J., Dymond, C. C., and Neilson, E. T. (2008b). Risk of natural disturbances makes future contribution of Canada's forests to the global carbon cycle highly uncertain. *Proceedings of the National Academy of Sciences USA*, 105, 1551–1555.

Lamarque, J.-F., Bond, T. C., Eyring, V., et al. (2010). Historical (1850–2000) gridded anthropogenic and biomass burning emissions of reactive gases and aerosols: Methodology and application. *Atmospheric Chemistry and Physics*, 10, 7017–7039.

Lamarque, J.-F., Kyle, G. P., Meinshausen, M., et al. (2011). Global and regional evolution of short-lived radiatively-active gases and aerosols in the Representative Concentration Pathways. *Climatic Change*, 109, 191–212.

Lamarque, J.-F., Dentener, F., McConnell, J., et al. (2013). Multi-model mean nitrogen and sulfur deposition from the Atmospheric Chemistry and Climate Model Intercomparison Project (ACCMIP): Evaluation of historical and projected future changes. *Atmospheric Chemistry and Physics*, 13, 7997–8018.

Lambert, F. H., Harris, G. R., Collins, M., et al. (2013). Interactions between perturbations to different Earth system components simulated by a fully-coupled climate model. *Climate Dynamics*, 41, 3055–3072.

Lawrence, D. M., and Slater, A. G. (2005). A projection of severe near-surface permafrost degradation during the 21st century. *Geophysical Research Letters*, 32, L24401, doi:10.1029/2005GL025080.

LeBauer, D. S., and Treseder, K. K. (2008). Nitrogen limitation of net primary productivity in terrestrial ecosystems is globally distributed. *Ecology*, 89, 371–379.

Le Quéré, C., Raupach, M. R., Canadell, J. G., et al. (2009). Trends in the sources and sinks of carbon dioxide. *Nature Geoscience*, 2, 831–836.

Le Quéré, C., Andres, R. J., Boden, T., et al. (2013). The global carbon budget 1959–2011. *Earth System Science Data*, 5, 165–185.

Levis, S. (2010). Modeling vegetation and land use in models of the Earth System. *WIREs Climate Change*, 1, 840–856.

Lewis, S. L., Lopez-Gonzalez, G., Sonké, B., et al. (2009). Increasing carbon storage in intact African tropical forests. *Nature*, 457, 1003–1006.

Lewis, S. L., Brando, P. M., Phillips, O. L., van der Heijden, G. M. F., and Nepstad, D. (2011). The 2010 Amazon drought. *Science*, 331, 554.

Liu, L., and Greaver, T. L. (2009). A review of nitrogen enrichment effects on three biogenic GHGs: The CO_2 sink may be largely offset by stimulated N_2O and CH_4 emission. *Ecology Letters*, 12, 1103–1117.

Liu, X., Zhang, Y., Han, W., et al. (2013). Enhanced nitrogen deposition over China. *Nature*, 494, 459–462.

Lombardozzi, D., Sparks, J. P., and Bonan, G. (2013). Integrating O_3 influences on terrestrial processes: Photosynthetic and stomatal response data available for regional and global modeling. *Biogeosciences*, 10, 6815–6831.

Lombardozzi, D., Bonan, G. B., and Nychka, D. W. (2014). The emerging anthropogenic signal in land–atmosphere carbon-cycle coupling. *Nature Climate Change*, 4, 796–800.

Lombardozzi, D., Levis, S., Bonan, G., Hess, P. G., and Sparks, J. P. (2015). The influence of chronic ozone exposure on global carbon and water cycles. *Journal of Climate*, 28, 292–305.

Luo, Y. (2007). Terrestrial carbon-cycle feedback to climate warming. *Annual Review of Ecology, Evolution, and Systematics*, 38, 683–712.

Luo, Y. Q., Randerson, J. T., Abramowitz, G., et al. (2012). A framework for benchmarking land models. *Biogeosciences*, 9, 3857–3874.

MacFarling Meure, C., Etheridge, D., Trudinger, C., et al. (2006). Law Dome CO_2, CH_4 and N_2O ice core records extended to 2000 years BP. *Geophysical Research Letters*, 33, L14810, doi:10.1029/2006GL026152.

Magnani, F., Mencuccini, M., Borghetti, M., et al. (2007). The human footprint in the carbon cycle of temperate and boreal forests. *Nature*, 447, 849–851.

Matthews, H. D. (2005). Decrease of emissions required to stabilize atmospheric CO_2 due to positive carbon cycle-climate feedbacks. *Geophysical Research Letters*, 32, L21707, doi:10.1029/2005GL023435.

Matthews, H. D. (2006). Emissions targets for CO_2 stabilization as modified by carbon cycle feedbacks. *Tellus B*, 58, 591–602.

McDowell, N. G., and Sevanto, S. (2010). The mechanisms of carbon starvation: How, when, or does it even occur at all? *New Phytologist*, 186, 264–266.

McDowell, N., Pockman, W. T., Allen, C. D., et al. (2008). Mechanisms of plant survival and mortality during drought: Why do some plants survive while others succumb to drought? *New Phytologist*, 178, 719–739.

McDowell, N. G., Fisher, R. A., Xu, C., et al. (2013). Evaluating theories of drought-induced vegetation mortality using a multimodel–experiment framework. *New Phytologist*, 200, 304–321.

Menzel, A., Sparks, T. H., Estrella, N., et al. (2006). European phenological response to climate change matches the warming pattern. *Global Change Biology*, 12, 1969–1976.

Mercado, L. M., Bellouin, N., Sitch, S., et al. (2009). Impact of changes in diffuse radiation on the global land carbon sink. *Nature*, 458, 1014–1017.

Metsaranta, J. M., Kurz, W. A., Neilson, E. T., and Stinson, G. (2010). Implications of future disturbance regimes on the carbon balance of Canada's managed forest (2010–2100). *Tellus B*, 62, 719–728.

Monnin, E., Steig, E. J., Siegenthaler, U., et al. (2004). Evidence for substantial accumulation rate variability in Antarctica during the Holocene, through synchronization of CO_2 in the Taylor Dome, Dome C and DML ice cores. *Earth and Planetary Science Letters*, 224, 45–54.

Morgan, P. B., Mies, T. A., Bollero, G. A., Nelson, R. L., and Long, S. P. (2006). Season-long elevation of ozone concentration to projected 2050 levels under fully open-air conditions substantially decreases the growth and production of soybean. *New Phytologist*, 170, 333–343.

Myneni, R. B., Keeling, C. D., Tucker, C. J., Asrar, G., and Nemani, R. R. (1997). Increased plant growth in the northern high latitudes from 1981 to 1991. *Nature*, 386, 698–702.

Nadelhoffer, K. J., Emmett, B. A., Gundersen, P., et al. (1999). Nitrogen deposition makes a minor contribution to carbon sequestration in temperate forests. *Nature*, 398, 145–148.

Negrón-Juárez, R., Baker, D. B., Zeng, H., Henkel, T. K., and Chambers, J. Q. (2010). Assessing hurricane-induced tree mortality in U.S. Gulf Coast forest ecosystems. *Journal of Geophysical Research*, 115, G04030, doi:10.1029/2009JG001221.

Nemani, R. R., Keeling, C. D., Hashimoto, H., et al. (2003). Climate-driven increases in global terrestrial net primary production from 1982 to 1999. *Science*, 300, 1560–1563.

Norby, R. J., and Zak, D. R. (2011). Ecological lessons from free-air CO_2 enrichment (FACE) experiments. *Annual Review of Ecology, Evolution, and Systematics*, 42, 181–203.

Ollinger, S. V., Aber, J. D., Reich, P. B., and Freuder, R. J. (2002). Interactive effects of nitrogen deposition, tropospheric ozone, elevated CO_2 and land use history on the carbon dynamics of northern hardwood forests. *Global Change Biology*, 8, 545–562.

Page, S. E., Siegert, F., Rieley, J. O., et al. (2002). The amount of carbon released from peat and forest fires in Indonesia during 1997. *Nature*, 420, 61–65.

Pan, Y., Birdsey, R., Hom, J., and McCullough, K. (2009). Separating effects of changes in atmospheric composition, climate and land-use on carbon sequestration of U.S. Mid-Atlantic temperate forests. *Forest Ecology and Management*, 259, 151–164.

Pan, Y., Birdsey, R. A., Fang, J., et al. (2011). A large and persistent carbon sink in the world's forests. *Science*, 333, 988–993.

Peylin, P., Law, R. M., Gurney, K. R., et al. (2013). Global atmospheric carbon budget: Results from an ensemble of atmospheric CO_2 inversions. *Biogeosciences*, 10, 6699–6720.

Phillips, O. L., Malhi, Y., Higuchi, N., et al. (1998). Changes in the carbon balance of tropical forests: Evidence from long-term plots. *Science*, 282, 439–442.

Phillips, O. L., Aragão, L. E. O. C., Lewis, S. L., et al. (2009). Drought sensitivity of the Amazon rainforest. *Science*, 323, 1344–1347.

Piao, S., Friedlingstein, P., Ciais, P., Zhou, L., and Chen, A. (2006). Effect of climate and CO_2 changes on the greening of the Northern Hemisphere over the past two decades. *Geophysical Research Letters*, 33, L23402, doi:10.1029/2006GL028205.

Piao, S., Ciais, P., Friedlingstein, P., et al. (2008). Net carbon dioxide losses of northern ecosystems in response to autumn warming. *Nature*, 451, 49–52.

Piao, S., Ciais, P., Friedlingstein, P., et al. (2009). Spatiotemporal patterns of terrestrial carbon cycle during the 20th century. *Global Biogeochemical Cycles*, 23, GB4026, doi:10.1029/2008GB003339.

Piao, S., Sitch, S., Ciais, P., et al. (2013). Evaluation of terrestrial carbon cycle models for their response to climate variability and to CO_2 trends. *Global Change Biology*, 19, 2117–2132.

Pinder, R. W., Bettez, N. D., Bonan, G. B., et al. (2013). Impacts of human alteration of the nitrogen cycle in the US on radiative forcing. *Biogeochemistry*, 114, 25–40.

Pongratz, J., Reick, C., Raddatz, T., and Claussen, M. (2008). A reconstruction of global agricultural areas and land cover for the last millennium. *Global Biogeochemical Cycles*, 22, GB3018, doi:10.1029/2007GB003153.

Pongratz, J., Reick, C. H., Raddatz, T., and Claussen, M. (2009). Effects of anthropogenic land cover change on the carbon cycle of the last millennium. *Global Biogeochemical Cycles*, 23, GB4001, doi:10.1029/2009GB003488.

Pongratz, J., Reick, C. H., Houghton, R. A., and House, J. I. (2014). Terminology as a key uncertainty in net land use and land cover change carbon flux estimates. *Earth System Dynamics*, 5, 177–195.

Prentice, I. C., Harrison, S. P., and Bartlein, P. J. (2011). Global vegetation and terrestrial carbon cycle changes after the last ice age. *New Phytologist*, 189, 988–998.

Ramankutty, N., and Foley, J. A. (1999). Estimating historical changes in global land cover: Croplands from 1700 to 1992. *Global Biogeochemical Cycles*, 13, 997–1027.

Randerson, J. T., Thompson, M. V., Conway, T. J., Fung, I. Y., and Field, C. B. (1997). The contribution of terrestrial sources and sinks to trends in the seasonal cycle of atmospheric carbon dioxide. *Global Biogeochemical Cycles*, 11, 535–560.

Randerson, J. T., Field, C. B., Fung, I. Y., and Tans, P. P. (1999). Increases in early season ecosystem uptake explain recent changes in the seasonal cycle of atmospheric CO_2 at high northern latitudes. *Geophysical Research Letters*, 26, 2765–2768.

Randerson, J. T., Hoffman, F. M., Thornton, P. E., et al. (2009). Systematic assessment of terrestrial biogeochemistry in coupled climate–carbon models. *Global Change Biology*, 15, 2462–2484.

Raupach, M. R., Canadell, J. G., and Le Quéré, C. (2008). Anthropogenic and biophysical contributions to increasing atmospheric CO_2 growth rate and airborne fraction. *Biogeosciences*, 5, 1601–1613.

Reichstein, M., Ciais, P., Papale, D., et al. (2007). Reduction of ecosystem productivity and respiration during the European summer 2003 climate anomaly: A joint flux tower, remote sensing and modelling analysis. *Global Change Biology*, 13, 634–651.

Reichstein, M., Bahn, M., Ciais, P., et al. (2013). Climate extremes and the carbon cycle. *Nature*, 500, 287–295.

Richardson, A. D., Black, T. A., Ciais, P., et al. (2010). Influence of spring and autumn phenological transitions on forest ecosystem productivity. *Philosophical Transactions of the Royal Society B*, 365, 3227–3246.

Sarmiento, J. L., Gloor, M., Gruber, N., et al. (2010). Trends and regional distributions of land and ocean carbon sinks. *Biogeosciences*, 7, 2351–2367.

Scheiter, S., Langan, L., and Higgins, S. I. (2013). Next-generation dynamic global vegetation models: Learning from community ecology. *New Phytologist*, 198, 957–969.

Schimel, D., Stephens, B. B., and Fisher, J. B. (2015). Effect of increasing CO_2 on the terrestrial carbon cycle. *Proceedings of the National Academy of Sciences USA*, 112, 436–441.

Schuur, E. A. G., Bockheim, J., Canadell, J. G., et al. (2008). Vulnerability of permafrost carbon to climate change: Implications for the global carbon cycle. *BioScience*, 58, 701–714.

Schuur, E. A. G., Abbott, B. W., Bowden, W. B., et al. (2013). Expert assessment of vulnerability of permafrost carbon to climate change. *Climatic Change*, 119, 359–374.

Schwartz, M. D., Ahas, R., and Aasa, A. (2006). Onset of spring starting earlier across the Northern Hemisphere. *Global Change Biology*, 12, 343–351.

Shevliakova, E., Pacala, S. W., Malyshev, S., et al. (2009). Carbon cycling under 300 years of land use change: Importance of the secondary vegetation sink. *Global Biogeochemical Cycles*, 23, GB2022, doi:10.1029/2007GB003176.

Shevliakova, E., Stouffer, R. J., Malyshev, S., et al. (2013). Historical warming reduced due to enhanced land carbon uptake. *Proceedings of the National Academy of Sciences USA*, 110, 16730–16735.

Sitch, S., Cox, P. M., Collins, W. J., and Huntingford, C. (2007). Indirect radiative forcing of climate change through ozone effects on the land-carbon sink. *Nature*, 448, 791–794.

Sitch, S., Huntingford, C., Gedney, N., et al. (2008). Evaluation of the terrestrial carbon cycle, future plant geography and climate–carbon cycle feedbacks using five Dynamic Global Vegetation Models (DGVMs). *Global Change Biology*, 14, 2015–2039.

Slater, A. G., and Lawrence, D. M. (2013). Diagnosing present and future permafrost from climate models. *Journal of Climate*, 26, 5608–5623.

Slayback, D. A., Pinzon, J. E., Los, S. O., and Tucker, C. J. (2003). Northern hemisphere photosynthetic trends 1982–99. *Global Change Biology*, 9, 1–15.

Smith, N. G., and Dukes, J. S. (2013). Plant respiration and photosynthesis in global-scale vegetation models: Incorporating acclimation to temperature and CO_2. *Global Change Biology*, 19, 45–63.

Sokolov, A. P., Kicklighter, D. W., Melillo, J. M., et al. (2008). Consequences of considering carbon–nitrogen interactions on the feedbacks between climate and the terrestrial carbon cycle. *Journal of Climate*, 21, 3776–3796.

Stephens, B. B., Gurney, K. R., Tans, P. P., et al. (2007). Weak northern and strong tropical land carbon uptake from vertical profiles of atmospheric CO_2. *Science*, 316, 1732–1735.

Sutton, M. A., Simpson, D., Levy, P. E., et al. (2008). Uncertainties in the relationship between atmospheric nitrogen deposition and forest carbon sequestration. *Global Change Biology*, 14, 2057–2063.

Thomas, R. Q., Canham, C. D., Weathers, K. C., and Goodale, C. L. (2010). Increased tree carbon storage in response to nitrogen deposition in the US. *Nature Geoscience*, 3, 13–17.

Thomas, R. Q., Bonan, G. B., and Goodale, C. L. (2013). Insights into mechanisms governing forest carbon response to nitrogen deposition: A model–data comparison using observed responses to nitrogen addition. *Biogeosciences*, 10, 3869–3887.

Thornton, P. E., Lamarque, J.-F., Rosenbloom, N. A., and Mahowald, N. M. (2007). Influence of carbon–nitrogen cycle coupling on land model response to CO_2 fertilization and climate variability. *Global Biogeochemical Cycles*, 21, GB4018, doi:10.1029/2006GB002868.

Thornton, P. E., Doney, S. C., Lindsay, K., et al. (2009). Carbon–nitrogen interactions regulate climate–carbon cycle feedbacks: Results from an atmosphere–ocean general circulation model. *Biogeosciences*, 6, 2099–2120.

Tucker, C. J., Fung, I. Y., Keeling, C. D., and Gammon, R. H. (1986). Relationship between atmospheric CO_2 variations and a satellite-derived vegetation index. *Nature*, 319, 195–199.

van der Molen, M. K., Dolman, A. J., Ciais, P., et al. (2011). Drought and ecosystem carbon cycling. *Agricultural and Forest Meteorology*, 151, 765–773.

van der Sleen, P., Groenendijk, P., Vlam, M., et al. (2015). No growth stimulation of tropical trees by 150 years of CO_2 fertilization but water-use efficiency increased. *Nature Geoscience*, 8, 24–28.

van der Werf, G. R., Randerson, J. T., Collatz, G. J., et al. (2004). Continental-scale partitioning of fire emissions during the 1997 to 2001 El Niño/La Niña period. *Science*, 303, 73–76.

van der Werf, G. R., Dempewolf, J., Trigg, S. N., et al. (2008). Climate regulation of fire emissions and deforestation in equatorial Asia. *Proceedings of the National Academy of Sciences USA*, 105, 20350–20355.

van der Werf, G. R., Randerson, J. T., Giglio, L., et al. (2010). Global fire emissions and the contribution of deforestation, savanna, forest, agricultural, and peat fires (1997–2009). *Atmospheric Chemistry and Physics*, 10, 11707–11735.

Vitousek, P. M., Porder, S., Houlton, B. Z., and Chadwick, O. A. (2010). Terrestrial phosphorus limitation: Mechanisms, implications, and nitrogen–phosphorus interactions. *Ecological Applications*, 20, 5–15.

Wenzel, S., Cox, P. M., Eyring, V., and Friedlingstein, P. (2014). Emergent constraints on climate-carbon cycle feedbacks in the CMIP5 Earth system models. *Journal of Geophysical Research: Biogeosciences*, 119, 794–807, doi:10.1002/2013JG002591.

Williams, C. A., Collatz, G. J., Masek, J., and Goward, S. N. (2012). Carbon consequences of forest disturbance and recovery across the conterminous United States. *Global Biogeochemical Cycles*, 26, GB1005, doi:10.1029/2010GB003947.

Wittig, V. E., Ainsworth, E. A., and Long, S. P. (2007). To what extent do current and projected increases in surface ozone affect photosynthesis and stomatal conductance of trees? A meta-analytic review of the last 3 decades of experiments. *Plant, Cell and Environment*, 30, 1150–1162.

Wittig, V. E., Ainsworth, E. A., Naidu, S. L., Karnosky, D. F., and Long, S. P. (2009). Quantifying the impact of current and future tropospheric ozone on tree biomass, growth, physiology and biochemistry: A quantitative meta-analysis. *Global Change Biology*, 15, 396–424.

Wolkovich, E. M., Cook, B. I., Allen, J. M., et al. (2012). Warming experiments underpredict plant phenological responses to climate change. *Nature*, 485, 494–497.

Wu, C., Chen, J. M., Black, T. A., et al. (2013). Interannual variability of net ecosystem productivity in forests is explained by carbon flux phenology in autumn. *Global Ecology and Biogeography*, 22, 994–1006.

Xia, J., and Wan, S. (2008). Global response patterns of terrestrial plant species to nitrogen addition. *New Phytologist*, 179, 428–439.

Yang, X., Thornton, P. E., Ricciuto, D. M., and Post, W. M. (2014). The role of phosphorus dynamics in tropical forests – a modeling study using CLM-CNP. *Biogeosciences*, 11, 1667–1681.

Zaehle, S. (2013). Terrestrial nitrogen–carbon cycle interactions at the global scale. *Philosophical Transactions of the Royal Society B*, 368, 20130125, doi:10.1098/rstb.2013.0125.

Zaehle, S., and Dalmonech, D. (2011). Carbon–nitrogen interactions on land at global scales: Current understanding in modelling climate biosphere feedbacks. *Current Opinion in Environmental Sustainability*, 3, 311–320.

Zaehle, S., Friedlingstein, P., and Friend, A. D. (2010a). Terrestrial nitrogen feedbacks may accelerate future climate change. *Geophysical Research Letters*, 37, L01401, doi:10.1029/2009GL041345.

Zaehle, S., Friend, A. D., Friedlingstein, P., et al. (2010b). Carbon and nitrogen cycle dynamics in the O-CN land surface model, 2: Role of the nitrogen cycle in the historical terrestrial carbon balance. *Global Biogeochemical Cycles*, 24, GB1006, doi:10.1029/2009GB003522.

Zaehle, S., Medlyn, B. E., De Kauwe, M. G., et al. (2014). Evaluation of 11 terrestrial carbon–nitrogen cycle models against observations from two temperate Free-Air CO_2 Enrichment studies. *New Phytologist*, 202, 803–822.

Zeng, H. C., Chambers, J. Q., Negron-Juarez, R. I., et al. (2009). Impacts of tropical cyclones on US forest tree mortality and carbon flux from 1851 to 2000. *Proceedings of the National Academy of Sciences USA*, 106, 7888–7892.

Zeng, N., Zhao, F., Collatz, G. J., et al. (2014). Agricultural Green Revolution as a driver of increasing atmospheric CO_2 seasonal amplitude. *Nature*, 515, 394–397.

Zhang, L., Jacob, D. J., Knipping, E. M., et al. (2012). Nitrogen deposition to the United States: Distribution, sources, and processes. *Atmospheric Chemistry and Physics*, 12, 4539–4554.

Zhang, Q., Wang, Y. P., Pitman, A. J., and Dai, Y. J. (2011). Limitations of nitrogen and phosphorous on the terrestrial carbon uptake in the 20th century. *Geophysical Research Letters*, 38, L22701, doi:10.1029/2011GL049244.

Zhang, Q., Wang, Y.P., Matear, R. J., Pitman, A. J., and Dai, Y. J. (2014). Nitrogen and phosphorous limitations significantly reduce future allowable CO_2 emissions. *Geophysical Research Letters*, 41, 632–637, doi:10.1002/2013GL058352.

Zhao, M., and Running, S. W. (2010). Drought-induced reduction in global terrestrial net primary production from 2000 through 2009. *Science*, 329, 940–943.

Zscheischler, J., Mahecha, M. D., von Buttlar, J., et al. (2014). A few extreme events dominate global interannual variability in gross primary production. *Environmental Research Letters*, 9, 035001, doi:10.1088/1748-9326/9/3/035001.

Nitrogen, Chemistry, and Climate

30.1 | Chapter Summary

The flows of reactive nitrogen (Nr) in the Earth system affect climate through several processes. Additional Nr increases the production of nitrous oxide (N_2O). Deposition of Nr onto terrestrial ecosystems increases carbon dioxide (CO_2) uptake and reduces methane (CH_4) consumption by soils. Increases in N_2O destroy stratospheric ozone (O_3). The NO_x and NH_3 emitted to the atmosphere are short lived and have little or no direct radiative effects. However, they react in the atmosphere to produce aerosols, and NO_x additionally forms tropospheric ozone and destroys CH_4, all of which are key radiative forcings. High concentrations of tropospheric ozone additionally damage plants and reduce CO_2 uptake. This chapter provides an overview of reactive nitrogen and its impact on climate. This includes the effect of nitrogen on terrestrial carbon storage, the direct radiative forcing from N_2O emissions, and the chemistry of Nr in the atmosphere as its affects CH_4, tropospheric and stratospheric ozone, and secondary aerosols. The chemistry of NO_x in the presence of carbon monoxide (CO) and volatile organic compounds (VOCs) to form ozone and increase the oxidation capacity of the troposphere by producing the hydroxyl radical (OH), and thereby consuming CH_4, is a key perturbation to tropospheric chemistry. While these processes affect climate, the net effect of anthropogenic Nr on climate is uncertain.

30.2 | The Nitrogen Cascade

The amount of reactive nitrogen cycling in the Earth system has increased since the preindustrial era due to anthropogenic activities. Reactive nitrogen (Nr) is all nitrogen compounds except N_2. This includes the gases ammonia (NH_3), nitrous oxide (N_2O), nitric oxide (NO), and nitrogen dioxide (NO_2); the inorganic ions ammonium (NH_4^+), nitrite (NO_2^-), and nitrate (NO_3^-); and organic compounds. The compound NH_x is the general expression for reduced nitrogen ($NH_x = NH_4^+ + NH_3$). The compound NO_x is the general expression for the nitrogen oxides ($NO_x = NO + NO_2$). The compound NO_y represents all oxidized nitrogen including NO_x, nitric acid (HNO_3), organic nitrates, and other compounds. Nitric acid is an oxidation product of NO_2 in the atmosphere.

In the absence of human activities, natural inputs of Nr are from lightning and biological nitrogen fixation, which convert atmospheric N_2 to a useable form. Natural gaseous losses are from nitrification and denitrification, which return NO, N_2O, and N_2 to the atmosphere, and ammonia volatilization, which is the loss of NH_3 to the atmosphere (Figure 20.12). Since the beginning of the industrial era, this balance has been altered by anthropogenic addition of Nr produced during industrial processes and used in agriculture. Anthropogenic sources of Nr include the manufacture and use of fertilizers for agriculture; the cultivation of legumes and other nitrogen-fixing crops;

Table 30.1 | Global nitrogen budget with (a) creation of Nr from N_2 by natural and anthropogenic sources, (b) natural and anthropogenic emissions of NO_x and NH_3, and (c) atmospheric deposition of NO_y and NH_x

Process	Flux (Tg N yr^{-1})	
(a) Sources of Nr		
Natural		
Biological N fixation, land	58	
Biological N fixation, ocean	160	
Lightning	4	
Natural total	222	
Anthropogenic		
Fossil fuel combustion	30	
Fertilizer production	100	
Industrial	24	
Biological N fixation	60	
Anthropogenic total	214	
(b) Atmospheric emissions	NO_x	NH_3
Natural		
Soils	7.3	2.4
Oceans	–	8.2
Lightning	4	–
Natural total	11.3	10.6
Anthropogenic		
Fossil fuel combustion	28.3	0.5
Agriculture	3.7	30.4
Biomass burning	5.5	9.2
Anthropogenic total	37.5	40.1
(c) Atmospheric deposition	NO_y	NH_x
Land	27.1	36.1
Ocean	19.8	17.0
Global	46.9	53.1

Source: From Ciais et al. (2013).

and fossil fuel combustion and industrial processes. Anthropogenic sources of Nr amount to over 200 Tg Nr yr^{-1}, equal to natural sources (Table 30.1). These processes emit NO_x and NH_3 into the atmosphere, which are re-deposited onto land and oceans as NO_y and NH_x through dry deposition of gases and aerosols and wet deposition in precipitation.

Soils and lightning emit a small amount of NO_x naturally to the atmosphere; however, human activities contribute three-quarters of the 49 Tg NO_x-N added to the atmosphere each year. The major anthropogenic sources of NO_x are fossil fuel combustion, biomass burning, and fertilizer usage (Table 30.1). Fossil fuel combustion accounts for 75 percent of the anthropogenic emissions. Once in the atmosphere, NO_x is oxidized to nitric acid (HNO_3) and organic nitrates in a matter of hours or days and forms ozone and secondary aerosols. NO_x is deposited onto land in a variety of forms, collectively termed NO_y.

Emissions of NH_3 provide a flux of Nr to the atmosphere that is as large as that from NO_x (51 Tg NH_3-N yr^{-1}). Natural sources emit a small amount of NH_3-N to the atmosphere, but agriculture is the largest source of NH_3 (Table 30.1). The vast majority (~80%) of the 51 Tg NH_3-N yr^{-1} emitted to the atmosphere originates from volatilization of livestock wastes, fertilizer use, and biomass burning. Volatilization is the loss of nitrogen through conversion of NH_4^+ to NH_3. In the atmosphere, NH_3 combines with H_2SO_4 (from SO_2 oxidation) and HNO_3 to produce ammonium sulfate and ammonium nitrate aerosols. Gaseous NH_3 and aerosols fall back onto land through dry and wet deposition processes. Total ammonium deposition is termed NH_x.

The additional Nr in the Earth system has numerous environmental consequences, collectively termed the nitrogen cascade (Galloway et al. 2003, 2008). Natural and anthropogenic NO_x and NH_3 emissions add 100 Tg N yr^{-1} to the atmosphere (Table 30.1). Atmosphere deposition of NO_y and NH_x returns a similar amount of Nr back to land and oceans. The increased Nr contributes to the acidification of soils, rivers, and lakes; accelerates the loss of biodiversity; disrupts the functioning of terrestrial, freshwater, and marine ecosystems; pollutes rivers and water supplies; contributes to the eutrophication of coastal ecosystems; and degrades air quality.

The anthropogenic addition of Nr also affects climate (Figure 30.1). Anthropogenic inputs of Nr onto land affect the concentration of greenhouse gases in the atmosphere. The most direct

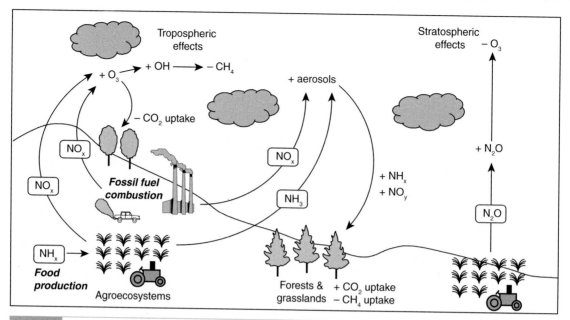

Fig. 30.1 Nitrogen and climate interactions in the nitrogen cascade. Fossil fuel combustion, food production, and other human activities add NO_x, NH_3, and N_2O to the atmosphere, with consequences for tropospheric and stratospheric ozone, aerosols, N_2O, CH_4 destruction, Nr deposition, and terrestrial CO_2 uptake. Adapted from Erisman et al. (2011). See also Galloway et al. (2003, 2008), Gruber and Galloway (2008), and Hertel et al. (2012).

effect is through increased emissions of N_2O. Fossil fuel combustion, Nr inputs in agricultural systems (through fertilizer usage and manure), biomass burning, and other human activities increase N_2O emissions to the atmosphere. Total anthropogenic sources of N_2O are of the same magnitude as natural terrestrial sources (Table 3.5). Deposition of NH_x and NO_y on land increases carbon storage in terrestrial ecosystems (Chapter 29). A secondary effect arises because of changes in CH_4 production and consumption by terrestrial ecosystems. Well-drained soils are a sink for atmospheric CH_4. Additional Nr deposition can reduce CH_4 consumption rates by microbes, thereby contributing to higher atmospheric CH_4. In wetlands, however, the additional Nr may increase CH_4 production by stimulating plant productivity. Other additional effects of increased Nr on climate are mediated through chemical reactions in the atmosphere.

30.3 | Atmospheric Chemistry

In the stratosphere, N_2O destroys ozone (O_3), and anthropogenic N_2O emissions are a key

ozone-depleting substance (Ravishankara et al. 2009). The N_2O in the stratosphere breaks down in the presence of sunlight:

$$N_2O + hv \rightarrow N_2 + O \qquad (30.1)$$

where hv denotes a single photon of radiation. The vast majority of N_2O loss (~90%) is by this reaction. The remainder is lost in the reactions:

$$N_2O + O \rightarrow 2NO \qquad (30.2)$$

and:

$$N_2O + O \rightarrow N_2 + O_2 \qquad (30.3)$$

The first of these reactions is critical to the ozone chemistry of the stratosphere. The NO produced from N_2O destroys ozone in the reactions:

$$\begin{aligned} NO + O_3 &\rightarrow NO_2 + O_2 \\ NO_2 + O &\rightarrow NO + O_2 \end{aligned} \qquad (30.4)$$

and the net reaction is:

$$O_3 + O \rightarrow 2O_2 \qquad (30.5)$$

These reactions destroy one molecule of ozone, but no molecules of NO or NO_2.

Reactive nitrogen is also critical to the chemistry of the troposphere. The gases NO_x and NH_3 are short lived in the atmosphere, in contrast with long-lived greenhouse gases such as CO_2, CH_4, and N_2O. However, they affect climate through several fast chemical transformations in the troposphere that produce ozone and the hydroxyl radical (OH), consume CH_4, and additionally form secondary aerosols (Figure 30.1). One key reaction is the formation of ozone. Tropospheric ozone is a greenhouse gas (Figure 8.13) and is a pollutant that harms human health and damages plants. Ozone is produced in the troposphere by the oxidation of carbon monoxide, CH_4, and nonmethane volatile organic compounds in the presence of NO_x. Emissions of these ozone precursors have increased since the preindustrial era. The resulting increase in ozone concentration also increases the abundance of OH and decreases CH_4.

Nitrogen oxides are a key regulator of tropospheric ozone in the cycling of NO_2 and NO. The photochemical destruction of NO_2 is the major source of tropospheric ozone. Ozone is formed when atomic oxygen (O) reacts with molecular oxygen (O_2). In the troposphere, the primary source of O is the photolysis of NO_2:

$$NO_2 + hv \rightarrow NO + O \tag{30.6}$$

This produces ozone in the reaction:

$$O + O_2 + M \rightarrow O_3 + M \tag{30.7}$$

where M represents another molecule that is not directly involved with, but is needed for, the reaction. However, ozone is consumed in reaction with NO:

$$NO + O_3 \rightarrow NO_2 + O_2 \tag{30.8}$$

which regenerates NO_2. This cycling is represented by:

$$NO_2 \xrightleftharpoons{hv+O_2} NO + O_3 \tag{30.9}$$

These reactions occur rapidly, within minutes during the daytime, and do not result in a gain of ozone. They do, however, quickly interchange NO and NO_2 and produce a steady-state amount of ozone, but this concentration is generally low.

Instead, the oxidation of carbon monoxide, CH_4, and nonmethane volatile organic compounds in the troposphere in the presence of NO_x explains the high concentration of ozone found in many regions.

Emissions of NO_x enhance the photochemical production of tropospheric ozone in the presence of carbon monoxide, CH_4, and nonmethane volatile organic compounds. In their absence, Eq. (30.9) shows that the oxidation of NO to NO_2 consumes the ozone produced in the photolysis of NO_2. However, chemical reactions that oxidize NO to NO_2 without consuming ozone will produce ozone. One means by which this occurs is the oxidation of carbon monoxide (CO) by the hydroxyl radical (OH).

Carbon monoxide is oxidized to CO_2 by OH:

$$CO + OH \rightarrow CO_2 + H \tag{30.10}$$

The atomic hydrogen (H) combines with O_2 to form the hydroperoxy radical (HO_2):

$$H + O_2 + M \rightarrow HO_2 + M \tag{30.11}$$

and this produces NO_2 in reaction with NO:

$$HO_2 + NO \rightarrow OH + NO_2 \tag{30.12}$$

The NO_2 then produces ozone in the presence of sunlight, from Eq. (30.6) and Eq. (30.7). The result is a system of connected cycles in which:

$$\begin{aligned} CO + OH &\xrightarrow{O_2} HO_2 + CO_2 \\ HO_2 + NO &\rightarrow OH + NO_2 \\ NO_2 &\xrightarrow{hv+O_2} NO + O_3 \end{aligned} \tag{30.13}$$

The net outcome is consumption of CO to produce CO_2 and O_3 in the presence of sunlight and NO_x ($CO+2O_2 \rightarrow CO_2 + O_3$). Neither OH nor HO_2 is consumed. Ozone is produced because the conversion of NO to NO_2 is accomplished by HO_2 rather than by O_3, as in Eq. (30.8).

Volatile organic compounds, in the presence of NO_x, also can produce ozone (Figure 30.2a). Their oxidation by OH forms organic peroxy radicals (RO_2). These function similar to HO_2 to produce NO_2:

$$RO_2 + NO \rightarrow RO + NO_2 \tag{30.14}$$

where here R denotes a hydrocarbon chain (e.g., CH_3). These and other oxidation products serve as

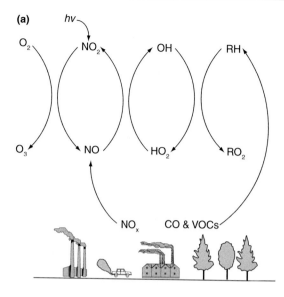

(a)

(b) Methane (CH$_4$)

HCHO is an O$_3$ precursor

Fig. 30.2 Processes by which Nr in combination with volatile organic compounds (VOCs) produce ozone. (a) General illustration of VOC–NO$_x$–O$_3$ chemistry and the oxidation of RH to RO$_2$ by OH. For methane, R = CH$_3$. (b) Specific illustration for methane. The cycle begins with the hydroxyl radical (OH) oxidizing methane. The OH radical is regenerated in reactions with two NO that produce two NO$_2$, each of which forms one O$_3$. The net result of the cycle is to oxidize one methane molecule and form two ozone molecules. Additional reactions yield formaldehyde (HCHO), which produces more ozone.

precursors to ozone formation. Volatile organic compounds (VOCs) are a broad class of organic chemicals that participate in atmospheric photochemical reactions. Methane is one such VOC, but there are many nonmethane VOCs. These are released into the atmosphere by anthropogenic and natural emissions. Anthropogenic sources are numerous and include gasoline combustion, industrial processes, and evaporation of liquid fuels, paints, varnishes, solvents, cleaning supplies, and many other household products. Many plants also emit VOCs.

Methane oxidation (Figure 30.2b) illustrates the reactions of hydrocarbons. (Hydrocarbons contain only carbon and hydrogen and are a broad class of VOCs.) Methane is destroyed in the troposphere when it is oxidized by the hydroxyl radical:

$$CH_4 + OH \rightarrow CH_3 + H_2O \qquad (30.15)$$

The product of this reaction (CH$_3$; methyl radical) itself combines near instantaneously with molecular oxygen to produce CH$_3$O$_2$ (methyl peroxy radical):

$$CH_3 + O_2 + M \rightarrow CH_3O_2 + M \qquad (30.16)$$

The methyl peroxy radical can undergo various reactions. Reaction with NO produces the methoxy radical (CH$_3$O) and NO$_2$:

$$CH_3O_2 + NO \rightarrow CH_3O + NO_2 \qquad (30.17)$$

This reaction is a critical step in ozone formation because it produces NO$_2$ without consuming ozone. The NO$_2$ forms ozone in the presence of sunlight, following Eq. (30.6) and Eq. (30.7). Some CH$_3$O combines with molecular oxygen to produce formaldehyde (HCHO) and the hydroperoxy radical:

$$CH_3O + O_2 \rightarrow HCHO + HO_2 \qquad (30.18)$$

The HO$_2$ radical is then available to convert NO to NO$_2$ from Eq. (30.12), which also regenerates the hydroxl radical. The net result is to oxidize one methane molecule and form two ozone molecules. Formaldehyde is itself a VOC that is oxided by OH and that also breaks down by photolysis, and the products of these reactions produce more ozone. The chemical transformations of other VOCs are complex, but for hydrocarbons in general involve oxidation by OH to yield organic peroxy radicals and secondary VOCs.

Biogenic VOC emissions from plants are a critical factor in ozone formation in regions polluted with NO_x (Chameides et al. 1988). In general, higher VOC concentrations lead to more ozone, but higher amounts of NO_x can create or destroy ozone. The exact outcome depends on the relative amount of VOCs and NO_x. Where NO_x is plentiful relative to VOCs (low VOC/NO_x ratio), the amount of VOCs limits ozone production, and ozone increases with more VOCs. However, higher NO_x itself reduces ozone. Where the concentration of NO_x is low relative to VOCs (high VOC/NO_x), such as in rural environments, the amount of NO_x limits ozone formation. At low NO_x levels, ozone formation is generally insensitive to VOCs.

The hydroxyl radical is the key oxidant in the troposphere and thereby controls many aspects of tropospheric chemistry (Prinn 2003). The major source of OH for oxidation reactions in the troposphere is from ozone, which breaks down in the presence of sunlight:

$$O_3 + h\nu \rightarrow O + O_2 \qquad (30.19)$$

Many oxygen atoms (O) recombine with O_2 to form ozone. Some, however, react with water vapor to produce OH:

$$O + H_2O \rightarrow 2OH \qquad (30.20)$$

The hydroxyl radical is a powerful oxidant in the troposphere. It reacts quickly with many chemicals (its lifetime in the atmosphere is on the order of seconds). Its oxidization of CO, CH_4, and nonmethane VOCs quickly destroys the chemicals. Emissions of NO_x, therefore, lead to CH_4 destruction by increasing the amount of the hydroxyl radical. The formation of ozone from NO_x produces OH, which destroys CH_4.

Reactive nitrogen has other consequences for tropospheric chemistry. The NO_x and NH_3 emitted to the atmosphere undergo chemical reactions that produce ammonium sulfate and ammonium nitrate aerosols. Ammonia (NH_3) produces ammonium nitrate (NH_4NO_3) or ammonium sulfate ((NH_4)$_2SO_4$) aerosols in a complex system of reactions. Ammonia and nitric acid (HNO_3) form aerosol ammonium nitrate:

$$NH_3(g) + HNO_3(g) \rightleftharpoons NH_4NO_3(s) \qquad (30.21)$$

Ammonium nitrate forms where NH_3 and HNO_3 concentrations are high, but sulfate concentrations are low. At high relative humidity the ammonium nitrate formed is an aqueous solution of NH_4^+ and NO_3^- instead of a solid. Emissions of NO_x are one source of nitric acid, and so NO_x produces aerosols. Sulfuric acid (H_2SO_4) is critical to aerosol formation. In an ammonia-poor environment, ammonium sulfate is produced from the reactions:

$$NH_3(g) + H_2SO_4(g) \rightleftharpoons (NH_4)HSO_4(s) \qquad (30.22)$$

$$NH_3(g) + (NH_4)HSO_4(g) \rightleftharpoons (NH_4)_2SO_4(s) \qquad (30.23)$$

In an ammonia-rich environment, the ammonia that does not react with sulfate additionally forms NH_4NO_3.

30.4 | Radiative Forcing

Reactive nitrogen from fossil fuel combustion and agricultural activities affects global climate by altering atmospheric chemistry, aerosols, and greenhouse gas concentrations. Various studies have reviewed the overall effects of anthropogenic Nr additions on climate (Galloway et al. 2008; Arneth et al. 2010; Erisman et al. 2011). In general, the net climate impact is the balance between positive radiative forcing from emissions of N_2O and production of tropospheric ozone and negative radiative forcing from aerosols, enhanced terrestrial carbon storage, and decreased atmospheric CH_4.

CO_2 – Nr deposited on forest ecosystems can increase plant production and carbon storage in nitrogen-limited ecosystems (Chapter 29). This results in less accumulation of anthropogenic CO_2 emissions in the atmosphere and thereby reduces the positive radiative forcing of CO_2.

CH_4 – The chemistry of VOC–NO_x–O_3 produces OH, which destroys CH_4. Methane is a powerful greenhouse gas with an atmospheric lifetime of 12 years. Its 100-year global warming potential is 28 times that of CO_2 (Myhre et al. 2013). By depleting atmospheric CH_4, emissions of NO_x reduce the positive radiative forcing of CH_4. This is the dominant effect of Nr on CH_4. Secondary effects from changes in CH_4

production and consumption in soils are poorly understood.

N_2O – The addition of fertilizer and manure to soils, fossil fuel combustion, biomass burning, and other anthropogenic activities increase the production of N_2O. Nitrous oxide is a strong greenhouse gas with an atmospheric lifetime of 121 years and a 100-year global warming potential 265 times that of CO_2 (Myhre et al. 2013). Additional N_2O is a positive radiative forcing.

Tropospheric ozone – Oxidation of CO, CH_4, and nonmethane VOCs in the presence of NO_x produces ozone, and emissions of NO_x, consequently, lead to the formation of tropospheric ozone. Tropospheric ozone provides a positive radiative forcing. An additional positive radiative forcing occurs because increased concentrations of ozone at the surface reduce plant productivity, which decreases the accumulation of anthropogenic CO_2 emissions in terrestrial ecosystems (Chapter 29). Higher NO_x emissions can, therefore, alter the global carbon cycle (Collins et al. 2010).

Stratospheric ozone – N_2O in the stratosphere destroys ozone. Stratospheric ozone has a small negative radiative forcing, and the loss of ozone is therefore a positive radiative forcing.

Aerosols – NO_x and NH_3 in the atmosphere form secondary aerosols such as ammonium sulfate, ammonium nitrate, and organic aerosols. These aerosols are a negative radiative forcing because they scatter solar radiation back to space and additionally alter cloud processes. The production of aerosols from anthropogenic Nr contributes to an enhanced negative radiative forcing. Aerosols also increase the proportion of solar radiation that is diffuse, which penetrates deeper into the plant canopy and stimulates terrestrial CO_2 uptake (Chapter 29). However, this process is small compared with aerosol–radiation and aerosol–cloud interactions.

Surface albedo – One poorly understood nitrogen–climate interaction may arise from the effect of foliage nitrogen on surface albedo (Ollinger et al. 2008; Hollinger et al. 2010). Across numerous temperate and boreal forests, canopy albedo increases with higher canopy nitrogen concentration. The relationship arises from a positive correlation between nitrogen concentration and canopy near-infrared reflectance. This suggests a possibly significant role of nitrogen in the climate system through solar radiation absorption. However, the biophysical mechanisms for this correlation are unclear and may be caused by changes in canopy structure (Wicklein et al. 2012; Knyazikhin et al. 2013; Ollinger et al. 2013). A global assessment of albedo in relation to nitrogen deposition shows a positive relationship between canopy albedo and nitrogen deposition, but this also reflects co-variation among forest canopy properties, environmental conditions, and albedo (Leonardi et al. 2015).

The net effect of Nr is thought to reduce the positive anthropogenic radiative forcing by a small amount. In an analysis by Erisman et al. (2011), increased N_2O and ozone damage to plants increase radiative forcing, but Nr deposition and NH_3 and NO_x emissions decrease radiative forcing. The negative radiative forcing of NH_3 arises from production of aerosols while the net negative radiative forcing of NO_x is a balance of a small positive radiative forcing from tropospheric ozone production and a larger negative radiative forcing from production of aerosols and a decrease in CH_4. The net influence of Nr on the global radiative balance is estimated to be -0.24 W m^{-2}, but the uncertainty is large and it may range from $+0.2$ to -0.5 W m^{-2}. This compares with a net anthropogenic radiative forcing of 2.3 W m^{-2} (Myhre et al. 2013).

Enhanced terrestrial carbon storage and increased N_2O emissions are two opposing consequences of anthropogenic addition of Nr. A review of published experimental studies concluded that increased N_2O emissions reduce, but do not completely offset, the positive benefits of carbon uptake (Liu and Greaver 2009). Regional analyses for Europe (Butterbach-Bahl et al. 2011) and the United States (Pinder et al. 2012) show that the negative radiative forcing from enhanced carbon uptake generally offsets the positive radiative forcing from N_2O emissions, at least on short timescales. However, global analyses indicate a contrary result – that the positive radiative forcing of N_2O emissions exceeds the negative radiative forcing from CO_2 sequestration (Zaehle et al. 2011).

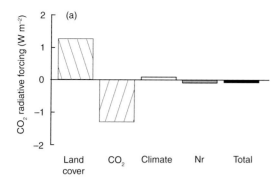

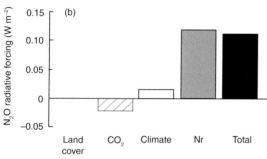

Fig. 30.3 Biogeochemical contribution of anthropogenic Nr and other global environmental changes to the present-day radiative forcing of (a) CO_2 and (b) N_2O. Shown are the individual contributions of land-cover change, CO_2 fertilization, climate change, and Nr additions as well as the total radiative forcing. Adapted from Zaehle et al. (2011).

Zaehle et al. (2011) used a global terrestrial biosphere model with coupled carbon–nitrogen biogeochemistry and nitrogen gas emissions to study the impact of anthropogenic Nr addition from fertilizer usage and atmospheric deposition on radiative forcing (Figure 30.3). In their simulations, anthropogenic Nr inputs account for about 20 percent of the carbon stored by terrestrial ecosystems between 1996 and 2005 and for most of the increase in N_2O emissions. They estimate that the increased carbon storage from nitrogen deposition has decreased the CO_2 radiative forcing by 96 mW m^{-2}. This is small compared with changes in CO_2 radiative forcing from land cover and CO_2 fertilization. Moreover, emissions of N_2O provide a positive radiative forcing of 125 mW m^{-2}, thereby offsetting the Nr reduction of CO_2 radiative forcing.

While these effects offset at the global scale, they vary regionally. Anthropogenic Nr increases carbon storage in the region 30–60° N, where most of the nitrogen deposition occurs. The effect on carbon storage is particularly large in the forests of eastern North America, central Europe, India, and China, where Nr addition accounts for up to 25 percent of the net carbon gain. Emissions of N_2O have increased in all regions of the world because of nitrogen deposition, with largest increases in the region 20–40° N as a result of agriculture.

Butterbach-Bahl et al. (2011) assessed the net radiative forcing of anthropogenic Nr in Europe. In that region, the net effect of anthropogenic Nr is a negative radiative forcing, thereby cooling climate, though this estimate is highly uncertain (Table 30.2). Increased N_2O in the atmosphere is a strong positive radiative forcing (17 mW m^{-2}). Production of tropospheric ozone from NO_x emissions augments the positive radiative forcing (2.9 mW m^{-2}), as does reduced land carbon uptake from higher surface ozone concentrations (4.4 mW m^{-2}). Decreased CH_4 uptake by soils is a minor radiative forcing (0.1 mW m^{-2}). The total positive radiative forcing is 24 mW m^{-2}. This warming is countered by a negative radiative forcing equal to –40 mW m^{-2}. The largest contributors to this cooling are enhanced carbon uptake from Nr deposition (–19 mW m^{-2}) and secondary atmospheric aerosols produced from emissions of NO_x and NH_3 (–16.5 mW m^{-2}). This aerosol radiative forcing is comparable to that from enhanced terrestrial carbon storage and is of similar magnitude, but opposite sign, to the positive radiative forcing from increased N_2O emissions. The reduction in atmospheric CH_4 from NO_x emissions produces a smaller negative radiative forcing (–4.6 mW m^{-2}). The net radiative forcing is a small cooling (–16 mW m^{-2}). However, the uncertainty is large, and the net effect could range from weak warming (15 mW m^{-2}) to strong cooling (–47 mW m^{-2}).

Pinder et al. (2012) assessed the climate effects of anthropogenic Nr in the United States and found that Nr has a cooling effect over a 20-year time frame (Figure 30.4). They used a different metric of climate change impact, defined as the global temperature potential calculated on a 20-year or 100-year basis and expressed in units of CO_2 equivalents. Emissions of N_2O produce

Table 30.2 | Global radiative forcing (mW m^{-2}) attributed to European anthropogenic Nr emissions

Forcing	Effect	Estimate	Minimum	Maximum
CO_2	Nr deposition increases land CO_2 uptake	−19	−30	−8
CH_4	Higher NO_x emissions decrease CH_4 lifetime	−4.6	−6.7	−2.4
	Nr deposition decreases soil CH_4 uptake	0.13	0.03	0.24
N_2O	N_2O emissions increase	17	14.8	19.1
Tropospheric ozone	Higher NO_x emissions increase ozone	2.9	0.3	5.5
	Ozone reduces land CO_2 uptake	4.4	2.3	6.5
Aerosols	Aerosol–radiation	−16.5	−27.5	−5.5
Net	All processes	−15.7	−46.8	15.4

Note: Aerosol–cloud interactions were not estimated.
Source: From Butterbach-Bahl et al. (2011).

a large warming, and this warming is slightly enhanced by ozone damage to plants. Greater carbon sequestration from nitrogen deposition, aerosol formation, and NO_x impacts on CH_4 and ozone decrease temperature. The net cooling of these forcings exceeds the warming of N_2O. The NO_x–O_3–CH_4 chemistry produces the largest temperature decrease. Pinder et al. (2012) estimate a small climate impact from aerosols relative to the nitrogen deposition carbon sink and N_2O emissions. Overall, combustion emissions produce cooling (mostly from NO_x) while agricultural emissions produce warming (mostly from N_2O). The uncertainty in these estimates is large, and the net effect may be zero. The lifetime of aerosols and other pollutants in the atmosphere is short so that their cooling effect is negligible on a 100-year time frame. Instead, the 100-year effect of Nr is warming, rather than cooling, because N_2O is long lived in the atmosphere.

30.5 | Reactive Nitrogen in the Twenty-First Century

Anthropogenic production of Nr has increased over the industrial era (Figure 2.10), and the amount of Nr in the Earth system is likely to increase in the future. Global anthropogenic fertilizer usage is projected to be 90–190 Tg N yr^{-1}, depending on population growth, food demand, improvements in agricultural efficiency, and other factors (Erisman et al. 2008). This is up to twice the current usage. Livestock production is also expected to grow (van Vuuren et al. 2011). As a result, global N_2O emissions will continue to increase (Figure 2.11e). A depiction of a future Earth without climate change policy intervention (RCP8.5) describes large population growth, modest gains in energy efficiency, high fossil fuel energy consumption, and large increases in agricultural land to meet the food demand. Emissions of N_2O increase substantially because of fertilizer use and intensification of agricultural production. A modest climate policy intervention scenario (RCP6.0) also describes increased N_2O emissions over the twenty-first century as global cropland area increases to feed a growing population. Emissions decline only with policies that promote low radiative forcing (RCP4.5, RCP2.6). However, even in these scenarios the atmospheric concentration of N_2O still increases over the twenty-first century (Figure 2.11f).

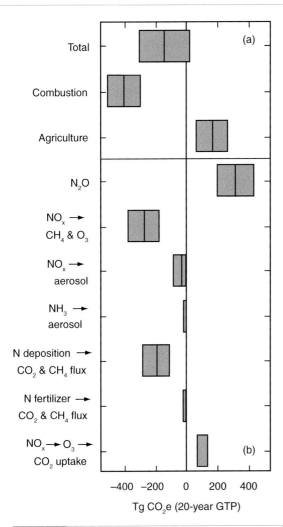

Fig. 30.4 Climate impacts of Nr emissions in the United States expressed in Tg CO_2 equivalents on a 20-year global temperature potential basis. The width of the bar shows the range and the black line is the best estimate. (a) Total forcing and the contribution from combustion and agriculture. (b) Individual forcing components. Adapted from Pinder et al. (2012).

near present-day throughout the twenty-first century (Figure 2.12b). This arises because of increased NH_x deposition and decreased NO_y deposition (Ciais et al. 2013). Nitrogen deposition is projected to decline in North America and Europe and increase in Asia. These projected changes in Nr deposition have low confidence because of uncertainty in Nr emissions, atmospheric transport, and the processes resulting in deposition.

Factors that will determine future Nr emissions include air pollution control measures, agricultural development, and climate change mitigation policy (van Vuuren et al. 2011). Economic activities that emit Nr are likely to grow in the future. Emissions of N_2O and NH_3 depend mostly on agricultural activities (fertilizer use, livestock production). The trajectory of agriculture in the future depends on population growth, socioeconomic growth, and demand for food and fiber. Emissions of NO_x depend on trends in energy production and transportation.

Anthropogenic alterations to the nitrogen cycle are driven by the same processes that influence CO_2 emissions – population growth, increasing energy usage, greater demand for food and fiber, and land-use change to feed the growing population. Policy interventions and socioeconomic developments that limit CO_2 emissions will also reduce the amount of Nr in the Earth system. Moreover, environmental policies that target the influence of Nr emissions on air quality and water quality may also enable management of the excess Nr in the Earth system for climate change mitigation. Such policies include (Galloway et al. 2008): reducing fossil fuel combustion; technological improvements to reduce NO_x emissions in fossil fuel combustion; decreasing fertilizer usage by increasing the nitrogen-use efficiency of crops and by better managing fertilizer application (timing, place, and rate of application) to reduce N_2O emissions; management of livestock manure to reduce N_2O emissions and NH_3 volatilization; and similar management of human sewage treatment. Nitrogen management of agricultural systems presents many tractable solutions, provided incentives are in place to encourage their adoption (Robertson and

While energy consumption is expected to increase in the future, NO_x emissions could decline because of air pollution control policies, improvements in fuel efficiency, and changes in energy systems (van Vuuren et al. 2011). Emissions of NH_3 are expected to increase because of growth in livestock production (van Vuuren et al. 2011). Global nitrogen deposition is expected to remain relatively constant at a rate

Vitousek 2009). Climate change mitigation policies must balance the multiple effects of intervention. For example, high nitrogen use to grow biofuels could cancel the CO_2 benefit of biofuels by contributing to N_2O emissions and higher tropospheric ozone.

Much of our understanding of the climate effects of anthropogenic Nr comes from synthesis of field studies in response to nitrogen addition or greenhouse gas inventory analyses. The representation of climate–nitrogen interactions in Earth system models is still incomplete. Observational estimates of the carbon gain in forests with nitrogen addition are highly variable, though it is generally accepted that nitrogen enrichment stimulates plant growth and decreases decomposition rates so as to increase terrestrial carbon storage and provide a negative radiative forcing (Chapter 29). Field experiments show that low nitrogen availability can restrict plant productivity increases with CO_2 enrichment (Chapter 20). Global terrestrial biosphere models with carbon–nitrogen biogeochemistry simulate less carbon gain with elevated atmospheric CO_2 concentrations compared with carbon-only models (Chapter 29). However, there is substantial variability among models in both their response to nitrogen enrichment and the nitrogen downregulation of CO_2 fertilization, and there is little consensus on how to model key aspects of carbon–nitrogen biogeochemistry (Zaehle and Dalmonech 2011). Nor do models fully replicate observed changes in carbon and nitrogen cycles with elevated CO_2 (Zaehle et al. 2014). Moreover, the full radiative forcing of anthropogenic Nr including chemistry–climate interactions is lacking in the current generation of Earth system models. Better understanding and modeling of nitrogen in the Earth system is a key requirement for simulating anthropogenic climate change over the coming century. A complete understanding of the effects of increased Nr on climate requires a multidisciplinary integration of many biogeochemical and ecological processes with an understanding of their effects on atmospheric chemistry, composition, and radiative forcing. Management of Nr for climate change mitigation requires an understanding of socioeconomic needs and drivers of nitrogen use.

30.6 | Review Questions

1. Discuss the benefits and adverse effects of anthropogenic Nr for climate.

2. Compare the effects of anthropogenic Nr addition on N_2O emissions and terrestrial carbon storage.

3. How would a decrease in NO_x emissions to improve air quality affect climate?

4. Discuss trends in anthropogenic Nr emissions that will affect climate.

5. What nitrogen management strategies could be used to mitigate climate change over the coming century?

30.7 | References

Arneth, A., Harrison, S. P., Zaehle, S., et al. (2010). Terrestrial biogeochemical feedbacks in the climate system. *Nature Geoscience*, 3, 525–532.

Butterbach-Bahl, K., Nemitz, E., Zaehle, S., et al. (2011). Nitrogen as a threat to the European greenhouse balance. In *The European Nitrogen Assessment*, ed. M. A. Sutton, C. M. Howard, J. W. Erisman, et al. Cambridge: Cambridge University Press, pp. 434–462.

Chameides, W. L., Lindsay, R. W., Richardson, J., and Kiang, C. S. (1988). The role of biogenic hydrocarbons in urban photochemical smog: Atlanta as a case study. *Science*, 241, 1473–1475.

Ciais, P., Sabine, C., Bala, G., et al. (2013). Carbon and other biogeochemical cycles. In *Climate Change 2013: The Physical Science Basis. Contribution of Working Group I to the Fifth Assessment Report of the Intergovernmental Panel on Climate Change*, ed. T. F. Stocker, D. Qin, G.-K. Plattner, et al. Cambridge: Cambridge University Press, pp. 465–570.

Collins, W. J., Sitch, S., and Boucher, O. (2010). How vegetation impacts affect climate metrics for ozone precursors. *Journal of Geophysical Research*, 115, D23308, doi:10.1029/2010JD014187.

Erisman, J. W., Sutton, M. A., Galloway, J., Klimont, Z., and Winiwarter, W. (2008). How a century of ammonia synthesis changed the world. *Nature Geoscience*, 1, 636–639.

Erisman, J. W., Galloway, J., Seitzinger, S., Bleeker, A., and Butterbach-Bahl, K. (2011). Reactive nitrogen in the environment and its effect on climate change. *Current Opinion in Environmental Sustainability*, 3, 281–290.

Galloway, J. N., Aber, J. D., Erisman, J. W., et al. (2003). The nitrogen cascade. *BioScience*, 53, 341–356.

Galloway, J. N., Townsend, A. R., Erisman, J. W., et al. (2008). Transformation of the nitrogen cycle: Recent trends, questions, and potential solutions. *Science*, 320, 889–892.

Gruber, N., and Galloway, J. N. (2008). An Earth-system perspective of the global nitrogen cycle. *Nature*, 451, 293–296.

Hertel, O., Skjøth, C. A., Reis, S., et al. (2012). Governing processes for reactive nitrogen compounds in the European atmosphere. *Biogeosciences*, 9, 4921–4954.

Hollinger, D. Y., Ollinger, S. V., Richardson, A. D., et al. (2010). Albedo estimates for land surface models and support for a new paradigm based on foliage nitrogen concentration. *Global Change Biology*, 16, 696–710.

Knyazikhin, Y., Schull, M. A., Stenberg, P., et al. (2013). Hyperspectral remote sensing of foliar nitrogen content. *Proceedings of the National Academy of Sciences USA*, 110, E185–E192.

Leonardi, S., Magnani, F., Nolè, A., Van Noije, T., and Borghetti, M. (2015). A global assessment of forest surface albedo and its relationships with climate and atmospheric nitrogen deposition. *Global Change Biology*, 21, 287–298.

Liu, L., and Greaver, T. L. (2009). A review of nitrogen enrichment effects on three biogenic GHGs: The CO_2 sink may be largely offset by stimulated N_2O and CH_4 emission. *Ecology Letters*, 12, 1103–1117.

Myhre, G., Shindell, D., Bréon, F.-M., et al. (2013). Anthropogenic and natural radiative forcing. In *Climate Change 2013: The Physical Science Basis. Contribution of Working Group I to the Fifth Assessment Report of the Intergovernmental Panel on Climate Change*, ed. T. F. Stocker, D. Qin, G.-K. Plattner, et al. Cambridge: Cambridge University Press, pp. 659–740.

Ollinger, S. V., Richardson, A. D., Martin, M. E., et al. (2008). Canopy nitrogen, carbon assimilation, and albedo in temperate and boreal forests: Functional relations and potential climate feedbacks. *Proceedings of the National Academy of Sciences USA*, 105, 19336–19341.

Ollinger, S. V., Reich, P. B., Frolking, S., et al. (2013). Nitrogen cycling, forest canopy reflectance, and emergent properties of ecosystems. *Proceedings of the National Academy of Sciences USA*, 110, E2437.

Pinder, R. W., Davidson, E. A., Goodale, C. L., et al. (2012). Climate change impacts of US reactive nitrogen. *Proceedings of the National Academy of Sciences USA*, 109, 7671–7675.

Prinn, R. G. (2003). The cleansing capacity of the atmosphere. *Annual Review of Environment and Resources*, 28, 29–57.

Ravishankara, A. R., Daniel, J. S., and Portmann, R. W. (2009). Nitrous oxide (N_2O): The dominant ozone-depleting substance emitted in the 21st century. *Science*, 326, 123–125.

Robertson, G. P., and Vitousek, P. M. (2009). Nitrogen in agriculture: Balancing the cost of an essential resource. *Annual Review of Environment and Resources*, 34, 97–125.

van Vuuren, D. P., Bouwman, L. F., Smith, S. J., and Dentener, F. (2011). Global projections for anthropogenic reactive nitrogen emissions to the atmosphere: An assessment of scenarios in the scientific literature. *Current Opinion in Environmental Sustainability*, 3, 359–369.

Wicklein, H. F., Ollinger, S. V., Martin, M. E., et al. (2012). Variation in foliar nitrogen and albedo in response to nitrogen fertilization and elevated CO_2. *Oecologia*, 169, 915–925.

Zaehle, S., and Dalmonech, D. (2011). Carbon–nitrogen interactions on land at global scales: Current understanding in modelling climate biosphere feedbacks. *Current Opinion in Environmental Sustainability*, 3, 311–320.

Zaehle, S., Ciais, P., Friend, A. D., and Prieur, V. (2011). Carbon benefits of anthropogenic reactive nitrogen offset by nitrous oxide emissions. *Nature Geoscience*, 4, 601–605.

Zaehle, S., Medlyn, B. E., De Kauwe, M. G., et al. (2014). Evaluation of 11 terrestrial carbon–nitrogen cycle models against observations from two temperate Free-Air CO_2 Enrichment studies. *New Phytologist*, 202, 803–822.

Aerosols, Chemistry, and Climate

31.1 | Chapter Summary

Aerosols affect the radiative balance of the atmosphere by absorbing and scattering radiation, by altering cloud albedo, and through precipitation. Their primary radiative effect is to increase planetary albedo, and aerosols have a negative radiative forcing. However, black carbon (commonly called soot) is an absorbing aerosol that heats the atmosphere, and deposition on snow and ice decreases surface albedo (a positive radiative forcing). Aerosols also have indirect effects by altering biogeochemical cycles. Mineral aerosols affect climate directly by altering the radiative balance of the atmosphere and indirectly by fertilizing ecosystems. Dust emissions may initiate a positive land–atmosphere feedback that enhances drought. Fires influence climate through emissions of long-lived greenhouse gases, organic and black carbon aerosols, and short-lived reactive gases. These latter emissions produce ozone, alter the oxidation capacity of the troposphere through the hydroxyl (OH) radical, and thereby affect the concentration of methane (CH_4). The net radiative forcing of fires is the balance of these biogeochemical emissions and also biogeophysical effects from changes in surface albedo and energy fluxes. It is estimated the fires provide a negative radiative forcing. They may also decrease precipitation in a positive feedback whereby biomass burning promotes drought and greater susceptibility to fire. Plants emit numerous biogenic volatile organic compounds (BVOCs). Oxidation of these BVOCs in the presence of nitrogen oxides (NO_x) forms ozone. Emissions of BVOCs also reduce the oxidation capacity of the atmosphere (OH), decreasing the atmospheric sink for CH_4 and increasing its lifetime in the atmosphere. Chemical transformations also produce secondary organic aerosols. Emissions of BVOCs are thought to provide a negative radiative forcing, but this is likely to diminish in the future because of human activities. Chemistry–climate interactions from short-lived climate forcers (NO_x, BVOCs, ozone, CH_4, and secondary organic aerosols) are now recognized as being significant and comparable in magnitude to other climate forcings. The aerosol effects of dust, fire, and BVOCs, as well as their chemistry–climate interactions, are important feedbacks with climate change.

31.2 | Aerosol Sources

Aerosols are small solid and liquid particles suspended in the atmosphere. They have a variety of sizes and compositions, typically range in size from a few nanometers to tens of micrometers, and are commonly seen as dust, smoke, or haze. These aerosols directly affect climate by absorbing or scattering atmospheric radiation and have other indirect effects such as altering clouds and precipitation by acting as cloud condensation nuclei or ice nuclei (Table 31.1). The balance among these processes depends on the

Table 31.1 Key aerosols, their sources, tropospheric lifetime, and climate properties

Aerosol species	Sources	Lifetime	Climate properties
Sulfate	Marine and volcanic emissions; fossil fuel combustion	~1 week	Light scattering; cloud condensation nuclei (CCN)
Nitrate	Oxidation of NO_x	~1 week	Light scattering; CCN
Black carbon	Combustion of fossil fuels, biofuels, and biomass	1 week–10 days	Light absorption; CCN
Organic	Combustion of fossil fuels, biofuels, and biomass; terrestrial and marine ecosystems	~1 week	Light scattering; CCN
Primary biogenic particles	Terrestrial ecosystems	1 day–1 week	CCN and ice nuclei
Mineral dust	Wind erosion	1 day–1 week	Light scattering and absorbing; ice nuclei
Sea salt	Wave breaking	1 day–1 week	Light scattering; CCN

Source: From Boucher et al. (2013).

type of aerosol, their abundance, and numerous environmental factors, but the net effect of aerosols is to cool climate. However, aerosols have numerous other climate influences. By decreasing solar radiation at the surface, aerosols reduce the energy available for evapotranspiration. High aerosol concentrations may have decreased evapotranspiration and increased runoff over the twentieth century in polluted regions of North America and Europe (Gedney et al. 2014). Aerosols additionally alter biogeochemical cycles; e.g., by changing the ratio of direct and diffuse radiation aerosols affect gross primary production, and by depositing nutrients to marine and terrestrial ecosystems. These biogeochemical effects may be as important as aerosol–radiation and aerosol–cloud interactions (Mahowald 2011; Mahowald et al. 2011). Changes in emissions of dust and fire aerosols and in the formation of secondary organic aerosols from biogenic emissions are expected to be important feedbacks with climate change (Carslaw et al. 2010).

The lifetime of aerosols in the troposphere is typically on the order of days to weeks. They are removed by wet deposition in precipitation and by dry deposition from turbulent motions and gravitational settling. However, volcanic aerosols can remain in the stratosphere for several years. Plumes of desert dust and smoke can extend for several hundreds of kilometers and travel long distances. In this manner, desert dust from Africa travels across the Atlantic Ocean to the Caribbean and the Amazon, Asian dust and anthropogenic aerosols cross the Pacific Ocean to North America, and black carbon from wildfires falls on snow and ice in the Arctic.

Aerosols are broadly categorized as: inorganic (sulfate, SO_4^{2-}; nitrate, NO_3^-; ammonium, NH_4^+; sea salt, NaCl); organic (containing carbon–carbon bonds); black carbon produced by incomplete combustion of fossil fuels and biomass (e.g., diesel engines during transportation, wildfires, wood and coal burning); mineral dust; and primary biogenic particles. They are further distinguished as primary or secondary aerosols. Primary aerosols originate from direct emissions of particulate matter into the atmosphere (Table 31.2). Natural sources of primary aerosols include sea salt from ocean spray, desert dust, black carbon from wildfires, organic carbon from wildfires, and primary biogenic particles.

Table 31.2 | Annual sources of aerosols

Sources	Amount (Tg yr^{-1})
Sea salt	10,100
Mineral dust	1600
Primary biogenic particles	110
Primary organic aerosols	95
Biomass burning	54
Fossil fuel	4
Biogenic	35
Black carbon	10
Biomass burning	6
Fossil fuel	4
Sulfates	200
Biogenic	57
Volcanic	21
Anthropogenic	122
Nitrates	18
Secondary organic aerosols	28
Biogenic	25
Anthropogenic	3

Note: Annual source estimates have large uncertainty (Boucher et al. 2013).

Source: From Andreae and Rosenfeld (2008) and Mahowald et al. (2011).

This latter form of aerosols includes pollen, fungal spores, bacteria, and plant debris. Secondary aerosols are produced from gaseous emissions as products of chemical reactions in the atmosphere in a gas-to-particle conversion. For example, sulfur dioxide (SO_2) emitted by volcanoes or during fossil fuel combustion produces the secondary inorganic aerosol sulfate. Sulfate aerosols also form when dimethyl sulfide (CH_3SCH_3) is produced in the oceans by phytoplankton. This is the largest natural emission of sulfur gas to the atmosphere, where it is converted to sulfate aerosols. Ammonia (NH_3) and nitrogen oxides (NO_x) produce ammonium nitrate and ammonium sulfate aerosols (Chapter 30). Secondary organic aerosols result from emissions of biogenic volatile organic compounds such as isoprene and monoterpenes from vegetation, which undergo chemical transformations to produce aerosols, and from chemical

emissions during wildfires. Anthropogenic aerosols arise from fossil fuel combustion, which emits black carbon and organic carbon, SO_2, and NO_x; from biomass burning (black carbon and organic carbon); and from dust created by overgrazing and deforestation.

There is large geographic variability in aerosol mass concentration and composition (Figure 31.1). Urban locations have considerably higher concentrations than rural locations. Several types of aerosols are common throughout the world (Boucher et al. 2013). Organic aerosols comprise a substantial fraction of aerosol mass (with diameter less than 10 μm) in many locations. Sulfate is about 10–30 percent of aerosol mass in much of the world, and the concentrations of ammonium and nitrate are less (4% and 6%, on average). Black carbon is less than 5 percent of aerosol mass in most areas, though it can be larger (12%) in urban Europe, urban Africa, and regions of South America and Asia where combustion sources are prevalent. Mineral dust is a large component of aerosol mass in South Asia and China (35%). Sea salt is dominant at remote oceanic locations (50–70% of aerosol mass).

Aerosols are extensive in many regions of the world, as a brownish haze in the atmosphere over India and Southeast Asia, dust and haze plumes extending from Asia east over the Pacific Ocean, and biomass burning and dust plumes extending from North Africa west over the subtropical Atlantic Ocean (Ramanathan et al. 2001). Aerosols can also be found in seemingly remote, pristine locations of the world. One such area is the Amazon (Martin et al. 2010; Pöschl et al. 2010). Biomass burning within the basin generates numerous particles, as do primary biogenic particles from pollen, spores, bacteria, and plant debris. Copious emissions of biogenic volatile organic compounds and their photochemical oxidation in the atmosphere generate a high abundance of secondary organic aerosols. Marine aerosols are imported into the region from the Atlantic Ocean. Winds transport dust and biomass burning particles across the Atlantic from Africa. In the wet season (December–March), the heavy rainfall removes these aerosols through wet deposition and limits the occurrence of fire. However, large

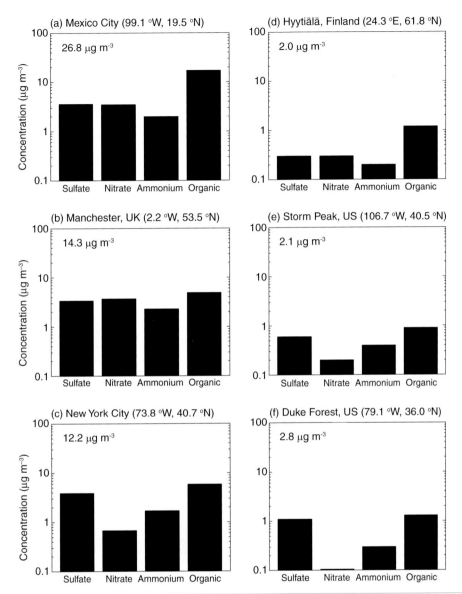

Fig. 31.1 Mass concentration of submicron sulfate, nitrate, ammonium, and organic aerosols in urban (a–c) and rural (d–f) locations. Data for New York City and Manchester are for summer. The number in each panel is the total concentration. Data from Jimenez et al. (2009). See also Zhang et al. (2007) and Boucher et al. (2013).

areas are covered by smoke from biomass burning during in the dry season (June–September). A high abundance of aerosols emitted or formed within the Amazon, which serve as nuclei for clouds and precipitation, suggests an active biogeochemical coupling between the biosphere and atmosphere that sustains the hydrologic cycle (Pöschl et al. 2010).

31.3 Radiative Forcing

Aerosols affect climate through the absorption and scattering of radiation in the atmosphere and by additionally altering clouds and precipitation (Ramanathan et al. 2001; Andreae and Rosenfeld 2008; Boucher et al. 2013; Rosenfeld et al. 2014).

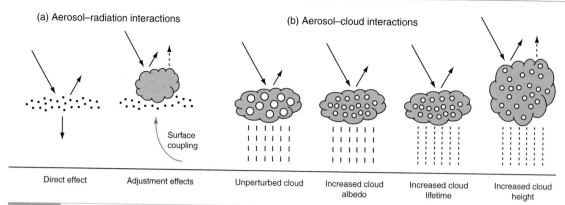

Fig. 31.2 Radiative feedbacks associated with aerosols. (a) Aerosol–radiation interactions due to the direct effects from scattering and absorption of solar radiation and also cloud adjustments. The small black dots represent aerosol particles. The solid black arrows represent solar radiation, and the dashed arrows represent longwave radiation. Adjustment effects include aerosol radiative effects on the surface energy budget, atmospheric profile, and cloudiness. (b) Aerosol–cloud interactions. The white circles represent cloud droplets. The dashed lines represent rainfall. The unperturbed cloud has large droplets and plentiful rainfall. The perturbed cloud has more and smaller droplets. This increases cloud albedo and suppresses drizzle thereby increasing cloud lifetime. Cloud height can also increase. Adapted from Forster et al. (2007) and Boucher et al. (2013).

The former is termed aerosol–radiation interactions (Figure 31.2a). The primary radiative effect of aerosols is to increase planetary albedo. Sulfate aerosols reflect incoming solar radiation back to space, as do mineral dust, sea salt, and many other aerosols. Scattering aerosols have a negative radiative forcing that cools planetary climate. In contrast, black carbon and organic aerosols absorb solar radiation in the atmosphere, as also does mineral dust. Absorbing aerosols have a positive radiative forcing, particularly over bright surfaces. The presence of aerosols in the troposphere initiates many rapid adjustments within the physical climate system. For example, both scattering and absorbing aerosols reduce the solar heating of the land surface so that less energy is available to evaporate water and heat the atmospheric boundary layer. Other rapid adjustments occur because absorption of solar radiation by black carbon and dust heats the troposphere, which lowers relative humidity, promotes re-evaporation, and reduces cloud liquid water. This reduces low level cloud formation and allows more solar radiation to reach the surface. Tropospheric heating also increases atmospheric stability by warming the air relative to the surface, and the increased stability can inhibit convection.

Aerosols also alter cloud microphysics, the radiative properties of clouds, and the amount and lifetime of clouds. The net effect of these is termed aerosol–cloud interactions (Figure 31.2b). One mechanism by which aerosols affect clouds is by acting as cloud condensation nuclei and ice nuclei. These are particles on which water vapor condenses into cloud droplets and on which ice crystals form. Greater aerosol concentration can increase cloud condensation nuclei and produce more cloud droplets. For the same amount of liquid water in the cloud, cloud droplet size decreases. More droplets of smaller size increases cloud albedo (the cloud albedo effect). Aerosols can also increase the lifetime of clouds by inhibiting rainfall because the smaller droplets are slower to coalesce into raindrops (the cloud lifetime effect). Aerosols can increase cloud height. Whereas clouds primarily reflect solar radiation and increase planetary albedo, high clouds additionally reduce the emission of longwave radiation to space because the cloud tops are cold. This compensates for their albedo effect, and high clouds warm climate. Additionally, aerosols deposited onto snow and ice decrease surface albedo.

The effects of aerosols on climate are quantified in terms of the radiative forcing due to

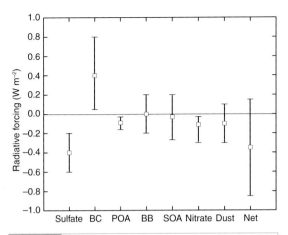

Fig. 31.3 Global annual mean radiative forcing (1750–2011) from aerosol–radiation interactions for sulfate aerosols, black carbon (BC) aerosols from fossil fuel and biofuel, primary organic aerosols (POA) from fossil fuel and biofuel, biomass burning (BB) aerosols (black carbon and organic), secondary organic aerosols (SOA), nitrate aerosols, and mineral aerosols (dust). Also shown is the net radiative forcing. This is calculated without allowing for rapid adjustments in atmospheric temperatures, water vapor, and clouds and is slightly smaller than that shown in Figure 8.13. Data from Boucher et al. (2013) and Myhre et al. (2013).

aerosol–radiation interactions, aerosol–cloud interactions, and changes in surface albedo due to deposition of black carbon on snow and ice (Boucher et al. 2013; Myhre et al. 2013). The net effect is a negative radiative forcing that cools global climate (Figure 8.13). The radiative forcing from aerosol–radiation interactions is -0.45 W m^{-2} (with an uncertainty range of -0.95 to 0.05 W m^{-2}). The aerosol–cloud interactions radiative forcing is similar (-0.45 W m^{-2}; -1.2 to 0.0 W m^{-2}). The total radiative forcing is -0.9 W m^{-2} (-1.9 to -0.1 W m^{-2}), excluding black carbon on snow. However, the regional forcing can be much larger in locations having high concentrations of anthropogenic aerosols.

The radiative forcing from aerosol–radiation interactions is the net forcing from sulfate aerosols, black carbon aerosols from fossil fuel and biofuels, primary organic aerosols from fossil fuel and biofuels, black carbon and organic aerosols from biomass burning, secondary organic aerosols, nitrate aerosols, and mineral aerosols. Figure 31.3 partitions the net aerosol–radiation

interactions radiative forcing into these aerosol sources. Black carbon from fossil fuel and biofuels is a large positive radiative forcing. Sulfate aerosols have the greatest negative radiative forcing. The net effect of biomass burning aerosols is small, but consists of large offsetting terms from organic aerosols and black carbon.

Dirty snow and ice from deposition of black carbon has a lower albedo than pristine snow and ice, which contributes to global warming (Hansen and Nazarenko 2004; Flanner et al. 2007, 2009). This black carbon deposition onto the surface contributes a small positive radiative forcing of 0.04 W m^{-2} with an uncertainty range of 0.02 to 0.09 W m^{-2} (Figure 8.13). This estimate comes from climate model simulations and is subject to uncertainties in black carbon emissions, atmospheric transport, the aerosol parameterization including deposition onto the surface, and the parameterization of snow albedo (Bond et al. 2013; Lee et al. 2013). However, all models show that the Arctic warms in response to black carbon deposition. The warming is generally greatest over land in spring, when the amount of incoming solar radiation is high.

31.4 | Mineral Aerosols

An estimated 1600 Tg yr^{-1} of mineral dust is mobilized by winds (Table 31.2). This dust is entrained into the atmosphere from dry soils with sparse vegetation and where soil properties allow erosion by wind. Key determinants of dust emissions are wind speed, soil moisture, vegetation cover, and topography (Ravi et al. 2011). Another critical feature is that dust emissions are not widespread throughout arid climates, but rather are concentrated in preferential source areas with strong mobilization potential. The precise way in which these factors control dust entrainment is not well understood for global models. Annual dust emissions simulated by global models vary by a factor of ten, from 500 to 4300 Tg yr^{-1} (Huneeus et al. 2011). Anthropogenic sources of dust from crop cultivation, deforestation, and overgrazing are important, though the magnitude of these

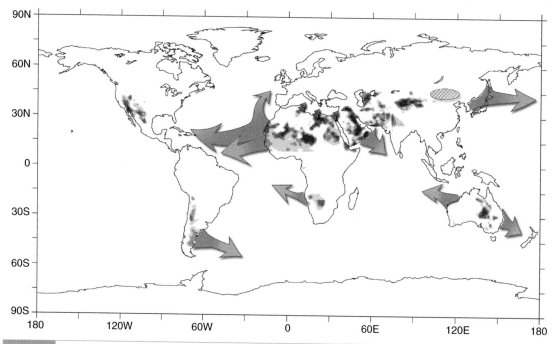

Fig. 31.4 Mineral aerosol source regions and atmospheric transport. Graphic provided courtesy of Joseph Prospero as in Mahowald et al. (2005).

sources is uncertain. A discernible increase in dust emissions from North Africa is evident due to human activities aver the past two hundred years (Mulitza et al. 2010). Anthropogenic activities are estimated to contribute as little as 10 percent of the global dust emissions (Tegen et al. 2004; Stanelle et al. 2014) to 14–60 percent (Mahowald and Luo 2003) and 30–50 percent (Tegen and Fung 1995). A more recent analysis estimated global dust emissions to be 1535 Tg yr⁻¹, of which anthropogenic sources contributed 24 percent (Ginoux et al. 2012). Principal dust source regions extend across the drylands and deserts of North Africa, the Middle East, central Asia, and China. In these regions, large plumes of dust are carried into the atmosphere and transported great distances (Figure 31.4). North Africa accounts for 55 percent of the global dust emissions (Ginoux et al. 2012).

Dust deposition is important to the ecology of the biosphere (Field et al. 2010; Ravi et al. 2011), and mineral dust additionally influences climate through a variety of radiative and biogeochemical feedbacks (Mahowald et al. 2011; Ravi et al. 2011). Dust aerosols are both

scattering and absorbing and so alter the radiative heating of the atmosphere. Over snow and ice, atmospheric dust decreases top-of-the-atmosphere albedo by reducing the backscattering of solar radiation to space that would otherwise occur because of the high albedo of snow and ice. At the surface, dirty snow from dust deposition increases the absorption of solar radiation by land, warming the surface. Mineral dust aerosols also carry nutrients such as iron and phosphorus (Jickells et al. 2005; Mahowald et al. 2005, 2008, 2009, 2011). Some of the nutrients deposited into the North Pacific are provided by dust from desert regions of Asia, and Saharan dust similarly deposits nutrients into the North Atlantic.

The Sahara Desert is a large source of dust, and plumes of desert dust can extend for several hundreds of kilometers into the Atlantic Ocean (Figure 31.5). Trans-Atlantic transport of dust from North Africa to South America, the Caribbean, and southeastern United States in plumes extending over several hundred kilometers in latitude at altitudes up to 5–7 km is common (Prospero et al. 1981, 1987, 1996; Prospero

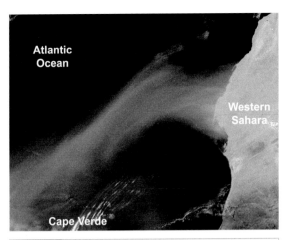

Fig. 31.5 Dust plume off the western coast of Africa extending several hundred kilometers into the Atlantic Ocean past Cape Verde, October 8, 2012. NASA image courtesy Jeff Schmaltz (NASA Goddard Space Flight Center, Greenbelt, Maryland) and provided by NASA's Earth Observatory.

and Nees 1986; Prospero and Lamb 2003). This transport and deposition of dust influences soil biogeochemistry. Saharan dust has contributed significant amounts of silicate clay and other minerals to soils of the Caribbean and eastern United States seaboard (Herwitz et al. 1996). Winds carry dust from North Africa to the Amazon (Prospero et al. 1981; Swap et al. 1992; Formenti et al. 2001; Martin et al. 2010). In these nutrient-poor soils, critical nutrients such as phosphorus and potassium are delivered in trace amounts by intermittent pulses of dust deposited during rainstorms (Swap et al. 1992; Okin et al. 2004). Thus, the productivity of parts of the rainforest is linked to events in sub-Saharan West Africa 5000 km distant. In particular, the concentration of dust in the Carribean (at Barbados) correlates with drought in sub-Saharan Africa (Prospero and Nees 1977, 1986; Prospero and Lamb 2003), though this relationship has broken down over the past few decades (Mahowald et al. 2009; Ridley et al. 2014).

Dust plays an important role in the hydrology of the western interior of the United States. This region is greatly dependent on the mountain snowpack for water. By decreasing snow albedo, dust causes earlier springtime snowmelt.

Grazing, agriculture, and other human activities have degraded the land and greatly increased dust emissions in this region over the past two centuries (Neff et al. 2008). Snow in the mountains melts earlier because of high dust deposition (Painter et al. 2007, 2010, 2012). Dust from far away regions also influences the regional hydrology. Long distance transport of dust and other particles from both Asia and the Sahara Desert have been detected in the Sierra Nevada Mountains of western United States and have been implicated in cloud formation and in enhancing precipitation (Creamean et al. 2013).

In desert regions, dust aerosols may decrease precipitation by cooling the surface and promoting subsidence, thereby reducing convection, and also by aerosol–cloud interactions that reduce drop size and suppress rainfall. Dust emissions may thereby initiate a positive land–atmosphere feedback that enhances drought (Rosenfeld et al. 2001; Ravi et al. 2011; D'Odorico et al. 2013). Dieback of vegetation during drought exposes dry soil and increases dust emissions, which in turn reduces rainfall and decreases soil moisture to prolong the drought. Such a feedback may have occurred during the 1930s Dust Bowl in North America and at other times (Cook et al. 2008, 2009, 2013) and may also operate in North Africa (Nicholson 2000; Prospero and Lamb 2003; Yoshioka et al. 2007; Marcella and Eltahir 2014). An additional feedback loop may be mediated through biogeochemical cycles, whereby increased soil erosion with loss of vegetation decreases soil fertility and further reduces vegetation cover (D'Odorico et al. 2013).

Over longer timescales, changing biogeography affects the amount of mineral dust in the atmosphere. Dust concentrations have varied over the past 800,000 years and were higher in glacial periods than in interglacials (Figure 8.2e). This could be due to expansion of unvegetated areas in high latitudes and central Asia as a result of increased aridity and lower atmospheric CO_2 concentration (Mahowald et al. 1999, 2006), though other factors such as high winds are also possible (McGee et al. 2010). Over the twentieth century, desert dust may have doubled over much of the world as a result

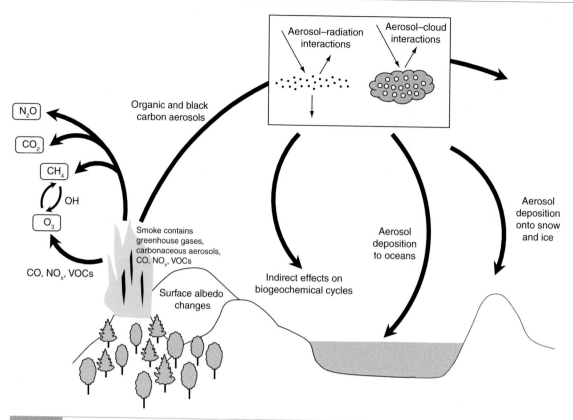

Fig. 31.6 Schematic illustration of the radiative effects of fire. Adapted from Ward et al. (2012). See also Carslaw et al. (2010) and Mahowald et al. (2011) for a review.

of climate change and anthropogenic land use (Mahowald et al. 2010). Vegetation changes with future climate change and anthropogenic land use are expected to affect dust emissions (Mahowald and Luo 2003; Tegen et al. 2004; Mahowald et al. 2006).

31.5 | Biomass Burning

Fires are a critical feature of the Earth system and influence air quality, atmospheric composition, and climate through a variety of biogeochemical and biogeophysical processes (Bowman et al. 2009). Figure 31.6 illustrates key processes: the emissions of greenhouse gases, which provide a positive radiative forcing; emissions of carbon monoxide (CO), oxides of nitrogen ($NO_x = NO + NO_2$), and volatile organic compounds, which are precursors to ozone formation and also affect the oxidation capacity of the troposphere through the hydroxyl (OH) radical; emissions of organic and black carbon aerosols; and changes in surface albedo.

The combustion of biomass during fires emits long-lived greenhouse gases (CO_2, CH_4, N_2O); many short-lived reactive gases including CO, NO_x, numerous nonmethane hydrocarbons, and other volatile organic compounds; and particulate matter (Crutzen and Andreae 1990; Andreae and Merlet 2001; Hoelzemann et al. 2004; Akagi et al. 2011; Wiedinmyer et al. 2011). This is commonly seen as smoke plumes, which can extend for great distances in the atmosphere (Figure 31.7). These emissions are quantified in terms of emission factors (g emitted per kg biomass burned). Estimates of emission factors vary among biome, type of fire, and chemical constituent. Table 31.3 gives representative values. Fire emissions are then calculated by:

$$E_i = A \times B \times F \times e_i \qquad (31.1)$$

where E_i is the emission of species i, A is the area burned, B is biomass (or fuel load), F is the fraction of the biomass burned, and e_i is the emission factor (Hoelzemann et al. 2004; Schultz et al. 2008; Wiedinmyer et al. 2011).

These emissions alter the chemistry and composition of the atmosphere, air quality, and climate. For example, satellite retrievals show high concentrations of CO, NO_2, formaldehyde (HCHO), and glyoxal (CHOCHO) are found in regions of the world with high biomass burning (Rosenfeld et al. 2014). These compounds undergo various chemical transformations in the atmosphere. In particular, NO_x, CO, and volatile organic compounds increase the amount of ozone in the troposphere (Chapter 30), and wildfires are a significant source of ozone (Jaffe and Wigder 2012). Biomass burning also affects the hydroxyl radical and its global oxidation capacity (Ward et al. 2012; Mao et al. 2013). The production of ozone from biomass burning emissions of NO_x increases the concentration of OH in the troposphere. Greater amounts of OH destroy CH_4 and decrease its lifetime. However, OH is also destroyed through its oxidation of CO and volatile organic compounds, both of which are emitted during fire. This increases the lifetime for CH_4.

Fires are a significant source of organic and black carbon aerosols in the atmosphere (Mahowald et al. 2011). These reduce solar radiation at the surface, thereby cooling the surface, but carbonaceous aerosols also increase

Fig. 31.7 Smoke plume from wildfires in southern California, October 26, 2003. NASA image courtesy Jacques Descloitres (NASA Goddard Space Flight Center, Greenbelt, Maryland) and provided by NASA's Earth Observatory.

Table 31.3 Biomass burning emission factors (g per kg dry matter burned) for various chemical constituents by fire type

Fire type	CO_2	CO	CH_4	NMHC	H_2	NO_x	N_2O	$PM_{2.5}$	OC	BC
Tropical deforestation	1626	101	6.6	7.00	3.50	2.26	0.20	9.05	4.30	0.57
Savanna	1646	61	2.2	3.41	0.98	2.12	0.21	4.94	3.21	0.46
Tropical woodland	1636	81	4.4	5.21	2.24	2.19	0.21	7.00	3.76	0.52
Extratropical forest	1572	106	4.8	5.69	1.78	3.41	0.26	12.84	9.14	0.56
Agricultural	1452	94	8.8	11.19	2.70	2.29	0.10	8.25	3.71	0.48
Peat	1703	210	20.8	7.00	3.50	2.26	0.20	9.05	4.30	0.57

Note: NMHC, nonmethane hydrocarbons. $PM_{2.5}$, particulate matter with diameter < 2.5 μm. OC, organic carbon. BC, black carbon.
Source: From van der Werf et al. (2010).

radiative heating of the troposphere. Black carbon deposition onto snow and ice provides a positive radiative forcing. Fires can have indirect effects on climate through biogeochemical cycles (Mahowald 2011; Mahowald et al. 2011). This occurs by depositing nitrogen and phosphorus onto terrestrial ecosystems and iron to marine ecosystems, by changing climate, and by altering diffuse radiation.

The net radiative forcing of fires is the balance of these biogeochemical emissions, as well as the biogeophysical effects from changes in surface albedo and energy fluxes. For example, wildfires in boreal forest emit CO_2 and CH_4, which provide a positive annual radiative forcing (Randerson et al. 2006). Ozone, black carbon deposited on snow and ice, and aerosols are an additional positive annual radiative forcing but are short lived. These are countered by an increase in surface albedo following loss of vegetation (a negative radiative forcing). Over long time periods, the negative radiative forcing from surface albedo exceeds the smaller positive biogeochemical radiative forcing (Table 23.1).

Ward et al. (2012) calculated the global radiative forcing from the various effects of fire depicted in Figure 31.6. They estimated a net negative radiative forcing in the preindustrial era of −1.02 W m⁻² and attributed that to greenhouse gases, aerosols, surface albedo, and climate–biogeochemical feedbacks (Table 31.4). These estimates have large uncertainty, but illustrate the various components of fire radiative forcing. The largest forcing is from aerosols, mostly aerosol–cloud interactions (−1.6 W m⁻²). This is augmented by the increase in surface albedo with fires (−0.2 W m⁻²) and additional negative radiative forcing from feedbacks with the carbon cycle (−0.22 W m⁻²). Fires emit CO_2 directly to the atmosphere during combustion, but also create a long-term carbon sink during vegetation regrowth. The net outcome of fires is to increase CO_2 concentrations, a positive radiative forcing of 0.83 W m⁻². The contribution of CH_4, N_2O, and tropospheric ozone is 0.16 W m⁻². Over the period 1850–2000, anthropogenic activities decreased the cooling effect of fires by 0.5 W m⁻².

The effects of aerosols on clouds and precipitation are complex and vary depending on the

Table 31.4 | Global radiative forcing of fires in the preindustrial era (1850)

Effect	Radiative forcing (W m⁻²)
Greenhouse gases	
CO_2	0.83
CH_4	0.06
N_2O	0.03
Tropospheric ozone	0.07
Aerosols	
Aerosol–radiation interactions	0.10
Aerosol–cloud interactions	−1.60
Indirect BGC effects	−0.09
Surface albedo	
Land surface	−0.20
Deposition onto snow/ice	0.00
Carbon cycle feedback	−0.22
Net	−1.02

Source: From Ward et al. (2012).

type of cloud (Rosenfeld et al. 2008, 2014). Black carbon and organic carbon aerosols are emitted in large quantities during biomass burning. The aerosols contained in smoke serve as cloud condensation nuclei and ice nuclei, alter cloud microphysics, and by absorbing radiation change the radiative heating of the troposphere. Over the Amazon, smoke plumes have been observed to suppress cloud cover and reduce precipitation (Andreae et al. 2004; Koren et al. 2004). The heavy smoke diminishes convection by warming the troposphere, decreasing radiative heating of the land, reducing evapotranspiration, and stabilizing the lower troposphere. This suppresses the formation of convective clouds and precipitation. A further reduction of precipitation arises from a decrease in cloud droplet size. These changes in atmospheric heating and cloud dynamics may additionally affect large-scale atmospheric circulations.

Black carbon aerosols absorb solar radiation in the atmosphere and simultaneously heat the troposphere but cool the surface by reducing downwelling solar radiation. One region where black carbon aerosols are abundant is over India and Southeast Asia. The changes in atmospheric heating and surface cooling can impact, and perhaps weaken, the monsoon (Ramanathan and Carmichael 2008; Bond et al. 2013). One such mechanism to reduce precipitation originates with fires. The occurrence of fires in tropical forests and peatlands in equatorial Asia during drought, by emitting carbonaceous aerosols, may create a positive feedback whereby biomass burning increases drought (Tosca et al. 2010). This region experiences drought during El Niño events, leading to increased fire emissions and large regional smoke clouds over Sumatra, Borneo, and the surrounding ocean. Climate model simulations show that fire aerosols (organic carbon and black carbon) reduce solar radiation at the surface, cooling the land and ocean, but warm the troposphere. These changes suppress convection and decrease precipitation throughout the region.

Climate model simulations also suggest an influence of fire aerosols on global temperature and precipitation (Tosca et al. 2013). Black carbon and organic carbon aerosols emitted by fires reduce surface solar radiation and decrease global surface temperature by 0.13°C. These aerosols lower temperature by more than 0.5°C in regions of South America, Africa, and equatorial Asia that experience extensive burning. The presence of aerosols decreases precipitation in these regions, likely due to local aerosol effects on boundary layer processes as previously discussed, but also from changes in the Hadley circulation. The surface cooling in combination with greater tropospheric heating increases equatorial subsidence and weakens the Hadley circulation. This decreases precipitation over tropical South America, Africa, and equatorial Asia.

These studies of fires and precipitation in Amazonia, Southeast Asia, and throughout the tropics suggest that smoke aerosols can decrease precipitation. Indeed, a positive feedback has been proposed among fire, rainfall, and the mixture of trees and grasses that maintains the savanna biome by promoting drought and more fire (Beerling and Osborne 2006). Recurring fires

that kill trees help maintain the grass cover. These fires may themselves promote drought that additionally favors C_4 grasses with high water-use efficiency.

31.6 Biogenic Volatile Organic Compounds

Plants actively exchange hundreds of species of organic chemicals collectively known as biogenic volatile organic compounds (BVOCs) in trace amounts with the atmosphere, many of which are commonly recognized as scents and odors (e.g., in pines and eucalyptus). Measurements over an orange grove in California identified 494 compounds actively exchanged between plants and the atmosphere (Park et al. 2013). Excluding methane, major BVOCs are isoprene (C_5H_8); monoterpenes, consisting of two isoprene units ($C_{10}H_{16}$); and sesquiterpenes with three isoprene units ($C_{15}H_{24}$) (collectively called isoprenoids). These are nonmethane hydrocarbons (organic compounds containing only hydrogen and carbon, excluding CH_4). Other hydrocarbons include ethene (C_2H_4) and propene (C_3H_6). Another broad class of BVOCs is oxygenated organic compounds (organic compounds that contain oxygen; e.g., methanol, CH_3OH; ethanol, C_2H_5OH; acetone, CH_3COCH_3; acetaldehyde, CH_3CHO). The BVOCs produced by plants have many physiological and ecological functions including those related to protection from abiotic and biotic stressors (e.g., heat stress, ozone damage, wounds), scents to attract pollinators, and as airborne signals to deter herbivores, attract predators of herbivores, and activate herbivore defense mechanisms (Sharkey et al. 2008; Laothawornkitkul et al. 2009; Loreto and Schnitzler 2010; Peñuelas and Staudt 2010). About 1–2 percent of the annual carbon uptake in net primary production is returned to the atmosphere by BVOC emissions (~1 Pg C yr⁻¹; Guenther et al. 1995, 2012).

The concentration of isoprene, monoterpenes, sesquiterpenes, and other BVOCs in the troposphere is low (a few parts per trillion to several parts per billion) and their lifetime is short (minutes to hours), but they undergo chemical

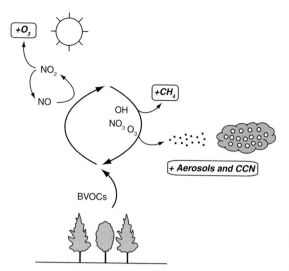

Fig. 31.8 Effects of BVOC emissions on atmospheric chemistry and climate through the photochemical production of tropospheric ozone (O_3); the oxidation of BVOCs by the hydroxyl radical (OH), the nitrate radical (NO_3), and ozone resulting in an increase in atmospheric CH_4; and the formation of secondary organic aerosols that affect climate by scattering radiation and serving as cloud condensation nuclei (CCN). Adapted from Peñuelas and Staudt (2010).

transformations that affect atmospheric chemistry and climate (Figure 31.8). Chief among these are oxidation by the hydroxyl radical (OH), the nitrate radical (NO_3), and ozone (O_3) in matters of minutes to hours that yield numerous compounds (Atkinson 2000; Fuentes et al. 2000; Atkinson and Arey 2003). These reactions are important chemistry–climate feedbacks that decrease the concentration of OH, form secondary organic aerosols and ozone, and increase the lifetime of methane (Andreae and Crutzen 1997; Laothawornkitkul et al. 2009; Pacifico et al. 2009; Peñuelas and Staudt 2010). In the troposphere, isoprene, monoterpenes, and sesquiterpenes are rapidly oxidized by OH. This decreases the concentration of OH, which is the major atmospheric sink for CH_4. Enhanced BVOC emissions, by reducing OH concentrations, increase the lifetime of CH_4. Isoprene and monoterpenes additionally react in the presence of sunlight and high levels of NO_x from fossil fuel combustion to form tropospheric ozone (Chapter 30). In a low NO_x environment, however, these BVOCs

react with ozone directly, leading to ozone destruction. Organic aerosols comprise a significant fraction of aerosol mass concentration (Figure 31.1), and secondary organic aerosols account for a large portion of these. Volatile organic compounds are a chief natural precursor of secondary organic aerosols. Some of these emissions originate from anthropogenic sources and biomass burning. Plant emissions of BVOCs (e.g., isoprene, monoterpenes) also produce secondary organic aerosols in reactions with OH, nitrate, and ozone.

Isoprene is the most abundant BVOC. It is produced by many tree species, including oaks, poplars, and eucalyptus, but its emission is very dependent on the particular species (e.g., Keenan et al. 2009). The amount emitted increases with greater sunlight and warmer temperatures up to some temperature optimum, is inhibited at high atmospheric CO_2 concentrations, decreases with prolonged drought stress, and varies with ozone damage (Grote and Niinemets 2008; Laothawornkitkul et al. 2009; Pacifico et al. 2009; Peñuelas and Staudt 2010). Isoprene emission varies with leaf traits, is higher in shade intolerant, early successional species than in shade-tolerant species, and increases with high photosynthetic capacity, short leaf lifespan, and low leaf mass per unit area (Harrison et al. 2013). Global emission inventories are obtained from models (Guenther et al. 1995, 2006, 2012; Arneth et al. 2008). The yearly production is about 450–750 Tg yr^{-1}, or 400–660 Tg C yr^{-1} (Emissions are reported as either chemical mass or the mass of carbon. Carbon is ~88% of the molecular mass of isoprene and monoterpenes.) Emissions of monoterpenes are considerably less, about 35–160 Tg yr^{-1} (30–140 Tg C yr^{-1}). A principal species is α-pinene, found in many conifer trees and especially pine and which is noticeable as a distinct scent. Other common monoterpenes include β-pinene (which has a pine-like scent) and limonene (citrus scent). Tropical and temperate broadleaf forests are significant emitters of isoprene, while boreal forests are strong emitters of monoterpenes.

Global BVOC emissions sum to about 1000 Tg yr^{-1} (Guenther et al. 2012). Isoprene comprises one-half of the total emissions (Table 31.5).

Table 31.5	Global annual BVOC emissions
Compound	Emissions (Tg yr^{-1})
Isoprene	535
Monoterpenes	
α-pinene	66
Other	96
Sesquiterpenes	29
Methanol	100
Acetone	44
Ethanol	21
Acetaldehyde	21
Ethene	27
Propene	16
Other	51
Total BVOC	1007

Source: From Guenther et al. (2012).

Monoterpenes and sesquiterpenes comprise 16 percent and 3 percent, respectively. Almost one-half of the total monoterpene emissions is α-pinene. Other significant emissions include oxygenated BVOCs (methanol, 10%; acetone, 4%; ethanol, 2%; acetaldehyde, 2%), ethene (3%), and propene (<2%). Collectively, these species comprise almost one-quarter of BVOC emissions. Tropical trees emit ~80 percent of global isoprene and monoterpene emissions and about one-half of other BVOC emissions. Temperate forests account for about 10 percent each of global isoprene, monoterpene, and other BVOC emissions. Monoterpene emissions exceed isoprene emissions in temperate and boreal conifer forests.

The formation of secondary organic aerosols from BVOCs is complex, and their precise radiative forcing is uncertain. However, studies find that biogenic secondary organic aerosols may provide a negative radiative forcing. This is particularly evident in boreal forests at northern high latitudes. Conifer forests are strong emitters of monoterpenes. Atmospheric measurements find a high amount of aerosols over boreal forests originating from monoterpenes (Tunved et al. 2006). The formation of aerosols over conifer forests correlates closely with seasonal variation in forest productivity, likely from monoterpene emissions (Kulmala et al. 2004). Model simulations show that boreal forest BVOC emissions exert a negative radiative forcing on climate from cloud–albedo interactions (Spracklen et al. 2008; Scott et al. 2014). Measurements in Finland also find a negative radiative forcing from biogenic aerosol formation over boreal forests (Kurtén et al. 2003; Lihavainen et al. 2009).

Emissions of BVOCs are expected to increase with a warmer planet, creating a negative climate feedback due to increased aerosol concentrations (Carslaw et al. 2010). However, the complex response to other agents of global change must also be considered, and leaf-scale responses may not manifest at regional or continental scales (Sharkey and Monson 2014). For example, inhibition of isoprene emissions by elevated atmospheric CO_2 concentrations counters the expected increase in emissions with a warmer climate (Arneth et al. 2007; Heald et al. 2009). Anthropogenic land use and land-cover change are also critical determinants of BVOC emissions. Over the twentieth century, there has been a reduction in global isoprene emissions because the increase with warming has been offset by decreases in emissions with elevated atmospheric CO_2 and because of land-use changes that replaced forests (high BVOC emitter) with low BVOC-emitting croplands and pasturelands (Lathière et al. 2010; Unger 2013).

Future uses of land will play an important role in BVOC emissions and chemistry–climate interactions involving ozone and secondary organic aerosols. A key land cover transition is loss of forests with an increase in croplands and pasturelands. Tropical deforestation, for example, reduces isoprene emissions (Lathière et al. 2006). Changes in land cover over the twenty-first century have a greater effect than warming and elevated CO_2 on isoprene emissions (Guenther et al. 2006; Heald et al. 2008; Chen et al. 2009; Wu et al. 2012; Tai et al. 2013; Squire et al. 2014). In contrast, tree planting for carbon sequestration may increase emissions, as does the expansion of northern temperate and boreal forests with a warmer climate (Lathière et al. 2005, 2006; Heald et al. 2009; Wu et al. 2012). Some biofuel crops (e.g., tropical oil palm

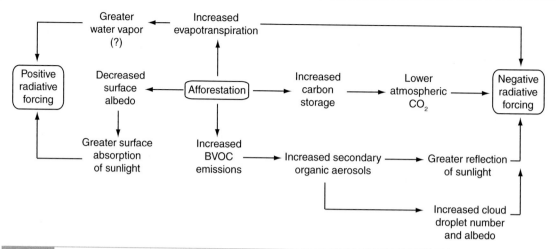

Fig. 31.9 Biogeophysical (albedo, evapotranspiration) and biogeochemical (carbon, aerosols) processes by which increased tree cover in the boreal forest affects climate. The positive radiative forcing from increased water vapor is uncertain and pertains to deciduous trees. Adapted from Spracklen et al. (2008).

and short rotation plantations of eucalyptus, poplar, or willow in mid-latitudes) are strong emitters of isoprene, and cultivation of these crops increases emissions (Ashworth et al. 2012; Hardacre et al. 2013). Replacement of cropland and grassland in Europe with biofuels increases isoprene emissions and leads to higher ozone, and this can offset policies to improve air quality by limiting ozone precursors (Ashworth et al. 2013).

The southern region of the United States and the eastern seaboard are a principal region of isoprene and monoterpene emissions. These BVOCs contribute to the high ozone of the southeastern region (Chameides et al. 1988) and also to secondary organic aerosols that form a regional cooling haze (Goldstein et al. 2009). In this region of the United States, harvesting, timber management, and forest succession substantially impact BVOC emissions through changes in forest leaf area and community composition (Purves et al. 2004).

Changes in the disturbance regime may also affect BVOC emissions. Fires emit large quantities of BVOCs during biomass combustion, which impact regional ozone formation and atmospheric chemistry (Ciccioli et al. 2014). Pine beetle infestations in coniferous forests of British Columbia and western United States increase monoterpene emissions when trees are under attack, which may increase the amount

of secondary organic aerosols (Amin et al. 2012; Berg et al. 2013), though the increases are small compared with large interannual variability and tree mortality decreases monterpene emissions in the long run (Berg et al. 2013).

Emissions of BVOCs in a changing environment may initiate feedbacks that affect climate change. One such negative feedback arises from increased forest productivity in a warmer, CO_2-enriched world. The productivity-driven increase in BVOC emissions enhances the production of secondary organic aerosols, which cool temperatures (Kulmala et al. 2004). Such a feedback among the biosphere, aerosols, and climate, in which warmer temperatures enhance BVOC emissions and aerosol formation, has been observed (Paasonen et al. 2013). Another feedback may arise because of enhancement of forest productivity from diffuse radiation generated by the high aerosol loading (Kulmala et al. 2014).

Studies of BVOC emissions and aerosols in boreal forests show that these forests influence climate not just by albedo, evapotranspiration, and carbon storage, but also through aerosol formation (Figure 31.9). Indeed, an analysis of a 40-year-old Scots pine stand in southern Finland shows that biogenic aerosols augment the negative radiative forcing from large carbon sequestration in this fast growing forest and counter the positive radiative forcing from surface albedo (Kurtén et al. 2003). A broad biogeophysical and biogeochemical understanding of the role of boreal forests in the Earth system and its effect on climate is required (Spracklen et al. 2008).

31.7 Biosphere Mediated Chemistry–Climate Feedbacks

Chemistry–climate feedbacks related to trace gas emissions, chemical reactions in the atmosphere, and aerosol formation (as described here and in Chapter 30) are increasingly recognized as being important for climate simulation (Arneth et al. 2010). It is not just the long-lived greenhouses gases (e.g., CO_2, CH_4, N_2O) that are important. Short-lived climate forcings from tropospheric chemistry related to NO_x, BVOCs, ozone, CH_4, and aerosols are also significant.

Unger (2014) used a global model to examine the chemistry–climate consequences of historical cropland expansion between the 1850s and 2000s. The loss of forests and increase in croplands reduces global BVOC emissions by ~35 percent. This substantially lowers the amount of ozone, CH_4, and secondary organic aerosols in the atmosphere. The decreases in ozone and CH_4 are a negative radiative forcing of -0.13 W m^{-2} and -0.06 W m^{-2}, respectively, while the reduction in aerosols is a positive radiative forcing equal to 0.09 W m^{-2}. The net radiative forcing is -0.11 W m^{-2}, comparable to that from land-use changes in surface albedo (-0.15 W m^{-2}) or from CO_2 emissions (0.17 W m^{-2}) over the same period. These estimates are likely to be revised with additional studies, but they suggest that altered emission of BVOCs from anthropogenic uses of land has an effect on global climate that is of similar magnitude to other more commonly assessed effects of land use.

Biosphere mediated chemistry–climate feedbacks are also important for simulation of past climates millions of years ago. Biogenic production of isoprene and monoterpenes, emissions of NO_x, CO, black carbon, organic carbon and other materials during wildfire, and emissions of NO_x from lightning and soil helped maintain high atmospheric concentrations of CH_4, N_2O, and tropospheric ozone in the warm climate of the mid-Pliocene 3 million years ago (Unger and Yue 2014). Biogenic reactive gas emissions were similarly critical to the high atmospheric concentrations of CH_4, N_2O, and tropospheric ozone in the early Eocene 55 million years ago and the late Cretaceous 90 million years ago (Beerling et al. 2011).

Biogenic chemistry–climate feedbacks for global climate simulation are an emerging research frontier. In particular, the climate consequences of land use and land-cover change are mediated through biogenic emissions of reactive gases and atmospheric chemistry in addition to biogeophysical and carbon cycle processes. Tropical deforestation, for example, causes changes in soil NO_x and plant BVOC emissions that affect tropospheric ozone and the hydroxyl radical. The chemistry-mediated effects of deforestation are a complex response to changes in these biogenic emissions, dry deposition, boundary layer processes, and atmospheric chemistry (Ganzeveld and Lelieveld 2004; Ganzeveld et al. 2010). Meteorology within the canopy and in the atmospheric boundary layer above the canopy affect biogenic emissions of BVOCs, NO_x, and other reactive gases, dry deposition within the canopy, and the chemistry of the canopy air. As a result, there can be sharp gradients of ozone and NO_x in the canopy. How to represent this in-canopy chemistry for climate simulations is a key research need, though such models have been developed at the stand scale (Ganzeveld et al. 2002a,b; Forkel et al. 2006; Boy et al. 2011; Wolfe and Thornton 2011; Saylor 2013).

31.8 Review Questions

1. Contrast the radiative effect of scattering aerosols (e.g., dust) over a surface covered by fresh snow and a vegetated surface. Over which surface are aerosol–radiation interactions likely to be greatest? Compare this with the effect of absorbing aerosols (e.g., black carbon).

2. Describe how mineral aerosols affect climate.

3. Describe the effects of fire on climate. What are the consequences of more fires in a warmer and drier world?

4. Describe how BVOCs affect climate. What processes regulate BVOC emissions in a future Earth and what climate feedbacks do these future emissions create?

5. Describe the effects of increases in NO_x, CO, CH_4, and nonmethane VOCs on tropospheric ozone and OH.

6. Reduced emission of NO_x is one means to decrease tropospheric ozone. What are the likely consequences for OH and CH_4 of future anthropogenic emissions that restrict NO_x? How will higher future emissions of carbon monoxide affect this outcome?

7. Our current understanding of tropical deforestation emphasizes changes in surface energy fluxes (biogeophysics) and carbon emissions. Contrast these with the chemistry–climate effects of deforestation.

8. Describe the processes by which the presence of boreal forests provides a positive and a negative radiative forcing.

31.9 | References

Akagi, S. K., Yokelson, R. J., Wiedinmyer, C., et al. (2011). Emission factors for open and domestic biomass burning for use in atmospheric models. *Atmospheric Chemistry and Physics*, 11, 4039–4072.

Amin, H., Atkins, P. T., Russo, R. S., et al. (2012). Effect of bark beetle infestation on secondary organic aerosol precursor emissions. *Environmental Science and Technology*, 46, 5696–5703.

Andreae, M. O., and Crutzen, P. J. (1997). Atmospheric aerosols: Biogeochemical sources and role in atmospheric chemistry. *Science*, 276, 1052–1058.

Andreae, M. O., and Merlet P. (2001). Emissions of trace gases and aerosols from biomass burning. *Global Biogeochemical Cycles*, 15, 955–966, doi:10.1029/2000GB001382.

Andreae, M. O., and Rosenfeld, D. (2008). Aerosol–cloud–precipitation interactions, Part 1: The nature and sources of cloud-active aerosols. *Earth-Science Reviews*, 89, 13–41.

Andreae, M. O., Rosenfeld, D., Artaxo, P., et al. (2004). Smoking rain clouds over the Amazon. *Science*, 303, 1337–1342.

Arneth, A., Miller, P. A., Scholze, M., et al. (2007). CO_2 inhibition of global terrestrial isoprene emissions: Potential implications for atmospheric chemistry. *Geophysical Research Letters*, 34, L18813, doi:10.1029/2007GL030615.

Arneth, A., Monson, R. K., Schurgers, G., Niinemets, Ü., and Palmer, P. I. (2008). Why are estimates of global terrestrial isoprene emissions so similar (and why is this not so for monoterpenes)? *Atmospheric Chemistry and Physics*, 8, 4605–4620.

Arneth, A., Harrison, S. P., Zaehle, S., et al. (2010). Terrestrial biogeochemical feedbacks in the climate system. *Nature Geoscience*, 3, 525–532.

Ashworth, K., Folberth, G., Hewitt, C. N., and Wild, O. (2012). Impacts of near-future cultivation of biofuel feedstocks on atmospheric composition and local air quality. *Atmospheric Chemistry and Physics*, 12, 919–939.

Ashworth, K., Wild, O., and Hewitt, C. N. (2013). Impacts of biofuel cultivation on mortality and crop yields. *Nature Climate Change*, 3, 492–496.

Atkinson, R. (2000). Atmospheric chemistry of VOCs and NO_x. *Atmospheric Environment*, 34, 2063–2101.

Atkinson, R., and Arey, J. (2003). Gas-phase tropospheric chemistry of biogenic volatile organic compounds: A review. *Atmospheric Environment*, 37(S2), S197–S219.

Beerling, D. J., and Osborne, C. P. (2006). The origin of the savanna biome. *Global Change Biology*, 12, 2023–2031.

Beerling, D. J., Fox, A., Stevenson, D. S., and Valdes, P. J. (2011). Enhanced chemistry–climate feedbacks in past greenhouse worlds. *Proceedings of the National Academy of Sciences USA*, 108, 9770–9775.

Berg, A. R., Heald, C. L., Huff Hartz, K. E., et al. (2013). The impact of bark beetle infestations on monoterpene emissions and secondary organic aerosol formation in western North America. *Atmospheric Chemistry and Physics*, 13, 3149–3161.

Bond, T. C., Doherty, S. J., Fahey, D. W., et al. (2013). Bounding the role of black carbon in the climate system: A scientific assessment. *Journal of Geophysical Research: Atmospheres*, 118, 5380–5552, doi:10.1002/jgrd.50171.

Boucher, O., Randall, D., Artaxo, P., et al. (2013). Clouds and aerosols. In *Climate Change 2013: The Physical Science Basis. Contribution of Working Group I to the Fifth Assessment Report of the Intergovernmental*

Panel on Climate Change, ed. T. F. Stocker, D. Qin, G.-K. Plattner, et al. Cambridge: Cambridge University Press, pp. 571–657.

Bowman, D. M. J. S., Balch, J. K., Artaxo, P., et al. (2009). Fire in the Earth System. *Science*, 324, 481–484.

Boy, M., Sogachev, A., Lauros, J., et al. (2011). SOSA – a new model to simulate the concentrations of organic vapours and sulphuric acid inside the ABL – Part 1: Model description and initial evaluation. *Atmospheric Chemistry and Physics*, 11, 43–51.

Carslaw, K. S., Boucher, O., Spracklen, D. V., et al. (2010). A review of natural aerosol interactions and feedbacks within the Earth system. *Atmospheric Chemistry and Physics*, 10, 1701–1737.

Chameides, W. L., Lindsay, R. W., Richardson, J., and Kiang, C. S. (1988). The role of biogenic hydrocarbons in urban photochemical smog: Atlanta as a case study. *Science*, 241, 1473–1475.

Chen, J., Avise, J., Guenther, A., et al. (2009). Future land use and land cover influences on regional biogenic emissions and air quality in the United States. *Atmospheric Environment*, 43, 5771–5780.

Ciccioli, P., Centritto, M., and Loreto, F. (2014). Biogenic volatile organic compound emissions from vegetation fires. *Plant, Cell and Environment*, 37, 1810–1825.

Cook, B. I., Miller, R. L., and Seager, R. (2008). Dust and sea surface temperature forcing of the 1930s "Dust Bowl" drought. *Geophysical Research Letters*, 35, L08710, doi:10.1029/2008GL033486.

Cook, B. I., Miller, R. L., and Seager, R. (2009). Amplification of the North American "Dust Bowl" drought through human-induced land degradation. *Proceedings of the National Academy of Sciences USA*, 106, 4997–5001.

Cook, B. I., Seager, R., Miller, R. L., and Mason, J. A. (2013). Intensification of North American mega-droughts through surface and dust aerosol forcing. *Journal of Climate*, 26, 4414–4430.

Creamean, J. M., Suski, K. J., Rosenfeld, D., et al. (2013). Dust and biological aerosols from the Sahara and Asia influence precipitation in the western U.S. *Science*, 339, 1572–1578.

Crutzen, P. J., and Andreae, M. O. (1990). Biomass burning in the tropics: Impact on atmospheric chemistry and biogeochemical cycles. *Science*, 250, 1669–1678.

D'Odorico, P., Bhattachan, A., Davis, K. F., Ravi, S., and Runyan, C. W. (2013). Global desertification: Drivers and feedbacks. *Advances in Water Resources*, 51, 326–344.

Field, J. P., Belnap, J., Breshears, D. D., et al. (2010). The ecology of dust. *Frontiers in Ecology and the Environment*, 8, 423–430.

Flanner, M. G., Zender, C. S., Randerson, J. T., and Rasch, P. J. (2007). Present-day climate forcing and response from black carbon in snow. *Journal of Geophysical Research*, 112, D11202, doi:10.1029/2006JD008003.

Flanner, M. G., Zender, C. S., Hess, P. G., et al. (2009). Springtime warming and reduced snow cover from carbonaceous particles. *Atmospheric Chemistry and Physics*, 9, 2481–2497.

Forkel, R., Klemm, O., Graus, M., et al. (2006). Trace gas exchange and gas phase chemistry in a Norway spruce forest: A study with a coupled 1-dimensional canopy atmospheric chemistry emission model. *Atmospheric Environment*, 40, S28–S42.

Formenti, P., Andreae, M. O., Lang, L., et al. (2001). Saharan dust in Brazil and Suriname during the Large-Scale Biosphere-Atmosphere Experiment in Amazonia (LBA) – Cooperative LBA Regional Experiment (CLAIRE) in March 1998. *Journal of Geophysical Research*, 106D, 14919–14934.

Forster, P., Ramaswamy, V., Artaxo, P., et al. (2007). Changes in atmospheric constituents and in radiative forcing. In *Climate Change 2007: The Physical Science Basis. Contribution of Working Group I to the Fourth Assessment Report of the Intergovernmental Panel on Climate Change*, ed. S. Solomon, D. Qin, M. Manning, et al. Cambridge: Cambridge University Press, pp. 129–234.

Fuentes, J. D., Lerdau, M., Atkinson, R., et al. (2000). Biogenic hydrocarbons in the atmospheric boundary layer: A review. *Bulletin of the American Meteorological Society*, 81, 1537–1575.

Ganzeveld, L., and Lelieveld, J. (2004). Impact of Amazonian deforestation on atmospheric chemistry. *Geophysical Research Letters*, 31, L06105, doi:10.1029/2003GL019205.

Ganzeveld, L. N., Lelieveld, J., Dentener, F. J., Krol, M. C., and Roelofs, G.-J. (2002a). Atmosphere–biosphere trace gas exchanges simulated with a single-column model. *Journal of Geophysical Research*, 107, doi:10.1029/2001JD000684.

Ganzeveld, L. N., Lelieveld, J., Dentener, F. J., et al. (2002b). Global soil-biogenic NO_x emissions and the role of canopy processes. *Journal of Geophysical Research*, 107, doi: 10.1029/2001JD001289.

Ganzeveld, L., Bouwman, L., Stehfest, E., et al. (2010). Impact of future land use and land cover changes on atmospheric chemistry-climate interactions. *Journal of Geophysical Research*, 115, D23301, doi:10.1029/2010JD014041.

Gedney, N., Huntingford, C., Weedon, G. P., et al. (2014). Detection of solar dimming and brightening effects on Northern Hemisphere river flow. *Nature Geoscience*, 7, 796–800.

Ginoux, P., Prospero, J. M., Gill, T. E., Hsu, N. C., and Zhao, M. (2012). Global-scale attribution of anthropogenic and natural dust sources and their emission rates based on MODIS Deep Blue aerosol products. *Reviews of Geophysics*, 50, RG3005, doi:10.1029/2012RG000388.

Goldstein, A. H., Koven, C. D., Heald, C. L., and Fung, I. Y. (2009). Biogenic carbon and anthropogenic pollutants combine to form a cooling haze over the southeastern United States. *Proceedings of the National Academy of Sciences USA*, 106, 8835–8840.

Grote, R., and Niinemets, Ü. (2008). Modeling volatile isoprenoid emissions – a story with split ends. *Plant Biology*, 10, 8–28.

Guenther, A., Hewitt, C. N., Erickson, D., et al. (1995). A global model of natural volatile organic compound emissions. *Journal of Geophysical Research*, 100D, 8873–8892.

Guenther, A., Karl, T., Harley, P., et al. (2006). Estimates of global terrestrial isoprene emissions using MEGAN (Model of Emissions of Gases and Aerosols from Nature). *Atmospheric Chemistry and Physics*, 6, 3181–3210.

Guenther, A. B., Jiang, X., Heald, C. L., et al. (2012). The Model of Emissions of Gases and Aerosols from Nature version 2.1 (MEGAN2.1): An extended and updated framework for modeling biogenic emissions. *Geoscientific Model Development*, 5, 1471–1492.

Hansen, J., and Nazarenko, L. (2004). Soot climate forcing via snow and ice albedos. *Proceedings of the National Academy of Sciences USA*, 101, 423–428.

Hardacre, C. J., Palmer, P. I., Baumanns, K., Rounsevell, M., and Murray-Rust, D. (2013). Probabilistic estimation of future emissions of isoprene and surface oxidant chemistry associated with land-use change in response to growing food needs. *Atmospheric Chemistry and Physics*, 13, 5451–5472.

Harrison, S. P., Morfopoulos, C., Dani, K. G. S., et al. (2013). Volatile isoprenoid emissions from plastid to planet. *New Phytologist*, 197, 49–57.

Heald, C. L., Henze, D. K., Horowitz, L. W., et al. (2008). Predicted change in global secondary organic aerosol concentrations in response to future climate, emissions, and land use change. *Journal of Geophysical Research*, 113, D05211, doi:10.1029/2007JD009092.

Heald, C. L., Wilkinson, M. J., Monson, R. K., et al. (2009). Response of isoprene emission to ambient CO_2 changes and implications for global budgets. *Global Change Biology*, 15, 1127–1140.

Herwitz, S. R., Muhs, D. R., Prospero, J. M., Mahan, S., and Vaughn, B. (1996). Origin of Bermuda's clay-rich Quaternary paleosols and their paleoclimatic significance. *Journal of Geophysical Research*, 101D, 23389–23400.

Hoelzemann, J. J., Schultz, M. G., Brasseur, G. P., Granier, C., and Simon, M. (2004). Global Wildland Fire Emission Model (GWEM): Evaluating the use of global area burnt satellite data. *Journal of Geophysical Research*, 109, D14S04, doi:10.1029/2003JD003666.

Huneeus, N., Schulz, M., Balkanski, Y., et al. (2011). Global dust model intercomparison in AeroCom phase I. *Atmospheric Chemistry and Physics*, 11, 7781–7816.

Jaffe, D. A., and Wigder, N. L. (2012). Ozone production from wildfires: A critical review. *Atmospheric Environment*, 51, 1–10.

Jickells, T. D., An, Z. S., Andersen, K. K., et al. (2005). Global iron connections between desert dust, ocean biogeochemistry, and climate. *Science*, 308, 67–71.

Jimenez, J. L., Canagaratna, M. R., Donahue, N. M., et al. (2009). Evolution of organic aerosols in the atmosphere. *Science*, 326, 1525–1529.

Keenan, T., Niinemets, Ü., Sabate, S., Gracia, C., and Peñuelas, J. (2009). Process based inventory of isoprenoid emissions from European forests: Model comparisons, current knowledge and uncertainties. *Atmospheric Chemistry and Physics*, 9, 4053–4076.

Koren, I., Kaufman, Y. J., Remer, L. A., and Martins, J. V. (2004). Measurement of the effect of Amazon smoke on inhibition of cloud formation. *Science*, 303, 1342–1345.

Kulmala, M., Suni, T., Lehtinen, K. E. J., et al. (2004). A new feedback mechanism linking forests, aerosols, and climate. *Atmospheric Chemistry and Physics*, 4, 557–562.

Kulmala, M., Nieminen, T., Nikandrova, A., et al. (2014). CO_2-induced terrestrial climate feedback mechanism: From carbon sink to aerosol source and back. *Boreal Environment Research*, 19 (suppl. B), 122–131.

Kurtén, T., Kulmala, M., Dal Maso, M., et al. (2003). Estimation of different forest-related contributions to the radiative balance using observations in southern Finland. *Boreal Environment Research*, 8, 275–285.

Laothawornkitkul, J., Taylor, J. E., Paul, N. D., and Hewitt, C. N. (2009). Biogenic volatile organic compounds in the Earth system. *New Phytologist*, 183, 27–51.

Lathière, J., Hauglustaine, D. A., De Noblet-Ducoudré, N., Krinner, G., and Folberth, G. A. (2005). Past and future changes in biogenic volatile organic compound emissions simulated with a global dynamic vegetation model. *Geophysical Research Letters*, 32, L20818, doi:10.1029/2005GL024164.

Lathière, J., Hauglustaine, D. A., Friend, A. D., et al. (2006). Impact of climate variability and land use changes on global biogenic volatile organic compound emissions. *Atmospheric Chemistry and Physics*, 6, 2129–2146.

Lathière, J., Hewitt, C. N., and Beerling D. J. (2010). Sensitivity of isoprene emissions from the terrestrial biosphere to 20th century changes in atmospheric CO_2 concentration, climate, and land use. *Global Biogeochemical Cycles*, 24, GB1004, doi:10.1029/2009GB003548.

Lee, Y. H., Lamarque, J.-F., Flanner, M. G., et al. (2013). Evaluation of preindustrial to present-day black carbon and its albedo forcing from Atmospheric Chemistry and Climate Model Intercomparison Project (ACCMIP). *Atmospheric Chemistry and Physics*, 13, 2607–2634.

Lihavainen, H., Kerminen, V.-M., Tunved, P., et al. (2009). Observational signature of the direct radiative effect by natural boreal forest aerosols and its relation to the corresponding first indirect effect. *Journal of Geophysical Research*, 114, D20206, doi:10.1029/2009JD012078.

Loreto, F., and Schnitzler, J.-P. (2010). Abiotic stresses and induced BVOCs. *Trends in Plant Science*, 15, 154–166.

Mahowald, N. (2011). Aerosol indirect effect on biogeochemical cycles and climate. *Science*, 334, 794–796.

Mahowald, N. M., and Luo, C. (2003). A less dusty future? *Geophysical Research Letters*, 30, 1903, doi:10.1029/2003GL017880.

Mahowald, N., Kohfeld, K., Hansson, M., et al. (1999). Dust sources and deposition during the last glacial maximum and current climate: A comparison of model results with paleodata from ice cores and marine sediments. *Journal of Geophysical Research*, 104D, 15895–15916.

Mahowald, N. M., Baker, A. R., Bergametti, G., et al. (2005). Atmospheric global dust cycle and iron inputs to the ocean. *Global Biogeochemical Cycles*, 19, GB4025, doi:10.1029/2004GB002402.

Mahowald, N. M., Muhs, D. R., Levis, S., et al. (2006). Change in atmospheric mineral aerosols in response to climate: Last glacial period, preindustrial, modern, and doubled carbon dioxide climates. *Journal of Geophysical Research*, 111, D10202, doi:10.1029/2005JD006653.

Mahowald, N., Jickells, T. D., Baker, A. R., et al. (2008). Global distribution of atmospheric phosphorus sources, concentrations and deposition rates, and anthropogenic impacts. *Global Biogeochemical Cycles*, 22, GB4026, doi:10.1029/2008GB003240.

Mahowald, N. M., Engelstaedter, S., Luo, C., et al. (2009). Atmospheric iron deposition: Global distribution, variability, and human perturbations. *Annual Review of Marine Science*, 1, 245–278

Mahowald, N. M., Kloster, S., Engelstaedter, S., et al. (2010). Observed 20th century desert dust variability: Impact on climate and biogeochemistry. *Atmospheric Chemistry and Physics*, 10, 10875–10893.

Mahowald, N., Ward, D. S., Kloster, S., et al. (2011). Aerosol impacts on climate and biogeochemistry. *Annual Review of Environment and Resources*, 36, 45–74.

Mao, J., Horowitz, L. W., Naik, V., et al. (2013). Sensitivity of tropospheric oxidants to biomass burning emissions: Implications for radiative forcing. *Geophysical Research Letters*, 40, 1241–1246, doi:10.1002/grl.50210.

Marcella, M. P., and Eltahir, E. A. B. (2014). The role of mineral aerosols in shaping the regional climate of West Africa. *Journal of Geophysical Research: Atmospheres*, 119, 5806–5822, doi:10.1002/2012JD019394.

Martin, S. T., Andreae, M. O., Artaxo, P., et al. (2010). Sources and properties of Amazonian aerosol particles. *Reviews of Geophysics*, 48, RG2002, doi:10.1029/2008RG000280.

McGee, D., Broecker, W. S., and Winckler, G. (2010). Gustiness: The driver of glacial dustiness? *Quaternary Science Reviews*, 29, 2340–2350.

Mulitza, S., Heslop, D., Pittauerova, D., et al. (2010). Increase in African dust flux at the onset of commercial agriculture in the Sahel region. *Nature*, 466, 226–228.

Myhre, G., Shindell, D., Bréon, F.-M., et al. (2013). Anthropogenic and natural radiative forcing. In *Climate Change 2013: The Physical Science Basis. Contribution of Working Group I to the Fifth Assessment Report of the Intergovernmental Panel on Climate Change*, ed. T. F. Stocker, D. Qin, G.-K. Plattner, et al. Cambridge: Cambridge University Press, pp. 659–740.

Neff, J. C., Ballantyne, A. P., Farmer, G. L., et al. (2008). Increasing eolian dust deposition in the western United States linked to human activity. *Nature Geoscience*, 1, 189–195.

Nicholson, S. E. (2000). Land surface processes and Sahel climate. *Reviews of Geophysics*, 38, 117–140.

Okin, G. S., Mahowald, N., Chadwick, O. A., and Artaxo, P. (2004). Impact of desert dust on the biogeochemistry of phosphorus in terrestrial eco-systems. *Global Biogeochemical Cycles*, 18, GB2005, doi:10.1029/2003GB002145.

Paasonen, P., Asmi, A., Petäjä, T., et al. (2013). Warming-induced increase in aerosol number concentration likely to moderate climate change. *Nature Geoscience*, 6, 438–442.

Pacifico, F., Harrison, S. P., Jones, C. D., and Sitch, S. (2009). Isoprene emissions and climate. *Atmospheric Environment*, 43, 6121–6135.

Painter, T. H., Barrett, A. P., Landry, C. C., et al. (2007). Impact of disturbed desert soils on duration of mountain snow cover. *Geophysical Research Letters*, 34, L12502, doi:10.1029/2007GL030284.

Painter, T. H., Deems, J. S., Belnap, J., et al. (2010). Response of Colorado River runoff to dust radiative forcing in snow. *Proceedings of the National Academy of Sciences USA*, 107, 17125–17130.

Painter, T. H., Skiles, S. M., Deems, J. S., Bryant, A. C., and Landry, C. C. (2012). Dust radiative forcing in snow of the Upper Colorado River Basin, 1: A 6 year record of energy balance, radiation, and dust concentrations. *Water Resources Research*, 48, W07521, doi:10.1029/2012WR011985.

Park, J.-H., Goldstein, A. H., Timkovsky, J., et al. (2013). Active atmosphere–ecosystem exchange of the vast majority of detected volatile organic compounds. *Science*, 341, 643–647.

Peñuelas, J., and Staudt, M. (2010). BVOCs and global change. *Trends in Plant Science*, 15, 133–144.

Pöschl, U., Martin, S. T., Sinha, B., et al. (2010). Rainforest aerosols as biogenic nuclei of clouds and precipitation in the Amazon. *Science*, 329, 1513–1516.

Prospero, J. M., and Lamb, P. J. (2003). African droughts and dust transport to the Caribbean: Climate change implications. *Science*, 302, 1024–1027.

Prospero, J. M., and Nees, R. T. (1977). Dust concentration in the atmosphere of the equatorial North Atlantic: Possible relationship to the Sahelian drought. *Science*, 196, 1196–1198.

Prospero, J. M., and Nees, R. T. (1986). Impact of the North African drought and El Niño on mineral dust in the Barbados trade winds. *Nature*, 320, 735–738.

Prospero, J. M., Glaccum, R. A., and Nees, R. T. (1981). Atmospheric transport of soil dust from Africa to South America. *Nature*, 289, 570–572.

Prospero, J. M., Nees, R. T., and Uematsu, M. (1987). Deposition rate of particulate and dissolved aluminum derived from Saharan dust in precipitation at Miami, Florida. *Journal of Geophysical Research*, 92D, 14723–14731.

Prospero, J. M., Barrett, K., Church, T., et al. (1996). Atmospheric deposition of nutrients to the North Atlantic Basin. *Biogeochemistry*, 35, 27–73.

Purves, D. W., Caspersen, J. P., Moorcroft, P. R., Hurtt, G. C., and Pacala, S. W. (2004). Human-induced changes in US biogenic volatile organic compound emissions: Evidence from long-term forest inventory data. *Global Change Biology*, 10, 1737–1755.

Ramanathan, V., and Carmichael, G. (2008). Global and regional climate changes due to black carbon. *Nature Geoscience*, 1, 221–227.

Ramanathan, V., Crutzen, P. J., Kiehl, J. T., and Rosenfeld, D. (2001). Aerosols, climate, and the hydrological cycle. *Science*, 294, 2119–2124.

Randerson, J. T., Liu, H., Flanner, M. G., et al. (2006). The impact of boreal forest fire on climate warming. *Science*, 314, 1130–1132.

Ravi, S., D'Odorico, P., Breshears, D. D., et al. (2011). Aeolian processes and the biosphere. *Reviews of Geophysics*, 49, RG3001, doi:10.1029/2010RG000328.

Ridley, D. A., Heald, C. L., and Prospero, J. M. (2014). What controls the recent changes in African mineral dust aerosol across the Atlantic? *Atmospheric Chemistry and Physics*, 14, 5735–5747.

Rosenfeld, D., Rudich, Y., and Lahav, R. (2001). Desert dust suppressing precipitation: A possible desertification feedback loop. *Proceedings of the National Academy of Sciences USA*, 98, 5975–5980.

Rosenfeld, D., Lohmann, U., Raga, G. B., et al. (2008). Flood or drought: How do aerosols affect precipitation? *Science*, 321, 1309–1313.

Rosenfeld, D., Andreae, M. O., Asmi, A., et al. (2014). Global observations of aerosol-cloud-precipitation climate interactions. *Reviews of Geophysics*, 52, doi:10.1002/2013RG000441.

Saylor, R. D. (2013). The Atmospheric Chemistry and Canopy Exchange Simulation System (ACCESS): Model description and application to a temperate deciduous forest canopy. *Atmospheric Chemistry and Physics*, 13, 693–715.

Schultz, M. G., Heil, A., Hoelzemann, J. J., et al. (2008). Global wildland fire emissions from 1960 to 2000. *Global Biogeochemical Cycles*, 22, GB2002, doi:10.1029/2007GB003031.

Scott, C. E., Rap, A., Spracklen, D. V., et al. (2014). The direct and indirect radiative effects of biogenic secondary organic aerosol. *Atmospheric Chemistry and Physics*, 14, 447–470.

Sharkey, T. D., and Monson, R. K. (2014). The future of isoprene emission from leaves, canopies and landscapes. *Plant, Cell and Environment*, 37, 1727–1740.

Sharkey, T. D., Wiberley, A. E., and Donohue, A. R. (2008). Isoprene emission from plants: Why and how. *Annals of Botany*, 101, 5–18.

Spracklen, D. V., Bonn, B., and Carslaw, K. S. (2008). Boreal forests, aerosols and the impacts on clouds and climate. *Philosophical Transactions of the Royal Society A*, 366, 4613–4626.

Squire, O. J., Archibald, A. T., Abraham, N. L., et al. (2014). Influence of future climate and cropland expansion on isoprene emissions and tropospheric ozone. *Atmospheric Chemistry and Physics*, 14, 1011–1024.

Stanelle, T., Bey, I., Raddatz, T., Reick, C., and Tegen, I. (2014). Anthropogenically induced changes in twentieth century mineral dust burden and the associated impact on radiative forcing. *Journal of Geophysical Research: Atmospheres*, 119, 13526–13546, doi:10.1002/2014JD022062.

Swap, R., Garstang, M., Greco, S., Talbot, R., and Källberg, P. (1992). Saharan dust in the Amazon Basin. *Tellus B*, 44, 133–149.

Tai, A. P. K., Mickley, L. J., Heald, C. L., and Wu, S. (2013). Effect of CO_2 inhibition on biogenic isoprene emission: Implications for air quality under 2000 to 2050 changes in climate, vegetation, and land use. *Geophysical Research Letters*, 40, 3479–3483, doi:10.1002/grl.50650.

Tegen, I., and Fung, I. (1995). Contribution to the atmospheric mineral aerosol load from land surface modification. *Journal of Geophysical Research*, 100D, 18707–18726.

Tegen, I., Werner, M., Harrison, S. P., and Kohfeld, K. E. (2004). Relative importance of climate and land use in determining present and future global soil dust emission. *Geophysical Research Letters*, 31, L05105, doi:10.1029/2003GL019216.

Tosca, M. G., Randerson, J. T., Zender, C. S., Flanner, M. G., and Rasch, P. J. (2010). Do biomass burning aerosols intensify drought in equatorial Asia during El Niño? *Atmospheric Chemistry and Physics*, 10, 3515–3528.

Tosca, M. G., Randerson, J. T., and Zender, C. S. (2013). Global impact of smoke aerosols from landscape fires on climate and the Hadley circulation. *Atmospheric Chemistry and Physics*, 13, 5227–5241.

Tunved, P., Hansson, H.-C., Kerminen, V.-M., et al. (2006). High natural aerosol loading over boreal forests. *Science*, 312, 261–263.

Unger, N. (2013). Isoprene emission variability through the twentieth century. *Journal of Geophysical Research: Atmospheres*, 118, 13606–13613, doi:10.1002/2013JD020978.

Unger, N. (2014). Human land-use-driven reduction of forest volatiles cools global climate. *Nature Climate Change*, 4, 907–910.

Unger, N., and Yue, X. (2014). Strong chemistry–climate feedback in the Pliocene. *Geophysical Research Letters*, 41, 527–533, doi:10.1002/2013GL058773.

van der Werf, G. R., Randerson, J. T., Giglio, L., et al. (2010). Global fire emissions and the contribution of deforestation, savanna, forest, agricultural, and peat fires (1997–2009). *Atmospheric Chemistry and Physics*, 10, 11707–11735.

Ward, D. S., Kloster, S., Mahowald, N. M., et al. (2012). The changing radiative forcing of fires: Global model estimates for past, present and future. *Atmospheric Chemistry and Physics*, 12, 10857–10886.

Wiedinmyer, C., Akagi, S. K., Yokelson, R. J., et al. (2011). The Fire INventory from NCAR (FINN): A high resolution global model to estimate the emissions from open burning. *Geoscientific Model Development*, 4, 625–641.

Wolfe, G. M., and Thornton, J. A. (2011). The Chemistry of Atmosphere-Forest Exchange (CAFE) Model – Part 1: Model description and characterization. *Atmospheric Chemistry and Physics*, 11, 77–101.

Wu, S., Mickley, L. J., Kaplan, J. O., and Jacob, D. J. (2012). Impacts of changes in land use and land cover on atmospheric chemistry and air quality over the 21st century. *Atmospheric Chemistry and Physics*, 12, 1597–1609.

Yoshioka, M., Mahowald, N. M., Conley, A. J., et al. (2007). Impact of desert dust radiative forcing on Sahel precipitation: Relative importance of dust compared to sea surface temperature variations, vegetation changes, and greenhouse gas warming. *Journal of Climate*, 20, 1445–1467.

Zhang, Q., Jimenez, J. L., Canagaratna, M. R., et al. (2007). Ubiquity and dominance of oxygenated species in organic aerosols in anthropogenically-influenced Northern Hemisphere midlatitudes. *Geophysical Research Letters*, 34, L13801, doi:10.1029/2007GL029979.

Urbanization

32.1 | Chapter Summary

Urban land uses represent an alteration of the natural landscape. The vast tracts of impervious roads, sidewalks, driveways, parking lots, roofs, and walls alter the surface energy budget and the hydrologic cycle. The most prominent characteristic of the urban climate is the urban heat island, by which air temperature in cities can be several degrees warmer than that of rural landscapes. The heat island arises due to reduced emission of longwave radiation by the city surface, much of which is trapped by tall buildings; the low surface albedo of cities; reduced latent heat flux and increased sensible heat flux because of greater impervious surface area; storage of heat in urban materials during the day that is released at night; and from anthropogenic heat sources within the city. Cities also generate more runoff compared with rural landscapes because of greater impervious surface area. Vegetated parks within cities ameliorate the urban heat island by reducing impervious surface area and allowing for infiltration and evaporation of rainfall.

32.2 | Urban Morphology

Large cities have a distinct physical morphology. Table 32.1 illustrates this for 10 cities in the United States. Single-family residential housing is the dominant land use in all cities, comprising 49–78 percent of total land area. Apartment housing is generally modest (about 5% of area) except in Baltimore and Philadelphia, where row houses are abundant. Industrial zones are 10–22 percent of total land area, and commercial zones comprise 7–17 percent of area. Building heights are lowest in single-family residences (one to two stories) and highest in residential apartment and commercial zones, which range from one-story retail centers to multi-story skyscrapers. The combined surface area of roofs and walls ranges from 72 km^2 in Pittsburgh to 213 km^2 in Philadelphia. Residential housing is the single biggest contributor to total surface area, accounting for 68–88 percent of total wall and roof area. Expressed as a percentage of the corresponding ground area, single-family residential walls and roofs cover about 40–50 percent of the corresponding ground area, though this number is as high as 74 percent in Baltimore. Walls and roofs generally cover a similar or smaller fraction of ground area in industrial and commercial zones.

Urban vegetation has a distinct pattern that reflects human values, preferences, and choices. This is seen in an examination of urban forest cover in relation to population size and ecoregions. In a study of eight cities in the northern hardwood forest region of the United States with populations ranging from 2000 to 200,000 people, tree cover was found to range from 18 percent to 38 percent of total area (Halverson and Rowntree 1986). Tree cover decreased as population increased due to the larger area devoted to

Table 32.1 | Morphology of 10 large cities in the United States by land use

	Area					Wall and roof area (%)				Mean height (number of stories)			
	Total (km²)	R-SF (%)	R-A (%)	IZ (%)	CZ (%)	R-SF	R-A	IZ	CZ	R-SF	R-A	IZ	CZ
Pittsburgh	158	68	6	17	10	41	114	30	66	2	7	3	7
Boston	200	56	7	21	17	51	98	19	38	2	6	2	4
Sacramento	208	72	2	10	16	43	11	13	30	1	2	2	4
Cincinnati	226	62	6	22	10	42	33	43	34	2	2	2	4
Baltimore	228	49	21	14	16	74	130	39	32	2	5	3	4
Philadelphia	323	53	20	13	14	55	119	51	48	2	7	3	4
Denver	346	78	3	12	7	42	43	22	29	1	4	2	3
Seattle	391	73	1	15	10	32	105	43	31	2	3	2	3
Atlanta	440	74	5	12	8	39	40	27	38	1	2	2	5
Houston	493	67	3	21	10	42	42	25	63	2	2	2	5

Note: Total land area is divided into residential single family (R-SF), residential apartment (R-A), industrial (IZ), and commercial (CZ) zones. Wall and roof area is the combined surface area expressed as a percentage of land area in the land-use zone. Data from Ellefsen (1990/9). His 17 urban zones were reduced to 4 zones: residential single family (A3, Dc3, Do3); residential apartment (A2, Dc2, Do2); industrial (A4, Dc4, Do4); commercial (A1, A5, Dc1, Dc5, Dc8, Do1, Do5, Do6).

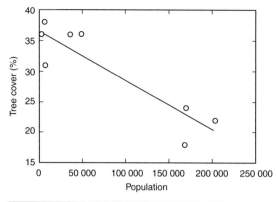

Fig. 32.1 Percentage of land covered by trees in relation to population for eight cities in the northern hardwood forest region of the United States. Data from Halverson and Rowntree (1986).

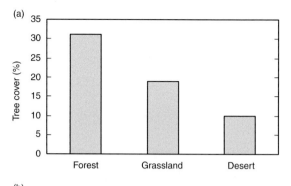

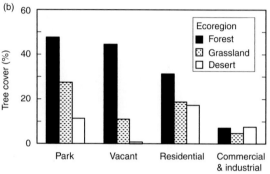

Fig. 32.2 Tree cover for 58 cities in forest, grassland, and desert ecoregions of the United States. (a) Percentage of land covered by trees in relation to ecoregion. (b) Percentage of land covered by trees in relation to urban land use and ecoregion. Data from Nowak et al. (1996).

transportation, commercial, and industrial uses (Figure 32.1). Another study related tree cover in 58 cities in the United States to the surrounding natural vegetation (Nowak et al. 1996). Tree cover as a percentage of total land area ranged from 0.4 percent to 55 percent. Tree cover was highest in cities located in forested ecoregions, lower in cities in grassland ecoregions, and lowest in desert regions (Figure 32.2a). Within a city, land use was the primary determinant of tree cover, and each land use had a characteristic structure and function that influenced tree cover (Figure 32.2b). Tree cover for park and vacant urban land was highest in forest ecoregions and lowest in desert ecoregions, reflecting the decreased availability of water. Differences in tree cover were not as large for residential land use (though forested regions had higher cover than other regions), reflecting a desire by homeowners for trees regardless of environment.

32.3 | The Urban Heat Island

Cities are often warmer than surrounding rural areas, especially at night. This phenomenon is known as the urban heat island (Landsberg 1981; Oke 1982, 1995; Arnfield 2003; Stewart and Oke 2012). This warming occurs because of greater absorption of solar radiation and trapping of longwave radiation within the city, storage and release of heat by buildings and paved surfaces, low vegetation cover and low evapotranspiration, and from anthropogenic heat sources within the city.

The development of Columbia, Maryland, a planned community along the Baltimore–Washington, D.C. corridor, provides a case study of how urbanization alters temperature. In 1968, when the population of Columbia was 1000 people, the maximum warming compared with surrounding rural area was 1°C throughout much of the town (Figure 32.3). A small business center with office buildings and parking lot was a local heat island of up to 3°C. By 1974, the population was 20,000 people, and the heat island was larger. Most of the town was more than 2°C warmer than the surrounding rural land. A central commercial and residential district was 5–7°C warmer. This trend in Columbia mirrors regional population

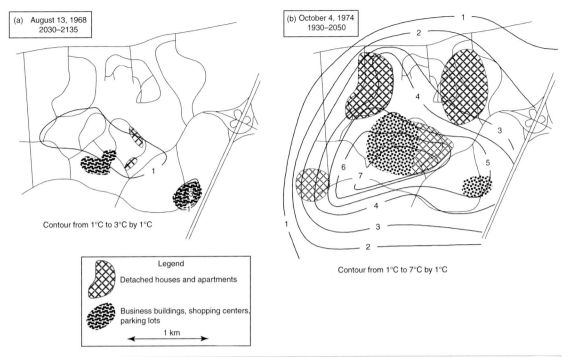

(a) August 13, 1968
2030–2135

Contour from 1°C to 3°C by 1°C

(b) October 4, 1974
1930–2050

Contour from 1°C to 7°C by 1°C

Legend

Detached houses and apartments

Business buildings, shopping centers, parking lots

1 km

Fig. 32.3 Evening air temperature for Columbia, Maryland. Temperature is the departure from a rural location. (a) 1968 when the population was 1000. (b) 1974 when the population was 20,000. Adapted from Landsberg and Maisel (1972) and Landsberg (1979).

growth and urban warming throughout the Baltimore–Washington, D.C. corridor between 1950 and 1979 (Viterito 1989).

In general, there is a relationship between urban warming, defined as the difference in temperature between a city and surrounding rural area, and population size. The maximum difference in urban and rural temperatures at any time increases as a logarithmic function of population (Figure 32.4a). Large cities of 100,000 to 1,000,000 people can be 8–12°C warmer than rural areas. For North American and European cities, Oke (1973) related the maximum instantaneous temperature difference ($\Delta T_{u-r(\max)}$) and population (P):

$$\text{North America: } \Delta T_{u-r(\max)} = 2.96 \log_{10}(P) - 6.41$$
$$\text{Europe: } \Delta T_{u-r(\max)} = 2.01 \log_{10}(P) - 4.06$$
$$(32.1)$$

Large cities in North America are a few degrees warmer than comparable European cities.

The formation and intensity of an urban heat island depends on weather conditions. The urban–rural temperature difference, when it exists, is generally largest at night, especially during clear, calm conditions. It is smallest with cloudy and windy conditions. Data collected in and near St. Louis, Missouri, illustrate the diurnal cycle (Figure 32.5a). On average, the urban site was warmer than the rural site (25.5°C versus 23.8°C). This difference was small during the day and larger at night, with a greatest difference of 3–4°C between 2300 and 0400 hours.

The dependence of the urban heat island with population shown in Figure 32.4a is the largest instantaneous temperature difference, which usually occurs at night with clear sky and calm wind. Karl et al. (1988) developed for cities in the United States relationships of population with daily mean temperature, daily minimum temperature, and daily maximum temperature averaged for all days in the year and for each of the four seasons. These relationships have the form:

$$\Delta \bar{T}_{u-r} = aP^{0.45} \qquad (32.2)$$

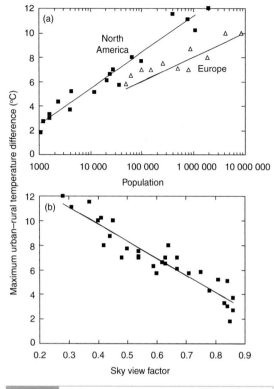

Fig. 32.4 Maximum air temperature difference between urban and rural areas in relation to (a) population and (b) sky view factor. Data are for 18 cities in North America, 11 cities in Europe, and 2 cities in Australia/Asia. Data from Oke (1981).

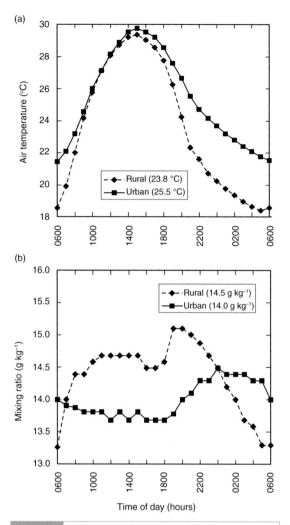

Fig. 32.5 Diurnal cycle of (a) air temperature and (b) humidity for urban and rural sites in St. Louis, Missouri. Data are for an average summer day from 1972 to 1975. Numbers in parentheses are daily means. Adapted from Semonin (1981). Changnon (1981) describes the St. Louis study.

where a equals 1.82×10^{-3} (daily mean), 3.61×10^{-3} (daily minimum), and -3.9×10^{-4} (daily maximum) when averaged over the year. Although the maximum urban–rural temperature difference at any time can be quite large, this temperature difference is much smaller when averaged over the year. The mean daily temperature of a city in the United States of one million people is about 1°C warmer than its rural counterpart. Most of this warming occurs at night, when the average daily minimum temperature is 1.8°C warmer. Daytime maximum temperature is relatively unchanged.

Vegetation phenology derived from satellite data provides an assessment of the urban heat island. Several studies have shown that leaf emergence in spring occurs several days earlier in urban areas than in rural areas while leaf senescence in autumn occurs later (White et al. 2002; Zhang et al. 2004). In eastern North America, for example, Zhang et al. (2004) estimated that springtime green-up occurs seven days earlier compared with surrounding rural areas and the onset of dormancy occurs eight days later. The urban heat island is particularly extensive along the corridor from Washington, D.C.–Philadelphia–New York City–Boston. In much of this region, spring green-up occurs 6–10 days earlier than adjacent rural areas.

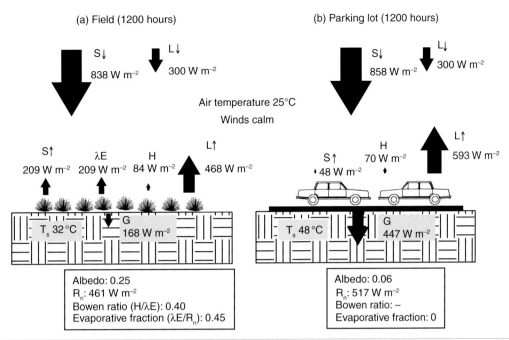

(a) Field (1200 hours)

$S\downarrow$
838 W m^{-2}

$L\downarrow$
300 W m^{-2}

Air temperature 25°C
Winds calm

$S\uparrow$
209 W m^{-2}

λE
209 W m^{-2}

H
84 W m^{-2}

$L\uparrow$
468 W m^{-2}

T_s 32°C

G
168 W m^{-2}

Albedo: 0.25
R_n: 461 W m^{-2}
Bowen ratio ($H/\lambda E$): 0.40
Evaporative fraction ($\lambda E/R_n$): 0.45

(b) Parking lot (1200 hours)

$S\downarrow$
858 W m^{-2}

$L\downarrow$
300 W m^{-2}

$S\uparrow$
48 W m^{-2}

H
70 W m^{-2}

$L\uparrow$
593 W m^{-2}

T_s 48°C

G
447 W m^{-2}

Albedo: 0.06
R_n: 517 W m^{-2}
Bowen ratio: –
Evaporative fraction: 0

Fig. 32.6 Surface energy fluxes measured at noon at (a) a field and (b) a nearby parking lot in Columbia, Maryland. Arrows are proportional to the size of the flux ($S\downarrow$, incoming solar radiation; $L\downarrow$, atmospheric longwave radiation; $S\uparrow$, reflected solar radiation; $L\uparrow$, emitted longwave radiation; λE, latent heat flux; H, sensible heat flux; G, soil heat flux; T_s, surface temperature; R_n, net radiation). Data from Landsberg and Maisel (1972).

32.4 | Urban Energy Fluxes

The energy balance of a city is quite distinct from that of rural locations (Landsberg 1981; Oke 1982, 1988, 1995; Arnfield 2003). In heavily polluted areas, incoming solar radiation can be reduced by 10–20 percent compared with rural regions. This, however, is compensated by increased downward longwave radiation from the gases and aerosols in the air. Anthropogenic heat sources from motor vehicles, power plants, industrial processes, and heating systems augment radiative heating. Anthropogenic heat sources can be considerable and in some cases may be comparable in magnitude to net radiation. The net energy impinging on the system is balanced by the sensible and latent heat returned to the atmosphere and heat stored in the urban system. Studies of the energy budget of cities routinely find less latent heat flux and greater sensible heat flux with urbanization. Storage heat flux can be a significant part of the

urban surface energy budget (Grimmond and Oke 1999).

The energy balance of an asphalt parking lot illustrates the manner in which urban surfaces alter energy fluxes. Figure 32.6 shows surface energy fluxes for a parking lot and nearby field at noon on a warm day. Both sites received the same incoming longwave radiation and similar solar radiation during the measurement period. However, the field had a higher albedo (0.25) than the parking lot (0.06), and the radiative forcing on the parking lot was 181 W m^{-2} more than that impinging on the field. For both surfaces, the vast majority of this energy was returned to the atmosphere as longwave radiation. Because it was hotter, the parking lot emitted more longwave radiation than the field. Of the 461 W m^{-2} net radiation on the field, 45 percent was returned to the atmosphere as latent heat, 36 percent was stored in the ground, and 18 percent was dissipated as sensible heat. In contrast, the parking lot had no latent heat flux. Sensible heat flux was reduced relative to the

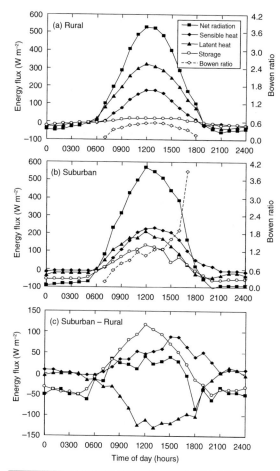

Fig. 32.7 Diurnal cycle of net radiation, sensible heat, latent heat, heat storage, and Bowen ratio (ratio of sensible heat to latent heat) at Vancouver, B.C., on an average summer day for (a) a rural site, (b) a suburban site, and (c) the suburban–rural difference. Data from Cleugh and Oke (1986).

diurnal cycle at rural and suburban locations measured on a typical summer day. During daytime, suburban–rural differences in net radiation were generally small compared with greater sensible heat, decreased latent heat, and increased heat storage at the suburban site than the rural location. The largest differences were a decrease in latent heat flux of greater than 100 W m^{-2} and a corresponding increase in heat storage. The suburban site also had a larger Bowen ratio than the rural site. The maximum Bowen ratio at the rural location was 0.6 at midday whereas the midday Bowen ratio was 1.2 at the suburban site. Late afternoon values were in excess of 1.8 at the suburban site. At night, turbulent fluxes were negligible and radiation was lost to the atmosphere (i.e., net radiation was negative). The loss of radiation was compensated for by release of heat stored during the day. The suburban site lost more radiation than the rural site (about 50 W m^{-2}) and had correspondingly larger release of stored heat.

Figure 32.8 shows the same data, but averaged over daylight hours. The rural site of managed grassland had a higher albedo than the suburban site (0.20 versus 0.13). Because more solar radiation was absorbed at the surface, the suburban site had a slightly higher net radiation flux than the rural site. Most of the net radiation at the rural site (66%) was returned to the atmosphere as latent heat; 30 percent was returned as sensible heat; only 4 percent was stored in the ground. In contrast, only 34 percent of the net radiation at the suburban site was dissipated as latent heat; 44 percent was returned as sensible heat; and 22 percent was stored in the urban fabric. Over the course of the day, the suburban site had 50 percent higher sensible heat flux and 46 percent lower latent heat flux compared with the rural site. These differences in energy fluxes were related to the impervious surface area at the suburban site (25% building, 11% paved) and reduced greenspace (64%).

32.5 | Urban Canyons and Radiation

The physical basis for the development of a heat island is understood through the concept of an

field, perhaps because the lower surface roughness reduced the efficiency with which heat was carried away from the surface. Instead, 86 percent of the net radiation at the surface was stored in the ground. As a result of the higher radiative forcing and greater heat storage, the parking lot surface was 16°C hotter than the field.

The urban energy budget has been studied in detail at Vancouver, British Columbia (Yap and Oke 1974; Oke 1979; Kalanda et al. 1980; Oke and McCaughey 1983; Cleugh and Oke 1986; Oke and Cleugh 1987; Grimmond 1992; Roth and Oke 1995). Figure 32.7 illustrates the

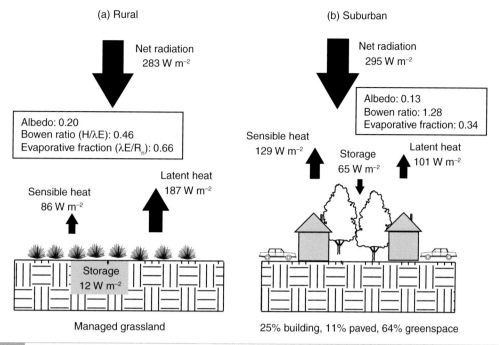

(a) Rural

Net radiation
283 W m^{-2}

Albedo: 0.20
Bowen ratio (H/λE): 0.46
Evaporative fraction (λE/R$_n$): 0.66

Sensible heat
86 W m^{-2}

Latent heat
187 W m^{-2}

Storage
12 W m^{-2}

Managed grassland

(b) Suburban

Net radiation
295 W m^{-2}

Albedo: 0.13
Bowen ratio: 1.28
Evaporative fraction: 0.34

Sensible heat
129 W m^{-2}

Storage
65 W m^{-2}

Latent heat
101 W m^{-2}

25% building, 11% paved, 64% greenspace

Fig. 32.8 As in Figure 32.7, but averaged over daylight periods. Arrows are proportional to the size of the flux.

urban canyon, in which the cityscape is represented by a street lined by buildings on either side. The geometry of the canyon, especially the height of buildings relative to the width of the street, affects the radiative balance of the city by reducing the emission of longwave radiation from the city surface and by decreasing surface albedo.

Radiative trapping within an urban canyon occurs due to changes in the sky view factor. The sky view factor is the proportion of the viewing hemisphere occupied by sky. Tall buildings block some of the sky so that a point on the street at the center of an urban canyon is exposed to only a portion of the sky (Figure 32.9a). The fraction of the sky seen is the sky view factor. The sky view factor for an infinitely long street lined by buildings of uniform height is:

$$\psi_{sky} = \cos(\alpha) \qquad (32.3)$$

where α is the angle defined by street width (W) and building height (H) as $\tan(\alpha) = 2H/W$. For a street that is as wide as the buildings are tall (i.e., $H = W$), $\alpha = 63.4°$ and the sky view factor is $\psi_{sky} = 0.45$. That is, the center of the

street is exposed to only 45 percent of the sky. The remainder $(1 - \psi_{sky})$ is the wall view factor. As H/W increases, buildings block a greater portion of the sky from the street; the sky view factor decreases while the wall view factor increases (Figure 32.10). Similar behavior is seen for a point at the center of a building wall (Figure 32.9b). The wall is exposed to the sky, the opposing wall, and the road. As H/W increases, less of the sky is viewed and more of the canyon (opposing wall and road) is viewed.

As H/W increases, an increasing portion of the cold sky is blocked by the warmer urban buildings and roads. More of the longwave radiation at the street surface comes from the surrounding building walls and less comes from the sky. Additionally, the magnitude of longwave emission to space is proportional to the street's exposure to the sky. As H/W increases, more of the longwave radiation emitted by the street surface is absorbed by the surrounding buildings and less escapes to the atmosphere. This is illustrated in Figure 32.11, which shows that the net longwave radiation loss to the atmosphere from a road decreases with greater H/W. The

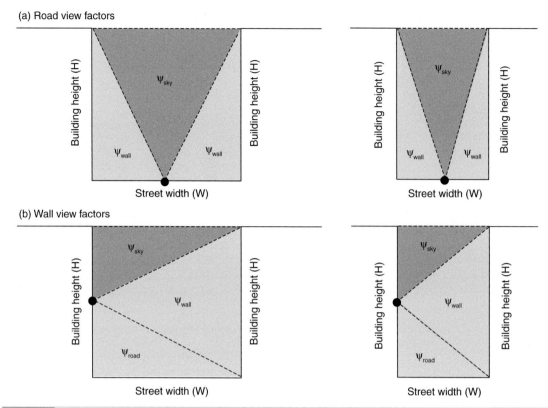

Fig. 32.9 Urban canyon and associated view factors. Panels show a street lined by buildings on each side. View factors are for (a) a point at the center of the street and (b) a point at the center of a building wall. ψ_{sky} is the sky view factor, ψ_{wall} is the wall view factor, and ψ_{road} is the road view factor. View factors sum to one for each canyon. Panels depict a wide (left) and narrow (right) canyon. Harman et al. (2004) provide mathematical equations for these view factors.

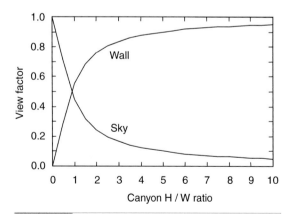

Fig. 32.10 Sky and wall view factors in relation to the ratio of building height (H) to street width (W) for a point on a street at the center of an infinitely long, symmetrical urban canyon.

absorption and emission of longwave radiation by building walls behave similarly. As the canyon becomes narrower or buildings taller, more of the longwave radiation on a wall comes from the opposing wall and road and less from the sky; more of the longwave radiation emitted by walls is absorbed by the street and opposing walls and less escapes to the sky. In contrast, the net longwave radiation of horizontal roofs above the canyon height is independent of canyon height or width.

One reason, therefore, why the heat island is largest at night with clear skies and little wind relates to the trapping of longwave radiation by tall buildings in the urban canyon. With these meteorological conditions, surface cooling occurs mostly due to radiative exchange with the atmosphere. Rural sites lose more longwave

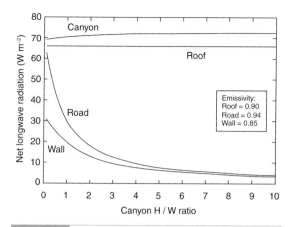

Fig. 32.11 Net longwave radiation for surfaces in an urban canyon as a function of building height (H) and street width (W). In this figure, net longwave radiation is the radiation emitted by the surface minus incoming atmosphere radiation ($L\uparrow - L\downarrow$, positive to the atmosphere). Atmospheric longwave radiation is $L\downarrow = 340\ \text{W m}^{-2}$, and the temperature of each surface is 19°C. Net longwave radiation for the canyon is the sum of road and wall fluxes after converting the wall fluxes to per unit ground area using the height to width ratio. Adapted from Oleson et al. (2008).

radiation than do urban sites and therefore cool more rapidly after sunset. Studies of the energy balance of urban canyons confirm the importance of building height to street width in determining trapping of radiation within the canyon and warming (Nunez and Oke 1976, 1977; Arnfield and Mills 1994a,b; Eliasson 1996; Arnfield and Grimmond 1998).

Surface and air temperature data collected in and near an apartment complex on a clear, calm, summer afternoon illustrate the urban canyon effect at a small scale (Figure 32.12). Late in the afternoon, the courtyard surface was 19°C warmer than air in the courtyard. In contrast, an adjacent grass lawn and wooded area outside the courtyard were less than 2°C warmer than the air. As the Sun set, all three sites cooled, but even at 2115 hours the surface temperature of the courtyard (30°C) was several degrees warmer than that of the lawn (23°C) or woodland (25°C). The air within the courtyard was about 1°C warmer than the external air.

At the urban scale, there is a strong relationship between the urban–rural temperature difference and sky view factor in which the magnitude of urban warming decreases as more of the sky is seen (Figure 32.4b). Oke (1981) related the instantaneous temperature difference ($\Delta T_{u-r(\max)}$) and view factor:

$$\Delta T_{u-r(\max)} = 15.27 - 13.88\,\psi_{sky} \tag{32.4}$$

The generality of this relationship is evident in the fact that European cities are generally cooler than North American cities of the same size. When related to view factor, however, all cities regardless of location show the same relationship between urban warming and view factor. Thus, differences between cities in Europe and North America might be attributed to the denser and taller buildings in North America.

Similar trapping of solar radiation occurs within the urban canyon and contributes to the low surface albedo of cities. During the day, tall buildings along a narrow street create more opportunity for radiative trapping within the canyon. This occurs when solar radiation reflected by a surface impinges on other surfaces in the canyon, being partially absorbed and re-reflected. Tall buildings block a portion of the sky so that some of the solar radiation reflected by the street is trapped within the canyon. The net effect is that more solar radiation is absorbed than would be expected from the reflectivity of the surface material, and the overall albedo of an urban canyon declines markedly as H/W increases (Figure 32.13). The albedo of the city as a whole, however, depends on the albedo of rooftops above canyon height in addition to the canyon albedo. Increasing the albedo of roofs can increase the overall city albedo.

32.6 | The Urban Canopy Layer

In studying the urban climate, it is necessary to distinguish the urban canopy layer, which lies below rooftop level and comprises microclimates created by buildings, roads, and vegetation, from the urban boundary layer above

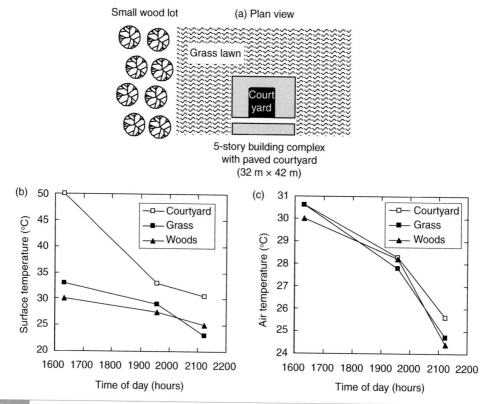

Fig. 32.12 Surface and air temperature measured in a small courtyard and adjacent grass lawn and woods on a clear, calm, summer afternoon. (a) Plan view of courtyard and adjacent grass lawn and woods. (b) Surface temperature measured in late-afternoon and evening. (c) Air temperature. Data from Landsberg (1970) and Landsberg (1981, p. 73, p. 85, p. 86).

rooftop level that is an integration of the microclimates of the urban canopy layer over a large area (Oke 1976, 1995; Arnfield 2003). The effect of urbanization differs in these two layers. Although cities as a whole warm the urban boundary layer, shade from buildings in the urban canopy layer can create cooler local temperatures than found in open areas. A city may have reduced latent heat flux compared with rural landscapes, but residential lawns within the city can have large latent heat fluxes (Oke 1979; Suckling 1980; Peters et al. 2011). In addition, it is necessary to distinguish surface and air temperature (Voogt and Oke 1997). Surface temperature is the temperature at which a surface emits longwave radiation and is strongly controlled by solar radiation. The urban–rural difference in surface temperature is generally greatest during the day whereas the air temperature difference is greatest at night.

Within a city there is substantial variation in temperature related to topography, proximity to water bodies, the density of development, the amount of vegetated cover, and type of building materials. Numerous local climate zones can be found within a city, and such zones can be defined based on building height and street width, sky view factor, impervious surface fraction, vegetation cover, and other factors (Stewart and Oke 2012; Stewart et al. 2014). Urban structure – the size, shape, and orientation of buildings and streets – is an important determinant of temperature within a city. For example, the height of buildings and the orientation of streets create complex patterns of sunlight and shade over the course of a day that affect air and surface temperature (Arnfield 1990; Ruffieux et al. 1990; Nichol 1996). The urban cover – particularly impervious area and vegetated area – determines surface wetness and

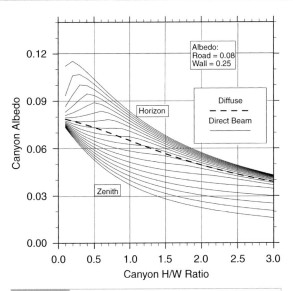

Fig. 32.13 Direct beam and diffuse albedo of an urban canyon as a function of building height (*H*) to street width (*W*). Direct beam albedo is shown for solar zenith angles from 0° to 85° in increments of 5°. In this example, the road has an albedo of 0.08 and the wall has an albedo of 0.25. Adapted from Oleson et al. (2008).

the partitioning of energy into latent and sensible heat. The urban fabric – the types of materials used in construction of buildings, streets, parking lots, and other urban surfaces – determines properties such as albedo, heat capacity, thermal conductivity, and wetness. A study of summer and autumn air temperature in Lawrence, Kansas, found land use (e.g., residential, commercial, industrial, park) accounted for 17–25 percent of the variance in measured air temperature (Henry et al. 1985; Henry and Dicks 1987). The type of surface material (e.g., asphalt, concrete, brick, gravel, grass) accounted for a similar amount of temperature variance.

32.7 | Urban Parks

One important component of the urban canopy layer is vegetation and parks. Numerous studies have demonstrated the role of vegetation in ameliorating the urban heat island. The cooling effect of parks is particularly evident in surface temperature. High-resolution aerial or satellite measurements within cities show large differences in surface temperature related to urban land use, with buildings and impervious areas having the warmest surfaces and parks having lower temperature (Carlson et al. 1981; Vukovich 1983; Roth et al. 1989; Nichol 1996; Spronken-Smith and Oke 1998; Rotach et al. 2005).

The presence of parks produces a discernible signal in air temperature, as shown by a study of nighttime air temperature during summer in Washington, D.C. (Figure 32.14). On the particular night studied, the average temperature obtained along a transect from the Northwest section of the city southeast through Rock Creek Park, the 16th Street business district, and through the open parks of the Mall was 23°C. However, temperature varied greatly depending on location. A large wooded area of Rock Creek Park had air temperature as low as 20°C. The downtown business district had air temperature of about 25°C while temperature on the nearby large open parks of the Mall was about 1°C cooler. Even during the day, Rock Creek Park heated more slowly than the commercial districts and was typically 1–2°C cooler at midday.

Similar cooling has been found in other cities (Bowler et al. 2010). Parks in Vancouver are typically 1–2°C cooler than surrounding cityscapes, and similar cooling is seen in Sacramento, California (Spronken-Smith and Oke 1998). Air temperature in a 15 ha (1 ha = 10,000 m²) park in Stockholm on a summer day was 0.5–0.8°C less than built-up areas during the day and 2°C cooler at sunset (Jansson et al. 2007). A study of three parks in Göteborg, Sweden, documents both the cooling of urban parks and the extension of the park microclimate into the surrounding built-up area (Upmanis et al. 1998). The parks ranged in size from 2 to 156 ha. Maximum cooling compared with surrounding area occurred at night and was 6°C for the large park compared with about 2°C for the smallest park. Cooling from the largest park extended 1000 m from the park boundary whereas the influence of the smallest park was on the order of 30 m. In general, the nocturnal air temperature difference between urban parks and surrounding built

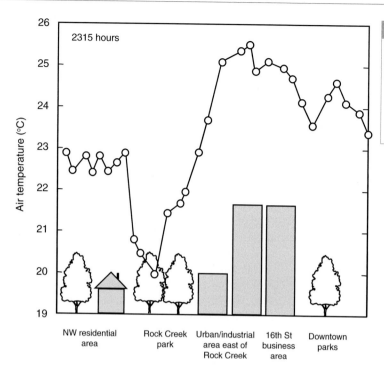

Fig. 32.14 Air temperature measured on a summer night in Washington, D.C., along a transect from the northwest residential area southeast through Rock Creek Park into the 16th Street business district and the grass parks of the Mall. Adapted from Landsberg (1981, p. 233).

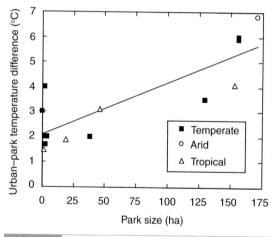

Fig. 32.15 Influence of park size on the urban–park temperature difference. Data are for nine temperate parks, four tropical parks, and one arid city. 1 ha = 10,000 m². Adapted from Upmanis et al. (1998).

areas increases with park size (Figure 32.15). Midday measurements in Rotterdam on a summer day found that the downtown was 1.2°C warmer than the surrounding rural area, but a city park was 4.0°C cooler than the downtown

(Heusinkveld et al. 2014). Within the city, evening temperature decreased with greater fractional vegetation cover. Cooler temperature in parks compared with urban street canyons arises because of greater latent heat flux in parks, especially when wet, and because of greater radiative loss at night as a result of higher sky view factors.

The influence of parks can be seen in surface energy fluxes measured within cities. Offerle et al. (2006) measured energy fluxes over several summer days in the central business district and in industrial, suburban residential, and rural areas of Łódź, Poland (Table 32.2). The central business district site consisted of dense urban attached buildings with an average height of 10 m and H/W ratio of 0.8. The industrial area had long, narrow buildings with lower height and smaller H/W ratio. The residential site had lower building density and higher vegetation cover compared with the two urban sites. Single-family detached houses were the dominant building type along widely spaced streets with H/W ratio of 0.3. An open grass field in the city served as a rural site. Buildings and

Table 32.2 Characteristics of four sites in Łódź, Poland, and seven sites in Basel, Switzerland, where surface energy fluxes were measured

Site	Description	Building height (m)	H / W	Cover (%) Vegetation	Building	Impervious
Łódź						
CBD	Urban commercial	10	0.8	24	35	41
IND	Industrial	9	0.4	39	17	44
RES	Single-family residential	8	0.3	78	10	13
RUR	Grassland	0	0	100	0	0
Basel						
U1	Urban residential	15	1.3	16	54	30
U2	Urban residential/ commercial	13	0.8	31	37	32
U3	Urban commercial	19	0.7	0	100	0
S1	Single-family residential	8	0.6	53	28	19
R1	Rural grassland	0	0	91	2	7
R2	Rural agricultural	0	0	98	0	2
R3	Rural grassland	0	0	94	1	5

Source: Basel, from Christen and Vogt (2004). Łódź, from Offerle et al. (2006).

other impervious surfaces (e.g., streets) comprised most of the source area for the measured fluxes in the central business district and industrial area. In contrast, the vegetated fraction increased to three-quarters or more of the source area in the residential and rural sites. Daytime sensible heat flux ranged from a high of 44 percent of net radiation at the central business district to a low of 22 percent of net radiation at the rural site (Figure 32.16). Similarly, the Bowen ratio decreased from 1.8 at the central business district to 0.4 at the rural site. Across the four sites, the ratio of sensible heat flux to available energy decreased with greater vegetation cover and increased with greater H/W. The Bowen ratio similarly decreased with greater vegetation cover and increased with greater H/W.

A comparison of surface energy fluxes measured during summer at seven urban and rural locations in Basel, Switzerland, found similar dependences of fluxes on vegetation cover (Christen and Vogt 2004; Rotach et al. 2005). Three urban sites in the city center were characterized by tall buildings, large H/W, low vegetation cover, and high cover by buildings and impervious surfaces (Table 32.2). In contrast, a residential neighborhood consisting mostly of two- to three-story single-family housing had greater vegetation cover, lower building and impervious surface cover, and lower H/W. Three rural sites had greater than 90 percent vegetation cover. Daytime sensible heat flux was about 50 percent of net radiation at the city center, but less than 30 percent of net radiation at the rural sites. Across the seven sites, the Bowen ratio and the ratio of sensible heat flux to available energy decreased with greater vegetation cover and increased with greater H/W (Figure 32.16).

Similar relationships have been found in comparisons of surface fluxes between cities. A study of the energy budget in three cities in the United States located in arid climates found that the Bowen ratio decreased as irrigated greenspace within the cities increased (Grimmond and Oke 1995). In a comparison of 10 sites in seven North American cities, latent heat flux increased and

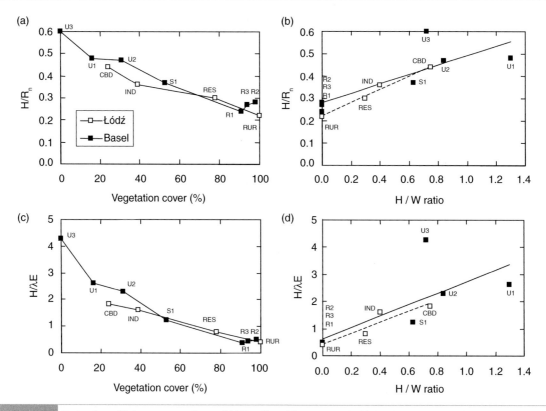

Fig. 32.16 Ratio of sensible heat to net radiation (H / R_n, a–b) and Bowen ratio ($H / \lambda E$, c–d) in relation to vegetation cover (left) and H / W (right) for four sites in Łódź, Poland, and seven sites in Basel, Switzerland. Labels indicate particular sites (Table 32.2). Lines in the right panels show trends for Łódź (dashed line) and Basel (solid line) in relation to H / W. Data from Christen and Vogt (2004) and Offerle et al. (2006).

Bowen ratio decreased with greater vegetation cover (Grimmond and Oke 2002). These studies suggest that vegetation cover and conversely impervious area are key indicators of urban climates. Indeed, historical trends for 51 urbanizing watersheds in eastern United States show decreased watershed evaporation and increased watershed sensible heat with greater urban development (Dow and DeWalle 2000).

The effects of vegetation extend to microclimate and can be seen in residential landscapes. Residential lawns, for example, can have large latent heat fluxes, particularly when irrigated (Oke 1979; Suckling 1980; Peters et al. 2011). In the semiarid climate of Colorado, proximity to irrigated greenspace affects local microclimates within suburban residential landscapes (Bonan 2000). Air temperature measurements across a suburban development show that in this

climate dry, native grass landscapes are warmer than irrigated greenbelts and irrigated residential lawns during summer. Hard materials such as flagstone patio and rock mulch have a higher midday surface temperature (> 50°C) than an irrigated grass lawn (30°C); dry, native grass is intermediate (~40°C).

32.8 | Urban Energy Balance Models

Numerical models of the energy balance and aerodynamics of urban areas give insights to the physical causes of the urban heat island. Various modeling approaches have been devised to mathematically represent the urban energy budget and turbulent transfer, and urban surfaces are being included in atmospheric models. Such models simulate surface energy fluxes, urban temperature, and other characteristics of urban climates (Grimmond et al. 2010, 2011).

The simplest urban model is to represent the energy balance of a city by a bulk formulation similar to that for soil, foliage, or plant canopies:

$$(1-r)S\downarrow +L\downarrow +F = L\uparrow[T_s] \\ + H[T_s] + \lambda E[T_s] + G[T_s]$$

(32.5)

where $S\downarrow$ and $L\downarrow$ are the incoming solar and longwave radiation, respectively, and r is surface albedo. Anthropogenic energy fluxes (F) from vehicular traffic, building heating and air conditioning, and industry are additional sources of energy. The fluxes of emitted longwave radiation ($L\uparrow$), sensible heat (H), latent heat (λE), and heat storage in the urban volume (G) depend on surface temperature (T_s). As with a leaf, plant canopy, or soil, Eq. (32.5) is solved for the surface temperature that balances the energy budget, but using parameters appropriate for urban surfaces. Important surface properties are the albedo of cities, surface roughness, the thermal properties of buildings and paving materials, and the amount of evaporating surface (i.e., vegetation) in the city. This one-dimensional energy budget formulation has been used to study urban climates (Myrup 1969; Ross and Oke 1988; Todhunter and Terjung 1988; Best 2005; Best et al. 2006).

Mathematical models of the energy balance of an urban canyon have been devised (Mills 1993, 1997; Mills and Arnfield 1993; Arnfield 2000), and a widely used class of urban models for climate simulation simplifies the complexity of a cityscape into a single idealized urban canyon defined by its H/W (Figure 32.17). These models represent the urban canyon as a single, mixed volume of air. An idealized roof, wall, and street comprise the canyon, with a separate energy budget for each surface. This allows for distinction of the various construction materials in a city and the unique thermal environment of the various city surfaces. The canyon floor can be divided into impervious and pervious (greenspace) portions. Such models provide simplified treatments of radiation, energy fluxes, and turbulence within cities while representing the important role of building height and street

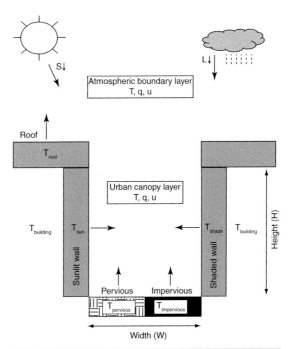

Fig. 32.17 Depiction of an urban canyon model. The canyon is represented by sunlit and shaded walls, a pervious and impervious canyon floor, and a roof. The model solves for the temperature of these surfaces, and the associated energy fluxes determine the temperature (T) and humidity (q) of the urban canopy layer. Flux calculations depend on the wind speed (u) in the canyon. The urban canopy layer itself interacts with the atmospheric boundary layer. See Oleson et al. (2008) and Oleson (2012) for model details.

width in determining radiative exchange in the canyon. The canyon concept provides a simplified parameterization of urban land cover that can be included in atmospheric numerical models (Masson 2000; Masson et al. 2002; Lemonsu et al. 2004; Oleson et al. 2008, 2011; Demuzere et al. 2013). These models can also include explicit treatment of vegetation and green roofs (Lemonsu et al. 2012; de Munck et al. 2013). The models are being used to examine changes in the temperature and heat stress within cities in response to climate change (McCarthy et al. 2010; Fischer et al. 2012; Oleson 2012).

32.9 | Rainfall

Though the urban heat island is the most prominent signature of cities on climate, urbanization can also alter regional precipitation (Shepherd 2005; Collier 2006; Mahmood et al. 2014). Numerous studies have found increased rainfall and increased frequency of severe weather associated with the urban heat island. For example, a study of St. Louis, Missouri, found summer rainfall on the leeward side of the city was greater than in other sectors (Figure 32.18). Moreover, storms resulting in over 25 mm of rain were 50 percent more frequent over urban areas than suburban or rural areas. Despite methodological difficulties in establishing causality, it is likely the St. Louis precipitation anomaly is caused by the city itself. Numerical modeling studies confirm the importance of the urban heat island in initiating precipitation around St. Louis (Rozoff et al. 2003).

Summer rainfall increases of 10–20 percent and increased thunderstorm activity have been observed in many cities (Table 32.3). Analyses of satellite-derived estimates of rainfall find similar increases in precipitation associated with cities (Shepherd et al. 2002), and case studies for Atlanta, Georgia, reveal rainstorms initiated by the urban heat island (Bornstein and Lin 2000; Dixon and Mote 2003; Diem and Mote 2005; Mote et al. 2007). However, industrial pollution can also inhibit rainfall. The small pollution particles emitted into air can inhibit water droplets from coalescing into larger droplets to create rain. Consequently, precipitation from certain types of clouds can be substantially reduced downwind from large urban and industrial areas (Rosenfeld 2000; Givati and Rosenfeld 2004).

Evidence of urban influences on rainfall also comes from reports of a weekly cycle in precipitation (Sanchez-Lorenzo et al. 2012). Differences in precipitation between the end and beginning of the workweek were first observed in 1929 (Ashworth 1929). For example, near-coastal ocean areas along the Atlantic seaboard of the United States receive more rain on weekends than on weekdays (Figure 32.19). This corresponds to the observed weekly cycle of air pollution in which lowest concentrations of carbon monoxide and ozone, two common urban pollutants, occur early in the week. Weekend storms may be enhanced by pollution, which

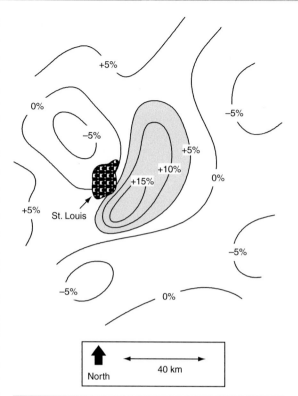

Fig. 32.18 Average summer rainfall near St. Louis, Missouri, expressed as a percentage of urban rainfall. Adapted from Changnon (1981, p. 6).

Table 32.3 Maximum increases in summer rainfall and thunderstorm activity for nine cities in the United States expressed as a percentage of rural value

	Rainfall	Thunderstorms
Cleveland	27	38
Detroit	25	–
Chicago	17	42
St. Louis	15	25
New Orleans	10	27
Washington, D.C.	9	36
Houston	9	10
Tulsa	0	0
Indianapolis	0	0

Source: From Changnon (1981, p. 4). See also Changnon (2001).

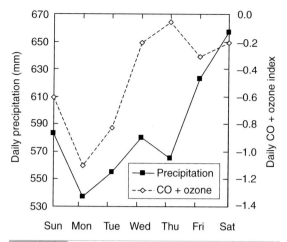

Fig. 32.19 Weekly cycle of air pollution (carbon monoxide and ozone) and precipitation along the Atlantic coast of the United States. Data from Cerveny and Balling (1998).

provides condensation nuclei around which rain drops form. Other weekly cycles have also been detected in temperature, lightning, and severe storms (Sanchez-Lorenzo et al. 2012). Since it is unlikely that natural processes vary on a seven-day cycle, the weekly cycles of precipitation and temperature in urbanized regions have been taken as indicators of human influences, primarily in the form of pollutants and heat from fuel combustion, on weather and climate. However, there is debate over the statistical methods used to identify these cycles and their physical explanation (Daniel et al. 2012; Sanchez-Lorenzo et al. 2012).

32.10 | Urban Hydrology

Urban landscapes generate more surface runoff compared with rural landscapes because of extensive roofs, streets, sidewalks, parking lots, and other impervious surfaces. The hard impervious surfaces and compacted soils typical of cities hinder infiltration, and constructed drainage systems quickly convey this water to rivers. The net effect is that urbanization increases the peak discharge rate, the speed of the runoff, and the total volume of runoff. A study of historical streamflow for 39 urban watersheds in the

United States found increases in annual streamflow in proportion to cumulative changes in population density (DeWalle et al. 2000).

The United States Soil Conservation Service provides a simple means to examine the effect of urbanization on runoff. This methodology relates runoff to soil type, land use, land cover, and soil moisture through a variety of curve numbers using Eq. (10.7). For a given amount of rainfall, runoff decreases as the curve number decreases (Figure 10.5a). In urban settings, the curve number is a composite of the curve number for pervious surfaces and the curve number for impervious surfaces (typically taken as $CN = 98$). Increasing impervious area results in a larger composite curve number, but the exact value depends on the means by which the runoff is conveyed to the drainage system. For urban surfaces that discharge directly into the drainage system, the composite curve number is:

$$CN_{composite} = CN_{pervious} + f_i \left(98 - CN_{pervious}\right) \quad (32.6)$$

where $CN_{pervious}$ is the curve number for the pervious surface and f_i is the fraction of the landscape that is impervious.

Figure 32.20 illustrates an application of the Soil Conservation Service method to estimate runoff in various settings. Urban surfaces greatly increase the amount of runoff compared with rural landscapes. For example, a forest generates only 13 mm of runoff in a 100 mm rainstorm. A residential setting that is 50 percent impervious generates 50 mm of runoff in the same storm. A business district where the surface area is 90 percent impervious generates 83 mm of runoff.

The effect of urbanization on runoff is greatest for small storms. Larger floods brought about by intense or prolonged rainfall are less affected because even in rural settings the soil moisture deficit is satisfied, the saturated zone expands, and runoff is rapidly conveyed to stream channels. This is illustrated by Figure 32.21, which relates the size of urban floods relative to rural floods to impervious area and recurrence interval. Recurrence interval, also called the return period, is the average time between events. It is the inverse of the annual probability of

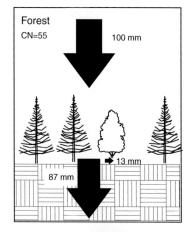

Forest
CN=55
100 mm
13 mm
87 mm

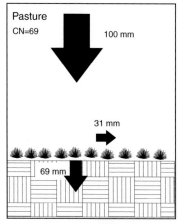

Pasture
CN=69
100 mm
31 mm
69 mm

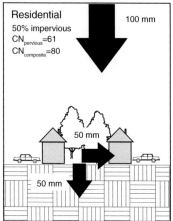

Residential
50% impervious
$CN_{pervious}=61$
$CN_{composite}=80$
100 mm
50 mm
50 mm

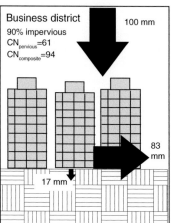

Business district
90% impervious
$CN_{pervious}=61$
$CN_{composite}=94$
100 mm
83 mm
17 mm

Fig. 32.20 Infiltration and surface runoff in response to 100 mm of rainfall over 24 hours for forest, pasture, residential, and commercial lands using the Soil Conservation Service method. Also shown are the appropriate curve numbers.

occurrence. In general, small storms are more frequent (have a low recurrence interval) than large storms. The ratio of urban-to-rural runoff decreases from 10–20 for storms with recurrence intervals less than one year (i.e., small, frequent storms) to less than 2 with recurrence intervals greater than ten years (i.e., large, rare storms). Greater impervious area increases runoff for a given recurrence interval. For example, runoff during a 100-year storm doubles in magnitude as the impervious area increases from 10 percent to 45 percent.

32.11 | Review Questions

1. How does tree cover on vacant land in cities of the United States vary by ecoregions (forest, grassland, desert)? What is a likely cause of this variation? Does tree cover in residential areas show a similar pattern? Explain why.

2. You live in a small town in eastern United States with a population of 10,000. What is the maximum heat island that you experience? You move to a nearby large city of 500,000. How much greater is the maximum heat island effect? How much longer would you expect the growing season to be at your new urban home?

3. In a city of 100,000 people in the United States, how much warmer is daily mean temperature averaged over the year compared with surrounding rural areas? Contrast this with a city of 1 million people.

4. An architect is designing an urban renewal project that consists of shops, office space, and apartments along a street that is 25 m wide. One design

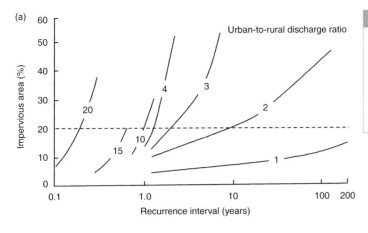

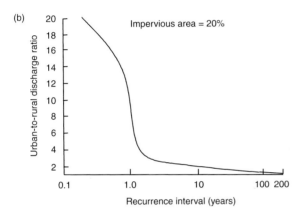

Fig. 32.21 Ratio of discharge after urbanization to discharge before urbanization in relation to impervious area and flood recurrence interval. (a) Contour lines show the urban-to-rural discharge ratio. (b) Discharge ratio in relation to recurrence interval for 20 percent impervious area. Adapted from Hollis (1975).

calls for buildings 10 m tall. Another design calls for buildings 30 m tall. What is the sky view factor from the center of the street in both designs? Which design allows better cooling at night? Why?

5. The mayor of a large city wants to reduce the heat island effect and proposes to build a city park to create a cooler local climate. The proposed park size is 50 ha. What magnitude of cooling should the mayor expect?

6. A commercial builder proposes to convert an existing grassland area to a shopping mall and assures city planners that runoff from a 50-mm rainstorm will not be more than twice that of the original grassland based on the Soil Conservation Service runoff calculation. Three alternative plans are proposed that vary in the amount of directly connected impervious area

(10%, 20%, 30%). Which plan meets the design requirement? The grassland has a curve number $CN = 69$.

7. Urban planners in a metropolitan area want to build a large city park with grass lawns and abundant trees. Discuss three ways in which the park will affect the climate of the city.

8. Constructing gardens in raised beds on building rooftops has been proposed as a means to mitigate the deleterious effects of the cityscape. How might a rooftop garden affect the climate and water balance of a city?

9. A landscape architect is designing a residential lawn in suburban Denver, Colorado (semiarid climate). What elements can be included in the design to create cool summer microclimates while minimizing water usage?

32.12 | References

Arnfield, A. J. (1990). Street design and urban canyon solar access. *Energy and Buildings*, 14, 117–131.

Arnfield, A. J. (2000). A simple model of urban canyon energy budget and its validation. *Physical Geography*, 21, 305–326.

Arnfield, A. J. (2003). Two decades of urban climate research: A review of turbulence, exchanges of energy and water, and the urban heat island. *International Journal of Climatology*, 23, 1–26.

Arnfield, A. J., and Grimmond, C. S. B. (1998). An urban canyon energy budget model and its application to urban storage heat flux modeling. *Energy and Buildings*, 27, 61–68.

Arnfield, A. J., and Mills, G. M. (1994a). An analysis of the circulation characteristics and energy budget of a dry, asymmetric, east–west urban canyon, I: Circulation characteristics. *International Journal of Climatology*, 14, 119–134.

Arnfield, A. J., and Mills, G. M. (1994b). An analysis of the circulation characteristics and energy budget of a dry, asymmetric, east–west urban canyon, II: Energy budget. *International Journal of Climatology*, 14, 239–262.

Ashworth, J. R. (1929). The influence of smoke and hot gases from factory chimneys on rainfall. *Quarterly Journal of the Royal Meteorological Society*, 55, 341–350.

Best, M. J. (2005). Representing urban areas within operational numerical weather prediction models. *Boundary-Layer Meteorology*, 114, 91–109.

Best, M. J., Grimmond, C. S. B., and Villani, M. G. (2006). Evaluation of the urban tile in MOSES using surface energy balance observations. *Boundary-Layer Meteorology*, 118, 503–525.

Bonan, G. B. (2000). The microclimates of a suburban Colorado (USA) landscape and implications for planning and design. *Landscape and Urban Planning*, 49, 97–114.

Bornstein, R., and Lin, Q. (2000). Urban heat islands and summertime convective thunderstorms in Atlanta: Three case studies. *Atmospheric Environment*, 34, 507–516.

Bowler, D. E., Buyung-Ali, L., Knight, T. M., and Pullin, A. S. (2010). Urban greening to cool towns and cities: A systematic review of the empirical evidence. *Landscape and Urban Planning*, 97, 147–155.

Carlson, T. N., Dodd, J. K., Benjamin, S. G., and Cooper, J. N. (1981). Satellite estimation of the surface energy balance, moisture availability and thermal inertia. *Journal of Applied Meteorology*, 20, 67–87.

Cerveny, R. S., and Balling, R. C., Jr. (1998). Weekly cycles of air pollutants, precipitation and tropical cyclones in the coastal NW Atlantic region. *Nature*, 394, 561–563.

Changnon, S. A., Jr. (1981). *METROMEX: A Review and Summary*, Meteorological Monographs, Volume 18, Number 40. Boston: American Meteorological Society.

Changnon, S. A. (2001). Assessment of historical thunderstorm data for urban effects: The Chicago case. *Climatic Change*, 49, 161–169.

Christen, A., and Vogt, R. (2004). Energy and radiation balance of a central European city. *International Journal of Climatology*, 24, 1395–1421.

Cleugh, H. A., and Oke, T. R. (1986). Suburban–rural energy balance comparisons in summer for Vancouver, B.C. *Boundary-Layer Meteorology* 36, 351–369.

Collier, C. G. (2006). The impact of urban areas on weather. *Quarterly Journal of the Royal Meteorological Society*, 132, 1–25.

Daniel, J. S., Portmann, R. W., Solomon, S., and Murphy, D. M. (2012). Identifying weekly cycles in meteorological variables: The importance of an appropriate statistical analysis. *Journal of Geophysical Research*, 117, D13203, doi:10.1029/2012JD017574.

de Munck, C. S., Lemonsu, A., Bouzouidja, R., Masson, V., and Claverie, R. (2013). The GREENROOF module (v7.3) for modelling green roof hydrological and energetic performances within TEB. *Geoscientific Model Development*, 6, 1941–1960.

Demuzere, M., Oleson, K., Coutts, A. M., Pigeon, G., and van Lipzig, N. P. M. (2013). Simulating the surface energy balance over two contrasting urban environments using the Community Land Model Urban. *International Journal of Climatology*, 33, 3182–3205.

DeWalle, D. R., Swistock, B. R., Johnson, T. E., and McGuire, K. J. (2000). Potential effects of climate change and urbanization on mean annual streamflow in the United States. *Water Resources Research*, 36, 2655–2664.

Diem, J. E., and Mote, T. L. (2005). Interepochal changes in summer precipitation in the southeastern United States: Evidence of possible urban effects near Atlanta, Georgia. *Journal of Applied Meteorology*, 44, 717–730.

Dixon, P. G., and Mote, T. L. (2003). Patterns and causes of Atlanta's urban heat island-initiated precipitation. *Journal of Applied Meteorology*, 42, 1273–1284.

Dow, C. L., and DeWalle, D. R. (2000). Trends in evaporation and Bowen ratio on urbanizing watersheds in eastern United States. *Water Resources Research*, 36, 1835–1843.

Eliasson, I. (1996). Urban nocturnal temperatures, street geometry and land use. *Atmospheric Environment*, 30, 379–392.

Ellefsen, R. (1990/91). Mapping and measuring buildings in the canopy boundary layer in ten U.S. cities. *Energy and Buildings*, 15/16, 1025–1049.

Fischer, E. M., Oleson, K. W., and Lawrence, D. M. (2012). Contrasting urban and rural heat stress responses to climate change. *Geophysical Research Letters*, 39, L03705, doi:10.1029/2011GL050576.

Givati, A., and Rosenfeld, D. (2004). Quantifying precipitation suppression due to air pollution. *Journal of Applied Meteorology*, 43, 1038–1056.

Grimmond, C. S. B. (1992). The suburban energy balance: Methodological considerations and results for a mid-latitude west coast city under winter and spring conditions. *International Journal of Climatology*, 12, 481–497.

Grimmond, C. S. B., and Oke, T. R. (1995). Comparison of heat fluxes from summertime observations in the suburbs of four North American cities. *Journal of Applied Meteorology*, 34, 873–889.

Grimmond, C. S. B., and Oke, T. R. (1999). Heat storage in urban areas: Local-scale observations and evaluation of a simple model. *Journal of Applied Meteorology*, 38, 922–940.

Grimmond, C. S. B., and Oke, T. R. (2002). Turbulent heat fluxes in urban areas: Observations and a local-scale urban meteorological parameterization scheme (LUMPS). *Journal of Applied Meteorology*, 41, 792–810.

Grimmond, C. S. B., Blackett, M., Best, M. J., et al. (2010). The international urban energy balance models comparison project: First results from phase 1. *Journal of Applied Meteorology and Climatology*, 49, 1268–1292.

Grimmond, C. S. B., Blackett, M., Best, M. J., et al. (2011). Initial results from Phase 2 of the international urban energy balance model comparison. *International Journal of Climatology*, 31, 244–272.

Halverson, H. A., and Rowntree, R. A. (1986). Correlations between urban tree crown cover and total population in eight U.S. cities. *Landscape and Urban Planning*, 13, 219–223.

Harman, I. N., Best, M. J., and Belcher, S. E. (2004). Radiative exchange in an urban street canyon. *Boundary-Layer Meteorology*, 110, 301–316.

Henry, J. A., and Dicks, S. E. (1987). Association of urban temperatures with land use and surface materials. *Landscape and Urban Planning*, 14, 21–29.

Henry, J. A., Dicks, S. E., and Marotz, G. A. (1985). Urban and rural humidity distributions: Relationships to surface materials and land use. *Journal of Climatology*, 5, 53–62.

Heusinkveld, B. G., Steeneveld, G. J., van Hove, L. W. A., Jacobs, C. M. J., and Holtslag, A. A. M. (2014). Spatial variability of the Rotterdam urban heat island as influenced by urban land use. *Journal*

of Geophysical Research: Atmospheres, 119, 677–692, doi:10.1002/2012JD019399.

Hollis, G. E. (1975). The effect of urbanization on floods of different recurrence interval. *Water Resources Research*, 11, 431–435.

Jansson, C., Jansson, P.-E., and Gustafsson, D. (2007). Near surface climate in an urban vegetated park and its surroundings. *Theoretical and Applied Climatology*, 89, 185–193.

Kalanda, B. D., Oke, T. R., and Spittlehouse, D. L. (1980). Suburban energy balance estimates for Vancouver, B.C., using the Bowen ratio-energy balance approach. *Journal of Applied Meteorology*, 19, 791–802.

Karl, T. R., Diaz, H. F., and Kukla, G. (1988). Urbanization: Its detection and effect in the United States climate record. *Journal of Climate*, 1, 1099–1123.

Landsberg, H. E. (1970). Micrometeorological temperature differentiation through urbanization. In *Urban Climates: Proceedings of the WMO Symposium on Urban Climates and Building Climatology, Brussels, October 1968, Volume I*. Geneva: World Meteorological Organization, pp. 129–136.

Landsberg, H. E. (1979). Atmospheric changes in a growing community (the Columbia, Maryland experience). *Urban Ecology*, 4, 53–81.

Landsberg, H. E. (1981). *The Urban Climate*. New York: Academic Press.

Landsberg, H. E., and Maisel, T. N. (1972). Micrometeorological observations in an area of urban growth. *Boundary-Layer Meteorology*, 2, 365–370.

Lemonsu, A., Grimmond, C. S. B., and Masson, V. (2004). Modeling the surface energy balance of the core of an old Mediterranean city: Marseille. *Journal of Applied Meteorology*, 43, 312–327.

Lemonsu, A., Masson, V., Shashua-Bar, L., Erell, E., and Pearlmutter, D. (2012). Inclusion of vegetation in the Town Energy Balance model for modelling urban green areas. *Geoscientific Model Development*, 5, 1377–1393.

Mahmood, R., Pielke, R. A., Sr., Hubbard, K. G., et al. (2014). Land cover changes and their biogeophysical effects on climate. *International Journal of Climatology*, 34, 929–953.

Masson, V. (2000). A physically-based scheme for the urban energy budget in atmospheric models. *Boundary-Layer Meteorology*, 94, 357–397.

Masson, V., Grimmond, C. S. B., and Oke, T. R. (2002). Evaluation of the Town Energy Balance (TEB) scheme with direct measurements from dry districts in two cities. *Journal of Applied Meteorology*, 41, 1011–1026.

McCarthy, M. P., Best, M. J., and Betts, R. A. (2010). Climate change in cities due to global warming and urban effects. *Geophysical Research Letters*, 37, L09705, doi:10.1029/2010GL042845.

Mills, G. M. (1993). Simulation of the energy budget of an urban canyon – I: Model structure and sensitivity test. *Atmospheric Environment*, 27B, 157–170.

Mills, G. (1997). An urban canopy-layer climate model. *Theoretical and Applied Climatology*, 57, 229–244.

Mills, G. M., and Arnfield, A. J. (1993). Simulation of the energy budget of an urban canyon – II: Comparison of model results with measurements. *Atmospheric Environment*, 27B, 171–181.

Mote, T. L., Lacke, M. C., and Shepherd, J. M. (2007). Radar signatures of the urban effect on precipitation distribution: A case study for Atlanta, Georgia. *Geophysical Research Letters*, 34, L20710, doi:10.1029/2007GL031903.

Myrup, L. O. (1969). A numerical model of the urban heat island. *Journal of Applied Meteorology*, 8, 908–918.

Nichol, J. E. (1996). High-resolution surface temperature patterns related to urban morphology in a tropical city: A satellite-based study. *Journal of Applied Meteorology*, 35, 135–146.

Nowak, D. J., Rowntree, R. A., McPherson, E. G., et al. (1996). Measuring and analyzing urban tree cover. *Landscape and Urban Planning*, 36, 49–57.

Nunez, M., and Oke, T. R. (1976). Long-wave radiative flux divergence and nocturnal cooling of the urban atmosphere, II: Within an urban canyon. *Boundary-Layer Meteorology*, 10, 121–135.

Nunez, M., and Oke, T. R. (1977). The energy balance of an urban canyon. *Journal of Applied Meteorology*, 16, 11–19.

Offerle, B., Grimmond, C. S. B., Fortuniak, K., and Pawlak, W. (2006). Intraurban differences of surface energy fluxes in a central European city. *Journal of Applied Meteorology and Climatology*, 45, 125–136.

Oke, T. R. (1973). City size and the urban heat island. *Atmospheric Environment*, 7, 769–779.

Oke, T. R. (1976). The distinction between canopy and boundary layer urban heat islands. *Atmosphere*, 14, 268–277.

Oke, T. R. (1979). Advectively-assisted evapotranspiration from irrigated urban vegetation. *Boundary-Layer Meteorology*, 17, 167–173.

Oke, T. R. (1981). Canyon geometry and the nocturnal urban heat island: Comparison of scale model and field observations. *Journal of Climatology*, 1, 237–254.

Oke, T. R. (1982). The energetic basis of the urban heat island. *Quarterly Journal of the Royal Meteorological Society*, 108, 1–24.

Oke, T. R. (1988). The urban energy balance. *Progress in Physical Geography*, 12, 471–508.

Oke, T. R. (1995). The heat island of the urban boundary layer: Characteristics, causes and effects. In *Wind Climate in Cities*, ed. J. E. Cermak. Dordrecht: Kluwer, pp. 81–107.

Oke, T. R., and Cleugh, H. A. (1987). Urban heat storage derived as energy balance residuals. *Boundary-Layer Meteorology*, 39, 233–245.

Oke, T. R., and McCaughey, J. H. (1983). Suburban–rural energy balance comparisons for Vancouver, B.C.: An extreme case? *Boundary-Layer Meteorology*, 26, 337–354.

Oleson, K. (2012). Contrasts between urban and rural climate in CCSM4 CMIP5 climate change scenarios. *Journal of Climate*, 25, 1390–1412.

Oleson, K. W., Bonan, G. B., Feddema, J., Vertenstein, M., and Grimmond, C. S. B. (2008). An urban parameterization for a global climate model, Part I: Formulation and evaluation for two cities. *Journal of Applied Meteorology and Climatology*, 47, 1038–1060.

Oleson, K. W., Bonan, G. B., Feddema, J., and Jackson, T. (2011). An examination of urban heat island characteristics in a global climate model. *International Journal of Climatology*, 31, 1848–1865.

Peters, E. B., Hiller, R. V., and McFadden, J. P. (2011). Seasonal contributions of vegetation types to suburban evapotranspiration. *Journal of Geophysical Research*, 116, G01003, doi:10.1029/2010JG001463.

Rosenfeld, D. (2000). Suppression of rain and snow by urban and industrial air pollution. *Science*, 287, 1793–1796.

Ross, S. L., and Oke, T. R. (1988). Tests of three urban energy balance models. *Boundary-Layer Meteorology*, 44, 73–96.

Rotach, M. W., Vogt, R., Bernhofer, C., et al. (2005). BUBBLE – an urban boundary layer meteorology project. *Theoretical and Applied Climatology*, 81, 231–261.

Roth, M., and Oke, T. R. (1995). Relative efficiencies of turbulent transfer of heat, mass, and momentum over a patchy urban surface. *Journal of the Atmospheric Sciences*, 52, 1863–1874.

Roth, M., Oke, T. R., and Emery, W. J. (1989). Satellite-derived urban heat islands from three coastal cities and the utilization of such data in urban climatology. *International Journal of Remote Sensing*, 10, 1699–1720.

Rozoff, C. M., Cotton, W. R., and Adegoke, J. O. (2003). Simulation of St. Louis, Missouri, land use impacts on thunderstorms. *Journal of Applied Meteorology*, 42, 716–738.

Ruffieux, D., Wolfe, D. E., and Russell, C. (1990). The effect of building shadows on the vertical temperature structure of the lower atmosphere in downtown Denver. *Journal of Applied Meteorology*, 29, 1221–1231.

Sanchez-Lorenzo, A., Laux, P., Hendricks Franssen, H.-J., et al. (2012). Assessing large-scale weekly cycles in meteorological variables: A review. *Atmospheric Chemistry and Physics*, 12, 5755–5771.

Semonin, R. G. (1981). Surface weather conditions. In *METROMEX: A Review and Summary*, ed. S. A. Changnon, Jr. Boston: American Meteorological Society, pp. 17–40.

Shepherd, J. M. (2005). A review of current investigations of urban-induced rainfall and recommendations for the future. *Earth Interactions*, 9, 1–27.

Shepherd, J. M., Pierce, H., and Negri, A. J. (2002). Rainfall modification by major urban areas: Observations from spaceborne rain radar on the TRMM satellite. *Journal of Applied Meteorology*, 41, 689–701.

Spronken-Smith, R. A., and Oke, T. R. (1998). The thermal regime of urban parks in two cities with different summer climates. *International Journal of Remote Sensing*, 19, 2085–2104.

Stewart, I. D., and Oke, T. R. (2012). Local climate zones for urban temperature studies. *Bulletin of the American Meteorological Society*, 93, 1879–1900.

Stewart, I. D., Oke, T. R., and Krayenhoff, E. S. (2014). Evaluation of the "local climate zone" scheme using temperature observations and model simulations. *International Journal of Climatology*, 34, 1062–1080.

Suckling, P. W. (1980). The energy balance microclimate of a suburban lawn. *Journal of Applied Meteorology*, 19, 606–608.

Todhunter, P. E., and Terjung, W. H. (1988). Intercomparison of three urban climate models. *Boundary-Layer Meteorology*, 42, 181–205.

Upmanis, H., Eliasson, I., and Lindqvist, S. (1998). The influence of green areas on nocturnal temperatures in a high latitude city (Göteborg, Sweden). *International Journal of Climatology*, 18, 681–700.

Viterito, A. (1989). Changing thermal topography of the Baltimore–Washington corridor: 1950–1979. *Climatic Change*, 14, 89–102.

Voogt, J. A., and Oke, T. R. (1997). Complete urban surface temperatures. *Journal of Applied Meteorology*, 36, 1117–1132.

Vukovich, F. M. (1983). An analysis of the ground temperature and reflectivity pattern about St. Louis, Missouri, using HCMM satellite data. *Journal of Climate and Applied Meteorology*, 22, 560–571.

White, M. A., Nemani, R. R., Thornton, P. E., and Running, S. W. (2002). Satellite evidence of phenological differences between urbanized and rural areas of the eastern United States deciduous broadleaf forest. *Ecosystems*, 5, 260–277.

Yap, D., and Oke, T. R. (1974). Sensible heat fluxes over an urban area – Vancouver, B.C. *Journal of Applied Meteorology*, 13, 880–890.

Zhang, X., Friedl, M. A., Schaaf, C. B., Strahler, A. H., and Schneider, A. (2004). The footprint of urban climates on vegetation phenology. *Geophysical Research Letters*, 31, L12209, doi:10.1029/2004GL020137.

33

Climate Intervention and Geoengineering

33.1 | Chapter Summary

Geoengineering is the purposeful intervention in climate through various methods and technologies applied at a scale sufficiently large enough to alter regional or global climate. Such approaches are broadly grouped into solar radiation management and CO_2 removal and on land include: increasing the albedo of the land surface; agricultural management to reduce greenhouse gas emissions; large-scale afforestation and reforestation to enhance the terrestrial carbon sink; and use of biofuels to reduce anthropogenic CO_2 emissions. In cities, building design practices such as white roofs and green roofs provide a means to lessen the urban heat island. Managing terrestrial ecosystems for CO_2 removal has received much scientific study, but biogeophysical climate consequences (chiefly through albedo and evapotranspiration) must be considered in addition to the biogeochemical benefits. Future land-use choices have varying biogeophysical and biogeochemical outcomes for climate over the twenty-first century, and in some cases these counteract one another. Land-use choices additionally alter the cycling of reactive nitrogen in the Earth system, emissions of mineral dust and biomass burning aerosols, and emissions of biogenic volatile organic compounds, each of which has significant chemistry–climate feedbacks. The climate consequences of land use must be studied in an integrative, interdisciplinary framework to prevent unintended outcomes of climate intervention policies. Moreover, many biogenic emissions affect air quality, and air pollution control and climate change must be treated in a common policy framework. To balance the climate and socioeconomic outcomes of land use, Earth system models are beginning to include socioeconomic decision making.

33.2 | Adaptation, Mitigation, and Geoengineering

A changing climate has various impacts on human systems including food availability, water resources, health and disease, and socioeconomic stability, as well as disrupting the functioning of natural systems. Many approaches are available to respond to the environmental and socioeconomic challenges posed by anthropogenic climate change. In scientific parlance, *adaptation* is the adjustment to climate change (Allwood et al. 2014). In the context of human systems, adaptation refers to intervention to moderate or avoid harmful effects or to exploit beneficial outcomes of climate change. *Mitigation* describes human intervention to reduce the sources or enhance the sinks of greenhouse gases or to reduce the emissions of other substances that contribute to climate change (Allwood et al. 2014). In this meaning, mitigation pertains only to emissions. *Geoengineering*, or climate engineering, is

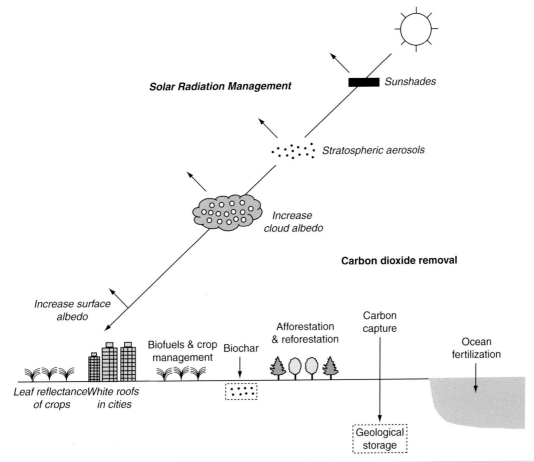

Fig. 33.1 Overview of various climate geoengineering methods to mitigate anthropogenic climate change by solar radiation management to reflect sunlight (italic text) and carbon dioxide removal (semi-bold text). Adapted from Lenton and Vaughan (2009) and Vaughan and Lenton (2011).

a deliberate modification of Earth's climate to alleviate the impacts of climate change (Allwood et al. 2014). Geoengineering describes a broad set of methods and technologies applied to the atmosphere, ocean, and biosphere at a scale sufficiently large enough to alter regional or global climate. In fact, however, there is broad overlap among adaptation, mitigation, and geoengineering methods (Boucher et al. 2014). For example, the use of white roofs and green roofs in cities is an adaptation to a warmer climate that is effective precisely because it alters climate, albeit at a local scale. Farming practices (e.g., no-till management) can simultaneously be an adaptation that sustains crop productivity in a warmer world, mitigate climate change by reducing greenhouse gas emissions, and engineer a cooler climate by higher surface albedo. Whether considered adaptation, mitigation, or geoengineering, they have the same intent – a human intervention in the cycling of energy, water, chemicals, and other substances to lessen the detrimental impacts of anthropogenic climate change.

The geoengineering of climate encompasses techniques to mitigate greenhouse gas emissions, large-scale adaptation of human systems (e.g., cropland, cities) to reduce anthropogenic global warming, and other direct interventions in the climate system (Figure 33.1). One aspect of geoengineering is solar radiation management (Crutzen 2006; Lenton and Vaughan 2009;

Vaughan and Lenton 2011; Boucher et al. 2013, 2014; Caldeira et al. 2013). This is an increase in planetary albedo to reduce the amount of solar radiation absorbed by the planet and to counter the effects of anthropogenic greenhouse gas emissions. Possible means to achieve this include positioning sunshades or mirrors in space, brightening marine clouds by injecting sea salts into the atmosphere, or injecting sulfate aerosols into the stratosphere.

For example, a 4–5 percent reduction in solar insolation at the top of the atmosphere mostly counters the planetary warming from a quadrupling of atmospheric CO_2, but the effects vary regionally and there are associated reductions in precipitation that require tradeoffs between temperature and precipitation mitigation (Kravitz et al. 2013, 2014; Tilmes et al. 2013). Moreover, solar insolation reduction produces complex changes in temperature and precipitation extremes (Tilmes et al. 2013; Curry et al. 2014). The use of planetary mirrors or shades, sea salts, and sulfate aerosols to modify solar radiation each produces planetary cooling to mitigate greenhouse gas warming, but the effectiveness of these methods varies and they have quite different effects on precipitation (Niemeier et al. 2013). Solar radiation management may also produce unintended consequences with undesirable outcomes. For example, injection of sulfate aerosols alters atmospheric chemistry and may lead to ozone depletion (Tilmes et al. 2008; Pitari et al. 2014).

Another geoengineering technique is to increase the albedo of the land surface. A higher albedo of cropland reduces the absorption of solar radiation and cools surface climate (Ridgwell et al. 2009; Singarayer et al. 2009; Doughty et al. 2011; Irvine et al. 2011; Caldeira et al. 2013). However, the cooling is regional (e.g., in areas of North America and Eurasia with widespread croplands), and the required increase in leaf reflectance to achieve this may not be possible. Such changes in surface albedo, while cooling climate, also alter the hydrologic cycle and can reduce precipitation because of greater subsidence over land (Bala and Nag 2012). Another possibility is to artificially increase the albedo of

desert surfaces by laying highly reflective material over the ground (Irvine et al. 2011). The albedo of city surfaces can also be increased by using white surfaces for building roofs, and such techniques are effective to decrease air temperature within a city (Section 33.7). This type of cropland management or urban planning is technically not mitigation, because it does not directly reduce greenhouse gas emissions; and it is seen more as an adaptation to climate change than as geoengineering because its effects on climate are localized (Boucher et al. 2014).

Carbon dioxide removal is another broad category of geoengineering, in this case removing CO_2 from the atmosphere by increasing natural carbon sinks or creating new carbon sinks (Lenton and Vaughan 2009; Vaughan and Lenton 2011; Caldeira et al. 2013; Ciais et al. 2013; Boucher et al. 2014). One means is by direct industrial carbon capture from air and subsequent storage. Iron fertilization of oceans has been proposed to increase ocean carbon uptake. Several methods of carbon capture and storage involve terrestrial ecosystems by enhancing the land carbon sink. Afforestation and reforestation remove carbon from the atmosphere. Biochar (charcoal) produced during pyrolysis is a stable, recalcitrant material which if buried in soils may be a long-term carbon sink. Bioenergy produced from biomass (e.g., forests, switchgrass, miscanthus, sugarcane) also captures carbon and reduces fossil fuel emissions. Avoided deforestation is an additional land management practice. It does not remove carbon from the atmosphere, but it does reduce carbon emissions from land use. In this context, carbon dioxide removal is both mitigation (to reduce emissions) and geoengineering (by deliberately intervening in the climate system).

Whether it is considered adaptation, mitigation, or geoengineering, there are many local, regional, and global methods to reduce the damaging impacts of anthropogenic climate change at the scale at which people and natural systems perceive climate. Many of these solutions involve terrestrial ecosystems and recognition of their vital role in regulating climate through biogeophysical and biogeochemical processes. Management of forests, grasslands, and agricultural lands to mitigate greenhouse gas emissions has received considerable attention (Smith et al. 2008, 2014). The most straightforward pertains to afforestation, reforestation, and avoided deforestation to manage the carbon cycle. This builds upon principles of carbon cycle–climate coupling to determine the anthropogenic CO_2 emissions compatible with a climate stabilization policy in light of terrestrial and oceanic carbon sinks (Chapter 29). Enhanced terrestrial sinks require less reduction in fossil fuel emissions to achieve a stabilization target. However, anthropogenic land use and land-cover change also alter the physical state of the land and affect climate by modifying surface albedo, surface roughness, and the partitioning of net radiation into sensible and latent heat fluxes (Chapter 28). The addition of nitrogen fertilizers in croplands has numerous consequences through the nitrogen cascade (Chapter 30). Human modification and management of the biosphere also alters the emissions of biogenic volatile organic compounds and other chemical exchanges with the atmosphere that affect aerosols and other short-lived climate forcings (Chapter 31). A comprehensive understanding of ecosystem management to mitigate climate change must address the multiple influences of ecosystems on climate in an integrative manner.

33.3 | Climate Services and Terrestrial Ecosystems

Much of the study of the purposeful intervention in Earth's climate through afforestation, reforestation, and cultivation of biofuels has focused on the radiative forcing from changes in surface albedo and carbon storage. The simple story is that the low surface albedo of forests warms climate while their large carbon storage cools climate. This is particularly evident in boreal and northern temperate forests, where the presence of trees masks the high albedo of snow. High rates of evapotranspiration further promote cooling, and this is particularly evident in tropical forests.

Figure 33.2 summarizes this understanding using a measure of the climate regulation value of ecosystems (Anderson-Teixeira et al. 2012). This combines the biogeophysical effects of ecosystems on the surface energy budget and the biogeochemical emissions of greenhouse gases into an index of CO_2 equivalents relative to bare ground. Biogeophysical climate values manifest through the effects of vegetation on net radiation and latent heat flux. In all the ecosystems considered, vegetation removal increases albedo, thereby decreasing net radiation, but decreases latent heat flux. The evaporative cooling of ecosystems is a positive climate service while surface warming from high net radiation (low albedo) is a negative climate service. Biogeochemical processes are greenhouse gas fluxes (CO_2, N_2O, CH_4) and include both annual fluxes of the intact ecosystem and also the emissions that would occur upon clearing the land. Forests, in particular, contain large stores of carbon that would be released to the atmosphere with deforestation. Removal of this carbon from the atmosphere is a large positive climate service.

Natural ecosystems generally have larger net climate regulation values than agroecosystems, primarily because of differences in greenhouse gases (Figure 33.2). Biogeophysical climate services are generally minor compared with biogeochemical services in natural ecosystems. In these ecosystems, the evaporative cooling provided by intact vegetation is smaller than the warming provided by their low albedo (high net radiation) so that the net biogeophysical climate service is negative in most natural ecosystems. The exceptions are tropical evergreen forest and tropical savanna. The large positive climate regulation value of tropical evergreen forests points to the importance of protecting these forests, while, in this analysis, the climate value of boreal evergreen forest is minor because the biogeophysical and biogeochemical services offset. In contrast, the biogeochemical climate services of

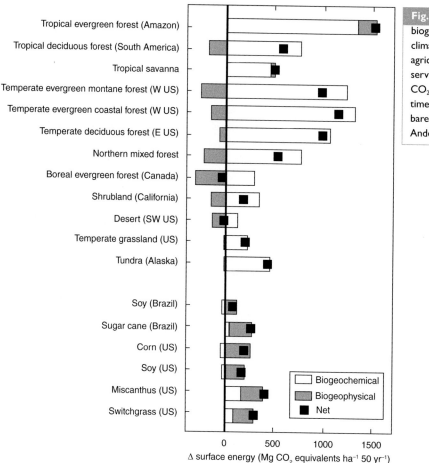

Fig. 33.2 Biogeochemical, biogeophysical, and net climate services of natural and agricultural ecosystems. Climate services are given in terms of CO_2 equivalents over a 50-year time frame and are relative to bare ground. Adapted from Anderson-Teixeira et al. (2012).

agroecosystems are small. Their net positive climate service arises mostly from strong evaporative cooling.

33.4 | Climate Outcomes of Land-Use Pathways

Climate model simulations of historical land use and land-cover change show that the biogeophysical land-use forcing of climate, while small globally, is in North America and Eurasia of similar magnitude, but of opposite sign, to greenhouse gas climate change (Figure 28.18). A similar conclusion pertains to possible pathways of land use over the twenty-first century. For example, Feddema et al. (2005) simulated climate change for the twenty-first century using two different depictions of future socioeconomic conditions. These scenarios provided the greenhouse gas concentrations used in the simulation. Associated with each scenario are socioeconomic trends that drive fossil fuel emissions and land-cover change, similar to the representative concentration pathways shown in Figure 2.12. In these simulations, the two extremes are represented by the A2 and B1 scenarios (Figure 33.3). The A2 scenario has the highest fossil fuel CO_2 emission, which increases over the twenty-first century, and the largest expansion of agriculture to support a burgeoning global population. Most land suitable for agriculture is converted to cropland by 2100. The B1 scenario represents the opposite extreme, with the lowest CO_2 emission, loss of farmland, and net reforestation due to declining global population and farm abandonment in the latter part of the century.

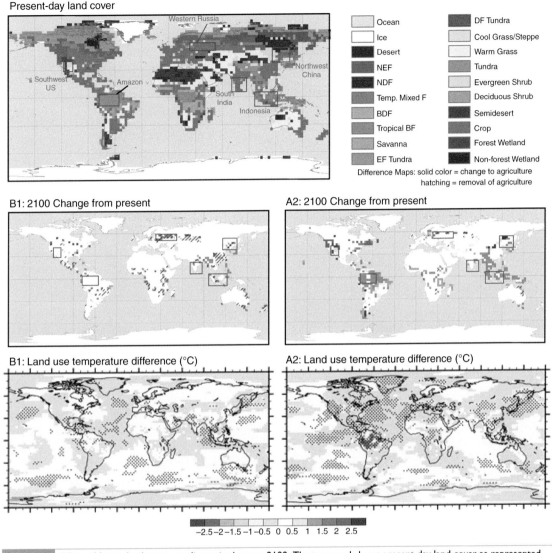

Fig. 33.3 Effect of future land cover on climate in the year 2100. The top panel shows present-day land cover as represented in a climate model (Labels – B, broadleaf; N, needleleaf; E, evergreen; D, deciduous; and F, forest). The middle panels show land-cover change at 2100 for the B1 and A2 scenarios. The bottom panels show boreal summer (June–August) temperature differences due to land-cover change in the B1 and A2 scenarios. The data were calculated by subtracting the greenhouse gas forcing from a simulation including land-cover change and greenhouse gas forcings. Stippling indicates statistically significant differences. Adapted from Feddema et al. (2005). See color plate section.

Simulations were performed with greenhouse gas forcing only and with both greenhouse gas forcing and land-cover change. The A2 scenario has high CO_2 emission, and the model simulates a 2°C warming of planetary temperature in the absence of land-cover change. Agricultural expansion in the A2 scenario leads to additional warming in boreal summer (June–August) in the Amazon but cooling that helps mitigate warming in mid-latitudes (Figure 33.3). The B1 is a low greenhouse gas scenario, which produces a 1°C warming. The associated land-use temperature change is much smaller in this scenario, which depicts net loss of farmland by 2100.

Sitch et al. (2005) also examined the climate impacts of the A2 and B1 land-use scenarios,

Table 33.1 | Contribution of land-cover change to global annual mean temperature change over the twenty-first century

SRES scenario	Atmospheric CO$_2$ at 2100 (ppm)	Temperature change (°C)
A2		
With land-cover change	957	2.7
Without land-cover change	830	2.4
B1		
With land-cover change	592	1.7
Without land-cover change	572	1.5
A1B		
With land-cover change	847	2.6
Without land-cover change	789	2.4
B2		
With land-cover change	677	2.1
Without land-cover change	631	1.8

Source: From Sitch et al. (2005).

as well as the A1B and B2 scenarios, which are intermediate and have moderate expansion of agricultural land by 2100. Their model included coupled carbon cycle–climate feedbacks. Two climate simulations were performed for each scenario: with fossil fuel emission and anthropogenic land-cover change; and with fossil fuel emission but cropland held constant at 1990 values (thereby eliminating land-cover change). Atmospheric CO$_2$ was simulated interactively as the balance between fossil fuel emission, the net atmosphere–ocean carbon flux, and the net atmosphere–land carbon flux of managed and natural vegetation. In the first experiment, land-cover change alters surface energy fluxes and the hydrologic cycle (biogeophysical processes) and alters carbon storage (biogeochemical processes). In the latter experiment, changes in land cover are removed so that only fossil fuel CO$_2$ emission drives climate change.

Table 33.1 summarizes results. Atmospheric CO$_2$ in 2100 varies greatly among the four scenarios, but land-cover change increases atmospheric CO$_2$ in all scenarios compared with simulations without land-cover change. This increase ranges from 127 ppm in A2 to 20 ppm in B1. All scenarios show an increase in global annual mean temperature in 2100 compared with 2000, ranging from 1.7°C in B1 to 2.7°C in

A2. Land-cover change contributes 0.2–0.3°C to this warming. The importance of biogeophysical effects of land-cover change is highlighted by the similar warming found in the A2 and A1B simulations despite a 110 ppm lower atmospheric CO$_2$ in A1B (847 ppm) compared with A2 (957 ppm). This is due, in part, to greater biogeophysical cooling in A2 as a result of deforestation in northern latitudes.

The biogeophysical effects of land-cover change can be compared with the biogeochemical effects from carbon emissions. To examine this, Sitch et al. (2005) repeated the simulations for A2 and B1 but with only the biogeophysical or biogeochemical effects of land-cover change represented (Figure 33.4). In both scenarios, carbon loss from land-cover change causes biogeochemical warming. The temperature increase is greatest in A2 (0.25-0.5°C) because of widespread deforestation. The B1 biogeochemical warming is smaller (0.1–0.25°C) because of less tropical deforestation and because of temperate reforestation. The two scenarios also differ in biogeophysical effects. The widespread expansion of agriculture in A2 leads to a cooling of 0.1–0.25°C throughout much of the world. In B1, biogeophysical processes lead to warming, primarily because of temperate forest regrowth. The biogeophysical contrast between A2 and

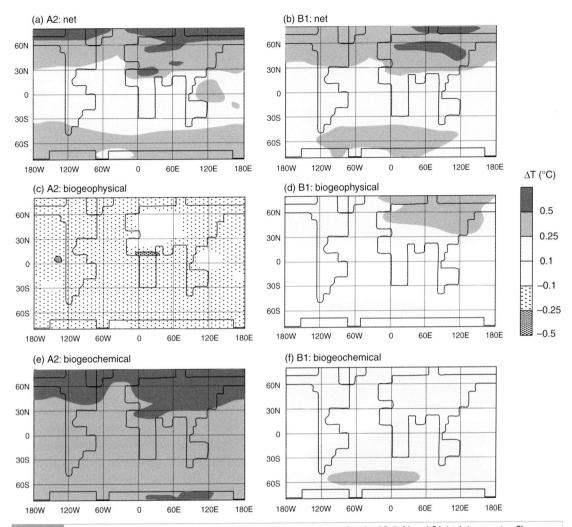

Fig. 33.4 Difference in annual mean temperature due to land-cover change for the A2 (left) and B1 (right) scenarios. Shown are (a, b) the net effect, (c, d) the biogeophysical effect, and (e, f) the biogeochemical effect. Adapted from Sitch et al. (2005).

B1 is similar to results of Feddema et al. (2005) shown in Figure 33.3, but the simulations of Sitch et al. (2005) additionally highlight biogeochemical effects. In A2, the biogeochemical warming offsets the biogeophysical cooling. The B1 net warming is similar to A2, but for different reasons: moderate biogeophysical warming from temperate reforestation augments weak biogeochemical warming from tropical deforestation.

More recent simulations using representative concentration pathways point to a similar conclusion. Davies-Barnard et al. (2014b) used an Earth system model to simulate the biogeophysical, biogeochemical, and net temperature change arising from the land-cover changes in RCP2.6, RCP4.5, and RCP8.5. The world depicted by RCP2.6 and RCP8.5 shows continued deforestation and an increase in cropland during the twenty-first century; in contrast, forests expand in RCP4.5 to sequester carbon and meet climate stabilization goals (Figure 2.12). The different depictions of deforestation, afforestation, and agricultural land have different consequences for climate (Table 33.2). In particular, the afforestation in RCP4.5 enhances twenty-first century climate warming because of biogeophysical processes (i.e., lower

Table 33.2 Change in annual mean temperature (°C) over land at the end of the twenty-first century due to land-cover change			
Scenario	Biogeophysical	Biogeochemical	Net
RCP2.6	−0.02	0.04	0.015
RCP4.5	0.19	−0.08	0.11
RCP8.5	−0.04	0.04	0.0035

Note: Temperature is averaged over the period 2070–2100.
Source: From Davies-Barnard et al. (2014b).

surface albedo). The biogeophysical warming outweighs the biogeochemical cooling from carbon sequestration. In contrast, biogeochemical warming from carbon emissions dominates the net temperature response in RCP2.6 and RCP8.5.

33.5 | Forest Management

Forests provide a large positive climate service (Figure 33.2), and their preservation and management has been highlighted for their climate benefits (Bonan 2008; Chapin et al. 2008; Jackson et al. 2008; Anderson et al. 2011). Forest management to enhance carbon uptake is one means to mitigate anthropogenic CO_2 emissions. Prominent in this is the widespread planting of trees in forest plantations on abandoned cropland, pastureland, or marginal land to sequester vast amounts of carbon. Other programs promote forest management to decrease land-use emissions and preserve carbon sinks. However, forests in mid- to high latitudes are thought to warm climate, primarily because of their lower albedo compared with cropland or grassland. Additional changes in climate occur with differences in evapotranspiration, but these are less certain. By decreasing surface albedo, temperate and boreal afforestation and reforestation could warm climate and diminish the reduction in global warming resulting from carbon sequestration. In contrast, tropical afforestation, reforestation, and avoided deforestation cool climate through combined biogeophysical and biogeochemical outcomes. Model simulations suggest this is indeed the case (Betts 2000; Claussen et al. 2001; Gibbard et al. 2005; Schaeffer et al.

2006; Bala et al. 2007; Betts et al. 2007; Bathiany et al. 2010; Arora and Montenegro 2011).

Arora and Montenegro (2011) used an Earth system model to compare the climate consequences of five afforestation scenarios. Areas of the world that are presently occupied by cropland but which could potentially support forests were allowed to be afforested. One scenario was 100 percent global afforestation, in which all crops were replaced by trees. A 50 percent global afforestation scenario restricted tree planting to only 50 percent of the global crop area. Other scenarios restricted afforestation to boreal, northern temperate, or tropical latitudes. The afforestation occurred over a 50-year period, from 2011 to 2060.

Atmospheric CO_2 increases over the twenty-first century in the control simulation without afforestation as a result of fossil fuel emissions, and the planet warms. Afforestation increases the land carbon uptake over the twenty-first century and reduces atmospheric CO_2 concentration compared with the control simulation. The 100 percent global afforestation simulation has the largest decrease (93 ppm). The reduction in atmospheric CO_2 is one-half as much in the 50 percent global afforestation scenario. Temperature decreases in both scenarios. Complete (100%) and partial (50%) global afforestation reduce planetary warming by 0.45°C and 0.25°C, respectively, over the period 2081–2100 compared with the control simulation (Table 33.3). The temperature reduction with complete afforestation is mostly uniform globally and results from biogeochemical effects (Figure 33.5). The global biogeophysical effect of afforestation averages to no temperature

Table 33.3 Impact of afforestation on atmospheric CO_2 and global annual mean temperature

Scenario	Afforested area (million km²)	Atmospheric CO_2 (ppm)	Temperature (°C)		
			Net	Biogeophysical	Biogeochemical
100% global afforestation	20.2	−93	−0.45	0.00	−0.45
50% global afforestation	10.1	−45	−0.25	−0.01	−0.24
50% boreal afforestation	2.0	−9	−0.04	0.01	−0.05
50% northern temperate afforestation	4.7	−20	−0.11	0.04	−0.15
50% tropical afforestation	2.7	−20	−0.16	−0.07	−0.09

Note: Temperature is the difference of the afforestation simulations compared with a control simulation and is for the period 2081–2100. Atmospheric CO_2 is for 2100.
Source: From Arora and Montenegro (2011).

change, but is a prominent warming in northern high latitudes, where the warming from the lower albedo is important and initiates loss of sea ice. With partial global afforestation, biogeophysical warming is more dominant because of the smaller drawdown in CO_2, and albedo warming dominates the net temperature response in parts of Eurasia.

The net climate effect of afforestation in these simulations is the balance between surface changes in albedo and evapotranspiration and changes in the land carbon sink. Modeling studies show a distinctly different net effect of boreal, temperate, and tropical forests on climate. The presence of boreal forests may warm climate because the positive radiative forcing from low surface albedo offsets the negative radiative forcing from carbon storage. In contrast, tropical forests cool climate because high rates of evapotranspiration augment the biogeochemical cooling. Similar regional differences are seen with afforestation (Figure 33.5). Biogeophysical changes arising from boreal and northern temperate afforestation enhance warming, while tropical afforestation reduces warming because of greater evapotranspiration (Table 33.3). When normalized by the area

of land afforested, the temperature decrease is three times more in the tropics (–0.06°C per million km²) than in the boreal and northern temperate regions (–0.02°C per million km²). This suggests that tropical afforestation, reforestation, and avoided deforestation are effective forest management policies to mitigate climate change, as seen also in Figure 33.2.

These model simulations highlight the need to consider biogeophysical (albedo, evapotranspiration) effects in addition to the carbon benefits of afforestation. The albedo of forests varies with stand age, canopy structure, and species composition, and several analyses illustrate forest management of surface albedo, with resulting climate benefits. In temperate conifer and deciduous forests, for example, forest thinning can alter summer canopy albedo by reducing leaf area index, though the changes are modest – up to 0.02 in the visible and 0.05 in the near-infrared wavebands (Otto et al. 2014). The albedo of deciduous forests in boreal regions is generally higher than in conifer forests, both in summer and winter, and forest management that favors deciduous species over conifers can provide climate benefits (Bright et al. 2014). The higher summertime evapotranspiration

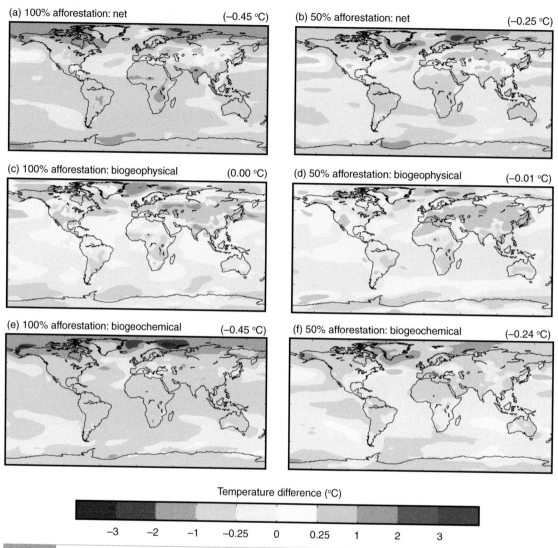

(a) 100% afforestation: net   (−0.45 °C)

(b) 50% afforestation: net   (−0.25 °C)

(c) 100% afforestation: biogeophysical   (0.00 °C)

(d) 50% afforestation: biogeophysical   (−0.01 °C)

(e) 100% afforestation: biogeochemical   (−0.45 °C)

(f) 50% afforestation: biogeochemical   (−0.24 °C)

Temperature difference (°C)

−3   −2   −1   −0.25   0   0.25   1   2   3

Fig. 33.5 Difference in annual mean temperature due to 100 percent afforestation (left) and 50 percent afforestation (right). Temperature difference is the change over the period 2081–2100 compared with a control simulation without afforestation. Shown are (a, b) the net effect, (c, d) the biogeophysical effect, and (e, f) the biogeochemical effect. The numbers in parentheses are the global average. Adapted from Arora and Montenegro (2011) and provided courtesy of Vivek Arora. See color plate section.

of deciduous forests further augments the climate mitigation potential. Forest management may have as large an impact on climate as does land-cover change (Luyssaert et al. 2014).

33.6 | Agricultural Management

The use of land to grow food and fiber and to raise livestock releases significant amounts of CO_2, CH_4, and N_2O. Application of nitrogen fertilizer increases N_2O emissions. Livestock add CH_4 to the atmosphere, and their manure releases N_2O and CH_4. Rice cultivation is a source of CH_4. Mechanized farming, the manufacturing of fertilizers and other chemicals, and other food production processes release additional greenhouse gases. These emissions can be reduced by agricultural management practices (Smith et al. 2008, 2014). Such options include: increasing crop yield; increasing soil carbon storage; management of nitrogen fertilizer to reduce N_2O emissions; management of livestock and feedlots to reduce emissions of CH_4 and N_2O; and production of biofuels. For example, increasing crop yields lessen the need to clear forests and other lands to grow crops. More efficient use of fertilizers reduces emissions from excess application.

Agricultural practices to increase crop yield or store carbon in soil may affect climate through biogeophysical processes. The addition of biochar to soil sequesters carbon in soil, but also decreases soil albedo (Genesio et al. 2012; Verheijen et al. 2013). Crop cultivars determine specific phenological events such as planting date, vegetative and reproductive growth periods, and time to harvest. These influence yield, but also affect the seasonality of energy fluxes in temperate croplands (Sacks and Kucharik 2011). An earlier planting date of maize (corn), for example, increases latent heat flux and decreases sensible heat flux in June. A shorter time period from maturity to harvest increases net radiation in October by exposing the low albedo of soil compared with brown foliage.

No-till agriculture is a means to grow crops with reduced disturbance of soil by decreasing the use of plows, cultivators, or other mechanization in soil preparation for planting or after harvesting. It improves the soil by promoting infiltration and by increasing organic matter retention, and it may increase crop productivity. It has been advocated to store carbon in soil and reduce land-use carbon emissions (Smith et al. 2008, 2014), though the actual benefit is still being debated (Powlson et al. 2014; Pittelkow et al. 2015). In addition, elimination of tillage increases surface albedo when plant residues are left on the ground. Measurements in a winter wheat field, for example, show a daily albedo of ~0.2 following harvest with conventional tillage and an albedo of ~0.3 without tillage (Davin et al. 2014). The increased albedo from no-till farming cools climate (Lobell et al. 2006; Davin et al. 2014). Crop residues also decrease soil evaporation, which by itself results in a modest warming, but the net biogeophysical outcome of no-till farming is to cool climate (Davin et al. 2014).

The planting and harvesting of biofuels can mitigate anthropogenic CO_2 emissions, but the biogeophysical changes associated with biofuels may alter climate. Climate models simulations show that the introduction of biofuels in midwestern and central United States decreases temperature during the growing season (Georgescu et al. 2011; Anderson et al. 2013). Georgescu et al. (2011) found a regional cooling of up to 1°C during the growing season by replacing annual crops with perennial grasses such as switchgrass and miscanthus. This cooling occurs from increased evapotranspiration, but also due to higher albedo. A key consideration is the deeper rooting depth of biofuels compared with annual crops and their longer period with leaf emergence. The higher water-use efficiency of switchgrass and miscanthus relative to crops (e.g., maize) is also important (VanLoocke et al. 2012).

Analysis of land-use transitions in the tropical savanna region of Brazil provides a detailed analysis of climate change from cultivation of biofuels. Much of this region has been converted to cropland and pastureland, but Brazil has an expanding production of sugarcane for biofuel. These land uses differ in their albedo and evapotranspiration and have different climate impacts. Analysis of remotely-sensed surface temperature, albedo, and evapotranspiration shows that conversion of native savanna to cropland and pastureland warms surface temperature regionally by 1.5°C (Loarie et al. 2011). This warming occurs because of decreased evapotranspiration, which counters the cooling from increased albedo. The introduction of sugarcane alleviates some of these changes. Conversion from cropland and pastureland to sugarcane cools the region by 0.9°C. This cooling occurs from the combined effects of greater evapotranspiration and higher albedo with the transition to sugarcane. Climate model simulations show that conversion of existing land cover (with a high crop and pasture mosaic) to sugarcane decreases temperature by about 1°C during the growing season (Georgescu et al. 2013a). However, the climate effects of sugarcane are not distributed equally throughout the year. Temperature increases by about the same amount after harvest, when evapotranspiration decreases relative to the existing landscape. When averaged over the year, the seasonal differences offset one another.

33.7 | Urban Planning and Design

Cities, too, can be planned, designed, and adapted for climate change. More than one-half of the world's population lives in urban settings (Seto et al. 2014). There are 23 cities with more than 10 million inhabitants, 449 with more than 1 million people, and nearly 1000 cities with populations of 500,000 or more. About two-thirds of the world's population is expected to be living in urban areas by 2050. In the context of climate change mitigation, urban planning and design are seen as means to reduce greenhouse gas emissions (Seto et al. 2014). The kinds of towns, cities, and urban settings that develop over the coming decades will determine the use of fossil fuels and emission of greenhouse gases. Future growth in urban population and urban area will necessitate management of the physical form and infrastructure of cities (extent, density, design of streets and buildings, land use mix, transportation, movement) to reduce emissions of CO_2 and other greenhouse gases. In the context of adaptation, it is also recognized that urban planning and design provide an opportunity to lessen excessive urban temperatures through use of green spaces, porous pavements, white and green roofs, and other practices (Revi et al. 2014). Parks have a lower temperature than the surrounding cityscape (Chapter 32) and provide a planning tool to lessen the urban heat island (Bowler et al. 2010). Similarly, use of light colored and permeable pavements instead of dark, impervious asphalt surfaces can reduce temperature by reflecting solar radiation and through evaporative cooling (Li et al. 2013).

Increasing the albedo of city surfaces (e.g., roofs, roads) is widely-studied as a means to reduce the high temperatures found in cities. A study in New York City compared the temperature of white and black roofs (Gaffin et al. 2012). The white roofs had an albedo of about 0.65 and decreased peak daytime surface temperature during summer by 24°C and average daily temperature by 5–7°C compared with black roofs. Modeling results show that a higher city albedo can directly reduce summer air temperatures (Akbari et al. 2009, 2012; Menon et al. 2010; Oleson et al. 2010; Akbari and Matthews 2012; Jacobson and Ten Hoeve 2012; Georgescu et al. 2013b; Li et al. 2014). Additionally, cooler building temperatures reduce energy consumption by air conditioning. This is a direct cost savings. Furthermore, it lessens the waste energy produced in air conditioning, which is a contributor to the urban heat island.

Many such modeling studies do not explicitly represent cities, but rather increase the albedo of the land surface to mimic a whitening of urban surfaces (e.g., Menon et al. 2010; Akbari et al. 2012). In contrast, Oleson et al. (2010) used a global climate model that explicitly represents cities as an urban canyon comprised of streets, walls, roofs, and greenspace (Figure 32.17). They increased the albedo of roofs to 0.9 to mimic a highly reflective surface. The effective increase in surface albedo varied with the fraction of the city surface covered by roofs. For one particular location (New York City), the increase in roof albedo from ~0.3 in the control simulation to 0.9 decreases the summer urban heat island by 0.5°C. When averaged over all urban areas worldwide, increasing the albedo of city roofs decreases the daily maximum temperature in cities by 0.6°C. At high northern latitudes in winter, higher roof albedo is not as effective at reducing the heat island because of low solar radiation, the high albedo of snow on roofs, and because of increased energy usage to heat buildings. In these regions, the benefits from a reduction in the summertime heat island must be weighed against increased heating costs in winter.

Rooftop gardens are another means to alleviate high temperatures within cities (Revi et al. 2014). These so-called green roofs operate through evaporative cooling of the surface, by storing heat in the soil substrate and insulating the building from temperature extremes, and by shading from the plant canopy, each of which lowers the internal building temperature and may also lower air temperature above the roof (Simmons et al. 2008; Getter et al. 2011;

Table 33.4 Summer (June–August) temperature change (°C) in cities due to urban growth over the twenty-first century and with white and green roof adaptation strategies

Region	Urban growth	White roofs	Green roofs
California	1.3	−1.5	−0.2
Arizona	0.9	−0.5	−0.2
Texas	1.2	−1.2	−0.5
Florida	0.8	−0.4	−0.2
Mid-Atlantic	1.2	−1.8	−1.2
Chicago/Detroit	1.1	−1.4	−0.9

Note: Temperature is the difference from the control simulation with present-day cities.
Source: From Georgescu et al. (2014).

Coutts et al. 2013; Li et al. 2014). The effectiveness of green roofs varies with design criteria that determine evapotranspiration rates (e.g., substrate depth, vegetation composition, rainwater retention). The hydrological performance of the rooftop garden influences the surface energy balance and the partitioning of net radiation into sensible and latent heat fluxes and so determines the rooftop microclimate. The thermal benefit of green roofs requires maintaining actively transpiring vegetation.

Climate model simulations show the influence of future urban expansion over the twenty-first century in the United States on climate and the effectiveness of white and green roofs to alleviate urban warming (Georgescu et al. 2014). The study contrasted the climate simulated with present-day cities to that simulated with an urban growth scenario to accommodate a projected population of 690 million people in 2100. Without any intervention, city growth increases summer air temperature by 1–2°C in large urban regions throughout the country, separate from greenhouse gas warming. White roofs and green roofs alleviate this warming, but their usefulness varies by region (Table 33.4). White roofs are a more effective adaptation strategy than are green roofs, but the magnitude of the difference relates to prevailing climate. In Florida, for example, white roofs provide an additional 0.2°C cooling compared with green roofs; in California the additional cooling is 1.3°C. The effectiveness of the adaptation strategies also varies seasonally. During

winter, the cooling from white roofs enhances energy demand to heat buildings.

Perhaps because climate affects human health and comfort in so many ways, consideration of climate in the built environment has received much attention. There have been many treatments by designers and architects of the urban environment, its relationship with climate, and how to incorporate climate into urban planning and design (Aronin 1953; Olgyay 1963; Givoni 1976, 1998; Robinette 1983; Lowry 1988; Brown and Gillespie 1995). Cooling breezes and shade are needed during hot, overheated periods; the warmth of sunlight and protection from winds are needed during cold, underheated periods. Use of vegetation to conserve energy and to create thermally pleasant environments by blocking cold winter winds, shading hot summer sun, and by evaporative cooling is well documented. Precipitation can also be more effectively managed and stored by vegetated landscapes that promote infiltration in contrast to hard surfaces that promote runoff. Building materials and construction can be adapted to reduce energy use and mitigate greenhouse gas emissions (Lucon et al. 2014).

We have developed a modern lifestyle independent of climate and the natural environment. Buildings, cities, and landscapes are designed without regard to unique regional environments and ecosystems; we no longer know where in the world we live. This contrasts greatly with less technological eras. For example, colonists settling Virginia and Massachusetts during the

1600s quickly adapted building design techniques brought with them from England to provide relief from the hot, humid Virginia summer and to accommodate the cold New England winter (Fitch 1948, 1966, 1972). In contrast, now we use technology to modify the prevailing climate for our needs. This is evident in our homes and commercial buildings, which are heated in winter and cooled by air conditioning in summer, and our residential yards, which may be irrigated to supplement rainfall.

Topographic, edaphic, and ecological features of the landscape create noticeable climate changes. Orographic precipitation creates a stark contrast between moist and arid climates on the windward and leeward slopes of mountains. Northeast slopes in the Northern Hemisphere are typically a few degrees cooler than southwest slopes, particularly during summer afternoons. Air temperature typically cools by 1°C with every 100 m gain in elevation. Cold air may collect at night in low-lying sites while a slightly higher location basks in relative warmth. Oceans and large lakes moderate the seasonal variation in temperature. Sea breezes provide relief on a hot summer day while inland cities may suffer from a sweltering heat wave. A person only has to stand under a tree on a hot summer day to realize the shade beneath the canopy is cooler than open spaces where one is directly exposed to the Sun's rays. Windbreaks formed by trees conserve energy by reducing heat loss from strong winter winds. By understanding the science of ecology and climatology, we can take advantage of natural landscape features and make climate work for us.

33.8 | The Coupled Climate–Natural–Human System

Our current understanding of the climate services of ecosystems is based on surface albedo, evapotranspiration, and carbon storage. It emphasizes the positive climate benefits of tropical rainforest preservation, the possibly harmful effects of afforestation in boreal and northern forests, and the uncertain climate services of temperate forests. However, biosphere–atmosphere interactions and ecosystem management for climate services are much more complex than this simple understanding. Greenhouse gases are well-mixed in the atmosphere and influence global climate, while biogeophysical processes are minor at the global scale but have large regional impact. Biogeophysical processes influence climate more immediately than does the carbon cycle so that a timescale must be considered in policy actions. Slow rates of long-term carbon accumulation in forests may in the short-term be offset by more rapid biogeophysical effects.

A broad understanding of terrestrial ecosystem functioning in the Earth system is needed to lessen the risk of unintended consequences of policy actions. An integrated understanding of land-use and climate requires considering more than albedo, evapotranspiration, and carbon to include other greenhouse gases, reactive nitrogen, biogenic volatile organic compounds, and aerosols. Increased fertilizer usage, for example, may be required in a future Earth to sustain food and fiber yield or for biofuel production. Such an outcome would increase N_2O emissions and have other climate consequences through the nitrogen cascade (Figure 30.1). Interactions among terrestrial ecosystems, climate change, and atmospheric chemistry can dampen or enhance climate change, particularly through emissions of biogenic volatile organic compounds (Figure 31.8). These emissions may increase with warmer temperatures, enhancing the production of secondary organic aerosols, or their emissions may by inhibited by elevated atmospheric CO_2. Anthropogenic land use and land-cover change will also greatly influence future emissions. Future reforestation and afforestation and the cultivation of biofuels are expected to increase isoprene emissions, with subsequent effects for ozone, methane, aerosols, and the oxidation capacity of the troposphere.

Tropospheric ozone, NO_x, and aerosols are key atmospheric pollutants that affect climate and ecosystem functioning. Ozone and fine particulate matter (less than 2.5 μm in diameter) also cause serious human health problems. The climate influences of these pollutants is complex; some warm climate, others cool climate, and they have indirect consequences from biogeochemical cycles. They are short lived in the atmosphere (days to weeks) and their concentration can be regionally localized, in contrast with CO_2, CH_4, and N_2O, which are long-lived greenhouses gases and well-mixed in the atmosphere. However, it is becoming clear that short-lived climate forcings are significant within the Earth system. Air pollution control and climate change must be treated in a common policy framework so as to assess the human health and ecosystem benefits gained by limiting pollutant emissions while also preventing unintended climate consequences (Prinn et al. 2007; Arneth et al. 2009; Unger 2012).

Land use takes place in the context of other environmental changes: higher atmospheric CO_2 that simulates plant productivity but reduces stomatal conductance; a warmer climate with less snow and more droughts; increased nitrogen deposition that enhances terrestrial carbon uptake but has other negative climate consequences; higher ozone concentration that damages stomata; and increased aerosols that decrease solar radiation at the surface but increase the fraction that is diffuse. These forcings have different consequences for ecosystem functioning (Huntingford et al. 2011). Biosphere–atmosphere interactions must be considered in terms of the multitude of forcings over the past and coming century. Moreover, the climate services of ecosystems depend on the prevailing climate. Evaporative cooling is less important in a climate with hot, dry summers. The albedo difference among forests, grasslands, and croplands is greatest in winter and spring in snowy climates. Forest masking of snow albedo becomes less important in a warm climate with no snow (Pitman et al. 2011) or if land-use change occurs on land without much snow (Pongratz et al. 2011).

Afforestation, reforestation, croplands, and biofuels are key human uses of land that influence climate through the carbon cycle, emissions of other greenhouse gases, changes in land surface biogeophysics (albedo, evapotranspiration), and other processes. The exact climate outcome of these is still uncertain. Models show highly divergent responses to land-use and land cover change (Chapter 28) and in carbon cycle–climate feedbacks (Chapter 29). Moreover, widespread afforestation may have unexpected consequences by changing the planetary energy balance and altering the Hadley circulation (Swann et al. 2012), as shown in Figure 28.20. Nonetheless, it is increasingly recognized that land-use decisions do affect climate, particularly at the regional scale, and that an understanding of future Earth requires study of the interactions and feedbacks among socioeconomic needs, human decision making, land use, and climate, because climate affects the way in which people use land to grow food and fiber (van Vuuren et al. 2012; Hibbard and Janetos 2013; Arneth et al. 2014; Rounsevell et al. 2014).

Future land use to grow food and fiber that does not exacerbate, and perhaps lessens, anthropogenic climate change is a key aspect of the coupled climate and human system. The yield of wheat, rice, and maize in temperate and tropical regions is expected to decrease with future climates, and adaptations such as changes in cultivar, planting date, and irrigation are needed to alleviate the decline (Challinor et al. 2014). Climate change may necessitate still more transformational adaptations such as crop relocation or changes in farming system (e.g., dryland, irrigated). Food production must increase substantially to feed a growing global population in a changing climate, but at the same time the environmental footprint of agriculture must decrease to avoid planetary harm (Foley et al. 2011).

Earth system models currently prescribe land-use pathways as forcings, not feedbacks, and so decouple land-use and climate. Economic factors (e.g., crop yield, market pricing, financial incentives for carbon storage) can drive different land-use decisions with vastly different climate outcomes (Hallgren et al. 2013; Davies-Barnard et al. 2014a). A full two-way interactive coupling between climate and land use in which human uses of land are simulated interactively in response to climate, similar to terrestrial biosphere models, is elusive. Initial such efforts illustrate that the effects of land use driven by socioeconomic considerations have an important role in determining the outcomes of climate policies over the twenty-first century (Voldoire et al. 2007; Jones et al. 2013).

33.9 | Review Questions

1. Describe a set of climate model experiments to examine the biogeophysical and biogeochemical effects of future land-cover change on climate and also their combined effects.

2. Contrast the climate change mitigation potential of boreal, temperate, and tropical forests. Describe this potential in terms of biogeophysical processes and the carbon cycle. Which forest has the greatest potential? Explain why.

3. How does consideration of biogenic volatile organic compounds affect the climate change mitigation potential of boreal, temperate, and tropical forests?

4. Devise forest management policies to lessen anthropogenic climate warming in the Brazilian Amazon, the United States, and Canada. Discuss socioeconomic and environmental consequences of these policies.

5. In a climate with mild winters and hot summers, is it preferable to build a house on a north- or a south-facing slope to promote energy conservation? Which is preferable in a climate with cold winters and mild summers?

6. The following table compares the climate of London with that of the Massachusetts Bay area (Boston) and the Tidewater region of Virginia (Norfolk). Discuss changes in climate faced by English settlers of Massachusetts and Virginia in the 1600s. How might these changes have affected building design?

	London	Boston	Norfolk
Average January temperature (°C)	4.4	−2.2	4.4
Days with minimum temperature <0°C	41	98	54
Average July temperature (°C)	16.7	22.2	25.6
Days with maximum temperature >32°C	1	13	32
Annual precipitation (mm)	584	1118	1143

7. Landsberg (1973) coined the phrase "the meteorologically utopian city." How would you design such a city for the region of the world in which you live? How would you address anthropogenic climate change?

33.10 | References

Akbari, H., and Matthews, H. D. (2012). Global cooling updates: Reflective roofs and pavements. *Energy and Buildings*, 55, 2–6.

Akbari, H., Menon, S., and Rosenfeld, A. (2009). Global cooling: Increasing world-wide urban albedos to offset CO_2. *Climatic Change*, 94, 275–286.

Akbari, H., Matthews, H. D., and Seto, D. (2012). The long-term effect of increasing the albedo of urban areas. *Environmental Research Letters*, 7, 024004, doi:10.1088/1748-9326/7/2/024004.

Allwood, J. M., Bosetti, V., Dubash, N. K., Gómez-Echeverri, L., and von Stechow, C. (2014). Glossary. In *Climate Change 2014: Mitigation of Climate Change. Contribution of Working Group III to the Fifth Assessment Report of the Intergovernmental Panel on Climate Change*, ed. O. Edenhofer, R. Pichs-Madruga, Y. Sokona, et al. Cambridge: Cambridge University Press, pp. 1249–1279.

Anderson, C. J., Anex, R. P., Arritt, R. W., et al. (2013). Regional climate impacts of a biofuels policy projection. *Geophysical Research Letters*, 40, 1217–1222, doi:10.1002/grl.50179.

Anderson, R. G., Canadell, J. G., Randerson, J. T., et al. (2011). Biophysical considerations in forestry for climate protection. *Frontiers in Ecology and the Environment*, 3, 174–182.

Anderson-Teixeira, K. J., Snyder, P. K., Twine, T. E., et al. (2012). Climate-regulation services of natural and agricultural ecoregions of the Americas. *Nature Climate Change*, 2, 177–181.

Arneth, A., Unger, N., Kulmala, K., and Andreae, M. O. (2009). Clean the air, heat the planet? *Science*, 326, 672–673.

Arneth, A., Brown, C., and Rounsevell, M. D. A. (2014). Global models of human decision-making for land-based mitigation and adaptation assessment. *Nature Climate Change*, 4, 550–557.

Aronin, J. E. (1953). *Climate and Architecture*. New York: Reinhold.

Arora, V. K., and Montenegro, A. (2011). Small temperature benefits provided by realistic afforestation efforts. *Nature Geoscience*, 4, 514–518.

Bala, G., and Nag, B. (2012). Albedo enhancement over land to counteract global warming: Impacts on hydrological cycle. *Climate Dynamics*, 39, 1527–1542.

Bala, G., Caldeira, K., Wickett, M., et al. (2007). Combined climate and carbon-cycle effects of large-scale deforestation. *Proceedings of the National Academy of Sciences USA*, 104, 6550–6555.

Bathiany, S., Claussen, M., Brovkin, V., Raddatz, T., and Gayler, V. (2010). Combined biogeophysical and biogeochemical effects of large-scale forest cover changes in the MPI earth system model. *Biogeosciences*, 7, 1383–1399.

Betts, R. A. (2000). Offset of the potential carbon sink from boreal forestation by decreases in surface albedo. *Nature*, 408, 187–190.

Betts, R. A., Falloon, P. D., Klein Goldewijk, K., and Ramankutty, N. (2007). Biogeophysical effects of land use on climate: Model simulations of radiative forcing and large-scale temperature change. *Agricultural and Forest Meteorology*, 142, 216–233.

Bonan, G. B. (2008). Forests and climate change: Forcings, feedbacks, and the climate benefits of forests. *Science*, 320, 1444–1449.

Boucher, O., Randall, D., Artaxo, P., et al. (2013). Clouds and aerosols. In *Climate Change 2013: The Physical Science Basis. Contribution of Working Group I to the Fifth Assessment Report of the Intergovernmental Panel on Climate Change*, ed. T. F. Stocker, D. Qin, G.-K. Plattner, et al. Cambridge: Cambridge University Press, pp. 571–657.

Boucher, O., Forster, P. M., Gruber, N., et al. (2014). Rethinking climate engineering categorization in the context of climate change mitigation and adaptation. *WIREs Climate Change*, 5, 23–35.

Bowler, D. E., Buyung-Ali, L., Knight, T. M., and Pullin, A. S. (2010). Urban greening to cool towns and cities: A systematic review of the empirical evidence. *Landscape and Urban Planning*, 97, 147–155.

Bright, R. M., Antón-Fernández, C., Astrup, R., et al. (2014). Climate change implications of shifting forest management strategy in a boreal forest ecosystem of Norway. *Global Change Biology*, 20, 607–621.

Brown, R. D., and Gillespie, T. J. (1995). *Microclimatic Landscape Design: Creating Thermal Comfort and Energy Efficiency*. New York: Wiley.

Caldeira, K., Bala, G., and Cao, L. (2013). The science of geoengineering. *Annual Review of Earth and Planetary Sciences*, 41, 231–256.

Challinor, A. J., Watson, J., Lobell, D. B., et al. (2014). A meta-analysis of crop yield under climate change and adaptation. *Nature Climate Change*, 4, 287–291.

Chapin, F. S., III, Randerson, J. T., McGuire, A. D., Foley, J. A., and Field, C. B. (2008). Changing feedbacks in the climate–biosphere system. *Frontiers in Ecology and the Environment*, 6, 313–320.

Ciais, P., Sabine, C., Bala, G., et al. (2013). Carbon and other biogeochemical cycles. In *Climate Change 2013: The Physical Science Basis. Contribution of Working Group I to the Fifth Assessment Report of the Intergovernmental Panel on Climate Change*, ed. T. F. Stocker, D. Qin, G.-K. Plattner, et al. Cambridge: Cambridge University Press, pp. 465–570.

Claussen, M., Brovkin, V., and Ganopolski, A. (2001). Biogeophysical versus biogeochemical feedbacks of large-scale land cover change. *Geophysical Research Letters*, 28, 1011–1014.

Coutts, A. M., Daly, E., Beringer, J., and Tapper, N. J. (2013). Assessing practical measures to reduce urban heat: Green and cool roofs. *Building and Environment*, 70, 266–276.

Crutzen, P. J. (2006). Albedo enhancement by stratospheric sulfur injections: A contribution to resolve a policy dilemma? *Climatic Change*, 77, 211–219.

Curry, C. L., Sillmann, J., Bronaugh, D., et al. (2014). A multimodel examination of climate extremes in an idealized geoengineering experiment. *Journal of Geophysical Research: Atmospheres*, 119, 3900–3923, doi:10.1002/2013JD020648.

Davies-Barnard, T., Valdes, P. J., Singarayer, J. S., and Jones, C. D. (2014a). Climatic impacts of land-use change due to crop yield increases and a universal carbon tax from a scenario model. *Journal of Climate*, 27, 1413–1424.

Davies-Barnard, T., Valdes, P. J., Singarayer, J. S., Pacifico, F. M., and Jones, C. D. (2014b). Full effects of land use change in the representative concentration pathways. *Environmental Research Letters*, 9, 114014, doi:10.1088/1748-9326/9/11/114014.

Davin, E. L., Seneviratne, S. I., Ciais, P., Olioso, A., and Wang, T. (2014). Preferential cooling of hot extremes from cropland albedo management. *Proceedings of the National Academy of Sciences USA*, 111, 9757–9761.

Doughty, C. E., Field, C. B., and McMillan, A. M. S. (2011). Can crop albedo be increased through the modification of leaf trichomes, and could this cool regional climate? *Climatic Change*, 104, 379–387.

Feddema, J. J., Oleson, K. W., Bonan, G. B., et al. (2005). The importance of land-cover change in simulating future climates. *Science*, 310, 1674–1678.

Fitch, J. M. (1948). *American Building: The Forces that Shape It*. Boston: Houghton Mifflin.

Fitch, J. M. (1966). *American Building. 1: The Historical Forces that Shaped It*. Boston: Houghton Mifflin.

Fitch, J. M. (1972). *American Building. 2: The Environmental Forces that Shape It*. Boston: Houghton Mifflin.

Foley, J. A., Ramankutty, N., Brauman, K. A., et al. (2011). Solutions for a cultivated planet. *Nature*, 478, 337–342.

Gaffin, S. R., Imhoff, M., Rosenzweig, C., et al. (2012). Bright is the new black – multi-year performance of high-albedo roofs in an urban climate. *Environmental Research Letters*, 7, 014029, doi:10.1088/1748-9326/7/1/014029.

Genesio, L., Miglietta, F., Lugato, E., et al. (2012). Surface albedo following biochar application in durum wheat. *Environmental Research Letters*, 7, 014025, doi:10.1088/1748-9326/7/1/014025.

Georgescu, M., Lobell, D. B., and Field, C. B. (2011). Direct climate effects of perennial bioenergy crops in the United States. *Proceedings of the National Academy of Sciences USA*, 108, 4307–4312.

Georgescu, M., Lobell, D. B., Field, C. B., and Mahalov, A. (2013a). Simulated hydroclimatic impacts of projected Brazilian sugarcane expansion. *Geophysical Research Letters*, 40, 972–977, doi:10.1002/grl.50206.

Georgescu, M., Moustaoui, M., Mahalov, A., and Dudhia, J. (2013b). Summer-time climate impacts of projected megapolitan expansion in Arizona. *Nature Climate Change*, 3, 37–41.

Georgescu, M., Morefield, P. E., Bierwagen, B. G., and Weaver, C. P. (2014). Urban adaptation can roll back warming of emerging megapolitan regions. *Proceedings of the National Academy of Sciences USA*, 111, 2909–2914.

Getter, K. L., Rowe, D. B., Andresen, J. A., and Wichman, I. S. (2011). Seasonal heat flux properties of an extensive green roof in a Midwestern U.S. climate. *Energy and Buildings*, 43, 3548–3557.

Gibbard, S., Caldeira, K., Bala, G., Phillips, T. J., and Wickett, M. (2005). Climate effects of global land cover change. *Geophysical Research Letters*, 32, L23705, doi:10.1029/2005GL024550.

Givoni, B. (1976). *Man, Climate and Architecture*, 2nd ed. New York: Van Nostrand Reinhold.

Givoni, B. (1998). *Climate Considerations in Building and Urban Design*. New York: Van Nostrand Reinhold.

Hallgren, W., Schlosser, C. A., Monier, E., et al. (2013). Climate impacts of a large-scale biofuels expansion. *Geophysical Research Letters*, 40, 1624–1630, doi:10.1002/grl.50352.

Hibbard, K. A., and Janetos, A. C. (2013). The regional nature of global challenges: A need and strategy for integrated regional modeling. *Climatic Change*, 118, 565–577.

Huntingford, C., Cox, P. M., Mercado, L. M., et al. (2011). Highly contrasting effects of different climate forcing agents on terrestrial ecosystem services. *Philosophical Transactions of the Royal Society A*, 369, 2026–2037.

Irvine, P. J., Ridgwell, A., and Lunt, D. J. (2011). Climatic effects of surface albedo geoengineering. *Journal of Geophysical Research*, 116, D24112, doi:10.1029/2011JD016281.

Jackson, R. B., Randerson, J. T., Canadell, J. G., et al. (2008). Protecting climate with forests. *Environmental Research Letters*, 3, 044006, doi:10.1088/1748-9326/3/4/044006.

Jacobson, M. Z., and Ten Hoeve, J. E. (2012). Effects of urban surfaces and white roofs on global and regional climate. *Journal of Climate*, 25, 1028–1044.

Jones, A. D., Collins, W. D., Edmonds, J., et al. (2013). Greenhouse gas policy influences climate via direct effects of land-use change. *Journal of Climate*, 26, 3657–3670.

Kravitz, B., Caldeira, K., Boucher, O., et al. (2013). Climate model response from the Geoengineering Model Intercomparison Project (GeoMIP). *Journal of Geophysical Research: Atmospheres*, 118, 8320–8332, doi:10.1002/jgrd.50646.

Kravitz, B., MacMartin, D. G., Robock, A., et al. (2014). A multi-model assessment of regional climate disparities caused by solar geoengineering. *Environmental Research Letters*, 9, 074013, doi:10.1088/1748-9326/9/7/074013.

Landsberg, H. E. (1973). The meteorologically utopian city. *Bulletin of the American Meteorological Society*, 54, 86–89.

Lenton, T. M., and Vaughan, N. E. (2009). The radiative forcing potential of different climate geoengineering options. *Atmospheric Chemistry and Physics*, 9, 5539–5561.

Li, D., Bou-Zeid, E., and Oppenheimer, M. (2014). The effectiveness of cool and green roofs as urban heat island mitigation strategies. *Environmental Research Letters*, 9, 055002, doi:10.1088/1748-9326/9/5/055002.

Li, H., Harvey, J. T., Holland, T. J., and Kayhanian, M. (2013). The use of reflective and permeable pavements as a potential practice for heat island mitigation and stormwater management. *Environmental Research Letters*, 8, 015023, doi:10.1088/1748-9326/8/1/015023.

Loarie, S. R., Lobell, D. B., Asner, G. P., Mu, Q., and Field, C. B. (2011). Direct impacts on local climate of sugar-cane expansion in Brazil. *Nature Climate Change*, 1, 105–109.

Lobell, D. B., Bala, G., and Duffy, P. B. (2006). Biogeophysical impacts of cropland management changes on climate. *Geophysical Research Letters*, 33, L06708, doi:10.1029/2005GL025492.

Lowry, W. P. (1988). *Atmospheric Ecology for Designers and Planners*. McMinnville, Oregon: Peavine Publications.

Lucon, O., Ürge-Vorsatz, D., Zain Ahmed, A., et al. (2014). Buildings. In *Climate Change 2014: Mitigation of Climate Change. Contribution of Working Group III to the Fifth Assessment Report of the Intergovernmental Panel on Climate Change*, ed. O. Edenhofer, R. Pichs-Madruga, Y. Sokona, et al. Cambridge: Cambridge University Press, pp. 671–738.

Luyssaert, S., Jammet, M., Stoy, P. C., et al. (2014). Land management and land-cover change have impacts of similar magnitude on surface temperature. *Nature Climate Change*, 4, 389–393.

Menon, S., Akbari, H., Mahanama, S., Sednev, I., and Levinson, R. (2010). Radiative forcing and temperature response to changes in urban albedos and associated CO_2 offsets. *Environmental Research Letters*, 5, 014005, doi:10.1088/1748-9326/5/1/014005.

Niemeier, U., Schmidt, H., Alterskjaer, K., and Kristjánsson, J. E. (2013). Solar irradiance reduction via climate engineering: Impact of different techniques on the energy balance and the hydrological cycle. *Journal of Geophysical Research: Atmospheres*, 118, 11905–11917, doi:10.1002/2013JD020445.

Oleson, K. W., Bonan, G. B., and Feddema, J. (2010). Effects of white roofs on urban temperature in a global climate model. *Geophysical Research Letters*, 37, L03701, doi:10.1029/2009GL042194.

Olgyay, V. (1963). *Design With Climate: Bioclimatic Approach to Architectural Regionalism*. Princeton: Princeton University Press.

Otto, J., Berveiller, D., Bréon, F.-M., et al. (2014). Forest summer albedo is sensitive to species and thinning: How should we account for this in Earth system models? *Biogeosciences*, 11, 2411–2427.

Pitari, G., Aquila, V., Kravitz, B., et al. (2014). Stratospheric ozone response to sulfate geoengineering: Results from the Geoengineering Model Intercomparison Project (GeoMIP). *Journal of Geophysical Research: Atmospheres*, 119, 2629–2653, doi:10.1002/2013JD020566.

Pitman, A. J., Avila, F. B., Abramowitz, G., et al. (2011). Importance of background climate in determining impact of land-cover change on regional climate. *Nature Climate Change*, 1, 472–475.

Pittelkow, C. M., Liang, X., Linquist, B. A., et al. (2015). Productivity limits and potentials of the principles of conservation agriculture. *Nature*, 517, 365–368.

Pongratz, J., Reick, C. H., Raddatz, T., Caldeira, K., and Claussen, M. (2011). Past land use decisions have increased mitigation potential of reforestation. *Geophysical Research Letters*, 38, L15701, doi:10.1029/2011GL047848.

Powlson, D. S., Stirling, C. M., Jat, M. L., et al. (2014). Limited potential of no-till agriculture for climate change mitigation. *Nature Climate Change*, 4, 678–683.

Prinn, R., Reilly, J. M., Sarofim, M., Wang, C., and Felzer, B. (2007). Effects of air pollution control on climate: Results from an integrated global system model. In *Human-Induced Climate Change: An Interdisciplinary Assessment*, ed. M. Schlesinger, H. Kheshgi, J. Smith, et al. Cambridge: Cambridge University Press, pp. 93–102.

Revi, A., Satterthwaite, D. E., Aragón-Durand, F., et al. (2014). Urban areas. In *Climate Change 2014: Impacts, Adaptation, and Vulnerability. Part A: Global and Sectoral Aspects. Contribution of Working Group II to the Fifth Assessment Report of the Intergovernmental Panel on Climate Change*, ed. C. B. Field, V. R. Barros, D. J. Dokken, et al. Cambridge: Cambridge University Press, pp. 535–612.

Ridgwell, A., Singarayer, J. S., Hetherington, A. M., and Valdes, P. J. (2009). Tackling regional climate change by leaf albedo bio-geoengineering. *Current Biology*, 19, 146–150.

Robinette, G. (1983). *Energy Efficient Site Design*. New York: Van Nostrand Reinhold.

Rounsevell, M. D. A., Arneth, A., Alexander, P., et al. (2014). Towards decision-based global land use models for improved understanding of the Earth system. *Earth System Dynamics*, 5, 117–137.

Sacks, W. J., and Kucharik, C. J. (2011). Crop management and phenology trends in the U.S. Corn Belt: Impacts on yields, evapotranspiration and energy balance. *Agricultural and Forest Meteorology*, 151, 882–894.

Schaeffer, M., Eickhout, B., Hoogwijk, M., et al. (2006). CO_2 and albedo climate impacts of extratropical carbon and biomass plantations. *Global Biogeochemical Cycles*, 20, GB2020, doi:10.1029/2005GB002581.

Seto, K. C., Dhakal, S., Bigio, A., et al. (2014). Human settlements, infrastructure, and spatial planning. In *Climate Change 2014: Mitigation of Climate Change. Contribution of Working Group III to the Fifth Assessment Report of the Intergovernmental Panel on Climate Change*, ed. O. Edenhofer, R. Pichs-Madruga, Y. Sokona, et al. Cambridge: Cambridge University Press, pp. 923–1000.

Simmons, M. T., Gardiner, B., Windhager, S., and Tinsley, J. (2008). Green roofs are not created equal: The hydrologic and thermal performance of six different extensive green roofs and reflective and non-reflective roofs in a sub-tropical climate. *Urban Ecosystems*, 11, 339–348.

Singarayer, J. S., Ridgwell, A., and Irvine, P. (2009). Assessing the benefits of crop albedo bio-geoengineering. *Environmental Research Letters*, 4, 045110, doi:10.1088/1748-9326/4/4/045110.

Sitch, S., Brovkin, V., von Bloh, W., et al. (2005). Impacts of future land cover changes on atmospheric CO_2 and climate. *Global Biogeochemical Cycles*, 19, GB2013, doi:10.1029/2004GB002311.

Smith, P., Martino, D., Cai, Z., et al. (2008). Greenhouse gas mitigation in agriculture. *Philosophical Transactions of the Royal Society B*, 363, 789–813.

Smith, P., Bustamante, M., Ahammad, H., et al. (2014). Agriculture, forestry and other land use (AFOLU). In *Climate Change 2014: Mitigation of Climate Change. Contribution of Working Group III to the Fifth Assessment Report of the Intergovernmental Panel on Climate Change*, ed. O. Edenhofer, R. Pichs-Madruga, Y. Sokona,

et al. Cambridge: Cambridge University Press, pp. 811–922.

Swann, A. L. S., Fung, I. Y., and Chiang, J. C. H. (2012). Mid-latitude afforestation shifts general circulation and tropical precipitation. *Proceedings of the National Academy of Sciences USA*, 109, 712–716.

Tilmes, S., Müller, R., and Salawitch, R. (2008). The sensitivity of polar ozone depletion to proposed geoengineering schemes. *Science*, 320, 1201–1204.

Tilmes, S., Fasullo, J., Lamarque, J.-F., et al. (2013). The hydrological impact of geoengineering in the Geoengineering Model Intercomparison Project (GeoMIP). *Journal of Geophysical Research: Atmospheres*, 118, 11036–11058, doi:10.1002/jgrd.50868.

Unger, N. (2012). Global climate forcing by criteria air pollutants. *Annual Review of Environment and Resources*, 37, 1–24.

VanLoocke, A., Twine, T. E., Zeri, M., and Bernacchi, C. J. (2012). A regional comparison of water use efficiency for miscanthus, switchgrass and maize. *Agricultural and Forest Meteorology*, 164, 82–95.

van Vuuren, D. P., Batlle Bayer, L., Chuwah, C., et al. (2012). A comprehensive view on climate change: Coupling of earth system and integrated assessment models. *Environmental Research Letters*, 7, 024012, doi:10.1088/1748-9326/7/2/024012.

Vaughan, N. E., and Lenton, T. M. (2011). A review of climate geoengineering proposals. *Climatic Change*, 109, 745–790.

Verheijen, F. G. A., Jeffery, S., van der Velde, M., et al. (2013). Reductions in soil surface albedo as a function of biochar application rate: Implications for global radiative forcing. *Environmental Research Letters*, 8, 044008, doi:10.1088/1748-9326/8/4/044008.

Voldoire, A., Eickhout, B., Schaeffer, M., Royer, J.-F., and Chauvin, F. (2007). Climate simulation of the twenty-first century with interactive land-use changes. *Climate Dynamics*, 29, 177–193.

Coevolution of Climate and Life

34.1 | Chapter Summary

It is well known that life depends on climate. That climate regulates the structure and functioning of terrestrial ecosystems is a foundational principle of geography and ecology. Anthropology shows, too, that climate was central in the development of human societies. We now know as well that life itself influences climate. Numerous biosphere–atmosphere feedbacks are evident at long paleoclimate timescales spanning tens of thousands and millions of years, but also at the shorter timescale of the past century. A physical and chemical understanding of climate has grown to a biological perspective that includes the biogeophysical and biogeochemical functioning of plants and terrestrial ecosystems. The chemical composition of the atmosphere and its temperature, water vapor, clouds, and heat transport are regulated in part by the biosphere. In addition, human societies, socioeconomic systems, and political systems have emerged over the past several centuries as dominant forces shaping the planet. To the physical, chemical, and biological understanding of climate is a new view that sees climate change over the coming centuries through a socioeconomic perspective and shaped by human actions. Life – the microbes and microorganisms in soil, the plants reaching skyward, and the people inhabiting the land – is a key factor that determines Earth's climate.

34.2 | Ecosystems, Humans, and Climate Change

The notion that plants and terrestrial ecosystems affect climate and planetary habitability is embodied in the concept of coevolution of climate and life. Numerous books have explored this topic (Budyko 1974, 1986; Schneider and Mesirow 1976; Lovelock 1979, 1988; Schneider and Londer 1984; Schneider et al. 2004), and it is seen in the profound influence plants had on the geologic history of the planet (Beerling 2007). It is one part of an emerging recognition that a physical and chemical understanding of climate must expand to a biological perspective that includes the biogeophysical and biogeochemical functioning of plants and terrestrial ecosystems.

This view, as detailed in this book, arises from the multitude of anthropogenic perturbations in the Earth system and their multidisciplinary consequences. Carbon dioxide is a greenhouse gas that is central to understanding natural and anthropogenic climate change. It also alters leaf physiology and ecosystem functioning, and it inhibits isoprene emissions. Methane and N_2O are important greenhouse gases, and they also affect atmospheric chemistry. Methane emissions can produce ozone and decrease the oxidizing capacity of the troposphere (OH radical) through NO_x–VOC–O_3 chemistry. Nitrous oxide destroys stratospheric ozone. Tropospheric ozone is another important greenhouse gas, and it additionally damages leaves, decreases plant productivity, and decreases the terrestrial carbon sink. Aerosols increase planetary albedo and provide a negative radiative forcing, and they have indirect radiative effects by altering biogeochemical cycles through, for example, fertilizing ecosystems or enhancing diffuse radiation. Additional reactive nitrogen in the Earth system affects the greenhouse gases CO_2, CH_4, N_2O, and ozone and produces aerosols. Emissions of BVOCs affect tropospheric ozone, OH, and CH_4 and produce secondary organic aerosols. Underlying all these is the central role of plants and terrestrial ecosystems and of human actions that disrupt the functioning of ecosystems.

Interactions of climate and life are evident in the numerous terrestrial feedbacks with climate, as shown in this book and summarized in Table 34.1. Climate–vegetation coupling is found in past climates, where the evolution of leaf size, shape, and stomatal functioning in response to high atmospheric CO_2 concentration altered climate. It is seen in climate-induced changes in biogeography that altered albedo, evapotranspiration, carbon storage, and reactive gas emissions on land over the Holocene and during earlier glacial cycles. Changes in the structure and functioning of terrestrial ecosystems over the coming century are likewise expected to influence the trajectory of climate change. The scientific debate is no longer do plants and terrestrial ecosystems influence climate and should these processes be included in models of Earth's climate. Rather, the research continues to demonstrate how much climate, from local to regional to global scales, does indeed depend on biospheric functioning.

Our understanding of Earth's climate has progressed even further beyond the physics, chemistry, and biology of the Earth system. It

Table 34.1 Key land–atmosphere interactions that affect climate

Process	Climate effect
Soil moisture	Wet soils increase evapotranspiration and create a cool, moist climate; dry soils amplify droughts and heat waves.
Snow	High albedo cools climate; snow cover affects precipitation in North America and Eurasia; loss of snow with warming decreases surface albedo (+ climate feedback); low thermal conductivity insulates soil.
Leaf phenology	Springtime leaf emergence in the Northern Hemisphere increases evapotranspiration and cools air.
Stomatal conductance	Lower stomatal conductance with elevated CO_2 decreases evapotranspiration and increases temperature; also increases leaf photosynthesis and water-use efficiency.
Carbon cycle	The terrestrial biosphere is a carbon sink from elevated CO_2, nitrogen deposition, and past land-use.
(a) Carbon–concentration feedback	Higher atmospheric CO_2 stimulates plant productivity (– feedback); soil nitrogen and phosphorus restrict CO_2 fertilization.
(b) Carbon–climate feedback	Climate change decreases terrestrial carbon storage (+ feedback).
Reactive nitrogen	May cool the climate of North America and Europe as the net balance of: increasing N_2O; fertilizing the terrestrial carbon sink; forming tropospheric O_3 and decreasing CH_4; producing aerosols; and reducing the terrestrial carbon sink by O_3 damage to plants.

Process	Climate effect
Aerosols	Absorb and scatter radiation and alter clouds; net effect is − radiative forcing; decreased surface solar radiation (solar dimming), but enhanced proportion of diffuse radiation stimulates plant productivity; fertilizes ecosystems.
(a) Mineral aerosols	Cool climate (− radiative forcing), but deposition on snow is + radiative forcing; fertilize tropical ecosystems; may be + feedback that enhances drought.
(b) Biomass burning	Emits CO_2 and other greenhouse gases; emits reactive gases that affect O_3 and atmospheric chemistry; emits absorbing aerosols that heat the troposphere, cool the surface, and decrease precipitation; black carbon deposition on snow is + radiative forcing; alters surface albedo and surface energy fluxes.
Biogenic volatile organic compounds	Produce ozone in the presence of NO_x; reduce OH and increase CH_4 lifetime; produce secondary organic aerosols that cool climate.
Land use and land-cover change	Changes surface albedo, roughness length, and Bowen ratio; source of CO_2, CH_4, N_2O, and Nr (NO_x, NH_x); alters BVOCs, mineral aerosols, and biomass burning aerosols; deforestation increases CO_2 emission (+ radiative forcing), increases albedo (− radiative forcing), and decreases BVOC emissions (− radiative forcing).
Tundra	Snow melt and permafrost carbon loss with a warmer climate is + climate feedback; woody shrub expansion with climate warming augments warming by decreasing surface albedo.
Boreal forest	Boreal forest warms climate because of low surface albedo; high evapotranspiration of deciduous trees may warm climate through greater atmospheric water vapor; monoterpenes from conifer forests form secondary organic aerosols, which cool climate; outcome of afforestation is a balance of lower surface albedo (+ radiative forcing) but higher carbon storage and aerosols (− radiative forcings).
(a) Wildfire	Boreal fires cool climate because of higher surface albedo following loss of forest. In boreal spruce forests, the positive biogeochemical radiative forcing exceeds the negative biogeophysical radiative forcing in the first year following fire, but the biogeophysical forcing is dominant over longer time periods.
(b) Pine beetle	Tree death increases surface albedo in winter and spring; summer evapotranspiration decreases; daytime temperature warms; the forest carbon sink decreases; BVOC emissions increase while under attack.
Temperate forests	Have low surface albedo, high evapotranspiration, and high carbon storage; are thought to cool climate.
Tropical rainforests	Cool climate through carbon storage and high evapotranspiration; maintain high precipitation; produce many aerosols that additionally affect climate; biomass burning may decrease precipitation.
North Africa	Soil moisture and surface albedo produce strong vegetation–climate coupling; vegetation increases precipitation, seen 6 kyr BP; loss of vegetation decreases precipitation.

is no longer convenient to distinguish natural and human systems when studying climate. The influence of people, through fossil fuel combustion, agriculture, deforestation, land clearing, and urbanization, on climate and atmospheric composition is evident. The climate of future Earth is not simply a physical, chemical, or biological problem. Climate change is a socioeconomic dilemma, driven by our actions that emit greenhouse gases, aerosols, and short-lived climate forcers, that alter the landscape, and that co-opt the functioning of terrestrial ecosystems. Indeed, one avenue of inquiry is to understand when in the course of human history we first altered the functioning of the Earth system (Ruddiman 2003, 2007, 2013).

34.3 | Climate and an Ecological Aesthetic

One of the means by which people influence climate is through our transformation of the biosphere. We need a certain amount of land to grow food and fiber, to build our cities, and to sustain our socioeconomic well-being. These actions alter climate, freshwater availability, and biogeochemical cycles. The scientific concepts of ecological climatology can be applied to design and manage the biosphere for human needs. There is a growing awareness of the goods and services provided by ecosystems and that they provide a natural solution to many environmental problems. An understanding of the science of ecology and climatology allows us to take advantage of natural landscape features and ecosystems functions to improve our planetary environment.

This is seen locally in the building of our communities and cities, where there is a similar ecological aesthetic within the landscape planning and design professions (McHarg 1969; Hough 1984, 1995; Spirn 1984; Bormann et al. 1993). This movement recognizes that the landscape is not only where we live, but also regulates climate, air quality, water resources, and supports plants, animals, and other living creatures that sustain the healthy functioning of ecosystems. It, too, emphasizes Earth as a system, with ecosystems as regulators of environmental health through flows of energy, water, nutrients, and biomass. Ecological design advocates a new design aesthetic stemming from ecological functions and services rather than the traditional design principles of form, composition, color, and texture. The goods and services supplied by ecosystems provide a natural solution to urban environmental problems. Identity, form, and aesthetics arise from natural processes and features of the land.

Similar concepts pertain to our global environment. Forests, for example, provide ecological, economic, social, and aesthetic services to natural systems and humankind, including refuges for biodiversity, provision of food, medicinal, and forest products, regulation of the hydrologic cycle, protection of soil resources, recreational uses, spiritual needs, and aesthetic values. Additionally, forests influence climate through exchanges of energy, water, CO_2, and other chemicals with the atmosphere. It is now understood that forests and human uses of forests provide important climate forcings and feedbacks, that climate change may adversely affect ecosystem functions, and that forests can be managed to mitigate climate change.

This represents a profound change in attitude towards forests. In the colonial era of the United States, as land was being cleared for settlement, forests were seen as a hostile wilderness that needed to be tamed and civilized (Williams 1989; Stegner 1990). Progress could only be achieved through the felling of trees and cultivation of land. The colonial leader and future president John Adams wrote in his diary (June 15, 1756) that prior to European settlement "the whole continent was one continued dismal wilderness, the haunt of wolves and bears and more savage men. Now, the forests are removed, the land covered with fields of corn, orchards bending with fruit, and the magnificent habitations of rational and civilized people." The clearing of forests and cultivation of land was thought to make the climate more

Fig. 34.1 "View from Mount Holyoke, Northampton, Massachusetts, after a Thunderstorm–The Oxbow" (Thomas Cole). Reproduced with permission of the Metropolitan Museum of Art (www.metmuseum.org).

habitable, a requirement to advance learning and the arts and to promote the health, well-being, and prosperity of Americans (Fleming 1998). A lively debate arose, and notable colonial politicians and scholars advocated the merits of deforestation to improve the climate. Thomas Cole's 1836 painting "View from Mount Holyoke, Northampton, Massachusetts, after a Thunderstorm–The Oxbow" conveys the views Americans at that time felt toward forests (Figure 34.1). The untamed forest on the left is a threatening wilderness while the pastoral farmland on the right is serene. In an increasingly technological world fraught with ensuing global environmental challenges, it is ironic that the role of terrestrial ecosystems – nature's technology – in regulating climate and mitigating climate change is becoming apparent.

34.4 Concluding Thoughts

Research in the geophysical and biological sciences over the past decades has shown that Earth's climate must be understood as an interdisciplinary system. The role of the world's oceans in affecting weather and climate is well known, and study of the atmospheric sciences requires knowledge of ocean sciences. Similarly, there is growing awareness that an understanding of Earth's biota and soils is necessary in the study of climate and climate change. People, too, have been recognized as important determinants of the climate on Earth through emissions of greenhouse gases and through land use and land-cover changes that alter the biogeochemical, hydrologic, and energy cycles that regulate climate. Greater understanding of Earth and its climate requires that all components of the

Earth system – physical, chemical, biological, socioeconomic – be considered.

Earth system models that couple terrestrial ecosystems and climate are abstractions of complex physical, chemical, and biological processes. Many ecological, biogeochemical, and hydrological processes are represented in detail, but nature continually surprises us with our lack of understanding. It will be many years, if ever, before the full complexity of ecosystems and all the possible feedbacks are included in Earth system models. However, the picture emerging now is one of terrestrial ecosystems as an important regulator of climate. The scientific study of climate and solutions to adapt to and mitigate anthropogenic climate change must be interdisciplinary and must account for the various terrestrial biogeophysical and biogeochemical feedbacks with climate. Moreover, it must recognize people as the major agents of planetary change. We are not independent of the environment; nor are our lifestyles independent of climate. Our individual actions and choices,

and collectively those of our communities and nations, do matter and do influence climate and planetary habitability. Like terrestrial ecosystems, the rich complexity of human socioeconomic systems cannot be fully represented in the models; but like terrestrial ecosystems, future climate will be determined by human actions.

Science provides the framework to understand the functioning of the Earth system and to identify pathways to ensure a healthy balance between human and natural systems that does not comprise planetary habitability for future generations. However, science by itself cannot enable the solutions. Sustainable development over coming generations requires balancing a scientific understanding of the Earth system with a cultural, religious, and political understanding of human societies. It requires a balance between the facts of science and the values of individuals, cultures, religions, and countries. Balancing these is essential in the quest to manage Earth's future.

34.5 | Review Questions

1. The following data from Lovelock (1979) describe Earth with and without life. Which planet represents Earth with life? What does this suggest about relationships between the biosphere and climate?

	Planet A	Planet B
Carbon dioxide	98%	0.03%
Nitrogen	1.9%	79%
Oxygen	trace	21%
Argon	0.1%	1%
Surface temperature	290°C ± 50°C	13°C
Surface pressure	60,000 hPa	1000 hPa

2. Based on a climate modeling study published in the early 1990s that showed loss of the boreal forest cools climate, an American radio personality at the time suggested that the Canadian and Eurasian boreal forest should be cut down to lessen climate warming. Discuss the pros and cons of this.

3. Do you think that the biosphere can, and should, be managed to alleviate anthropogenic climate change?

4. Human uses of land affect climate and in turn are shaped by climate. Discuss the pros and cons of including interactive socioeconomic systems in Earth system models.

5. Many science fiction movies set in the future depict cramped cities of towering buildings, monolithic infrastructure, and devoid of trees. What does this say about our values of, and relationship with, trees?

6. The world's population exceeds 7 billion people and is likely to exceed 10 billion in the future. How do we feed the growing population, provide social and economic prosperity, and maintain healthy oceans, terrestrial ecosystems, and wildlife?

7. Describe the future Earth that you would like to see and discuss how to achieve that.

34.6 | References

Beerling, D. (2007). *The Emerald Planet: How Plants Changed Earth's History*. Oxford: Oxford University Press.

Bormann, F. H., Balmori, D., and Geballe, G. T. (1993). *Redesigning the American Lawn: A Search for Environmental Harmony*. New Haven: Yale University Press.

Budyko, M. I. (1974). *Climate and Life*. New York: Academic Press.

Budyko, M. I. (1986). *The Evolution of the Biosphere*. Dordrecht: Reidel.

Fleming, J. R. (1998). *Historical Perspectives on Climate Change*. New York: Oxford University Press.

Hough, M. (1984). *City Form and Natural Process: Towards a New Urban Vernacular*. New York: Van Nostrand Reinhold.

Hough, M. (1995). *Cities and Natural Processes*. London: Routledge.

Lovelock, J. E. (1979). *Gaia: A New Look at Life on Earth*. Oxford: Oxford University Press.

Lovelock, J. E. (1988). *The Ages of Gaia: A Biography of Our Living Earth*. New York: Norton.

McHarg, I. L. (1969). *Design with Nature*. Garden City, New York: Natural History Press.

Ruddiman, W. F. (2003). The anthropogenic greenhouse era began thousands of years ago. *Climatic Change*, 61, 261–293.

Ruddiman, W. F. (2007). The early anthropogenic hypothesis: Challenges and responses. *Reviews of Geophysics*, 45, RG4001, doi:10.1029/2006RG000207.

Ruddiman, W. F. (2013). The Anthropocene. *Annual Review of Earth and Planetary Sciences*, 41, 45–68.

Schneider, S. H., and Londer, R. (1984). *The Coevolution of Climate and Life*. San Francisco: Sierra Club Books.

Schneider, S. H., and Mesirow, L. E. (1976). *The Genesis Strategy: Climate and Global Survival*. New York: Plenum Press.

Schneider, S. H., Miller, J. R., Crist, E., and Boston, P. J. (2004). *Scientists Debate Gaia: The Next Century*. Cambridge: MIT Press.

Spirn, A. W. (1984). *The Granite Garden: Urban Nature and Human Design*. New York: BasicBooks.

Stegner, W. (1990). It all began with conservation. *Smithsonian*, 21(1), 35–43.

Williams, M. (1989). *Americans and Their Forests: A Historical Geography*. Cambridge: Cambridge University Press.

Appendix – Species names

Abies (Fir)
Abies balsamea (Balsam fir)
Abies fraseri (Fraser fir)
Acer (Maple)
Acer pensylvanicum (Striped maple)
Acer rubrum (Red maple)
Acer saccharinum (Silver maple)
Acer saccharum (Sugar maple)
Aesculus octandra (Yellow buckeye)
Alnus sinuata (Sitka alder)
Betula (Birch)
Betula alleghaniensis (Yellow birch)
Betula lenta (Sweet birch)
Betula papyrifera (Paper birch)
Betula populifolia (Gray birch)
Carya (Hickory)
Carya glabra (Pignut hickory)
Carya tomentosa (Mockernut hickory)
Castanea (Chestnut)
Castanea dentata (American chestnut)
Castanea mollissima (Chinese chestnut)
Cornus florida (Dogwood)
Fagus grandifolia (American beech)
Fraxinus (Ash)
Fraxinus americana (White ash)
Fraxinus nigra (Black ash)
Fraxinus pennsylvanica (Green ash)
Gleditsia triacanthos (Honey locust)
Halesia monticola (Mountain silverbell)
Juniperus occidentalis (Western juniper)
Larix laricina (Tamarack)
Liquidambar styraciflua (Sweetgum)
Liriodendron tulipifera (Yellow poplar)
Picea (Spruce)
Picea glauca (White spruce)
Picea mariana (Black spruce)
Picea rubens (Red spruce)
Picea sitchensis (Sitka spruce)
Pinus (Pine)
Pinus banksiana (Jack pine)

Pinus echinata (Shortleaf pine)
Pinus elliottii (Slash pine)
Pinus ponderosa (Ponderosa pine)
Pinus pungens (Table mountain pine)
Pinus resinosa (Red pine)
Pinus rigida (Pitch pine)
Pinus strobus (Eastern white pine)
Pinus sylvestris (Scots pine)
Pinus taeda (Loblolly pine)
Pinus virginiana (Virginia pine)
Populus (Aspen, cottonwood, poplar)
Populus balsamifera (Balsam poplar)
Populus deltoides (Cottonwood)
Populus tremuloides (Quaking aspen)
Populus trichocarpa (Black cottonwood)
Prunus (Cherry)
Prunus pennsylvanica (Pin cherry)
Prunus pumila (Sand cherry)
Pseudotsuga menziesii (Douglas fir)
Quercus (Oak)
Quercus alba (White oak)
Quercus coccinea (Scarlet oak)
Quercus macrocarpa (Bur oak)
Quercus marilandica (Blackjack oak)
Quercus prinus (Chestnut oak)
Quercus rubra (Northern red oak)
Quercus velutina (Black oak)
Rhus glabra (Sumac)
Robinia pseudoacacia (Black locust)
Salix (Willow)
Thuja occidentalis (Northern white cedar)
Tilia (Basswood)
Tilia americana (American basswood)
Tilia heterophylla (White basswood)
Tsuga (Hemlock)
Tsuga canadensis (Eastern hemlock)
Tsuga heterophylla (Western hemlock)
Tsuga mertensiana (Mountain hemlock)
Ulmus americana (American elm)

Index

Printed in the United States
By Bookmasters